Mass units

pounds × 16	= ounces
ounces × 0.0625	= pounds
ounces × 28.35	= grams
grams × 0.0353	= ounces
pounds × 0.454	= kilograms
kilograms × 2.205	= pounds
short tons × 2000	= pounds
pounds × 0.0005	= short tons
long tons (metric tons) × 1000	= kilograms
kilograms × 0.001	= long tons (metric tons)
short tons × 0.907	= metric tons
metric tons × 1.10	= short tons
milligrams × 0.001	= grams
grams × 1000	= milligrams
micrograms × 0.000001	= grams
grams × 1,000,000	= micrograms
nannograms × 0.000000001	= grams
grams × 1,000,000,000	= nannograms

Energy units

horsepower × 745.7	= watts
Btu × 0.293	= watt-hours
kilowatt-hours × 3416	= Btu

One quad is one quadrillion Btu or 10^{15} Btu.
One pound of coal yields 13,000 Btu when burned.
One kilogram of coal yields 28,600 Btu when burned.
One metric ton of coal yields 28.6 million Btu when burned.
One pound of coal burned in a 40%-efficient electric power plant
    yields 1.52 kilowatt-hours of electrical energy.
One barrel of oil yields 5.5 million Btu when burned.
One billion barrels of oil yields 5.5 Quads when burned.
One ton of oil = approximately 300 gallons
                = approximately 7 barrels
One cubic foot of natural gas yields 1,032 Btu when burned.

Units used in water resources

million acre-feet × 1.23	= cubic kilometers
cubic kilometers × 0.813	= million acre-feet
million acre-feet per year × 0.893	= billion gallons per day
billion gallons per day × 1.12	= million acre-feet per year

Concentration units

milligrams/liter	= parts per million (ppm)
grams/liter × 1000	= parts per million (ppm)
ppm × 0.001	= grams/liter
micrograms/liter (μg/l)	= parts per billion (ppb)
grams/liter × 1,000,000	= parts per billion (ppb)
ppb × 0.000001	= grams/liter
nannograms/liter (ng/l)	= parts per trillion (ppt)
grams/liter × 1,000,000,000	= parts per trillion (ppt)
ppt × 0.000000001	= grams/liter

THE ENVIRONMENT

Issues and Choices for Society

THE ENVIRONMENT

Issues and Choices for Society

Third Edition

Penelope ReVelle
Essex Community College

Charles ReVelle
The Johns Hopkins University

Jones and Bartlett Publishers
Boston • Portola Valley

Editorial, Sales and Customer Services offices
Jones and Bartlett Publishers
20 Park Plaza
Boston, MA 02116

Library of Congress Cataloging-in-Publication Data

ReVelle, Penelope.
  The environment: issues and choices for society / Penelope ReVelle,
Charles ReVelle. — 3rd ed.
    p.   cm.
  Bibliography: p.
  Includes index.
  ISBN 0-86720-072-3
    1. Environmental protection. 2. Human ecology. I. ReVelle,
Charles. II. Title.
  TD174.R49 1988   87-28741
  363.7—dc19   CIP

ISBN 0-86720-072-3

Printed in the United States of America.
10 9 8 7 6 5 4 3 2 1

Cover photograph: Cottongrass and Mount McKinley by Ed Cooper.
Cover design: Rafael Millán
Production: Del Mar Associates
Copy Editor: Jackie Estrada
Text compositor: Thompson Type
Artists: Wayne Clark, Cindie Clark-Huegel, Richard Carter

Credits for the color inserts can be found on page 749.

To Parents and Children,
the two ends of the living arrow of which we are a part

Preface

hen we wrote the first and second editions of *The Environment: Issues and Choices for Society*, our goal was to produce a stimulating text for one-term introductory courses in environmental studies. The text was designed to accommodate students with various academic backgrounds. In addition, we wanted to write a book that would complement a variety of teaching approaches, including lecture, discussion, and case study. The response from users of the first two editions, professors and students alike, has been gratifying.

In this edition we have retained the major features that made the first and second editions so useful. However, we have made some significant changes. We have added new material throughout the text, updating all of the topics. And, recognizing that the problems of the environment occur on a global scale, we have included a new element: a series of "Global Perspectives." As Daniel Koshland wrote recently in *Science*, "It's time to take a global look at the policies and priorities that are dooming our ecosystem—we can win but only if we recognize [that] we all win or lose together, and no one can be excused from the game."★ Examples of international or global problems include desertification, the greenhouse effect, the use of pesticides in developing nations, and nuclear winter.

Concluding six of the eight parts of the book are special sections that examine global aspects of the environmental issues covered in those parts. For example, the United Nations initiative to bring safe drinking water to every human being on earth by 1990 is discussed in Global

Perspective III, which falls at the end of Part III ("Water Resources"), while the possibility of a worldwide nuclear winter is examined in the Global Perspective following Part IV ("Energy Resources"). Concern for these issues represents a broadening and maturation of the environmental movement. We want students to know that their concerns are the concerns of millions worldwide.

Organization of the Text

Most of the chapters contain two types of text materials. In the basic text portion of each chapter, information is presented at a level suitable for beginning college students. Here students are given a solid foundation in the topic and a background that will enable them to participate in class discussions. The second type of material consists of controversial issues set off from the general text. The controversy boxes provide examples of disagreements on environmental policy and contain capsule quotes from opposing parties. In these debates, sometimes surprising and other times predictable conflicts arise. Who is hero and who is villain is not always clear. Experts often disagree on environmental matters. In the controversy boxes, differing viewpoints are argued, popular and unpopular causes are debated, and the difficulties inherent in making fair, effective laws are illustrated. We hope this exposure will encourage students to sift facts and come to their own conclusions rather than take the word of any one expert. In doing so, students will discover a surprising fact: Many environmental questions do not have easy answers. There are only social agreements, sometimes expressed in laws. The agreements represent society's consensus at the moment and are open to change.

★D. Koshland, "Inexorable Laws and the Ecosystem," *Science* (July 3, 1987), p. 9.

Each chapter also begins with an outline of main headings and ends with questions, references, and suggestions for further reading on the subjects in the chapter. All readings are described briefly to help students choose the ones that will be most helpful on a particular topic.

New Topics

New topics have been introduced because environmental issues are dynamic; laws and regulations are open to change as new information becomes available and as political conditions change. Examples in the present revision include the EPA decision to ban all uses of asbestos over the next ten years; the ending of tax credits for energy conservation and renewable energy; and a bold new law to encourage control of erosion by farmers. Not only do laws and regulations change, but the urgency surrounding various problems may wax and wane over time. The problems of groundwater contamination, acid rain, and hazardous wastes have moved to the forefront of public attention in recent years.

Following is a brief rundown of some of the changes that have been made in this updated edition of the text.

The book's revised introduction provides historical vignettes of important environmentalists. Included are such well-known environmentalists such as Rachel Carson, John Muir, and Lois Gibbs, as well as some lesser-known figures whose lives have influenced our environment: William Green, whose efforts turned a fragile barrier island into a protected national seashore and wildlife refuge; Robert Betz, a modern-day Johnny Appleseed who is planting prairie rather than apples; and John Van Dyke, who did for the desert what Muir did for his beloved redwoods.

In Part I, "Humans and Other Nations That Inhabit the Earth," the ecology discussion has been augmented with new material on evolution, behavior, and competition. The clear and thorough treatment of biogeochemical cycles remains a central part of the ecology discussion. In the chapter on saving species, new material on plants has been added, and that theme is further developed in Global Perspective I. This Global Perspective focuses on a distinctly bioethical dimension of preserving plant and animal wildlife: the competition between the needs of people in developing countries and the saving of species. By popular demand, the discussion of human population has been expanded to

two chapters. The new population chapter explores how nations in different areas of the world face and deal with their population problems. The clear lesson is that no single answer is applicable everywhere and that answers and responses must be tuned to the social and political environment in which the population situation occurs.

In Part II, "Resources of Land and Food," the Global Perspective on desertification deals with the potential for damaging the land in the attempt to increase food production.

In Part III, "Water Resources," a new chapter explores the contamination of surface water and groundwater by hazardous wastes. Water resource problems and solutions have been updated with new information on the progress of the United States toward cleanup of its lakes and rivers.

In Part IV, "Conventional Sources of Energy," energy facts and figures have been extensively updated, with a recognition of our temporary breathing space in the global energy policy conflict. The time between crises is portrayed as an opportunity to create new and environmentally safe energy technologies that can carry us when conventional energy resources again are in short supply. The chapter on nuclear power examines the Chernobyl disaster and the potential hazards associated with nuclear technology. Global Perspective IV notes that the spread of nuclear technology inevitably brings with it the potential for the spread of nuclear weapons. The use of these weapons can in turn precipitate a full scale nuclear war. The recent concept of a nuclear winter following a nuclear war is investigated. The history of severely cold years following volcanic activity is presented as the conceptual base of the nuclear winter model.

Part V, "Air Pollution," includes a significant updating of facts and data on air pollution. Global Perspective V is divided into two parts. In the first part, acid rain is discussed as a world-wide phenomenon. Among topics examined are the problem of transnational transport of air pollution and the control problems it creates. The decline and destruction of forests in Europe and the United States and the link to air pollution are documented. In the second part, carbon dioxide buildup in the atmosphere and potential global warming are explored.

In Part VI, "Natural Sources of Power and Energy Conservation," we survey exciting new developments in natural sources of power. Among these are the onrush of wind power in the western states, particularly

California, as well as developments in photovoltaic energy from the sun, a technology that shows increasing promise as an energy source.

In Part VII, "Human Health and the Environment," the latest data are presented on the effects of smoking and on environmental carcinogens. Global Perspective VI examines the problem of pesticides in developing nations, where they contaminate foodstuffs and damage the health of agricultural workers. The United States exports pesticides to these nations, and the chemicals return to haunt us as toxic residues on our extensive imported food supply.

Instructional Package

We have prepared an instructional package that contains a variety of materials intended to help organize and enrich environmental courses. Components of the instructional package include:

- Instructor's Resource Manual—contains additional references, more controversies, suggestions for class projects, sources of environmental films, and transparency masters.
- Instructor's Test Bank—contains over 1,000 questions and answers of varying difficulty that correspond to the text chapter-by-chapter.
- Environmental Science Slide Set—contains 60 important illustrations and photographs highlighting key concepts from the text.
- Environmental Science Film Series—a collection of educational films on the subject.

Further information on the instructional package is available by contacting the publisher.

Acknowledgments

Many people influenced this book and made it possible. We are indebted to our good friend and agent John Riina, who recognized new possibilities in our ideas and prose. John steered us with a gentle hand toward creation of a new and better shape for our text and has always been available for advice and counsel.

The many suggestions and guidance of our editor for this third edition, Joe Burns, have been especially valuable to us. Jean-Francois, editor for the second edition, also provided useful ideas for this revision.

Julie Allen provided research assistance. She tracked down numerous articles and other references and penetrated federal agencies in search of information and photographs. Katharine Lacher, our good neighbor, aided us in proofreading the endless stream of galleys.

Dr. Bruce MacBryde very graciously reviewed the material on endangered plants. Along with Dr. Faith Campbell, he answered a variety of questions and helped us find other wildlife experts.

Our major sources for the controversy boxes were articles we found in scientific journals and in the popular press. In this area, two parties contributed particularly. Peter and Linda Rottmann called our attention to the fire at Baxter Park and the sharp debate in the *Bangor Daily News*. It was an archetype of a problem without an answer. And Donald W. Taube, University of California at Santa Barbara, mentioned his own interest in Palau, the Pacific paradise on the verge of becoming a supertanker port. His bibliography and articles supplemented our own, enabling us to develop the Palau story more deeply.

We would also like to thank the following professors who have reviewed this book over its three editions and contributed greatly to its quality:

Ruth Allen, American University
Richard Andren, Montgomery Community College
David Appenbrink, Chicago State University
David Ashley, Missouri Western State College
Bill Baker, Rockland Community College
William Battin, State University of New York, Binghamton
James Beddard, New Hampshire College
Bruce Bennett, Community College of Rhode Island
Robert Bertin, Miami University
Charles Biernbaum, College of Charleston
Thomas Bilbs, Mobile College
Richard Bonalewicz, Gannon University
Arthur C. Borror, University of New Hampshire
George Carey, Stonehill College
Don Collins, Montana State University
Arthur Driscoll, Westfield State College
N. DuBowsky, Westchester Community College
F.F. Flint, Randolph-Macon College
David Gates, University of Michigan, Ann Arbor

Harry Gershenowitz, Glassboro State College
Joseph Gould, Georgia Institute of Technology
Ted L. Hanes, California State University at Fullerton
Karen Harding, Ft. Steilacoom Community College
Jay Hatch, University of Minnesota
Tom Hellier, University of Texas, Arlington
Clyde W. Hibbs, Ball State University
Clyde Houseknecht, Wilkes College
Jerry Howell, Morehead State University
Eric Karlin, Ramapo College
Terry Keiser, Ohio Northern University
Michael Kelley, United States Military Academy
Phyllis Kingsburg, Drake University
Paul Knuth, Edinboro University
Kipp Kruse, Eastern Illinois University
Edward Lukacevic, Cuyahoga Community College
Daniel Mark, William Jewell College
R. J. McCloskey, Boise State University
Earl McCoy, University of South Florida
Edward McLean, Clemson University
John Meyers, Middlesex Community College
John Mikulski, Oakton Community College
Kenneth Moore, Seattle Pacific University
Steven Mueller, Brainard Community College
Norton Nickerson, Tufts University
John Pawling, Temple University
Robert Pearson, Northern Illinois University
John Peck, St. Cloud State University
Jon Pigage, Waubonsee Community College
G. Puttick, Harvard University
Dominic Roberti, St. Joseph's College
David Robinson, Albany State College
C. Lee Rockett, Bowling Green State University
Diane Schulman, Erie Community College
Donald Scoby, North Dakota State University
A. J. Slavin, Trent University

Barbara Smigel, Clark County Community College
George E. Stanton, Columbus College
Harold Stevenson, McNeese State University
Ellis Sykes, Albany State College
James Taylor, University of Alabama
Susanna Tak-Yung Tong, University of California, Los Angeles
Margaret Trussell, California State University at Chico
Irvine Wei, Northeastern University
Donald Winslow, Indiana University
Dennis Woodland, Andrews University

The Instructor's Manual owes much to the diligent and creative effort of Greg Lindsey, a colleague who is also involved in environmental education. His hands-on approach to environmental action with the Mc-Henry County Defenders is a model for local efforts.

Many people were kind enough to allow the use of their valuable photographs. We have acknowledged them in the photo captions. Special mention should be made, however, of the expert and patient help provided by Renaldo Reyes, librarian at the United Nations photo library, who aided us in locating many of the excellent color photos. We also were privileged to have the assistance of Michael Ritter of Joliet, Montana. We met Mike while stranded on a stalled Amtrak train in New Jersey. Mike furnished us with photos of modern mechanized agriculture from his farm in Montana.

Finally, Jeff Wright, an environmentally concerned colleague and fellow educator, helped us form the portion of the final chapter that deals with the opportunities we all have to contribute to the environmental movement. His ideas helped prevent us from preaching, for which we are all grateful.

Penelope and Charles ReVelle

Contents in Brief

Contents

THE ENVIRONMENT

Issues and Choices for Society

Introduction:
On Solving
Environmental
Problems

William Green: Rescuer of Assateague Island

John Muir: From Wilderness Prophet to
Political Force

Lois Gibbs: Not in *Anyone's* Backyard

John Van Dyke, Who Recognized the Beauty
and Value of Deserts

Robert Betz: Johnny Appleseed of the
Prairie

Rachel Carson: Prose-Poet for
Environmental Protection

ost of us would agree that environmental protection is a matter of primary importance to humans and to the future of the human species. Survey after survey of public opinion shows this concern to be strong and deeply rooted in Americans today.

Yet many of us also feel a sense of frustration, a feeling of "Where could I begin?" Some are afraid that they don't know enough or that they are just "amateurs," unlikely to have any significant effect whatever they do.

To counter these feelings, we'd like to share with you the following stories about people who have made a difference. These stories, which cover a span of more than a century, deal with people who came from all areas of the country and from all walks of life. Importantly, none of the people who figure in these accounts were "experts" in the kinds of problems with which they eventually had to grapple. Although they were amateurs in that sense, they had in common with you and me a concern for the environment, a love of natural beauty, and a determination to leave for their descendants a world as whole as the one they knew.

William Green: Rescuer of Assateague Island

Assateague Island National Seashore, a 37-mile-long barrier island off the coast of Maryland and Virginia, is home to a wide variety of seabirds and shorebirds. Mil-

lions of migrating waterfowl rest on the island's shallow ponds and sandy shores during their travels. The island also boasts the famous Chincoteague wild ponies, believed to be descended from horses that swam ashore from a shipwrecked Spanish galleon in the 1600s. The island is wild and beautiful, and it was saved, forever, for all of us and our children by an act of Congress.

The island was saved not because it was a famous wildfowl sanctuary, well known to the U.S. Congress, but because of the efforts of one determined citizen, William E. Green. Green was a heating contractor who fell in love with the area during visits in the 1940s. He and his wife bought 1,000 acres of land on neighboring Chincoteague Island and eventually moved to Chincoteague permanently.

When Green realized that developers intended to sell lots and begin construction of homes on Assateague Island, he began a campaign to save the island. His struggle would continue for the rest of his life and consume a major portion of his personal fortune.

In 1955 he made his first trip to Annapolis, the state capital of Maryland, to lobby for a public authority to govern Assateague Island. After that he went again every year until he died. He sent telegrams, made phone calls, attended committee meetings and hearings (invited or uninvited), and carried petitions for signatures. He tried every way he could think of to educate lawmakers to the unique value of Assateague Island as a wildlife refuge. He also tried to underscore how unsuitable the island was for homes and development by pointing out that much of the proposed area for development was underwater for at least part of the year. Green harangued local officials in person and in newspaper advertisements. In one advertisement he challenged all candidates for state or local office to state their position on Assateague's future. He then printed all the answers he obtained. He gained a reputation as a character and a pest, but he did not let that stop him. "I have to use all the tools open to me as a private citizen," he once said, "one of which is a big mouth."

In 1963, though he was ill with cancer, Green made one more trip to hearings in the state capital. His friends believe that after these hearings he was finally convinced that the island would be protected. He lived only six more months.

In 1965, two years after his death, Assateague Island was designated a national seashore. A phrase he had used in his newspaper ads to the citizens of his county was finally a reality: "[Assateague Island] is the birthright of your children, and their children and their children's children ad infinitum."

John Muir: From Wilderness Prophet to Political Force

John Muir was born in Scotland in 1838. When he was 10 years old, his family emigrated to Wisconsin to farm. Although his father wanted him to stay and work on the family farm after he was grown, Muir thought the world offered other possibilities. Muir was a clever inventor who already had several devices, including an automatic page-turner and an "alarm clock," to his credit. His "alarm clock" raised the bed to a vertical position, at a preset time, thus dumping the occupant out. Friends who recognized his keen mind and potential helped him to enroll at the University of Wisconsin, where he thought he might study to be either an engineer or a doctor.

After completing only three years of college, however, he set out in 1863 to explore the California wilderness. For almost 10 years he lived the life of a wilderness recluse, supporting himself with a variety of odd jobs. He was a shepherd and a guide, but he saved his best efforts for his independent research into the geology of the Sierras and the botany of the wilderness.

This brilliant but independent and solitary person seems an unlikely character to have become an environmental activist, someone later described as the man who made conservation a major political force in the

United States. Yet in his writings we can see that even as Muir lived this isolated and contemplative life, he was drawn by two strong forces to rejoin human society.

"Bless me, what an awful thing town duty is! I was once free as any pine-playing wind," he wrote in 1875, "and feel that I have still a good length of line. But alack! there seems to be a hook or two of civilization in me that I would fain pull out, yet *would not pull out . . ."*

One of these strong forces drawing him back to civilization was his intense desire to communicate to his fellow beings the sheer joy to be found in the wilderness:

> The Shasta woods are full of wild bees, and their honey is exactly delicious . . . and no wonder, inasmuch as it was in great part derived from the nectar bells of a huckleberry bog by bees that were left alone to follow their own sweet ways. . . .
>
> There is no daylight in towns, and the weary public ought to know that there is light here . . . Come all who need rest and light, bending and breaking with overwork, leave your profits and losses and metallic dividends and come a beeing.

The second and more powerful force at work to draw Muir into the crusade that was to dominate the rest of his life was the simple fact that his beloved wilderness was disappearing at an alarming rate.

Sheep ranchers were destroying huge portions of the Sierra wilderness by deliberately setting fires in order to clear out the forest undergrowth so their sheep might more easily pass to graze in the higher meadows. Sawmills were already making inroads among the giant sequoia trees, while entrepreneurs sought the largest specimens to exhibit at fairs and expositions.

> . . . it appears that all the destructible beauty of this remote Yosemite is doomed to perish like that of its neighbors, and our tame law-loving citizens plant and water their garden daisies without concern, wholly unconscious of loss.

Muir was unable to accept this loss quietly. He began writing articles to acquaint people with the wilderness they stood to lose. Although he never claimed to enjoy writing for its own sake (he described it as similar to the progress of a glacier, "one eternal grind"), and although he said he could not get used to public speaking, becoming physically ill before his lectures, he gained a reputation as a famous naturalist and lecturer.

Muir founded a local conservation club, which eventually became the Sierra Club. By the turn of the century, John Muir was clearly the country's leading spokesman for conservation of our natural resources. By 1914, when he died, Muir had made conservation an important political force—a force that it remains to this day.

John Muir's legacy to us all is that we are able to follow the advice he gave:

> Climb the mountains and get their good tidings. Nature's peace will flow into you as sunshine flows into trees. The winds will blow their own freshness into you, and the storms their energy, while cares drop off like autumn leaves.

Lois Gibbs: Not in *Anyone's* Backyard!

In these stories of citizens who worked for environmental improvement, we often see that it is a strong emotion that turns an ordinary citizen into an environmentalist. Nowhere is this more clear than in the story of Lois Gibbs, founder of the Citizens Clearing House on Hazardous Wastes, and before that president of the Love Canal Homeowners Association. Before *that*, however, she was a 22-year-old housewife, mother of two children, and self-confessed "shy person."

The emotion that turned Lois Gibbs into the sort of person who confronts complacent officials, who organizes public opinion, and who stubbornly campaigns for environmental improvement was anger. She was angry because her 5-year-old son had been forced to attend a school built on top of an abandoned chemical dump, Love Canal in New York State. The boy developed asthma and suffered from seizures after he began school. Convinced that the location of the school was responsible for her son's problems, Gibbs asked that he be transferred. However, school officials refused to allow him to transfer on the basis that other parents might then demand the same for their children.

A survey of her neighborhood showed Gibbs that other people were concerned about the effect of chemicals leaking from the old canal dumpsite. Not only were people worried about their children attending the school built over the dump, but those living in houses bordering the canal feared that exposure to unknown chemicals was making them ill. As far as she could see, her friends and neighbors did seem to be suffering from an excessive number of illnesses, miscarriages, and birth defects. Yet officials that she contacted seemed anxious to prevent knowledge about the dump and its effects from spreading. One chided her for stirring up trouble that was likely to hurt tourism at nearby Niagara Falls.

At this point, Lois Gibbs became angry enough to knock on doors and collect information supporting her contention that the area around the former chemical dump was an unhealthy place to live. Several reporters had written about the problems that seemed to be occurring around Love Canal. Chemicals and even whole drums of waste were migrating to the surface and into nearby cellars and swimming pools. But it was Lois Gibbs, with her door-to-door survey, who met with the people affected and who became the rallying force for community concern.

When the Love Canal issue was satisfactorily resolved, with the government moving and resettling families from the area, she went on to found Citizens Clearing House for Hazardous Wastes, an organization dedicated to helping other local groups fight similar dangers from hazardous wastes in their communities.

From a person who had once avoided public speaking at all costs and who had once worried that her lack of extensive schooling would hamper her in dealing with officials and politicians, she has become an articulate spokesperson for a principle in which she believes: "I believe that ordinary citizens, using the tools of dignity, self-respect, common sense, and perseverance, can influence solutions to important problems in our society."

John Van Dyke, Who Recognized the Beauty and Value of Deserts

John C. Van Dyke was an art professor, visiting in southern California, when he decided to explore the Arizona and New Mexico deserts. He suffered from a respiratory ailment, which may explain why he was first drawn to the clear, clean air of the deserts. Regardless of why he first went there, however, he soon fell in love with "the sandy wastes, the arid lands, the porphyry mountain peaks." He was one of the first white men to explore the area and was the first to write about his desert experiences for the general public. In this way he became part of the awakening environmental consciousness of the late 1800s and early 1900s.

Many before him had praised and seen the worth of forest slopes, flower-filled meadows, and waterfalls. Van Dyke could see the pure, harsh beauty and also the value of deserts—lands his contemporaries considered useless.

"How often have you and I," wrote Van Dyke, "found beauty in neglected marshes, in wintry forests, and in barren hillsides!"

"One begins by admiring the Hudson-River landscape and ends by loving the desolation of the Sahara," he continued, as if he could scarcely understand this transition himself. "The deserts should never be reclaimed." (By this he meant irrigated and then farmed.) "They are the breathing-spaces of the West and should be preserved forever." John Van Dyke was a person who realized that in order to preserve the environment, it is necessary to help people see the beauty and value of the natural world. He was the forerunner of all those who love deserts and work for their protection.

Robert Betz: Johnny Appleseed of the Prairie

Some people think Johnny Appleseed isn't around anymore, but they haven't met Robert Betz. Johnny's spirit lives on in this enthusiastic midwestern college professor.

Whereas Johnny Appleseed's ambition had been to introduce his beloved apple trees to the American wilderness, Robert Betz's dream is to restore a part of the glory that was once found all across our midwestern states—the broad expanses of tallgrass prairie.

Four hundred thousand square miles of Kansas, Nebraska, Oklahoma, North and South Dakota, Missouri, Iowa, Minnesota, Illinois, Ohio, and Wisconsin were once covered by the mixture of grasses, wildflowers, and other plants known as prairie. Migratory wildfowl rested by the prairie ponds, potholes, and sloughs, while the prairie birds—dickcissels, horned larks, and scissortails—sang from the grasses. The grass itself, dozens of different kinds, grew 8 feet tall on favorable sites and 3 to 4 feet high elsewhere. Nor was this all. A vast array of insects lived on the prairies, providing food for the birds and other small animals

living there. These in their turn were food for the larger hunters: hawks and badgers, bobcats and coyotes.

Of this tightly integrated living system less than 0.06 percent remains, in isolated patches. An overgrown cemetery remains in one area; a quarter section saved by a farmer for sentimental reasons exists in another. There are 93 acres on the outskirts of Chicago, next to a McDonalds, and a few larger parcels such as the Flint Hills in Kansas, where the land was not suitable for farming.

What almost all of these remnants lack, and what Robert Betz is trying to re-create, is space. The feeling of limitless expanse is one of the most characteristic features of the prairie landscape. He envisions 5,000 acres of prairie, undulating to the horizon, farther than the eye can see. This is the view our pioneer ancestors once looked on, and one that was subsequently plowed and paved almost into oblivion.

Betz, a biology professor at Northeastern Illinois University, coordinates a group of volunteers who collect seeds from remnants of prairie in the area surrounding Chicago. Since 1972 he has been supervising the planting of this seed on a square mile plot of land that covers the Fermi National Accelerator Laboratory.

The Fermi lab's particle accelerator is a giant, doughnut-shaped atom smasher buried beneath the surface of the ground. After the accelerator was built, when it was time to landscape the grounds, directors of the laboratory contacted a university group for help. Instead of conventional landscaping however, it was suggested that the land be used as another laboratory—one in which prairie would be re-created.

"I like wildness," Betz said in a *National Geographic* interview. "I like the idea of wildness. I'd like to see us restore areas as reserves for our grandchildren. The reserves would be kind of like Noah's ark. Then if our grandchildren ever wanted to expand the areas of wildness, they'd have these nuclei. That's what we really are trying to do now—save the nuclei."

Rachel Carson: Prose-Poet for Environmental Protection

Like other citizens, scientists can be drawn to environmental action in a field different from the one in which they usually work. This seems to be what happened to Rachel Carson, who was trained both as a biologist and as a writer. She was already a famous author when she began what she thought would be a magazine article about the effect human beings could have on the natural world. Carson's previous books, *Under the Sea Wind, The Sea Around Us*, and *The Edge of the Sea*, had all been best-sellers. In these books she wrote in a lyrical, almost poetic way about the oceans and shorelines, about the creatures that lived in these places and their lives. Her books remain some of the best examples of scientific writing: both accurate and a pleasure to read.

In her planned article Carson hoped to expand on her thesis that the natural world needs to be viewed as an interconnected whole rather than a collection of unconnected parts. She also wanted to explore further the relationship of humans to this natural world. Carson thought she might be able to use the example of DDT, a pesticide that had come into widespread use by the late 1950s, when she was writing. As she did the research for her article, however, she became more and more alarmed about what she saw as the growing problem with the use of pesticides in general. Pesticide usage was increasing each year, with seemingly little recognition of the dangers such chemicals could present to people or to animals. The article grew from its originally planned size into a full-scale book investigating the pesticide problem: *Silent Spring*. The book became a warning to the public about the dangers to which they were being unknowingly exposed and also a call to action.

Although she had planned to take only a few months to write the article, the book was 3 years in preparation. ". . . in the end I believe you will feel, as I do," she wrote her editor, "that my long and thorough preparation is indispensable to doing an effective job . . . Now it is as though all the pieces of an extremely complex jigsaw puzzle are at last falling into place."

Silent Spring was written as an exposé, much as Upton Sinclair wrote *The Jungle* in order to stimulate public outcry against conditions in the meat-packing industry. Nevertheless, the book serves as a vehicle for Carson's basic philosophy, as she restated it in a *CBS Reports* special on *Silent Spring*:

> (the natural world) is built of a series of interrelationships between living things, and between living things and their environment. You can't just step in with some brute force and change one thing without changing a good many others.

Public reaction to *Silent Spring* was extremely favorable, making Rachel Carson something of a household name, but industrial reaction was swift and bitter. The chemical industry began an attack that was based on both personal and scientific criticism in an attempt to discredit not only Carson's science but also the author herself. Some of the most vehement attacks were made by writers who admitted they had not even read *Silent Spring*.

Interestingly, much of Carson's argument has stood the test of time, although knowledge of the effects of pesticides on both humans and the natural environment was nowhere near as advanced when she was writing as it is today. Pesticides *are* hazardous substances, and their overuse can be both dangerous and ineffective. *Silent Spring* forcefully brought this knowledge to public attention and spurred the passage of important laws to control pesticide use.

None of the people we have been discussing brought to their cause all the knowledge they needed to solve the problems they had to tackle. All were willing to learn. They learned from books, from other people, from experience—whatever was necessary to achieve their aims. Their stories are an inspiration to those of us who wish to preserve our environment.

Their stories also contain a promise—the promise that we can have pure water, fresh air, uncrowded places of solitude, and a rich heritage of fellow species with which to share life. Moreover, these pioneers of a healthy and whole environment show us that we can help to realize this promise through our own efforts and energies. The rest of this book is dedicated to the knowledge needed to accomplish this aim.

There are some facts you will want to know. This book contains many of them. And there are issues and controversies that surface over and over again. You will want to know how people have argued these issues and resolved such controversies in the past. In the future, however, additional facts will be uncovered and new controversies may replace the old ones. What we hope will finally remain with you is a spirit: the spirit that you can make a difference.

References

Weimer, L. G. "The Irascible Savior of Assateague Island," *Sierra* (May/June 1986), p. 64.

Engberg, R., ed. *John Muir Summering in the Sierra*. Madison: University of Wisconsin Press, 1984.

Mighetto, L. "Muir Among the Animals," *Sierra* (March/April 1985), p. 69.

Gohen, M. P. *The Pathless Way*. Madison: University of Wisconsin Press, 1984.

Gibbs, L. M. *Love Canal: My Story*. Albany: State University of New York Press, 1982.

Van Dyke, J. *The Desert*. New York: Charles Scribner's Sons, 1904.

Revard, C. "To Make a Prairie," *National Parks* (March/April 1985), p. 21.

Farney, D. "The Tallgrass Prairie: Can It Be Saved?" *National Geographic* (January 1980), p. 37.

"Grasslands," *The Nature Conservancy News* (June/July 1986).

Hoien, M. "Native Struggle, a Fond Look at Prairie Life," *Free Environment News*, **72** (May/June 1984), 4.

Gartner, C. B. *Rachel Carson*. New York: Frederick Ungar, 1983.

Brooks, P. *The House of Life: Rachel Carson at Work*. Boston: Houghton Mifflin, 1972.

(Courtesy of F. C. Sunquist)

Humans and Other Nations That Inhabit the Earth

It is undeniable that there is too much of many things in the world today: too much mercury in the water, too much sulfur in the air, too many wrecked automobiles. But surely the most curious excess is people. How has it happened that human populations have grown so rapidly that food and job shortages have gripped some countries, that both wildlife and wilderness are threatened, and that cities have grown to unmanageable sizes?

There are many reasons. Throughout the history of humankind, children have been valued. They have provided hands to work on the farm, to help in the blacksmith and carpentry shops and in the kitchen. Children represent security for parents in old age because they provide for parents too aged and infirm to work. And, for different reasons to different people, children are desirable in and of themselves. Children are a form of immortality, both of genes and of values. Through their children, a people's physical characteristics and also their beliefs can survive into the future. In traditional African religion, people are immortal as long as their descendants remember them. All societies, from the most primitive to the most cultured, consider the rearing and teaching of children one of their most important tasks.

Children are not only a precious resource; they have also been a most fragile one. In the United States and other countries, before the development of modern medicine and sanitation, five or six children, on the average, were born to each family. Of these, perhaps only two or three lived to adulthood. Because the survival of a child was so uncertain, people customarily had large families. Among other considerations, this helped ensure that at least two children would live to care for the parents when they grew old.

In many societies, especially where most of the population is engaged in farming, the tradition of large families has remained. At the same time, however, modern medicine has made child survival much more certain. Within the past 50 years, the **death rate** (or the number of people who die, each year, per 1,000 people in the population) has been reduced in most societies by one-half or more. This reduction in the death rate is due only partly to the medical advances that help adults to live longer; it is due in greater part to the survival of infants and young children who now live to become adults. This means that, in many countries, the number of people who die each year is much smaller than the number of people who are born. And so the populations of these countries grow rapidly.

As the population of a country grows, many facets of life undergo stress. Food supplies must grow or shortages, malnutrition, and finally starvation will result. Job opportunities must increase or unemployment spreads. New housing must be constructed, new parks opened, more roads, hospitals, and schools built or unbearable crowding will occur. Even if construction needs are met, the stresses of city living are forced on many people.

How, and how well, a country meets the stresses of rapid population growth depends to a large extent on how technically advanced that country is. For instance, technically underdeveloped countries may find it difficult or impossible to feed, house, and educate a rapidly growing population. Technically developed countries, on the other hand, can often expand production of food, goods, and services, but cannot cope with increased pollution, vanishing wilderness, or the mushrooming of social problems in growing cities.

Population growth is thus an underlying cause of, or contributor to, many of the environmental problems we know by other names: food crisis; disappearance of wilderness; energy crisis; air and water pollution; urban sprawl.

We begin this part with two chapters that discuss the ecological terms and principles related to populations and their interactions in ecosystems. The following two chapters explore the problem of rapid growth in human populations. How has the population problem come about? What are the effects of rapid population growth in developed and undeveloped nations? What means do we have of controlling population growth? And finally, what is the outlook for the future?

The last chapter in this part considers wildlife resource problems because the human population is, of course, not alone on this planet. Human populations cannot grow without inevitably affecting other populations, both plant and animal, living on the earth. For instance, the Indian tiger, once the most coveted prize of the trophy hunter, is now gravely endangered. Although hunters used to be the most serious danger the tiger faced, now India's rapidly growing population is taking over the little habitat left to the tiger. Before World War II, India literally teemed with great herds of wildlife. Although hunting by wealthy Indian royalty and foreigners had some impact on the numbers of game, it was nothing compared to what happened after independence in 1947. According to Hari Dang, an ecologist: "Postwar exploitation—open season on all resources! Shoot everything, burn what's left, destroy the rest. It was our disaster period. You had the same thing in the American West—in the 1880s everybody started shooting, and the great herds were destroyed." But even worse was to come.

As the human population grew, competition with wildlife increased for food, water, living space, and forest products. John Putman of *National Geographic* tells of a bird-watching walk he took in India's Borivli National Park with Dr. Salim Ali, in 1976: "There was the rustle of palms in a light breeze. We spotted palm swifts, then a rufous-backed shrike, black-headed orioles, racket-tailed drongos. Then out of the forest came a column of tribesmen, each carrying branches. They filed by, looking at us only briefly. 'What is this! What is happening! An army coming out of the forest!' Dr. Salim Ali cried. Part of Borivli's habitat was disappearing before our eyes. 'What can you do? What should you do?' Dr. Salim Ali added. 'They must eat, and need wood for their cooking fires.' He shook his head: 'Population—it is at the root of every problem.'" (Quotes from *National Geographic*, September 1976). The Global Perspective that closes Part One highlights this conflict between human needs and the right to survival of these "other nations," the wildlife populations with whom we share the earth.

CHAPTER ONE

Lessons from Ecology: Structure in Ecosystems

Peregrine Falcons and the Definition of a Species

What Is a Species?/Populations and Subspecies

How Organisms Live Together

Habitats and Niches/Communities/Ecosystems

Abiotic Factors in Ecosystems

Light/Moisture/Salinity/Temperature/Oxygen Supply/ Fire/Soil Type/Abiotic Factors Working Together

Trophic Structure in Ecosystems

Food Chains/Carbon Isotopes and Trophic Structure/ Food Webs

Diversity in Ecosystems

Reasons for Diversity/Diversity and Stability/Peregrines and Diversity

A student of the environment uses knowledge from many other fields. Biology, law, sociology, mathematics, anthropology, and physics are only some of the subjects useful in understanding environmental problems. The discipline most closely tied to environmental studies, however, is probably ecology.

The word **ecology** is made up of the Greek words *oikos*, meaning "house," and *logos*, "to study." Thus *ecology* means, roughly, the study of where organisms live, or their environment. Perhaps a better way to put it is that ecology is concerned with the relationship between organisms and their environment.

Many ecological terms are used in environmental studies. The purpose of the ecology chapters, which appear in several main sections of this book, is to explain ecological principles that you will need in order to understand environmental problems.

In this first chapter, we examine the parts of natural ecosystems: both the living organisms and the non-living environment that make up these systems.

Peregrine Falcons and the Definition of a Species

Birds belonging to the species called peregrine falcons are spectacular hunters (Figure 1.1). Sailing with wings outspread, peregrines search for the smaller birds that are their prey. When a falcon spots something that looks as if it would make a meal, it plunges downward

Figure 1.1 A mature peregrine falcon grows to 15–20 in. (37–50 cm) in length and has a wingspread of up to 43 in. (108 cm). The birds are noted for their beautiful markings and for the spectacular dives they make in mid-air. (The Peregrine Fund)

at speeds over 200 mi/hr (320 km/hr) and snatches its unsuspecting dinner in mid-air.

Since the early 1960s there has been no natural population of peregrine falcons east of the Mississippi. The birds fell victim to poisons in their ecosystem, most likely DDT. But DDT, which was first used in this country in 1945, is now banned, and levels in the environment seem to be decreasing. For this reason, Tom Cade, a Cornell University ornithologist, thinks it is a good time to try to restore the Eastern peregrine falcon population. He breeds the birds in captivity and then trains them to live successfully in the wild areas in which they were once found (Figure 1.2).

During the course of this project, Cade and the U.S. Fish and Wildlife Service have come into conflict over the definition of a species. Cade breeds his falcons from European peregrine falcons as well as from native North American birds. Because the European birds live in fairly populous areas, Cade feels that European falcons have characteristics that would be valuable to falcons attempting to settle along the populous eastern coast of the United States. The U.S. Fish and Wildlife Service is uneasy about allowing the result of this cross-breeding, which might be called a foreign species, to be established in this country because an executive order prohibits the introduction of "exotic species" into the United States. Many scientists have come to Cade's defense by pointing out that all the different birds Cade uses are simply subspecies of peregrine falcons. Who is right?

What Is a Species?

An animal **species** is most often defined in terms of reproduction. That is, a species consists of a group of individuals that can successfully breed with one another; that share ties of common parentage; and that therefore possess a common **pool of genes**, or hereditary material. In most cases it is possible to tell species apart on the basis of their different appearance or behavior or physiological makeup. However, these differences in themselves do not define species. If two apparently similar groups of organisms cannot interbreed successfully when given the opportunity, the two groups make up two separate species. Similarly, if two different-seeming groups of organisms are capable of interbreeding, there can be a flow of genes between them, and they are thus members of the same species, no matter how different they are in appearance.

Figure 1.2 Young falcons compared to a more mature bird held by Tom Cade. Many peregrine falcons flew in North America until about the early 1960s, when they fell victim to pesticides such as DDT. Now that DDT is no longer used in the U.S., Cade, a Cornell University ornithologist, is breeding the birds in captivity and then reintroducing them to the wild. More than 200 birds have been successfully released. This is equal to as many birds as the whole eastern population of peregrines once raised in a year, before DDT came into use. (The Peregrine Fund)

We should keep in mind that by defining species, scientists are attempting to describe part of the natural world in an orderly way, and the natural world does not always fit neatly into scientific categories. Thus, there are exceptions to these rules for defining species. Remmert (1980) pointed out that two species that have developed in different parts of the world, but with similar requirements for food, climate, living space, and so on, can sometimes interbreed if they eventually meet

Figure 1.3 The grizzly bear is a prized trophy for hunters, but few are left in the 48 states. Some 1,000 are believed to live in small populations in wilderness areas of Idaho, Wyoming, and Montana. The Interior Department has listed grizzlies as "threatened," a category that indicates they need less protection than "endangered" animals. Large populations still exist in Canada (11,000–18,000) and Alaska (8,000–10,000), where hunting is still allowed. Proposed Katmai National Park in Alaska would provide increased protection for the Alaskan population. These four grizzlies are salmon fishing in the McNeil River just outside the proposed park. (National Park Service photograph by Keith Trexler)

in natural surroundings. This has been noted in several species of birds, fish, and insects.

In a similar way, plants cannot always be classified into species on the basis of whether they can interbreed. Plants as a group have more diverse reproductive methods and genetic systems than animals do.

What, then, is a species? If we remember that the definition of a species is not based on hard and fast rules, we can say that as a general rule, a species is a group of organisms that can breed successfully, that share ties of parentage, and that therefore possess a common pool of hereditary material.

An important part of the species concept is that, since different species are not able to interbreed, they will follow different evolutionary paths as they adapt to their environment. In this way the *adaptive zone* of a species, which is made up of the resources available to that species and the parasites and predators★ it encounters, becomes different from that of any other species.

Populations and Subspecies

A **population** is all the individuals of a given species that live together in the same location. For example, the 50 or so birds that make up the entire population of peregrine falcons west of the Rockies. One or more

populations of a particular kind of plant or animal make up a species (Figure 1.3).

In general, even though they all belong to the same species, the members of a particular population resemble each other more closely than they resemble members of other populations. The reasons are that pairing is more likely to occur between individuals within a population than between those in different populations, and that members of a population are all subject to similar environmental influences upon their direction of evolution.

In some cases, geographic conditions in which various populations of a species live are very different, or the barriers to travel between the localities are great. For example, populations of the same species may live on different islands or in different rivers or on opposite sides of mountain ranges. The populations may then be found to have some very different genetic characteristics (although not enough differences to prevent successful interbreeding if given the opportunity). Populations of a species that are unlikely to breed because of geographic factors and that show genetic differences are sometimes defined as *subspecies* or *races*. Because Cade's peregrine falcons seem to breed successfully (although he does use some complicated artificial insemination techniques), it has been argued that they are all members of one species—or at worst, subspecies—and that the executive order does not apply. The Fish and Wildlife Service appears to be unconvinced by this argument but has agreed to allow an

---

★Predators are generally larger than the organisms they eat, while parasites are smaller than their host organisms.

exception to the executive order so that the peregrine falcons can still be released.

How Organisms Live Together

Habitats and Niches

Before we go any further, two useful ecological terms need to be defined. One is habitat; the other is niche.

The physical surroundings in which an organism lives are called its **habitat**. This is where you would go to find that particular organism. For example, the habitat of bloodworms, or chironimids, is the mud at the bottom of lakes, while the habitat of deer mice is temperate zone woodlots. These habitats consist, of course, of a variety of physical factors, such as temperature, soil type, and moisture, but they also contain many other animals as well as plants. Thus, an organism's habitat includes living as well as nonliving elements.

Peregrine falcons prefer to build their nest, or eyrie, on a narrow ledge or a steep cliff. This habitat provides protection for the baby falcons from predators such as owls or raccoons. In at least one case, a skyscraper seems to fill a peregrine's habitat requirements. Scarlet, one of the peregrines released by Tom Cade, chose to roost on a 35-story building in Baltimore, to the delight of its inhabitants.

The term **niche** is somewhat more difficult to define, in part because different people use it in different ways. It can refer to an organism's particular habitat. But it can also mean the organism's ecological function—that is, what the organism does, such as what time of day or night it is active, how much sunlight it requires, and in what ways it acts as a **predator** (eats other organisms) or **prey** (is eaten by other organisms). The niche thus corresponds to an organism's way of life.

It is a general rule that two species cannot occupy the same niche forever. Competition between the species should cause a change in some aspect of the lifestyle of one of them, force one to move somewhere else, or eliminate one of them. Sometimes two species are found who do live in the same habitat and eat the same food. However, closer examination often proves that their niche is not the same because some other factor, such as predators or parasites, prevents either of the species alone from using all the available habitat or eating all that type of food.

A major problem that can arise when a new species is introduced into an area is that it may occupy the same niche as a native species. Competition between the new and native species may result in elimination of the native species.

Communities

Definition. Within a given environment, organisms are not grouped randomly; rather, populations of organisms characteristic of that particular type of environment are organized into a **community**. This community includes all the living organisms, both plant and animal (including microorganisms), interacting in that particular environment. The term *community* can be applied to a relatively small number of organisms, such as those that make up a small pond community, or it can be used to describe a much larger number, such as would be found in the forest communities into which Cade hopes to reintroduce the peregrine falcon. These large forest communities include populations of small bird species that are the prey of the peregrine falcons, as well as populations of many other species of large and small animals and populations of various species of trees and other plants.

Functions. The degree to which organisms in a community interact with one another varies. In some environments, such as the African Serengeti grasslands, interactions are so important that the various species could probably not survive without one another. In the grasslands zebra graze when the grass is tall, exposing some of the sheath portion of the grass, which is the food for wildebeest. Wildebeest crop the grass further, allowing small herbs, which are the food of Thompson's gazelles, to grow.

In a similar way, the presence of the grazing animals helps to maintain the grass itself. Grazing encourages continued growth of the grasses, while lack of grazing may allow forest species to grow up and take over. Forest species may also take over if fires, which keep the fire-sensitive forest species down, are controlled by humans.

In other communities, notably those in harsh or fluctuating environments such as floodplains or the Arctic, there seems to be less of this type of species interaction. The animals in floodplains are mostly transients; they leave the community during a flood and return when conditions improve for them. They therefore act less to shape or maintain the environment than do the grazing species on the Serengeti plains.

Ecosystems

In ecological terms, *community* refers to a system of living organisms. This living system and the non-living, or **abiotic**, components of the environment plus the ecological processes that take place there make up an **ecosystem** (Figure 1.4). The nonliving parts of an ecosystem include such things as soil, amount of rainfall, and sunlight.

Because of the interactions among the organisms in an ecosystem, we might say that an ecosystem adds up to more than the sum of its parts. That is, the system itself has certain ecological properties in addition to the characteristics of the individuals or species making up the community. These properties include the flow of materials or energy through ecosystems; such processes are the subject of Chapter 2.

Some systems are not self-sufficient but depend on neighboring communities for such inputs as organic food materials; an example might be a small stream ecosystem. Others are large and complete enough to require little more than energy from the sun to function as a unit. Examples of ecosystems include a meadow, a stream, a forest, a lake, or any other clearly defined part of the landscape. (See Figure 1.5.)

Abiotic Factors in Ecosystems

The abiotic factors in an environment influence the kind and numbers of organisms found there. Often when we speak of pollution we are concerned about substances or procedures that change these nonliving factors in an ecosystem and, in turn, affect the com-

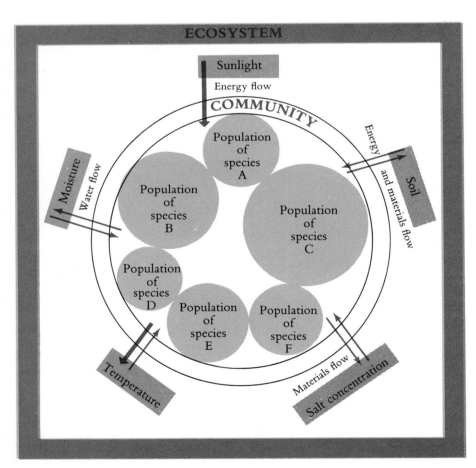

Figure 1.4 A diagrammatic representation of the relationship of population, community, and ecosystem. The populations (circles) of the various species living together in a specified location make up a community. Together with the nonliving features of the area (rectangles) and the ecological processes (arrows) that take place there, they make up an ecosystem. Many ecological processes take place in an ecosystem, such as energy flow and the cycling of water. Only a few are shown in the diagram. (All these processes are discussed in Chapter 2.)

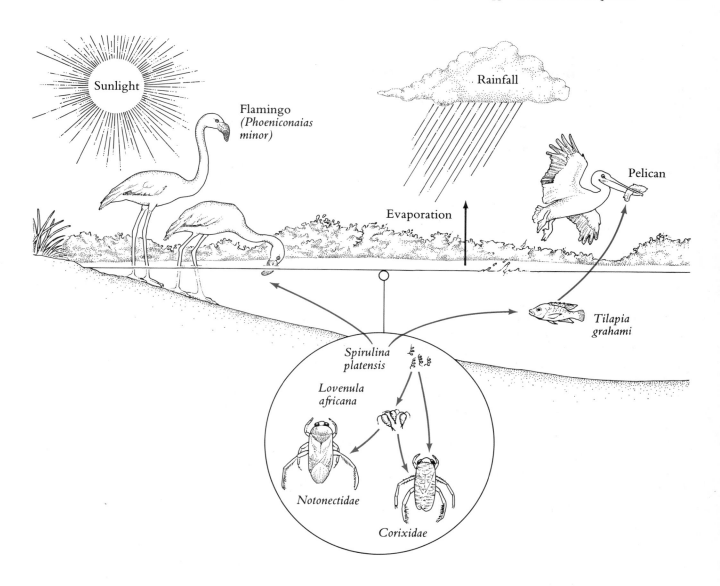

Figure 1.5 A simple ecosystem: Lake Nakuru, Kenya. Conditions are such that few species can live here. The living community in Lake Nakuru consists of populations of various species that can withstand the high concentration of sodium carbonate and the low concentration of hydrogen ions (pH 10.5). The main species found are blue-green algae (*Spirulina platensis*); flamingos (*Phoeniconaias minor*) and fish (*Tilapia grahami*), both of which eat the algae; and fish-eating birds such as pelicans and cormorants. Some of the important nonliving components of this ecosystem, in addition to sodium carbonate, are sunlight and rainfall. In more hospitable environments, such as the eastern deciduous forest ecosystem, many more species are found. (Adapted from H. Remmert, *Ecology*. New York: Springer-Verlag, 1980.)

munity of organisms living there. For instance, later we discuss pollutants that can change patterns of rainfall, increase or decrease environmental temperature, affect the oxygen or salt content of water, or change the concentration of inorganic plant nutrients in water. For this reason, we now examine more closely the important abiotic components of ecosystems and briefly consider how they affect the living community.

Light

Sunlight is one of the principal nonliving factors in an ecosystem because green plants use sunlight to produce organic material. This process is called **photosynthesis**. In addition to sunlight, photosynthesis requires water and carbon dioxide. The organic products of photosynthesis include a variety of sugars. Carbon dioxide is a chemical compound that can be described as "low energy," while sugars have a great deal of energy stored in their chemical bonds. Thus, in the process of photosynthesis the energy from sunlight is turned into, or stored as, chemical energy.

   In addition to organic materials, oxygen is produced during photosynthesis. Plants themselves use some of this oxygen, but they generally produce more than they need. All animal life, including human life, depends on this excess oxygen for respiration. We also depend on the sugars produced during photosynthesis for food, whether we eat the plants directly or whether we eat animals that eat plants.

   The process of photosynthesis can be summarized as follows:

$$\text{Sunlight} + \begin{array}{c}\text{carbon}\\\text{dioxide}\end{array} + \text{water} \xrightarrow{\text{green plants}} \begin{array}{c}\text{organic}\\\text{material}\\\text{(sugars)}\end{array} + \text{oxygen}$$

Or, using chemical formulas, photosynthesis can be written as the following equation:

$$6CO_2 + 6H_2O \xrightarrow[\substack{\text{green plants}\\\text{(chlorophyll, enzymes)}}]{\substack{\text{energy from}\\\text{sunlight}}} C_6H_{12}O_6 + 6O_2$$

However, this equation is only a summary. Photosynthesis actually consists of many separate reactions facilitated by a number of different enzymes (biological catalysts). A variety of organic materials are formed, here represented by the formula $C_6H_{12}O_6$.

   In most ecosystems, sunlight is present in suffi-cient amounts. Only in the depths of the ocean or of inland lakes or in caves does the lack of sunlight limit growth in an ecosystem. How the energy captured in photosynthesis is transferred within ecosystems is discussed in the next chapter.

Moisture

The amount of moisture in environments varies widely, from desert areas to lakes and oceans. All forms of life on earth require water to live, and the abundance and quality of water are major factors in determining what kinds of communities will develop in a given environment. In land environments, the amount of available moisture is a function of precipitation, humidity, and the evaporation rate. In water environments, the types of communities may also depend on the availability of water; however, in this case the availability of water means changes in water levels—that is, changes with the tides. Availability of water can also refer to differences in salt content, which affects the rate at which water enters or leaves organisms. In Chapter 9 we examine the different kinds of water environments and the specific communities that inhabit them. In this section we will look at the properties of water itself that have directly influenced the development of life as we know it.

Water and temperature. Water is unusual in that a relatively large amount of heat is needed to change its temperature or to change solid water (ice) to a liquid or liquid water to a gas (water vapor). For these reasons, temperature changes in water tend to occur slowly, and changes in water temperature occur more slowly than changes in air temperature. This is important for organisms living in water, since it gives them more time to adjust to temperature changes.

   Water reaches its greatest density at 3.94°C. That is, a given volume of water (for example, a 1 cm cube) weighs more at 3.94°C than at any other temperature. Its density decreases as the temperature decreases below this point. If you keep in mind that ice forms at 0°C, you can see that a given volume of ice at 0°C is lighter than the same volume of water at 3.94°C. This is why ice floats on cold water. This is an important property because it prevents lakes from freezing solid. The ice layer floats on top of the lake and insulates the water beneath it, allowing many aquatic creatures to survive during winter in the water below the ice.

Figure 1.6 During the summer in reservoirs such as this, which is used to generate hydropower, the water forms layers. The warm, oxygen-rich water floats on top of cooler bottom water, which may be low in oxygen. If, as is common practice, the cooler bottom water is released through the dam during power generation, stream communities below the reservoir may suffer from lack of oxygen in the water. (U.S. Department of Energy)

Warm water, being less dense than cold water, also floats on cold water. This is important in managing reservoirs (Figure 1.6) and also in determining the effects of pollutants on lakes, such as the phosphorus in detergents (Chapter 12).

Water as a solvent. Water is the most common solvent in nature. The amount and kinds of nutrients dissolved in water affect the growth of organisms. In a similar way, pollutants dissolved in water, even those that are only slightly soluble, affect organisms in land or water environments. For instance, *acid rain* is formed when sulfur oxides, produced by burning fossil fuels, dissolve in rain. Acid rain has reduced forest growth in Scandinavian countries and has caused entire populations of sport fish to disappear from some lakes in the Adirondack Mountains of New York State. (See Global Perspective V for more on acid rain.)

Salinity

Salt waters, such as the oceans, generally contain about 3.5% salt, or 35 parts of salt for every 1,000 parts of water. (Terms such as parts per thousand, parts per million, and parts per billion are explained inside the back cover of this book.) In contrast, fresh waters average 0.05% salt, or 0.5 parts per thousand. Most of the salt in the oceans is sodium chloride, but many other salts are present.

The salt content of water is one of the major factors determining what organisms will be found there. Freshwater organisms, both plant and animal, have a salt concentration in their body fluids and inside their cells higher than that of the water in which they live. Because substances tend to move from areas of higher concentration to areas of lower concentration, water tends to enter and salts tend to leave these organisms. Freshwater organisms have developed mechanisms or structural parts to cope with this situation. In addition, freshwater organisms have evolved so that they contain lower salt concentrations in their bodies than organisms found in salt water.

Some salt-water organisms (for example, marine algae and many marine invertebrates) have a salt concentration in their bodies or cells almost identical to that of ocean water. However, many marine organisms have body fluids with a lower salt concentration than the water in which they live. For these organisms, water tends to leave their bodies or cells and salts tend to enter. Their regulatory mechanisms must solve a different problem from that of freshwater organisms. Bony fish, for instance, have developed ways of excreting salt and retaining water. The main point is that the two environments, salt water and fresh water, provide different conditions for organisms to adapt to and thus are inhabited by different kinds of organisms.

In addition to salt and fresh waters, there are brackish waters, with intermediate salt concentrations. Such waters occur wherever salt and fresh waters meet—in estuaries, for instance, or where salt water intrudes on fresh groundwaters. Certain organisms are adapted, for all or part of their life cycles, to various intermediate salt concentrations.

Land-dwelling animals and plants tend to lose water to the atmosphere. In this respect they resemble many marine species because during their evolution they have also had to develop mechanisms to conserve water.

The kinds of water communities that develop in salt and fresh water are examined more closely in Chapter 9.

Temperature

Temperature has a profound effect on the growth and well-being of organisms. The biochemical reactions necessary for life are dependent on temperature. In general, chemical reactions speed up two to four times for a 10°C rise in temperature. Nonetheless, it is not possible to make sweeping generalizations about the effects of environmental temperature on the distribution of organisms because organisms have developed so many and varied mechanisms to deal with temperature changes.

Warm-blooded organisms, such as humans, are able to maintain a constant body temperature independent of the temperature of their environment. They are called *endothermic* or *homeothermic* organisms. Body warmth is a by-product of internal biochemical reactions that produce energy for the organism. In a cold environment, warm-blooded animals can retain this body warmth by insulation (blubber, feathers, fur, clothing). In a warm environment, the heat is lost by processes usually involving the evaporation of water (humans sweating from the skin; dogs panting and allowing their tongue to hang out). During very cold weather, when animals have difficulty finding enough food to burn to keep their body temperatures at a high level, some of them hibernate. During this time, their rate of energy use falls, so their body temperature falls as well.

The temperature of so-called cold-blooded (*ectothermic*) animals, as well as of plants and microorganisms, varies with that of the environment. However, even in this group of organisms there are a variety of mechanisms to adjust body temperature. Most of these mechanisms would be classified as behavioral methods of regulation. For example, bees can warm their hive by beating their wings. This is such an effective method that bees can live and reproduce in arctic regions. Many insects, snakes, and lizards warm themselves in the sun, taking up a position broadside to the sun's rays during the cool morning hours. Mosquito larvae develop quickly in the uppermost layers of ponds, where the sun's rays warm the water. When temperatures rise, many organisms take refuge in holes or burrows or under rocks. This helps them escape a lethal rise in body temperature or, in the case of desert organisms, prevents excessive use of precious water for cooling. During freezing weather, ectothermic animals and plants may produce antifreeze substances in their cells to prevent them from freezing. Many animals produce glycerol, while plants produce sugars such as hamamelose.

Photosynthesis does not depend on temperature as strongly as other reactions do because it is not just a biochemical reaction but also involves photochemical (light-driven) reactions. Thus photosynthesis is almost as effective in producing organic material in cold as in warm climates.

Most of the observations we can make about temperature and its effects on organisms are on a large scale. For instance, fewer types of organisms seem able to adapt to conditions in the arctic, where temperatures are far below the biological optimum. Even this seemingly obvious principle is complicated by the observation that besides the severity of temperature, the variability of conditions is also important. That is, fewer types of organisms are found where temperatures vary widely from day to night or season to season than where temperatures are more constant. However, one thing we can say with confidence is that organisms, during the course of their evolution, have developed mechanisms to deal with temperature as it is found in their environment, whether that is warm or cold, constant or fluctuating. Human actions that change these temperatures can have devastating effects on ecosystems. For example, in Denmark the brown weevil (*Hylobius abieties*) normally takes three years to develop. When the forest is **clear cut** (all trees cut, regardless of size or species), the sun warms the ground more than before and the weevil matures in two years. For this reason, the weevil does much greater damage to forests where clear cutting is allowed. In Chapter 15 we discuss the effects of *thermal pollution*: the addition of excess heat to water environments.

Oxygen Supply

Both plants and animals use oxygen in the process of respiration, whereby they obtain energy for growth and metabolism. Respiration consists of a series of biochemical reactions in which an energy-containing organic material, such as the sugar glucose, is broken down by biological catalysts called enzymes. The energy released is used to drive other reactions in the cell. If oxygen is available, the material is fully broken down to carbon dioxide, water, and energy. The summary of these reactions resembles photosynthesis in reverse:

$$\text{Oxygen} + \underset{\text{(e.g., glucose)}}{\overset{\text{organic}}{\text{material}}} \xrightarrow{\text{enzymes}} \underset{\text{dioxide}}{\overset{\text{carbon}}{}} + \text{water} + \text{energy}$$

Using chemical formulas, respiration can be summarized as:

$$6O_2 + C_6H_{12}O_6 \longrightarrow 6CO_2 + 6H_2O + \text{energy}$$

In land environments, oxygen is rarely in short supply. (Exceptions would be in some soils or on high mountaintops.) In water, on the other hand, the supply of oxygen may easily be a problem. The concentration of oxygen in water depends on the rate at which the gas diffuses into water, as well as on the rates at which it is produced by the plants living there and used by the plants and animals in the water.

In some lakes the supply of plant nutrients allows the growth of masses of algae, which die, sink to the bottom, and are decomposed by bacteria. This last process can use up all the oxygen in the water. Other desirable organisms cannot live in this oxygen-poor water. The addition of sewage to natural water environments results in the loss of much of the oxygen present. These effects are discussed more fully in Chapter 12, which focuses on organic wastes, dissolved oxygen, and eutrophication.

Fire

Fire can also be considered an abiotic factor that influences the types of communities in an ecosystem. Some environments are subject to regular natural cycles of fire. In southeastern pine forests and grassy savanna and steppe regions, periodic fires are a natural event.

Trees in forests where fires are regular may have thick bark that enables them to survive fires. The cones of some pines, such as the Jack pine (*Pinus banksiana*),

release seeds best when heated to a certain temperature. In this way, the seeds are sown at times when other plants that might compete for living space have been eliminated from the area. In fact, in the pine-spruce forests in northern Europe fire actually allows pines to grow. There are certain areas in these forests where the spruce have grown up in dense stands, crowding out the pines, which do not compete well for living space. Although the spruce are easily damaged by fire, once they form a dense stand it is difficult for fire to spread because the short spruce needles pack tightly on the ground and are resistant to fire. However, where pine and spruce are mixed, the forest floor accumulates loose piles of litter shed by the pines. Periodic fires in the mixed pine-spruce forest injure the spruce and allow pines to flourish.

In several cases, it has been shown that the vegetation growing up after a fire has more nutrients, such as phosphorus, potassium, calcium, and magnesium. Animals feeding on this vegetation may be better nourished. When humans prevent these natural fires, they are really causing changes in ecosystems that have come to depend on fire for periodic renewal.

Fire has now become an accepted part of forest management (see more in Chapter 30), although the public has been slower to accept this idea.

Soil Type

The type of soil found in an area is very important to humans because soils vary widely in their ability to support crops. The most useful soils from this point of view are the grassland and temperate forest soils. Other types, such as desert soil or the soil in tropical rain forests, are not as generally suited for raising crops. However, scientists looking at the effect of soil type on the kinds and distribution of organisms in an area have concluded that while soil type can have some effects on communities, the communities themselves have a profound effect on the type of soil in an area. To understand this it is necessary to realize that soil is not an entirely abiotic component of ecosystems but rather is a mixture of living and nonliving materials.

The nonliving part of soil is the finely divided particles produced by the action of weathering on the parent material of the earth's surface. Combined with this is organic material: organisms and their products, which grow, die, and become mixed in. Some soil scientists believe that the kind of soil that develops in a given region is usually governed by the climate of the region and the community that grows there, rather than by the type of parent rock. (Climate as a determining factor in soil type is discussed in Chapter 6.)

As an example of the effect of community on soil type, grassland soil is characterized by its dark color, a result of having a great deal of finely divided organic material called *humus*. The entire grass plant, including the roots, is short-lived. Each year, therefore, large amounts of organic material are added to the soil, where it is quickly decomposed to humus. This humus is then slowly decomposed by soil microorganisms to inorganic materials (mineralization). In a forest, on the other hand, the leaf litter and roots are slow to decay to humus, but the humus is then quickly mineralized. Thus, little humus can accumulate in forest soil.

From this discussion you might get the impression that rather than soil affecting an ecosystem, the reverse is true: an ecosystem determines its soil type. To an extent, this is so, but the process takes a long time. In geologically young regions, soil type is more likely to be determined by parent material or by some other condition, such as water level or topographic features such as hills.

Exceptions also occur where climatic conditions are extreme, such as in the desert. Here, small differences in soil composition can make a large difference in the type of community that develops. Soil types and their suitability for crop production are discussed in Chapter 6, where the different land environments are examined more closely.

Abiotic Factors Working Together

Undoubtedly, many other abiotic components of ecosystems could be mentioned. However, those factors already discussed are generally agreed to be the most important ones.

Although they are discussed separately, it is important to note that abiotic factors act together. Temperature, for instance, almost always acts in combination with moisture and wind. To predict how particular temperatures will affect the kinds of organisms found in a given environment, we need to know about these other factors as well. (Together, these factors describe the climate of an area. See Chapter 19 for a more complete discussion of climate.) Likewise, the development of soil type involves the interaction of climate, the organisms that grow up in a given area, and the breakdown of parent rock.

Figure 1.7 In Montana's Rocky Mountains a 5-year program has begun to reintroduce peregrine falcons. Here a researcher watches the development of young falcons placed on a cliff-top eyrie and guards against golden eagles, which prey on the young birds. (The Peregrine Fund, William R. Heinrich)

Trophic Structure in Ecosystems

Attempts are being made to reestablish peregrine falcons not only in the east but also in parts of the west from which they have disappeared. For instance, the State of Montana is attempting to reestablish populations of peregrine falcons in abandoned eyries along the face of the Montana Rocky Mountains. This area contains a large enough prey base of smaller birds, such as cowbirds, meadowlarks, and killdeer, for the peregrines to feed on. However, while they are growing and learning to fend for themselves, the peregrines must be protected from predators such as golden eagles, great horned owls, coyotes, and raccoons (Figure 1.7).

In any particular community, the organisms are

linked together in a pattern of preying and being preyed upon—of eating and being eaten. This pattern is the **trophic structure** of the community: the way in which the various organisms in the community obtain their nourishment. Such patterns are often described as **food chains** or **food webs**.

Food Chains

Food chains can be diagrammed in a relatively simple, straight-line fashion. Creatures in the chain generally feed on only one or a few species and are preyed on by one or a few other species. Food chains are found where conditions are so harsh that few animals or plant species are able to survive—for instance, on the arctic tundra (Figure 1.8) or in the high carbonate concentrations in Lake Nakuru (Figure 1.5).

The food chain illustrated in Figure 1.8 is a grazing food chain. There are several distinct trophic levels in the chain. Organisms at the bottom of grazing food chains are **producers**, plant species that convert sunlight into food energy by photosynthesis (the grasses, sedges, and lichen in Figure 1.8). Creatures that eat producers are **herbivores**, or green-plant eaters (such as the caribou in Figure 1.8), and they are called *primary consumers*. Next in line are the meat eaters, or **carnivores**, which eat the primary consumers. These are the *secondary consumers* (men and wolves in Figure 1.8). Longer food chains include *tertiary consumers* that eat the secondary consumers, and so on.

At another trophic level are the **decomposers**, or **detritus** feeders. This group of organisms is extremely important because it feeds on dead organic matter, breaking it down into inorganic and organic materials.

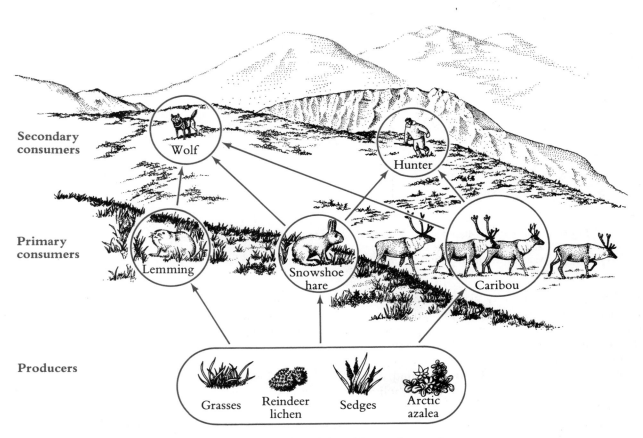

Figure 1.8 Arctic food chains are simple and short because few species are able to live in the harsh and unpredictable arctic climate. One arctic food chain is made up of a few arctic tundra plants; the caribou, lemmings, and snowshoe hares that eat the plants; and wolves and humans.

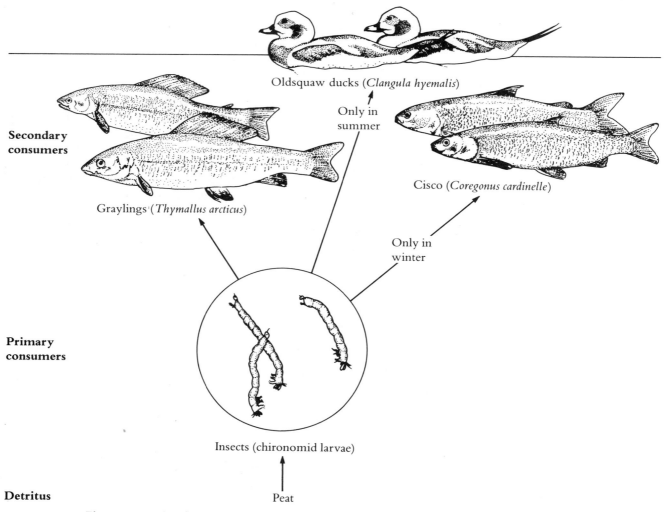

Figure 1.9 A detritus-based arctic food chain. Peat, formed from organisms that decayed thousands of years ago, is the carbon source for insect larvae, which in turn are consumed by arctic graylings, ducks, and whitefish. (Data from Schell, 1983.)

These materials may be used by plants and animals or may inhibit or stimulate other organisms in the ecosystem. Fungi and bacteria are common decomposers, but animal species may also be important. In the freshwater lakes of the Alaskan tundra, especially interesting examples of detritus-based food chains are found. Into these lakes is washed peat formed from plants that lived and died 8,000 to 12,000 years ago. Carbon from the peat is incorporated into insect larvae, which are eaten by consumer species such as arctic graylings and old-squaw ducks (Figure 1.9). In this case there is a delay of thousands of years between primary production and its use by consumers.

Carbon Isotopes and Trophic Structure

Carbon exists in living organisms in three different forms, called isotopes. These isotopes differ in the number of **neutrons** they have in their nucleus and are known as ^{14}C, ^{13}C, and ^{12}C. These isotopes also differ in that ^{12}C and ^{13}C are stable over time while ^{14}C is radioactive and decays or changes into another element, nitrogen. The loss of ^{14}C to nitrogen is just equalled by the formation of new ^{14}C high above the earth in the upper atmosphere, so a balance exists among the three isotopes of carbon. Important information about what a species is eating can be obtained by looking at the ratio of these three carbon isotopes in organisms.

The original ratio of ^{14}C to ^{12}C in an organism will be the same as that of the environment in which the organism lives. This is because carbon is exchanged rapidly and continuously between an organism and its environment. Plants, for instance, absorb carbon dioxide and incorporate it into their tissue during photosynthesis. Animals absorb carbon from their food and excrete it as wastes or carbon dioxide. Thus, until the moment it dies, an organism will have the same $^{14}C/^{12}C$ ratio as its environment.

Once an organism dies, however, the situation changes. Now there is no longer a free exchange of carbon between the dead organism and its environment, and ^{14}C slowly begins to decay or change into nitrogen. Thus, less and less ^{14}C compared to ^{12}C is found in decaying matter as time goes on. Peat, which was formed from organisms that died 8,000 to 12,000 years ago, has a much lower $^{14}C/^{12}C$ ratio than organic matter formed last week. This is how scientists were able to detect the fact that the insect larvae in tundra pools were feeding on peat carbon rather than the carbon in present-day phytoplankton species (Schell, 1983). Further, they could tell that the insect larvae were in their turn food for graylings, ducks, and whitefish. The ratio of the other carbon isotope, ^{13}C, to ^{12}C gives another important piece of information. Arctic land plants and arctic salt-water plants (phytoplankton) have differences in their photosynthetic processes that lead to a difference in the $^{13}C/^{12}C$ ratio found in their tissues. Thus, the examination of the $^{13}C/^{12}C$ ratio in an organism may give information about whether the organism is feeding on land plants or on salt-water phytoplankton species.

Now you can see that if many types of plants are present in a given environment and an organism feeds on a variety of plant species or on the detritus from plants with both types of photosynthetic pathways, the $^{13}C/^{12}C$ ratio will not yield much information: the signal will be confused. In the arctic, however, where relatively simple food chains are common, this type of investigation can be fruitful.

In the same study that determined peat to be a major carbon source in tundra lakes, the organisms in the near-shore region of the Beaufort Sea were also examined (Schell, 1983). This near-shore region receives large inputs of peat as well as detritus from modern land plants, from rivers flowing from the tundra, and from erosion of the shore itself. In this area, unlike in the tundra lakes, the organisms all showed modern $^{14}C/^{12}C$ ratios, indicating that they were not using peat as a carbon source.

Furthermore, the $^{13}C/^{12}C$ ratio of consumers was similar to that of salt-water phytoplankton, which indicated that these primary-producer species were the main carbon source utilized by consumers in the Beaufort Sea (Figure 1.10). (In contrast, species that live and graze on the tundra itself, lemmings and caribou, have $^{13}C/^{12}C$ ratios similar to land plants.) One further interesting feature of this study concerned migratory species: Oldsquaw ducks that winter in marine regions and then fly to the tundra lakes for the summer breeding season, and species of fish that winter in the freshwater streams and lakes and migrate back to salt water to breed in the spring. The isotope ratios in these organisms changed with the season, reflecting the fact that they fed on a food chain based on salt-water phytoplankton when in the Beaufort Sea and on a detritus-based food chain utilizing peat when in or near the freshwater tundra lakes.

Food Webs

In contrast to harsh arctic regions, which have relatively few species, many ecosystems in temperate and tropical climates, where a large number of species interact with one another, have a trophic structure best described as a *food web*. Food webs have more organisms and interconnections than food chains. Figure 1.11 shows an example of only part of a food web from a small stream in South Wales. Food webs are more complicated than food chains and are more difficult to sort into the trophic levels of producer, consumer, and so on. One reason is that food webs involve so many species, that we cannot always be sure we know all of the creatures in a particular food web. Also, some spe-

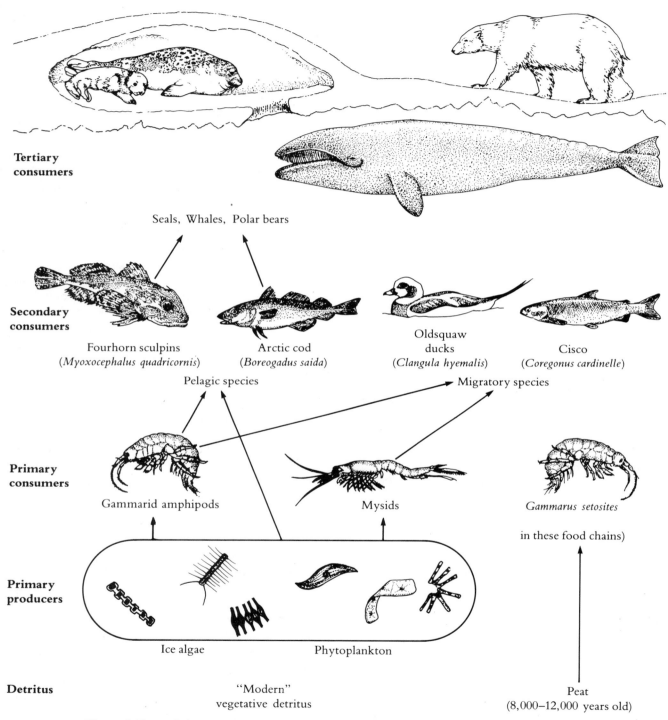

Tertiary consumers

Seals, Whales, Polar bears

Secondary consumers

Fourhorn sculpins
(*Myoxocephalus quadricornis*)

Arctic cod
(*Boreogadus saida*)

Oldsquaw ducks
(*Clangula hyemalis*)

Cisco
(*Coregonus cardinelle*)

Pelagic species

Migratory species

Primary consumers

Gammarid amphipods

Mysids

Gammarus setosites

in these food chains)

Primary producers

Ice algae

Phytoplankton

Detritus

"Modern" vegetative detritus

Peat
(8,000–12,000 years old)

Figure 1.10 Food chains in the near-shore region of the Beaufort Sea. Ice algae and phytoplankton are the primary producers in the ecosystem and are the main carbon source for consumers. (Data from Schell, 1983.)

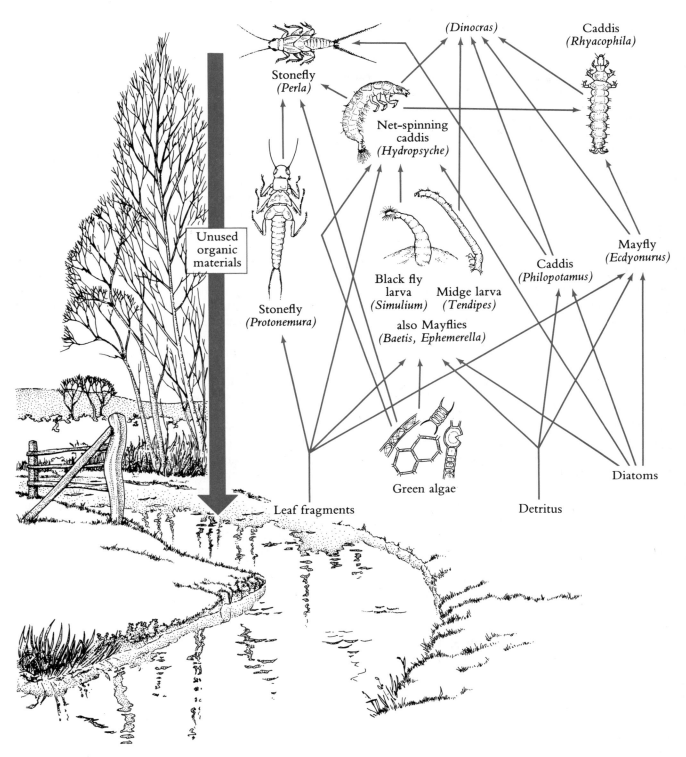

Stonefly
(Perla)

(Dinocras)

Caddis
(Rhyacophila)

Net-spinning
caddis
(Hydropsyche)

Stonefly
(Protonemura)

Unused
organic
materials

Mayfly
(Ecdyonurus)

Black fly
larva
(Simulium)

Midge larva
(Tendipes)

Caddis
(Philopotamus)

also Mayflies
(Baetis, Ephemerella)

Green algae

Detritus

Diatoms

Leaf fragments

Figure 1.11 Part of the food web in a small stream community in South Wales. Note that the net-spinning caddis is both a primary consumer (eats green plants) and a secondary consumer (eats primary consumers). The complexity of most food webs is emphasized by the fact that this illustration shows only some of the insects involved. Many other creatures—for example, small fish, frogs, and turtles—are a part of this web. (Adapted from J. R. Jones, *Journal of Animal Ecology,* **18** [1949] 142. Used by permission of Blackwell Scientific Publ. Ltd.)

cies occupy more than one position in a web. That is, an organism may eat both green plants and such primary consumers as insects, making it both a primary consumer and a secondary consumer. The net-spinning caddis in Figure 1.11 illustrates just such a complication.

In the small stream food web, the decomposers feed on materials carried by water from upstream, such as leaves, feces, dead organisms, and so on. Many food webs, especially in water, are based on detritus. The large arrow on the side of Figure 1.11 indicates that all of the unused organic material produced at various levels of the food chain (feces, secretions, dead plants or animal bodies) is either broken down by the decomposers to basic plant nutrients or cycled back up the food web as decomposers themselves are eaten.

Diversity in Ecosystems

When ecologists speak of **diversity**, they are referring to two features of an ecosystem. The first is the number of species in a given area compared to the number of individuals. In other words, it is a way of measuring variety. Sometimes this is called the *species richness* of an ecosystem. The second feature is *dominance*. This term measures relative numbers of individuals present in each species. Two ecosystems may have the same number of species present, but in one there may be approximately equal numbers of individuals of each species while in the second most of the individuals may be of the same species. The first ecosystem is then considered to have greater diversity.

One reason why diversity is an important measurement is that human interference with a natural ecosystem often results in reduced diversity. Fewer kinds of species are found in a polluted ecosystem than in a similar ecosystem that is not polluted. Measurements of diversity in a given area over time can be an indication of the effects of pollution.

Reasons for Diversity

The more diverse the environmental conditions, the more different niches are possible and, assuming a long enough time has passed for evolutionary changes to have taken place, the more species are found. It should be emphasized, though, that there are basic biological

processes common to all organisms. These processes work best at certain optimum temperatures, light intensities, and so forth. The farther conditions are from these optima, the less easily organisms can adapt to the conditions. For this reason, the farther conditions are from biological optima, the fewer species are found. To put it another way, fewer species are able to use the available niches in harsh climates. Thus, fewer species are found in arctic regions, on icy mountaintops, and in deserts than in more moderate environments such as temperate or tropical forests.

These three factors—diversity of environmental conditions, harshness of environmental conditions, and the length of time a system has had to evolve—are not the only factors affecting diversity, but they appear to be major controlling influences. Other factors will be mentioned in the next chapter.

Human interference may also move conditions far-takes the form of reducing the diversity of natural conditions. Thus, clearing a mixed forest to plant a single variety of pines destined for a paper mill reduces the number of niches. The result is a lower diversity of animal and plant species in the new pine forest than in the old mixed forest.

Human interference may also move conditions farther from biological optima. For instance, the addition of sewage to a stream reduces the amount of oxygen available in the water. Only a few organisms are adapted to low oxygen concentrations. As a result, the diversity in the polluted stream is much lower than in a normal, well-oxygenated stream. It should be emphasized that the actual number of organisms in the polluted and nonpolluted habitat may be similar, but the organisms in the polluted habitat will mostly be members of one or a few species. It is the number of different species that is reduced.

Diversity and Stability

Because of their complicated interrelationships, food webs are often described as more stable than food chains. That is, food webs should remain in balance despite the loss of a species. According to this theory, in a temperate-region field community, for example, the loss of an animal such as a fox, which preys on another species such as rabbits, should not cause a rapid increase in the rabbit population because there are other predators who also eat rabbits. In a food chain, the loss

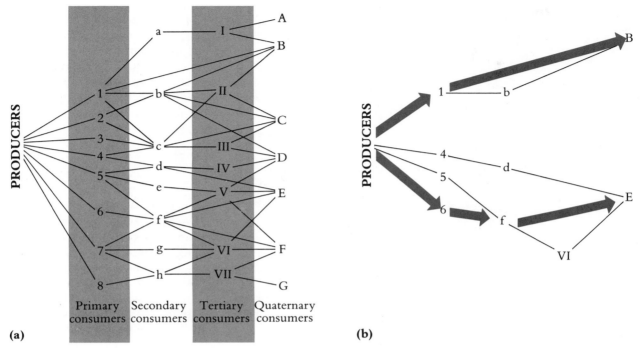

Figure 1.12 (a) A hypothetical food web, showing the flow of food or energy through an ecosystem from producers to quaternary consumers. (b) Another view of the same system, in which all pathways accounting for less than 0.001 of the total are left out. Now the significant structure more greatly resembles a set of connecting food chains. (Adapted from H. Remmert, *Ecology*. New York: Springer-Verlag, 1980.)

of a predator might lead to an explosive increase in its prey population. However, this point has not been proven. It should be remembered that because ecosystem webs are so complicated, no ecosystem has yet been completely examined, either in terms of all the creatures that may be a part of the system or in terms of how important the various pathways are. Possibly, most of the food energy in a web flows along only a few pathways, resembling food chains (Figure 1.12). In addition, the stability observed in communities that have many species could be due to other factors. For instance, mixed fields of crops do not suffer the massive outbreaks of insect pests often seen in fields with only one sort of crop. In some cases, this happens because insect pests locate the plants on which they feed by means of chemicals given off by the plants, and the insects have trouble locating these plants in a mixed field. The stability of the mixed-field ecosystem thus results from a sensory phenomenon rather than from

additional predator-prey relationships. As another example, tropical rain forests are among the most diverse ecosystems known, yet investigators believe them to be among those systems most easily disrupted by human actions.

We might say that very diverse ecosystems appear to have evolved under fairly constant conditions. Their higher diversity affords many checks and balances that act to keep the system constant as long as there is little outside interference. Changed environmental conditions and new species can have drastic effects on these delicately balanced systems. On the other hand, elastic systems—that is, those that can spring back after an outside influence—may well have evolved under more variable environmental conditions. For instance, the reed communities along lake shores, which are naturally subject to variable conditions such as water-level fluctuations, mowing, and ice movements, appear able to recover relatively quickly from pollution incidents.

Figure 1.13 Baby peregrines bred in captivity. In areas where DDT is found in the environment, peregrine falcons lay eggs with thin shells. These eggs are more likely to break before hatching than eggs with normal shells. This is believed to be the reason why there have been no natural populations of peregrine falcons in the eastern U.S. since the early 1960s. (The Peregrine Fund)

Peregrines and Diversity

Tom Cade's program to restore some natural diversity to our ecosystems by reestablishing the peregrine falcon population in the United States is an example of wildlife management techniques at their best and most imaginative. Not only has he developed ways of encouraging the birds to breed in captivity (Figure 1.13), but he and his staff also teach the newly fledged falcons to hunt for themselves in the wild. At about 4 weeks, when the birds can tear and eat their food, they are taken from the laboratory to old falcon nesting sites or to specially made towers. There they are protected 24 hours a day by attendants who keep away owls and raccoons that would eat the young birds. The peregrines return to the nest site to eat until they are skilled in catching their own meals.

Still, we do not have anywhere near enough researchers to undertake a similar program for other endangered creatures. There are also many reasons why it would not work in other cases. Clearly we must try to prevent species from becoming extinct rather than try to repair the damage once it is done.

In the following chapters, we explore first the dynamics of ecosystems and then the problems faced by the human species as well as those faced by wildlife populations. For, as the fate of the peregrine falcon was changed by human acts, first for ill and then for good, so are human fates and fortunes tied up with those of the other species on this planet.

Summary

A group of organisms that can sucessfully breed with each other, that share ties of parentage, and that therefore have a common pool of hereditary material is called a species. A population is a group of organisms of the same species living together in a particular area. The habitat of a species consists of the biological and physical surroundings in which a species or population is found. Similar but somewhat more comprehensive is the term *niche*, which also includes what a species does in its habitat: what it eats or what feeds on it, whether it is active day or night, and so on.

Within a given habitat, the species are organized into a community. There are interactions between the species in a community. Sometimes these interactions may actually help to preserve the habitat in a particular form. For instance, grazing communities prevent grasslands from being invaded by woodland species. The abiotic factors in an environment along with the community of organisms found there form an ecosystem. Light, oxygen supply, moisture, soil type, salinity, temperature, and fire are all examples of abiotic factors.

The trophic structure of a community describes how the organisms are bound together in a pattern of eating and being eaten. Producers, which form organic materials by the process of photosynthesis, and detritus, the dead organic residue matter from living organisms, are the basis of either simple food chains or complex food webs. Consumers utilize producers and detritus as sources for their organic materials.

Diversity is a measure of the species richness of an ecosystem and also a measure of the dominance of one species over others. Major influences on the diversity of organisms found in an ecosystem include the diversity of environmental conditions, the length of time the ecosystem has been in existence, and the harshness of environmental conditions. Diversity is not a guarantee of stability in ecosystems. In fact, ecosystems evolving under harsh conditions may have low diversity and be quite stable, while highly diverse systems may be very fragile in terms of environmental change.

Questions

1. How would you define a species? a population? a subspecies? What could be the purpose of an executive order prohibiting the introduction of exotic species into the country?
2. What is the difference between a community and an ecosystem?
3. List some of the abiotic factors in an ecosystem. For each, briefly explain its importance to the function of an ecosystem.
4. Diagram a simple food web that might exist in the area where you live. What are the producers? the consumers? the decomposers?
5. Differentiate between an organism's habitat and its niche.
6. What does an ecologist mean by diversity? Of what importance are measurements of diversity?

Further Reading

If you are interested in ecology, you may want to read some of the following materials.

Cade, T. J. *The Falcons of the World*. Ithaca, N.Y.: Cornell University Press, 1982.

Cargo, D. N., and B. H. Mallory. *Man and His Geologic Environment*. Reading, Mass.: Addison-Wesley, 1977.

Gates, D. M. *Biophysical Ecology*. New York: Springer-Verlag, 1980.

Levin, D. A. "The Nature of Plant Species," *Science*, **204** (April 27, 1979), 381.

In this article, Levin states: "For humans the environment has meaning only when its components can be interrelated in a predictive structure. We try to make sense out of nonsense and put the world into some perspective which has order and harmony." The species concept has caused a great deal of argument and confusion. This article is a thoughtful consideration of the problem, as well as a thought-provoking discussion of the human need to impose order where it is not always clear that order exists.

Levin, H. L. *The Earth Through Time*. Philadelphia: Saunders, 1978.

Odum, E. P. *Fundamentals of Ecology*. Philadelphia: Saunders, 1971.

Watt, K. *Principles of Environmental Science*. New York: McGraw-Hill, 1973.

Richardson, J. L. *Dimensions of Ecology*. New York: Oxford University Press, 1979.

References

Remmert, H. *Ecology*. New York: Springer-Verlag, 1980.

Schell, D. M. "C^{13} and C^{14} Abundances in Alaskan Aquatic Organisms," *Science*, **219** (March 4, 1983), 1068.

CHAPTER TWO

Lessons from Ecology: Dynamics of Ecosystems

The Flow of Materials in Ecosystems
Hydrologic Cycle/Carbon Cycle/Nitrogen, Phosphorus, and Sulfur Cycles/Biological Magnification

Energy Flow in Ecosystems
Productivity/Thermodynamic Laws and Ecosystems

Population Dynamics
Age Distribution Graphs/Population Growth Curves/ r and K Selection/Factors Determining Carrying Capacity

Evolution and Competition
Forces Driving Evolutionary Change/Competition and Evolutionary Strategies/Alternatives to Competition Theory: Predation/The Variable Environment Alternative/Punctuated Versus Gradual Evolution

In the first chapter we discussed the parts of ecosystems—that is, the living organisms as well as important nonliving or abiotic factors. Figure 2.1 (page 34) is a simplified picture of a pine forest ecosystem and Figure 2.2 (page 35) is a photograph of just such a pine forest. However, while both the diagram and the photograph are representative of this type of ecosystem, one central fact needs to be emphasized: Ecosystems are not like diagrams or photographs, static in time. Instead, growth and change are characteristics of ecosystems. The living and nonliving components act on each other and cause changes. Further, there is a flow of materials and energy through ecosystems. These ecosystem dynamics are the subject of this chapter.

The Flow of Materials in Ecosystems

A unifying feature of the flow of materials in ecosystems is that the flow is cyclic. That is, materials flow through systems in a largely circular manner: into the living organisms, back to the abiotic environment, and then into the living organisms again.

Hydrologic Cycle

The cycle with which we are most familiar is the **hydrologic cycle**. It consists of three distinct and continuing events: the evaporation of water; condensation and rainfall; and runoff. Water evaporates from the surfaces of lakes, ponds, streams, wetlands, rivers, and oceans. It also evaporates from soil and vegetation and is transpired by plants.* This water returns to the earth as

---

*Transpiration refers to the release of water vapor from the aboveground parts of plants. Evaporation and transpiration from plants together are sometimes referred to as "evapotranspiration."

rainfall. More water evaporates from the surface of the oceans than returns to the oceans as rain, however. On land, the opposite is true—less water evaporates from soils, vegetation, and surface waters than returns by rainfall (except for deserts, where evaporation is greater than rainfall). The balance is maintained by runoff, or water flowing from streams, lakes, rivers, and groundwaters to the oceans (Figure 2.3, page 36).

The hydrologic cycle can be compared to the process of distillation, in which water is vaporized by heating it in a flask. The water vapor leaves behind in the flask dissolved materials such as salt. This does not mean that rain is completely free of contaminants—rain and snow may become contaminated with gases and particles in the air. (For example, acid rain may form; see Global Perspective V.) However, rain is relatively pure. In some areas of the world, with few rivers and little water in the ground for people to utilize, rain furnishes drinking water. In such places, rainfall is collected in cisterns and stored for later use.

The processes in the hydrologic cycle are being modified by human activities. Some of these changes are intentional; others are accidental. For instance, rainfall has increased in industrial areas because water droplets condense more quickly around minute mineral particles in the air. As another example, runoff increases when vegetation is destroyed. Trees, grasses, and other plant covers capture and hold rainfall, allowing it to percolate down through layers of decaying organic materials and rocks in the soil. Too rapid runoff may lead to flooding. Also, where there is less water slowly percolating through the deep soil layers, the amounts of trace minerals dissolved from buried rocks decreases. These minerals are necessary for plant growth and may have to be added to fertilizer in some areas. On the other hand, the runoff of water via streams, rivers, and lakes to the oceans is interrupted

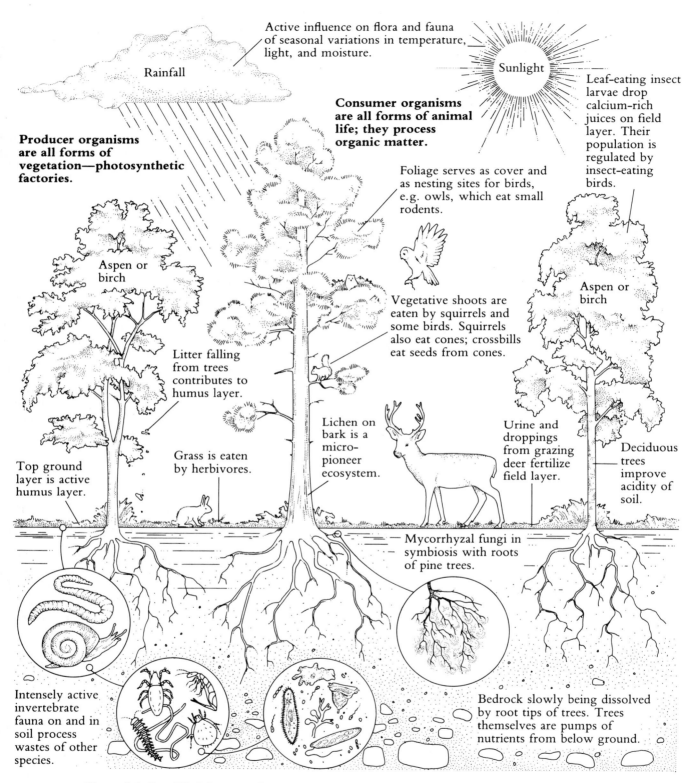

Active influence on flora and fauna of seasonal variations in temperature, light, and moisture.

Rainfall

Sunlight

Leaf-eating insect larvae drop calcium-rich juices on field layer. Their population is regulated by insect-eating birds.

Consumer organisms are all forms of animal life; they process organic matter.

Producer organisms are all forms of vegetation—photosynthetic factories.

Foliage serves as cover and as nesting sites for birds, e.g. owls, which eat small rodents.

Aspen or birch

Vegetative shoots are eaten by squirrels and some birds. Squirrels also eat cones; crossbills eat seeds from cones.

Aspen or birch

Litter falling from trees contributes to humus layer.

Lichen on bark is a micro-pioneer ecosystem.

Urine and droppings from grazing deer fertilize field layer.

Deciduous trees improve acidity of soil.

Top ground layer is active humus layer.

Grass is eaten by herbivores.

Mycorrhyzal fungi in symbiosis with roots of pine trees.

Intensely active invertebrate fauna on and in soil process wastes of other species.

Bedrock slowly being dissolved by root tips of trees. Trees themselves are pumps of nutrients from below ground.

Figure 2.1 Simplified diagram of some of the living and nonliving components of a pine forest ecosystem. (From Darling and Dasmann, in: *Impact of Science on Society*, vol. 19 no. 2, 1969. Courtesy of UNESCO.)

Figure 2.2 Pine forest ecosystem. Beside the pines, many other organisms are found in pine forest ecosystems, including otters, beavers, turtles, tree frogs, ospreys, and even bald eagles. Mosquitoes, deerflies, and poison ivy are also often found. (Larry E. Morse)

by humans at many points and for many purposes, some of which contaminate the water with chemical and biological wastes. These problems are described in greater detail in Part Three, Water Resources.

Carbon Cycle

All living creatures on earth consist mainly of water. About 80%–90% of living material is water. However, the most important structural material for life forms is carbon. This carbon, which forms the molecular backbone for all organic compounds in living creatures, comes from carbon dioxide in the atmosphere.

As explained in Chapter 1, plants use carbon dioxide during photosynthesis to manufacture some of the organic compounds necessary to life. Carbon from the atmosphere thus enters living systems through photosynthesis at the level of the ecosystem producers. The carbon moves through the ecosystem from one trophic level to another until it reenters the atmosphere during respiration or until the organisms in which it is contained die. The process of decomposition forms carbon dioxide again as dead material decays (Figure 2.4, page 37). This is only a part of the **carbon cycle**, however.

Most carbon involved in the cycle is in the oceans. This carbon (in the form of carbonates) controls, to a large extent, the amount of carbon dioxide in the air. Excess carbon dioxide in the atmosphere can dissolve in the oceans, forming carbonate and bicarbonate ions. In the opposite direction, the oceans can release carbon dioxide to the atmosphere. The oceans thus act to buffer, or keep constant, the carbon dioxide concentration of the atmosphere. Scientists believe that this mechanism kept the amount of carbon dioxide in the atmosphere relatively constant until industrialization began.

Human activities appear to be unbalancing the carbon cycle in several ways. An extra 5.8 billion tons of carbon dioxide are added to the atmosphere each year when fuels are burned at their present level of use. Some scientists believe that the rapid cutting of forests to satisfy human needs for farmland and wood products may be decreasing the amount of carbon dioxide used by plants.

Although some of the excess carbon dioxide from burning of fuels dissolves in the ocean, half or more remains in the atmosphere. Measurements show that the level of atmospheric carbon dioxide has increased steadily since the 1950s. The complete carbon cycle, as well as the possible effects of an increase in atmospheric carbon dioxide, is explored in Global Perspective V.

Nitrogen, Phosphorus, and Sulfur Cycles

Three other important materials cycle through ecosystems along with water and carbon: nitrogen, phosphorus, and sulfur. All three of these chemical elements are essential components of living matter.

Nitrogen cycle. Nitrogen is present in the atmosphere as nitrogen gas (N_2). The air we breathe is about 80% nitrogen. This nitrogen becomes part of biological matter almost entirely through the bacteria and algae that can "fix" atmospheric nitrogen into organic compounds and nitrates. Legumes, such as clover, al-

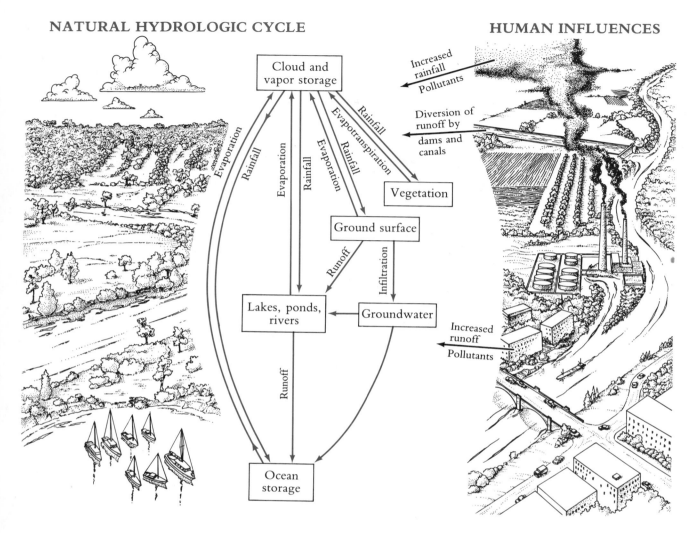

NATURAL HYDROLOGIC CYCLE

HUMAN INFLUENCES

Figure 2.3 The hydrologic cycle. Water evaporates from soil and surface waters and is transpired by vegetation. This water returns to the earth as rainfall. In addition, water moves from land areas to the ocean as runoff in streams, rivers, and groundwater flows. Humans influence the hydrologic cycle in many ways. For instance, humans divert runoff, decrease the amount of water held in soil by destroying natural vegetation, increase rainfall via air pollution, and pollute water.

falfa, soybeans, and locust trees, form little nodules in their roots where such nitrogen-fixing bacteria live. Farmers often fertilize their land naturally by growing these types of crops and then plowing them into the soil.

Nitrogen is lost from the **nitrogen cycle** (Figure 2.5, page 38) to deep-sea sediments, but this loss is just about balanced by additions from volcanoes. Humans interfere with the nitrogen cycle by producing large

quantities of nitrogen oxide gases, which are normally present in very small amounts. These gases contribute to the formation of smog (Chapter 21). In addition, excess nitrate from agricultural fertilizers can lead to overgrowths of algae, or eutrophication (Chapter 12).

Phosphorus cycle. The sea is a major part of a number of important cycles. One example is the hydrologic

NATURAL CARBON CYCLE

HUMAN INFLUENCES

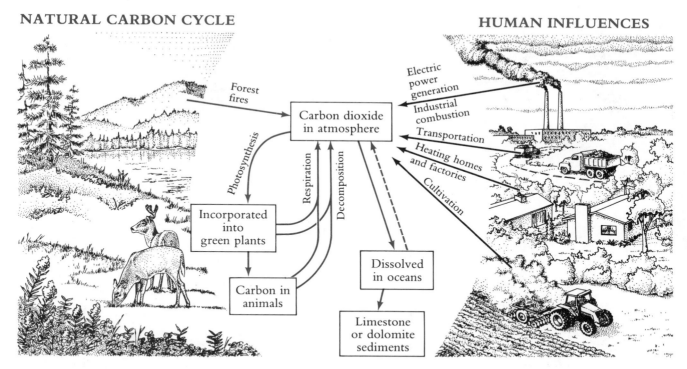

Figure 2.4 Part of the carbon cycle. (The complete cycle is discussed in Global Perspective V.)

cycle; another is the **phosphorus cycle**, shown in Figure 2.6 on page 39. Phosphorus enters the environment mainly from rocks or deposits laid down in past ages. Erosion and **leaching** gradually release the phosphorus. Some is used by biological systems, but much of it is washed into the sea, where it settles in the sediments. Phosphorus in shallow sediments is cycled by bottom dwellers through fish and then to birds, which deposit the phosphorus back on land as droppings or guano. But a portion of the phosphorus is not recycled; this is the amount buried in deep sediments.

Large quantities of phosphorus are used as fertilizer, and most is eventually lost as deep-sea sediments. At some time, though not in the near future, shortages could result. Phosphorus is an essential nutrient for plants and, along with nitrogen, is often in short supply compared to other nutrients in water. Phosphate, which enters natural waters from sewage, fertilizer runoff, and mining wastes, contributes to the problems of eutrophication (Chapter 12).

Sulfur cycle. In the **sulfur cycle**, sulfur is changed from one form to another by a variety of microorganisms in the soil and sediments (Figure 2.7, page 40).

Plants and animals require sulfur as sulfate (SO_4^{2-}) in order to build organic materials such as amino acids and proteins. In the deep sediments and soils, sulfur is changed to iron sulfides at the same time that phosphorus is changed from an insoluble to a soluble form. In this way, the two cycles interact.

Human activities, notably the burning of high-sulfur coal to generate electric power, are affecting the sulfur cycle by overloading the atmospheric pool of sulfur with sulfur dioxide, a compound normally present at very low levels. (The effects are described in Chapter 20.)

Water, carbon, nitrogen, phosphorus, and sulfur are not the only materials that cycle in ecosystems, but from the point of view of their significance to living organisms and their relationship to current environmental problems, they are the most important.

Biological Magnification

Some pollutants not only cycle in the environment but also tend to accumulate in living organisms. In these cases, the concentration of a pollutant found in living organisms increases as the pollutant is passed up the

NATURAL NITROGEN CYCLE   HUMAN INFLUENCES

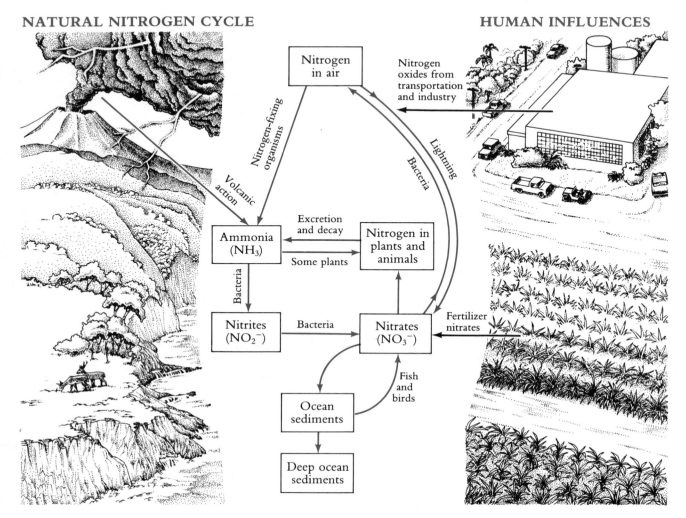

Figure 2.5 The nitrogen cycle. Plants and animals generally require nitrogen in the form of nitrates. A small amount of nitrate is formed from atmospheric nitrogen by lightning. However, most nitrogen enters living systems through bacteria and algae that can "fix" nitrogen gas from the air into ammonia. Biological matter is broken down into ammonia, which is cycled by bacteria through nitrites and back to nitrates. Some of the nitrates are lost to the cycle when they are buried in deep sea sediments, while nitrogen is added to the cycle by volcanic action. Certain bacteria can "denitrify" nitrates back to nitrogen in the air. Humans influence the nitrogen cycle by adding nitrates in fertilizers and nitrogen oxide gases from transportation and industry.

food chain. This happens when organisms take in a pollutant faster than they eliminate it. Mercury, for instance, may be present at relatively harmless concentrations in water or bottom muds but may be concen-

trated to lethal levels in shellfish growing in the water (see Chapter 14). Pesticides such as DDT are also affected this way: They may be almost undetectable in the water but present in greater and greater concentra-

NATURAL PHOSPHATE CYCLE   HUMAN INFLUENCES

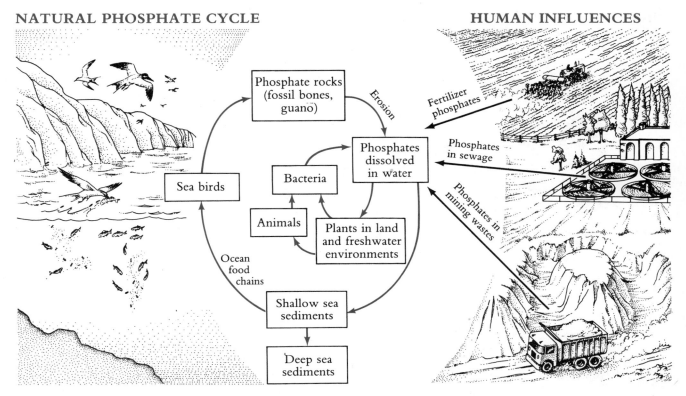

Figure 2.6 The phosphorus cycle. Erosion and mining release phosphates from rocks and other deposits to water environments. Phosphates are eventually deposited in shallow or deep sea sediments. From shallow sediments, some phosphate is recycled back to land environments.

tions at higher trophic levels. This phenomenon is called **biological magnification**. We will explain it more fully when we discuss pesticides in Chapter 7.

Energy Flow in Ecosystems

In contrast to the cyclic flow of materials we have been discussing, energy flow on earth is more like a one-way street. The energy in ecosystems comes from the sun and is finally lost as heat to the cosmic reaches of outer space.

The flow of energy through systems is of major concern to ecologists and should be of interest to everyone else, too. An understanding of energy flow, and the basic laws that govern it, leads to a better understanding of why natural systems work the way they do. Equally important, such an understanding helps us see

the limits of our ability to make changes in our environment without damaging it.

Figure 2.8 on page 41 is a simplified diagram of energy and materials flow in a stream ecosystem. The energy from the sun is the driving force. Green plants capture the sun's energy by photosynthesis and use this energy to produce organic materials. The energy stored in these materials serves as the energy source for the primary consumers that eat the plants and for the higher consumers that eat primary consumers, and so on. Even in ecosystems called "incomplete" because they lack the producers that capture sunlight, if we follow the energy source back far enough, it will be found to be the sun. For instance, in the deep sea bottom, where no light penetrates, photosynthesis cannot occur. Instead, the energy source is **detritus**. But this detritus originates from organic materials produced in

NATURAL SULFUR CYCLE   HUMAN INFLUENCES

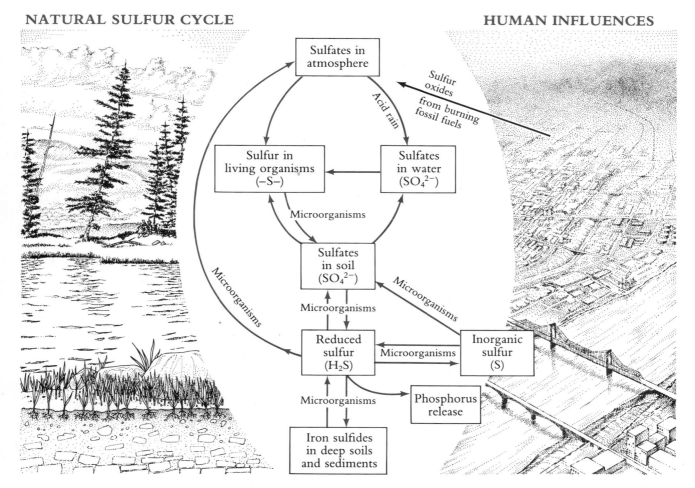

Figure 2.7 The sulfur cycle. There is a large pool of sulfur in soil and sediments and a smaller pool of sulfur in the atmosphere. Plants and animals use sulfur (as sulfates) from soil, water, and air. This sulfur is then incorporated into organic sulfur compounds, such as sulfur-containing proteins. Sulfur is changed from sulfates to reduced sulfur compounds and back again by a variety of microorganisms in the soil and sediments.

the upper layers of the ocean, where there is light and photosynthesis.

Productivity

The amount of organic matter produced in an ecosystem or by an individual in a given time is its **productivity**. Productivity is measured in a variety of ways. It is often measured as the *increase* in weight of biological matter for a given area over a given time period. (This includes all the new plant tissue, roots, leaves,

and so on as well as the increased weight of animals during the time specified.) Productivity may also be stated as a number of calories for a given time period, when measuring the energy captured by green plants and then transferred from one trophic level to another in an ecosystem.

In contrast, the **biomass** of an area or ecosystem is the *total* weight of all living material, plant or animal, in that area at a specified moment. For instance, the productivity of a forest ecosystem might be measured during a year's time. The biomass at the end of the year

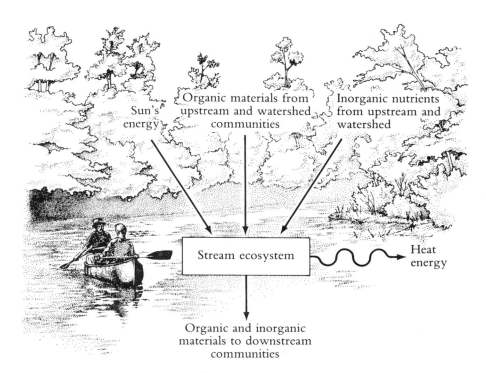

Figure 2.8 Energy and materials flow for a stream ecosystem. Living systems are open systems: materials and energy flow into the system from its surroundings and out again into the environment.

would include not only that year's productivity but also plant and animal tissue (such as tree trunks) produced in previous years.

The productivity of an ecosystem can be further divided into primary and secondary production. **Primary productivity** is the amount of organic material produced by green plants (and some bacteria) using the sun's energy. **Secondary productivity** is the organic material produced by organisms that are not photosynthetic—that is, fungi, most bacteria, and animals. These organisms consume organic matter from primary producers to gain energy and materials for their own bodies.

Ecological pyramids. Ecologists often draw **ecological pyramids** to show the amount of energy transferred from one trophic level to another in an ecosystem. Figure 2.9 on page 42 shows an imaginary and highly simplified system in which alfalfa plants in a field use sunlight to produce plant tissue, which is eaten by calves, which in turn are eaten by a boy. The first pyramid shows the biomass, or weight of living material, at each level. The second shows the yearly energy transfer in calories through the system.

These particular pyramids clearly illustrate how much primary productivity is necessary to provide the

yearly protein consumption of just one boy if his requirement is met by meat alone. The implications of this in terms of the world food supply are discussed in Chapter 8.

Availability of produced material. Several factors are not at all clear from illustrations such as Figure 2.9. In the first place, we should distinguish between gross and net productivity. **Gross productivity** is the total amount of organic material produced by plant photosynthesis in a given time period. Part of this, however, is used up by the plants themselves in respiration. The amount left is **net productivity**. This is actually what we measure, and it is the more important figure because only net productivity is available to the next trophic level.

Second, the available organic material is not usually completely consumed by the next trophic level. As a rule, primary consumers utilize only 5%–15% of the primary production. There are several reasons for this. One is that some of this primary production is simply not available. In a tropical rain forest, for instance, most of the primary production is in the leaves of the tree canopy high above the forest floor, largely unavailable to browsing animals. Another reason is that dry or cold seasons create a variation in primary productivity:

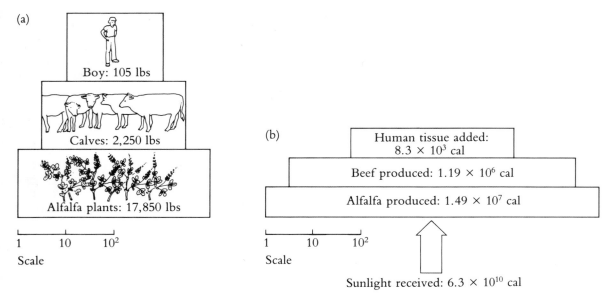

Figure 2.9 Ecological pyramids for an alfalfa–calf–boy system, calculated for 10 acres and 1 year. (a) The biomass of the living material at each trophic level; (b) the productivity in calories at each level. (Note that both pyramids are drawn using a logarithmic scale in order to accommodate the large values at the producer level.) (Adapted from Odum, *Fundamentals of Ecology*, 3rd ed. [Philadelphia: Saunders, 1971].)

When it is cold or dry, primary productivity drops. The number of animals that can live on primary production is governed by the availability of food at the worst times. There simply are not enough animals in an ecosystem to consume all primary production during periods of high productivity; however, this effect is partly offset by migrations of consumers into areas where productivity is high. Even that portion of the available food actually eaten by animals is not all turned into flesh to be eaten by the next trophic level. Some part of the food is not digestible, and part of the energy in the food that is digested must go to maintain normal vital processes. Finally, energy transfer is not 100% efficient because energy is lost as heat (this will be explained later). In general, only 1%–10% of the food eaten by animals is used to form new body materials.

All of these losses are summarized in Figure 2.10, which is a more detailed energy flow diagram than Figure 2.9. Such a diagram is sometimes called an *energy budget*, since it attempts to account for all of the energy entering and leaving a system. Even this complicated-looking diagram is not as complex as a real ecosystem. Several things are not shown in the diagram. For instance, small herbivores (plant eaters) probably migrate into or out of such ecosystems from other areas. Further, some of these small herbivores are eaten by secondary consumers, who are eaten by tertiary consumers, and so on. Note that all of the unused organic material eventually enters the food chains responsible for the decomposition of dead material. On land, these are mainly in the soil; in bodies of water, they are usually found toward the bottom.

Thermodynamic Laws and Ecosystems

Energy budgets follow the **first law of thermodynamics**, which states that energy can be changed from one form to another, but it cannot be created or destroyed. Thus, all the energy entering an ecosystem must balance with the amount remaining there plus the amount leaving the system, as is done in Figure 2.10. (Note that the amount of energy consumed by "other herbivores" does not quite balance the amount lost as heat and to decay processes. The remainder represents the "other herbivores" eaten by secondary consumers, which are not shown in the diagram.)

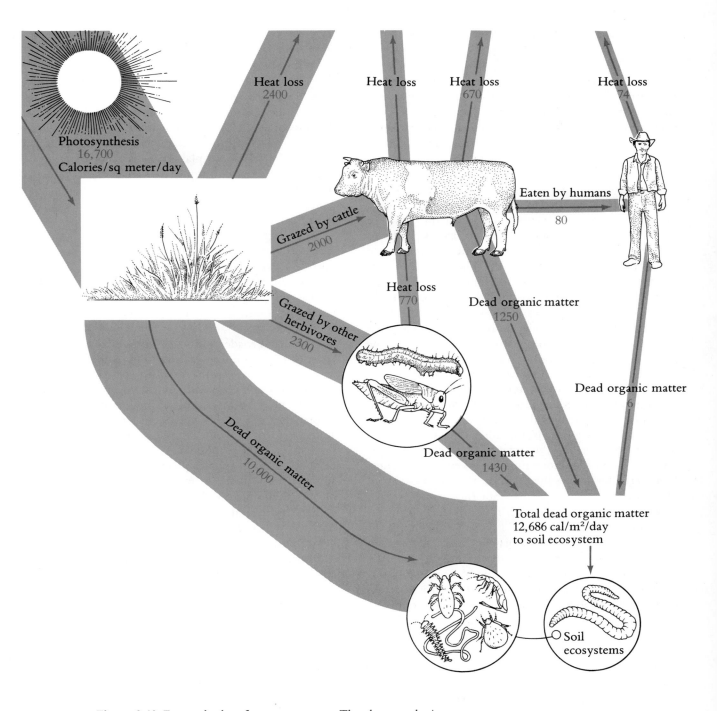

Figure 2.10 Energy budget for a cow pasture. The photosynthetic activity of the grasses captures 16,700 cal/m²/day. The energy flow through the system is shown. Note the large proportion of energy that enters detritus food chains in the soil. Note also the proportion of energy lost as heat at each stage of transfer. Eventually the 12,686 calories that enter soil food chains will also be lost as heat when the organic material containing this energy is completely broken down. Note that not all energy flow is accounted for in this simplified diagram—see text. (Adapted from H. Remmert, *Ecology* [New York: Springer-Verlag, 1980], p. 211.)

The first law of thermodynamics explains certain facts about the environment. For instance, it can help us understand limits on food production. Food contains stored chemical energy used by living organisms. When we eat food, it is broken down in our bodies as the stored energy is released. Some of the energy is captured and stored in body structure, some is lost as heat, and some powers the reactions essential to life.

In order to increase food production, especially in the developed world, farmers use great quantities of synthetic fertilizer. However, this fertilizer is produced by using fossil fuels, both as raw materials and to run the manufacturing process itself. That is, we are increasing the production of one kind of energy—the food energy in plant crops and the animals that eat them—by using up stores of high-energy fossil fuels. Similarly, synthetic foods cannot be manufactured without using some other energy resource to run the process—energy cannot be created.

This does not mean that no improvements can be made in the efficiency of food production. A major effort involves breeding plants in which more growth takes place in parts useful to humans as food, such as the kernels of a wheat plant rather than the stalk. In plant breeding, as in other sciences, however, ecological energy considerations must be taken into account. Remmert (1980) gives the example of attempts to breed plants that can fix nitrogen from the air rather than requiring soil nitrates (which in many farming areas are supplied in the form of expensive fertilizer). In terms of energy costs to the organism, fixing nitrogen costs more than the uptake of nitrate from the soil. This energy must come from some other biological process. Remmert speculates that if the experiment is successful, the plants will be found to be much less productive than the original variety, which could have the effect of cancelling out the cost savings in reduced fertilizer use.

While explaining ecological pyramids, we mentioned that energy transfer reactions are not 100% efficient—some energy is lost as heat. In this way, living systems obey the **second law of thermodynamics**. This law was first discovered by physicists attempting to turn heat energy into useful work using machines such as steam engines. They discovered that heat could never be converted 100% into useful work (more on this in Chapter 15). Similarly, in living systems, energy gained from breaking down energy-rich food substances is not all available for use in other reactions—some energy is always lost as heat.

This is one reason why few individuals (such as large carnivores) stand at the top of food chains compared to the vast numbers of individuals at the bottom. The loss of energy at each step in the food chain makes it impossible for more than a few individuals to be supported. This also explains why it is not energy efficient to eat from the top of food chains, as humans often do. When we eat meat or fish, we eat food that contains only a small part of the energy with which the food chain began.

The large amount of energy lost as heat during energy transfers in the food web must be made up from outside sources (sunlight, detritus) or the whole ecosystem will run out of energy and collapse. Materials such as minerals or water cycle in the environment, but energy does not. Energy flows into ecosystems and is lost as heat.

A much broader and sometimes more useful statement of the second law of thermodynamics is that all systems tend to become random, or disordered, on their own. For instance, if you dump a bag of marbles onto the floor, they will spread out spontaneously in a disorderly fashion. The release of air pollutants, such as those formed when coal is burned (Chapter 20), demonstrates the second law in action. The pollutants, released from a power-plant smokestack, disperse spontaneously in the atmosphere. We actually depend on this phenomenon: As the pollutants disperse, they become more dilute and less dangerous. Of course, this sort of system only works up to a point. As populations grow, power demands and pollutant levels increase, and dispersal or dilution of pollutants becomes a less and less effective solution.

On their own, the marbles we spill will not regroup and hop back into the bag to restore order in the room. We need to expend energy hunting for them and putting them back in the bag. The formation of order is a nonspontaneous process requiring energy, whether someone is cleaning up marbles from a floor or an organism is growing by the orderly arrangement of molecules.

When we first look at living organisms, it may seem that here is a system becoming more orderly spontaneously. A growing, living organism, considered alone, might seem to fit this description. However, at the same time that the organism is growing in an orderly fashion, biochemical reactions occurring inside it are producing heat. This heat, which is given off into the environment, increases the random motion of molecules in the organism's surroundings. Thus, the organism plus its environment is still tending spontaneously toward disorder.

Using the second law, we can project some of the effects of population growth. Increasing numbers of people, who themselves live and grow by the production of order, will cause increasing levels of disorder in their environment. Air pollution caused by increased power demands of growing populations is one example. Another is the erosion of soil from land that was once part of tightly knit forest ecosystems, after the forest has been cleared to plant crops or after the land has been turned over in the search for minerals.

Although the laws of thermodynamics are important in understanding relationships between organisms and the effects of organisms (including humans) on their environment, most people fail to take them into account. A study of how these natural laws operate with respect to power generation, for example (see Chapter 15), can help us find ways to minimize our impact on the environment.

Population Dynamics

Up to this point we have been considering individual parts of ecosystems and some of the processes that go on between these parts. Now we have enough information to consider how these processes affect the size and distribution of the various populations in ecosystems—that is, population dynamics.

Age Distribution Graphs

In many natural populations, if we compare the numbers of individuals at various ages, we obtain a graph resembling Figure 2.11(a). In such a case, younger individuals are present in much greater numbers than older ones. The death rate of the young organisms is very high compared to that in other age groups. Thus, the numbers of individuals that survive to enter the older age categories are relatively small. This is shown in the form of a bar graph in Figure 2.11(b).

The shape of an age distribution graph depends on the specific population under study. For instance, the female Pacific herring produces some 8,000 eggs each year, of which 95% hatch. Only 0.1%, however, live to maturity. A bar graph of this population would show a very broad base and a very narrow top. On the

other hand, a population distribution for species such as elephants and whales, which produce only one offspring—and that not every year—would not have as wide a base and would have a larger percent of the population in the adult age categories.

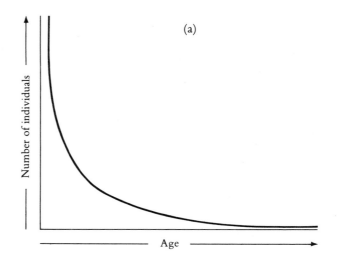

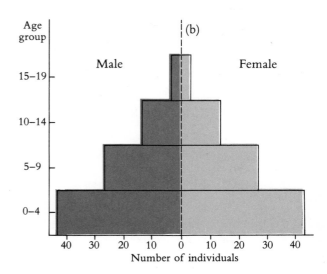

Figure 2.11 Age distribution graphs. (a) Generalized curve showing numbers of individuals in a population versus their age group. (b) Bar graph of this population type, characterized by high rates of birth and by high death rates in the younger age categories.

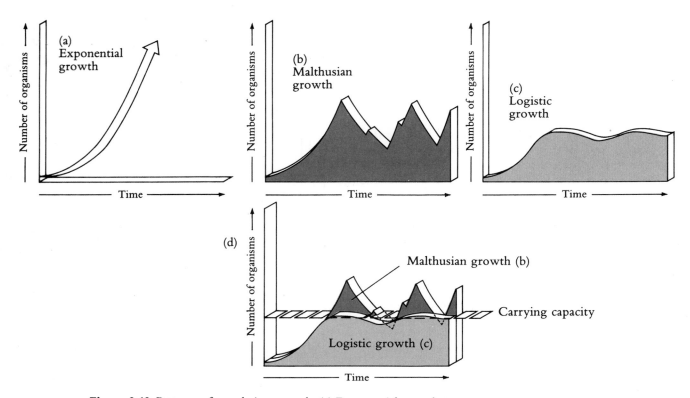

Figure 2.12 Patterns of population growth. (a) Exponential growth typical of organisms colonizing newly formed habitats. (b) Malthusian or irruptive growth pattern, in which population growth is limited by disastrous population crashes. (c) Logistic growth, in which the population gradually levels off. (d) Malthusian and logistic growth are superimposed on each other for comparison. Also shown is the carrying capacity of an environment, or the largest population that environment can support without being damaged. Note how the population showing Malthusian growth tends to overshoot the carrying capacity and then fall in a series of sharp population crashes. Populations growing logistically, however, level off smoothly around the carrying capacity of their environment.

Population Growth Curves

If a population of organisms grows under conditions in which essentially no death occurs before organisms become sexually mature, growth is exponential. A curve of such growth would resemble Figure 2.12(a). Exponential growth occurs in natural systems when organisms colonize a new habitat in which they have no competitors. However, it is obvious that this type of growth does not continue as the curve does, because otherwise such an organism would soon cover the earth completely.

Malthus recognized this fact in the late 1800s. He proposed that populations grow until they use up the resources available to them, and then growth is limited by catastrophes such as famine, disease, and violence. The growth curve thus levels off into a series of fluctuations, as shown in Figure 2.12(b).

Some animal populations do seem to follow this pattern of explosive growth and then collapse. In 1944, for instance, 29 reindeer were brought to St. Matthews Island in the Bering Sea. By 1963, there were 6,000 deer on the island. Immediately thereafter, however, the population suffered a catastrophic decline to less than 50 animals. Such growth is called irruptive or Malthusian.

Other populations do not seem to follow such a growth pattern but instead show a similar initial exponential increase followed by a smooth leveling off over a period of time, at a level often referred to as the carrying capacity of their environment. (Carrying capacity is discussed more fully shortly.) Organisms with simple life histories demonstrate this type of growth, called logistic growth, very well—see Figure 2.12(c). As life histories become more complex, however, no curve may fit a population's growth patterns very well, or a particular curve may fit only at certain periods of time.

r *and* K *Selection*

Habitats such as the Montana Rocky Mountains and the eastern deciduous forest, where attempts are being made to reestablish peregrine falcon populations, are relatively stable over long periods of time. That is, unless humans intervene, conditions do not change rapidly (with the notable exception of natural disasters). Most of the organisms in these ecosystems have adapted in ways that allow them to make the maximum use of their environment without harming it. In other words, these organisms are adapted to the capacity of the environment. A growth curve for such a population of organisms might resemble Figure 2.12(c), the logistic growth curve. There are other habitats, however, that develop rapidly and disappear again—for instance, seasonal ponds that fill with water only during rainy seasons. The growth curve for organisms colonizing such habitats commonly resembles Figure 2.12(a).

Organisms characteristic of stable habitats are often termed *K*-selected, where *K* is a symbol for the capacity of the environment. *K*-selected organisms are generally long-lived, have few offspring, and are successful in situations where they must compete for scarce resources. In general, trees and animals with long life spans are *K* strategists. A second type of organism, characteristic of fleeting environments, is called *r*-selected, where *r* refers to their high rate of reproduction. In addition to showing a high reproductive rate, these organisms generally develop rapidly. They are not good competitors, however, and can establish themselves only where there is little or no competition. Small animals and the plants that first colonize

habitats after a disaster (pioneer plants) are *r* strategists.

Of course, there is no neat division of all organisms into one or the other type. Many organisms are better classified somewhere between the two categories. Furthermore, even stable habitats have subdivisions that favor *r* strategists. The bodies of organisms that die in forests, for example, provide a habitat for *r* strategists (such as fly maggots) in an area where *K*-selected species are usually favored.

Factors Determining Carrying Capacity

Populations have certain space requirements that ensure such things as adequate food supply, space to seed or to raise young, the availability of reproductive partners, and a necessary minimum of stress-producing encounters. The **carrying capacity** of an ecosystem is the maximum-sized population for which the ecosystem can provide these requirements for an indefinite period of time. Carrying capacity is shown in graphic form in Figure 2.12(d). We might now ask, what are the factors that determine the carrying capacity of an environment for a particular species, and how do organisms in nature adjust to this capacity?

Density-dependent and density-independent factors. Most, if not all, factors that adjust population size and carrying capacity are density-dependent; that is, they may kill off 80% of a dense population but only 10% of a sparse one. An example would be a disease that spreads quickly in dense populations with many contacts between individuals but that does not make much headway in a sparse population whose individual members rarely come in contact.

Extreme climatic conditions sometimes act as density-independent factors in population regulation. For instance, a severe winter frost in the intertidal zone can cause rock surface to crumble away, removing snails that cling there without regard to the number of snails present. In many other cases, however, climatic factors act in a density-dependent manner. For example, during a flood, a larger percentage of individuals in a sparse population may find refuge (such as higher ground) than in a denser population.

Density-dependent factors controlling populations can work either by decreasing the birth rate in a population or by increasing the death rate.

Predator-prey relationships. Predator-prey relationships appear to play a part in controlling population size. The example given of a disease controlling a population size once it reaches a certain level can be considered a special case of predator-prey relations. The disease is the predator and the host the prey. However, each system must be studied to determine whether the predator is controlling the prey or vice versa. One famous example is the 9-year population cycle of alpine hares and the lynxes that prey on them. The alpine hare population peaks in approximately 9-year intervals, and each peak is followed by a peak in the lynx population. But as the lynx population peaks, the hare population crashes. This was originally interpreted as an example of a species (the lynx) exceeding the carrying capacity of its environment by consuming too much of its food supply, following which the lynx population collapses and the whole cycle starts again. However, in regions where the lynx has become extinct, the hare population still cycles. Whatever the reason for this cycling, it is clear that the hare population was controlling the lynx and not vice versa.

On the other hand, predators much smaller than their prey (some scientists would then call them parasites) may be able to control the prey population. For instance, a European leaf beetle imported into this country successfully controls the introduced weed St. John's wort (*Hypericum perforatum*).

In areas of great species diversity, several predators may keep the population of a prey species much smaller than it would be under the same environmental conditions with fewer predators. Several species with similar ecological needs can then occupy the same area. This may be a partial explanation for the richness of species on coral reefs and in rain forests.

Food supply and food competition. On the face of it, food supply and the competition for food appear to be a likely limiting factor for populations. However, field researchers have not been able to show that population density varies in any consistent way with food supply. Probably food supply acts with other factors to limit populations.

Abiotic factors. The density of a population tends to decrease as abiotic factors such as climate or soil type become less and less optimal for the species in question. As an example, soil type has been found in some cases to be related to the quality of the food available. In

Scotland, basaltic soils rich in minerals support a much greater population of hares and red grouse than do mineral-poor granitic soils. Even though the amount of primary production on both soils is similar, the plants are richer in minerals on the basaltic soil and so provide a higher-quality food for wildlife.

Autoregulation. In some cases, population size appears to be regulated by factors other than those already mentioned. Scientists have described this situation as *autoregulation*. The stress caused by overcrowding may be a factor in autoregulation. Stress is normally increased by crowding because there are more aggressive meetings with other organisms. Autoregulation can take several forms; for example, parental care of offspring can improve or deteriorate depending on density, or mass migrations out of a habitat may occur at peak population densities.

Sometimes an actual physical basis for autoregulation behavior is known. That is, some biochemical reaction to crowding affects births or deaths or triggers behavioral patterns. As an example, crowding in tree shrews (*Tupaia glis*) causes an increase in blood urea concentrations. Offspring mature more slowly or not at all. Mothers no longer protect their young by marking them with a glandular secretion, and the young are then eaten by other tree shrews. Population size is kept constant in these ways, even if food is present in excess. If crowding is increased artificially—for instance, by decreasing the size of the cage in which the population is kept—blood urea concentrations rise even higher and some of the animals die. In the field, lemmings have been shown to have similar blood urea concentrations during peak population densities.

The factors we have mentioned cannot give us a complete explanation of either the carrying capacity of an environment or how populations adapt to that capacity. In fact, the science of ecology has not advanced to a point at which it could give a complete explanation. However, there are fascinating questions raised by the insights ecology does give us. Considering mechanisms by which animal populations appear able to adjust to the carrying capacity of their environment leads us to wonder what mechanisms might or might not adjust human populations to the carrying capacity of our environment. How do we determine this capacity? Will we recognize it before we have exceeded it?

Evolution and Competition

Even a short study of the fossil record reveals that the very existence of a species is a dynamic phenomenon. That is, a species is not a static entity, shaped in a particular mold and assured of the same niche for eternity. Rather, fossil history shows a pattern of change in species as they adapt to changing conditions. If species are for some reason unable to adapt, the fossil record shows clearly that they cease to exist; in other words, they become extinct. From fossil evidence it appears that species exist, on average, less than 10 million years. In fact, about 99% of the species that have ever existed are believed to have suffered extinction.

Despite this record, some 1.5 million species have been identified in today's world. Even this number, however, is thought to be only a fraction of the mind-boggling number of total species that exist on earth—perhaps 5 to 10 million. In the following sections we will consider briefly some of the ideas about how evolution has led to such a great diversity of organisms. We must understand something about this natural process before we examine the effects humans have on the dynamic diversity of natural communities.

Evolution is one of the most exciting areas in biology today. Theories are being devised, torn apart, and revised. There is a great deal of argument, and as yet there are few points on which everyone agrees. Information on this topic comes from a variety of sources. One source is the field of genetics, in which the hereditary material (DNA or genes) of organisms living today is studied. Another source is the field of paleontology, in which fossils illustrating past evolutionary change are unearthed and examined. A third source is the field of ecology, in which the relationships between species alive today are investigated and relationships between species from the past are deduced.

Forces Driving Evolutionary Change

The most commonly held beliefs about the evolution of a diversity of life forms involve two major factors. The first is the adaptation of species to use new habitats. As climates have changed through the eons, as continents have drifted apart and come together, as mountains have risen or worn away, new habitats must have become available for organisms to colonize, and old ones must have become much less hospitable than they once were. These pressures, acting through natural se-

lection of the fittest organisms, could have caused species to change in character over long periods of time. In addition, when geologic changes cause separation of populations of the same species, new species can develop. On the other hand, the same factors of climatic or geologic change must cause some species, unable to live under the new conditions, to become extinct.

But, there is a second factor implicit in ideas about evolution: competition. If environmental change is the force that sets the stage for species to form, competition is the theme of the play itself. It is competition within a species that determines that the fittest individuals will survive to reproduce and that thus determines the direction of a species' evolution. It is competition between species that determines which species will survive to reproduce and which will become extinct. Furthermore, competition between species can be viewed as the force driving them to use new niches in order to avoid extinction. Thus, competition is seen as a major determinant of community structure today as well as an evolutionary force.

Climatic change. Although the idea of environmental change as a driving force in the evolution of species has great logical appeal, real fossil evidence for it is only now being collected. This is partly because in the fossil record extinction of species and the appearance of new species are hard to distinguish from the migration of various species into or out of an area. One location where fossil evidence for climate-caused evolution is fairly strong is in sediments from the Siwalik Hills in Pakistan, laid down from 1 to 18 million years ago. Here, scientists can see the sudden disappearance about 7 million years ago of a whole related group of apes. This disappearance coincides with the beginning of a drier, cooler climate.

Competition. Many ecologists believe that predation and competition for resources such as space or food are the two main forces that determine what species are found in particular communities. How well a species meets the challenge of being preyed upon is a factor that influences how abundant that species is in a community. Similarly, how well a species competes with the other species present in a community determines whether it will gather enough of the resources it needs to survive and reproduce.

Predation is sometimes said to be less important

than competition because predation pressure tends to lessen as a species becomes less abundant. (The predator itself becomes scarce as a result of starvation, or it turns to other prey.) Unlike predation, however, competition between two species can be unremitting and can lead to the loss or exclusion of the weaker species from a community. This is called the *principle of competitive exclusion.* According to this theory, two organisms cannot live in the same habitat and use the same resources—that is, occupy the same niche—indefinitely. Eventually one, the better competitor, will eliminate the other. This principle can actually be demonstrated in the laboratory with single-celled organisms, such as paramecia in a jug of growth medium. However, because of the complexity of natural ecosystems, in the real world it is not so easy to show that a species has disappeared as a result of competition as opposed to some other cause.

Competition theory—the body of evidence and conjecture that attributes the structure of communities and pressure for evolutionary change to competition between individuals and between species—has been a major part of ecological thought for many years. Recently, however, some scientists have questioned whether competition between species is really that common or that important in natural communities.

Competition and Evolutionary Strategies

The driving force or evolutionary "push" that competition exerts might work either directly or indirectly. A species could evolve directly toward an improved competitive position by becoming the stronger competitor for a given resource. For instance, a plant might evolve in the direction of improved absorption of essential nutrients such as dissolved phosphate in the soil. This sort of strategy has obvious physiological limits, however. The plant will eventually approach a maximum rate of movement of chemicals.

A more indirect method of improving competitive position would be to begin using a different resource or part of a resource that is not presently being used. In this strategy, competing species in a given habitat evolve so that they use different spaces, different resources, or the same resources but at different times. For instance, several bird species may continually occupy and find food in the same trees, but one species will occupy the topmost branches, another the middle ranges, and another the bottommost branches. Bats

and swallows would ordinarily compete directly for insects, but swallows feed on the insects until dusk while bats feed only at night.

Relief from competition is also obtained when organisms evolve ways to "mark" a particular resource space as their own. Certain trees species excrete chemicals that prevent the growth of other species or individuals nearby. Symbiosis, in which two species living in close touch provide each other with a material or a service, can be viewed as a way of improving the competitive advantage of one or both partners. The tree that provides a home in its roots for nitrogen-fixing bacteria receives in return a necessary nutrient that improves its competitive position in a nitrogen-poor environment.

In all these ways, then, competition is believed to exert an evolutionary pressure on organisms to change, to diversify, in fact to become new species in different habitats or to develop a different set of ecological requirements in the same habitat (a new niche). If a subpopulation of a species is separated by some geographical barrier, such as mountains or bodies of water, it may change (evolve) so that it becomes specialized to the conditions found in a particular subpart of its environment. If it can no longer breed with the original members of the species, a new species has arisen.

Competitionists point to the finches of the Galapagos Islands as proof of their theory. There are 14 species found in that habitat and all 14 are believed to have evolved from one ancestral species. The birds differ in food habits—all the way from seed eating to blood drinking—and in body characteristics such as beak size and shape. Certain pairs of closely related finches are never, or almost never, found on the same island. This is considered to be evidence for the principle of competitive exclusion. In one case where closely related finches do coexist, they occupy different parts of the island. Furthermore, one of the species has shifted from eating mainly seeds, which it does on islands where it exists by itself, to eating mainly insects. Competitionists believe this to be an example of a shift in the use of resources caused by competition between species.

Alternatives to Competition Theory: Predation

As neat as this theory is, not all ecologists subscribe to it. That is, most ecologists believe that competition exists, but not all believe it is the major force in either the present structure of communities or in evolutionary change. In fact, many believe that predation is more important than competition.

Several field studies seem to show that for some species, predators do keep populations well below the level at which they would need to compete for resources. A classic study involved predatory starfish (Schoener, 1982). When scientists removed starfish from intertidal regions, there was a decrease in the number of species the starfish had been preying on. This seemed to be due to competitive exclusion of some of the prey species by others. Such competition had been kept down by the starfish, which had prevented the prey species from using all their available resources. Thus in this community, competition had an effect only when predation was absent.

The Variable Environment Alternative

Other ecologists believe that environmental change, which occurs often and without regard to any biological interaction (examples would include floods, hurricanes, cold spells, drought), acts to keep populations well below the level at which competition will occur. These scientists note that it is true that periods of scarcity of resources (such as food) may occur, thus causing competition among species. However, they argue that the more common periods of relative plenty that follow scarcity obscure any evolutionary effects. That is, any characteristics that might have given a competitive advantage to certain individuals during the period of scarcity would no longer represent an advantage during the normal times in which resources are abundant compared to population size. These characteristics would just become part of the normal variation in characteristics within a species.

Examination of natural communities to determine whether competition is occurring (or whether it occurred in the past, leading to the exclusion of some species) is a field of study currently fraught with controversy. It appears that the amount of competition between two species can vary from year to year in some areas, while competition can be a relatively constant fact of life in others. Even in the same environment, abiotic factors such as climate and fire can keep some species well below the level at which competition would occur, while at the same time competition is taking place between other species. In other words, the complexity of the situation regarding competition is pretty much what one might expect from the complexity of the natural world!

Punctuated Versus Gradual Evolution

Basic to the theory of competition as an evolutionary driving force is the idea that natural communities are in equilibrium. That is, amounts of resources compared to the numbers of organisms remain in balance from season to season or year to year. Whether great or small, all of the available resources are used by a dynamic balance of the species present. These species are in constant and unremitting competition with each other for resources. Further, competition theory predicts a gradual evolutionary change, a fine honing of competitive edges, resulting from this constant pressure.

In some environments, however, the fossil record does not support such gradual changes. Instead, scientists studying the fossils theorize that species remain the same for millions of generations even in the face of environmental change such as cooling of the climate or decreases in rainfall. These scientists believe that evolution does not take place by gradually accumulated changes in a species that, bit by bit, help it adapt to a changing environment. Rather, they feel evolutionary change is abrupt. New species form suddenly and remain the same for a long time. This is called *punctuated evolution*. Such a view of communities and evolution has interesting implications for ecology and environmental science. The idea of communities as perfectly balanced groups of species perfectly adapted to their environment may not always be the case. Natural communities may not be "a perfect and fragile blend of irreplaceable species" (Bakker, 1985).

On the other hand, the concept of equilibrium implies a natural balance of forces that act to correct a push in the population from one direction or another. Equilibrium systems are in balance and tend to remain in balance. If natural systems are often not in equilibrium, natural communities may have less resilience to environmental change than some people have thought. Until we understand more about evolution and what drives species toward extinction or toward better adaptation to their environment, we must be very careful about the environmental pressures we exert on natural systems.

Summary

Ecosystems are not fixed but rather dynamic or changing. Materials such as water, carbon, nitrogen, phosphorus, and sulfur tend to cycle in ecosystems, while energy flow is in one direction. Productivity is a measure of the organic material produced by a system or an individual in a given time period. Ecological pyramids, which show the productivity of various levels in the trophic structure, illustrate that energy transfer from one level to another is quite small. This is partly explained by the second law of thermodynamics, which states that energy transfers are never 100% efficient. The first law of thermodynamics, which holds that energy can neither be created nor destroyed, reminds us that all the energy in a particular system can be accounted for and also that schemes for increasing the production of food energy will always require an input of energy from some other source, such as fossil fuels.

Populations can be characterized by their growth curves and age distribution curves. A population that grows until it exceeds the carrying capacity of its habitat and then crashes is said to exhibit Malthusian growth. A population that grows until it levels off at the carrying capacity of its environment is said to exhibit logistic growth. Stable environments characteristically support *K*-selected organisms, which are good competitors adapted to the carrying capacity of their environments. Fleeting environments encourage *r*-selected organisms, which have higher reproductive rates but are poor competitors.

The carrying capacity of an environment, or the size of populations that can be supported indefinitely, is determined by density-independent factors, such as some climatic disasters; by density-dependent factors, such as disease, that act more strongly as a population increases in density; by abiotic factors, such as soil quality; and by autoregulation.

Species are not fixed entities. They can be shown to have changed, become extinct, or arisen over time due to environmental change and the interaction between organisms and species. Competition between individuals and between species appears to be and to have been an important factor, but not the only factor, in shaping community structure and evolutionary adaptation to environmental change.

Questions

1. Illustrate the cyclic nature of the movement of materials in the ecosystem, using water, carbon, nitrogen, phosphorus, or sulfur. How does the flow of energy differ from such cyclic processes?
2. How do the first and second laws of thermodynamics operate in ecosystems?
3. Explain what is meant by exponential, Malthusian, and logistic growth. What is the relation of these various growth patterns to the carrying capacity of an environment?
4. What forces shape community structure and influence evolutionary change?

Further Reading

The following readings all deal with competition theory and evolution.

Lewin, R. "Finches Show Competition in Ecology," *Science* **219** (March 25, 1983), 1411.
Lewin, R. "Santa Rosalia Was a Goat," *Science* **221** (August 12, 1983), 636.
Lewin, R. "Punctuated Equilibrium Is Now Old Hat," *Science* **231** (February 14, 1986), 672.
May, R. M. and J. Seger. "Ideas in Ecology," *American Scientist* **74** (May–June 1986), 256.

References

Bakker, R. T. "Evolution by Revolution," *Science 85* (November 1985), 80.
Remmert, H. *Ecology*. New York: Springer-Verlag, 1980.
Schoener, T. "The Controversy over Interspecific Competition," *American Scientist* **70** (Nov.–Dec. 1982), 586.

CHAPTER THREE

Human Population Problems

Growing and Changing Populations

*Population Growth Through History/Developed Versus
Developing Nations/Birth Rates, Death Rates, and the
Demographic Transition/Components of Rapid Population
Growth/Causes of Rapid Population Growth—Solving
the Mystery/Population Profiles/Population Momentum/
Total Fertility Rate*

Effects of Rapid Population Growth on
Health and Welfare in Developing Nations

*Food Production/Health Consequences/Educational
Consequences/Environmental Consequences*

Population Growth and Economic
Development in Developing Nations

*Population Growth as a Block to Economic Development/
Social Inequalities and the Reforms Needed for
Development*

Population Growth in Developed Nations

Resources and Technology/Effects of Crowding

Rural to Urban Migration

*Advantages of Cities/Disadvantages of Cities/Reversing
the Trend*

Migration Between Countries

Benefits of Immigration/Immigration to the United States

CONTROVERSIES:
3.1 *Malthus, Population, and Resources*
3.2 *Immigration*

Growing and Changing Populations

The rapid growth of human populations, sometimes called the population "explosion," appears in some ways to be a mystery. Families in poorer countries do not always respond to poverty by having fewer children. Instead, they often want many children. In some of these poorer countries the population will double in the next 20 years and may approach five or six times its present size in the next half-century. Such increases can only lead to further misery for these people. Meanwhile, in wealthier countries, families that could well afford to feed and clothe many children purposely limit their family size. As a result, populations in these countries grow only slowly or not at all.

How has this situation come about, and why does it persist? As in any real mystery, part of the explanation can be discovered by careful examination of the facts. Another part must be deduced from past history, and filling in the final pieces of the puzzle calls for an imaginative leap on the part of the investigator.

Population Growth Through History

Figure 3.1 shows the growth of the world's population through time. There were probably no more than 10 million humans in the world until people learned, about 8000 B.C., to domesticate animals and to grow and store their own food. After this, the human population grew more quickly.

Experts estimate that there were perhaps 300 million people in the world by the year 1. This population doubled to about 600 million over the next 1,500 years—a steady but slow increase interrupted by war, famine, and plague. Then, just before the start of the Industrial Revolution, a tremendous growth spurt began.

In the eighteenth and nineteenth centuries the world population grew at an estimated rate of 0.5% a year. In only 150 years, from 1750 to 1900, the world population doubled to 1.7 billion. But faster rates were yet to come. In only 30 years, from 1950 to 1984, the world population, achieving growth rates of 2% per year, doubled once again from 2.5 billion to almost 4.8 billion.

Developed Versus Developing Nations

The areas of the world in which population growth is taking place have shifted with time. From the middle 1700s well into the 1900s, populations grew in the countries we now call the **developed nations**, such as the United States and Great Britain. Since 1950, however, the major portion of world population growth has been taking place in the **developing nations**.

In general, the terms *developed* and *developing* refer to whether a country is industrialized (developed) or not industrialized (developing). Because industrialization leads to a higher standard of living, "developed versus developing" usually (although not always) means wealthy versus poor as well.

If a country's total economic production—called its gross national product (GNP) or gross domestic

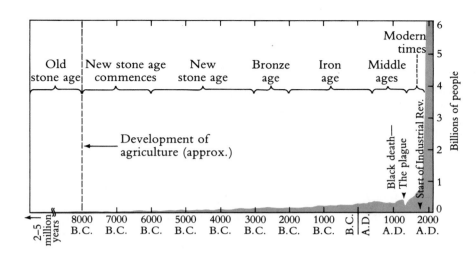

Figure 3.1 World population growth through history. There has been a steady increase in the human population since 8000 B.C., when agriculture became a significant way of obtaining food. However, the current enormous population growth spurt began much more recently, just before the start of the Industrial Revolution. (From Elaine M. Murphy, *World Population: Toward the Next Century* [Population Reference Bureau, Washington, D.C., 1981].)

product (GDP)—is divided by the population of the country, a number is obtained, the per capita GNP, that is a rough indicator of the economic status of people in that country. For countries classified as low income, the per capita GNP is less than $410 per year. India, for instance, had a per capita GNP of $260 in 1985. In comparison, the U.S. per capita GNP was $14,000 in 1985 (see Table 3.1).

In Table 3.1 countries are divided into developing or developed on the basis of how industrialized they are. However, a third grouping is necessary to accommodate oil-exporting countries in which a relatively nonindustrialized population still enjoys many of the same benefits found in developed nations. This is due, of course, to the large income derived from oil.

Several qualifications must be given about such per capita figures. In the first place, the income in any country is not evenly distributed. Some people benefit by a much greater share of the GNP than others. Thus, there can be, and are, very poor people living in countries with a high per capita GNP. Nevertheless, when differences in GNP are as great as those shown in Table 3.1, it is clear that vast differences in living standards must exist as well.

A second qualification is that the absolute difference in living standards between developing and developed countries is not quite as great as the difference in GNP. Not only incomes but also prices are lower in developing nations. Even if adjustments are made for this, however, the income gap between Indians and U.S. citizens was still estimated to be $8,000 per year in 1980.

Table 3.1 Per Capita GNP in 1985 for Selected Countries (in U.S. dollars)

	Per capita GNP
Developing countries: Low-income economies	
Mali	$ 150
Zaire	160
Uganda	220
India	260
China	290
Senegal	380
Developing countries: Middle-income economies	
Mexico	2,240
Egypt	700
Brazil	1,890
Singapore	6,620
Israel	5,360
Developed countries	
Sweden	12,400
United Kingdom	9,050
Switzerland	16,390
United States	14,090
Canada	12,000
High-income oil-exporting economies	
Kuwait	18,180
Saudi Arabia	12,180
Qatar	21,170

Source: Population Reference Bureau, *1986 World Population Data Sheet.*

Birth Rates, Death Rates, and the Demographic Transition

In the developed nations, **death rates**—the number of persons out of every thousand who die each year—began to fall in the middle of the eighteenth century. This decline was attributable to a variety of medical and sanitary improvements that we will discuss in the next section. The drop in death rates continued to the middle of the 1900s. Because fewer people out of every thousand were dying each year, populations in these countries began to grow. Such growth did not continue indefinitely however, because **birth rates**—the number of births each year per 1,000 people—also began to decline.

Those countries in which both the birth and death rates have fallen are said to have undergone the **demographic transition**. Most industrial nations have already undergone this transition. Further changes in birth and death rates in these countries in the near future are likely to be small.

It would be easy to conclude that birth rates in developed nations fell *because* people became aware of the decline in death rates. In other words, people could see that more children would survive to adulthood and so chose to have fewer children. However, there is evidence that birth rates actually began to fall *before* the medical and sanitary advances that lowered death rates. Further, this seeming paradox is an important clue to why rapid population growth continues in developing countries today.

Components of Rapid Population Growth

In countries all over the world, death rates are being lowered by public health measures, by improved medical care, and by improvements in nutrition.

Diseases such as malaria and yellow fever, which are spread by mosquitoes, are prevented by the use of insecticide sprays and by draining and filling the swamps where these insects breed. Cholera, typhoid, and other intestinal diseases are prevented by chlorinating and filtering public water supplies. The pasteurization of milk can prevent the spread of bacterial diseases through this most common of childhood foods. General food sanitation, made possible by refrigeration and a pure water supply, also contributes to reducing disease.

Antibiotics save many children who might once have been victims of diseases such as pneumonia and scarlet fever. A vaccine that prevents tuberculosis and a drug that cures it are available. Vaccinations can also prevent most children from ever having smallpox, diphtheria, tetanus, whooping cough, polio, and measles. Finally, better nutrition and care for pregnant women have improved both the health of mothers and the chances that their babies will survive.

All these medical and public health advances have combined to cause a dramatic drop in the death rate among infants and small children, not only in developed countries but in developing countries as well. But even though the odds on child survival have increased in developing countries, and even though more of a couple's children survive to adulthood, there has not always been a corresponding drop in the birth rate. That is to say, many parents still have as many children as they did when it was necessary to have five or six children to be sure of raising two to adulthood.

Components of the growth rate. In developing countries, the birth rate is likely to be as high as 30 to 45 births per 1,000 people in the population each year, while the death rate may be 15 per 1,000 per year. In contrast, among the developed countries birth rates have fallen to 10 to 15 per 1,000 people per year while death rates range from 6 to 13 per 1,000 per year (Figure 3.2).

If we subtract the death rate from the birth rate, we obtain the **rate of natural increase**, a number that indicates how fast a population is growing. In countries classified as developing, this rate may range from 15 to 30 people per year for every 1,000 already in the population. For developed countries, on the other hand, the rate is much lower—in some cases, zero.

The actual **growth rate** of a country includes not only the rate of natural increase but also the amount of migration into or out of the country. This is called the **net immigration rate**, or immigration minus emigration. The growth rate is thus the rate of natural increase combined with the net immigration rate (Figure 3.3).

In some developed countries with low birth rates, immigration may be a major part of the growth rate. In the United States, for instance, net immigration accounted for almost one-third of population growth in 1981.

Immigration can have a significant impact in some undeveloped countries as well. In Nigeria, for instance, the government issued a decree in 1982 that all aliens

Developed countries

Developing countries

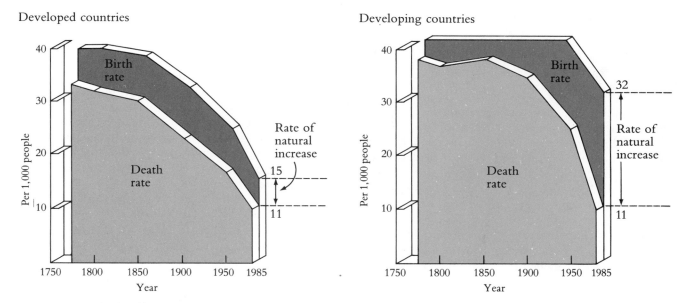

Figure 3.2 World birth and death rates. The rates of birth and death are shown for both developed and developing countries. By subtracting the death rate from the birth rate, we obtain the rate of natural increase. (From Elaine M. Murphy, *World Population: Toward the Next Century* [Washington, D.C.: Population Reference Bureau, 1981].)

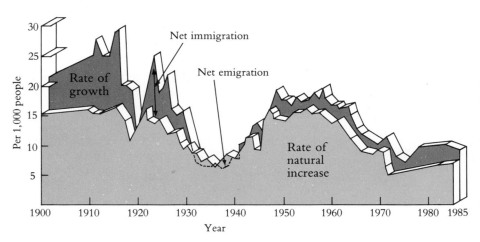

Figure 3.3 Growth rate and rate of natural increase. The growth rate for a country depends on both the natural rate of increase (birth rate minus death rate) and the rate at which people immigrate (move in) or emigrate (move out). These rates are shown for the United States from 1900 to 1985. At the present rate of growth, the population will double in about 87 years. (Data from Population Reference Bureau.)

were to leave the country. Hundreds of thousands of workers had entered Nigeria from countries such as Ghana to work in jobs generated by Nigeria's oil boom. When falling world oil prices dried up the job market, the Nigerian government determined that Nigerians should fill all remaining jobs. (Immigration problems are further discussed later in this chapter.)

A comparison of growth rates. Growth rates vary, even among the developed countries. In Denmark, Sweden, the Federal Republic of Germany, and Austria, populations are hovering near a zero growth rate. In Italy, Poland, Canada, and the United States, birth rates still exceed death rates. As a group, however, the developed nations have a growth rate of about 0.6%

per year contrasted with an average of 2% per year for the developing nations.*

Causes of Rapid Population Growth— Solving the Mystery

Understanding the cause of high population growth rates in developing nations requires understanding the value of children in these countries compared to the value of children in developed nations. In developed societies today, most of the benefits of raising children could be called psychological. The pleasures of parenting, the gratification of passing on values to another generation, the fulfilling of other people's (such as grandparents') expectations—none of these bring any economic reward to the parents. Most of these psychological rewards are gained by having one or two children (although a third child may be wanted if the first two are the same sex). These factors, combined with the high cost of raising children in Western society today and with the availability of effective contraceptive methods, have made the two-child family more and more the accepted norm.

The situation is different in most developing nations, where children may increase a family's wealth or provide parents with more leisure or with old-age security. In parts of rural Africa, labor is a scarce resource and land is not. The family that has many working members can farm more land and become wealthier. As Jones (1970) noted of Mali:

. . . except for recently introduced plows and carts, the physical factors a man needs to produce food are readily available. Land is abundant, "capital goods"—hoes and oxen—are cheap, and the wood and iron for making them are abundant. The crucial factors of production are the family's labor, of course, and "technology."

While in Uganda:

. . . extensive labor is needed for herding . . . A moderately prosperous herd owner with say 100 to 150 cattle, 100 sheep and goats and a few don-

keys needs about six herd boys ranging from 6 to 25 years of age to maintain a herd by himself. A man with many cattle but few sons must herd together, and share the yield of his stock, with a man who has few cattle and many sons. (Dyson-Hudson and Dyson-Hudson, 1970)

Further, in some parts of Africa—for example, Burkina Faso and the Sudan—private ownership of land does not exist. A person owns the crops grown on land, but no one can own the land itself. Land-use priorities are often given out according to the number of people in a family who can work on the land.

African and Asian families are often not as child-centered as families in the West. Parents may be expected to eat more than a proportionate share of food and to use more of the family resources for clothing and so on than are the children.

Besides providing labor to free the parent from household or field chores so the parent can do other work, working children allow the parent to have some leisure time. Younger children even allow older children more freedom. In developing nations children can be useful at as young an age as five (Figure 3.4). If they cannot do an adult's labor in the fields, they can help take care of younger brothers and sisters and do some of the household chores. In developing countries such as Bangladesh, marketing and preparing meals may be a full day's work for one person in the family. Children can be trained to help with meal preparation chores. Cain (1977) estimated that a man living alone in Bangladesh would spend 90% of his working time on household chores—tasks that women and children would otherwise do.

Children also provide a sense of security against old age or ill health, especially where there are no government old-age programs. In at least one sense, children are better than pensions or government social security, since the benefits they promise will not be eaten away by inflation. Even when parents must invest resources in children during the years of child rearing, they can expect to reap some return when the children are grown. This enables parents to even out their own lifetime consumption: they spend more when they are young, vigorous, and able to work, and they draw on this investment when they are old or infirm. In Africa, parents who can expect their children to support them when they are old tend to respond to arguments that fewer children are needed as medical care improves.

*In 1981 the United States had a population growth rate of 1%, and the rate of natural increase was close to the average for developed countries. However, net immigration, combined with natural increase, pushed the U.S. growth rate to one of the highest for any developed nation.

Figure 3.4 Child worker. This Venezuelan boy is one of what the International Labour Organization estimates as at least 52 million child laborers in the world today. (UNICEF photo)

Thus, they have fewer children. However, African parents living in cities, who seem to feel less certain that the old values will hold, tend to have even more children because it is then more likely that at least a few will provide for the parents' old age.

Other factors that affect population growth rates include the way in which property is passed from generation to generation and the availability of educational opportunities. In many developing countries the definition of family differs from that in the West. In parts of Africa, a father, his sons, and their wives and children may live together, or at least work together, under the father's direction. The economic unit of one man, his wife, and their children is not necessarily the usual one. In fact, in parts of Africa a man may have more than one wife. In such situations, property rarely belongs to a husband and wife together and may not be passed down from father to son. In parts of Africa where land is not held privately or where a father's property is not divided among his sons, parents do not feel the need to limit family size to avoid splitting inherited land among too many heirs.

In some Indian states, the problems of inheritance are seen as the problems of the next generation. The father has the right and the duty to work for the maximum benefits for himself and the family he heads. This may very well mean more children now to provide more labor and produce more goods in the present, with no thought as to how the children will fend for themselves in the future.

Parents can make choices about how they will educate their children in developed and in certain developing nations. They may educate all of them, splitting the family's resources; they may educate only the brightest, sending that one child as far up the educational ladder as possible; or they may educate none of them. In countries that have enough economic development and in which social classes are not rigid, parents may see a benefit in producing only a few, well-educated children. These children can be expected to make something of themselves and add to the family's prestige and wealth. Further, in most of these countries, parents cannot benefit from their children's labor because of laws against child labor and because children are often required to be in school. In such a climate, parents limit family size in order to gain extra income by working themselves.

In poorer countries, such as Bangladesh, parents may not have this choice; their resources are often so limited that not even one child can be educated, or schools may not be available. In such a situation, the only wealth available to a family is a large number of children.

To summarize, in many developing nations, children provide parents with wealth (labor), with leisure, and with security in old age. These factors, which promote large families, are largely absent in industrialized societies. In addition, social factors that act to limit family size in developed countries, such as education opportunities for children and the inheritance of property from a father to his children, are often missing in developing countries.

Population Profiles

Before we go on to look at the consequences of rapid population growth and what can be done to change its course, it is useful to look at a particular way of representing the population of a country: the **population profile**. The males and females in a population can be represented as groups called *age cohorts*. These cohorts

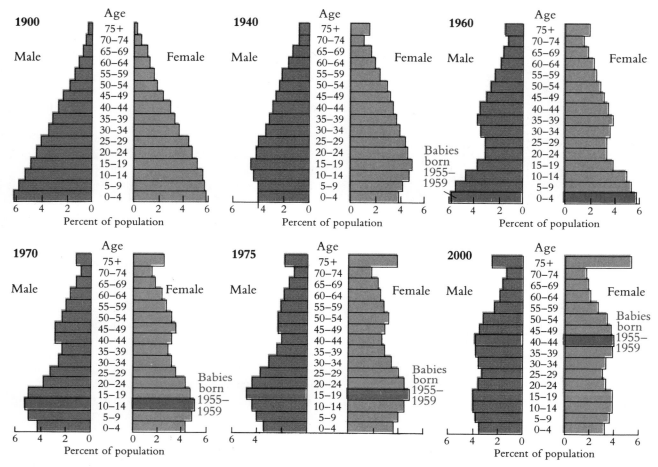

Figure 3.5 Population profiles for the population of the United States at various times. From these graphs you can see how a large group of babies born during one period (1955–1959) swells the size of age groups in later years. (Source: Population Reference Bureau, 1975.)

consist of all the males and females born in the same 5-year period.★ For instance, all the girl or all the boy babies born in the United States from 1955 through 1959 would constitute one age group. In Figure 3.5 the U.S. population is shown as a series of profiles over time. The 1955–1959 cohort of men and women first appears as the bottom bars on the population profile for 1960. In the 1970 profile, these people, then 10 to 14 years old, are represented by the third bar up from the bottom of the pyramid, and so on for the later decades.

Pinched profiles of developing nations. Certain features of populations show up well in population profiles. For example, many developing countries show a characteristic "pinched profile." In this case, a decreased death rate in younger portions of the population, brought about by better health care, has swollen the younger age groups compared to the older ones. The profile thus has a wide base and a narrow top (Figure 3.6). Compare this profile of India in 1970 to the one for India in 1951 (Figure 3.7), before modern medical advances were introduced in many parts of that country. The increased size of childhood age groups was the result of better infant and child survival as health care improved.

★Sometimes the population may be divided into 1-year age groups, but 5-year groups are more commonly used.

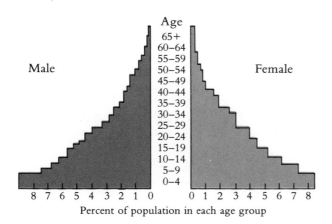

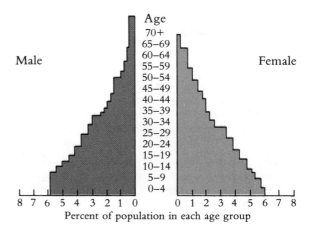

Figure 3.6 Population of India in 1970—an example of a pinched profile. (From *Population Bulletin*, Population Reference Bureau, 1970.)

Figure 3.7 Population of India in 1951. (From W. S. Thompson and D. T. Lewis, *Population Problems*, 5th ed. [New York: McGraw-Hill, 1965]. Used with permission.)

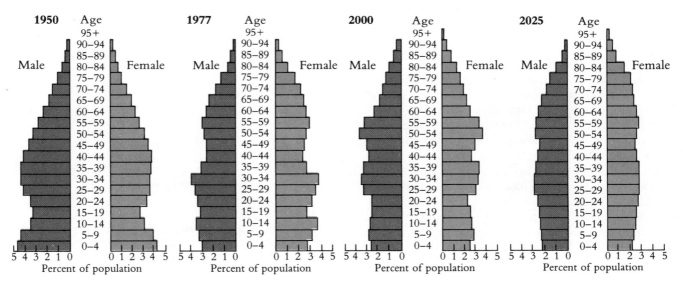

Figure 3.8 Population of Sweden, 1950–2025 (projected). (From Population Reference Bureau, *Population Bulletin*, No. 35 [June 2, 1980], page 28.)

Profiles of developed nations. Figure 3.8 shows a series of profiles for Sweden from 1950 to 2025 (projected). In this case, the profiles are approaching an obelisk shape, which is characteristic of countries that have achieved replacement-level fertility. That is to say, each couple has only two children, on average, and so only replaces themselves. In such a situation, the distribution among the age groups becomes more uniform.

Population Momentum

One more feature of population growth that can be seen well in these bar graphs is **population momentum**, the growth that can take place in a country's population even if each couple has, on an average, no more than two children. Momentum can be understood by considering people in terms of age ranges:

those 0–15 years who are not yet old enough to have children, those 15–45 who are now in their childbearing years, and all those older than 45, who are no longer likely to have children. In terms of a country's future population growth, the two important groups are those 0–15, who will have children in the future, and those 16–45, who are having their families now. People in all the older age groups can cause no further increase in the population. In most developing countries, the large numbers of children now 0–15 years old are likely to marry and have at least two children per couple. As these large numbers of parents and their children grow older, they will swell the older age groups, which are now relatively small.

Because changes in the near future are the result of the age structure existing today, population reduction programs started right now would have a 40- to 60-year lag time before they could result in stable populations. This would be the case even if programs were immediately and successfully implemented. (Actual experience shows that population reduction programs are rarely successful immediately but take a number of years to become accepted.)

In Figure 3.5, which showed the age structure of the U.S. population from 1900 to 2000 as a series of population profiles, the large group of babies born in 1955–1959 forms a bulge. The females in this age group will be in their childbearing years through the 1990s. Even if their fertility rate remains at less than replacement level (1.8 for instance), the large actual number of children born to these women will cause a growth in the U.S. population.

In Figure 3.9, two overlapping pyramids for the U.S. population, one using actual data for 1960 and the other using data projected by the U.S. census for 60 years from now, show how the U.S. population will take 60 to 70 years to stabilize, or reach the state of **zero population growth**, even though the fertility rate is now at slightly less than two children per woman.

Total Fertility Rate

The birth rate and death rate defined and used at the beginning of this chapter are often referred to as the *crude birth rate* and *crude death rate.* The adjective *crude* points out that there could be something misleading in these numbers. One problem is that the crude death rate does not take into account the age structure of a

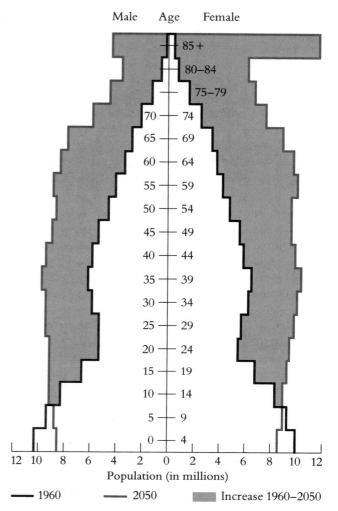

Figure 3.9 The two overlapping "population pyramids" give a vivid picture of the changes in U.S. population size and composition projected to the year 2050. Because both pyramids are drawn to the same scale, they give a comparable picture of overall size in both years. In 1960, as the "baby boom" began to wane, U.S. population was 179 million and the pyramid had a wide base as a result of prior years' high fertility. By 2050, continued low fertility should cause the pyramid to take on a relatively uniform appearance. (From Population Reference Bureau, *Population Today*, February 1984, p. 5.)

population. For instance, the crude death rate for Sweden in 1985 was 11 deaths per 1,000 in the population, while for Mexico it was 6. This might seem to indicate that Mexicans are receiving better health care. In fact, the difference arises from the fact that Mexico's population has proportionately more people in the younger age groups. Only 4% of Mexicans are in the over 65 age group and thus subject to the likelihood of dying from old age, while 17% of Swedes are in this group. Actually, the life expectancy in Sweden is 75 years, while in Mexico it is about 66.

The crude birth rate is less subject to this particular kind of misinterpretation because the proportion of women in childbearing age groups is similar in most countries. Both Mexico and Sweden had 23% of their population in the 15–49 age group in 1980. Problems do arise, however, in the few cases in which the proportion of women of childbearing age differs between countries, or cases in which more women in the age group 15–49 in a particular country are in the subgroup 20–29, the most fertile period of a woman's life.

A more sensitive measure of a country's fertility than crude birth rate is the **total fertility rate (TFR)**: the total number of live births a hypothetical average woman would be expected to have during her reproductive lifetime. This figure is based on the fertility rates current when the calculation is made. Because this hypothetical woman will be part of several different age groups over her lifetime, the TFR is derived by combining all the fertility rates for the different age groups in the population. Whether the number of women in the childbearing age group is large or small does not affect the TFR, which measures only how many children each woman can be expected to have. For this reason, TFR gives much earlier warning of changes in reproductive behavior than the crude birth rate does. The crude birth rate for a country may remain high or continue to increase as long as there are much larger numbers of women in the childbearing age groups than in the older age groups. TFR, however, tells whether the couples reproducing now are choosing to lower their family size toward replacement levels or are continuing to have more than the average two children per couple.

Effects of Rapid Population Growth on Health and Welfare in Developing Nations

The types of problems caused by rapid population growth are different for developing countries than for developed ones.

Food Production

In developing countries, the most pressing need is often simply to produce enough food to feed the rapidly growing numbers of people. Government programs must be heavily weighted in this direction to prevent starvation.

The three graphs in Figure 3.10 show India's rate of population growth from 1950 to 1984, the increased production of food grains in India during that period, and the actual amount of grain produced per person over the same time span. While India managed a significant increase in food production, the population increase used up all but a small part of the gain, and during bad years there was no improvement.

Investments in food production are also important for the stability of a developing country's government. A hungry population is a discontented one, and discontent can flare into revolution and overthrow of those in power. This possibility sometimes leads governments to act in ways that are not economically wise but that will help them stay in power. For instance, food prices are often kept at a low level in large cities. However, this policy results in low prices for the farmer, which in turn discourages increases in food production. Farmers will not want to raise more food if they do not receive a reasonable price for their efforts. Low farm prices also mean that farmers cannot invest in modern farming methods that would increase their crop yields. Thus, the countries that most need more food may, in some ways, be hindering greater food production. In this manner, population pressures can force a government to slant its policies in ways that are not in the country's long-term best interest. The world food situation is covered in more detail in Chapter 8.

Health Consequences

Because so much money must be spent on food production in developing nations, little is left for health programs. Among women who have many children, the rates of maternal illness and death are higher than in developed countries, and the babies born to these women tend to suffer from malnutrition (Figure 3.11). The lack of sufficient food in childhood leads to poor health and decreased mental ability.

A developing nation's annual investment in health activities must cover medicines, rents, equipment, salaries of physicians and nurses, and so on. Yet in most

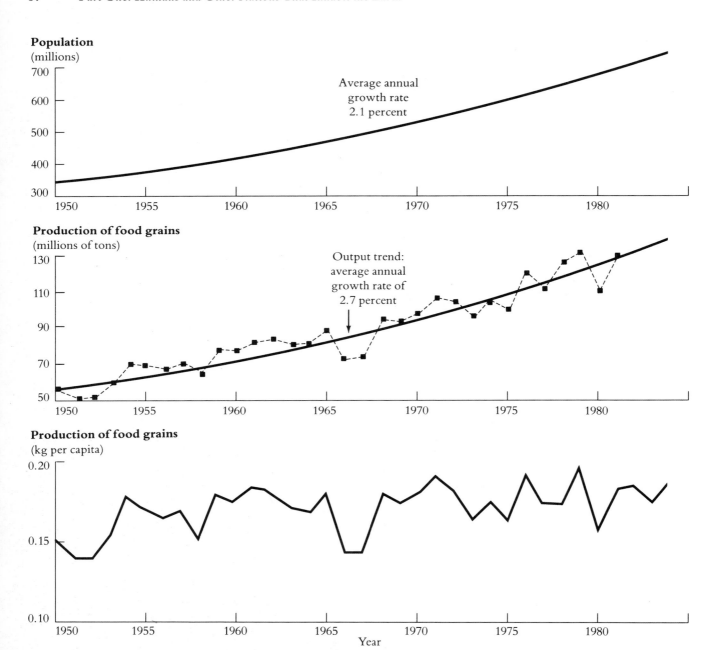

Figure 3.10 Population growth and production of food grains in India, 1950–1984. Although grain production increased almost 250% over the period, population growth resulted in barely a 20% increase in food-grain production on a per person basis. (From The World Bank, *World Development Report 1984* [New York: Oxford Press, 1984], p. 93.)

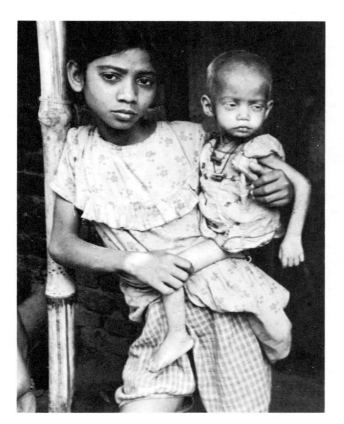

Figure 3.11 Severe malnutrition affects about 10 million children under age 5 in developing countries. (UNICEF/ Abigail Heyman)

developing countries, the total annual investment in health activities, when divided by the population, comes to only $1–$2 per person per year. Obviously, not much health care can be obtained for this sum.

Ill health has its effect on productivity. Burdened by ailments of many sorts, individuals are not likely to be able to pull themselves and their families out of poverty by their own labor. Thus, the expanding populations of developing nations diminish the health resources available to each person and so reduce each person's potential to succeed.

Educational Consequences

Education is also likely to be slighted in developing nations, as governments are often unable to increase educational services at the same rate at which the population is growing (Figure 3.12).

Developing countries have made tremendous gains in primary education for their young people over the past 20 years. Improvements in enrollment figures have been recorded in almost all countries. However, government spending on education in developing countries, as a whole, has become a significant burden— about 4% of the gross national product. Predicted increases in school-age children will make programs difficult to maintain (Figure 3.13). In fact, recessions in the world economy have already put severe strains on the ability of developing nations to sustain educational spending. In some cases, the amount spent per student has dropped in real terms, even though there is not much margin for decrease. In Malawi and Kenya, class sizes are often greater than 60 students. Whereas Scandinavian countries spend over $300 per year per student for classroom supplies, countries such as Bolivia, El Salvador, and Malawi spend less than $2 per student. It is difficult to see how improvements in class size or improvements in the materials for each student can be achieved if populations continue to grow.

As increasing numbers of young people go out to look for jobs each year, their low educational levels and the scarcity of jobs force them into unskilled positions at best. People who accept these low-paying jobs have limited upward social mobility and little opportunity to improve their economic lot even by hard work.

Environmental Consequences

To feed increasing numbers of people, developing countries usually attempt to increase farm outputs. However, increased agricultural production carries environmental costs. The clearing of large forest tracts and the soil erosion that accompanies intensive agriculture have led to landslides, floods, and the silting of reservoirs needed for hydroelectric power. Animal habitats are being endangered as more and more land is turned over to agriculture or the gathering of fuel wood. Agricultural pesticides and fertilizers have polluted waters and pose new health problems for agricultural workers.

In some environmentally fragile areas, the combination of population growth and drought has turned large areas into virtual desert (see Chapter 7). Erosion, pollution, loss of species, and loss of land have all resulted from unchecked population growth.

We catalog these events because they matter to us.

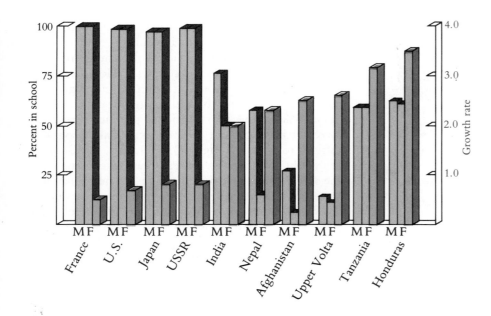

Figure 3.12 School enrollment in ten selected countries compared to growth rates. The graph shows the percent of girls and boys ages 6–11 enrolled in school compared to the rate of natural increase for each country. (Data from 1982 World Population Data Sheets, Population Reference Bureau, and UNESCO Statistical Yearbook, 1981.)

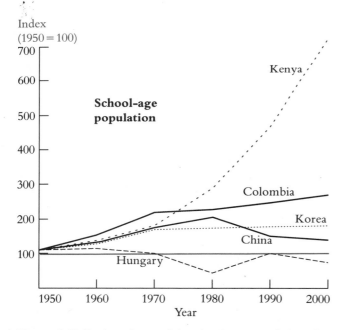

Figure 3.13 Projected growth in school-age populations for several countries. (From The World Bank, *World Development Report 1984* [New York: Oxford Press, 1984].)

Our perspective is influenced, however, by the mountain of plenty on which we stand. Well-fed, well-clothed, well-educated, we can afford the luxury of seeking quality in our environment.

It is not hard to understand why wise land-use policies, pollution control, and protection of endangered wildlife are sometimes low priorities in developing nations. Where children and parents are perpetually hungry, often sick, and without adequate clothing, where there is little hope for people to better themselves, the immediate future of the environment does not seem to matter very much. Until people's primary needs are met, they will have little enthusiasm about seeking a quality environment.

Nonetheless, certain elements of the environment are of long-term importance to people everywhere. Without good agricultural methods, such as proper irrigation techniques, careful use of pesticides, and control of erosion, the productivity of the soil will eventually be lost and food production will fall. Without adequate treatment of water supplies and wastewaters, diseases will continue to ravage and weaken the population and periodic epidemics of chemical poisoning will occur. (See, for instance, Agarwal et al. [1982] on environmental problems in India.)

These are but a few of the potential long-term problems. The point is, even in developing nations where the most pressing needs are the most basic human ones—food, shelter, and good health—consideration must be given to the environment. For unless the environment remains healthy, it cannot long support a human population.

Population Growth and Economic Development in Developing Nations

Economic development is one of the major aims of governments in developing nations. It is through economic development that governments hope to raise the living standards of their people.

Population Growth as a Block to Economic Development

In the Western nations, economic development took place at the same time as the discovery of the health and sanitation measures that led to rapid population growth. In these countries, industrialization provided jobs to absorb the increasing labor force that resulted from population growth, and incomes rose steadily. Within the same time frame, birth rates in these countries began to fall. Eventually, a point was reached in most of the developed nations where birth rates and death rates balanced each other. Today these nations are approaching a state in which their populations are not growing at all (zero population growth).

To many people there seems a clear relationship: Increasing economic development leads to a lowering of the population growth rate. Therefore, to achieve decreases in population growth in developing countries, the first step must be economic development. However, looked at from another point of view, there is a great deal of evidence that population growth itself, at the level taking place in developing countries today, is blocking their economic development.

According to the analysis of experts at the World Bank, economic development alone would not bring down birth rates quickly enough to prevent serious consequences in developing nations. Moreover, rapid population growth may in some ways make economic development impossible. There are several reasons for this opinion.

In the first place, population growth in developing countries is both quantitatively and qualitatively different than it was in the now-developed countries 200 years ago. In the second place, rapid population growth causes governments to spend their limited capital in ways that do not speed economic progress. In the third place, rapidly growing populations place stresses on resources and social institutions, thereby hindering economic progress.

How the situation differs in developing nations today. Even during the time of their most rapid growth, Western nations never experienced the growth rates now prevalent in the developing nations. In Europe population growth peaked at about 1.5% in the early 1800s and declined from there. In North America, relatively high fertility combined with heavy immigration raised population growth rates to more than 2% in the 1800s, but the rates had declined well below 2% by the early 1900s. In contrast, growth rates are near to or exceed 3% in many developing countries today. An additional difference is that a large part of that increase is taking place in rural areas. In Africa and Asia rural populations are growing at well over 2% per year, whereas in developed countries rural populations grew at a rate of less than 1% during the 1800s.

This rural growth is a problem because population growth is less easily absorbed in rural areas, where most jobs are in agriculture. Land suitable for farming is a scarce commodity in many countries. Further, modernization of agriculture tends to decrease the number of jobs available rather than increase it (see Chapter 8 for more on this subject).

Population growth and the capital needed for development. Rapid population growth forces governments to choose between making investments that would improve the long-term economic welfare of their people and spending money on the short-term health and food requirements of the people. For instance, education through at least the primary level has become a prerequisite for economic improvement in the modern world. But, as populations grow, the number of children who must be educated increases. If a government attempts to continue spending the same per capita amount on schooling, larger and larger sums of money must be allocated to education each year. This is a form of **capital widening**. In contrast, slower population growth would free some of this capital for improvements in education or job training. This would be **capital deepening**, an increase in available capital per person. Countries that are faced with food shortages and that must take emergency actions to prevent starvation or the spread of epidemic diseases may be forced to spend their limited resources on food and medical care and decrease the amount spent on education.

A second example of capital widening involves growth of the labor force in developing countries. To increase productivity per worker and thus increase incomes, capital must be invested in roads, energy sources, farm machinery, parts, and factories. If it is not, each additional worker will produce less with the reduced land and capital available to him or her. Wages will fall in relation to profits, while rents and income inequalities will increase. Table 3.2 shows how much investment could have been available for each new worker in selected countries in 1980 (if all available investment were allocated to new workers). Rapid population growth in developing countries leading to increases in the size of their labor force will widen these differences even more in the future.

In short, the human capital (an educated, skilled labor force) and the physical capital to invest in the physical requirements of manufacturing are difficult or even impossible to accumulate when a country's population is growing rapidly.

Population growth and the resources needed for development. Because there is little advanced technology in developing countries, there is little demand for the kind of resources that fuel an industrialized society. The one serious exception is materials used to produce fertilizers. Developing nations need large amounts of fertilizer to produce food, but production of fertilizers requires nitrates (which are produced from natural gas), phosphates, and energy. For this reason, high world oil prices affect not only the developed parts of the world but also developing countries such as India.

Increasing populations put pressure on scarce resources such as land and water and may make wise use of these resources impossible. In many developing countries today, large areas of tropical forest land and steep mountainous slopes are being cleared to grow food for increasing populations. Erosion of fertile topsoil often follows, leaving land unfit for cultivation in the future. Similarly, overgrazing of semiarid lands can turn them into deserts. In such cases, the welfare of future generations is being mortgaged to serve the needs of the present.

In developing countries that possess an abundance of natural resources (such as Brazil, Ivory Coast, and Zaire), more people could theoretically be supported

Table 3.2 Investment Per Potential New Worker in Selected Countries, 1980

Country	Gross domestic investment per potential new worker*
Bangladesh	$ 1,090
Ethiopia	1,530
Nepal	1,260
Rwanda	1,660
Kenya	4,700
Egypt	8,960
Thailand	10,660
Colombia	10,100
Korea	29,850
Brazil	40,360
Japan	535,040
Australia	219,350
France	461,340
Germany	481,330
United States	188,990

Source: The World Bank, *World Development Report 1984* (New York: Oxford Press, 1984).

*In 1980 dollars, if all available investments were allocated to new workers.

in the long term. However, rapid population growth will hinder the investment in roads, public services, and training programs that would allow use of these resources.

Developing countries whose natural resources are already scarce (such as Bangladesh, Borundi, China, Egypt, India, Java, Kenya, Malawi, Nepal, and Rwanda) must look to investment in manufacturing and agricultural modernization to improve their economic picture. Again, this will be difficult if increasing numbers of people must be fed and cared for.

Social Inequalities and the Reforms Needed for Development

As the population grows in a developing country, the per capita income (the Gross National Product divided by the number of people) generally decreases. There are two ways of interpreting this particular fact. Some experts emphasize the points we have been making: Lower per capita income decreases savings, cuts down on the amount of capital available for investment, and prevents people from buying manufactured goods, thus retarding industrialization. Others point out that in many developing countries, the capital and most of the income is in the hands of a privileged few anyway. This has led to the development of two groups: one rich and well fed, one poor and ill nourished. Furthermore, the rich are likely to remain rich and privileged, while the poor and the children of the poor have little hope of escaping their fate. These experts argue that unless population programs are combined with some form of redistribution of wealth, population reduction among the poorer people in a developing nation will not help improve their economic status. In the next chapter we discuss this issue further with reference to several South American countries.

When governments do attempt to make changes, rapid population growth can cause difficulties in instituting the social and economic reforms that can lead to increased economic development. For instance, social and political pressures to employ large numbers of young people have contributed to the growth of a large government workforce in many developing nations. Yet such a use of scarce resources is not always in the long-term best interests of a developing country. Egypt's guarantee of a job to all college graduates tends to siphon funds away from the uneducated who might otherwise benefit from government spending and toward those who are already relatively well off.

Population Growth in Developed Nations

Resources and Technology

In developed countries, technology and the scarcity of resources are primary issues raised by population growth. The energy crisis has made many people aware of the depletion of such resources as natural gas and oil. Shortages are also predicted by the turn of the century in many minerals, among them copper, tin, magnesium, and zinc.

Looking back in history, it is hard to find instances in which we have actually run out of some critical resource. What commonly happens is that scarcity causes a price rise that eventually forces a switch to some other resource. Thus, a scarcity of wood in England in past centuries led to the burning of coal. Whether this mechanism, which economists call *substitution*, can be counted on to continue to work in the face of increasing population growth is not clear. For instance, will tech-

nology provide us with solar or fusion power before we run out of uranium, oil, and gas? We are at least gambling with the possibility of future resource shortages. Not everyone agrees with this argument, however—see Controversy 3.1.

The example of the switch from wood to coal fires in England brings up another problem. By the 1900s, coal fires were responsible for a level of air pollution that made London almost uninhabitable at certain times of the year. Air pollution incidents, such as the London fog of 1952 that caused 4,000 deaths, were primarily due to pollutants derived from the burning of coal.

Pollution is one of the most serious problems faced in developed countries. We need more fertilizer and pesticides to grow ever increasing quantities of food. We need more petroleum to produce the gasoline required to get increasing numbers of people to ever more distant places of work. The result is increasing levels of smog. More people need more housing and more farm products. Housing and farming use up forested lands, leading to increased soil erosion. More people generate more sewage and solid wastes, which either foul natural waters or demand technological solutions for their disposal. And more people consume more products whose manufacture generates chemical wastes. Although technology may be able to solve some of these problems, more and more of our money and effort go to maintaining the quality of life. Like the Red Queen in *Through the Looking Glass*, we run in order to stay in the same place.

Effects of Crowding

Population growth limits not only our choices about spending society's money but also our personal choices. For example, whole areas of the United States (such as the Eastern seaboard) are becoming urban. Recreational areas are threatened by throngs whose presence may even destroy the appeal of the area (Figure 3.14). Furthermore, as population increases, available open space decreases or becomes more costly. Wilderness turns into overcrowded parks, and parks degenerate into outdoor slums.

Scientists who have studied animals under crowded conditions have reported some results worth noting. For instance, rats in crowded cages begin to lose their normal patterns of social behavior. They neglect their young and sometimes resort to cannibalism. Some rats become overly aggressive, while others

CONTROVERSY 3.1

Malthus, Population, and Resources

[P]opulation growth in most countries will surely be halted substantially below [these] levels . . . either because of humane and voluntary measures taken now, or because of the old Malthusian checks . . .
Robert McNamara

[I]f resources are not fixed, then the Malthusian doctrine of diminishing returns does not apply.
Julian L. Simon

Take some time to consider what the future would be like if populations approach the levels predicted by the World Bank (moderate population projection):

> In the next 45 years, Bangladesh will have nearly tripled and will have 259 million people jammed into an area, alternately swept by flood and drought, the size of the state of Wisconsin—and Kenya, will have quintupled. . . .
>
> The total population of the developing countries as a group, 3.3 billion in 1980, will rise to over 7 billion by 2025, and to over 8.5 billion by 2050. A century from now, the world's population will total about 11 billion.

In an address to the 1984 World Population Conference, Robert S. McNamara presented these statistics and went on to say:

> In the end, population growth in most countries will surely be halted substantially below [these] levels. . . . That will happen either because of the old Malthusian checks, or perhaps even more likely, in tomorrow's world, it will occur as a result of coercive government sanctions and the recourse by desperate parents to both frequent abortion and clandestine infanticide.

What will that world be like? Will such huge populations finally fall victim, as McNamara fears, to the plagues and starvation predicted by Malthus so long ago? Or will we be subject to governmental edicts about how many children we can have?

Not everyone agrees that these are the only possibilities. Might not the human mind, ever ingenious when confronted with an immovable barrier (here the finite limit to the amount of land, water, and air on earth), make an end-run around the barrier or a quantum leap over it? One proponent of this idea, Julian Simon (1984) writes:

> In the short run, an additional person—baby or immigrant—reduces the standard of living. Another consumer causes the fixed stock of goods to be divided among more people. And as Malthus argued, more workers result in less output per worker.
>
> However, if resources are not fixed, then the Malthusian doctrine of diminishing returns does not apply. Given some time to adjust to shortages with known methods and new inventions, people create additional resources.
>
> What is extraordinary about the creation of new resources is that a shortage due to population

or income growth usually leaves us better off than if the shortage had not arisen, thanks to the resulting new techniques. (For example, plastics began as a substitute for elephant ivory in billiard balls.) Raw materials have actually become less scarce. Throughout history, prices of raw materials have fallen relative to consumer prices—despite the common sense that if one begins with an inventory and uses some up, there will be less left.

Is Simon right when he draws on historical precedent to say that "Given some time to adjust to shortages with known methods and new inventions, people create additional resources"? Has McNamara left anything out of his consideration of the problem of rapid population growth? Has Simon?

Sources: Robert S. McNamara, World Population Conference, 1984. Julian Simon, quoted in *Population Today* (May 1984), p. 8.

withdraw from the community. Sexual behavior becomes abnormal. Monkeys also show some of this same behavior when crowded together.

In the wild, animal populations occasionally experience rapid growth, and the needs of the expanded population may exceed the food supply or living space available. In such cases, starvation, disease, and reduced fertility lead to a decline in the animal population. Some scientists think the mass migration of lemmings is a response to rapid population growth.

Can the results of animal studies be applied to human populations? Probably not directly. Nevertheless, the implications of the changes that occur in animal communities under stress from rapid growth and crowding warrant consideration.

Perhaps humans can survive under relatively crowded conditions. On the other hand, we may prefer not to see the changes such crowding will cause. If we wish to influence the future, we must make decisions about population growth. Otherwise, the choice will have been made for us.

Rural to Urban Migration

Although the rapid growth of the world's population in general is a major cause for concern, another equally far-reaching phenomenon is occurring in this century that affects both the environment and how people interact with one another: the way in which society is organizing itself. We are in the midst of transition from a mainly rural society to a mainly urban society. Although at one time most people lived on farms, soon most will live in cities. In the early 1900s, only one-quarter of the world's population lived in cities. By the year 2000, it is believed that 60% will live in cities (Figure 3.15). Furthermore, the cities themselves are

Figure 3.14 Crowded campground in South Tyrol. In developed countries, rapid population growth leads to overcrowding of recreational and wilderness areas. In a real sense, some of our freedom of choice is lost. (Werner H. Muller/Peter Arnold, Inc.)

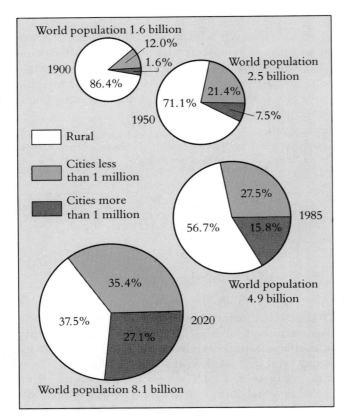

Figure 3.15 Patterns of world urbanization. (From Population Reference Bureau, *World Population: Towards the Next Century*, 1985.)

getting larger. There were only two cities with more than 10 million people in 1950, but eleven had reached that level by 1985. If such trends continue, there will be twenty-five megalopolises by the year 2000 (Table 3.3).

Advantages of Cities

Cities have certain advantages. They allow for greater division of labor, providing more jobs. Cities offer more specialized occupations for people to find work—for example, taxicab or rickshaw driver, police officer, firefighter—jobs that may not even exist in the hinterlands. Even crime and begging provide ways to sustain life in the cities. On average, 75% of migrants to cities are economically better off then they were in rural areas.

While cities provide more job opportunities for growing populations, the number of jobs in rural areas is decreasing. In both developed and developing nations, fertilizers, pesticides, and modern farming equipment (when available) make it possible for fewer people to till the same area of land. Employment opportunities are then fewer and people must move to the cities to support themselves.

Another advantage of cities is that the more concentrated population can be more easily provided with health and welfare services. In sub-Saharan Africa, for instance, only 10% of the rural population has a safe water supply, whereas 66% of the urban population

Table 3.3   Ten Largest Cities in the World, 1950, 1984, and 2000

1950	Population (in millions)	1984	Population (in millions)	2000	Population (in millions)
1. New York–N.E. New Jersey	12.3	1. Mexico City	18.1	1. Mexico City	26.3
2. London	10.4	2. Tokyo–Yokohama	17.2	2. Sao Paulo	24.0
3. Rhine-Ruhr	6.9	3. Sao Paulo	15.9	3. Tokyo-Yokohama	17.1
4. Tokyo-Yokohama	6.7	4. New York–N.E. New Jersey	15.3	4. Calcutta	16.6
5. Shanghai	5.8	5. Shanghai	11.8	5. Greater Bombay	16.0
6. Paris	5.5	6. Calcutta	11.0	6. New York–N.E. New Jersey	15.5
7. Greater Buenos Aires	5.3	7. Greater Buenos Aires	10.9	7. Seoul	13.5
8. Chicago–N.W. Indiana	4.9	8. Rio de Janeiro	10.4	8. Shanghai	13.5
9. Moscow	4.8	9. Seoul	10.2	9. Rio de Janeiro	13.3
10. Calcutta	4.6	10. Greater Bombay	10.1	10. Delhi	13.3

Source: Population Reference Bureau, 1985.

does. Studies in India show that substantial economies of scale are found in supplying water and electricity to cities of at least 150,000 people. However, whether a point is reached at which cities are too large to economically provide such services is not yet clear.

Disadvantages of Cities

In developing nations, poor people who move to cities to find work often find conditions of extreme squalor. The services needed in cities—housing, traffic control, sewerage, water supplies—are difficult to provide when cities double in size in only 10 years. Administrators may have neither the money nor the skills to cope with such growth. In the major cities of India, many people must live on sidewalks, making their home on nothing more than the pavement (Figure 3.16). The water and sanitary conditions of these wretchedly poor people may be worse than in the rural areas from which they came. Cardboard and tar-paper shacks seem a step up to people who have so little. Sanitation conditions are a modest improvement in communities of shacks, but they remain extremely primitive, with no running water and only ground disposal of waste. Intestinal diseases and tuberculosis are rampant in these unsanitary and crowded conditions.

In the wealthier, developed countries, poor people who move from rural areas to the cities may find noise, dirty air, violence-ridden schools, crime, readily available drugs, and, too often, ugly surroundings.

These problems of cities in the developing and developed world are due in large part to the speed of the transition from rural to urban society. Laws, managerial skills, and customs have not evolved quickly enough to cover situations in which enormous numbers of people are crowded together into cities. Perhaps new laws and methods will help solve some of these problems. As an example, wise land-use programs may help to redirect the rural-urban migration. Such programs may preserve open space, limit "urban sprawl," prevent strip development, or confine industrial operations.

Reversing the Trend

A number of countries have attempted to slow the growth of cities. China has successfully kept the proportion of its urban population constant over the past

Figure 3.16 Sidewalk dwellers. These typical sidewalk-squatter dwellings in Bombay contrast starkly with the modern high-rise apartment buildings that house a fortunate few. (UNICEF)

30 years using a system of restrictions in movement and a resettlement program.

In Korea, schemes to attract population and industry to small towns through economic incentives have been successful. Elsewhere, however, such schemes have had no visible impact on the growth of major cities. In fact, the World Bank has concluded that in economic terms, the money used in urban-rural resettlement plans would be better spent to intensify production in already settled areas. This view is admittedly a purely economic one, however. Indonesia's resettlement program, which relocates families from populous Java to the country's sparsely populated other islands, has dealt mainly with families from the poorest strata of society. These families have enjoyed a better standard of living once resettled. In purely economic terms, the money spent on relocation might have brought greater returns in other programs, but in terms of economic justice, the government considers the program a success.

Migration Between Countries

In the eighteenth and nineteenth centuries **emigration** provided something akin to a safety valve for population growth in Europe, absorbing some 10%–20% of the increase in population. Today, migration between countries is reaching historic proportions if we include legal **immigration**, refugee movements, and illegal (and often temporary) movements into and out of countries by workers in search of jobs. Yet for developing countries today, emigration cannot provide relief from problems of overpopulation, for two major reasons. First, the size of growing populations in developing countries makes it unlikely that any significant fraction can be absorbed by the more developed countries. Only 0.2% of the population growth in India was siphoned off by emigration between 1970 and 1980, and that figure is less than 1% for Africa or Asia as a whole. The World Bank points out that even if 700,000 immigrants from developing nations were allowed into the developed countries each year between 1985 and 2000, this flow would account for only 2% of the projected population growth in these countries, while it would account for 20%–40% of the population increase in the host countries. A larger migration from developing to developed countries in the near future is not only of little use, it appears unlikely. Second, the receiving countries place severe restrictions on the number of immigrants they will accept. Host countries are concerned about foreign workers taking jobs from natives. They also fear social changes or political tensions caused by large immigrant groups. Thus, host countries not only restrict numbers of immigrants but also try to insure control over the type of immigrants allowed so that particular needs in their country will be met (skilled versus unskilled laborers, professionals versus nonprofessionals).

Benefits of Immigration

Given the fact that immigration is not a solution for rapid population growth, it still offers certain benefits for both sending and receiving countries. For receiving countries, immigration provides a flexible labor supply—one that can be turned to the particular needs of the country. For sending countries, emigration of particular groups can ease job pressures at home, raising wages for those who remain. This has been true for construction workers in Pakistan and agricultural workers in Yemen Arab Republic. Emigrants themselves often send home money, which helps not only their own families but also their government's trade deficit. Such monies provided almost the same amount of foreign exchange as exports for Bangladesh and Burkina Faso in 1980. Many governments recognize this benefit and have designed taxation and incentive policies that encourage emigrants to send money home.

Many countries are justifiably concerned about emigration of skilled professionals and technicians, sometimes called the "brain drain." Countries that invest heavily in education of such people lose out when they cannot tax their income once they are trained or when they do not gain the benefit of the socially valuable services of professionals such as doctors and nurses.

Immigration to the United States

The United States accepts almost twice as many immigrants as all other countries combined, close to a half million people a year in recent years. To this figure should be added the number of refugees, who are not counted as part of normal immigrant quotas. The number of refugees accepted each year is determined by the President in consultation with Congress. On average, the United States has accepted about 200,000 refugees per year over the past decade. There is a further category to consider however, and that is illegal immigration. It is not clear how many illegal immigrants enter the United States each year, but the number may be substantial, perhaps a half million people. According to the Census Bureau, about 50% or 60% of the illegal aliens come from Mexico. Altogether, the total number of legal and illegal immigrants plus refugees now accounts for somewhere between one-fourth and one-half of U.S. population growth.*

Opinions about the economic effects of immigration, especially illegal immigration, vary. Many people

---

*These figures are challenged by a 1985 report from the National Academy of Sciences, which estimates the total illegal alien population at only 2 to 4 million, compared to earlier estimates of 3.5 to 6 million made by the Census Bureau. The NAS report suggests that this population has changed little since 1977.

feel that unskilled immigrants take jobs away from U.S. citizens and use social services for which they are not paying a fair share. A blue-ribbon government panel, The Interagency Task Force on Immigration Policy, reported in 1979 that the situation is complex. In times of high unemployment, immigrants may take jobs from U.S. citizens, but at other times they fill jobs in which pay is too low and working conditions are too poor to attract U.S. citizens. On the other hand, the availability of cheap labor acts to prevent wages in these jobs from rising to a level where they would attract U.S. workers. However, if raises did occur, some manufacturers would go out of business and some goods and services might become unavailable or unaffordable to the majority of people.

Data are scant on the use of government services (welfare, schooling, health care) by illegal immigrant families. Those few studies that have been done indicate that most illegal immigrants do in fact pay income and social security taxes and hospitalization premiums, through employer withholding channels. Further, few appear to use government services. In one study, 27% were found to have used hospitals or clinics, less than 4% had children in school or collected unemployment benefits, and only 0.5% were on welfare (Population Reference Bureau, 1982).

Immigrants are concentrated in certain states. In 1980, only six states accounted for 70% of the 4.5 million permanent resident aliens (those intending permanent settlement in the United States). Such concentration can impose a severe burden on those states for social services (schools, hospitals) to immigrants. Because of the concentration of illegal as well as legal immigrants in only a few states, it is possible that financial and social burdens fall unevenly on state or local governments. Florida officials estimate they were forced to spend $150,000 in refugee services between 1980 and 1982 when the United States accepted large numbers of Cuban and Haitian refugees. On the other hand, in a landmark 1982 court case, the state of Texas was unable to support the claim that it was spending a disproportionate amount on educating children of illegal aliens. Investigations showed that such children constituted less than 1% of the entering school population in 1980. The Supreme Court ruled that denial by a state of education or other provisions of the 14th Amendment to any person within its jurisdiction is unconstitutional.

Overall, many employers, middle-class consumers, illegal immigrants themselves, and U.S.

workers whose industries might disappear without illegal immigrant labor all benefit from illegal immigration, while unskilled U.S. workers stand to lose in some ways by continued illegal immigration. Legal immigrants, on the other hand, appear to consist only one-third of males, according to a Labor Department study. Apparently, two-thirds of legal immigrants are women and children and thus represent less of a threat to the unskilled labor market.

Of equal concern with the economic effects of immigration are the social and political effects on U.S. society. Figure 3.17 illustrates how the origins of immigrants have changed. The proportion of the immigrant population that is of Hispanic origin (Mexican-American, Puerto Rican, Cuban, and so on) has risen through the years. Although the earlier European immigrants usually adopted American customs and language quickly, parts of the recent Hispanic immigrant population have desired to retain their own language and customs. This has raised fears of separatist difficulties, such as those between French-speaking Quebec and the rest of Canada. However, since the Hispanic immigrants themselves come from a variety of cultural backgrounds and, in fact, share only a common language, these fears may be unjustified.

It is clear, however, that the composition of the U.S. population is slowly changing in response to the preponderance of Hispanic immigrants, both legal and illegal. Hispanics are the fastest growing U.S. minority both because of their immigration levels and their relatively high fertility level (2.5 children per woman compared to 1.8 for the United States as a whole). Table 3.4 illustrates how the composition of the U.S. population will change under two different assumptions. One is that the total of legal and illegal immigration will be kept to 500,000 persons per year; the other

Table 3.4 Composition of the U.S. Population Under Different Immigration Levels

Population groups	Percent of total in 1980	Percent of Total in 2020	
		Annual net immigration (500,000)	Annual net immigration (1,000,000)
White non-Hispanic	79.9	69.5	64.9
Black	11.7	14.3	14.0
Hispanic	6.4	11.1	14.7
Asian and other	2.0	5.0	6.4

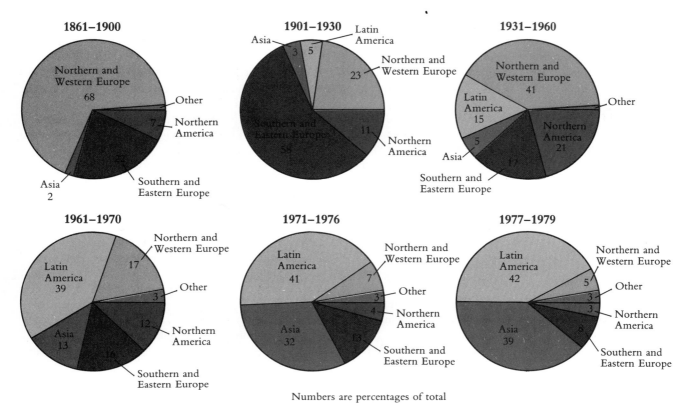

Numbers are percentages of total

Figure 3.17 U.S. immigrants by region of birth, 1861–1979. (From L. F. Gouvier, *Immigration and Its Impact on U.S. Society*, Population Trends and Public Policy Series #2, Population Reference Bureau, 1981.)

is that it will remain nearer the current level of 1 million persons per year. At current levels of immigration and assuming that the fertility of all racial/ethnic groups will have dropped to 1.8 children per woman, Hispanics would be the largest U.S. minority by 2020 (see Controversy 3.2).

Summary

When a country's death rates and birth rates have both declined to similar low levels, it has undergone the demographic transition from the older type of high-birth-rate–high-death-rate society to the more modern low-birth-rate–low-death-rate society. However, if death rates decline due to modern sanitary and health advances without a corresponding decrease in the birth rate, the result is rapid population growth. Birth rates may not decline if, besides the more personal satisfac-

tion they bring parents, children have an economic value to the parents as well. In some countries children can add to the family's income by working, can justify additional land allotments, may free parents from simple household and child-care tasks, and represent a source of security for the parents' old age. A lack of educational opportunities also contributes to high birth rates, as may certain social patterns such as the way property is distributed or inherited.

Population pyramids illustrate the comparative sizes of the various age groups in a population. In developing countries, the population of children is much greater than the population of older people. This is due to decreasing death rates, especially among infants and children, that have not been matched by decreased birth rates. Population pyramids for these countries have a wide base and a narrow top. In contrast, population pyramids in developed countries are approaching

CONTROVERSY 3.2

Immigration

Most Americans recognize the value of immigrants.
Lawrence Fuchs, former executive director, Select Commission on Immigration and Refugee Policy

To me, it's a matter of common sense: the unemployed in this state and across the nation will never get jobs as long as we continue to take in twice as many immigrants as the rest of the world combined.
Governor Richard D. Lamm, Colorado

The doctrinaire liberal view tends to equate immigration control with discrimination, even racism. And I resent that. Immigration can and should be controlled.
Edward Abbey, Arizona author and environmentalist

Certainly most Americans recognize the value of past immigration: most of us are the descendants of immigrants. And as Lawrence Fuchs (1986) points out, "All research shows that immigrants are extremely productive. They contribute to economic growth. They enrich our cultural and social life. Accepting immigrants into the United States also shows the world how confident we are about our ideals and institutions."

But has the situation changed today? Do immigrants threaten to take jobs from U.S. citizens? Does it make a difference that the cultural mix of immigrants has changed radically over the past 100 years? Do you favor stricter controls on immigration, or do you think we should take in more immigrants? What about refugees? Does the United States have a historic mission to accept refugees? Will we benefit or lose if we continue to accept large numbers of refugees? Are you in favor of the new U.S. immigration law that allows illegal aliens resident since 1982 and illegal alien farm workers who have worked at least 90 days in the year prior to May 1986 a chance to become citizens?

Sources: L. Fuchs, quoted in P. Murphy and P. Cancellier, *Immigration: Questions and Answers*, Population Reference Bureau, 1982. D. Lamm and E. Abbey quoted in W. Schmidt, "Colorado Governor Seeks to Halt Illegal Aliens," *New York Times*, August 1, 1983, p. A-9.

---

an obelisk shape because converging birth and death rates tend to result in the same numbers of people in all the various age groups. Population momentum refers to the fact that even where successful population control programs are in effect, the populations of most developing countries will increase greatly over the next 20 to 40 years because the large numbers of children now age 0–15 are likely to marry and have at least two children per couple. This will swell all of the older age groups (which are now relatively small).

Total fertility rate is a measure of the number of children a hypothetical woman will have over her lifetime. It is a more sensitive measure of reproductive behavior than crude birth rate.

Rapid population growth makes it difficult for a country to feed, educate, and provide health care for increasing numbers of people. Environmental degradation is often found as larger numbers of people try to live off land that cannot support them. There is much evidence that population growth itself is a block to economic development. Available funds must often be spent on short-term needs—medical care, food, disaster relief—rather than on long-term investments such as roads, industrial development, and job training that would provide a better standard of living for people in the future.

In developed countries population growth problems include absorption of immigrants into the mainstream, possible scarcities of resources, and crowding (which limits choices).

The world is tending to become an urban rather than a rural society. Advantages associated with living in cities include easier delivery of health care and social services and greater economic opportunities. Disadvantages include crowding, violence, and crime.

Immigration is a growing phenomenon in the world today as people move in search of better economic opportunities and escape from political persecution. Between one-fourth and one-half of U.S. population growth is now due to immigration rather than natural increase. Still, immigration cannot provide the safety valve it once did for overpopulation. There are not enough unpopulated countries to absorb the excess of people. However, immigration offers advantages to both host countries (which receive needed labor) and sending countries (which often receive funds sent home).

Questions

1. Define the terms *birth rate, death rate, rate of natural increase,* and *fertility rate.*
2. Although the U.S. fertility rate has reached replacement level or lower, the population is still growing. At current birth, death, and immigration rates, the population will double in about 90 years. Explain why the U.S. population will continue to grow.
3. What is meant by the *demographic transition*? Why do you think this transition has occurred in developed nations but not yet in developing ones?
4. List the problems caused by rapid population growth in developing nations. Compare them to the problems caused by rapid population growth in developed nations.
5. Some experts have called high fertility a result of poverty as well as a cause of poverty. What does this mean?
6. Can you support the proposition that developing countries should spend some of their scarce funds on protecting the environment?

Further Reading

Dumond, D. E. "The Limitation of Human Population: A Natural History," *Science,* **187** (February 28, 1975), 713.

Faucett, J. T., et al. "The Value of Children in Asia and the United States: Comparative Perspectives," *Papers of the East-West Population Institute,* No. 38, July 1974.

Ware, Helen. "The Economic Value of Children in Asia and Africa: Comparative Perspective," *Papers of the East-West Population Institute,* No. 50, April 1978.

These papers are available in single copies from the East-West Population Institute, East-West Road, Honolulu, Hawaii 96822.

World Development Report 1984. New York: Oxford University Press, 1984.

This is probably the most comprehensive and up-to-date report on population issues, although the slant is heavily, and understandably, economic.

Merrick, T. "World Population in Transition," *Population Bulletin,* **41**(2) (April 1986).

A good summary of the social factors involved in rapid population growth and methods to slow such growth.

Population Reference Bureau, 2213 M Street N.W., Washington, D.C. 20037.

Publishes a monthly newsletter as well as many position papers, teaching aids, wall posters, and population data sheets. Memberships, which entitle the owner to almost all of the publications (a real wealth of information), are available to teachers and students at reduced rates. The writing is clear and at a level intended for the general public.

McNamara, R. "Time Bomb or Myth: The Population Problem." Address to the World Population Conference, 1984.

Good summary of current world population issues and current U.S. policy, which is, at the moment, in conflict with much of accepted theory of population growth control.

Immigration Statistics: A Story of Neglect. Washington, D.C.: National Academy Press, 1985.
The NAP doesn't agree with illegal immigration statistics published by the U.S. Immigration and Naturalization Service.

References

Jones, W. "The Food Economy of the Ba Bugu Ijoliba, Mali." In P. F. McLouglin (Ed.), *African Food Production Systems.* Baltimore: Johns Hopkins University Press, 1970.

Dyson-Hudson, R. and N. Dyson-Hudson. "The Food Production System of a Semi-Nomadic Society: The Kari-mojong, Uganda." In P. F. McLouglin (Ed.), *African Food Production Systems.* Baltimore: Johns Hopkins University Press, 1970.

Cain, M. "The Economic Activities of Children in a Village in Bangladesh." Paper presented at the IUSSP International Population Conference, Mexico City, 1977.

Agarwal, A., et al. *The State of India's Environment.* Washington, D.C.: International Institute for Environment and Development, 1982.

U.S. Population: Where We Are, Where We're Going. Washington, D.C.: Population Reference Bureau, June 1982.

Simon, Julian. Quoted in *Population Today*, May 1984, p. 8.

Fuchs, L. Quoted in L. F. Bouvier and R. W. Gardner. "Immigration to the U.S.: The Unfinished Story," *Population Bulletin*, **41**(4) (November 1986).

CHAPTER FOUR

Limiting Rapid Population Growth

Components of the Problem

The Value of Factors Other Than Economic Development/ The Proximate Determinants of Fertility/The Ethics of Population Control

Methods of Limiting Population Growth

Providing Birth Control Information/Instituting Measures That Improve People's Social and Economic Well-being/ Providing Incentives and Disincentives

Time Scale and Control of Rapid Population Growth

Population Momentum/The Speed at Which Different Policies Take Effect

Experience of Population Programs in Four Regions

China: The One-Child Family/Brazil: An Unequal Distribution of Wealth/India: Social Factors and the Limits to Coercive Measures/The Sub-Sahara: Human Dignity and Population Control

Outlook for Population Growth in the Future

Problems Developing Countries Must Solve/ZPG and Other Problems Developed Countries Must Solve/ Population Projections

Components of the Problem

Two important lessons have been learned in the past 20 years of population research. The first is that reductions in population growth are possible for countries at all levels of economic development. It is not necessary to progress through the traditional stages of industrialization before population growth can be controlled. The second is that the mix of methods that will be successful in a particular country will vary, just as beliefs and social customs vary from country to country.

The Value of Factors Other than Economic Development

At the 1974 World Population Conference, a popular saying among delegates from developing nations was "Development is the best contraceptive." The examples of Korea, Singapore, and Hong Kong, all developing nations that were rapidly industrializing and at the same time experiencing fertility declines in the 1960s, seemed to prove the truth of this slogan. Now, however, we have other examples that do not fit this picture. Birth rates have fallen faster in Thailand and Colombia than in Brazil and Venezuela, despite the fact that these last two countries have been industrializing more rapidly, have had better economic growth, and have achieved higher average incomes. What is the reason for this? A closer look at the four countries shows that incomes and social services are distributed more evenly in Thailand than in Brazil or Venezuela. Further family planning services are more available in both Colombia and Thailand. Thus, other factors besides economic development are able to affect fertility.

Untangling the various factors that affect fertility is both a difficult and an important task. To design successful population programs, we must understand how individuals make decisions about family size and how various programs affect these decisions.

The Proximate Determinants of Fertility

Fertility, or the number of children a family will actually have, is determined by the parents' age at marriage, their use of contraception, and their decisions about breastfeeding and abortion. These are called the *proximate determinants* of fertility. These behaviors are the result of decisions the family makes on how many children they want, whether they will educate the children, whether the mother will work outside the home, and other savings and consumption decisions.

To go one step further back, the decisions a family reaches in these matters depend on the socioeconomic climate in which the family lives (see Figure 4.1). When a government wishes to affect fertility, it can attempt to influence the proximate determinants. However, it can also alter the socioeconomic climate. Although such government actions will be indirect, they can have the strongest and most lasting possible effects on fertility.

The Ethics of Population Control

When a government attempts direct control over the proximate determinants of fertility, ethical issues often arise. Forced contraception, forced abortion, and insistence on breastfeeding are generally felt to be unacceptable interference with personal liberties. For

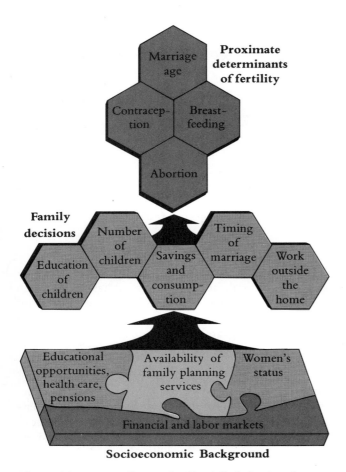

Figure 4.1 Factors affecting fertility. The behaviors that determine fertility, called the proximate determinants of fertility, are the result of family decisions. These decisions crystalize from the socioeconomic climate in which the family lives. Governments can have their strongest and most lasting effect on fertility with policies that change this socio-economic climate: restrictions on marriage age and child labor, provisions for health care and universal education, improvements in the status of women, old age pensions and incentives for small families, promotion of breastfeeding, and provision of contraceptive services.

instance, suppose a government decided to try to take active control of individual reproduction. In such a situation, a license might be issued to allow a couple to have a child, or forced sterilization programs might be carried out, as they were in parts of India in the 1970s.

There are a number of valid objections to these policies. Who would decide, and on what basis, which couples deserve to have children? The tenets of individual freedom might be lost in such programs. On the practical side, the technical capability does not yet exist

to carry out such a policy. There is no magic chemical that can be put in water to prevent fertility without other serious effects. Not enough doctors and nurses are available for forced, large-scale contraceptive or sterilization procedures. Thus, from a purely practical standpoint, for a family limitation program to succeed it must be made to seem desirable to everyone. (See Controversy 4.1 for another view about this issue.)

Only one of the determinants of fertility—marriage age—is commonly legislated. This is usually for social and health reasons, to prevent childbearing by young girls whose health would suffer. On the other hand, in at least one country, China, the allowed marriage age has been raised to decrease fertility (because the span of married women's childbearing years will then be decreased.)

Another possible means of population control that should immediately be ruled out is allowing mortality to rise or remain high when measures exist to prevent this. Most people would agree that a government's role should include improving the lives of its people. It is for this reason that population control programs are begun: to prevent the miseries that can accompany rapid population growth. Failure to reduce mortality when it is possible works counter to this goal. (Not everyone subscribes to this belief. See Controversy 8.1, page 193.)

Methods of Limiting Population Growth

Providing Birth Control Information

When a government wishes to limit the growth of its population, the first policy it usually adopts is to provide people with birth control information and materials. In many cases, abortions are made legal and thus easier for the poor to obtain. Although several religious organizations object to both abortion and the distribution of birth control information, there is usually also a great deal of popular support for this type of policy. In most situations, people will at least tacitly agree that couples should be able to limit their family size if they wish to do so. For instance, in mostly Roman Catholic Colombia, despite initial opposition by the Church, wide government distribution of contraceptives has helped reduce the birth rate from 3.4% to 2%.

A policy of providing access to information and materials on birth control can be justified on grounds

CONTROVERSY 4.1

The Ethics of Limiting Family Size

Any choice and decision with regard to the size of the family must irrevocably rest with the family itself . . .
United Nations Universal Declaration of Human Rights (1967)

Freedom to breed will bring ruin to all.
Garrett Hardin

Many difficult decisions are involved in the control of population growth. One of the most difficult is whether couples should have the complete freedom to decide how many children they wish to have. That is, should governments set and enforce upper limits of family size?

In 1967, the United Nations passed the Universal Declaration of Human Rights, which stated:

The Universal Declaration of Human Rights describes the family as the natural and fundamental unit of society. It follows that any choice and decision with regard to the size of the family must irrevocably rest with the family itself, and cannot be made by anyone else.

In the past few years, however, a few nations have decided that this concept is unworkable. Thus, in Bangladesh, Singapore, China, and India, governments have begun to use both economic threats and bonuses designed to limit population size.

Demographer Garrett Hardin believes that the decision to limit family size cannot be left to the individual. He says that such a decision is similar to one that faced cattle owners in eighteenth-century England. At that time, pastureland was held in common, so there was a great advantage to a herdsman to add more cattle to his herd, if he could. If everyone did that, however, the pasture would be overgrazed and thus ruined.

Hardin argues that, in a sense, our welfare-oriented society is like a commons. Individuals can choose to use up more than a fair share of resources, both social and environmental, by having more than their fair share of children.

Hardin (1968) wrote:

Perhaps the simplest summary of this analysis of man's population problems is this: the commons, if justifiable at all, is justifiable only under conditions of low-population density. As the human population has increased, the commons has had to be abandoned in one aspect after another.

First we abandoned the commons in food gathering, enclosing farm land and restricting pastures and hunting and fishing areas. These restrictions are still not complete throughout the world.

Somewhat later we saw that the commons as a place for waste disposal would also have to be abandoned. Restrictions on the disposal of domestic sewage are widely accepted in the Western world; we are still struggling to close the commons to pollution by automobiles, factories, insecticide sprayers, fertilizing operations, and atomic energy installations.

. . . I believe it was Hegel who said, "Freedom is the recognition of necessity."

The most important aspect of necessity that

we must now recognize, is the necessity of abandoning the commons in breeding. No technical solution can rescue us from the misery of overpopulation. Freedom to breed will bring ruin to all. At the moment, to avoid hard decisions many of us are tempted to propagandize for conscience and responsible parenthood. The temptation must be resisted, because an appeal to independently acting consciences selects for the disappearance of all conscience in the long run, and an increase in anxiety in the short.

The only way we can preserve and nurture other and more precious freedoms is by relinquishing the freedom to breed, and that very soon.

Do you agree with Hardin? Should the decision on family size be made by each couple, or should governments have a voice in it? If you feel that population growth should be slowed, should appeals to conscience be tried first? What did Hardin mean when he said "an appeal to independently acting consciences selects for the disappearance of all conscience in the long run"? Suppose such appeals don't work—can you think of economic measures, such as the use of taxes, that could influence family size? Suppose neither appeals nor economic measures are effective—are penalties justifiable?

Source: Garrett Hardin, "The Tragedy of the Commons," *Science*, **162** (December 13, 1968), 1243.

---

of humanitarianism. Fewer children will, in general, be better fed and cared for, since a family's resources will not be spread too thin. These children may even be healthier, since repeated, closely spaced pregnancies have been shown to be harmful both to mothers and to babies (see Figure 4.2). Furthermore, if only wanted pregnancies occur, mothers have no need to resort to abortion, which, like any operation, carries some risk. For this reason, many antiabortion groups support contraceptive programs. The cost to society of unwanted children who may become wards of the state in one way or another is also reduced.

There is clearly still an unmet need for contraceptive services in many developing countries today. By unmet need we refer to the percentage of women who want no more children yet who are exposed to the risk of pregnancy and are not using a contraceptive method. Figure 4.3 illustrates the size of this unmet need for three developing countries.

Couples wishing to use birth control must solve several problems. Financial costs may be a significant problem for couples in many developing countries. Medical assistance must be found for a variety of methods, while methods such as sterilization have time-related costs. Furthermore, social pressures may work against contraceptive use. Government programs can attack all of these problems, helping to increase contraceptive use. Box 4.1 notes the major forms of contraception available for family planning programs today, including their advantages and disadvantages. Material on the human reproductive system and a more detailed

description of contraceptive methods can be found in Belcastro (1986).

Instituting Measures That Improve People's Social and Economic Well-being

Access to birth control services alone is not enough to insure a reduction in population. As previously mentioned, there are a number of reasons why couples may choose to have more children than is considered ideal for society as a whole. A massive program of public education and contraceptive distribution in Pakistan from 1965 to 1980 failed almost completely, because of social factors. The failure was partly administrative: information about the program failed to reach a large part of the population. But in addition, a demand for contraceptives was not generated in this society.

Two other types of policies are available to governments: those that improve the social and economic well-being of people in general, which as a by-product spur a reduction in population growth, and those that offer specific rewards and incentives for individual families who reduce fertility and economic punishments for those who do not.

Old-age security programs. Recall from earlier discussion that children represent security for parents in old age. If old-age pensions, such as the social security program of the United States, are created in developing nations, parents may be able to give up the idea that they need many children. Bangladesh, for instance, is

Infant mortality per 1,000 live births, Peru

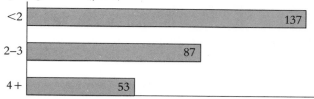

Spacing interval (years)

<2	137
2–3	87
4+	53

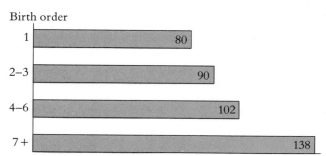

Birth order

1	80
2–3	90
4–6	102
7+	138

Maternal mortality per 100,000 live births, Matlab, Bangladesh

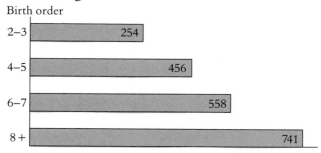

Birth order

2–3	254
4–5	456
6–7	558
8+	741

Figure 4.2 Mortality of infants and mothers compared to birth order and spacing. Infants and mothers both suffer higher death rates as more and more children are born into a family. (From The World Bank, *World Development Report 1984* [New York: Oxford University Press, 1984], pp. 128–129.)

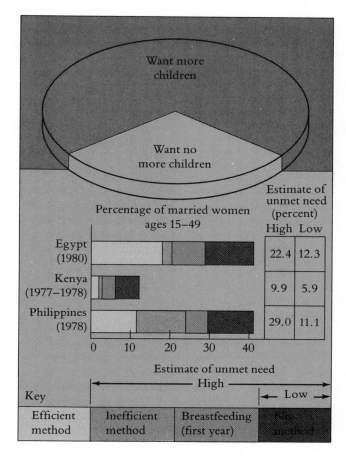

Figure 4.3 Contraceptive use and the unmet need for contraception for selected countries, 1977–1981. The low estimate of unmet need is the percentage of women using no contraception. The high estimate is the percentage using an inefficient method (withdrawal, rhythm, and the like), breastfeeding (within a year of a birth), or no method. (From The World Bank, *World Development Report, 1984* [New York: Oxford University Press, 1984], p. 131.)

planning to institute a program of social security benefits. To emphasize the desirability of small families, these benefits are to be awarded only to parents of two or fewer children.

Reductions in infant and child mortality. Another reason many parents have more than two children is to ensure that they will be able to raise at least two children to adulthood. Africa, the region with the world's highest birth rate, also has the world's highest death

rate.* Continuing reductions in infant and child diseases in developing nations will pave the way for a reduction in births.

Child labor and child education laws. In many developing countries, children may be sent to work and so represent to the parents the possibility of additional

---

*In 1982 the birth rate was 46 and the death rate near 17 per 1,000 per year.

Box 4.1 Birth Planning Technology

Several methods of birth control have been practiced throughout human history—abstinence, abortion, prolonged breastfeeding, and coitus interruptus (withdrawal)—but with uncertain effectiveness, and psychological and health damage. Contraceptive research in the past 30 years has made possible a much greater variety of more effective methods. Combined estrogen and progestin oral contraceptives (the "pill") and various intrauterine devices (IUD) were the first major breakthroughs in the late 1950s and early 1960s. Since then other methods have been developed: injectable contraceptives effective for two to three months; more effective copper and hormone-releasing IUDs; menstrual regulation (vacuum aspiration of the uterus within seven to fourteen days of a missed period); male sterilization; simplified female sterilization by laparoscopy and minilaparotomy; low-estrogen pills with fewer side effects; and a progestin-only "mini-pill." Barrier methods, such as the condom, diaphragm, and spermicides, have also been improved.

In 1980 the most commonly used methods of birth control worldwide were sterilization and the pill. Among developed countries the pill is the most used method, but sterilization has gained in popularity in the United States and in Great Britain, where it accounts for about a quarter of total use among married couples of childbearing age. The major exceptions to this pattern are Spain, Italy, and the Eastern European countries (except Hungary), where withdrawal, rhythm, and abstinence are still the most prevalent methods.

Among developing countries, sterilization is the most common modern method in Bangladesh, El Salvador, India, Korea, Nepal, Pakistan, Panama, Sri Lanka, Thailand, and Tunisia. The pill is the most favored method in Egypt, Jordan, Syria, much of Latin America, Malaysia, and Indonesia. Injectable contraceptives are widely used in Jamaica (11% of eligible women), Thailand (7%), Trinidad and Tobago (5%), and Mexico (3%); this method is convenient to use for rural women and, unlike the pill, does not interfere with lactation. Both the World Health Organization and IPPF have approved injectables—legal in more than 100 countries—but greater use in developing countries is partly constrained by the method's limited availability. The United States, the major contraceptive donor worldwide, cannot donate injectables because U.S. assistance policy prohibits supplying drugs not approved for domestic use.

Despite the greater variety of contraceptive methods now available, all have shortcomings:

Effectiveness. Under the ideal conditions of controlled studies in developed countries, existing methods can be highly effective in preventing pregnancy: nearly 100% for sterilization, the pill, and injectables; 98% for the IUD; and as much as 97% for the condom and the diaphragm after one year of use. But outside these controlled studies, some methods can be significantly less effective owing to incorrect or inconsistent use. In the United States, one in 100 couples using the pill will have a pregnancy within one year, more than two couples using the IUD, twelve using the condom or diaphragm, and twenty using rhythm. In the Philippines more than three women out of 100 using either the IUD or the pill and thirty-three using rhythm will become pregnant within a year. The motivation of couples to prevent pregnancy is important in the effectiveness of contraceptives. Couples who want no more children are likely to use methods more effectively than those who are spacing births.

Side effects. Physical side effects are a main reason that people switch, or stop using, contraceptives. For some methods, the long-term health risks of prolonged use are unknown. Methods such as the IUD and injectables, which alter bleeding patterns—by causing spotting between periods, increased or decreased flow, or amenorrhea—may be culturally unacceptable or restrict the activities of users.

Inconvenience. Barrier methods (condom, diaphragm, spermicides) have to be used each time couples have intercourse. In households in developing countries, pills and diaphragms are difficult to store and condoms difficult to dispose of.

Reversibility. Sterilization is highly effective but rarely reversible. Injectables are completely reversible but delay the return to fertility for several months.

Acceptability. To some couples, abortion and sterilization are religiously or culturally unacceptable; some may regard only abstinence or rhythm as acceptable.

Delivery. Sterilization (of both men and women) requires skilled medical or paramedical staff, who are often scarce in developing countries. The IUD, injectables, and the pill require medical backup for treatment of complications and side effects. Programs that promote the condom, pill, and spermicidal foam require a good network of supply points.

No single method of contraception is appropriate to the needs of all people, nor is there one that is completely safe, reversible, effective, and convenient. Nor is such an "ideal" method likely to be developed in the next 20 years. Family planning programs will have to rely on a mix of existing methods and a few new ones whose development is already well advanced.

Research is being concentrated in two areas: improving the safety, convenience, and life span of existing methods, such as the IUD, pill, injectables, and female barrier methods; and developing new methods, such as a monthly pill to induce menstruation, long-lasting biodegradable hormonal implants for women, nonsurgical chemical sterilization for men and

women, a male "pill," and an antipregnancy vaccine for women. Some of these new methods—such as the hormonal implant (in the arm), improved IUDs, the vaginal sponge, cervical cap, and diaphragms that release spermicide—may be widely available in the near future. Others, such as new male methods and an antipregnancy vaccine, require much more research and are unlikely to be marketed before the end of this century.

Compared with the past few decades; the pace of technological development is slowing. Worldwide funding for contraception-related research was $155 million in 1979, but has been declining in real terms since 1972–1973. About 30% of the total is spent on contraceptive development and safety studies; the rest goes to training and basic research on human reproduction. Some 72% of the total was spent in the United States. Over 80% of the total was financed by the public sector; industry's share shrunk from 32% in 1965 to less than 10% in 1980. Special testing and regulatory requirements, combined with product-liability problems, have lengthened the time between product development and marketing, increased the cost of developing new products, and made the future profitability of research more uncertain for private firms.

Source: The World Bank, *World Development Report 1984* (New York: Oxford University Press, 1984), p. 132.

family income. The passage and enforcement of laws prohibiting child labor can make additional children less desirable by removing their income potential. Parents must then feed and clothe each child for a longer time during which the child cannot work. Such laws benefit children by allowing them time for schooling and, depending on what work they would have done, possibly improving their health (Figure 4.4). The state of Kerala in India has a per capita income below the average for all of India, yet it devotes 39% of its budget to education and 16% to health and family planning services. Almost all children in Kerala attend at least the primary school grades. The result? Birth rates in Kerala declined from 37 per thousand in 1966 to 25 per thousand by 1978.

Education, besides helping the children themselves and society as a whole, has several benefits related to

population control. Educated people tend to marry later and to have fewer children. This is related to the fact that education allows people to move upward in economic and social status. Upward mobility is hindered by large numbers of children to feed, clothe, and care for. Educated women tend to limit their families to allow themselves the satisfaction of a career and additional family income.

Improvements in the status of women. Improvement in the status of women may act as a brake on population growth. Given the opportunity to take advantage of new or better job openings, women often limit their family size to reduce their child-care responsibilities. In the United States, the educational achievements of women, as well as men, have been steadily

Figure 4.4 Young, nimble-fingered children toiled long hours at low pay in the nineteenth-century factories and mines of today's more developed countries. (U.S. Library of Congress; photo by Lewis Hines)

growing. In addition, the percentage of women who work has been increasing. These factors have probably contributed to the downward trend in the U.S. birth rate. On the other hand, in parts of Africa and Asia where many family members can help care for young children or where uneducated and unskilled babysitters are easy to find, an increase in the number of women in the workforce does not have much influence on the birth rate.

Indirect influences on marriage age. Additional ways to influence population growth are to control housing availability and to require a period of military or other national service for all young people. These actions tend to increase the age at which marriage occurs, reducing the span of years in which a couple can conceive.

Promotion of breastfeeding. Breastfeeding has a definite contraceptive effect. In fact, up to 10 years ago it was responsible for more contraception in developing countries than family planning programs. The

contraceptive effect of breastfeeding is not lasting, however; it decreases with time after childbirth. Furthermore, although failure to menstruate as a result of breastfeeding is a reasonable guarantee of contraception, about 7% of women become pregnant without resuming menstruation. Despite the fact that breastfeeding is well recognized as contraceptive, the trend appears to be toward less and shorter periods of breastfeeding in developing countries. This is partly due to women working outside the home, especially if the workplace is far from the home. But it is also due to feelings that bottle-feeding is more modern or that breastfeeding is difficult or does not provide enough food for the child.

A decrease in breastfeeding not only increases the risk of pregnancy (if no other contraceptives are used) but also may endanger the health of the baby. In countries without a safe water supply and where mothers cannot read or understand formula-making directions, bottle-fed babies suffer much more diarrhea and malnutrition.

Government programs pointing out the health benefits of breastfeeding and discouraging the use of formula in hospitals and clinics can increase the percentage of mothers breastfeeding their babies. In Papua New Guinea a campaign to discourage advertising of infant formula, combined with restrictions on formula use and distribution in hospitals, increased the proportion of breastfed children under age 2 from 65% to 88% in only 2 years. The additional benefit of breastfeeding, over and above child health improvements, is, of course, a "free" contraceptive effect.

Raising of incomes. There is no question that raising of incomes decreases fertility in the long run. In the short run, however, raising incomes for poor families can increase fertility. This occurs in areas where marriages are often delayed by the costs of setting up a household or by the need for a dowry. Higher income allows earlier marriage in such cases.

Over the long term, however, fertility decreases with increasing income (see Figure 4.5). Children's incomes become less important, and education of children becomes both possible and a better investment. Furthermore, money is available to save and invest for old age.

The various population control methods that affect the socioeconomic climate in which people live all allow individuals some freedom of choice in the number

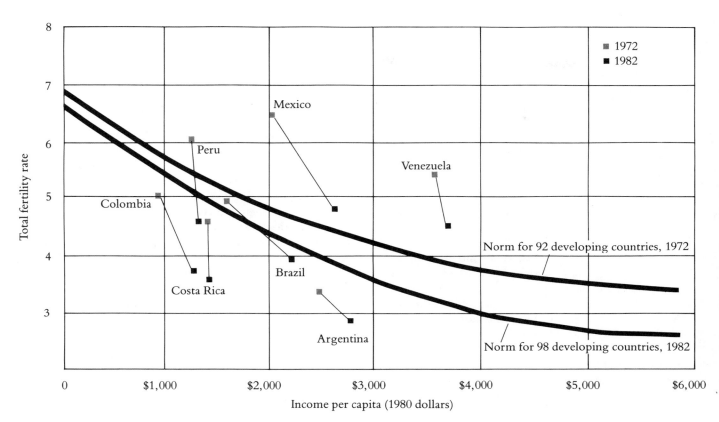

Figure 4.5 Fertility in relation to income, selected developing countries in Latin America and the Caribbean, 1972 and 1982. (From The World Bank, *World Development Report 1984* [New York: Oxford University Press, 1984], p. 171.)

of children they plan to have. These methods open other areas of satisfaction, and they also make people aware of the additional costs that children bring. (In 1982, it was estimated that each child cost a middle-income American family $85,000 in direct costs and another $55,000 in lost opportunity costs, mostly income the mother went without while taking care of the child.)

Through such mechanisms, population growth might be reduced without coercion. Evidence shows that, under the right circumstances, people become aware that having more than two children is not in their own best interest. Countries offering the greatest possibility of upward movement in economic and social status have shown the most dramatic decreases in birth rate. Thus, the ideal of bringing population under control may be achieved by means that, at the same time, help people reach a satisfying standard of living.

Providing Incentives and Disincentives

The second group of policies for controlling population growth consists of rewards for families with few children and penalties for those with more than a designated number. Although such schemes are in effect in over 30 countries, it is difficult to determine how much of an effect they have had, because they are usually accompanied by social change and an increase in family planning services.

Incentives and disincentives have included such things as educational or other bonuses for having few children and fines for having more than a specified number of children. The main objection to such policies is that children in the larger families, who had no control over whether they were born, suffer along with the parents who are subject to penalties or forgo rewards. Another objection is that verification that couples are in fact complying with the limits on children can be difficult. Further, the cost of the rewards may be high, and money might be better spent on other investments.

Incentives are transfer payments. That is, people are paid not to have children who, if they were born, would require government money in other ways (education, health care, and so on). In this sense, incentives do not use up government resources but rather use them in a different way. But, unless parents put money aside for such things as old-age care, which the child would have been expected to provide, the transfer will not be successful.

Despite these drawbacks, such policies are, in fact, being followed in many countries and are planned in others. Government employees in certain Indian states are not eligible for housing loans and land grants if they have more than three children. Singapore levies increasing fines for the birth of a third, fourth, fifth, or sixth child. Maternity leave is not allowed for the birth of a third or subsequent child, and hospital fees rise with the number of children in a family. On the other hand, sterilization after two children is rewarded with education and employment for the children and extra vacation time for the parent. Foreigners who wish to marry Singapore nationals must agree to be sterilized after the birth of their second child.

China has instituted the most extensive set of incentives and disincentives of any country in its attempt to achieve a fertility level of only one child per family. The specific example of China is discussed later in this chapter.

Time Scale and Control of Rapid Population Growth

Population Momentum

As mentioned in Chapter 3, programs designed to reduce population growth have one very serious built-in problem: It takes some 10 to 20 years before the results of any such program are seen. The reason is that fertility reduction programs have an effect only on the number of babies currently being born; they can obviously have no effect on the group of children, ages 0–15 years, already born. These children will grow to adulthood and marry and have children themselves during the next 10 to 20 years.

In most countries, the size of the 0–15 age group is larger than that of any other group in the population. If the people in the 0–15 age category turn out to have only enough children to reproduce themselves, they will still swell the size of their country's population. Thus, most countries would continue to grow in population for a time even if birth control programs immediately reduced the birth rate to about 2.1 to 2.3 children per couple, the birth rate that provides for each couple to reproduce themselves. This phenomenon is called *population momentum*.

The fact that population programs take such a long time to make their effects felt is a great handicap to effective action. It takes 5 to 6 years before decreased school enrollments are seen, 15 years before a reduction in the labor force occurs, and 15 to 19 years before stabilization begins in the amount of food needed. A government may thus seem to be spending a great deal of money for many years with very little obvious effect. Because of this time gap, it is important to try to separate birth control programs from politics and politicians, as political movements and ideologies have a way of changing or going in and out of favor in less than 10 or 20 years.

The Speed at Which Different Policies Take Effect

Besides the inherent problem of population momentum, various fertility control methods differ in the time it takes for their effects to be seen. The long-term effect of raising incomes is clearly to lower fertility, although the short-term effect may be just the opposite. Other methods that tend to require a long time to be effective include educational programs, reduction in infant and child mortality, and improvements in the status of women. In contrast, promoting later marriage, encouraging breastfeeding, and making contraception easier to obtain all have more immediate effects.

Experience of Population Programs in Four Regions

The main problem in devising ways for reducing population growth is, of course, the fact that the issue is not some abstract scientific problem solvable by dollars and technology. Instead, it is a problem that touches people directly. We can program computers to provide projections of population, but only people's behavior, plans, and aspirations can determine what the future will be. People's needs and feelings affect how they react to incentives and plans. People also exert a degree of control over policies that affect them. In the United States, for instance, population control policies must be seen as desirable to the majority of people before they have any chance of being put into action.

The experience of population control programs in four countries or regions is instructive because it illustrates not only successes versus failures but also how population control programs must be tailored to the specific values and sociopolitical system of each country.

China: The One-Child Family

The most ambitious experiment in controlling rapid population growth is taking place in China today. One billion people, almost a quarter of the world's population, live in China. (India is second to China in population, with 762 million people, the Soviet Union is third, and the United States fourth.) Population growth in China was not considered a problem until relatively recently. China had a **pronatalist** policy (in favor of population growth) until 1957. By 1971, however, Chinese leaders clearly felt that population growth was making it difficult to achieve their goal of a better living standard for all Chinese people. Late marriage was legislated, and a spacing of 3 to 7 years between births was encouraged, as was a maximum of two children per family. Local governments were charged with the responsibility of seeing that fertility goals were met through incentives, disincentives, and social pressure on individual families.

In 1978, however, Chinese leaders learned that the situation was worse than they had realized. Better counting procedures revealed that there were almost 10% more people in China than had been thought. In addition, a huge group of young people was about to enter their reproductive years. The population was close to 1 billion and growing, while new studies of land and water resources showed that no more than 1.2 billion could be supported at a reasonable standard of living. Thus began a campaign to convince each family to have no more than one child. The goal of this campaign is to stop the growth of the Chinese population

at 1.2 billion people by the end of the century and thereafter to allow it to shrink toward 800 million.

Couples are encouraged to sign a pledge to have only one child. In return, they receive a certificate entitling them to a variety of benefits, depending on where they live. Examples of benefits include: free medical care and schooling for the only child, cash bonuses or work points usable for extra food, old-age pension benefits, preferential housing treatment, and job preference for the child when grown up. If the couple then has another child, the couple must not only surrender all privileges but must pay fines amounting to a substantial portion of the family's income.

While the campaign has been extremely successful in urban areas, it has not fared as well in rural areas, where 75% of the population lives (see Figure 4.6). Furthermore, there is a large group of young people in China now coming into their childbearing years (see Figure 4.7). This group of young people will severely test the Chinese system of population control. If the one child per family system is not successful and total fertility falls only to 2.1 children per woman, the population will grow to 1.63 billion in 50 years. On the other hand, if the policy *is* successful, the population will peak at 1.2 billion in 2007 and then decline. However, this can bring other problems. Figure 4.8 shows what China's population pyramid would look like in 2032 if a fertility rate of 1.0 were achieved by 1992. The problem with such an age structure is that most of the population would be concentrated in the older, nonworking age groups and would need to be supported by a much smaller labor force. If, on the other hand, the one child per family policy is maintained for only about 20 years, through the 1990s, and then is changed to a replacement level policy, such economic problems would be largely avoided.

The new Chinese policy of responsibility, in which rural families are assigned a certain quota of farm production and can then keep the excess they produce, works against the one-child campaign, since more children enable a family to produce more. In some cases, larger families are even given more land to start with. In addition, few Chinese are currently covered by old-age security systems—perhaps 15% in urban areas and 1% in rural brigades. Children thus remain the main source of security for old age.

A major feature of the Chinese system of population control is that local authorities are given re-

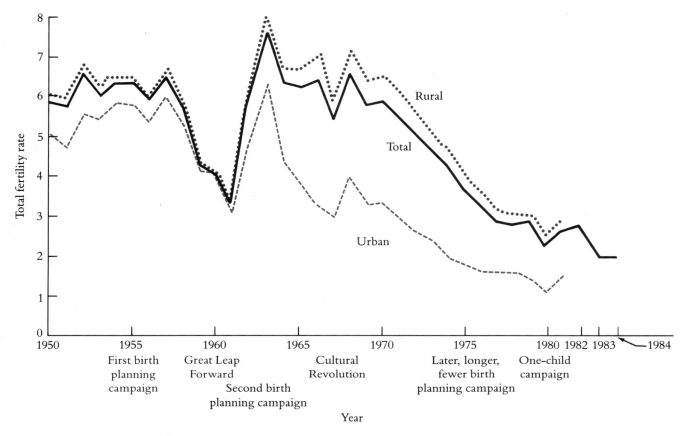

Figure 4.6 Urban and rural fertility rates in China. In 1950 there was only about a 10% difference between rural and urban fertility rates. Rapid declines in urban fertility rates combined with much slower declines in rural areas have resulted in rural rates that are now twice urban rates.

sponsibility for meeting stated goals. Although the government itself does not advocate coercion in the strict sense, social pressure is used extensively to encourage families to conform to the one-child policy. There have been rumors of forced abortion and sterilization by overzealous local officials. Another kind of problem has appeared in reports of sex ratios at birth. Normally about 105 males are born for each 100 females. In rural areas, for births of children after the second child the sex ratio is 112.4 males for each 100 females. Thus, in 1981 there appeared to be some 60,000 girls missing in the group of about 1 million children born third, fourth, or fifth in their family. This is thought to be evidence that a substantial number of girl babies were unreported or done away with. In areas where sons are still considered essential for parents' security in old age, female infanticide is a possibility.

Chinese psychologists are also concerned about the effects on children who grow up without siblings, uncles, aunts, or cousins. It remains to be seen whether such radical changes as the one-child program are possible or whether a more moderate program will eventually be adopted. (One possibility is the recently proposed two-child family with a 10-year space between the births of the two children. In this plan it is assumed that 30% of couples will still choose to have only one child, due to the economic incentives the government offers to such couples.) The sheer size of the Chinese population makes these issues of fertility control important not only to the Chinese but to the world as a whole.

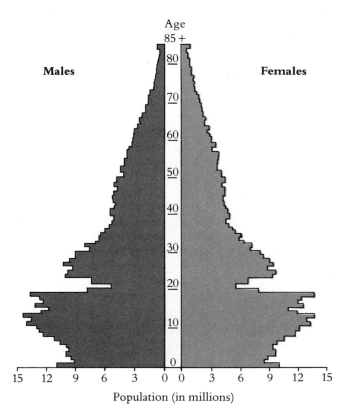

Figure 4.7 Population pyramid for China in one-year age groups.★ From the top to about age 20, China's pyramid resembles a country with rapidly slowing growth and might suggest a country on a rapid demographic transition. Below age 20, the pyramid gives China a most unusual demographic appearance. Despite the recent declines in Chinese fertility, those below 20 account for nearly half of the population. Looking at the pyramid, it seems clear that the smaller birth cohorts of recent years (the age groups between 1 and 10) result at least in part from the drastically reduced numbers of women in their 20s, the most fertile years. (From Population Reference Bureau, *Population Today*, March 1985, p. 5; data from the 10 percent Sample Survey of the 1982 Third National Census of China, as reported in *Beijing Review*, January 16, 1984.)

★Because of the Chinese belief in astrology, everyone knows the year and date of his or her birth. In many developing countries, one-year age groupings are inaccurate because people do not pay much attention to their age and tend to report it as the nearest 5- or 10-year interval. In Bangladesh's 1974 census, for example, 2.8 million people reported they were 30 years old and only 124,209 said they were 31.

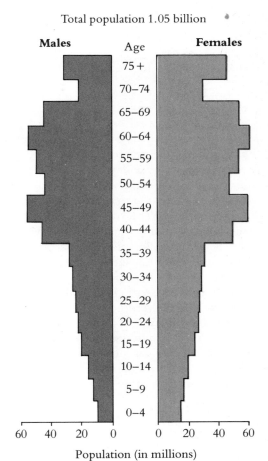

Figure 4.8 Projected Chinese population in 2032 if fertility rate reaches 1.0 by 1992. (From H. Yuan Tien, *Population Today*, April 1984, p. 7.)

Brazil: An Unequal Distribution of Wealth

In many Latin American countries, income and population are unequally distributed. The poorest 20% of households in Brazil have one-third of the children and only 2% of the income. The richest 20% of households have 8% of the children and 64% of the income. These differences are much greater than those found elsewhere in the world, even in such countries as India.

Until 1974, Brazil's government was pronatalist, on the theory that Brazil's vast land resources could support a much larger population. In addition, the Catholic church's official position against any but natural contraceptive methods (such as rhythm) inhibited the spread of modern birth control methods.

In this climate (lack of government support of birth control policies and widely unequal distribution of wealth), even a doubling of per capita real incomes between 1972 and 1982 failed to make a large impact on fertility. On the other hand, in Colombia, a country in which income is as poorly distributed as it is in Brazil but where government support of birth control policies is strong, fertility has decreased more than in Brazil despite a much smaller increase in per capita income.

Per capita income almost doubled in Brazil between 1970 and 1980 compared to only a 50% increase in Colombia and Mexico. Yet fertility fell by less than 20% in Brazil while decreasing by one-third in Mexico and Colombia. School enrollments have increased significantly in Colombia, and the labor force is growing more slowly than in Brazil, phenomena that should help to decrease the inequalities in income distribution.

Meanwhile, official government policy in Brazil appears to be changing in favor of a more active family planning role. Still, population is expected to grow at 2%–3% a year in Brazil as in most South American countries, and unless very different policies are initiated, the population will double in the next 30 years.

India: Social Factors and the Limits to Coercive Measures

It is probably misleading to consider India as a single country in terms of population growth experience. Differences in background and customs have led to differing successes in population growth control in the various regions of India. In Uttar Pradesh, infant mortality was 171 per thousand in 1978, only about 1 in 10 rural women were literate, and fertility was 5.6 children per woman. In Kerala, on the other hand, infant mortality was 47 per thousand, 75% of rural women were literate, and fertility was the lowest in India: 2.7 children per woman. The apparent explanation for such differences is found less in economic differences between the two areas, which are small, than in differences in social development, such as availability of health services, women's education, and family planning services.

Similarly, in another South Asian state, Sri Lanka, where per capita income was only $320 in 1982, fertility had fallen to 3.4 children per woman from 5.5 in 1960.

The fall was the result of decreased infant mortality, virtually 100% school enrollment of primary age girls, and increased contraceptive availability. Some South Asian countries, such as Bangladesh, have instituted incentive programs to encourage population limitation; however, it is not at all clear that a strong program similar to one in a central planned economy such as China is possible in South Asian countries such as India (see Controversy 4.2).

The Sub-Sahara: Human Dignity and Population Control

The region known as sub-Saharan Africa, which includes Ethiopia, Kenya, Sudan, Burkina Faso, Tanzania, Nigeria, and Zaire (Figure 4.9), is the only region of the world in which fertility has not begun to fall. It is also the poorest region of the world. Life expectancy is 10 years lower than in other low-income countries. Some of these countries have abundant natural resources but not the skills or physical capital to benefit from them. A larger number have neither skills and capital nor resources. Education reaches less than two-thirds of primary school children, and modern health care is available to only a small part of the population.

The current food situation in sub-Saharan Africa is probably the most serious in the world. Drought has worsened the food crisis caused by too many people attempting to grow food in a fragile semiarid land. Population control policies are found in only a few countries. In part this is because competing religious and tribal factions are suspicious that such policies will be used for the political advancement of one group or another. (In some countries census results have not even been published for political reasons.) Population reduction programs are also seen as a requirement of foreign aid programs rather than being of local concern (see Controversy 4.3). Furthermore, social pressures in Africa are still strongly in the direction of large families. Recent surveys among women in six African countries indicate a desire for six to nine children.

Lack of medical care has certainly contributed to the problem. In some regions of Kenya, people do not give their children permanent names until they are confident the children will survive the hazardous years of infancy. It is difficult to see how any real progress in the adoption of birth control methods can occur until child survival is improved. Further, extremely high

Figure 4.9 The countries of Africa.

CONTROVERSY 4.2

Family Planning in India— A Lesson on the Meaning of Freedom

Among the nations of the developing world, India has had one of the fastest-growing populations. At 300 million just after World War II, India doubled its population by the mid-1970s. Efforts to limit population have been underway there since the early 1950s. In fact, India was the first nation to create a national program to limit population through family planning. Even into the 1970s, however, the results were barely noticeable. In 1976, it was estimated that only 17.5 million couples out of 103 million in the reproductive age groups were using contraceptives.

India's program up to that time was traditional: Sterilization or birth control methods and devices were made available to those who asked for them. Although no one was forced to seek sterilization, incentives were offered. For a time, a man could obtain cash or a transistor radio (a coveted article) in return for having a vasectomy.

In April 1976, a new policy was adopted emphasizing sterilization by vasectomy and, in many ways, compelling men to submit to the operation. Indira Gandhi, prime minister of India, was quoted as saying, "We must act decisively and bring down the birth rate. . . . We should not hesitate to take steps which might be described as drastic. Some personal rights have to be kept in abeyance for the human rights of the nation: the right to live, the right to progress." The government's goal: a reduction in the birth rate from 35 per thousand to 25 per thousand by 1984. It is of use to note that the Indian government had declared a state of emergency in June 1975 and that a number of civil rights, including free speech, had been either suspended or decreased since that time. The intensive program of sterilization was begun in this political climate. Although the government did not mount a national campaign of compulsory sterilization, individual states were encouraged to do so.

The state of Maharashtra passed laws calling for compulsory sterilization for the father of three living children and compulsory abortion of a pregnancy leading to a fourth child. Incentive payments to those who submitted to vasectomies were a part of the law, as were payments to informers. In India's capital, Delhi, and in the states of Punjab and Haryana, laws were passed withdrawing vital government benefits and services from married men with two children who did not submit to vasectomy. Loss of subsidized housing, loss of free medical care and loans, and even loss of employment were possible if a man did not comply with the sterilization laws.

Widespread abuses, in which people—especially poor people—were compelled to submit, were reported as the state governments attempted to fill sterilization quotas. Riots broke out in Northern India over the issue. In the final months of Mrs. Gandhi's rule, sterilizations numbered a million per month. Then, in early 1977, Mrs. Gandhi's government fell, as the voters sent her and her Congress Party from office. It is not clear whether her government fell because of her near-dictatorial rule and her suspension of civil liberties or because of the highly visible, highly controversial program of sterilization she promoted. Many regarded the sterilization program as the largest factor in her defeat. The government that succeeded hers, led by Prime Minister Desai, abandoned the use of force and punishment for failure to be sterilized, presumably because the issue remained highly charged politically. In 1977–1978, sterilizations fell to 11% of the previous year's total. Although she was reelected in January 1980, Mrs. Gandhi afterward kept a low profile on population control efforts.

This episode raises some basic questions about government birth control policies. Kaval Gulhati asked, in an article in *Science* (1977):

> Should they stand by and wait for economic development and family planning programs to motivate contraception? Or should they take the destiny of the people in their hands, and force a fertility decline?

Gulhati asked this question before Mrs. Gandhi's government was rejected. In light of her government's defeat, the question might be expanded to ask:

> Can a government successfully undertake a compulsory program of population control and itself survive?

rates of infertility are seen in the region (an average of 12% in African countries as compared to 2%–3% in other developing countries). Fear of infertility discourages couples from using contraceptives and even from delaying or spacing births. Social as well as economic development is vital to slowing population growth in these countries. Education for women as well as a reduction in infertility and in infant and child mortality is needed.

Programs that recognize and respect the traditional methods of population control are more likely to succeed in Africa than ones that are designed along the lines of Western cultural values. For instance, in sub-Saharan Africa, certain taboos, such as one on intercourse during the long traditional period of breastfeeding, have evolved. These taboos increase child survival by increasing spacing between the children. Such taboos have been disappearing along with economic development and changing customs. Programs that not only explain the use of reversible contraceptives to encourage child spacing but also encourage traditional patterns of long breastfeeding are essential.

Outlook for Population Growth in the Future

Late in the 1970s, it became clear that population growth rates in developing countries were decreasing. In fact, two trends are occurring. A small but real decrease is occurring in world population growth rates, and a general and important decline is occurring in the fertility of women in the less developed countries.

According to the U.S. Bureau of the Census, the rate of world population growth fell from 1.98% per year in the period 1965–1970 to 1.88% in the period 1975–1977 and to 1.7% in 1982. However, the decline in the growth rate for developing countries has not been evenly distributed. It is almost entirely accounted for by the 50% fall in the Chinese birth rate. In middle-income economies such as Hong Kong, Korea, Thailand, Singapore, and Indonesia, birth rate declines have also been significant. In fact, birth rates have declined in all areas except sub-Saharan Africa (Figure 4.10).

CONTROVERSY 4.3

Africa: Population Control or Genocide?

No programme will work in Africa if it is imported.
Professor Mungai, University of Nairobi

We are fighting the common belief that the white
man is trying to limit the number of our people.
Joyce Nkausu, Zambia Family Planning Association

Africa as a region has the highest birth rate and also
the highest death rate in the world. Moreover, Western
concerns with population reduction are not generally
shared by African government leaders. Louis Ochero
wrote in the Population Reference Bureau's *Newsletter*
of October 1981:

> The advocacy of family planning as synony-
> mous with or contributing to population control,
> considered by the advocates as a basic requirement
> for economic development, was almost univer-
> sally rejected in black Africa. In some areas these
> were viewed as thinly veiled efforts at genocide,
> or at best an unrealistic approach to the real issue
> of development.

> The obstacle to the spread of family planning
> practices in Africa is not unconnected with their
> foreign origin. Most of the personnel propagating
> family planning in Africa are either expatriate or
> foreign-financed. This has generated a suspicion
> that family planning is foreign-supported for po-
> litical and even racial reasons. . . .

> There is a great deal more to the control of
> population growth through reducing the birth
> rate than only the conception control which is dis-
> pensed at the family planning clinics. Experience
> has shown that the birth rate of the traditional so-
> cieties often falls effectively only under the impact
> of urban industrial development, education, im-
> provement of health, and the lowering of the
> death rate. Propaganda for birth control or gov-
> ernment-sponsored family planning programs
> does not seem to have played an important role in

> lowering the birth rate of the developed nations
> either. The unjustifiably exaggerated emphasis on
> family planning only and its identification with
> population growth control has annoyed and irri-
> tated the intelligentsia in the developing countries
> and has aroused opposition to family planning.
> The less intellectual fail to see the reasons for
> "downgrading" their existing traditional methods
> of child spacing.

> The most important factors which have caused
> the lowering of the birth rate in the advanced
> countries have been the same as have been respon-
> sible for the raising of the standard of living of the
> people living there. Industrialization, urbaniza-
> tion, rural development, the spread of education,
> better health services, raising the age of marriage
> for women can be specially mentioned. No
> amount of exhortation for population control can
> take the place of these factors.

What do you think Ochero meant when he said
there is a suspicion in Africa that family planning is
"foreign-supported for political and even racial reasons"? Do you think this is an accurate perception of
the policies of developed nations toward family planning aid to undeveloped nations?

Do you think wealthier countries should increase
their aid to poorer countries, such as those in Africa,
whose populations are growing rapidly? If so, what
kind of aid should be given?

Do you think developing countries can lower their
fertility rates in the same way the developed countries
did a century ago?

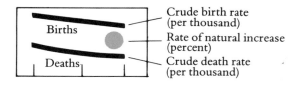

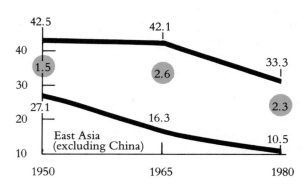

East Asia (excluding China)

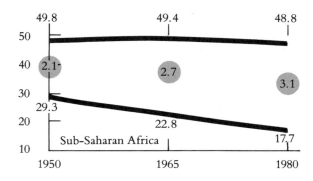

Sub-Saharan Africa

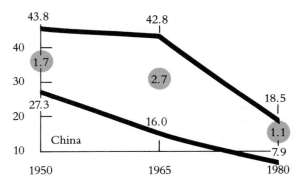

China

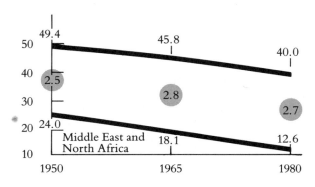

Middle East and North Africa

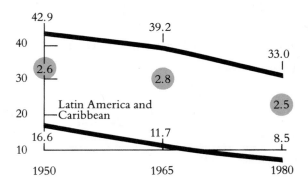

Latin America and Caribbean

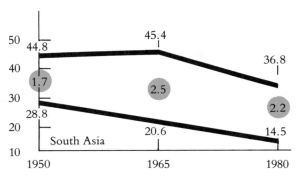

South Asia

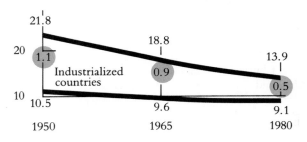

Industrialized countries

Figure 4.10 Birth and death rates for regions of the world, 1950–1980. (From The World Bank, *World Development Report 1984* [New York: Oxford University Press, 1984], p. 64.)

Crude birth and death rates depend on the age structure of a population. If a large percentage of a population consists of women of childbearing age, the crude birth rate tends to be high. This is the situation in developing countries today. The real hope that population growth is slowing comes from looking at the total fertility rate for each country. Because the total fertility rate measures the number of children each woman of childbearing age is expected to have, it can reveal changes in reproductive behavior that aren't confused by the size of various groups in the population. On a worldwide basis, the average total fertility rate fell from 4.6 in 1968 to 3.6 in 1987. Fully 80% of the world's population is in countries with declining fertility rates. In a study of 113 less developed countries, 95 nations had seen fertility declines from 1968 to 1975 (U.S. Census Bureau, 1979). Table 4.1 lists the decline in fertility in a number of larger less developed nations.

Because of the age structure in most developing nations (a large group of children yet to enter their childbearing years), birth rates and growth rates are expected to remain high in most developing countries for some time to come. The growth rates caused by lowering mortality will not play as large a role in the future as they have in the past. This is partly because much of the fall in mortality has already been accomplished. But it is also because mortality declines are not as significant at low birth rates as they are at high birth rates. When fertility is high, saving a new baby's life means that child will add many more people to the population when he or she comes to reproductive age. When fertility is low, children can be expected only to replace themselves. Similarly, increasing the life expectancy to 80 of someone who now expects to live to age 60 adds only one person to the population, since the years being added are not reproductive years.

Using the new estimates of fertility, world population growth is expected to level off in 2110 at 10.5 billion people. Faster or slower declines in fertility could mean leveling off at as few as 8 billion or as many as 14.2 billion.

Problems Developing Countries Must Solve

Statistics such as these provide hope that world population is coming under control. This by no means erases the problem of population growth. On the contrary, it appears to show that rapid population growth is a problem that more effort and money can affect. Studies from the cooperative international effort known as the World Fertility Studies (WFS) show that the mean number of children wanted by women in many developing nations is still well above replacement level (which is 2.1 to 2.5, depending on a country's death rate). Table 4.2 shows some of these family size preferences.

In some countries, especially in Africa, an effective government family planning policy has not been devised. An unmet need for contraceptive services still exists in these and many other countries. However, in some developing countries where fertility had previously been declining rapidly, the decline has stalled somewhere above replacement level. In Tunisia, for instance, fertility fell from 7.1 to 5.7 between 1966 and 1976 but only continued down to 5.1 by 1981. Some of this stall appears to have been due to dissatisfaction with available contraceptives. Tunisians are concerned about the health risks of modern contraceptives and note that family planning services are not always easy to obtain. In addition, women's status still depends in a large part on having children; outside work is viewed as unimportant.

In Costa Rica, fertility fell until the late 1970s, when it coincided with the desired family size of four children. Further declines are unlikely until desired family size decreases. Examples of economic and social

Table 4.1 Decline in Fertility in Some Less Developed Nations

Country	Fertility (live births per woman) 1968	1985	Percent decline	Population in 1987 (millions)
Bangladesh	6.98	6.2	11	107.1
Colombia	6.54	3.1	43	29.9
India	5.67	4.3	24	800.3
Indonesia	6.46	4.2	35	174.9
Mexico	6.59	4.0	39	81.9
Pakistan	6.84	6.1	4	104.6
Peoples Republic of China	4.20	2.4	43	1,062.0
Thailand	5.86	3.5	40	53.6

Source: 1987 World Population Data Sheet, Population Reference Bureau.

Table 4.2 Mean Number of Children Desired in Selected Countries

Country	Mean number desired
Kenya	6.8
Jordan	6.3
Sierra Leone	6.1
Paraguay	5.1
Costa Rica	4.7
Dominican Republic	4.6
Guyana	4.6
Mexico	4.4
Philippines	4.4
Malaysia	4.4
Panama	4.2
Pakistan	4.2
Fiji	4.2
Venezuela	4.2
Bangladesh	4.1
Colombia	4.1
Indonesia	4.1
Jamaica	4.0
Nepal	3.9
Peru	3.8
Sri Lanka	3.8
Thailand	3.7
Haiti	3.6
Republic of Korea	3.2
New Zealand	3.0
Turkey	3.0
Spain	2.8
Taiwan	2.8
Great Britain	2.6
Czechoslovakia	2.4
Belgium	2.3
Japan	2.2
Hungary	2.1

Source: Mary M. Kent and Ann Larson, *Family Size Preferences: Evidence from the World Fertility Surveys* (Washington, D.C.: Population Reference Bureau, 1982).

changes that could decrease the desired family size include more work opportunities for women outside the home, an increase in the number of jobs requiring a longer education period and provision of a widespread old-age pension system.

In Korea, a preference for sons may be contributing to a stall in fertility declines, while in Sri Lanka continued dependence on less effective traditional methods of birth control is probably a factor. Further progress in each of these countries will require studies of the exact causes of fertility stalls to determine effective policies.

ZPG and Other Problems Developed Countries Must Solve

In developed countries there are two main concerns as populations approach zero population growth (ZPG). Some countries (including West Germany, Hungary, and Denmark) have gone past the level of replacement fertility and actually have negative growth rates. The fear that such declines could cause a nation to "die out" or at least lose international prestige and power has led some Eastern European countries to tighten previously liberal abortion laws and to increase pronatalist incentives, such as tax deductions and educational allowances for children. However, there is no evidence that these policies have been successful in offsetting the high social and economic costs that lead parents to choose to have fewer children.

The second problem of concern to developed countries is the "aging" of the population that accompanies increasing life expectancy and decreasing birth rates. Figure 4.11 illustrates this trend in the United States. Low fertility rates eventually produce populations with higher proportions of the elderly than in the past. Eventually, at ZPG, the number of people in all age groups (except the very old) will approach the same percentage. For most countries, people in the working age groups will then be supporting more retired people than in past times. This is partially offset by the fact that there will be fewer children to be supported per family, but because older people consume about twice as much as children, problems will arise.

In Sweden, for instance, where population growth is at zero and there is a comprehensive welfare system, costs for support of the elderly have been rising much faster than other welfare costs. Proposals designed to help solve these problems include increasing the retirement age and changing tax policies to encourage older people to remain productive longer. Families might also be helped to care for their elderly members so that costly institutionalization is not necessary. Flexible work arrangements in which some "retirement" could be taken at younger ages, for education or retraining, might also help.

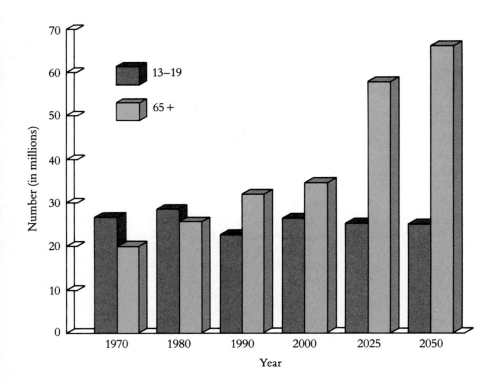

Figure 4.11 Number of teenagers and elderly in the United States, 1970–2050. (From *Population Today*, January 1985, p. 5.)

Population Projections

The World Bank has prepared a series of projections of population through the twenty-second century. The projections differ by how fast death rates (mortality) and fertility decline. Figure 4.12 shows how population can be expected to grow in developing nations. Even in the best case, the population of developing nations is expected to more than double by 2050.

The graph illustrates how population growth depends critically on how fast fertility declines and much less on how fast death rates decline. Rapid declines in both death rates and fertility rates appear difficult but not impossible to achieve. For India, fertility would need to reach 2.2 children per woman by 2000 compared to 4.8 today. Brazil would need to reach 2.1 compared to 3.9 today.

Some countries have set their goals for lowering fertility rates or crude birth rates to levels that are even lower than those used in the World Bank rapid fertility decline projections (see Table 4.3). Whether these countries can achieve their goals, and whether China and perhaps other countries can successfully sustain a policy of below-replacement fertility, could change the course of world population growth. Moreover, success could substantially decrease suffering in the world of the future. Such efforts deserve our respect and help.

Tell people where you live that we are working to surmount these problems. That our people are helping themselves and that we value your assistance. Together, together we will get the task done.*

Summary

Industrialization and economic development may eventually lower population growth rates. However, lower growth rates have also been achieved by means of social welfare policies: universal education, easily accessible birth control services, improvements in the status of women, better health care, and old age security plans.

---

*Village elder, Diomga Village, Burkina Faso, quoted in *A Shift in the Wind* (20) (1985), The Hunger Project, P.O. Box 789, San Francisco, CA 94101.

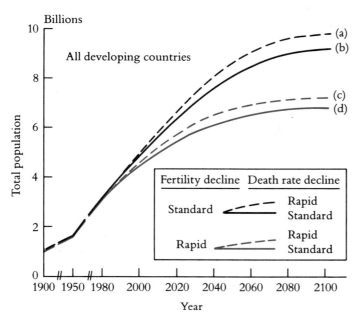

Figure 4.12 Possible future trends in fertility and death rates. The top two curves illustrate the population growth that could be expected with standard fertility declines combined with (a) rapid declines in death rates or (b) standard declines in death rates. The bottom two curves represent the alternative situation in which population grow is the result of rapid fertility declines combined with (c) rapid declines in death rates or (d) standard declines in death rates. Note that rapid reductions in fertility rates have more of an effect on the total population size than do changes in the speed at which death rates decline. (From World Bank, *World Development Report* [New York: Oxford University Press, 1984], p. 75.)

Table 4.3   Population Policy Aims in Selected Countries Compared to World Bank Projections for Those Countries

Country	Target year	Country's goal		World Bank projection*	
		Total fertility rate	Crude birth rate	Total fertility rate	Crude birth rate
Bangladesh	2000	2.5		2.8	23
Indonesia	1990	2.7	22	2.9	24
Korea	1988	2.1		2.2	20
Nepal	2000	2.5		2.9	24
Pakistan	1988		36	5.2	38
Thailand	1986	2.6		3.0	25
Ghana	2000	3.3		3.2	27
Tunisia	2001		22	2.2	20
Haiti	2000		20	2.4	22
Mexico	1988		25	3.6	29

Source: Adapted from The World Bank, *World Development Report 1984* (New York: Oxford University Press, 1984).

*World Bank projection given is for rapid scenario of fertility decline.

When the socioeconomic climate is altered so that families can increase their economic welfare by having fewer children, evidence is accumulating that they will do so.

Future programs need to be tailored to the specific needs of each country. Some populations still lack adequate birth control services; others need health care programs to reduce infant and maternal mortality and infertility. In other countries, these needs have been met. Further progress may depend on government incentives and disincentives. These should increase the cost of children to a family or generate opportunities that cannot be enjoyed if parents choose to have large numbers of children.

Current world population projections predict a doubling of world population to 8 billion between 2025 and 2050. However, the delay or failure of population programs in developing countries could mean that even this projection could be exceeded. Conversely, the success of more ambitious government policies, such as China's "one-child" plan, could foretell a faster decline in fertility rates and a lower final total for the world's population.

Questions

1. Population reduction policies that penalize families with more than a certain number of children can be viewed as unfair to innocent children. Are policies that reward small families similarly unfair? Explain, using examples such as housing loans, employment and education bonuses, old-age social security, and so on.
2. How do economic development and population growth affect each other?
3. Why is the status of women (education, employment opportunities) an important factor in fertility levels?
4. What major problem does a country face when it has stopped growing? Why?

Further Reading

Bouvier, L. *Planet Earth 1984–2034: A Demographic Vision.* Washington, D.C.: Population Reference Bureau, 1984.
A fascinating vision of what the future may be like if present population trends continue.
Day, L. H. "What Will a ZPG Society Be Like?" *Population Bulletin*, **33** (June 1978).
ReVelle, R., A. Khosla, and M. Vinovskis. *The Survival Equation.* Boston: Houghton Mifflin, 1971.
A thoughtful, humanitarian approach to rapid population growth and possible solutions to the problems it causes. Well worth reading for the arguments against coercion in birth control programs and the arguments that, under the right circumstances, all peoples can see that control of rapid population growth is in their best interest.

"World Population and Fertility Planning Technologies: The Next 20 Years." Washington, D.C.: Congress of the United States, Office of Technology Assessment, 1981. This report examines the various available contraceptive methods as well as prospects for new methods.
van de Kaa, D. "Europe's Second Demographic Transition," *Population Bulletin*, **42** (1) (March 1987).
As European countries move from zero-growth to below-replacement-level fertility, they seem to be undergoing a second demographic transition. This article examines the causes and consequences of such a trend.

References

Hardin, G. "The Tragedy of the Commons," *Science*, **162** (December 13, 1968), 1243.
Belcastro, P. A. *The Birth Control Book.* Boston: Jones and Bartlett, 1986.
Coale, A. J. *Rapid Population Change in China.* Washington, D.C.: National Academy Press, 1984.
Gulhati, K. "Compulsory Sterilization: The Change in India's Population Policy," *Science*, **195** (March 25, 1977), 1300.
Curtin, L. B. "Status of Women: A Comparative Analysis of 20 Developing Countries." *Reports on World Fertility Survey*, No. 5. Washington, D.C.: Population Reference Bureau, June 1982.
Kent, M. and A. Larsen. "Family Size Preferences: Evidence from the World Fertility Surveys." Washington, D.C.: Population Reference Bureau, 1982.
World Bank. *World Development Report 1984.* New York: Oxford University Press, 1984.
Lightbourne, R., et al. "The World Fertility Survey: Charting Global Childbearing." Washington, D.C.: Population Reference Bureau, March 1982.
Robinson, W. C., et al. "The Family Planning Program in Pakistan: What Went Wrong?" *International Family Planning Perspectives*, **7** (3) September 1981.
U.S. Census Bureau. *World Population 1977: Recent Demographic Estimates for the Countries and Regions of the World.* Washington, D.C.: Census Bureau, January 1979.

CHAPTER FIVE

Protecting Wildlife Resources

Is a Little Fish Worth More Than a Big Dam?

In the free-running waters of the Little Tennessee River lived a tiny fish called the snail darter (Figure 5.1). When this small member of the perch family was first discovered in 1973, all the known snail darters in the world lived in this one place.

In the same year that the snail darter was discovered, Congress passed the Endangered Species Act. An **endangered species** has so few living members that it is in danger of becoming extinct in the near future. The act states, in part, that actions of federal government agencies may not "jeopardize the continued existence of endangered species and threatened species or result in the destruction or modification of habitat of such species which is determined to be critical."

In 1966, seven years before anyone even knew the snail darter existed, Congress had authorized the Tennessee Valley Authority (TVA) to build the Tellico Dam and reservoir across the Little Tennessee River. Builders had half-completed the dam before the snail darter was discovered. The Tellico Dam was three-quarters finished in 1975 when the Secretary of the Interior listed the snail darter as an endangered species.

Now, the snail darter cannot breed in the still waters of a reservoir (it needs free-running water). Thus, by destroying the breeding grounds of the snail darter, completion of the $116 million Tellico Dam threatened to destroy the entire newly discovered population at a blow and would violate the Endangered Species Act. Several environmental groups sued to

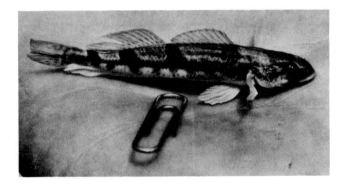

Figure 5.1 Snail darter. This endangered fish stood in the way of completion of the $116 million Tellico Dam. The controversy over the fish and the dam led the Tennessee Valley Authority to rethink the whole idea of placing a dam on the Little Tennessee River, which provided one of the few remaining stretches of good cold-water fishing in the region. The controversy also caused Congress to rethink the Endangered Species Act, which protects even tiny fish from large construction projects. (NYT Pictures)

have construction stopped, and the case went all the way to the Supreme Court. In 1978, with the dam 90% complete, the Court ruled that the project did indeed violate the Endangered Species Act and so must be either stopped or changed.

Was the protection of a small population of 3 in. (7.5 cm) fish really what Congress had in mind when it passed the Endangered Species Act? As Holden (1977) put it: "Undoubtedly many members of Congress were thinking thoughts of brown-eyed creatures and soaring winged things when they cast their vote, and now are finding themselves confronted with a Pandora's box containing infinite numbers of creeping things they never dreamed existed."

What is the value of a species? Why should we attempt to save species from extinction? Can we say that some species are more worth saving than others?

Scientists estimate that between 5 and 10 million species exist in the world, but ecologists have so far discovered and described only some 1 to 1.5 million. However, the discovery of new species is becoming a race against extinction. In prehistoric times, one species was lost perhaps every 1,000 years. Today, one species becomes extinct every year. But over the next 20 years, as many as a million species may be lost, most in the tropical forests.

The Value of Species

Wildlife as a Harvestable Resource

Even in highly developed countries such as the United States, wildlife species provide humans with food and fuel in the form of fish, nuts, berries, and wood for heating. The value of fresh foods from wild organisms in the United States is estimated at $2.8 billion per year. The use of wood for home heating increased by over 50% in the 1970s. In Vermont, for example, over half the homes now use wood as their primary heat source.

In developing countries, wild sources of food and fuel are even more important. Fish protein provides 10% of the animal protein in human diets worldwide. In many developing countries, wood is the only source of heat for cooking or heating.

Ecosystem Services

In addition to the easily quantified value of wild species as food and fuel, wildlife species provide the planet with a set of what might be termed *ecosystem services*. Plants produce oxygen that humans and other animals need for respiration. In addition, plants and microorganisms remove air and water pollutants, recycle nutrients, and moderate climate. While some of these services might be carried out by technological processes (the removal of phosphates in runoff water by a coastal wetland area could be accomplished at much greater cost by a sewage treatment plant), others are clearly irreplaceable.

Practical Value in Medicine, Agriculture, and Industry

Lost species can be viewed as lost opportunities. Animals and plants provide us with drugs, with foods, and with raw materials for industry. Twenty-five percent of prescription drugs dispensed in the United States contain plant extracts that cannot be synthesized. These drugs include the tranquilizer reserpine as well as a variety of antibiotics, pain killers, and drugs used to treat heart disease and high blood pressure. Vincristine, a drug extracted from a tropical periwinkle, is used in the successful treatment of Hodgkins disease, a cancer that strikes 5,000–6,000 Americans each year. However, only 5,000 plant species have been investigated for useful drugs. Scientists feel that another 5,000 usable drugs could be found among the 500,000 species of plants believed to grow in the world (Figure 5.2).

Figure 5.2 Howell's gum weed. This rare plant is a close relative of one used by Indians as a sedative, expectorant, and cure for poison ivy. (Courtesy of Bob Kiesling)

Agricultural researchers have found uses for a number of organisms. For instance, an important technique in farming is the use of biological controls, which involves using one species to prevent another from harming crops. As an example, certain wasps can successfully prevent the sugar-cane borer from destroying whole fields of cane. Another technique of modern agriculture involves the cross-breeding of various plant species to develop crops with higher yields or resistance to disease, drought, or heat. (In some parts of the world, the spread of modern hybrid grain varieties threatens the survival of traditional varieties, many of which have useful characteristics. This problem is discussed again in Chapter 8.)

Many plants produce chemicals that are natural insecticides (insect killers) or herbicides (weed killers). Still other plants provide waxes, lubricating oils, resins, oils for perfumes and flavorings, and dyes. The list could go on. And these are only the useful plant and animal products discovered so far. Many other agricultural, drug, and manufacturing substances await discovery. (In a few cases organisms have become too useful to humans for their own good. See Controversy 5.1.)

Value as Part of the Gene Pool

The science of genetic engineering is in its infancy, yet it is already clear that in the future scientists will be able to transfer, from one plant to another, desirable genes for certain characteristics. Examples might include disease or drought resistance, insect resistance, drug-producing ability, and higher protein content (Figure 5.3). Reducing the number of species in the world reduces the size of the gene pool. Every time we allow a plant or animal to become extinct, we run the risk of losing a possibly helpful organism or gene.

Value as Part of Food Webs

The loss of a particular species or group of species may have far-reaching effects on the community in which the species live. Complex food webs are common in temperate and tropical climates; however, since only a relatively few webs have been thoroughly studied, we usually lack the knowledge to predict the effects of the extinction of any particular plant or animal species. Many rare insects, snails, and birds depend on a particular kind of plant for food or as a place to live. If the plant becomes extinct, the animal is also likely to be-

Figure 5.3 A kind of wild corn (*Zea diploperennis*) was recently discovered on a hillside in Mexico. This wild corn is a perennial—that is, it doesn't need to be planted each year, as cultivated corn must. In addition, the corn is resistant to various viruses and grows well in wet soil. These are all desirable characteristics to transfer to cultivated corn by traditional plant-breeding techniques or by genetic engineering. The new species has so far been found only in this one location. The discovery was made just before the hillside was to be plowed up. (World Bank Photo: Larry Daughters)

CONTROVERSY 5.1

Endangered Species Versus Human Health Benefits

Surely it is not beyond our scientific ingenuity to find
alternative methods.
F. B. Orlans

It is not consistent with the genius of the American people
to restrict the progress of scientific knowledge.
A. S. Packard, Jr. and E. D. Cope

The use of animals in scientific research was once opposed mainly on the grounds of possible pain and cruelty. Scientists went to great lengths to assure the general public that the animals used in experiments never felt pain.

Now, however, new ethical concerns are being raised. Are research animals housed in such a way that social and behavioral needs are met? That is to say, normally social animals like chimpanzees should not be kept in individual, isolated cages because this would be a form of mental cruelty.

To go even further, should an animal with a dwindling population be used in research at all, even if humans stand to benefit greatly? N. Wade (1978) wrote:

> . . . production of the [hepatitis] vaccine may well pose a fatal conflict between the interests of mankind and those of chimpanzees. Chimps are the only species, other than man, in which the safety of the vaccine can be tested . . . if the chimpanzees are protected—the species is already classified as threatened—it may prove impossible to safety test and hence to manufacture the vaccine. Yet even in developed countries, where the disease is comparatively rare, hepatitis B takes a heavy toll. In the United States 15,000 cases were reported in 1976.

The true incidence was probably 150,000 according to the Center for Disease Control, of which probably about 1500 cases ended in death. . . .

> . . . officials deny that their chimpanzees would be captured inhumanely. "The method of capture is generally by locating a group of chimpanzees, surrounding them with a number of people and chasing them. The juveniles would usually tire first and these were captured by hand," a Merck official told the Federal Wildlife Permit Office. . . .

> ". . . Totally impossible unless you had big nets," says Jane Goodall. "Utterly fanciful. . . . Given the sort of habitat where wild chimpanzees are found, no human being could keep up with a wild chimpanzee, much less run it to the ground. . . . I can only conclude that someone is seeking to conceal the actual but less humane method of capture used—that is, shooting the mother to recover the young, which is the standard method used in Africa."

F. B. Orlans (1978) added:

> . . . a way must be sought to solve this conflict in a manner that is not detrimental to the chimpan-

zees. In the past, alternative methods of producing other vaccines (notably that for polio) have been found so that animal lives are spared . . . The ethical concerns for elimination of inhumane killing (in Wade's words, "to capture a chimpanzee: first shoot the mother") and for preservation of this dwindling species of animal are overriding.

Do you feel that a clear human need should outweigh the need to preserve an animal species?

Sources: N. Wade, "New Vaccine May Bring Man and Chimpanzee into Tragic Conflict," *Science,* **200** (June 2, 1978), 1027; F. B. Orlans, Letter to the Editor, *Science,* **201** (July 7, 1978), 6.

come extinct. In another case, a predator that normally keeps some pest under control might be lost. As an example of this effect, in some areas sprayed with DDT, red spider mites have gone out of control and damaged crops. This occurred because DDT killed the ladybugs that usually eat the mites, leaving the mites (which were not affected by the DDT) free to reproduce in enormous numbers.

Wolves are threatened by humans in part because their role in food webs is not understood. Wolves kill animals like deer for food, but they tend to select the sick, the weak, and the old. They can thus help keep a herd of deer healthy and of a size proper to the available food supply. Humans, hunting competitors of the wolf, pride themselves in taking the finest deer, thus lowering the quality of a herd. Another example of human alteration of food webs is described in the caption for Figure 5.4.

Concern about endangered plant species has come much more slowly than concern about endangered animal species; yet the two are so intimately related that they cannot be conserved separately. Many examples are known of animals driven to the brink of extinction because a particular plant on which they feed or under which they shelter became scarce. Peter Raven, of the Missouri Botanical Garden, has estimated that for every plant species that becomes extinct, 10 to 30 species of insects, higher animals, and other plants may also face extinction.

The red-cockaded woodpecker is threatened with extinction because it nests only in mature longleaf and loblolly pines. In many areas these mature trees are being replaced by young trees grown for pulp. Maturation of the longleaf pine itself appears to depend on a ground cover of wiregrass (*Aristida stricta*), which en-

courages fires that help the pine seedlings germinate and grow. Wiregrass grows, slowly, from stolons rather than from seed. Often, an area of wiregrass and longleaf pine will not grow back when logged, apparently because the slow-growing wiregrass cannot compete well with faster-growing species.

Intrinsic Value of Species

Besides all the practical reasons we might give, there are philosophical arguments in favor of preserving as many species as possible. Any species lost is gone forever. If we fail to do what is in our power to prevent these losses, we make a choice not only for ourselves but also for future generations. We are saying that they will never see the same living creatures we can see; they will never enjoy the diversity we enjoy. It may not even be a question of enjoyment—having evolved in the midst of such diversity, humans may require it to maintain their own mental health.

All these reasons, of course, consider other species only from the viewpoint of their usefulness to humans. Henry Beston (1928) wrote:

Remote from universal nature, and living by complicated artifice, man in civilization surveys the creature through the glass of his knowledge and sees thereby a feather magnified and the whole image in distortion. We patronize them for their incompleteness, for their tragic fate of having taken form so far below ourselves. And therein we err and greatly err. For the animal shall not be measured by man. In a world older and more complete than ours they move finished and complete, gifted with extensions of the senses we

Endangered Species

a California condor

b African elephant

c Tropical periwinkle
(*Catharanthus roseus*)

The developed nations, such as the U.S. and Western Europe, and the developing nations as well, are losing wildlife species at an alarming rate. The major cause is loss of wildlife habitat as growing human populations reach out for more farmland, more forest wood, and more recreational space.

(**a**) When this was written, there was only one California condor left in the wild. Perhaps by now there are none. This picture was taken at the San Diego Zoo, where a captive breeding program is being carried out in hopes of one day reintroducing the condor into its former habitat. (**b**) Large animals such as the condor and the elephant require large ranges in which to find enough food to survive. In Asia the elephant is endangered because its habitat is increasingly being converted to farmland. In Africa the elephant is hunted for its ivory tusks as well.

In addition to animal species, many plants are endangered because of loss of habitat. (**c**) The tropical periwinkle is a representative of a group of plants in danger because of the loss of tropical forests. One of the periwinkles produces a compound that is used to treat leukemia.

All the news is not bad, however. (**d**) and (**e**) Hunting restrictions and protected reserves have saved the buffalo while wildlife management techniques, such as removal of competing species, have been successful in bringing back puffins to islands off the coast of Maine. Unfortunately, we do not know how to save all endangered species by management techniques. For most creatures, the best hope lies in protecting them in their natural habitat.

d Buffalo (American bison)

e Puffin

Food Production

a U.S. mechanized farming

b Cattle ranching in Montana

c Plowing in developing nation

d Cattle herding in Kenya

Forging ahead to meet the challenge of food production for rapidly growing populations, agricultural technologists have had notable successes. Farmers in the U.S. have coaxed more than 300 bushels of corn from a single acre in one growing season. In the Indian State of Kerala almost miraculous increases in wheat yields followed the introduction of modern agricultural technology. Yet, in both developed and developing countries, there are prices to pay for such rapid forward strides. (**a**) Mechanized farming in the U.S. Although growing a single crop over a large area makes it easier and cheaper for the farmer to plant, cultivate and harvest, single crop fields are more subject to erosion and to invasion by pests. (**b**) Feeding cattle in Montana. Large farms are more efficient than small ones. Many owners of small family farms find they cannot run their farms at a profit and so must find other ways of making a living. This Montana rancher is filling feed troughs for his 8,000 head of cattle by truck. (**c**) Farmer plowing in Indonesia. (**d**) Young herders in Kenya. In developing nations, small farmers often cannot afford the outlay for fertilizer or seed that would enable them to farm more successfully. Furthermore, mechanization decreases the number of jobs, forcing people to move to overcrowded cities.

have lost or never attained, living by voices we shall never hear. They are not brethren, they are not underlings; they are other nations, caught with ourselves in the net of life and time, fellow prisoners of the splendour and travail of the earth.

How Species Become Endangered

Hunting

The reason that comes most quickly to mind for the disappearance of species is probably hunting. Hunting has contributed to the loss of a number of animals, especially vertebrates (animals that have a backbone). In certain well-managed wildlife populations, hunting need not harm the population—in fact, it can contribute to its welfare, most notably in cases where a population threatens to grow too large for its habitat. Unregulated hunting, however, has contributed to the loss of species. The buffalo of the American plains was hunted almost to extinction in the 1800s. Trainloads of hunters came for the sport, often carrying home no more than a buffalo head to mount as a trophy. In Africa, game officials have stopped or limited the hunting of many big game species lest these animals cease to exist except in zoos. (See Figure 5.5 and Global Perspective I.)

Habitat Alteration

Hunting is not the main problem faced by most endangered species. The majority are threatened with a loss of their **habitat**, the area in which they grow, breed, seek food, and find shelter. As human populations grow, they require more houses, roads, and shopping centers. Forests are cut down; marshes, estuaries, and bays are filled in; and land is overturned in the search for coal. All of these processes reduce the land or food supply available to various animals and plants. In a sense, humans are increasing their own habitat at the expense of the habitats for other creatures.

In some cases habitat destruction is a result of game-management procedures such as burning or flooding, done to make areas more attractive to game species. Populations of elk, pronghorn antelope, white-tailed deer, and mule deer have all increased greatly as a result of such management techniques. In the process, however, the habitat has become unsuitable for many other, nongame species.

Many endangered plants are living links with past

Figure 5.4 Sea otters were almost wiped out by fur trappers in the eighteenth and nineteenth centuries. Now, due in part to laws such as the Marine Mammal Protection Act, sea otter populations are recovering. In fact, they are threatening to avenge themselves in the process, if not on humans, then on species dear to our gastronomic heart: abalone, Pacific lobster, and crabs. From a few individuals discovered near Monterey, California in 1938, the otters have increased to almost 2,000 animals, ranging along a 150-mile (240 km) stretch of coast. Unfortunately, this same stretch of coast is also famous for seafoods such as abalone, a shellfish marketed for $8–10 a pound. Commercial fishermen are demanding that otter herds be limited in size to prevent further depredation of the profitable fishing industry. On the other hand, ecological studies show that the sea otter is a vital member of the shore community. By feeding on species such as sea urchins, otters protect seaweeds such as kelp from being overgrazed. In turn, kelp beds are at the bottom of the food webs that sustain species such as harbor seals and bald eagles. (Dr. Daniel Costa, Joseph M. Long Marine Laboratory, University of California, Santa Cruz)

Figure 5.5 African elephant herds have decreased dramatically in the past 10 years, due mainly to poachers seeking ivory tusks. In 1970, approximately 5 million elephants lived in Africa; by 1980, there were only 1.3 million. In Uganda's Kabalega National Park, poachers reduced a herd of 9,000 elephants to 160 terrified beasts who kept moving night and day, unable to find refuge. In Asia, the elephant is also endangered—only about 15,000 remain. Here the problem is loss of habitat as increasing areas are farmed. (Judy Rensberger/NYT Pictures)

ages, remnants of species that flourished in an earlier age and climate. They now exist in special niches along river banks, in bogs, marshlands, and barrens. Others are found on isolated mountain faces or in valleys or areas where glaciers never reached. These plants are rare precisely because they are adapted to grow in their present location. They can only be preserved if their habitat is protected.

Tropical Forest Loss

Although destruction of all types of habitats is occurring, the problem is most critical in tropical rain forests. Each year, an area equal in size to Great Britain is clear-cut or otherwise degraded. At present rates of destruction, in 20 to 30 years these forests will not exist in their present form. Yet two-thirds of the 5–10 million organisms believed to exist on earth are found in the tropics, most in forests.

Human population growth has often been blamed for much of the loss of tropical forests. Population increases in developing countries can lead to increased wood gathering and "slash-and-burn" agriculture by native peoples. In this type of agriculture, farmers cut down trees and raise crops for a few years. Then, as the soil becomes depleted of nutrients, the farmers move on to new plots and cut down more trees. Some experts feel the blame is misplaced, however, stating that such subsistence farming is responsible for only 10%–20% of forest loss (Figures 5.6 and 5.7). Large-scale cattle-raising projects and military road building in Brazil, as well as the demand for tropical woods from Brazil,

Africa, and Southeast Asia, destroy a much greater portion of tropical forests (see Global Perspective I).

Competition with Introduced Species: The Special Case of Islands

A number of species have become endangered or are already extinct because they were unable to compete with new species introduced into their habitat. Island ecosystems are especially fragile in terms of species loss due to competition with introduced species.

Of the 161 birds that have become extinct in their native habitat since the year 1600, 149 lived on islands. On some islands—for instance, the Hawaiian Islands—many species have evolved fairly recently. These islands exhibit a variety of different habitats due to differences in soil, rainfall, and elevation. During the evolution of native plants, competition from other species was different from what happened on the continents, allowing many new species to survive. In Hawaii, 97% of the native species are endemic (found nowhere else). Unfortunately, many of these species have been unable to compete with plant species or survive predation by animal species introduced in modern times. Imported mouplan sheep threaten both the mamane tree and the honeycreeper, a bird dependent on the tree as a food source. Mongooses were brought to the American Virgin Islands to control rats, but instead they attacked other native species. Of the native plant and animal species of the Hawaiian Islands, 36% are in danger and more than 10% may already be extinct.

Figure 5.6 Slash-and-burn agriculture in Fiji. In this photograph the forest has been cleared for a garden plot. Forest waste is left lying on the ground. This practice helps to stabilize the soil and prevent erosion. A mixture of crops is usually planted, mimicking to some extent the natural diversity in a tropical forest. (Photo courtesy of S. Siwatibau)

Figure 5.7 The same plot as in Figure 5.6, but 18 months later. Yams and taro have been harvested, leaving yaqona, bananas, paw-paws, and taro suckers. (Photo courtesy of St. Siwatibau)

Pesticides and Air Pollution

Many habitats that are otherwise undisturbed are poisoned by air pollutants, acid rain, or pesticides. Pines in the mountains near Los Angeles are injured by smog from the city. The large-scale use of pesticides in agriculture places further stress upon many endangered species. For instance, birds in the raptor group, which includes hawks and falcons, are affected by the use of DDT. About 20–30 years ago, these birds began laying eggs with very thin shells, so thin that they cracked before the chicks could hatch. Thinning is believed to have been due to DDT (Grier, 1982). The use of DDT is now banned in the United States, largely because of the effects it has on certain bird species.

As part of a pest-control program in the American West, attempts were made to kill coyotes, foxes, and wolves by using poisoned baits. These methods had a severe effect on populations of endangered species, among them the bald eagle, that also took the bait (Figure 5.8).

Collection of Plants

Certain plants, especially cacti, orchids, and carnivorous plants, are so desirable that they have been collected almost to extinction. Dealers in Texas and Mexico dig up huge piles of cactus and truck them to market, selling them to collectors and for use in southwestern landscaping schemes. Half the cacti go as far afield as Europe and the Far East.

Living creatures must certainly change as environmental conditions change. Species unable to adapt to new conditions die out and new species evolve to take their place. Dinosaurs and flying reptiles no longer live on the earth, but other creatures have arisen that did not exist when the dinosaurs did. Humans, however, are speeding up the rate of change to the point where species cannot evolve quickly enough to replace those that are lost. The hard truth is that one-half of the world's extinct mammals died out within the last 50 years.

Protecting Wildlife Resources

The protection of wildlife resources can be approached in a variety of ways. Laws can be written to protect

Figure 5.8 Eight-week-old bald eagle chick. According to the Audubon Society's 1985 Christmas bird survey, five of America's endangered bird species are recovering their numbers, mainly due to the decrease in the amount of DDT in the environment. Bald eagles, gyrfalcons, merlins, prairie falcons, and northern goshawks are now seen in greater numbers than 10 to 15 years ago. The eagle, a symbol of national pride, was endangered not only by pesticides but also by loss of habitat and hunting by farmers who believed eagles killed their livestock. Not long ago it was feared that bald eagles would be extinct by the end of the century. Programs in which young eaglets are taken from nests and raised at research stations before reintroduction into the wild have helped to restore the species, as have stiff laws against cutting down trees in which eagles nest or killing the birds themselves. (Craig Koppie, U.S. Department of the Interior, Fish and Wildlife Service)

species or to enhance their survival on an individual basis. This is the way the U.S. Endangered Species Act works. Samples of species can also be collected in zoos, botanical gardens, or seed banks to insure that we have representative examples of all living organisms. For many species, however, this second approach does not appear feasible. Special requirements for habitat or the population size for breeding may be too difficult to meet in captivity. A third approach is to set aside biological reserves containing entire ecosystems. In this way not only obviously threatened species but also the organisms tied to them in complex food webs can be saved. (See Controversy 5.2.)

Preserving Species Individually

Among the first federal laws dealing with wildlife were those that taxed hunting and fishing equipment and that required permits for these sports. The tax money collected has been used to buy land for wildlife refuges. Hundreds of millions of dollars have been raised for this purpose (a fact that must be taken into account by those who oppose hunting). It has been suggested that horticultural items could be taxed to provide a similar fund for plant preservation.

In 1966, Congress passed the Endangered Species Act, which was designed not only to protect wildlife but to determine the extent of the problem of disappearing wildlife. It directed that a list be made of endangered species, including estimates of the number of remaining individuals of a species as well as the range over which the species can be found.

In 1973, this law was made much stronger by a series of amendments. The new law recognized that we may want to protect species that face extinction in the United States, though not worldwide. It also established a new category called "threatened species." **Threatened species** are those not now endangered but whose populations are heading in that direction. By recognizing this fact early, perhaps we can do more to save them. A further important change was that a new category—endangered plants—was added to the endangered species list. In addition, the new amendments directed that federal agencies could not undertake projects that threaten endangered species or their habitat. Although this particular provision drew little notice when the amendments were passed, it became the basis for the conflict between the snail darter and Tellico

CONTROVERSY 5.2

The Most Important Environmental Problem: Extinction of Species

Despite concern in the U.S. over pollution, it is about the least important aspect of environment.
Lee M. Talbot

Lee M. Talbot, an ecologist at the Council of Environmental Quality, has pointed out that pollution "is about the least important aspect of environment" because it is, in most cases, reversible. But changing land use, such as leveling forests or filling in wetlands, eradicates entire habitats and causes some species to be lost to the world forever. Plants and animals that may now be regarded as dispensable may one day emerge as valuable resources (Talbot, 1974).

Russell Train, President of the World Wildlife Fund, agrees:

> I have spent most of my time over the past several years working on a variety of pollution problems—air, water, and chemical among others. As I review these efforts, I am struck by the fact that the real "bottom line" is the maintenance of life on this earth. Time is running out rapidly on the natural systems of the earth, and particularly on the survival of species. The loss of genetic diversity which threatens everywhere and the resulting biological impoverishment of the planet have grave implications for our long-term future.

> We need nothing less than a comprehensive program worldwide to preserve and protect representative ecosystems. . . .

> We human beings are relative newcomers on the face of the earth, but we now possess the power of life or death over our fellow creatures. While the scientific and economic arguments for the maintenance of species are compelling, it seems to me that we have an overriding moral responsibility to help preserve the other forms of life with which we share the earth (1978, p. 324).

Do you agree that loss of habitats of wild animals and plants is the most serious environmental problem we face? If not, what do you think is the most serious problem?

Sources: Lee M. Talbot, in *Science*, **184** (May 10, 1974), 646; Russell Train quoted from "Letters to the Editor," *Science*, **201** (July 28, 1978), 324.

Dam. As written, the law allowed no weighing of benefits versus costs if the extinction of a species was involved. (See Controversy 5.3.)

In 1978, as a result of the snail darter controversy, the act was again amended to make it more flexible when in conflict with government projects. Now a committee must first decide whether those in charge of a project have considered all reasonable alternatives. If conflict still exists, another committee, composed of the Secretaries of Interior, Agriculture, and the Army, the chairman of the Council of Economic Advisors, and representatives of the state in which the project is located will rule on whether a disputed project gives benefits that clearly outweigh preserving an endangered species. At its first meeting, in January 1979, this committee ruled against the completion of the Tellico Dam. Although millions of dollars of public money had already been spent on the project, this was ruled not sufficient reason to allow a species to be exterminated, even though the species involved had no sport or commercial value. The committee noted further that the threat to the snail darter pointed to other environmental problems that completion of the Tellico Dam would cause, problems that only came to public attention because of the fight over the snail darter. Many acres of fertile farmland would be covered with water when the dam was completed, decreasing a vital agricultural resource. In addition, a recreational resource would be lost: the last free-flowing stretch of the Little Tennessee River. Third, land of historical value, the ancestral homeland and grave sites of the Cherokee Indians, would be flooded. This case illustrates how endangered species, even seemingly insignificant ones such as the snail darter, act as barometers for environmental problems in general.

Jimmie Durham, a Cherokee Indian leader, noted in testimony before a House committee that many people were making fun of the snail darter. "I would like to ask why it is considered so humorously insignificant," he asked. "Because it is little, or because it is a fish?"

Although it seemed that the snail darter and the stretch of free-flowing river in which it lived were now safe, this was not so. In September 1979, supporters of the dam tacked onto an energy-development bill an amendment authorizing completion of the dam "notwithstanding the provisions of" the Endangered Species Act. This bill was signed by the President, to the

dismay of many environmentalists.★

In 1982, the Endangered Species Act was reauthorized. In addition, several amendments designed to improve procedures for the protection of endangered plants were added. Before 1982, only about 70 U.S. plant species had been listed as endangered (compared to about 225 animal species), despite studies showing that almost 3,000 U.S. plant species are in danger (Figure 5.9).

Several factors have combined to make the listing of plants a long and time-consuming procedure. Plant taxonomy is not as well advanced as animal taxonomy, nor has as much field work been done. Thus, it is not always clear whether plant species are rare or just little known. Further, before the 1982 amendments, the Endangered Species Act required that a plant's critical

Figure 5.9 Endangered Boott's rattlesnake root. Unlike endangered animals, which are protected by the Endangered Species Act from being "taken" at any place in their natural habitat, endangered plants are protected only where they occur on federal lands. Furthermore, it is only illegal to "take" plants. Vandalism, uprooting, or trampling endangered plants is not, strictly speaking, illegal. (Larry Master, Heritage Task Force)

★Since 1979, three small populations of snail darters have been found in remote locations by TVA zoologists. In 1984 the snail darter was demoted from endangered to threatened. "That's not to say it's on the rebound," said a spokesman, "it just means we're more informed on where it is" (*The New York Times*, August 8, 1974.)

CONTROVERSY 5.3

Do We Need Flexibility in Laws Protecting Endangered Species?

No compromise is possible when the problem is stated in terms of the question "Do organisms have the right to exist?"

Wayne Grimm, National Museum of Canada

Are you going to do anything to get the snail darter off our backs?

Unidentified Alabama Congressman to Interior Secretary Cecil Andrus

When the Endangered Species Act halted construction of the Tellico Dam because it appeared that a completed dam would wipe out the endangered snail darter, many congressmen began to feel that the act was "inflexible." That is, the act had no provision for considering the value of a project compared to the value of an endangered species. People began thinking of cases in which, they felt, a project or action could have more value to humans than the continued existence of a species. Senator William Scott of Virginia (1978) argued:

> Suppose a bird of some endangered species was in front of an intercontinental ballistic missile. . . . They could not release that missile. To me that would be a ridiculous offense. . . . Any commander worth his salt . . . would go ahead and release the missile, but he would be disobeying the law and would be subject to a fine of $20,000 and imprisonment for up to a year.

Others argued that the act was actually working well. They pointed out that the Tellico Dam was the only project ever halted by the law and one of only three to go to the courts at all. (Some 5,000 possible problems were solved by consultation with the Fish and Wildlife Service.)

But aside from questions about how well the original act worked is the question of whether flexibility is desirable. Do you feel there are instances in which a project could have more significance to humankind than the survival of an endangered species? Can you think of an example? Or do you think that no species should become extinct because of a construction project, however beneficial to humans? Can you justify this view?

Sources: Wayne Grimm and Alabama Congressman, quoted in *Science*, **196** (June 24, 1977), 1427–28. William Scott, quoted in *Science*, **201** (August 4, 1978), 427.

habitat be defined when it was listed as endangered. The *critical habitat* includes the area in which the plant is found, any areas to which it could be moved if necessary, and a buffer zone if needed. Field work to determine critical habitat has not been done for most plants in jeopardy. Finally, several analyses were required before a species could be listed. An economic analysis of the effect of setting aside a critical habitat was required by the Endangered Species Act, and economic and regulatory analyses were required by other acts and executive orders (Regulatory Flexibility Act, Paperwork Reduction Act). Staff and budget cuts at the Fish and Wildlife Service have slowed progress in this kind of analysis for both plants and animals.

The 1982 renewal specifically addressed several of these problems, noting that the only criteria for listing a species should be biological—that is, it is either in danger of extinction or it is not. Economic and other criteria are not to be used. Further, critical habitat is now to be defined as well as possible at the time of listing. This is an area in which citizen participation can be very effective. Several states have private conservation groups devoted to plant protection. Almost one-half the states have groups such as the California Native Plant Society, or botanical clubs such as those in Michigan and Maine, that work to determine the rarity and distribution of their state's plant species. Federal funds and advice are sometimes available through state governments to local society botanists. Their work helps in assembling the data needed before a plant can gain official endangered-species status.

Local garden clubs are encouraged by the Fish and Wildlife Service to "adopt-a-plant" (by protecting its habitat in the wild rather than by moving it to a garden). The FWS is willing to provide help in choosing those plants most in need of attention.

The Endangered Species Act does not control the effects of nongovernmental projects. Private citizens are not bound by law to consider the effects of projects, such as housing developments, on endangered species. Several states do have programs to encourage wildlife conservation on private lands, providing cost-sharing, free or low-cost materials, and free labor. Most of these programs involve farmland and are meant to conserve animal rather than plant species.

The 1973 amendments to the Endangered Species Act served one further purpose, which was to ratify the Convention on International Trade in Endangered Species of Wild Fauna and Flora (CITES). The treaty sets up a system of permits for both exporting and importing threatened and endangered species, or products made from them. Trade in nearly extinct species is practically prohibited, while strict controls are set for other endangered or threatened species. Other countries, including Canada, have signed CITES.

Wildlife Management Techniques

A variety of special techniques have been developed to preserve species in danger of extinction or to increase the range of animals considered highly desirable (i.e., those that people like to hunt). In some cases, animals may be transferred from their natural habitat to a similar area where they were not previously found. This has been done mainly with nonendangered game species such as Canada geese. The wild turkey, which has been introduced in a number of areas, now occupies more territory than it did during Colonial times.

When the judgment has been made that a species will not survive on its own, even if given a fair chance, eggs may be collected and hatched in captivity, or breeding programs can be instituted at zoos. The animals can in some, but not all cases, be successfully reintroduced into the wild. Sea turtles, which by instinct run to the sea after hatching and later return to their birthplace to lay eggs, never seem to get their bearings right if hatched in captivity. They swim off into dangerous waters and fail to return to suitable beaches for successful egg laying. On the other hand, about half of the whooping cranes alive today were hatched and reared in captivity.

In some cases management procedures on preserves are so successful that limited hunting can again be allowed. One hundred years ago, the American bison lived in herds so huge it sometimes required several hours to pass by a herd. Fifty years ago there were only a few hundred bison left. Within the past few years, bison numbers have increased enough to again allow some hunting.

Refuges and Reserves

U.S. refuge system. In the United States during the early 1900s, Congress began to set aside areas of wildlife habitat, or refuges, to help protect endangered wildlife. The development of the National Wildlife Refuge System is explained in the general context of land preservation in Chapter 30.

Plant species, especially, can best be saved by setting aside part of their natural habitat as a preserve. A few individuals of a species in botanic gardens are not enough to ensure reproduction and species survival. The first refuge intended to save endangered plants was purchased in 1980. This is the Antioch Dunes in California, home of the endangered Contra Costa wallflower and the Antioch Dunes evening primrose. A number of animals are also protected in refuges. The trumpeter swan, for instance, flourishes in Red Rocks Lake Refuge in Montana.

How big do reserves need to be? Many wildlife experts point out that refuges must be large areas, measured in thousands of square kilometers. Smaller areas may not be able to support certain species, often those most endangered. For example, large predators such as wolves or the big cats must roam vast areas to find food. In addition, larger reserves are able to buffer species from border pressures, such as pollution or human disturbance.

Certain research done on islands has a bearing on how big parks or reserves must be. The size of an island seems to influence how many different species can exist there. Ecologists E. O. Wilson (1984) and Robert MacArthur found in a study of animals on Pacific islands that a doubling of the area of an island is not accompanied by a doubling of the number of species found. Rather, an island ten times as big is needed to support twice the number of species. The reason this research applies to parks or reserves is that such areas are more and more becoming like natural islands in the midst of a sea of human-influenced habitat destruction.

Following the rule of island biogeography, if 90% of a natural habitat is destroyed and 10% is set aside in a reserve or park, we might expect to save no more than half the original number of species found in the area. By this reasoning, if we save only as much of the Amazonian tropical forests as is now found in parks or preserves, we may well be unable to save two-thirds of the half-million species found there.

It is not yet certain that the island theory applies to parks. However, studies now being done in tropical forests by ecologists, such as Thomas Lovejoy of the World Wildlife Fund, seem to show that it does. Lovejoy gives the example of a 25-acre preserve that lost its far-ranging piglike animals called peccaries. In an unanticipated chain reaction, ten species of frogs also

disappeared because they needed the moist wallows created by the peccaries.

In addition, wildlife experts are concerned about the size of reserve needed to preserve genetic diversity within a particular species. As the population of a particular species becomes smaller and smaller, animals have fewer choices in breeding. As a result, the offspring become more and more alike in their genetic composition.

In evolutionary terms this is not a good thing. A population in which all the individuals contain almost the same genes is vulnerable to any changes in the environment. Without a range of different characteristics among members of the population, there is no longer the possibility that some members will be able to stand a change in the environment or disease better than the average individual. There is always the danger that climatic change, disease, or competition from a new species could wipe out the whole population.

Furthermore, studies on endangered animals such as the cheetah, which has little natural genetic variability, show that there is a higher mortality of young born in the wild or in zoo breeding programs. This is apparently due to the large number of birth defects that can show up whenever close relatives breed. Small reserves that can support only small populations of a species (especially the larger mammals) force this kind of genetic uniformity.

With all these factors in mind, the United Nations Educational, Scientific and Cultural Organization (UNESCO) has begun to identify "biosphere reserves" or "ecological reserves," a network of protected samples of the world's major ecosystem types. The reserves must be large enough to support all the species living there, buffer them from the outside world, and protect their genetic diversity. In this way, the reserve allows both growth and evolution and acts as a standard against which human effects on the environment can be measured.

Besides laws establishing preserves, stronger laws are necessary to limit the spraying of pesticides near game preserves or near the other habitats of endangered species.

The Economics of Preserving Endangered Species Worldwide

Most people would agree that other creatures have a right to survive on the earth. People rarely intentionally set out to wipe out other species. Yet, as a number of experts have shown, our economic system is set up in such a way that we tend to do just that.

Species, like air and water, are, in a sense, a common resource. That is, we all stand to benefit from having a wide variety of plants and animals on the planet. None of us, however, "owns" any particular wild species and so no one is directly responsible for the survival of any particular species. Individuals can easily see the benefits gained from hunting tigers, capturing apes, or building a housing development on some other creature's habitat. Much harder to keep in mind are the benefits, to all of us, of having a great variety of species, since these benefits (medical uses, agriculture) are long term or less visible (aesthetic or moral). In other words, the short-term, visible benefits go directly to the individuals involved, while the losses are mainly long term and are spread over society as a whole. As a result, we are as wasteful of the resource of species as we have been of air and water, other resources that no one owns* (Figure 5.10).

Because these sorts of effects are not an intended result of people's actions, economists call them *externalities*, or spillover effects. People who undertake more economic activities are likely to cause greater spillover effects. People in developed countries use the most raw materials, which are often obtained by disturbing the habitat of creatures in less developed countries, and are, in this sense, most responsible for loss of species.

Several laws have been passed to decrease the effects of U.S. actions on endangered species worldwide. Under the National Environmental Policy Act (NEPA), U.S. government agencies are required to examine their actions abroad for possible ecological harm. International public organizations, such as the U.S. Agency for International Development and United Nations agencies, could be required to write environmental impact statements for their projects. Even if it were decided, because of great benefits to humans, to go ahead with plans that would damage habitats of other species, the costs would be recognized and weighed, avoiding much unnecessary damage. The United States has signed the Convention on Nature Protection and Wild-

Figure 5.10 Devil's hole pupfish (*Cyprinodon diabolis*). There are 12 species of pupfish in various locations around the U.S. Important to science because of their ability to withstand extremes of temperature and salt concentration, the fish have run afoul of humans in several instances. In 1976, the U.S. Supreme Court ruled that the Endangered Species Act prohibited ranchers from pumping so much underground water for their ranches that the water level was lowered in Devil's Hole, home of the endangered Nevada pupfish. The fish have been marooned there since the last glacier receded, leaving much of Nevada desert country. Not so lucky, California's Tecopa pupfish was recently declared extinct. Thirty years ago, builders of a bathhouse diverted waters from the thermal pools and springs in which the fish lived. Unable to adapt to life in the new swift stream, the fish finally died out. The bathhouse is no longer in use. (U.S. Fish and Wildlife Photo/Jack E. Williams)

life Preservation in the Western Hemisphere. Because of this convention, special plans were developed during Pan American Canal negotiations by the United States and Panama to protect endangered vegetation on Barro Colorado Island. This topic is explored further in Global Perspective I, following this chapter.

In the final analysis, the fate of other species is a barometer for the fate of the human species. If the outlook is not sunny for the survival of other creatures on this planet, surely human survival is in for a stormy time (Figure 5.11):

I heard the song
Of the world's last whale
As I rocked in the moonlight

---

*These principles were first clearly applied to environmental problems by Garrett Hardin in his "Tragedy of the Commons," *Science*, **162** (December 13, 1968), 1243.

Figure 5.11 Humpback whale. This whale is "breaching," or leaping out of the water. See Global Perspective I following this chapter for a discussion of the plight of this endangered giant. (Ken Balcomb/World Wildlife Fund)

And reefed the sail.
It'll happen to you
Also without fail
If it happens to me
Sang the world's last whale.★

We would do well to be guided by the maxim attributed to Aldo Leopold:

The first requirement of intelligent tinkering is to save all the pieces.

Summary

Wild plants and animals represent harvestable resources of food, fuel, and building materials. Natural communities also provide ecological services such as oxygen production and reduction of air and water pollution. Individual species manufacture chemicals useful as drugs, pesticides, flavorings, and a variety of other industrial raw materials. Additional agricultural uses for some species are found in programs for bio-logical control or species improvement through cross-breeding. All species are important complements of the food webs of which they are part. Finally, species have an importance and a right to exist in and of themselves.

In a variety of ways, this right to exist is being threatened. Hunting and collecting, habitat destruction, competition from introduced species, and pollution all threaten wildlife resources. Species can be protected individually by laws, such as the Endangered Species Act, which attempt to discover the needs and preserve the habitats of specific endangered species. Examples or seeds of endangered species can be collected in zoos, seed banks, and botanical gardens. The most comprehensive and most likely to succeed solution would be to save whole ecosystems in parks or reserves. It is important, however, that such reserves be large enough to fully protect an ecosystem and allow it to function normally.

Questions

1. How would you explain the value of preserving a wide variety of species, both plant and animal?
2. What is meant by the terms *extinct, threatened, endangered* when they are applied to species?

3. Briefly list the main reasons that plant and animal species become extinct or endangered. Which is the most important reason?
4. Do you believe hunting should be permitted? Give your reasons for or against and any special requirements you would like to see enforced.
5. How does the fact that wildlife species are usually viewed as a common resource, in the same way that water or air are considered common resources, contribute to the rapid loss of species today?
6. Recount some of the ways in which attempts are currently being made to preserve species from extinction. What further measures could be taken?

Further Reading

"Biological Diversity." Washington, D.C.: The Global Tomorrow Coalition, Natural Resources Defense Council, 1985.

A good review of the causes of and possible solutions to species extinction.

Canadian Nature Federation, 75 Albert Street, Ottawa, Ontario, K1P 6G1.

Information on endangered species in Canada and Canadian wildlife laws can be obtained from this organization.

U.S. Fish and Wildlife Service. *Endangered Species Technical Bulletin*, vol. 6, No. 1. Washington, D.C.: U.S. Department of the Interior, 1981.

This bulletin (and the September 27, 1985 Federal Register) lists those plants considered in danger in the United States and defines the problems they face. If you want to know which plants in your area need protection, this is the place to look. A similar list of vulnerable animal species was published in the Federal Register, December 30, 1982. More information is available from: The Director (OES), U.S. Fish and Wildlife Service, Department of the Interior, Washington, D.C. 20240.

Morse, L. E. and M. Henifin, eds. "Rare Plant Conservation: Geographical Data Organization." New York: New York Botanical Garden, 1981.

Citizens interested in helping with studies necessary before plants can gain official endangered species status can get an idea of the type and quantity of needed data from this publication.

Audubon Wildlife Report 1987, ed. by Roger L. DiSilvestro. Orlando, FL: Academic Press, 1987.

This volume features a comprehensive compilation of federal wildlife conservation activities.

"Wild Life Resources," in *The State of the Environment*. Paris, France: Organization for Economic Cooperation and Development, 1985.

Although written from the point of view of the developed countries, this chapter contains a good summary of the value and the plight of the world's wildlife resources.

Wilson, E. O. *Biophilia*. Cambridge: Harvard University Press, 1984.

A famous ecologist explains his fascination with the natural world and why it is important to all of us to preserve the variety of species we now enjoy.

Myers, N. "The End of the Lines," *Natural History*, **94**(2) (1985), 2.

The effect humans are having on the natural course of evolution is explored in this provocative article.

References

Bolandrin, M. F., et al. "Natural Plant Chemicals: Sources of Industrial and Medicinal Materials," *Science*, **228** (June 7, 1985), 1154.

C. Holden, "Endangered Species: Review of Law Triggered by Tellico Impasse," *Science*, **196** (June 27, 1977), 1427.

Wade, N. "New Vaccine May Bring Man and Chimpanzee into Tragic Conflict," *Science*, **200** (June 2, 1978), 1027.

Orlans, F. B. Letter to the Editor, *Science*, **201** (July 7, 1978), 6.

Beston, H. *The Outermost House*. New York: Holt, Rinehart and Winston, 1928.

Grier, J. W. "Ban of DDT and Subsequent Recovery of Reproduction in Bald Eagles," *Science*, **218** (December 17, 1982), 1232.

Talbot, L. M. Quoted in C. Holden, "Scientists Talk of the Need for Conservation," *Science*, **184** (May 10, 1974), 646.

Train, R. Letter to the Editor, *Science*, **201** (July 28, 1978), 324.

Scott, W. Quoted in *Science*, **201** (August 4, 1978), 427.

MacBryde, B. "Why Are So Few Endangered Plants Protected?" *American Horticulturist*, October/November 1980, p. 29.

O'Brien, S. J., et al. "Genetic Basis for Species Vulnerability in the Cheetah," *Science*, **227** (March 22, 1985), 1428.

The World's Tropical Forests: A Policy, Strategy and Program for the United States. Washington, D.C.: U.S. Interagency Task Force on Tropical Forests, 1980.

A GLOBAL PERSPECTIVE I

Human Needs Versus Animal Rights

Three Cases

*Eskimos and the Seal Boycott/Africa's Big Game Animals/
Tropical Forests*

Looking for Solutions

*Endangered Sea Mammals Versus Endangered Eskimo
Culture/Bowhead Whales/Economic Solutions and
International Cooperation in Africa/A Workable
Compromise in Nepal/Proposals to Stem Tropical Forest
Loss/Different Viewpoints*

CONTROVERSY:
I.1 ***The Ethics of Hunting Whales***

In my father's day, there were many wild animals around
here. It is a very good thing that we chased them away.
Joseph Mkomba, *farmer in Western Kenya's Kerio Valley*

Three Cases

Eskimos and the Seal Boycott

In Canada's Baffin islands, a group of arctic Eskimos,
the Inuit, once carried on a thriving trade in sealskins.
Now most families must accept welfare payments. The
reason is a successful boycott of seal furs. The boycott
was not called against these Eskimos but rather to protest the clubbing of baby harp seals in Newfoundland;
however, the effects are felt by the Eskimos as well as
the Newfoundland peoples. The price for a harp seal
skin has fallen from $33 in 1977 to as little as $4 today.
There is no longer any reason for the hunters to go out
for seal. Alcoholism and drug use seem to be increasing
in the villages. "The sealskin boycott has cost [the Es-
kimos] even more in self-respect. We always liked to
think that our men were better hunters than other
men," an Inuit told Meeka Kilabuk (1985).

Africa's Big Game Animals

In Africa, some conservationists fear that the last herds
of wild game, and the last of the big cats that prey on
them, will be gone by the end of this century. Yet, as
the quote at the opening of this Global Perspective illustrates, the primary concern of the African farmers is
not the loss of wildlife but protection of crops and
family from marauding beasts. The troublesome fact is
that the wildlife resources that conservationists wish to
protect are largely found in those developing countries or undeveloped areas of the world where the na-

tive peoples themselves are engaged in a struggle for
survival.

This is no accident, of course. Civilization has
brought both riches and ease to most of the population
of the developed countries. At the same time, civilization has destroyed much of the habitat for wildlife in
those same countries. Precisely because they are undeveloped, the poorer countries in Africa and southwest Central America still have much natural wildlife
habitat.

Add to this the fact that the tropical rain forest
habitat found in many of these countries is home to a
staggering richness of species, and the stage is set for a
conflict between the rights of poor but growing human
populations and the rights of wild plant and animal
species. Again and again, around the world these
poignant conflicts are played out. Whose rights are paramount: our fellow human beings struggling to feed
their families and live with dignity in the old ways, or
the wild species, struggling perhaps for their very existence on the earth?

Tropical Forests

Tropical rain forests, home to half of all the species now
in existence, are disappearing at an alarming rate. In
most cases the forests are first logged for valuable tropical hardwoods. Then, once the logging roads are in,
settlers follow and clear the remaining trees in order to
farm or ranch. But little tropical forest is suitable for
agriculture. In as little as 2 years or as much as 7 years,

Figure I.1 Tropical forest destruction. A major portion of the nutrients in tropical ecosystems are in the plants and animals. Thus, when forests are cleared, only impoverished soil remains. Left without cover, the soil is subject to massive erosion, with resultant flooding and loss of productive capacity. (George Seddon)

soil fertility is exhausted. Farms and ranches are abandoned to erosion and flooding. If current rates of destruction were to continue, in 20 years there would be no tropical forests at all, outside of the negligible portion in parks or reserves.

After viewing areas where loggers had *clear-cut* tropical forests, Nicholas Guppy (1984) wrote: "Visiting such areas, it is hard to view without emotion the miles of devastated trees, of felled broken and burned trunks, of branches, mud and bark crisscrossed with tractor trails. Such sights are reminiscent of photographs of Hiroshima, and Brazil and Indonesia might be regarded as waging the equivalent of thermonuclear war upon their own territories" (Figure I.1).

This is the point of view of a biologist from a developed country, however. From the point of view of the developing country itself, these same tropical forests are a resource wealth, sometimes the only wealth that it has. "People are starving on one side of the Andes," says Mariano Prado (1983), who is developing a resort on the other, tropical forest side of the mountains, "and here there are so many riches." The Peruvian government is trying to shift some of the desperately poor population along the coast into the relatively unpopulated interior forests.

The question, then, is, how can the rights of native people and native wildlife species be reconciled? In the following sections, we look first at agreements we in the United States have hammered out with native Alaskan Eskimos. Next we examine compromises suggested for or worked out in developing nations.

Looking for Solutions

Endangered Sea Mammals Versus Endangered Eskimo Culture

In 1972, the U.S. Congress passed the Ocean Mammal Protection Bill. This law protects arctic fur seals, walruses, and whales, among other ocean mammals, from commercial hunting, which has reduced some species almost to extinction. However, during hearings before the bill was passed, it became clear that a total prohibition of the hunting of sea mammals or the sale of products made from sea mammals would threaten something else that was endangered: native Eskimo culture.

The following testimony is excerpted from hearings from the Subcommittee on Oceans and Atmosphere of the Committee on Commerce of the United States Senate in the spring of 1972 (before the Ocean Mammals Protection Bill was passed):

Statement of John Henry: I don't have no written statements. . . . So, I'm going to make my own—what people thinks. As people thinks, I'm Eskimo, just like Indian or Aleut, all same. We are from Norton Sound on the coast area, where we could hunt on the sea and on the ground. We have two ways to hunt. To feed ourself from the sea and from the ground. Even we don't have no money. Sometimes we have hard times. We can't purchase from the ground, we can't purchase from the sea. But we are Eskimo, we can't write a check to purchase something that we could buy. Just sit down and write a check to purchase something for our children. . . .

Sea mammals. It's our living and I born on sea mammals. I born, we didn't have much money. My dad didn't have much money. He hunt and hunt, just the same as like other peoples in the other villages. He try to support us from the hunting, that's all. . . . No other resource we have, just from his hunting. Just like other villages does. And ancestor, from our ancestor I think we born like them. I will be like that, because my ancestors I know. I was born without money. I will be an ancestor for my own folks without money. . . .

Statement of Myrtle Johnson: I am a resident of Nome, and was born and raised in the village of Golovin. . . . I will draw from people I know personally to paint an imaginary picture based upon the truth.

John and Mary are in their mid-50s. John works part-time seasonally in casual labor, and they care for four foster children plus one grandchild. John is a successful seal hunter. He also catches Beluga. If he had a chance, he would join a crew and hunt black whale as well. This would bring him a share of meat and muktuk—something his family otherwise must buy at $3 per pound or receive as a gift that will obligate him to return in some other form.

John's family depends heavily upon sea mammal meat and oil all year long. The spring greens are stored in fresh seal oil, some of this being sent to family and friends living in the cities. The meat is dried, frozen, or in other ways preserved for the time when it is not so plentiful. . . .

The skins will end up as mukluks, parkas, parka-pants, and vests for the family, as well as surplus skins or garments to be sold or traded to others. Both raw and stretched, skins are a source of income and add to the internal welfare system of their village. . . .

Beef and pork, seldom carried at a village store, must be priced 25% to 50% higher than at the outlets with jet flight service. Thus, the importance of local meat harvest cannot be underestimated. Without the meats and oils, many families would be without this basic food in their diet. . . .

Caucasians speak of bread as the staff of life. For coastal Eskimos the seal represents the single basic food staple with all the meanings others may associate with bread. . . .

To take away our Eskimo bread is to deny the men and women of the villages the right to work to meet their needs as they see them and can mean the final end of our way of life before we are fully adapted to the modern world.

As a result of the hearings, the final act allowed exemptions for subsistence hunting (hunting for food or clothing) and the sale of the native handicrafts★ (Figure I.2).

Bowhead Whales

In 1976, even the Eskimo right to subsistence hunting was called into question.

For centuries, humans have hunted whales for oil and meat. In the 1700s and 1800s, right and bowhead whales were the main targets. These were called "right" whales because they float after they are killed and because they are slower than other species. Now these whales, along with the blue whale, the humpback, and the gray whale, are near extinction.

The great blue whale is the largest animal now living on the earth. Although they swim in the oceans, as fish do, whales and their smaller relatives, the dolphins, are mammals. Whales are warmblooded and suckle their young. They are also social animals, living in herds and possibly even in families. Within the past few years, scientists have realized that whales can "talk" to each other. Some beautiful records have been made of the singing sounds they make under water.†

Whaling was once a dangerous occupation. The small crew of men, with their harpoons and in their

---

★When the bill was renewed in 1981, the subsistence exemption was changed to one for rural users, rather than specifically mentioning Eskimos.

†One record of these sounds is *Songs of the Humpback Whale*, Capitol Records, Inc., Hollywood and Vine Streets, Hollywood, Calif.

Figure I.2 Eskimo craftsmen carve and sew materials from ocean mammals into useful and beautiful objects. This is an example of a parka sewn from sealskins. (Dobbs Collection/ Alaska Historical Library)

Figure I.3 Whaling with a hand-thrown harpoon in a small boat, as is still done in some areas, is a difficult and dangerous business. The whale might overturn the boat, with tragic results for the whalers. (Courtesy American Museum of Natural History, Dr. F. Rainey, photographer) In contrast, gigantic modern whaling ships, such as this Soviet factory ship, ensure there is little danger to the whalers. They are frighteningly efficient at killing whales, however. (Matt Herron/Black Star)

fragile boat, and the huge whale were relatively well matched. Mostly the men won, but sometimes the whale did, too. Modern methods, however, have tipped the balance far in favor of men (Figure I.3).

Modern non-Eskimo whaling fleets consist of fast killer boats, which kill the whales with harpoon guns, and large factory ships, which process the whales at sea. Whales are located by sonar or small planes, then driven by high-frequency sound waves until they are exhausted. A harpoon carrying a grenade, which explodes within the whale, is used to kill them. With these methods, modern whaling fleets have killed more whales in this century than whalers did in the sixteenth, seventeenth, eighteenth, and nineteenth centuries combined. Within the past 50 years, the numbers of blue, humpback, and gray whales have been reduced from hundreds of thousands to a few thousand. The California gray whale, whose breeding grounds are now pro-

tected by the Mexican government, has begun to recover in numbers, perhaps even to reach a population size near that of the 1800s. The bowhead whale, on the other hand, has been protected for 35 years but does not seem to be increasing in numbers.

About 30 nations belong to the International Whaling Commission, which sets yearly whaling quotas (the numbers of each type of whale that may be caught in the various oceans). Nations that have whaling fleets, such as Japan, and those that no longer have fleets, such as the United States, Canada, and South Africa, send delegates to the Commission meetings. Over the past few years, quotas have been slowly reduced as new scientific data have come to light on the population sizes and reproductive ability of whales. However, quotas have also declined because, in a number of cases, whalers have simply been unable to find enough whales to fill them.

For many years, conservationists have worked to pass a 10-year moratorium on all whaling, allowing time for the study of whales and their survival needs, before they are unintentionally hunted to extinction. Conservation groups such as the Animal Welfare Institute have conducted a vigorous campaign to support the moratorium. In addition, they urge that people boycott goods from Japan until the government agrees to stop whaling (see Controversy I.1). In 1982, the IWC finally approved a moratorium, to begin in 1985–1986. However, long before this, in 1946, the IWC treaty noted that the bowhead whale was the most endangered whale species. The commission at that time forbade all but subsistence hunting.

United States Eskimos had been catching about 10 to 15 bowheads per year since 1946. However, in the 1970s, the catch increased. Further, the number of whales struck but lost steadily increased (Table I.1). It is estimated that at least half of these later die. The IWC was concerned about the effect this increased hunting would have on an already severely endangered species. In 1976, the commission called for a total ban on subsistence hunting of the bowhead.

Why did Eskimo hunting of the bowhead whale increase? The answer is tied up with economics and the impact of outsiders on native culture. R. Storro-Patterson (1978) wrote:

> Correlated with the increase in the Eskimo harvest of bowhead whales is the growing number of whaling crews:
>
> Number of Whaling Crews—Spring Hunt
>
> | 1971 | 1972 | 1973 | 1974 | 1975 | 1976 |
> |------|------|------|------|------|------|
> | 25 | 27 | 45 | 46 | 75 | 86 |

The rise in the number of whaling crews may in turn be significantly related to the availability of jobs and money for Eskimos. Traditionally, to be a whaling captain was not only a matter of great prestige, but great difficulty. One had to either inherit the equipment, marry to obtain it, or gain sufficient wealth to purchase it. The latter option was seldom possible. Recently, however, such construction projects as the Alaskan pipeline, and oil drilling operations, as well as the Alaskan Native Claims Settlement have changed this. An ambitious Eskimo can save $9,000 for an "outfit" plus another $2,000 for provisions for a whaling crew.

The Eskimos were outraged by the ban, as were many other people who sympathize with the native Alaskans' difficult fight to survive and preserve their culture in a relatively hostile natural environment:

> "Hunting the bowhead is what keeps our communities together. Our people depend on the bowhead, first for food and, just as importantly, for the survival of our culture," said Dale B. Stotts, an official of Alaska's North Slope Borough, which includes a number of whaling towns and villages along the coast of the Beaufort Seas. (Rennsberger, 1977a)

W. H. DuBay (1977) wrote, in defense of the Eskimo hunters:

> . . . People outside Alaska don't seem to realize that the Eskimos are the residents of the Arctic and that, were it not for their aggressiveness in protecting their Arctic homeland from those who would destroy it, there would be no effective environmental safeguards at all operating in the U.S. or Canadian Arctic.

Table I.1   Alaskan Eskimo Bowhead Harvest

	1973	1974	1975	1976	1977[a]
Landed	37	20	15	48	26
Killed, lost	0	3	2	8	2
Struck, lost	10	28	26	37	77
	47	51	43	93	105

Source: R. Storro-Patterson, "The Bowhead Issue," *Oceans* (January 1978), p. 63.

a. The 1977 figures represent the spring hunt only.

CONTROVERSY I.1

The Ethics of Hunting Whales

Whaling is an affront to human dignity.
Sir Peter Scott

In a word, we are dissatisfied with the meeting. We are trying to avoid taking extreme actions.
Shigeru Hasui, *managing director, Japanese Whaling Association*

Most of us would agree that it would be terribly wrong to hunt whales to extinction. But to go a step further, should we "harvest" them at all?

At the 1982 meeting of the International Whaling Commission, a moratorium on commercial whale hunting was passed. The moratorium, which was delayed until 1988, displeased several members of the IWC, notably Japan.

Whale products are used to manufacture fertilizers, transmission fluids, and animal feeds, although other materials could be used to make all of these items. Whale meat is eaten in Japan and the Soviet Union. The Japanese claim that although whale meat forms only a small part of the Japanese diet (possibly as little as 1% of the protein), it is eaten mainly by the poor, who do not have many other choices.

On the other side is the evidence that whales are able to communicate with each other, that at least some whales travel in family groups, and that whales have large brains.

How should we balance these factors, as we deal with this "other nation" with which we inhabit the earth? Should there be any commercial whaling?

Do you think the United States should continue to ask for an exemption for Alaskan Eskimos to hunt endangered bowhead whales, even though this makes it more difficult for the United States to support a total moratorium on whaling at IWC meetings?

Sources: Peter Scott, quoted in *The New York Times*, June 8, 1982, p. A25; Shigeru Hasui, statement at the 1982 meeting of the International Whaling Commission.

. . . The issue is not just the best way to manage the preservation of the bowhead whale, but also the great question of subsistence hunting rights of American native peoples and the basic human right to eat what you have to eat in order to survive in your own environment.

Environmental groups were torn between their sympathy for the natives and the need to protect whales. Conservationists felt that if the United States filed a formal objection to the IWC decision, which would cancel the ban, other nations, such as Japan or Russia, would feel free to file objections to quotas the IWC had set on other whale species:

"For the first time, this has put the United States in the position of being the affected party in restricting whaling," said Dr. Roger Payne, a whale specialist with the New York Zoological Society. "If the United States files an objection to the I.W.C. position, this country could lose all its credibility in whale conservation."

"The original exemption on the bowhead was to guarantee what are known as the aboriginal rights of the Eskimos," Dr. Payne said. "The Eskimos aren't aborigines any more and along with modern hunting weapons there have to go some controls." (Rensberger, 1977b)

The U.S. government finally decided to ask that the Eskimos be given a quota on the number of bowheads they could catch. At an emergency meeting in December 1977, the IWC voted a quota of 12 whales landed or 18 struck, whichever came first.

The Eskimos objected that this meant that some people must go hungry in the coming year. Nonetheless, they agreed to obey the quota and set up a self-governing system to do so.

Perhaps the necessary balance has been found for the moment. It seems unfair that the Eskimos, as they attempt to protect their cultural heritage, must shoulder one more burden, that of nurturing back to health a whale population plundered by commercial whalers in the beginning of this century. But it also seems unfair that the bowhead whale, so close to extinction, should not have all possible chances to recover. "Surely the whales are also a 'special interest group' and need an advocate to defend their right to survive" (Ellis, 1980). As a *New York Times* editorial noted, "if Eskimo cul-

ture needs the bowhead, it must be saved; killing off the few that remain would not long satisfy either stomach or spirit."

Economic Solutions and International Cooperation in Africa

"It's crazy to try to educate an African to your point of view [on wildlife conservation] unless he's living in the same circumstances you are" (Parker, 1982). This lesson has been a bitter one for many conservation organizations attempting to influence African policies on wildlife conservation. In a country where large portions of the people don't have enough to eat and very few people have a living standard close to the average for the developed world, wildlife conservation simply isn't a priority issue.

There is a growing feeling that wildlife conservation in Africa will be possible only if it can be made to pay. The development of tourism based on visits to parks and reserves is one possibility. If people are willing to pay enough to see wild animals in their natural habitats, it makes sense to set those habitats aside. The area that is now set aside in this way is too small to preserve any significant portion of Africa's wild species, however. In most cases, the preserves are only gathering places for large numbers of animals at certain times of the year. At other times the animals disperse over a much wider area outside the preserves. Increasing the size of the preserves is not usually possible. Even if the large sums of money needed to purchase the areas were available, many of these areas are already settled by human populations.

Another possibility is to "crop" the wild animals—that is, to allow them to live and grow on ranches and in reserves without artificial feeding or watering and then to harvest surplus animals and sell the meat. It is not clear yet whether this is economically feasible or ecologically desirable. What *is* clear, however, is that the best solutions so far have been devised by people born and raised in the country concerned. Perhaps this is because they have the sensitivity to local customs and concerns, without which it is difficult to work out a plan acceptable to all factions.

A recent success is worth noting. The Amboseli region in Kenya has been used by humans as well as animals since the late Pleistocene. For the past 400 years, the Masai, nomadic cattle herders, have dominated the region. As in the rest of Africa, in recent years the Masai population has been growing rapidly. As a result, the tribes have been looking for additional pasture lands for increasing numbers of cattle. But, at the same time, increasing tourism has encouraged the Kenyan government to plan to set aside the entire Amboseli area as a park. To protest the park plan, the Masai began to kill large numbers of rhinoceroses.

David Westein, an African-born ecologist, has devised a plan whereby 150 square miles, constituting the most important area for wildlife conservation and tourism in the Amboseli region, will be protected from human intrusion. In the surrounding area, wildlife will coexist with cattle. There will, however, be some cost to the Masai ranchers in animals lost to predators and in land unusable for grazing. The government will pay in money and in services (health care, water development schemes) for these losses. So far the compromise appears to be working.

Tanzania, which has the largest wildlife population in Africa, has asked that other countries contribute to the cost of guarding wildlife from poaching. Needed equipment, such as surveillance helicopters for game wardens, is beyond the reach of many developing countries. That such measures can help is shown by the success of Project Tiger to preserve the Bengal tiger in India. This program, supported by international conservation groups, has increased the tiger population by 65%. Key features of the project include preserves, surrounded by buffer areas, in which humans are not allowed, and payments made to farmers who lose stock to the tigers.

A Workable Compromise in Nepal

In another case, Hemanta Mishra, ecologist with the National Parks and Wildlife Department in Nepal, has increased local acceptance of the Royal Chitwon National Park by meeting with local people to discover and address their concerns. One result has been that villagers are now allowed to enter the park once a year to collect elephant grass they need for thatching roofs. "By allowing the local people limited access to a resource central to their livelihood, the proposal illustrates one principle of conservation in terms that villagers can understand," says Mishra.

Proposals to Stem Tropical Forest Loss

Organizations such as the World Bank and the U.N. Food and Agricultural Organization have given a great deal of thought and money to attempts to stem the tide of loss of the world's tropical forests. The World Bank financed $1,154,900 in reforestation programs between 1968 and 1980. Yet it is not clear that any significant impact has been made on the problem.

One factor is that much larger sums are spent on agricultural projects. Given a choice between an agricultural versus a reforestation program, governments usually chose an agriculture program because of its immediate relief of human food needs. Another factor is that such loans as the World Bank grants may actually encourage logging. A country may decide that it can profit by selling its mature forest and then obtaining a loan to replant it. The loan is repaid by cutting and selling off more forest.

An interesting proposal by Guppy (1984) would involve an organization of timber-producing countries (OTEC) modeled in some ways on the successful OPEC oil cartel. According to Guppy, tropical forest woods are now grossly undervalued in the world market. As an example, when a forest is cut it is worthwhile to carry out only about 10% of the trees found there. Another 55% are irreparably damaged, and the remaining 35% are left standing. Yet many of the trees not sold are excellent woods with the potential for use and export. Market prices simply don't justify the cost of transporting them out. Because tropical forest woods bring so little on the world market, projects to maintain these lands as forest do not compete well with agricultural projects, hydroelectric dams, or other development schemes. A cartel, by artificially raising the price of tropical forest woods on the world market, could make the forests themselves more valuable as forest. In addition, part of the increased price could be set aside for reforestation projects.

Whether this is the answer to tropical forest loss is not certain. It does, however, meet an important criterion: It does not cause the entire burden of saving endangered species to fall on those members of the human race least in a position to bear it—the citizens of the developing countries.

In October 1985 an international task force organized by the World Resources Institute published a 56-country plan to halt tropical forest destruction. The

Figure I.4 Tree nursery in India. Several demonstration tree nurseries have been set up in India to encourage reforestation and to provide practical business experience for women. (Photo courtesy U.N. Environment Program)

report, funded by the World Bank and the United Nations Environment Plan, recognized that forest preservation must take its place beside agricultural and industrial development as an important goal in developing nations. The report also recommended more local participation and more participation by women (who traditionally gather and use many forest resources in developing nations) in schemes to preserve forests. (See Figure I.4.)

Different Viewpoints

Some conservationists feel that human impact criteria are laudable, but secondary to the most important issue. That, they say, is saving species from extinction. Other experts point out that any solution that sacrifices human needs to animal or plant welfare simply will not work. This remains a central point of debate about conserving wildlife resources.

Those of us in the developed countries, for the most part well fed and well clothed, with our own welfare secure, are in a position to consider the welfare of the entire planet. We have today a unique chance to influence the evolution and future of those "other nations," as Henry Beston termed other species, that share time and space with us. We must find the wisdom to meet this challenge.

Questions

1. Explain why efforts to conserve endangered species so often seem to be complicated by the plight of native peoples.
2. Do you think solutions based on the economic value of wild species are a good idea? What if it turns out that other uses of the land are more valuable?
3. What problems can you see arising if an "OTEC" cartel were formed? Do you think it's a good idea?
4. Are genuine human needs always paramount in a conflict between native peoples and wildlife species?

Further Reading

Graves, William. "The Imperiled Giants," *National Geographic* (December 1976), p. 752.

This beautiful pictorial essay, and the following essay by V. B. Scheffer, explore the life history of whales as well as their often ill-fated relationship with humans.

Mowat, F. *People of the Deer.* New York: Pyramid, 1968.

Mowat, F. *The Desperate People,* 2 vols. Toronto: McClelland and Steward Ltd., 1975.

These three books tell of the vanishing people, the Ihalmiut, who were the Eskimos of Canada's inland region. Mowat describes the delicate balance between humans and wildlife, which once enabled the Ihalmiut to live and flourish in a harsh climate. The books will give you an understanding of the plight of native peoples and endangered animals in an increasingly technological world.

Plumwood, V., and R. Routley. "World Rainforest Destruction—The Social Factors," *The Ecologist*, **12**(1) (1982), 4. The social, political, and economic factors involved in rain forest destruction are thoroughly explored in this article. Its conclusions are not the same as those in the *National Geographic* article on India and wildlife (below).

Putnam, John J. "India Struggles to Save Her Wildlife," *National Geographic* (September 1976), p. 299. India provides one of the best examples of how an expanding human population exerts intolerable pressures on wildlife populations in developing countries. The almost unresolvable conflicts between human and animal needs are highlighted in this perceptive article.

Elias, Thomas S. "Rare and Endangered Species of Plants—The Soviet Side," *Science*, **219** (January 7, 1983), 19. In 1980 the USSR passed an endangered animal protection law. Although some 2,000 plant species are in need of some protective measures and about 200 are in danger of extinction, plants are not yet protected by a specific law in the USSR. This paper contrasts U.S. and USSR approaches to the protection of endangered species.

Animal Welfare Institute, P.O. Box 3650, Washington, D.C. 20007. This conservation organization is one of the most active in campaigns to save whales. If you write to them, they'll put you on a mailing list for current information and will also send directions for staging your own "Save the Whales" campaigns.

Lewis, D. M. and G. Kaweche. "The Luangwa Valley of Zambia: Preserving Its Future by Integrated Management," *Ambio* (December 1985), p. 362. All of the problems of wildlife management and protection are evidenced in this Zambian Valley: poaching, agricultural encroachment, watershed deterioration, and conflicts between human and wildlife needs.

References

Prado, M. Quoted in *The New York Times*, August 11, 1983.

Kilabuk, M. Quoted in *The New York Times*, August 8, 1985.

Guppy, N. "Tropical Deforestation: A Global View," *Foreign Affairs Quarterly* (Spring 1984), p. 928.

Tropical Forests: A Call for Action. World Resources Institute, 1985.

Storro-Patterson, R. "The Bowhead Issue," *Oceans* (January 1978), p. 64.

Rensberger, B. *The New York Times*, October 5, 1977a.

Dubay, W. H. Letter to the Editor, *The New York Times*, November 15, 1977.

Rensberger, B. *The New York Times*, September 29, 1977b.

Ellis, R. Letter to the Editor, *The New York Times*, November 11, 1980.

Parker, I. "African Born White," *New York Times Magazine*, September 12, 1982.

Shiva, V. et al. "Reforestation in India: Problems and Strategies," *Ambio* (December 1985), p. 329.

(WFP photo by T. Page)

PART TWO

Resources of Land and Food

he United States is truly blessed with soil resources. This country contains the largest area of prime farmland in the world, some 346 million acres. Prime farmland is land with the proper combination of soil characteristics, climate, and water supply for agriculture. With proper management, such land can produce high crop yields year after year. This land, as well as pastureland, rangeland, forest, and non-prime cropland, makes up a huge and valuable resource. It is, of course, valuable to humans for food production, but it is also a basic part of wildlife habitats and recreational resources.

Yet the soil that forms the very basis of these resources is threatened in several ways. The main threat to agricultural lands is soil **erosion**, the loss of soil to winds or water flow. In certain areas, **salinization**, the accumulation of salts in soil, is a problem. Salinization often occurs when irrigation water evaporates, leaving behind the salts it contains. For this reason, it is a problem in dry climates such as the American southwest, where a great deal of irrigation water is used. In other areas, the structure of the soils has been hurt by traditional farming methods that compact soil with heavy equipment, or by recreational vehicles run over delicate desert soils.

Acid rain leaches needed minerals from both agricultural and forest soils, especially in the northeast. Finally, we are paving over some 3 million acres (1.2 million hectares) a year of our soil, almost one-third of it prime farmland, as cities and suburbs grow to accommodate a growing U.S. population.

The most tragic loss of soil and land resources is the process known as **desertification**. All over the world are areas where groundwater tables are declining as people pump water out faster than it can be naturally replaced, or where erosion or salinization is destroying soil resources. In some of these places, especially the semiarid regions of the African Sahel and the American southwest, vegetation has vanished.

Although it may appear that desert is creeping along and engulfing more and more land, this is not what actually happens. *Desertification* is a term applied to the severe degradation of land, to the point where it resembles a desert. It results from overuse by humans, sometimes in combination with variations in climate such as droughts. Desertification is the ultimate human insult to land and soil resources.

Like desertification, other soil resource problems are not unique to the United States. Erosion, for instance, is a serious problem in many countries today, especially in areas where tropical forests have been cleared for crops. Similarly, acid rain affects forests and croplands in a number of European countries.

In this part of the text we examine some of these threats to soil resources, and thus to food production. Chapter 6 explains soil structure and processes of soil formation. It also describes various natural plant communities. In the next chapter we discuss erosion, a major soil problem. An equally important environmental problem associated with food production is pesticide use, covered in Chapter 7.

In Chapter 8 the world food situation and prospects for the future are reviewed. We first look at the scope of the food problem. What is the extent of the food supply crisis? How can farmers, especially those in developing countries, where population growth is most rapid, grow more food? But these questions deal with only part of the story. Social and political realities control, to a large extent, how much food is actually produced and where it goes (Figure A). In Chapter 8 we attempt to give a feeling for these additional factors that affect our ability to provide enough food for all the world's people. Then we summarize the optimistic and the pessimistic views of future world food supplies. This last chapter also deals with the issue of world grain reserves, because here may lie the solution to both unnecessarily high food prices and catastrophic famine.

Water resources, which along with soil resources determine an area's food-producing ability, are covered in the next part, in Chapter 10, while acid rain and its

Figure A Feeding the world's people involves many problems—some technical, but many economic, social, and political as well. These schoolchildren are fed in a special government program to improve nutrition that also attracts children to school. (WFP photo/P. Morin)

effects on soil fertility are noted in Global Perspective V. The last two chapters in the book, 29 and 30, detail and summarize problems of urbanization and recreational use of land resources.

CHAPTER SIX

Lessons from Ecology: Land Habitats and Communities

Succession and Climax

A Tall-Grass Prairie Park/Succession/The Climax Concept/Succession and Climax in Natural Ecosystems/ Diversity and Succession/Energy Use and Succession/ Climax Community Characteristics

World Biomes

Tundra/Forests and Grasslands/Deserts

Soils and Ecosystems

Soil Types/Soil Formation/Soil Type and Agriculture

The Relationship of Climate, Soil, and Biomes

Climate/Climatic Change

Human Influences on Succession

Succession and Climax

A Tall-Grass Prairie Park

When the first American pioneers moved westward, they found vast expanses of grassland: the American prairies. Their covered wagons, moving through seemingly endless seas of waving grasses, were called prairie schooners. These prairies once covered more than a million square miles (2.6 million km²), from southern Canada all the way to the Gulf of Mexico, and from the Eastern forests to the Rockies (Figure 6.1). Prairie grasses range in size from short grasses a few inches (10 cm) high to tall grasses such as big blue stem, which grows taller than a human. At one time, tall-grass prairie covered 400,000 square miles (1 million km²), but now barely 1% of that remains. The rest, along with 50% of the short-grass prairies, has been farmed or paved over to make roads, housing developments, and shopping centers.

Tall-grass prairie forms a stable ecosystem composed of characteristic grasses: big bluestem, Indian, little bluestem, and switch grass, which grow to heights of 3–8 feet (1–2.4 m). Many wildflowers are mixed in with the grass, and 80 species of mammals, including deer, bobcats, coyotes, and badgers, find a home there (Figure 6.2). In addition, 300 bird species and well over 1,000 kinds of insects inhabit the prairie. However, once the grass is plowed up, it may be difficult or impossible to reestablish the system in any reasonable length of time. Several conservation groups are working to establish a Tall-Grass Prairie nature preserve in the Flint Hills of Oklahoma before there is nothing left to save. Other groups are working privately to reestablish the prairie ecosystem in areas where it has been plowed up and turned to agricultural uses (Figure 6.3).

Succession

The prairie grasslands are part of an ecosystem that is characteristic of that given area and also stable over time. That is, the kinds and proportions of the various plants and animals do not change much. Not all communities have this sort of stability.

Consider the fate of an abandoned farmer's field in a temperate climate such as the northern United States. The first year after the farmer gives up the field, it fills with annual weeds—the kinds of weeds that grow up each year, produce seed, and die. Perennial weeds, which are dormant during the winter months, start to appear the second year. Eventually, small shrubs and trees grow among the weeds and then take over the field. As the shrubs and trees grow larger and begin to shade the ground, the weeds that once grew in the field no longer find conditions suitable for growth. The kinds of animal species found in the field change as the vegetation alters. This rather orderly process is called **succession**. During succession, each community of plants and animals actually changes the environment in which it lives so that conditions favor a new commu-

Figure 6.1 The American prairie grasslands once occupied a million square miles across the entire mid-section of the U.S. Tall-grass prairie, where grass grew more than 6 feet (1.8 m) tall, covered the easterly portion and blended into a midregion of mid-height and mixed-height grasses. Farthest west was the short-grass prairie. (From Save the Tall-grass Prairie, Inc.)

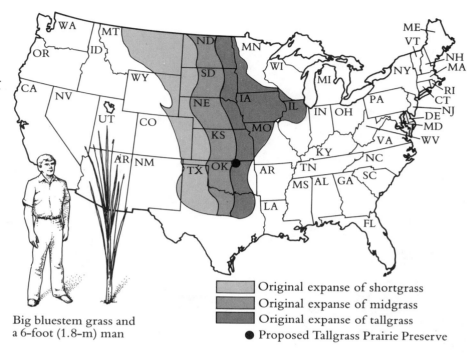

Big bluestem grass and a 6-foot (1.8-m) man

Original expanse of shortgrass
Original expanse of midgrass
Original expanse of tallgrass
● Proposed Tallgrass Prairie Preserve

nity, one usually involving larger species. The changes may involve such factors as the temperature and acidity of the soil or the amount of water and sunlight available. For example, pines may be the first trees to grow in a field. Pine branches eventually shade the ground, however, and pine seedlings do not do well in shade. Deciduous tree seedlings, such as oaks and hickories, which can grow in shade, may then take over the forest floor and supplant the pines themselves.

Succession occurs not only in abandoned fields but in all kinds of habitats. For instance, after a volcano erupts, tiny plants called algae begin to colonize the cooled lava. Over the years, a series of communities will replace each other until communities similar to those inhabiting the area before the volcano erupted appear. This succession, which starts from bare rock, is called *primary succession*. Primary succession also occurs in ponds and lakes (Figure 6.4). The term *secondary succession* is used to describe the process by which a community that had been removed from a particular site is reestablished. For instance, when a field is cleared for farming, a natural community is removed. If the field is later abandoned, and the original community begins to reestablish itself, that is called secondary succession.

Figure 6.2 A coyote roams through tall-grass prairie in spring, before the grass has reached its full height. (U.S. Fish and Wildlife Service/E. P. Haddon) "There was only the enormous, empty prairie, with grasses blowing in waves of light and shadow across it, and the great blue sky above it, and birds flying up from it and singing with joy because the sun was rising. And on the whole enormous prairie there was no sign that any other human being had ever been there."—From Laura Ingels Wilder, *Little House on the Prairie* (New York: Harper and Row, 1935).

(a)

(b)

Figure 6.3 Restoring the prairie. (a) Volunteer collects seeds of prairie grasses and wildflowers to plant on 650 acres of land at the Fermi laboratory in Illinois. This is the land that covers the Fermi labs' new subterranean proton accelerator ring. Laboratory officials from Fermi had approached the Morton Arboretum in Lisle, Illinois for help in landscaping their new site. However, Robert Betz, a Northern Illinois University professor and prairie restoration enthusiast,

suggested that the labs restore the original prairie ecosystem instead. So far over 300 acres have been restored. The Fermi lab prairie is now the largest restored prairie in the U.S. (b) Annual spring burn. In order to mimic a natural prairie condition that encourages the growth of prairie grasses, the experimental prairie is intentionally burned each March. (Courtesy of Fermi National Accelerator Lab, Batavia, Ill.)

The Climax Concept

In many areas, a community finally appears that is not supplanted by another one, as long as major climate changes do not occur. This last community does not change conditions to make them unsuitable for itself. Such a community may survive, perhaps for centuries, until other factors (such as climate) change or until disease or human activities are introduced. This relatively stable community is called the **climax community**. Major areas of the world have characteristic, or regional, climax communities. Regional climax communities are called **biomes**. They appear to be de-

termined mainly by temperature and rainfall within a given area, although other factors, such as soil type and soil age, can also influence the kind of vegetation that grows. We will take up the topic of climate and biomes again later in this chapter.

Succession and Climax in Natural Ecosystems

This concept of succession leading to a climax vegetation is a clear and straightforward idea. Unfortunately, it does not perfectly fit natural ecosystems. Thus, systems occur in which two or more climaxes appear to

changes in ecosystems, lest the natural and human changes add together and have an unexpectedly greater result. For instance, forest trees such as beech, spruce, and maple do not produce seeds every year but rather at irregular intervals depending on the climate. Passenger pigeons were probably dependent on seeds of this type. Some ecologists believe that large flocks of passenger pigeons were the result of a series of good seed years. The birds may have become extinct because humans killed so many birds that flocks were too small to survive the following series of poor seed years.

A second problem with the climax concept is that succession sometimes appears to linger at a particular stage rather than advance to the climax. Intermediate succession communities can be very vigorous and sometimes resist replacement by climax species. In some areas of Central Europe, for instance, dry grasslands do not move toward the expected climax beech forest. Regular mowing by humans or grazing by animals, which removes the shoot tips of small trees, halts succession at the grassland stage.

As another example, when tropical rain forests are cleared, only a poor secondary growth forest returns, not the original rain forest. This is due in large part to the fact that nutrients in tropical rain forests are tightly locked into cycles within the living organisms. There is no slow cycle of decay in the soil and eventual reuse of nutrients, as exists in temperate-region forests. Thus, when tropical forests are logged, most of the ecosystem nutrients are removed in the tree trunks, leaving little for a new forest to grow up on. If the tropical rain forest climax regenerates at all, it is on a time scale that has little relevance for humans.

Nevertheless, if the concepts of succession and climax are not applied with too much strictness, they can be useful in describing events in natural systems.

Diversity and Succession

Generally, during succession the number of species increases in each successive community. The first communities that colonize a lava flow, sand dune, or field are simple: they contain few species. As more plants appear, niches arise for more species of insects and other animals. These creatures, in turn, feed on plants and create pressures for species of resistant plants to flourish, and so on, through the various stages of succession. In this way, diversity increases as succession goes forward. However, diversity of species is not

Figure 6.4 Lake succession. With time, lakes and ponds may begin to fill up with sediments and plant debris. The waters become more and more shallow and less and less clear. (This is the process of eutrophication.) Eventually, the lake or pond fills so completely that it becomes a bog or marsh. Following this, a succession of land plants appears until the normal climax for the region establishes itself. Succession starting from an aquatic environment, such as a lake or pond, or from a terrestrial environment, such as an abandoned field, tends to lead to the same climax vegetation in a given area, although the path the succession takes may be very different. This supports the idea that the climate of a given area determines what the climax vegetation will be. The pond in the photo is a good distance along the way to becoming a bog. The margins are filled with spongy accumulations of plant material. Only a small area of open water remains. (The Nature Conservancy—Bruce Lund)

alternate. An example is Neusiedler Lake on the border between Austria and Hungary, which seems to appear and disappear in long cycles, perhaps 80 years or more. One theory holds that reeds in the lake grow until they occupy the entire lake (which is shallow). At this point the reeds use up the entire amount of water in the lake. The dry lake bed can no longer support the reed ecosystem, which collapses, only to reappear after the lake fills again.

The main point is that there may be long cycles of change even in climax ecosystems. It is important to take this into account when humans are planning

greatest in the climax community. Rather, diversity seems to peak sometime before the climax community appears.

Energy Use and Succession

Another community measure that increases during succession is energy use. As different groups of plants succeed each other, taller species grow up. The amount of biomass, or the weight of living material, both plant and animal, increases. Thus more and more biomass is produced for the same energy input (sunlight) as succession progresses. However, energy use, too, peaks before the climax community occurs.

Climax Community Characteristics

In climax communities, other factors seem to be more important than those leading to species diversity and increasing biomass. An example is the increasing size of individuals, which gives species the ability to store nutrients or water against seasonal shortages. This and other factors lead to increased competition between species and a loss of species in the climax community.

World Biomes

The following sections describe the major communities of plants and animals, or biomes, that are characteristic of different climatic regions.

Tundra

In the far north, the top few inches of soil thaw only in the summer. Below this thin layer, the soil remains permanently frozen year round. The frozen layer is called **permafrost**, and the whole area is known as the **tundra biome**. Roots cannot penetrate the icy layer, and so vegetation is limited to mosses, lichens (such as reindeer moss), grasses, and small woody plants (Figure 6.5). Overall, growth on the tundra is very slow. For this reason, when the tundra is disturbed, as by tire tracks, the scars do not heal. In the summer, the tracks fill with water and mud and in the winter they freeze to ice. This particular problem was of much concern during the planning of the Alaskan pipeline, which crosses so many miles of tundra.

Forests and Grasslands

Below the tundra is an area called the **taiga**, or **northern coniferous forest biome**. Although this biome is cold in winter, when summer comes the soil thaws completely—there is no permafrost (Figure 6.6).

Farther south are several biomes, depending on rainfall and temperature patterns. In areas with 30–60 inches (75–150 cm) of rainfall, moderate temperatures, and definite winter and summer seasons, the **deciduous forest biome** occurs. This includes most of the eastern United States, central Europe, and parts of Asia (Figure 6.7).

Where rainfall is too sparse to support trees, a **grasslands biome** develops. Grasslands need 10–30 inches (25–75 cm) of rain a year, compared to forests, which require 30–60 (75–150 cm). The taller grasses generally need more rainfall and deeper, richer soils than the short or mid-height grasses. Grasslands also occur in tropic areas where rainfall may total 40–60 inches (100–150 cm) per year. However, in these areas the rain comes all in one season, followed by a dry season.

Areas of the tropics where little seasonal variation in temperature occurs but much rain falls (60–90 inches or 150–225 cm a year) are generally covered by tropical rain forests. Many species of both plants and animals live in these forests. In fact, **tropical rain forest biomes** (and coral reefs) contain the greatest diversity of life found in any biome; that is, the largest number of different species per square meter are found there (Figure 6.8).

Deserts

In areas where rainfall is 10 inches (25 cm) per year or less, **desert biomes** occur. Only organisms adapted to long periods of drought and widely varying day-night temperatures can survive these conditions. Because there is little or no vegetation to moderate the sun's energy, days on the desert tend to be very hot, but as soon as the sun goes down, night temperatures plummet (Figure 6.9).

The biomes described here are the major biomes found in the world. Many subtypes exist, as do a variety of biomes covering smaller areas. (See Figure 6.10 for a map of world biomes.)

Figure 6.5 Arctic tundra biome. (Pro-Pix/Monkmeyer Press Photo Service) "Across the northern reaches of this continent there lies a mighty wedge of treeless plain, scarred by the primordial ice, inundated beneath a myriad of lakes, cross-checked by innumerable rivers and riven by the rock bones of an older earth. They are cold bones into which an eternal frost strikes downward five hundred feet beneath the thin skin of tundra bog and lichens which alone feel warmth under the long summer suns; and for eight months of the year this skin itself is wrinkled by the frosts and becomes part of the cold stone below. . . .

"Yet of all things that it may be, it is not barren. During the brief arctic summer it is a place where curlews circle in a white sky above the calling waterfowl on icily transparent lakes. . . . It is a place where minute flowers blaze in microcosmic revelry, and where the thrumming of insect wings assails the greater beasts, and sets them fleeing to the bald ridge tops in search of a wind to drive the unseen enemy away. And, not long since, it was a place where the caribou in their unnumbered hordes could inundate the land in one hot flow of life that rose below one far horizon, and reached unbroken past the opposite one." From Farley Mowat, *The Desperate People* (Boston: Little, Brown, 1959, p. 21).

(a)

(b)

Figure 6.6 Northern coniferous forest biome. This biome is made up of conifers such as spruce, fir, and tamarack, along with some deciduous trees such as birch. Bears, wolves, moose, squirrels, lynx, and many other mammals are common, as are many species of birds in the summer. (a) A spruce-pine forest in a mountainous region. (b) Alaskan taiga. (U.S. Department of the Interior, Bureau of Land Management)

Figure 6.7 Deciduous forest biome. In different areas, the mixture of deciduous trees (trees that shed their leaves each year) making up the forest will vary. For instance, in the north central U.S., the forests are predominantly beech and maple, while in the west and south, the forests are composed mainly of oak and hickory. Deer, bear, squirrels, and foxes roam the woodlands, while many bird species—such as woodpeckers, thrushes, and titmice—find a home here. In the fall, the leaves turn color and then fall to the ground. (Natural Park Service photo)

Figure 6.8 Tropical rain forest biome. Typically, three layers of vegetation occur. A sprinkling of tall trees, which lose their leaves during dry seasons, pokes up through a canopy of broadleaved evergreens and plants. Below this is a third layer of plants that flourish wherever there is a hole in the canopy. Lower levels of the forest are humid and have a constant temperature. Climbing vines and plants growing on other plants are common. So many species are found in tropical rain forests that it may be hard to find two trees of the same species over an area of several acres. Animals and insects are likewise numerous and diverse. (World Wildlife Fund—U.S./Russel A. Mittermeier)

Figure 6.9 Desert biome. Desert plants have adapted to survive long periods without water. Some have few, small, leathery leaves, which can be dropped during the dry season to reduce water loss—examples are sagebrush and mesquite. Other plants can store quantities of water in their tissues—these are the cacti and the euphorbias. A third group includes the annuals, which spring up, flower, and set seed in the short period after a rain. Desert animals have similarly adapted to the conditions of their environment. Many are active only in the night periods, burrowing into the ground during the day to remain cool. (National Park Service Photo: Richard Frear)

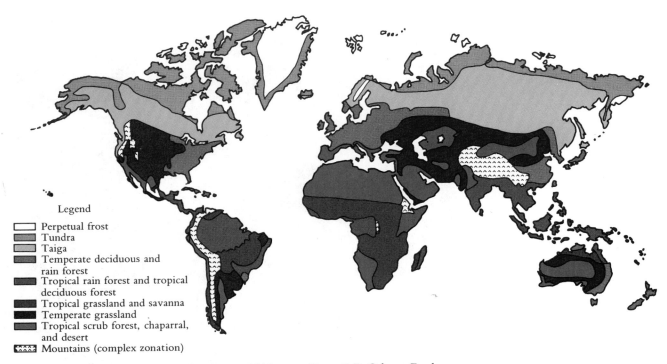

Figure 6.10 Map of major world biomes. (From E.P. Odum, *Fundamentals of Ecology*, 3rd ed. [Philadelphia: Saunders, 1971].)

Soils and Ecosystems

Soil Types

If we dig a trench vertically down into a particular kind of soil, along the sides of the trench we can see several layers, or horizons. The arrangement of these horizons is called the **soil profile** (Figure 6.11). This profile differs for soils found in various biomes.

The top layer, or A horizon, varies in depth from less than an inch (2.5 cm) in tropical rain forests to several feet in some grasslands. This layer, often called **topsoil**, contains plant roots, fungi, microorganisms, and a wide variety of soil insects and other burrowing animals. Also found here are dead and decaying parts of plants and animals. In land ecosystems, this is where the chief turnover of organic matter occurs. Here all the unused organic materials from the various trophic levels are recycled and broken down, first to humus and eventually to inorganic materials. **Humus** is an

organic substance that is broken down relatively slowly. It is not a plant food; however, it helps retain water in soil and to keep soil loose, or friable. These are important qualities for soil fertility.

Inorganic substances, formed from decomposition in the topsoil, filter down into the second soil layer, or subsoil. Finally, we come to the third layer, or parent material, on which the process of soil formation began.

Soil Formation

Many soil scientists believe that the type of soil that eventually forms in an area is dependent only on the climate of the region. Although plant and animal materials and the parent rock contribute the substances from which soil is formed, climate determines the process of soil development. According to this theory, the rock from which the soil is originally derived is not important except in early stages of soil formation, when it has an effect on the type of vegetation that

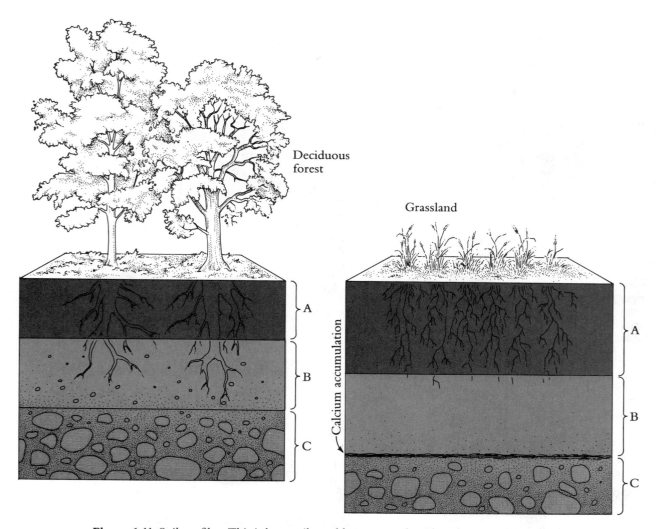

Figure 6.11 Soil profiles. This is how soil would appear on the sides of a trench cut into the ground. The upper layer, or A horizon, is topsoil and contains plant and animal debris, which is being turned into humus by soil microorganisms. In the B horizon, most of the organic material has been changed into its inorganic components. The third, or C, horizon is parent material—that is, soil characteristic of the rock from which it was formed. Climate and vegetation act on parent rock material over the ages to form the soil characteristic of a given area.

arises. In strong support of this theory is the fact that world climate maps (Figure 6.12) closely match maps of world soil types (Figure 6.13).

Temperature and precipitation are the two climatic factors of greatest importance in soil formation. As an example of how climate affects soil type, consider hot,

humid climates. The warm temperatures and moisture speed the processes of decay, while heavy rainfall causes soluble materials to leach rapidly out of the topsoil layers. You would expect the result to be a topsoil having little organic material and not much in the way of soluble plant nutrients either. This is exactly what is

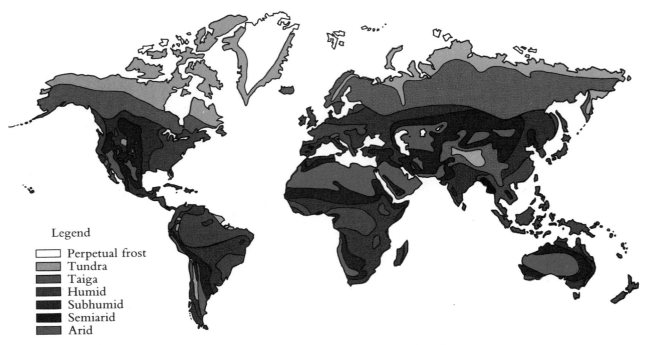

Figure 6.12 Map of major world climates. (U.S. Department of Agriculture Yearbook, 1941.)

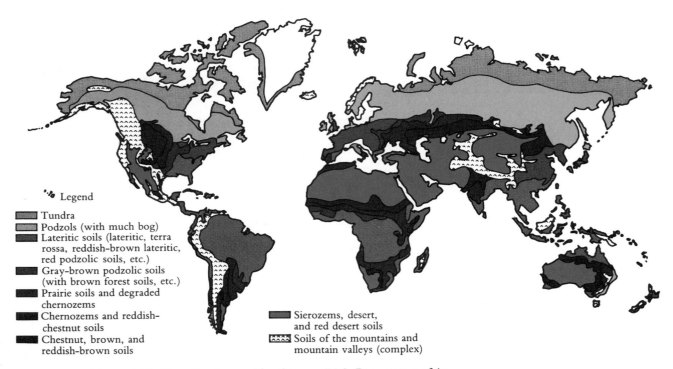

Figure 6.13 Map of major world soil types. (U.S. Department of Agriculture Yearbook, 1941)

found in hot, humid tropical rain forests. Some of these soils are so poor in humus and minerals that they are almost white.

Soil formation is a slow process, taking place over thousands of years in temperate regions. Some temperate-region soils are millions of years old. The tragic processes of erosion (see Chapter 7), which allow soil to wash away, are not naturally reversible within the length of human life spans.

Soil Type and Agriculture

Grassland soil is typically black and rich in finely divided, organic humus. Plant nutrients such as calcium, magnesium, and potassium are abundant. Such soils are valuable for agriculture. Temperate soil forests have less humus, and nutrients tend to leach out more easily, but these soils, too, can be successfully farmed if fertilizers and lime are added. Other soil types are not at all as suitable for agriculture.

Desert soils tend to be coarse and high in salts or lime, since little water is available to leach them out. Further, evaporation draws salts up to the surface where they can form a crust, called hardpan or caliche. In the tropics, with high temperatures and abundant rainfall, soils have little humus. Organic material rapidly decomposes to inorganic materials and is absorbed by a mat of plant roots on or close to the surface. Other materials, such as silica, are rapidly leached out of tropical soils by the heavy rains, leaving high concentrations of aluminum, magnesium, and iron oxides in the soil. In some areas, when tropical rain forests are cut down and the soil is laid bare, the iron enrichment of certain soil layers can cause the formation of **laterite**. The result is a soil so hard that it has been cut up for use as building blocks. Some of these blocks have lasted for 400 to 500 years in Southeast Asian temples. Obviously such soils can no longer be cultivated.

The Relationship of Climate, Soil, and Biomes

Climate

The world's climate can be described in various ways, but the most useful appears to be a system using temperature and precipitation. These two factors have perhaps the greatest effect on the distribution of soil types,

vegetation, animal life, and human activities. In Chapter 1 we noted that the severity and constancy of temperature and rainfall determine the kinds of communities that grow up in a region. Earlier in this chapter we discussed how climate (principally temperature and precipitation) determines the soil type that develops in a region. Because of these interactions, it is not surprising that a map of world climate corresponds in many major details to a map of world biomes or to a map of soil types around the world (Figures 6.10, 6.12, and 6.13).

Several factors control the climate for a given region. Most important is the amount of sunlight or solar radiation. Figure 6.14 summarizes some of the effects of solar radiation on climate. If this were the only controlling factor, however, the earth's climate would be a series of bands running horizontally around the globe, corresponding to the amount of solar radiation at different latitudes.

A glance at Figure 6.12 shows this is not the case. A major factor modifying the influence of sunlight is the distribution of land and water. Land heats and cools more quickly than water; thus, continental climates tend to be more extreme than marine climates. Similarly, the presence of mountains affects the amount of precipitation in a region, while cold or warm ocean currents cool or heat nearby coastal areas.

A more detailed description of the factors controlling climate is unfortunately beyond the scope of this text. Climate, weather, and air pollution episodes are considered further in Chapter 19.

Climatic Change

Although it is possible to identify different climatic regions in the world, these climates should not be regarded as unchangeable. Over periods of time, climatic conditions vary. You are probably familiar with the fact that over the past 2 million years, ice sheets have periodically advanced toward the equator and then receded again. But world climate varies on a much shorter time scale than the one on which ice ages are measured. For instance, it appears that global temperatures have been dropping over the past 30 years ($0.5°C$ since the 1940s), as part of a cooling trend that some scientists think may continue and others think will soon end.

Various theories have been proposed on the causes of natural changes in the earth's climate. These theories

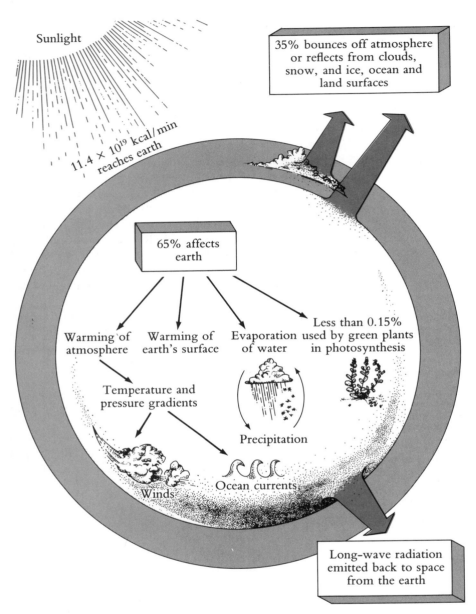

Sunlight

11.4 × 10¹⁹ kcal/min reaches earth

35% bounces off atmosphere
or reflects from clouds,
snow, and ice, ocean and
land surfaces

65% affects
earth

Warming of
atmosphere

Warming of
earth's surface

Evaporation
of water

Less than 0.15%
used by green plants
in photosynthesis

Temperature and
pressure gradients

Precipitation

Winds

Ocean currents

Long-wave radiation
emitted back to space
from the earth

Figure 6.14 Solar radiation and the earth's energy budget. Of the solar radiation reaching the earth and its atmosphere, about one-third is reflected away immediately by clouds, snow and ice, and water or land surfaces. The remaining two-thirds is absorbed by the earth and its atmosphere. Of this amount, less than 0.15% is used by green plants in photosynthesis. The rest causes water to evaporate and also warms the earth's surface and atmosphere. This warming is uneven over the globe, leading to temperature and pressure gradients. Along with the rotation of the earth, these gradients cause winds and ocean currents. All of the radiation absorbed by the earth and its atmosphere (except for a minute amount stored as fossil fuels) is eventually reradiated into space, mainly as long-wave radiation in the infrared range.

involve such phenomena as drifting continents, variations in the earth's orbit, variations in solar energy (sunspots), or volcanic activity. Although it is not possible to choose among these various theories at the present time, it is important to understand that climate does change naturally on both long and short time scales. Thus, the effects of human influences on climate may either add or subtract from the severity of natural changes. Some of the ways in which humans are influencing the earth's climate are discussed in Part Five.

Human Influences on Succession

Our discussion of succession ignored two factors that may have a strong influence on the type of vegetation in a specific area. One factor is, of course, human activities. Wherever we live, we humans turn the land to our own uses, cutting down or burning those species for which we have little use and encouraging those species we need. The way humans manage forests for the production of wood provides an example.

Figure 6.15 Desertification. Overgrazing strips land of its protective vegetation. Dust storms can then pick up soil particles and blow them away. This is a way in which once-fertile areas can come to resemble deserts. (USDA–Soil Conservation Service)

We noted earlier that climax communities do not produce the greatest biomass per unit of energy input. Productivity is not as great in climax ecosystems as in some earlier stages of succession. In order to increase production of food or wood, humans will try to keep an area at a younger successional stage. Younger stages, however, are less balanced and less able to maintain themselves than an older climax stage. Large inputs of energy and ingenuity are needed to maintain them. For instance, in some areas, paper companies plant vast tracts of forest land with softwood tree species that are fast-growing and suitable for pulp. Such stands are much more subject to severe insect damage than the normal, mixed climax community. For this reason, quantities of pesticides must be used to protect the trees. Manufacturing and applying pesticides, of course, requires energy. Pesticides can also have many adverse effects on the environment, some of which are detailed in Chapter 7.

Grasslands are used by humans to graze cattle or to grow corn and wheat. Up to a point, these are reasonable uses. It is not strange that the American farmer has done well farming ploughed grassland, since the two plant communities, grains and grasses, involve similar ecosystems. Grasslands, however, can be overgrazed (Figure 6.15). In tropical grasslands and in tropical rain forests, most of the plant nutrients are cycling in the plants themselves and do not spend much time in the soil. Thus, removal of too much vegetation by overgrazing removes almost all the nutrients, leaving soils that no longer support plant life (See Global Perspective II).

Similar to the problem of overgrazing is that of overuse of slash-and-burn farming in areas of tropical rain forest. In this type of agriculture, farmers cut and burn the natural vegetation before planting their crops. Such plots can be used for only a few consecutive years and then must lie fallow for 10 or even 30 years to regain their fertility. As populations increase in these areas, farmers are tempted to reuse a plot too soon.

A special influence humans have on vegetation is the result of strip mining. If the stripper is not careful to save the topsoil and spread it out on top again after he fills in his mining trenches, a wasteland can result. The top layers of soil, which contain the nutrients cycling from dead plants and animals to living ones, are the most important from the point of view of supporting growth. When the strip miner turns the layers over, the deeper nutrient-free layers are turned uppermost. This, in addition to the acidity produced by coal wastes, gives rise to a soil in which almost nothing can survive. Only imported topsoil or the slow, centuries-long processes of soil formation and washing away of acid will restore the original vegetation to land treated in this manner. Erosion, as described in Chapter 7, can produce similar results because the topsoil is washed away by rains. These are examples of how growing human populations, with their increasing needs for food, fiber, and mineral resources, can unbalance their environment.

On the positive side, people have turned deserts into gardens by adding the missing factor—water. Here again there are problems, however. The rapid loss of water by evaporation leaves behind minerals. Unless this is accounted for in irrigation schemes, minerals can accumulate until nothing will grow.

The other factor to which little attention has been given is fire. Many ecosystems are adapted to the natural occurrence of periodic fires. African grasslands, for instance, are composed of fire-tolerant species because fires are common during the long dry season. The topic of fire and forest management is covered in Chapter 30.

Summary

In a given area, an orderly series of communities tend to follow one another until a more stable or climax community is reached. This process is called primary succession if it starts on bare rock or in water, and secondary succession if it begins on abandoned agricultural land. Major climatic regions of the world have characteristic climax communities determined mainly by temperature and rainfall and, to a lesser extent, by soil type. During succession, diversity and biomass tend to increase. Both peak before the climax is reached, however, when competitive interaction becomes more important. Among the major world biomes are tundra, forests, grasslands, tropical rain forests, and deserts.

Soils consist of several vertical layers. The uppermost, or A horizon, is topsoil, a layer rich in biomass. The B horizon, or subsoil layer, receives inorganic substances from decomposition in the topsoil. The lowermost layer, or C horizon, is parent rock from which the soil was originally formed.

Because soil is composed not only of rock but also of the organic substances contributed by the plant and animal community in and on the soil, the type of soil that develops in a region is related to biome type. It is also related to climate, because climate influences the biome type. World maps of soil type, climate, and biomes illustrate this relationship. World climates are not fixed but appear to change in both long and short time frames.

Grassland soils and temperate forest soils are best suited for agriculture. Tropical rain forest soils and desert soils can be farmed if time is allowed for the ecosystem to recover between crops (and if water is added to desert ecosystems). Intensive agriculture in tropical and desert areas will lead to the formation of wastelands.

Humans influence succession by maintaining ecosystems in young, highly productive stages by removing more nutrients from ecosystems than nature can return or by rearranging the soil layers so the nutrient-poor layers are uppermost.

Questions

1. Describe what is meant by succession.
2. What is the main characteristic of a climax community? What are some of its other characteristics compared to earlier stages of succession?
3. What are the major factors influencing which climax community grows up in an area?
4. UNESCO has a program to identify and preserve examples of all the world biomes. What is a biome? Of what possible use could it be to preserve an example of each type?
5. What kinds of ecosystems have soils that are best for agriculture? Why?
6. Explain how clearing forest land for farming is an example of increasing food production at the cost of unbalancing the environment.

References

Lutgens, F. K. and E. J. Tarbuck. *The Atmosphere*, 2nd ed. Englewood Cliffs, N.J.: Prentice-Hall, 1982.

Kerr, R. A. "Climate Control: How Large a Role for Orbital Variations," *Science*, **201** (July 14, 1978), 144.

Levin, H. L. *The Earth Through Time*, Philadelphia: Saunders, 1978.

CHAPTER SEVEN

Soil Erosion and Pesticides Usage

Soil Erosion

Kinds of Water Erosion/Environmental Effects/Effects of Wind Erosion/Methods of Controlling Erosion/ Encouraging Soil Conservation

Pesticides and Food Production

Early Use of Pesticides/Biological Magnification/ Persistence/Resistance/Soil Microbes and Pesticides/The Organic Farming Solution/Integrated Pest Management and Biological Controls/Pest Control Experts

CONTROVERSIES:
7.1 ***Organic Gardening Economics***
7.2 ***Who Should Pay for IPM?***

he production of enough food to feed the world's growing population is without doubt one of the most important problems facing the human race today. This problem, however, is complicated by two other environmental problems. One is the loss of fertile cropland due to soil erosion. Scientists estimate that the four major food-producing countries (the United States, the Soviet Union, China, and India) may lose as much as 13.2 billion tons of soil from their croplands every year. The second problem is the environmental pollution that results from the toxic chemicals used to kill pests on crops.

Both erosion and pesticide pollution are also key issues with respect to the fastest spreading new agricultural technique, minimum tillage. In this new group of farming methods, erosion is markedly reduced. However, minimum tillage requires the use of large quantities of herbicides—pesticide chemicals that kill weeds.

In this chapter we detail the causes of erosion and possible solutions to the problem. We then investigate the environmental effects of the increasing use of herbicides and pesticides in the United States today. The special problems resulting from pesticide use in developing countries are examined in Global Perspective VI.

Soil Erosion

Storm clouds gather and a brisk wind stirs the poplars in the hedgerow. A tiny drop of rain falls on the freshly plowed earth. Then another drop, and another, and the storm begins in earnest. Each drop hits the earth like a miniature meteor—soil particles spray in a circle around the drop and a small crater is formed. As more droplets reach the earth, their arrival rate begins to exceed the rate at which the water can be absorbed into the soil. Puddles form, then rivulets. The looser and finer soil, the best soil, is picked up by the rivulets and carried downslope. The velocity of the rivulets increases; more and more soil is carried in suspension. Rivulets join to form temporary streams, which flow in gullies, and the rapidly flowing water, en route to a true stream, cuts sharply into the earth.

This event is not isolated or uncommon—it is any rainstorm on a plowed field. Multiplied by thousands of farms or millions of acres, it results in as much as 4 billion tons of soil being carried off each year by water erosion in the United States, about three-quarters of it from farmland. Wind may scour away another billion tons of soil. A report by the Soil Conservation Service in December 1980 estimated that as much as one-half of all cropland in the United States was being eroded so fast that its productivity would suffer. Such losses as these must have consequences. What are the effects of erosion on the land and water? How can erosion be slowed?

Kinds of Water Erosion

Soil scientists divide water erosion into three components: sheet, rill, and gully erosion. In sheet erosion, surface water moves downslope in a wide flow; it may roll soil particles with it, bump them along, or com-

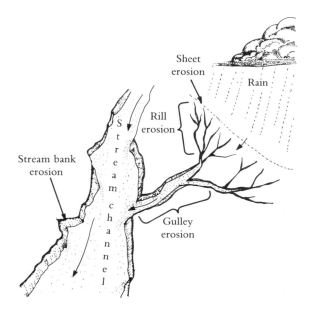

Figure 7.1 Types of erosion. Raindrops detach small particles of soil from the ground and carry them along as the water runs down a slope. Small rills or channels eventually form that may deepen and widen into gullies. Along the banks of the stream into which gullies flow, soil may also erode. (Adapted from M. F. Walter, and F. N. Swader, "Erosion from Agricultural Land," *Agricultural Engineering Bulletin*, **422** [1976].)

pletely suspend them in the water, depending on the size of the particles and the velocity of the water. Since sheet erosion occurs over wide areas in a field, it may not be noticeable at first, and awareness may dawn only after much damage has been done.

In rill erosion, in contrast to sheet erosion, the water joins into rivulets (Figures 7.1 and 7.2). Flowing at higher velocities through narrow miniature valleys, the water has greater erosive power. However, because rivulets are small and can be plowed up with the rest of the soil and smoothed over, they are easy to miss in farming. You may be able to see large rills forming in a hill of fresh earth at a construction site.

Gully erosion is not so easy to miss as sheet or rill erosion; the rivulets join into such large flows that tractors cannot smooth them over (Figure 7.3). Gullies resemble ordinary streams with soil banks, except that water flows in them only after rainstorms; at other times they are dry. The banks of the gully are usually only loose soil, which the water erodes away. Earth overhangs may be exposed by the water that undercuts the soil banks; since they are structurally weak, it is common for the overhangs to collapse into the gully, accelerating the erosion process.

Wind also erodes soil, lifting and transporting the finer particles and organic matter. Such erosion leaves

Figure 7.2 Severe rill erosion in the state of Washington. A fine tilling of the summer's cover crop, combined with poor growth in the fall of the wheat crop, left this steep slope in a highly eroded condition. The county crew that came to clear the road was shooed off by the farmer, who told them, "I'll haul the soil back up the hill." (USDA, Soil Conservation Service; Photographer: Fred Wetter)

Figure 7.3 Gully erosion in Montana. This gully cuts deeply through a hay meadow. The farmer will have to do significantly more plowing to prepare his land—now in two fields instead of one. (USDA, Soil Conservation Service; Photographer: D. J. Anderson)

behind the less desirable coarser soils, which, because they retain water less well, are themselves more likely to erode.

All these forms of erosion carry away soil nutrients and plant residues, essential components of a productive earth. Pesticides and even disease-causing organisms can be absorbed onto the soil particles and can later contaminate waters into which the soil is carried.

Environmental Effects

It is important to remember that erosion is a natural process. It has been observed for centuries; it even has beneficial effects. The rich delta lands built up at the mouths of the Mississippi, the Rhine, and the Nile all exist because soil was carried off the land in enormous quantities. The richness of the river deltas is a two-edged sword, however. If the delta soils are rich, the soil that remains on the land upstream must have been robbed of some of its value. This, in fact, is one of the principal concerns we have about soil erosion: the decreasing productive capacity of the soil.

Although erosion is a natural process, the massive extent to which it is presently occurring is the result of human activities. The farming of more land (only in this century made productive by irrigation), the grazing of cattle, the construction of buildings and highways, the activities associated with mining—all disturb the soil and accelerate erosion. In addition, the pavement that now overlays so much of America contributes to faster erosion. Because water cannot be absorbed by paving materials, it collects and may run swiftly off the pavements. Its increased volume and velocity will erode and suspend soil, carrying it into streambeds.

Rate of soil formation versus soil loss. Soil is formed in a slow and continuous process. The rate of new soil formation is about 1 inch (2.5 cm) of topsoil every 100–1,000 years. However, the rate varies widely, depending on climate, vegetation, soil type, and land use.

Soil erosion is measured in tons of soil lost per acre (metric tons/hectare, or t/ha). To understand what rates of soil loss mean, we need to know a few facts. About 6 inches (15 cm) of topsoil is usually cultivated in modern agriculture. This 6 inches weighs approximately 1,000 tons/acre (2245 t/ha). If soil erodes at about 15 tons/acre/year (33 t/ha/year), about 0.10 inch

(0.25 cm) of soil is lost each year and the whole plow layer in 60–70 years. Experts estimate that, depending on soil types, only 1–5 tons of soil can be lost per acre each year (2.2–11 t/ha) without decreasing productivity in the long term.

Loss of topsoil in the United States and worldwide. The 3 billion tons of soil eroded from U.S. farmland each year come from individual farms, where erosion could be more or less than some average figure. From sloped land that is unprotected and has been freshly plowed, under an intense rain, annual erosion losses could reach 60–100 tons per acre (132–220 t/ha). On a national basis, cropland erosion losses average about 4–5 tons per acre (9–11 t/ha) per year, but averages are deceiving. Only 6% of the nation's cropland accounts for some 43% of the soil erosion (Table 7.1). The Soviet Union may be losing even more soil than the United States as Soviet leaders try to reduce grain imports by bringing marginal, more easily eroded land into cultivation.

Such losses create numerous problems. The productive capacity of the soil may decrease, depending on the depth to which the topsoil extends and the character of the subsoil. The loss of 6 inches (15 cm) of topsoil will make most cropland much less productive, if not unsuitable for farming. For example, losses of this magnitude in Tennessee have reduced corn yields by 42%. These decreases in productivity may occur even if the farmer tries to make up for lost nutrients by adding more fertilizer. Loss in productivity is due not only to loss of the more fertile topsoil but also to changes in the soil structure. When topsoil has been carried off, the remaining soil may not absorb water as well. Hence, more runoff is likely to occur, making less water available to the crop. Since the subsoil is generally less permeable, less water can be stored between rains, making the likelihood of damage from droughts greater.

The less desirable subsoil remaining is also more difficult to till; rills make plowing more difficult yet; and gullies may block plowing entirely, resulting in the complete loss of portions of the land. Not only is the land lost, but the fields are cut into smaller blocks by the gullies; hence, the time and effort to cultivate an area increase. It is akin to having to mow the grass on an acre of land (about 200 feet by 200 feet or 61 m by 61 m) in blocks that are 10 feet (3 m) on a side, instead of making cuts that are 200 feet (61 m) long. A final

Table 7.1 Serious Erosion Problems in the United States

Area and location	Severity of problem
The Palouse (hilly area covering parts of Washington, Oregon, and Idaho)	Runoff from rain and snow causes erosion rates of 50–100 tons/acre (110–200 t/ha)
Southeastern Idaho	Erosion rates of up to 16 tons/acre (35 t/ha) on slopes
Texas Blackland Prairie	Gently rolling, highly erodable land; losses higher than national average
Southern Mississippi Valley	Row crops show losses of 20 tons/acre (44 t/ha)
Corn Belt states	Some of the highest average rates in the country: Iowa—9.9 tons/acre (22 t/ha); Illinois—6.7 tons/acre (15 t/ha); Missouri—10.9 tons/acre (24 t/ha)
Aroostook County, Maine	Potatoes grown on slopes of up to 25%; upper 2 feet (0.6 m) of soil gone

Source: America's Soil and Water Trends. Washington, D.C.: U.S. Department of Agriculture, Soil and Water Conservation Service, December 1980.

effect of water erosion on farmland is the damage to seedlings. These immature plants may be dislodged and carried off, or they may be buried by sediment.

Effects on streams, reservoirs, and aquatic life. Soil particles carried in suspension and bumping along the bottom of streambeds will eventually settle when the velocity of the stream decreases sufficiently. These sediments accumulating on the bottom of the stream channel use up part of the room in the channel. Thus, more water will be coursing down a stream channel than it is able to handle. Compounding the problem is the fact that the volume of runoff is further increased by the volume of the soil particles carried in suspension. Thus flooding by streams flowing through badly eroding areas is more likely.★

Not only are stream channels filled up by sediments, but reservoirs fill up as well. These structures,

★One extreme case of erosion has been documented on the Yellow River (Hwang Ho) in China. At flood stage, the flow in that river has been as much as 50% sediment by weight. The yellow color of the river is due to the color of the soil particles eroded from the uplands.

which are used to store water for irrigation, for hydropower, and for water supply, bring the flowing waters to a standstill. When this occurs, most of the sediments, which were in suspension, drop down. The buildup of sediments in reservoirs decreases the quantity of water available on a reliable basis. Large reservoirs are actually designed with extra capacity to hold the sediments. Of course, this added capacity drives up the cost of the reservoir. A 1968 survey of nearly 1,000 reservoirs in the United States showed about 40% of them filling at over 3% per year. These were most often smaller reservoirs, less than 100 acre-feet.

Besides using up the capacity of rivers and reservoirs, sediments carried in suspension affect aquatic life. Sediments may bury the habitat of bottom feeders. The turbid water allows less sunlight to enter, hindering the growth of aquatic plant species.

Finally, we must list among the effects of water erosion the poor water quality caused by the substances carried into our streams in runoff. Fertilizers and animal wastes use up vital oxygen in the water and destroy the habitat of aquatic life. Pesticides also enter the water through erosion. In fact, the Commission on Pesticides of the Department of Health, Education, and Welfare estimated that erosion is the most common route of pesticide entry into water.

Effects of Wind Erosion

Wind erosion has its own set of effects. Of course, its effect on the productivity of the soil is similar to the impact of water erosion. The lighter, more granular soils, with high organic content, are scoured away by the wind, leaving a coarser, less absorbent subsoil exposed. The impact of wind erosion on human activities is far different from that of water erosion, however.

Dust storms, when the winds lift and carry soil particles on a massive scale, are fearful indeed. One episode is recorded in which nearly 1,300 tons of particles were airborne per cubic mile of air (455 t/km³). Annual soil losses due to wind erosion have been measured at levels up to 7 tons per acre. People and livestock have difficulty breathing in dust storms and develop respiratory and eye ailments. Crops, especially seedlings, cannot survive such conditions. In the 1930s, the Midwest went through such a severe series of dust storms and blowing soil that the Southern Great Plains were nicknamed the "Dust Bowl" (Figure 7.4). Dust storms seem to come in bunches. For instance, there

were 120 storms recorded near Dodge City, Kansas, in 1936–1937. Another cluster of storms struck the area about 20 years later. Apparently some combination of wind, weather, and soil conditions triggers these storms.

Although the storms are distinct and noteworthy events, blowing soil is common. Its effects are the same as those of dust storms, though slower and less dramatic. Scour of plant leaves by soil particles may reduce crops. If plant roots are uncovered, the crop will not survive. Soil is stripped of its richer portions.

Methods of Controlling Erosion

The most important single activity that brings about soil erosion is farming. When soil is turned over, exposing a surface with no cover—indeed, with no protection whatsoever—the possibilities for erosion are greatly increased. One set of alternatives to control erosion, then, concentrates on how soil is tilled.

The fastest-spreading soil conservation technique today is undoubtedly **reduced tillage** farming. This is actually a group of methods in which some sort of crop cover is left on the soil at all times. The soil is never plowed and left bare to erosive wind and rain. Herbicides are applied to kill weeds or the previous crop. New crop seeds are often planted directly into the old stubble using special equipment. For instance, rye may be planted in a field in the fall and then killed by an herbicide in the spring before corn is planted into the residue of the rye crop. This residue reduces soil erosion by decreasing the impact of raindrops and slowing the flow of water across the field (Figure 7.5).

Reduced tillage farming can decrease soil erosion to almost zero. In the past 10 years, the number of acres in the United States planted by reduced tillage has increased dramatically from about 35 million to almost 100 million (14 to 40 million hectares). (Compare this to a total U.S. cropland of 413 million acres [165 million hectares].)

Developing countries also might profit from this technique. This is especially true in tropical climates, where the topsoil layer is very thin and quickly eroded. Although reduced tillage is now primarily used by large farmers in developing nations, small farmers could benefit from the reduced plowing labor involved. The energy involved in plowing fields with hand tools often limits the amount of land a farmer can plant.

Figure 7.4 Dust storm, Cimarron County, Oklahoma, 1936. A farmer and his two small children run for the shelter of their dugout during a duststorm in this famous photo by Arthur Rothstein. In the 1930s, the Midwest suffered a series of severe dust storms caused by winds blowing soil off fields during dry weather. So many dust storms hit the Southern Great Plains that the area became known as the "Dust Bowl."

Reduced tillage farming does require some skill. The farmer used to conventional tillage may need advice and encouragement getting started. *Farm Digest* (April 1982) summed it up: "For the farmer looking for a more efficient method, interested in using his head more than his back, concerned with conserving soil and fuel—conservation tillage is an option that's looking better and better" (p. 28).

Is reduced tillage farming the environmentally sound way to farm then? Maybe but it's not entirely clear yet. Large quantities of herbicides must be used in these methods. The long-term effect of such use on the environment is still to be determined. "Before we shift 90% of the nation's cropland into a farming technique that relies on increasing use of pesticides, we should take a hard look at where those pesticides are likely to end up and what their effects will be" (Speer and Paulson, 1984).

Figure 7.5 Reduced tillage farming: soybeans growing in grain stubble. Reduced tillage farming, in which crops are planted into the residue of a previous crop, has several advantages. Soil erosion is reduced almost to zero. In addition, water is conserved, as runoff and evaporation decrease. Less expensive machinery is needed and about 10–20% less energy is used, because the tractor makes only one or two passes over a field. This last feature also helps reduce the packing down of soil by heavy machinery. On the negative side, insects and plant diseases may be more common because the nonplowed seedbed provides a more favorable habitat. Soil temperatures are lower because of the plant cover. This is a negative feature in some areas, such as the northern U.S., but can be an advantage in the tropics. In general, greater management ability on the part of the farmer is needed, at least until the reduced tillage technique is completely adapted to the particular farming area. (USDA—Soil Conservation Service; Photographer: L. C. Harmon)

Crop rotation. On plowed land, the method of plowing and the type of crops planted can be planned to reduce erosion. When crops are rotated, by planting cash crops such as wheat alternately with grasses or legumes, the soil gains protection from erosion and may be nourished as well. Unfortunately, the high price of wheat in the late 1970s drew many farmers away from wheat-fallow rotations and wheat-sorghum-fallow rotations* to continuous planting of

_____
*A fallow crop is one that nourishes the soil.

wheat; no intermediate plantings were used to firm the soil and restore its nutrients. The problem of making farmers act in their long-term interests as opposed to short-term interests is a basic one that we shall address in a moment.

Contour plowing, terracing, and shelterbelts. An important factor in the severity of erosion is the slope of the soil: The greater the slope, the more severe erosion is likely to be (other factors such as rainfall and soil types remaining the same). This has led to the development of soil management practices such as contour plowing or terracing. Such methods are designed to interrupt, divert, and absorb the flow of water in its path down a field (Figure 7.6).

Two other major influences on the extent of erosion—the frequency and amount of rainfall—are not generally controllable by human activity.

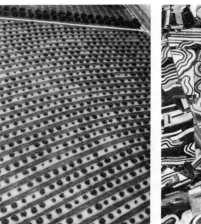

Figure 7.6 Soil conservation or modern art? The aerial view of a field of peach trees (left) in New Bridgeton, New Jersey resembles nothing so much as a tufted bedspread. Yet the pattern actually reflects good soil conservation practice. Between the rows of peach trees are strips of clover sod, which simultaneously enrich the soil and help to anchor it in place. (USDA, SCS photo by Clarence Deland) (Right) One of the first Soil and Water Conservation Districts in the U.S. was founded in Lancaster County, Pennsylvania, in 1937. From a wasteful pattern of straight-line furrows that climbed up hillsides, the pattern of cultivation evolved to its present state. This aerial view shows furrows following the contours of the hills and strip cropping (growing alternate strips of the main crop and a ground cover, such as hay or grass), two soil conservation practices that check erosion and hence preserve the soil. (USDA, SCS photo by Don Schuhart)

Methods involving the use of vegetation can protect the soil not only from water erosion but from wind erosion as well. Protection against wind erosion, however, may also involve planting trees such as willow or poplar, two fast-growing species. Arranged in long lines, called shelterbelts, the trees decrease the wind speed downwind from the belt. Earth banks, wooden fences, and rock walls may also be useful in checking wind erosion.

A model soil conservation program in Kenya, which includes strip cropping, tree planting, and drains, has improved farm incomes and shown that soil conservation is possible for small-scale farming in developing nations.

Encouraging Soil Conservation

One would think that soil conservation is in the farmer's best interests. It may be, but only in the long run. In the short run, the farmer must survive to plant another year. We indicated earlier that crop rotations in growing wheat may be abandoned in the quest for immediately higher profits. Soil conservation measures that involve restructuring the land cost money. And land taken out of production is land that will not produce a crop and an income. Furthermore, the loss in fertility of the soil due to erosion can temporarily be overcome by liberal doses of fertilizer.

The U.S. Department of Agriculture (USDA) has sponsored soil conservation programs since the middle 1930s, when the Dust Bowl era dramatized how very fragile our soil resource is. Studies have shown, however, that such programs were not working. A 1977 study by the U.S. General Accounting Office investigated conditions and practices on 283 farms, chosen on a random basis. The farms, which were in the Midwest, Great Plains, and Pacific Northwest, consisted of 119 farms that had had conservation plans prepared by the Soil Conservation Service of the USDA and 164 farms that had not. Of the 283 farms, 84% were losing soil by erosion at an annual rate of 5 tons per acre (11 t/ha) or more; experts believe a loss of 5 tons per acre each year cannot be sustained without a loss of productivity, even by deep soils. Of the 119 farms with conservation plans, fewer than half were actually using them.

The 1985 farm bill may help to solve some of these problems. The bill was designed to reduce the number of acres U.S. farmers cultivate by paying them *not* to grow crops on up to 40 million acres of erosion-prone grasslands and fragile wetlands. Furthermore, farmers who destroy such lands will be ineligible for government farm programs the following year. Farmers already cultivating fragile and erodable lands have 5 years to develop a soil and water conservation plan and another 5 years to put the plan into effect.

Pesticides and Food Production

Each year an estimated half of the world's critically short food supply is consumed or destroyed by insects, molds, rodents, birds and other pests that attack foodstuffs in fields, during shipment and in storage. (Brody, 1974)

Experts believe that if the insects and diseases that attack crops such as cereal grains in the field could be controlled, some 200 million extra tons of grain would be available each year. This amount of grain would feed 1 billion people. Moreover, if we could stop the pests, such as rats and insects, that eat or destroy food once it is harvested, there could be as much as 25% more grain available to the world's hungry people, without any increase in the amount of food actually grown (Figure 7.7). Experts say that in parts of India, as much as 70% of stored foods is lost to pests each year.

The challenge, and it's not a simple one, is to solve these pest problems without harming people or the environment. In India, stored grain is often illegally treated with the pesticide DDT to kill insects. As a result, people in India have more of this poisonous and carcinogenic chemical in their body fat than people in any other country in the world.

The use of pesticides creates many environmental problems. This section focuses on the use of pesticides in growing food. The problem of pesticides in water is covered in Chapter 14, while the effects of DDT on wildlife were noted in Chapter 5. In addition, the possible effects of pesticides in food on human health are considered in Chapter 28, and the special problems related to the use of pesticides in developing nations are noted in Global Perspective VI.

Early Use of Pesticides

Any material used to kill pests is called a **pesticide**, and usually any organism that competes with humans for

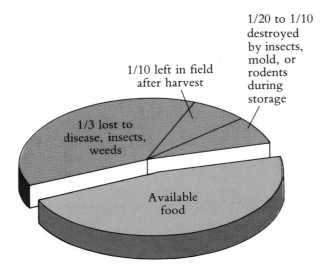

1/20 to 1/10 destroyed by insects, mold, or rodents during storage

1/10 left in field after harvest

1/3 lost to disease, insects, weeds

Available food

Figure 7.7 Crop loss in the field and in storage. In the U.S., more than one-third of the crops grown are lost to insects, disease, and rodents in the field and during storage. One-third are lost in the field, another 10% are left in the field after harvest, and 5%–10% are lost during storage. In other parts of the world, these figures are even higher. On average, one-half of the world's food crop is lost to disease, insects, and rodents during growing, shipping, and storage.

food, fiber, or living space is called a pest. The use of chemicals to kill pests is not a new idea. For centuries, farmers have used minerals such as arsenic, lead, and mercury or natural plant substances such as pyrethrum (obtained from a daisylike plant) to kill insects and other pests. However, in 1945 a new era in pest control began, when DDT came into widespread use to control the lice and fleas that plagued the armies of World War II. DDT was the first widely used synthetic (laboratory-produced) pesticide. In the years that followed, many other synthetic pesticides were introduced. In a number of cases, spectacular increases in crop yields resulted where the new pesticides were used. By 1967, half of all of the pesticides used were, like DDT, in the chemical family of chlorinated hydrocarbons.

Slowly, however, evidence began to accumulate that the synthetic pesticides were a mixed blessing.

Biological Magnification

Problems with pesticides arose in three main areas. In the first place, certain pesticides* tend to accumulate in

---

*Mainly, chlorinated hydrocarbons such as DDT and mercury-containing compounds.

living organisms. In some cases, pesticides not only accumulate to levels greater than are found in the environment but concentrations keep increasing as they move up food chains. This is the effect known as **biological magnification**.

DDT is an example of a pesticide that is magnified biologically. DDT is soluble in fat. When an organism ingests DDT—whether from water, from the detritus of plants that have been sprayed, or from insects that have eaten the plants—the DDT becomes concentrated in the organism's fatty parts. DDT is lost very slowly from these fatty tissues. If another creature in the food web eats the first organism, the consumer will be consuming a concentrated dose of DDT (Figure 7.8). Organisms at the top of their food chain (such as humans and predatory birds such as eagles and falcons) are eating foods with a much higher level of DDT than is generally present in the environment. One effect of such DDT levels on birds is that they lay eggs with shells that are much thinner than normal. These thin shells break easily and so are no protection for the developing chick inside the egg. In this way, the Eastern peregrine falcon failed to reproduce and so became extinct in the 1960s (see Chapter 1).

Persistence

A second area of concern is the length of time pesticides remain in the soil or on crops after they are applied. The chlorinated hydrocarbons, such as DDT, and pesticides that contain arsenic, lead, or mercury are known as **persistent** pesticides. This means that they are not broken down within one growing season by sunlight or bacteria. The half-life of DDT, for instance, may be as long as 20 years. (That is, after 20 years, only one-half of an amount of DDT used has been broken down to simpler compounds.) Mercury and arsenic are never really broken down—they are moved around or buried in muds.

Persistent pesticides can build up in the soil if a farmer applies them year after year. In some orchards where lead arsenate has been used to control insects, arsenic in the soil has built up to a level that kills the fruit trees. The long life of persistent pesticides is a major factor in the process of secondary contamination, whereby foods that were never treated with pesticides still become contaminated. For example, DDT sticks to soil particles after it is applied. In many areas, dust storms have picked up this contaminated dust and blown it, literally, around the world. Rain washes this dust out of the air in places where pesticides and even farming are practically unknown. Thus seals, penguins, and fish in the Antarctic all show traces of DDT in their fat. In the 1960s, when the use of DDT was at its height, 40 tons of DDT were dumped on England every year in rainfall. In this way, pastures and crops of all kinds received an unintended and unwanted treatment with DDT.

Resistance

Biological magnification and persistence are not the only problems associated with the use of pesticides. A third serious problem is that pests can become **resistant** to pesticides—the pesticide will no longer kill them. This can come about because of mutations occurring in some of the enormous numbers of insects hatching each season. **Mutations** are changes in an organism's inherited genetic material, the material that determines the organism's characteristics.

A pesticide may kill off most of the insects in a certain area. However, a few that have mutated so that they have slightly different characteristics than the others may survive. These few can repopulate the sprayed area, passing on their pesticide resistance to their offspring. In some cases, pesticides themselves may cause the development of resistance by stimulating the production in pests of enzymes that act to break down pesticides. Those insects that have been stimulated to produce the largest amounts of the enzymes will be most likely to survive and reproduce. In several generations, insects with sufficient enzyme to break down any reasonable amount of pesticide will appear. This sort of effect has led to the development of houseflies resistant to all of the major types of pesticides used.

The commonly used "broad-spectrum" pesticides, which kill a wide variety of insects, make the development of resistance simpler. They not only ensure that the newly resistant insects will have little competition for food but also kill off many parasites and predators of insect pests. In fact, predators may receive a higher dose of pesticide than their prey because of biological magnification. Resistant insects are thus given a good chance to establish themselves, free from normal controls. In a number of cases, pesticides have created new pests in this manner. For instance, the ladybug normally controls a type of citrus scale insect. If an orchard is sprayed with DDT, however, the ladybugs are killed while the scale insect is not. The scale insect

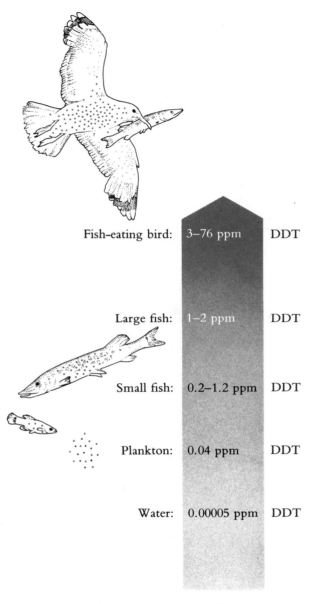

Fish-eating bird:	3–76 ppm	DDT
Large fish:	1–2 ppm	DDT
Small fish:	0.2–1.2 ppm	DDT
Plankton:	0.04 ppm	DDT
Water:	0.00005 ppm	DDT

Figure 7.8 Biological magnification. In 1967, fish-eating birds from a Long Island salt marsh estuary contained almost a million times more DDT than could be found in the water. At each step of the food chain, DDT was concentrated, as organisms consumed and absorbed more DDT than they were able to excrete. If you are not familiar with the term parts per million, see inside back cover. (Adapted from W. Keeton, *Biological Science*, 3rd ed. [New York: W. W. Norton, 1980].)

can then multiply unchecked and cause severe damage to the citrus crop.

In some cases, extensive use of pesticides has left farmers worse off than before. For instance, cotton has long been grown in Peru in the Cañete Valley. Before World War II, insects were controlled with arsenic and nicotine sulfate. Yields of cotton were about 470 bushels per acre (42.5 kl/ha). In 1949, crops were severely damaged by cotton bollworms and aphids, causing growers to try the new chlorinated hydrocarbon insecticides such as DDT and toxaphene. At first, it seemed to be a successful move. Yields nearly doubled. However, by 1955 the picture had changed. Not only had insect pests become resistant to synthetic insecticides, but new pests were appearing because the insecticides had killed off the beneficial insects that preyed on pests. Chemicals were unable to control insect populations and, by 1956, the cotton-growing industry had collapsed.

A similar sequence of events occurred in the Rio Grande Valley in Texas. By 1968, cotton pests were resistant to all available insecticides. Despite as many as 15–20 treatments with all possible combinations of pesticides, many fields were totally lost due to insect damage (but see Figure 7.9).

Even the successful public health uses of pesticides are threatened by insect resistance. In many countries, walls inside houses are sprayed with DDT and dieldrin. For months after the spraying, mosquitoes that land on the walls are killed. In this way, diseases like malaria and yellow fever, which are spread by mosquitoes, can be controlled. Worldwide, malaria is the largest single cause of death and debilitation. In the 1950s, campaigns to eradicate malaria were begun. Based mainly on spraying with chemical pesticides such as DDT, the campaigns were at first successful. However, the mosquitoes became resistant to DDT and then to propoxur, the spray that replaced DDT. Now cases of malaria are once more on the rise. In 1955, when eradication programs began, 250 million cases were reported and 2.5 million people died. By 1965, only 107 million cases were reported, but in 1976 the number rose to 150 million. Furthermore, the malarial parasite is becoming resistant to drugs used to treat the disease itself.

Between 1970 and 1980 the number of insects resistant to pesticides doubled from 224 to 428.

In line with the development of insect resistance to insecticides, weeds are becoming resistant to herbicides. In 1968, the first weed resistant to an herbicide

Figure 7.9 The boll weevil. Cotton growing has resumed in the Rio Grande Valley in Texas as a result of integrated pest management methods. Short-season cotton, which matures before boll weevil populations become large, cultural techniques such as destroying crop residues, and the use of small amounts of insecticides control pests without destroying beneficial insects or causing pest resistance to develop. In 1985, using a similar set of methods, USDA reported that it had reduced the boll weevil population in North and South Carolina to the point where economic damage to cotton was negligible. (USDA photo; Lewis Riley, photographer)

was discovered. Resistant weeds are rare in fields never treated with herbicides. However, as the use of herbicides has increased, a result of more farmers practicing reduced tillage farming, herbicide-resistant weeds are becoming more common.

Soil Microbes and Pesticides

A fourth problem has only recently been recognized. In the late 1970s, scientists trying to find out why certain pesticides worked well at first but slowly became ineffective discovered they weren't dealing with resistant weeds or insects but rather with soil microbes that break down pesticides. These microbes adapt to pesticide chemicals so they can break down, or use them, making the pesticides useless in controlling weeds or insects.

Studies have shown that although large quantities of herbicides are used in **minimum tillage** farming methods, the herbicides are generally bound to soil by crop residues. Because there is little soil erosion from fields farmed in this way, pesticide runoff is generally lower than from comparable fields farmed conventionally. However, the problems of resistant weeds and soil microbes that break down pesticides may lead farmers to use ever larger quantities of pesticides. If this occurs, minimum tillage farming could in the end cause greater environmental pollution with pesticides than conventional farming.

To summarize, synthetic chemical pesticides have not proven to be as easy to use nor as beneficial as was originally hoped. Insects and other pests such as rats have shown the ability to develop resistance to chemical pesticides. While pesticides that are persistent can accumulate from year to year in the soil and in the muds at the bottom of lakes and rivers, other pesticides are broken down by soil microorganisms before they can kill pests.

Further, fat-soluble pesticides absorbed from water, grass, or crops become concentrated in the milk and body fat of many animals, including humans, at the level of parts per million or parts per billion. (If you would like some comparisons to help you understand parts per million or parts per billion, see end papers.)

In Table 7.2, the toxicity and bioaccumulation of the major groups of pesticides in use today are compared. In order to say whether or not a pesticide is safe, a number of questions have to be asked. How poisonous is the pesticide to humans and to wildlife? How long does the pesticide persist in the environment? Does it accumulate in living organisms? Jerome Weber, an agricultural scientist at North Carolina State University, has developed a simplified rating scheme to determine pesticide safety on the basis of the answers to questions such as these (Table 7.2). According to this table, the all-around most dangerous pesticides are the chlorinated hydrocarbons and mercury compounds, such as ethylmercury chloride. It is easy to see why the EPA has forbidden or restricted the use of these chemicals in the United States.

Organophosphorus pesticides are very toxic to rats and humans and are therefore dangerous for the people who use them. But, since they do not accumulate in the environment or in living organisms, their total hazard is somewhat less than that of chlorinated hydrocarbons. Herbicides such as the triazines and organic acids

listed in the table are widely used in agriculture to kill weeds and to clear roadsides. These compounds score the lowest, generally 7 or lower.

Among the factors this rating scheme does not take into account are the potentials these chemicals have as mutagens (causing changes in heredity), carcinogens (causing cancer), or teratogens (causing birth defects). Such properties are more difficult and expensive to determine than the four factors considered but are equally important in the long run.

The Organic Farming Solution

If pesticides are now seen as a somewhat tarnished miracle, what might be a better strategy in growing food and fiber? One possibility is to apply no synthetic pesticides at all and to use only manure or special crops for fertilizer. This is, of course, organic farming. Once the only possible type of agriculture, organic farming became associated with health-food faddists and back-to-nature cultists after the development of synthetic pesticides and fertilizers. Today organic farming is often viewed as a fine ideal for the home gardener but not a reasonable goal for large-scale farmers. In 1971, Earl Butz, then Secretary of Agriculture, said in an interview, "Before we go back to an organic agriculture in this country, somebody must decide which 50 million Americans we are going to let starve or go hungry." Nonetheless, evidence is accumulating that in some situations, and when properly carried out, organic farming may be a match for what is now called conventional farming—that is, using synthetic pesticides and fertilizers (see Controversy 7.1).

The USDA estimates that if no pesticides at all were used in the United States, farmers would lose some 70% of their crops to pests, but other experts do not agree with this figure. Professor David Pimentel (1978) of Cornell University counters that the use of crop varieties resistant to pests, and other nonchemical techniques—such as changing the kinds of crops grown in different areas—could cut total losses to 16%. Certain crops, such as cabbage, potatoes, and some fruits, might suffer heavier losses.

Pimentel (1978) further reports that one study of organic farming in the corn belt seemed to show that yields of corn and soybeans were almost as good as on

(continued on page 168)

Table 7.2   Relative Toxicities of Selected Pesticides

Common name	Toxicity		Longevity[c]	Bioaccumulation[d]	Total
	To rat[a]	To fish[b]			
Chlorinated hydrocarbons					
Aldrin	3.2	3.9	4.0	3.1	14.2
DDT	2.7	3.7	4.0	4.0	14.2
Dieldrin	3.1	3.9	4.0	3.0	14.0
Endrin	3.5	4.0	4.0	2.8	14.3
Lindane	2.7	3.4	4.0	1.5	11.6
Mean value	3.0	3.8	4.0	2.9	13.7
Organophosphorus chemicals					
Dichlorvos	3.0	2.7	1.3	1.0	8.0
Disulfoton	3.9	3.3	1.1	1.0	9.3
Malathion	1.8	3.2	1.1	1.0	7.1
Parathion	3.6	3.3	1.3	1.0	9.2
Phorate	4.0	3.7	1.1	1.0	9.8
Mean value	3.3	3.3	1.2	1.0	8.8
Carbamates					
Carbaryl	2.1	2.4	1.1	1.0	6.6
Carbofuran	3.6	2.9	1.4	1.0	8.9
Mean value	2.8	2.6	1.2	1.0	7.6
Triazines					
Ametryn	1.8	2.4	1.2	1.0	6.4
Atrazine	1.7	2.0	3.6	1.0	8.3
Prometone	1.7	2.5	4.0	1.0	9.2
Mean value	1.9	2.3	2.5	1.0	7.7
Organic acids					
Dalapon	1.5	1.5	1.1	1.0	5.1
Dicamba	1.8	1.0	1.3	1.0	5.1
Picloram	1.1	2.4	3.6	1.0	8.1
2,4-D	2.4	1.4	1.1	1.0	5.9
2,4,5-T	2.5	2.8	1.4	1.0	7.7
Mean value	2.0	2.1	1.6	1.0	6.7
Miscellaneous					
Captan	1.0	3.0	1.0	1.0	6.0
Copper sulfate	1.7	3.0	4.0	1.0	9.7
Ethylmercury chloride	3.3	2.4	4.0	3.0	12.7

Source: Data used with permission of Dr. Jerome B. Weber, North Carolina State University.

a. How poisonous a pesticide is to rats (based on the LD50, the amount that will kill half of a group of test animals) is a good indication of how toxic it is to mammals in general. On a scale of 1 to 4, 4 is the most toxic.

b. How poisonous a pesticide is to fish (based on the LC50, the concentration that will kill half of a group of test fish) is a good indication of how toxic it is to other aquatic life. On a scale of 1 to 4, 4 is the most toxic.

c. How long a pesticide persists in the environment is indicated by how long it remains in the soil after it is applied.

readily broken down:	15 weeks or less
moderately broken down:	15–45 weeks
slowly broken down:	45–75 weeks
persistent:	75 weeks or longer

On a scale of 1 to 4, 4 is the most persistent.

d. Bioaccumulation can be measured by exposing a test species, such as oysters, to the pesticide for a length of time and then measuring how much accumulates in the oysters. On a scale of 1 to 4, 4 represents the highest accumulation.

CONTROVERSY 7.1

Organic Gardening Economics

You would never have believed it—we outyielded our neighbors by 100% or better on everything.
K. C. Livermore, Nebraska organic farmer

. . . The result of abandoning the use of pesticides? . . .
Silent Autumn.
ChemEcology, January 1978

Although most agriculturists remain unbelievers, a small but growing number of people believe that even large farms can be run, profitably, without synthetic pesticides and fertilizers. The high prices of fertilizer and pesticides, due partly to the energy crunch, has led many farmers to experiment with using fewer synthetic chemicals.

K. C. Livermore, a Nebraska farmer, grows alfalfa, oats, soybeans, and corn on 260 acres (104 hectares). He says:

We've done much better without chemicals. We hurt some at first when we switched over because we had to get the soil back in balance, get the poisons worked out of it. But in our fourth year there was a big turnaround and now we're outyielding our "chemical neighbors" by far.

Mr. Livermore says about insect and weed problems:

We don't have an insect problem like our chemical neighbors do. We don't have an altered plant. Our plants are natural and healthy. They pick up antibiotics from the soil, which turns insects away as nature intended. And we have insects, like ladybugs, which fight off the enemy insects. Ladybugs thrive on our farm.

Also, as soon as you get natural, healthy soil,

there isn't any weed problem. Nature puts in weeds to protect the soil. Weeds grow down in the soil and pick up trace minerals, and as they die they deposit these minerals on the soil's surface.

And when you have your soil in balance, weeds just don't grow as fast and you don't grow as many of them. Another thing is that when we used chemicals we had a clotty soil. Now it will run through your hands just like flour at times. Earthworms and other life in the soil are alive and can loosen it. It's easy to push the weeds right over when we cultivate.

Other farmers, such as Mike Shannon, who farms 30,000 acres (12,000 hectares) in San Joaquin, California, use a minimum of pesticides. His S–K ranch actually owns a pesticide supply company and a crop-dusting service, but has cut its own pesticide usage by two-thirds. "It costs money," he says. "We have the planes, but I'd rather not touch them." The S–K ranch is advised by Richard Clebenger, a man trained in integrated pest management—a strategy that attempts to use as little pesticide as possible. Clebenger admits, "There still are a lot of farmers . . . who can't sleep right unless they've given their fields a good spray."

Dow Chemical, on the other hand, has published a book, *Silent Autumn*, that describes the plagues of Biblical times as well as a variety of famines caused by uncontrolled pests in recent years. The book emphasizes the idea that careful use of pesticides is necessary to feed people in an ever more crowded world.

For the experimenting organic farmers, though, K. C. Livermore sums it up:

> We'd like to see this thing get turned around. . . . We'd like to see the wildlife and

the birds back here like it was in the 1940s and 50s. Is that a profitable way to farm? You bet it is. We use one-fourth less input and get as much or more back than anybody else. That should be real easy to calculate in your mind. . . .

Source: All quotes from *EPA Journal*, March 1978, pp. 24–25.

conventional farms (corn yields were about 10% lower and soybean yields about 5% lower). However, wheat yields were as much as 25% lower on organic farms. When energy cost savings on organic farms were taken into account, profits were almost the same for both kinds of farming.

Integrated Pest Management and Biological Controls

A second possibility is to use small amounts of synthetic pesticides in combination with a variety of other techniques to control pests. This solution is called IPM (for integrated pest management). The farmer practicing IPM chooses from a variety of pest control methods, including:

1. cultural practices, such as plowing, disposing of crop wastes at the end of the season, crop rotation;
2. planting resistant crops;
3. scouting for the actual presence of pests; and
4. selective use of pesticides.

Cultural techniques and resistant crops. Before synthetic pesticides were invented, farmers used many cultural practices to reduce insect damage. Some of these old techniques and some newly developed ones are very useful. A book published in 1860 lists this control for apple worms:

> The insect that produces this worm lays its egg in the blossom-end of the young apple. That egg makes a worm that passes down about the core and ruins the fruit. Apples so affected will fall prematurely and should be picked up and fed to swine. This done every day during their falling, which does not last a great while, will remedy the evil in two seasons. The worm that crawls from the fallen apple gets into crevices in rough bark,

and spins his cocoon, in which he remains till the following spring. (Walden, 1860, p. 22)

Plowing alone can destroy up to 98% of the corn earworm pupae that winter in the soil. Rotating the kinds of crops grown in a field each season, or mixing crops in a field rather than growing large fields of the same kind of crop year after year, can prevent the buildup of pests. Destroying crop residues after harvest destroys the winter home of pests such as the boll weevil. By growing certain kinds of hedgerows around fields, a farmer can increase the number of predatory insects in a field, leading to better control of the plant-eating insects that damage crops.

Some plant varieties have been bred so that they are resistant to major insect pests. In some cases, this has been accomplished by breeding plants that look or taste unattractive to pests. Wheat resistant to Hessian fly has been available for 30 years. As with pesticides, however, new insect mutants may develop and begin to attack the resistant species, making it necessary to again breed new crop varieties.

Monitoring and treatment. In integrated pest management, farmers must carefully monitor their crops to watch for the beginnings of insect attacks. Traps baited with a variety of substances lure insects and give the farmers early warning of a pest invasion (Figure 7.10). Only when something must be done to prevent serious crop damage do the farmers choose a pest control treatment. This need not be a chemical spray in all cases, since a number of methods of biological pest control are known.

Biological controls and use of natural biochemicals. Investigations of how insects reproduce and how they interact with each other have identified several

Figure 7.10 A scout collects insects from a trap in a field. This trap serves as an early warning of insect pests for farmers in the area. (USDA photo)

points at which humans can interfere in order to reduce insect damage to crops (Table 7.3). For example, insects produce chemical substances called pheromones. Some pheromones are sex attractants, which help insects find mates; other pheromones mark paths to food sources. If large amounts of a pheromone can be obtained, it can be used to bait traps. This has been done with the gypsy

Table 7.3 Biological Control Techniques

Control by other organisms
Parasites
Predators
Pathogens

Reproductive control
Release of sterile insects
Use of chemical sterilants
Release of incompatible pest strains

Hormones
Pheromones or other behavioral chemicals

moth, whose sex-attractant pheromone has been artificially synthesized. Traps baited with pheromone can be used either to catch insects and kill them with a poison or to give early warning of a pest's appearance in the field.

Insects and weeds are sometimes imported from other countries without their natural parasites, predators, or diseases. Such pests may then spread unchecked and cause great damage. Successful attempts have been made to control many of these undesirable immigrants, as well as some native pests, by introducing parasites, predators, and diseases. For instance, a tiny wasp, *Trichogramma*, controls cotton bollworms by parasitizing bollworm eggs. Two kinds of leaf-eating beetles have been able to keep Klamath weed, once a serious problem in California, under control. Another example is presented in Figure 7.11. In such cases, of course, it is necessary to determine in advance that the new species will not begin to eat crops or kill desirable insects if it runs out of pests.

Bacteria that attack only pest species also can be used as a control measure. Bacteria causing milky-spore disease will kill Japanese beetles and can be applied much like a synthetic insecticide.

Certain kinds of insects can be grown in large numbers and then sterilized by radiation or chemicals. Sterile males of these species, when released, mate in the wild with normal females, but the females then lay eggs that do not hatch. Screw worms have been eliminated from a Caribbean island, Caracoa, and controlled substantially in the southwestern United States by the release of sterile males. Mediterranean fruit flies are partially controlled in California and Florida by the release of sterile flies. However, not all insects can be grown artificially in large enough numbers, or sterilized easily enough, to use this method.

Economics of IPM. For a number of crops, integrated pest management is cheaper than conventional techniques, which involve spraying with chemicals at regular intervals. Even though farmers must pay to have their crops monitored, cotton, apples, and citrus fruits all can be grown more economically using IPM methods. The savings are mainly due to the use of less chemical spray, although yields are sometimes increased as well. In one example, 600 acres (243 hectares) of California tomatoes were being sprayed with chemicals four or five times a season, at a cost of $20–$30 per acre ($50–$75 per hectare). In spite of this, fruit

Figure 7.11 Gypsy moth caterpillar and parasitic fly. In this country, gypsy moths do much more damage than in their native Europe and Asia, where parasites keep their numbers low. However, none of the moth's natural enemies accompanied them here. Gypsy moths strip the leaves from millions of acres of trees (especially oaks) in the Northeast and central Michigan every year. The moth infestations seem to be spreading west and southward. At one time, 12 million acres (4.9 million hectares) a year were sprayed with DDT to control gypsy moths. The fly shown here is one of the parasites scientists are studying to find a biological control for the moths. (USDA photo; Murray Lemmon, photographer)

worms continued to damage the crop. A company specializing in IPM was able to control the fruit worms as well as reduce costs to $8–$10 an acre ($20–$25 per hectare), and during the second year of the program, no chemical sprays were necessary at all. As the costs of fertilizer and pesticides go up due to fossil-fuel shortages, the savings in such IPM programs will become more and more attractive.

In a similar way, the use of a biological form of pest control may prove much less expensive than repeated spraying with costly chemicals. Even in developing countries, biological control methods, such as the release of parasites, are being tried. Experts hope such methods will prove cheaper than chemicals and also less harmful to people and the environment in developing countries.

Pest Control Experts

Understandably, farmers are slow to change their farm practices. Their income depends on the crops they raise, and so they want to be sure of the effectiveness of a new method before they risk using it. Nonchemical pest control methods tend to be more complicated to use than chemical sprays, and they usually do not have the immediate and obvious "knockdown" effect that sprays have.

Farmers and experts who are used to reducing pest

populations as close as possible to zero may be uncomfortable with one of the basic ideas of IPM: that pest populations should be kept just below the level at which they cause economic injury, in order to maintain predators and parasites in the ecosystem. Furthermore, farmers get a good deal of advice from pesticide manufacturing companies, which are hardly disinterested parties.

These factors, as well as the fact that there are so many pest control techniques from which to choose, have led to the need for people specifically trained in pest control, as well as for "scouts"—people who monitor fields for the presence of insect pests. A number of small firms have already begun offering pest management services. The Department of Agriculture is attempting to work out certification programs to ensure that persons calling themselves pest control experts actually have the necessary training. Several university programs are being developed in pest control. They will eventually produce people knowing the advantages and disadvantages of chemical sprays, cultural practices, and biological controls. These trained specialists can then help to choose, for each situation, the combination of methods that will best solve a pest problem.

Unfortunately, budget cuts have reduced government efforts in the area of IPM. The Environmental Protection Agency now has no IPM program at all, and

CONTROVERSY 7.2

Who Should Pay for IPM?

I find it truly remarkable and ironic that we . . . are experiencing just as great if not a greater level of crop injury from pests as we did before DDT.

R. J. Prokopy, University of Massachusetts Department of Entomology

Different pests. Different times. One powerful answer. Pydrin

Pesticide ad, Farm Journal

Farmers need to get away from this notion of the automatic use of [pesticides]—We should consider the concept of integrated pest management.

Don Kuhlman, University of Illinois entomologist

Sometimes market forces don't work to provide the technology that is best from an environmental standpoint. A case in point seems to be integrated pest management (IPM). In the United States this method of pest control, which stresses cultural controls and scouting for the presence of pests, along with the use of a minimum amount of pesticides, got its start in universities with funding from EPA, USDA, and the National Science Foundation during the 1970s. However, budget cuts since 1981 have felled most IPM projects. The reasons are not just political. The new generation of synthetic pesticides, the pyrethroids, have an undeniable appeal as "magic bullets" that will solve pest problems simply and cheaply, much as DDT and its relatives promised in the 1940s and 1950s.

But many experts feel that these pesticides, like DDT, if used extensively will eventually become ineffective, due to insect resistance. The answer seems to be to use them in a program of integrated pest management, in which the pesticides are only one possible choice for a farmer who needs to control pests. How-

ever it is not in the best financial interest of a chemical company to give this sort of advice to farmers.

Some people have suggested that a tax on pesticides (say 2 cents a pound) could finance IPM research and educational activities. They insist this would transfer both the costs and the benefits to those who are most directly involved (farmers). Another possibility is to support IPM activities from general tax revenues, as was done in the 1970s, on the basis that we all benefit from a decreased use of pesticides. Still another possibility is to do nothing and allow market forces and natural processes to find whatever accommodation they can (this is basically what we are doing now).

Contrast these three positions. Which is most economical and for what reasons? Which is the best scientifically? Which is the most feasible from a political point of view? Which position do you support?

Source: All quotes from *EPA Journal*, March 1978, pp. 24–25.

the USDA's program is barely surviving. The chemical industry meanwhile is relying on a new generation of pesticides, synthetic copies of natural products such as the pyrethrin-like pesticides. (see Controversy 7.2.) Unless these new weapons are used in a managed program that recognizes how and why pests fight back, however, the same sequence of use, resistance, overuse, and finally uselessness may occur over and over.

Summary

Erosion and pesticide use are two important environmental problems associated with food production. Erosion, the loss of topsoil due to water or wind, is classed according to severity as sheet, rill, or gully erosion. New topsoil is formed slowly—at most, 1 inch every 100 to 1,000 years. Only 1 to 5 tons of soil per acre can be lost each year before a loss in productivity will result. In addition to lost productivity, erosion increases the chance of flooding from local streams, intensifies the effects of drought, makes plowing more difficult, decreases the capacity of reservoirs, and increases the organic matter and pesticide burdens of natural waters. Wind erosion also leads to decreased productivity and may result in dust storms. Erosion can be controlled by reduced or minimum tillage farming, crop rotation, contour plowing, terracing, and shelter belts.

The 1985 farm bill requires farmers on erosion-prone land to institute soil conservation methods or risk loss of government farm program benefits.

Pesticides are chemicals used to kill pests. The use of pesticides has led to a variety of problems. The chemicals may accumulate in living organisms by a process called biological magnification. If they do not break down within one growing season, they can persist and even accumulate in the environment. Pests often become resistant to pesticides that formerly controlled them. Furthermore, soil microorganisms can adapt to pesticides, breaking them down in the soil to useless by-products.

Alternatives to pesticide use include: organic farming, integrated pest management (IPM), and the use of biological control methods. In IPM, a variety of techniques are used, including cultural techniques, resistant crops, and monitoring for pest outbreaks, combined with limited pesticide use. Both IPM and the use of biological controls may prove more economical for a farmer than the use of chemical sprays. However, farmers must be informed about these alternatives and how they can benefit from them. This is the role of specially trained pest control experts.

Questions

1. Explain what erosion is and what its major causes are.
2. Summarize the effects erosion has on the environment.
3. What basic principles are involved in controlling erosion?
4. What are the major environmental problems associated with the use of pesticides?
5. Explain the basis of integrated pest management.
6. If you were a farmer, would you be an organic farmer, would you use integrated pest management, or would you use chemical pesticides whenever you felt you had an insect problem? Why?

Further Reading

Josephson, J. "Pesticides of the Future," *Environmental Science and Technology*, **17** (1983), 464A.
New directions in pest-control are predicted in this interesting article.

Phillips, R. E., et al. "No-Tillage Agriculture," *Science*, **208** (June 6, 1980), 1108.
No-tillage agriculture is a fast-spreading solution to soil erosion problems on cropland. The pros and cons are carefully detailed in this paper.

Graham, F., Jr. *Since Silent Spring*. New York: Fawcett, 1970.
The classic book *Silent Spring* by Rachel Carson began a public controversy over the effects of pesticides that continues today. Carson was eventually proven correct in many of her claims, although, as detailed in Graham's book, they were attacked violently when her book was first published.

Dover, M. "Getting off the Pesticides Treadmill," *Technology Review* (November/December 1985), p. 53.
This article is devoted to the methods and promise of integrated pest management.

Oelhaf, R. C. *Organic Agriculture*. New York: Halsted (Wiley), 1979.

"An Answer to Insecticide Failures," *Farm Digest* (May 1984), p. 3.
Read firsthand what farmers are learning about insect and weed resistance.

References

National Agricultural Lands Study. Washington D.C.: National Association of Conservation Districts, 1981.

Speer, L. and G. Paulson. Letter to the Editor. *The New York Times*, October 24, 1984.

Brody, J. E. "Pesticides," *The New York Times*, October 28, 1974.

Adkinson, P. L., et al. "Controlling Cotton's Insect Pests: A New System," *Science*, **216** (April 2, 1982), 19.

Lockeretz, W., et al. "Organic Farming in the Corn Belt," *Science*, **211** (February 6, 1981), 540.

Batra, S. "Biological Control in Agroecosystems," *Science*, **215** (January 8, 1982), 134.

Fox, J. L. "Soil Microbes Pose Problems for Pesticides," *Science*, **221** (September 9, 1983), 1029.

Walden, J. H. *Soil Culture*. New York: Saxton, Barker, and Co., 1860.

U.S. Department of Agriculture. *Report and Recommendations on Organic Farming*. Washington, D.C.: U.S. Government Printing Office, 1980.

Bottrell, D. G. *Integrated Pest Management*. Publication 041-011-0049-1. Washington, D.C.: U.S. Government Printing Office, 1980.

Larson, W. E., et al. "The Threat of Soil Erosion to Long-Term Crop Production," *Science*, **219** (February 4, 1983), 458.

Osuji, O. and O. Babalola. "Tillage Practices on a Tropical Soil," *Journal of Environmental Management*, **14** (1982), 343.

Benbrook, C. M. "New Tools and Policies in the 1985 Farm Bill," *Water Science and Technology Board Newsletter*, **3** (2) (March 1986).

Pimentel, D. (ed.). *World Food Pest Losses and the Environment*. AAAS Selected Symposium 13, 1978. Washington, D.C.: American Association for the Advancement of Science, 1978.

CHAPTER EIGHT

Feeding the World's People

How Much Food?

In this part of the world, we do not just think that people will die of hunger. We assume it.

Worker at a relief camp in Burkina Faso

Famine and Malnutrition

The worst famine in African history struck the continent in 1983–1985. The world was shocked by photographs and figures detailing the tragedy (Figure 8.1). A major drought, the most severe in modern history, had caused massive crop failures. This led inevitably to famine for populations that had little or no food in reserve for bad years; that had no money to buy food from the rest of the world; and that often lacked even a road or transportation system that could distribute the food aid finally sent by other nations.

As dreadful a crisis as the African famine was, it was but the acute form of a chronic problem. Such famines occur against a constant background of deaths due to lack of food. Some experts estimate that worldwide there are as many as 35,000 hunger-related deaths every day. The largest proportion of these deaths is among children under 5 years old.

In all parts of the world areas of chronic starvation exist where day after day, people eat fewer Calories or less protein than they need. (Calories are a measure of the heat or energy content in food. Human food requirements are measured in kilocalories, and the term is written with a capital c, "Calories.") Over the long term, this leads to both mental and physical crippling.

But problems arise in estimating how many people

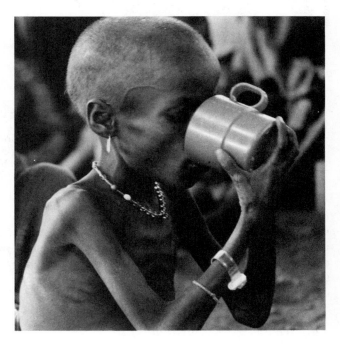

Figure 8.1 A small boy, victim of the drought in Ethiopia, drinks and gains some nourishment from newly delivered supplies. (U.N. photo by John Isaac)

are starving or malnourished and in defining what a reasonable standard of nutrition should be. In truth, people do not commonly die of starvation. In most cases, a person who has too little food is more likely to develop diseases than someone who eats well. Furthermore, the undernourished victim is more likely to die of diseases from which a well-nourished person could

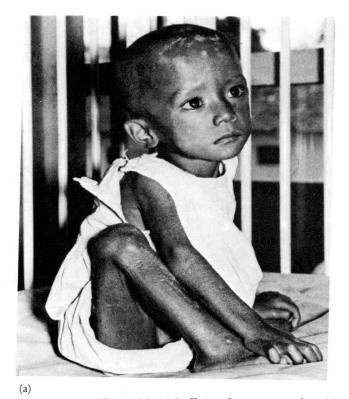

(a)   (b)

Figure 8.2 (a) Suffering from acute malnutrition, or kwashiorkor, a small boy sits on his hosptial bed in Djakarta. (UNICEF photo by Jack Ling) (b) Infantile marasmus. (UNICEF photo by Lynn Millar)

recover. Typhus, cholera, smallpox, plague, influenza, tuberculosis, and relapsing fever all commonly strike people weakened from lack of food. This means that the actual cause of death is listed as something other than starvation, even though starvation is clearly an underlying or major cause of death.

Children and Poor Nutrition

Children suffer from two diseases specially related to malnutrition. **Kwashiorkor** occurs when a child's diet may have enough Calories but too little protein. A major symptom of the disease is edema, or swelling of the abdomen (Figure 8.2). Kwashiorkor usually occurs after children are weaned. **Marasmus**, which involves a shortage of both protein and Calories, occurs when infants less than 1 year old are fed overdiluted formula from unsterile nursing bottles. Severe diarrhea almost always occurs in both diseases. Children with

kwashiorkor and marasmus have higher death rates and may also suffer permanent brain damage.

If mothers nurse their babies, infantile marasmus can be prevented because human milk provides needed protein for the child. In addition, because human milk is a completely sanitary food source, chances are less that the child will be infected with diarrhea-causing organisms. Even kwashiorkor is uncommon if a child is nursed through the second year or longer. Furthermore, nursing has an incomplete but significant contraceptive effect. Women who nurse are less likely to conceive again while nursing. This can have a positive effect in terms of population control where other birth control methods are not available or are unacceptable.

Unfortunately, in some developing countries companies that produce infant formulas have engaged in campaigns to sell parents on formula feeding of infants. In many of these countries, the cost of infant formulas is high compared to workers' incomes. In some cases, commercial formula, if fed at the proper strength, would cost one-fourth to one-third of a worker's income. Thus, the formula is often overdiluted. The high cost and the lack of necessary sanitary conditions for preparing formula mean that in many parts of developing countries a switch from breastfeeding to bottle-feeding or earlier weaning will cause infant and child health to suffer.

In 1984, the largest manufacturer of infant formula, the Nestlé Corporation, agreed not to continue advertising and distribution practices that might contribute to a shift from breastfeeding to bottle-feeding in developing countries. In return for this agreement, environmental action groups called off a consumer boycott of Nestlé products. If this agreement, along with active promotion of breastfeeding by the governments of developing nations, causes larger numbers of mothers to decide to breastfeed their babies, a significant amount of hunger-related disease will be ended.

Estimating Calorie Requirements

Estimates of the number of people receiving less food than they need are complicated by a lack of agreement on how many Calories are needed. For instance, people living in warmer climates may need fewer Calories. Even within a region, Calorie requirements probably vary by as much as 50% from person to person. Another problem is that nutritional estimates, based on what people eat in developed countries, may be too high, since people in these countries are generally regarded as overfed. In addition, economists who estimate the food available tend to take into account only those crops moving through the marketplace—for instance, cereal grains—or those animals that can be counted easily, such as beef cattle. This method probably underestimates how much people get to eat, at least in rural areas, since family gardens and other local sources of food do exist.

On the other hand, sometimes the food supply in an area has simply been divided by the number of people there. This method takes no account of the fact that income is unevenly divided, making purchasing power uneven. Starvation can exist in a land of plenty if some people cannot afford to buy food. This situation exists in all countries, developing or developed, including the United States. Even within a family, food distribution may be unequal. In some cultures men in general, or

Table 8.1 Hunger-Related Deaths[a]

	Early 1970s	Early 1980s
Annual hunger-related deaths in countries with an infant mortality rate over 50 per 1,000	15 million	13 million
Approximate number of hunger-related deaths among children under 5	11.3 million	9.8 million

Source: Adapted from R. L. Prosterman, "The Decline in Hunger Related Deaths," *The Hunger Project Papers* #1, May 1984, p. 10.

a. These numbers were calculated by R. L. Prosterman on the assumption that in developing countries 60% of deaths in the age groups 0–4 are due to hunger-related causes and that 30% of the deaths in the over 4 age groups are also due to hunger-related causes. He cites epidemological and social data to justify these assumptions.

the breadwinner in particular, eat first and the rest of the family shares what is left.

The latest Food and Agriculture Organization Survey estimates that 25% of the people in developing countries are malnourished (over 4 million people, mostly women and children).* These people lack one or more nutrients (such as protein) needed in a healthy diet. A smaller percentage are undernourished; that is, they do not get enough Calories to maintain their body weight with normal activity. Most undernourished people, of course, are also malnourished.†

The good news is that there is evidence that hunger-related deaths may be decreasing (Table 8.1). That is to say, efforts to improve the world food situation may be working, although very slowly, even in the face of rapid population growth in many areas.

The Food Crisis in Historical Perspective

Periodic famine has always existed. The concept of "food crisis" seems to imply something more. Are we likely to reach the point in terms of population growth where we simply cannot increase food production to meet the demand created by new mouths to feed? Might large segments of the world population starve to death in a nightmarish solution to overpopulation?

*This number does not include people living in Asian centrally planned nations—for example, China.
†See the reference by Poleman (1981) for a clear review of the problems, methods, and assumptions underlying such estimates.

Since World War II, just that sort of prediction has been made and then withdrawn several times. Perhaps something may be learned from past history of the world's food supply. After World War II, when public health measures reduced death rates in many undeveloped countries and population growth rates began increasing, many writers predicted that worldwide famine was on the way. This feeling was bolstered by Food and Agriculture Organization reports, which gave the impression that one-half to two-thirds of the world's population was malnourished. This was probably an overstatement. Statistics on food supplies were scanty and unreliable, and high values were used for the number of Calories required in a healthy diet.

Developing countries were actually making slow but steady progress toward feeding their populations until 1965–1966. In those years, India suffered two droughts that markedly decreased her food output. Because India is home to such a big chunk of the world's population (one out of every six people in the world is Indian), these crop failures loomed large in totals of world food production. Many people felt that the world famine had arrived.

Immediately after this, however, favorable weather and the introduction of high-yield grain varieties in India and other parts of Asia reversed the trend. World food production rose and a mood of optimism prevailed through 1971. The Green Revolution had begun. In the United States, the government paid farmers not to produce grain to prevent prices from falling too low. However, in 1972 the Russian grain crop failed. The Soviet Union and eastern European countries purchased 28 million tons of grain, reducing grain stocks in the United States and Canada to a 20-year low and raising grain prices all over the world. Demands from other countries also rose, in part due to rising incomes in countries such as Japan and in Western Europe that enabled more people to afford meat. Increased meat production means grain consumption by increasing numbers of cattle.

About this same time, the Peruvian anchovy fishing industry collapsed as a result of changes in the normal water temperature and overfishing. Harvests that had totaled 12.3 million metric tons in 1970 fell to 3 million metric tons in 1973. Visions of feeding the world from the oceans began to fade in the hard light of reality.

Then the weather demons struck again. In 1974, drought and early frosts caused the worst growing season in the United States in 25 years. Grain harvests were 20% below expected levels and grain reserves hit new lows. In addition, food prices hit new highs (Figure 8.3). World famine was again predicted.

There were certainly areas of famine in the 1970s and the 1980s, but the predicted world food crisis has not occurred. Many experts now agree that a world food crisis is unlikely in the near future. This does not mean we have no problems, however. Most of these problems revolve around the issue of fair distribution of available food.

Distribution and Wastage Problems

Even when there is a reasonable surplus of food, the world remains troubled by uneven distribution. One-half of the 1979 grain reserves were in one country, the United States.

While food production in developing countries has increased almost 120% since the 1950s and is now increasing at 3.3% per year, population growth has doubled. This growth is eating up much of the food production gains. In the developed nations, on the other hand, food production increased by 100% but population increased by only one-third.

One problem, then, is transportation: getting food surpluses to where the needy people are. Another problem involves decreasing the portion of crops lost to insects and disease in the field, in storage, and in transportation. As noted in Chapter 7, some 25% more grain could be available if such losses were prevented. The highest losses occur in developing countries, which can least afford them.

A third problem is damping the price fluctuations of grain on the world market. These fluctuations are caused mainly by bad weather, which decreases harvests in an unpredictable fashion. Because an increasing part of the world's grain harvests is spoken for in long-term contracts between the wealthier nations, any decreases in harvest seriously affect small, poor nations. These nations do not have the money or political clout to compete for grain when world supplies are small and prices high. Fourth is the problem of changing income distributions so that the countries and the people currently too poor to purchase enough food can afford a healthful diet. The social, political, and economic problems involved are enormous and will be discussed later. First, we look at the prospects for increasing the world's food supply in absolute terms.

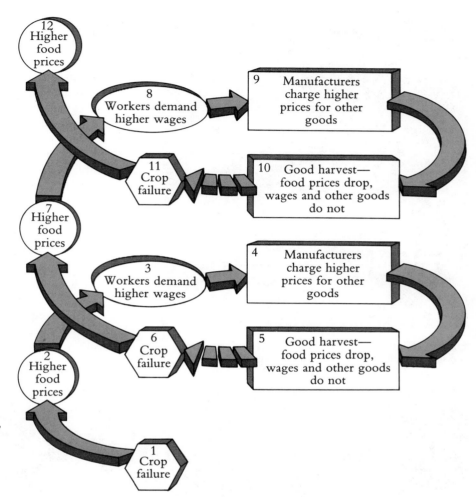

Figure 8.3 The "ratchet effect" of rising food prices. When food prices rise, they cause an increase in wage demands and also in the prices of other goods. When food prices drop, however, wages and other prices do not follow suit. Thus, if food prices lurch from high to low and back again, they have the effect of "jacking up" the cost of living.

How Can We Grow More Food?

Traditional Farming Versus the Modern American Farm

In the central highlands of the Philippine island of Luzon are the villages of a fierce mountain people, the Igorots. These people were once noted as head hunters, but their fame now rests on their outstanding ability to farm the steep mountain slopes of the island's central region. In the incredibly short period of a few hundred years, these people have built a series of terraces on the mountainside, by hand labor with only a few primitive tools (Figure 8.4). On these terraces, using no artificial fertilizers, they grow rice in amazingly high yields. A single hectare of land (2.47 acres) yields a whole year's supply of rice, their main food, to a family of five.

The high yields are due partly to the enormous amount of labor the Igorots put in. On the terraces, rice is grown by water culture. The terraces must be weeded continually to keep the water-retaining walls from crumbling. Another important factor appears to be nitrogen-fixing algae that grow in the water along with the rice. These algae provide nitrogen, which fertilizes the rice crop, while the rice provides carbon dioxide and some necessary shade for the algae. The Igorots have developed an environmentally sound and self-renewing agricultural system that supports them well. Along with the agricultural system, a social system has developed. Villages are organized into work groups that construct irrigation canals, terrace new slopes, plow the fields, and harvest crops, far more efficiently than solitary workers could (Figure 8.5).

Figure 8.4 Igorot rice terraces are marvels of engineering and skill, practiced by a people with no modern tools or equipment. (Photo by Charles Drucker)

From these work groups the whole social structure of the villages has grown:

> They sponsor rituals, and they help to resolve disputes; they form political parties and in times of trouble they become military units. They enter into virtually every phase of the communities' activities. Perhaps most importantly, they provide the Igorots with a stability of association, continuity in their personal relationships that in our own affluent, mobile society is becoming ever more rare and valued. The interdependence of the working groups has made the Igorots so culturally conservative that almost a century of contact with Western Society has resulted only in superficial changes in their way of life. (Drucker, 1978, p. 22)

The wet-culture of rice is a special type of farming. All over the world, farmers grow crops in a manner such as this, requiring much manpower but little or nothing of modern agricultural technology. Such crops are usually well adapted to the climate, the soil type, and the availability of water. But in many cases, these

Figure 8.5 Among the Igorots, work groups are much more than temporary labor gangs; they form the basis for society itself. Here work groups transplant rice seedlings. (Photo by Charles Drucker)

Figure 8.6 An American farm family in 1887 in front of their sod home. Although transformed by modern agricultural technologies, the family farm remains the characteristic unit in the Corn Belt and Wheat Belt. Nonetheless, the look of the farm has changed quite a bit since that time. In 1850 there were 1.5 million farms in the United States, with an average size of 196 acres (78 ha). By 1920, 6.5 million farms existed, averaging 149 acres (60 ha). By 1959, only 3.7 million farms were counted and the average size was 303 acres (121 ha). However, there are indications that this trend is now reversing, at least in the Northeast. The number of small farms is growing and may increase by 18–20% during the 1980s. (USDA photo)

farming operations are at or near the subsistence level. The farmer grows enough food for his own family and some to trade for the other goods needed to live. What hope does technology offer for bringing these farmers to a point where they can grow enough food for the rapidly growing urban populations in their countries? Before answering this question, it will be helpful to describe the modern American farm.

There appear to be two kinds of U.S. farms. Some 20% of the farms produce 75% of the food and fiber; the rest yield less than $20,000 per year in sales. Presumably these latter farms are worked only part-time. Profitable farms are big. This may mean 600–800 acres (240–320 hectares) in the Corn Belt, where one family can work this much land alone with the proper machinery and a small amount of help at harvest time (Figure 8.6). A dairy farm must have 40 cows per worker to see a profit because, again, a great deal of machinery is involved.

Still, except for certain types of farming, such as raising broiler chickens, large corporations are not farm owners. Most farming does not lend itself to corporate management, since managers would have to put in long hours and have knowledge of optimum farming techniques for each separate locality.★ Nor is farming as profitable, in terms of return on investment, as many other business ventures.

The next few sections detail some of the agricultural developments—such as synthetic fertilizers, irri-

★About 2.1% of farms are owned by corporations, but almost 90% of these are family corporations.

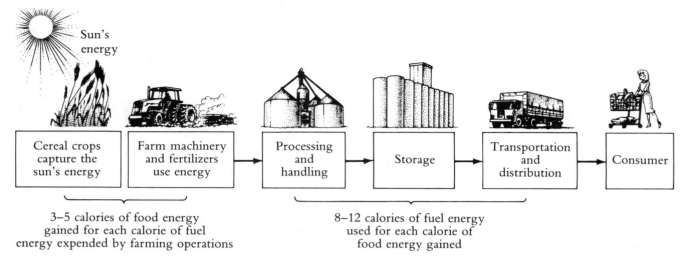

3–5 calories of food energy
gained for each calorie of fuel
energy expended by farming operations

8–12 calories of fuel energy
used for each calorie of
food energy gained

Figure 8.7 Energy costs for producing food in a developed country such as the U.S. Farming itself shows a net gain of energy because more of the sun's energy is captured by plants than is used by farming procedures. After this, however, processing, handling, storage, and distribution of food use up this energy gain and more, so that by the time food reaches the consumer, 5–7 Calories of fuel energy have been used to produce each Calorie of food energy.

gation schemes, and high-yield crops—that have increased U.S. food production to a point where North American farms produce most of the food exported in the world today. We will also examine the promise of similar techniques in developing countries.

Fertilizer, Energy, and Food Production

With few exceptions, the American farmer uses large amounts of artificial fertilizer. In fact, 30%–40% of increased U.S. productivity in recent years has been attributed to increased use of fertilizer. Nitrogen fertilizers, which account for half of all fertilizer in the world, are produced mainly from fossil fuels such as natural gas and coal. Energy is also needed for the manufacturing process itself. Shortages of fossil fuels have raised the cost of fertilizers over threefold in recent years, which, in turn, has contributed to the higher cost of food.

An especially unfortunate result of higher fertilizer prices is that poorer countries can now afford less fertilizer—yet these areas are precisely where it would do the most good. The reason is that as more and more fertilizer is added to a field, the increased crop yield per pound of fertilizer begins to decrease. The field is, so

to speak, saturated with fertilizer. Most large farms in developed countries, such as the United States, fall into this category. In contrast, farms in developing countries, where little or no fertilizer has been used in the past, show a much larger increase in yield for a smaller amount of fertilizer. In terms of the total world food supply, then, the distribution of fertilizer is uneven.

Energy is used in modern agriculture not only to produce fertilizer but also to run the machines that plant and harvest crops and in the after-harvest processes: handling, storage, processing, and distribution (Figure 8.7).

One further point should be made in relation to energy and agriculture. Most people prefer to eat meat if they can afford it. Yet, as the second law of thermodynamics predicts, meat is an inefficient way of obtaining food energy because of the energy transfers involved. As a general rule, an animal eats 3–10 pounds of grain to produce 1 pound of meat. A much more efficient use of food resources is for people to consume grain directly, as is now done in most developing nations (Figure 8.8).

Animals and animal products still have an important place in improving agricultural systems of developing nations, however. For instance, farmers can use

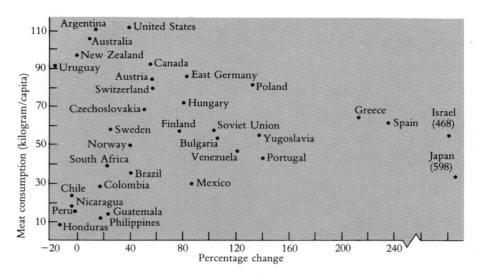

Figure 8.8 Change in meat consumption between 1961 and 1980. As people earn higher incomes, they spend more money on meat. The actual meat consumption per person is shown on the vertical axis, while the percent change from 1961 to 1980 is shown on the horizontal axis. Argentina and the U.S. lead in the amount of meat eaten per person, but the greatest changes have taken place in countries such as Israel and Japan. (From T. Barr, *Science*, **214** [December 4, 1981], 1087.)

animals to pull machinery such as plows or harvesting equipment. In many parts of the world, this means that a small family can tend the same amount of land as used to be tended by a larger family without animal help. This could lead to improved diet or excess crops for sale. The animal must be fed, of course, but it can graze on land that is not as suitable for human food crops (too steep, too cold, too rocky) or can be fed a second crop harvested at a different time from the main human food crop. In any case, animals and animal feed are generally easier to come by in developing nations than tractors and gasoline.

Animal products such as milk and meat can also provide the farmer with year-round income, as well as some small insurance against starvation in bad crop years.

Irrigation and Salinization

In many areas, notably the American southwest, fertile lands that receive little rainfall are successfully farmed after the development of irrigation systems. Some 15% of the world's farmland is irrigated, and 30% of the world's food is produced on these acres. Again, energy is a necessary component of the system. In most cases, fossil fuel is needed directly for pumping or draining systems.

In many areas, however, irrigation brings hazards along with benefits. In tropical countries, irrigation canals serve as a breeding place for the snails that carry the disease schistosomiasis. Although this disease is not fatal, it takes a great toll of strength from people living

in areas where the disease did not exist before irrigation (Figure 8.9).

Furthermore, irrigation water evaporates to some degree, leaving behind salts that were in the water. In some areas, including parts of the American west, this has left soils so salty that crops will no longer grow there. Soils can also become waterlogged if drainage

Figure 8.9 Irrigation project in Khuzestan, Iran. Canals such as this are good breeding spots for schistosomiasis, a disease spread by snails that live in slow-moving water. Children playing in the water are quickly infected. (Paul Hebert, 1975)

Figure 8.10 Salinization. In California's Imperial Valley, soil salinity levels on irrigated land pose a major problem to crop production. Areas of this cotton field show evidence of crop failure due to the high salt content in the soil. (USDA–SCS photo by Tim McCabe)

systems are not included in irrigation plans (Figure 8.10). Irrigation schemes may even have adverse social effects, such as causing overpopulation in newly irrigated areas.★

Some scientists have expressed concern that within the next hundred years, irrigation may become so widespread that the flow of major rivers into the sea may completely stop. The Nile and other rivers entering the Mediterranean are most likely to suffer that fate. The environmental effects of such an occurrence are unknown, but might include such things as decreased fishing, as the flow of nutrients from land into sea is stopped. Water as a scarce resource is discussed in Chapter 10.

The Supply of Arable Land

There is no shortage of land physically able to grow crops. In fact, the earth may have as much as twice again the amount of arable land as is now farmed. Some important qualifications must be made, however. In the

first place, most of the land that is easily cropped is already being farmed. Using the remaining land would require a great deal of energy and labor for such tasks as clearing forests, building irrigation systems, or transporting crops to distant markets. Furthermore, the land that is considered most fertile is generally already in cultivation. Some of the land not now being farmed is only marginally fertile. Thus, if new land is farmed, expected yields may not be as great as on present farmland. However, to offset this last problem, land can be improved by farming if the proper techniques and energy are available. Indian farmers cultivate almost the same number of acres as U.S. farmers but produce only 40% as much crops, partly because of a lack of fertilizers and machinery.

In many other countries, farmers keep their fields fertile by shifting cultivation every few years. When the land is allowed to lie fallow for a few years or a few decades, natural processes such as nitrogen fixation renew essential plant nutrients. In addition, populations of insect pests and weeds, which take a heavy toll of the crop, die down during the fallow period. These same ends can be achieved to a large extent if artificial fertilizers and pesticides are available. Apparently, even

---

★For more on this, see Worthington (1978).

tropical forest land can be farmed continuously if planters take care to prevent erosion and use suitable fertilizers or nitrogen-fixing crops.

In Africa, much of the fertile uncultivated land lies in areas infested with tsetse flies (which transmit trypanosomiasis, or sleeping sickness, to most breeds of cattle). Although the tsetse fly cannot yet be controlled over these large areas, conceivably it might be restrained in the future. Experts caution that clearing and planting or grazing this land would lead to severe erosion and desertification problems unless great care is taken.

Plans for clearing new lands for farming, however, put food production experts on a direct collision course with those concerned about preserving endangered plant and animal species. In South America, as well as in Africa, a large part of the land that could be farmed is now under forest cover and provides habitats for large numbers of animals and even larger numbers of plant species. The concept of preserving a wide diversity of species in the world comes into conflict with the vision of providing enough food for everyone. As human populations increase, so that more land is necessary for food production, other desirable uses, such as species protection, are edged out of the picture.

In the United States, many acres of fertile farmland fall victim to suburban development each year. This problem is examined in Chapter 29.

Fishery Resources

At one time humans looked to the oceans as a vast reservoir of food, waiting only for the technology to harvest and turn it into tasty dishes. However, reality, in the form of declining fish catches and even the complete collapse of some fisheries, has caught up with this notion.★ The seas hold a finite food resource. If we overexploit this resource, it appears that commercial enterprises will no longer be worthwhile and some species may be driven to extinction.

How much protein, then, could we harvest from the sea on a sustained basis—that is, year after year? Estimates vary from 60 million metric tons to 400 million tons, depending on the figures and assumptions

★The California sardine fishing industry collapsed in the 1950s, the North Sea herring industry in 1969, the West Africa/Namibia pilchard industry in 1970, and the Peruvian anchovy industry in 1972.

Table 8.2 Major Species Categories of World Marine Catch in 1976

Catch	Millions of metric tons
Diadromous fish (sturgeon, salmon, shad, etc.)	1.45
Marine fish	55.10
Crustaceans (lobster, shrimp, crab, etc.)	2.01
Molluscs (oysters, clams, squid, etc.)	3.05
Aquatic plants (brown, red, green seaweeds)	1.29

used. The current world catch of marine fish is 50–60 million metric tons per year (Table 8.2).

Increased yield estimates are sometimes based on the total biomass available in an area. This, however, includes species not now practical to catch (such as those too small) or use (such as those not accepted by consumers). Current fisheries usually exploit only one or a few species in an ecosystem. Increased yield estimates are also sometimes based on simple extrapolation of current trends, assuming that productivity limits would not apply for some time. This, however, ignores biological limits such as the population sizes needed for reproduction or interactions between populations in predator-prey systems, or the lack of a pollution-free environment.

In fact, the total world marine catch has remained relatively constant over the past 10 years, despite improved technology and increased fishing activity. This suggests that many currently harvested species are fully exploited and in some cases are decreasing in abundance. Rather than increasing the harvest, technology will be hard put to keep it from decreasing in the near future.

Although people have from time to time suggested harvesting other marine protein sources, such as plankton, current harvest and processing technology are lacking. Antarctic krill (Figure 9.9) might provide a significant harvest in the near future (perhaps 10 million metric tons per year). However, it is not known what effect harvesting krill or phytoplankton, which are very important parts of ocean food chains, could have on other fishery harvests. (See Chapter 9 for a comparison of productivity in ocean ecosystems and land ecosystems.)

The Green Revolution

High-Yield Grains

Perhaps the biggest hope technology has held out to farmers in developing nations is the promise of high-yield grains. New varieties of wheat and rice have been developed that far outproduce traditional varieties. Many new wheat and rice varieties are shorter in stature than older varieties and so channel more energy into seed production than into stem growth. Further, they often mature more quickly than older varieties, allowing a farmer to plant and harvest up to three crops a year. Using the new varieties, a farmer can grow more grain even without increasing the amount of land under cultivation. This development has become known as the **Green Revolution**.

In practice, however, difficulties remain. The new grains are very responsive to fertilizers—in fact, the best yields depend on the use of large amounts of fertilizer. In addition, the new varieties need more water than older varieties to reach their full potential, and in some cases they are more susceptible to disease and insects. All this means that a farmer must have the technical knowledge of how to use the new varieties and, often, the money to spend on necessary supplements such as fertilizers, pesticides, and irrigation mechanisms. Without these supplements, the new grain varieties may be no better and are sometimes worse than the older ones. Further, the intensive farming encouraged by the Green Revolution can cause serious erosion, salinization, and pesticide pollution unless farmers are educated to deal with such problems.

This is not to say the Green Revolution is without value. Even without optimum amounts of fertilizer, the new varieties often yield substantially more than older varieties. In a period of 6 years, from 1965 to 1971, farmers in the northwest area of India achieved spectacular increases in food grain production, in some cases averaging almost 10% more grain per year. A combination of high-yield grain, more fertilizer, and good rainfall led India to announce in 1971 that she soon expected to be self-sufficient in food. Although the lack of the monsoon rains in 1973 and 1974 caused a setback, India is now largely self-sufficient in cereal grains.

In short, however, the new varieties are best suited for certain environments where water, energy, edu-cation, and capital are not severely limited. Three-quarters of the rice and half of the wheat sown in developing nations still consists of locally adapted varieties.

Green Revolution II

New advances in agricultural production will need to take place in breeding crop varieties adapted to special local conditions—varieties that produce high yields on marginal lands. Agricultural researchers have developed new varieties that are tolerant of salt or aluminum, that can survive swamping during monsoons, or that have natural defenses against plant pests (Figure 8.11). Scientists hope that many of these advances will come from techniques such as genetic engineering or the use of chemicals to regulate plant growth. These relatively new kinds of biotechnology are sometimes labeled Green Revolution II.

Genetic engineering involves such things as the transfer of genes governing desirable traits between different species. Not long ago such feats were held to be still 10 years in the future. However, scientists have already been able to transfer genes responsible for traits such as resistance to certain herbicides to crop plants such as tomatoes, potatoes, and tobacco. Such resistance is important because it will enable farmers to use herbicides to kill weeds in a field without damaging the crop plants.

Commercially usable plants derived by gene-transfer methods are still a few years away, however. We may see results more quickly from techniques in which whole plants are regenerated from single cells in tissue cultures (Figure 8.12). Rapid screening of large numbers of plants for desirable traits is possible with tissue culture techniques.

Newly discovered plant-growth regulators may eventually be able to increase crop yields or, conversely, prevent weed growth. Other newly found chemicals seem to stimulate natural plant defenses against insect damage and so may eventually be used in reducing pesticide needs.

Fewer than 300 plants are commonly used by humans as food, and only about 20 plants account for most of the food consumed in the world. Of these, 8 (wheat, rice, corn, barley, oats, rye, sorghum, and millet) provide 60% of the world's Calories and half the world's protein. Other important crops include potatoes, soybeans, sugar cane, coconuts, and bananas.

Figure 8.11 Triticale—a new grain. On the right is triticale, a cross between wheat (left) and rye (middle). Triticale gives better yields than wheat, corn, or soybeans on marginal land, especially with acid soil. (USDA photo)

Leaf tissues are placed in enzyme solution to separate cells. Cell walls are removed, leaving "naked" protoplasts.

Protoplasts are exposed to selection pressure (e.g. high salt concentration or toxin).

Surviving protoplasts are treated to encourage cell division.

Protoplasts develop into calluses of undifferentiated cells. Calluses are cultured with hormones to promote cell differentiation.

New plant manifests desired trait.

Figure 8.12 Growing whole plants from single cells. Single cells are isolated from a plant and then treated so each one regenerates a whole new plant. In most cases, each cell will produce a plant exactly like the parent plant. (This is the process of cloning.) Orchids, Boston ferns, African violets, and oil palms are all propagated this way. When scientists use the method to develop new plant varieties, however, they look for those few cells that have some different characteristics than the parent plant. The number of these "different" cells can be increased by treating the cells with certain chemicals or ultraviolet light. Scientists can screen a group of cells to see whether any would develop into plants with useful characteristics, such as tolerance to high salt or high aluminum concentrations. Those single cells that survive and reproduce in such conditions in the laboratory are likely candidates for growing into whole plants that are tolerant to these conditions. (Adapted from Cooke, R., "Engineering a New Agriculture," *Technology Review*, May/June 1982, p. 24, drawing by Dan Collins.)

One unfortunate result of Green Revolution I was the virtual disappearance of local varieties of these major crop plants in some areas, when all the farmers switched to new hybrid varieties. As Green Revolution II gets underway, we can see even more clearly the need to preserve as many different species of plants as possible. For example, some of the traits that will be used to produce improved crop varieties by genetic engineering will probably be found in wild plants and locally adapted varieties of crop plants.

A number of countries have now established plant germ banks. These facilities store seeds or plant parts from local varieties of the important food crops or other possibly useful food plants. In this way, useful varieties won't be allowed to disappear as a result of most farmers' switching to one or a few "super" varieties of corn, rice, and wheat.

Social, Economic, and Political Aspects of Food Production

Technology offers farmers in developing countries a great deal of hope for increasing food production. Whether the farmers can take advantage of this help depends in large part on political and economic conditions. For instance, before small farmers in a developing country can take advantage of many kinds of improved technology, credit must be available so they can purchase seeds, pesticides, and other equipment. Large banks are usually wary of lending money to small farmers, however. Some of these farmers may turn to local moneylenders, friends, or relatives. The energy crunch is apparently making the credit problem worse, since prices for fertilizers, fuel, and related goods are rising. In the long run, price increases will be a curb on the success of the Green Revolution.

In the long run, too, land reform may be necessary in countries where a few wealthy landowners hold most of the political power. Such landowners are able to obtain new technological information for themselves fairly easily and see no need to make such information available to the small farmer. Yet the small farmer must have help, especially education in the technology of using high-yield grains, if the developing countries are to significantly increase agricultural production.

Another political restraint on agricultural production in the past has been that politicians have often tried to hold down food prices. This is often done to keep living costs down so that the large numbers of city dwellers will continue to vote for the politicians in power. The effect of this policy is to remove any incentive for farmers to grow more food than their family needs. Market prices simply remain too low to justify the work and investment involved in more crops. As an extreme example, a careful study in Mali showed that it cost a farmer 83 Malian francs to produce a kilo of rice, but the government paid the farmers only 60 francs per kilo. As a result, farmers smuggled rice across the borders into countries where they could get 108–128 francs per kilo for it.

In the Sahelian states (Mali, Niger, Upper Volta) it has been a practice to assure a job to any high school graduate. Such a policy has led to vast increases in the number of civil servants, whose salaries can be kept low by keeping food prices at low levels. This has acted to slow food production because the farmers see no reason to grow more crops at current prices.

The management of irrigation schemes, too, can have important effects on crop production. Managers are often political appointees, living in the cities, too far away from the scene to know how to manage the water resource effectively. Public water managers have tended to focus on avoiding disagreements about the proper distribution of water, rather than making hard decisions on what is really the most efficient way to use the available supplies. Furthermore, the water itself is usually undervalued. This is shown by the success of private water distribution schemes in countries such as India, even when they compete directly with public water supplies. That is, farmers are willing to pay higher prices for additional water from private sources even when some public water is available.

A Pessimist Looks at the Future

Population Growth May Outstrip Food Supply

As the history of the world's food supply shows, predictions about the ability of agriculture to feed the world are usually colored by the current stock of food on hand. When harvests are bad and grain surpluses fall, the pessimistic view prevails. Experts make calculations showing that we simply cannot boost agricultural production to a rate significantly higher than the population growth rate. When harvests are good and surpluses accumulate, the optimistic point of view catches hold and experts predict that a good diet for everyone is just around the corner.

Historically, pessimists can point to the fact that food production has just barely kept ahead of population growth. With so many of the world's people already undernourished, keeping up is not good enough. Population growth in developing countries continues, although there are signs that growth may be leveling off in some.

Practical Limits to a Green Revolution

Pessimists also warn us that high-yield grains themselves are susceptible to disease and insects. As these varieties become more common, the stage may be set for a crop failure similar to the potato blight that devastated seventeenth-century Ireland. A great variety of seed types in some measure protects us from widespread crop failure due to disease or insect pests. In semiarid regions, intensive agriculture without safeguards against erosion and the increasing use of irrigation that leaves salty residues in soil bring areas closer to desertification. This state, which has been loosely defined as human-induced barrenness of land, is often blamed on drought or the "march" of the desert to engulf nearby semiarid land. But desertification is, in fact, a result of human stresses. Natural events such as droughts are only the final straw. Unless great care is taken to prevent overgrazing, erosion, salinization, and compaction of the soil, the resultant desertlike area will no longer grow enough food for the human population dependent on it (see Global Perspective II).

In terms of increasing the actual supply of food, the Green Revolution is held back by the high price for energy, which prevents developing nations from obtaining the fertilizer and fuel necessary to grow high-yield grain varieties successfully. Even when farm machinery and the fuel to run it are available, problems will arise if people in rural areas are put out of work by machines (Figure 8.13).

In socioeconomic terms, as incomes rise in some countries and people can begin to afford meat, grain is diverted to feed meat-producing animals, and food purchasing power shifts further away from poorer countries. One result of the energy crisis has been to separate the developing countries into two groups: the wealthy OPEC countries, which are the oil exporters, and all the others, which have no oil. The oil-exporting countries are able to buy the food needed to upgrade the diet of their people. For this reason, they stand in a

Figure 8.13 A farmer in India irrigating his field. Agriculture in developing countries uses mostly human power and few machines. If farms are mechanized to increase agricultural production, will people be put out of work? Some experts think not. They point out that labor demands may actually increase when high-yield grains and machinery such as tractors, threshers, and harvesters are used because they enable farmers to farm more intensively, planting several crops a year. (William Borders/NYT Pictures)

somewhat different position with respect to possible food shortages than the other developing countries.

An Optimist Looks at the Future

What Technology Promises

Optimists support their point of view by noting that in most undeveloped countries the Green Revolution has barely started. New grain varieties and attention to crops such as the root vegetables, on which relatively little research has been done, promise great increases in yields. Both traditional plant-breeding techniques and the new biotechnology methods of Green Revolution II promise better grain varieties for even the people who must farm on land that is less than ideal. For instance, a new variety of sorghum has just been introduced in Africa that promises five times the yield of traditional varieties even without irrigation.

The United States itself could at least double agricultural production if the returns were great enough—and here lies a major if. If population growth can be slowed, if incomes can rise, then the historic problem of poor nourishment can be solved. In the simplest terms, people are undernourished because they are poor. With jobs and reasonable incomes, they could buy food, and, according to most experts, the world's capacity for producing food could probably meet this demand (Figure 8.14).

The paradox is that developing countries must find a way to provide more jobs in rural areas so that people there will not migrate to the cities (where there are no jobs either), while modernization of agriculture may reduce the number of people required to run farms. At the same time, agricultural and other development schemes are threatening to destroy the social and economic fabric on which present incomes depend.

One example of the problem of modernization is the Philippine government proposal to build four dams in the central mountains of Luzon. The dams are intended to provide hydroelectric power for a variety of Philippine development schemes. They would also flood as many as a dozen Igorot villages, submerging thousands of hectares of laboriously built rice terraces and forcing thousands of these mountain people to move—to cities where jobs are scarce, to a government resettlement area where they would have to begin building all over again, or to other Igorot villages

where overcrowding could destroy the carefully worked out social fabric.

Appropriate Technology

Rather than simply promoting a wholesale transfer of modern agricultural technology, many experts now recommend a careful study of what people in developing nations need and can use in the way of technology. This idea has come to be known as "appropriate technology." For instance, farm machines do not have to be labor-saving to be useful. In many developing countries, farmers must wait for the rainy season before plowing. Machines to help farmers plow the hard caked earth allow them to have their crops planted in time to take advantage of the monsoon rains.

Amulya Reddy, of the Indian Institute of Science, points out that technology in developing nations must meet several criteria. It must focus on what the people themselves need most; it must be environmentally sound; and most important, it must encourage self-reliance for those who use it. In terms of food production, this means that studies are needed to determine the actual practices and needs of farmers in developing nations. The technology then developed is more likely to be adopted and is more likely to improve production without causing severe environmental and social problems (Figure 8.15).

Reddy warns against trying to develop gadgets and devices for Indian villages while sitting in a lab in Cambridge, Massachusetts. They will never work, because one has to understand the whole social ecology as well as the physical circumstances of a village to know what people need and will use. Besides, the locals have to be involved every step of the way. "This is considered the obvious thing to do for the urban architect," notes Reddy, but somehow with poor people the idea of investigating their habits and preferences and doing test marketing falls by the wayside. "Then they don't like what we have done and we say they are stupid." (Holden, 1980)

There are really, then, two categories of food problems. First, can we actually produce enough food to provide an adequate diet for everyone? This question is one of technology, with certain added social and political features. The answer, for the near future, is prob-

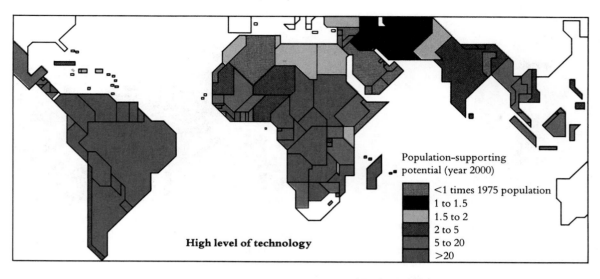

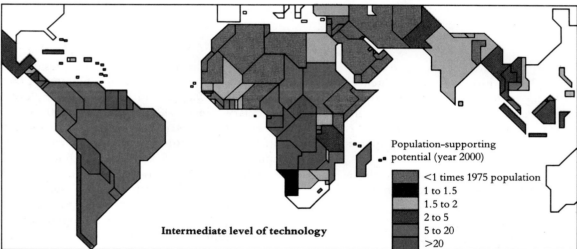

Figure 8.14 Food self-sufficiency. If technological advances and social changes were adopted, many developing nations would be capable of growing enough food to feed their entire population by the year 2000. These maps show two possible scenarios: the situation if an intermediate level of technology is achieved, and the situation if a high level of technology is achieved. At the intermediate level, a reasonable goal for most of the world, the only countries that could not grow enough food for their own needs would be 12 countries in Africa, 7 in Central America, 2 in Southeast Asia, and all the countries in Southwest Asia. If high tech agriculture could be adopted, only 4 countries in Africa (Western Sahara, Cape Verde, Ruwanda, and Mauritius) would not be able to feed themselves, along with 2 in Central America (Barbados and the Netherlands Antilles), 2 in Southeast Asia (Bangladesh and Singapore), and all except Iran, Iraq, and Syria in Southwest Asia. Some of the Southwest Asian countries are oil rich and along with Singapore can easily buy what food they need. (From M. Shah and G. Fischer, "Potential Population-Supporting Capacities of Lands in the Developing World," FAO/IIASA/UNFPA Technical Report of the Land Resources for Populations of the Future Project, Rome, Italy, 1984.)

Figure 8.15 In developing nations, new, "appropriate technology" solutions to farming problems hold out the hope of increasing yields without displacing workers. This human-powered tractor is used in Philippine rice paddies, along with an easily built and repaired rice thresher designed to thresh Philippine rice. (Courtesy of J. K. Campbell, Cornell University, Department of Agricultural Engineering)

ably yes. Second, can we ensure that everyone actually receives enough food? The problems in this category may be more difficult to solve. Based on social inequalities and economic variables, these problems and their solutions penetrate to the very core of whether humans, across the world, will care and take responsibility for each other (see Controversies 8.1 and 8.2).

One way that has been suggested for the short-term relief of famine and long-term relief from rising food prices is the establishment of grain reserves.

Are Grain Reserves a Solution?

Farm Economics and Political Solutions

. . . Let Pharaoh proceed to appoint overseers over the land, and take the fifth part of the produce of the land of Egypt during the seven plenteous years. . . . That food shall be a reserve for the land against the seven years of famine which are to befall the land of Egypt, so that the land may not perish through the famine. (Genesis 41:34, 36)

So spoke Joseph to the Pharaoh of Egypt some 3,600 years ago. In ancient China, too, we are told, the Confucians created a "constantly normal granary." In the modern era, Henry Wallace, Secretary of Agriculture

for President Franklin D. Roosevelt and presidential aspirant in the 1940s, was the leading advocate of grain reserves. Though Wallace had been urging an "ever-normal granary" since 1912, not until the 1930s, when farms were being destroyed by dust storms and drought, was he able to create a government agency that would store grain. Even then, the publicly announced objective behind government purchasing of grain stock was to stabilize farm incomes. The stabilizing of farm incomes was a goal for two reasons. First, the political power of farmers made a program of insured farm incomes important. Second, preventing farmers from going out of business because of losses in bad years did help make the supply of food more secure. Wallace's program used crop insurance, paid for by a contribution of a portion of the crop itself; limited direct payments; acreage allotments; and price-support loans that, if defaulted on, did not require repayment. Although farm incomes stayed up, output grew and the stocks of grain grew, bringing criticism of the program.

Criticism ended, however, when World War II broke out: The stocks of wheat, cotton, and corn became a vital military resource, as the United States undertook to aid its allies with badly needed contributions of food. By the close of the period of reconstruction after the war, all current U.S. farm production was

(continued on page 196)

CONTROVERSY 8.1

The Lifeboat Ethic

For posterity's sake we should never send food to any population that is beyond the carrying capacity of its land.
Garrett Hardin

This obscene doctrine . . .
Roger Revelle

It is just as obscene to let people die in the future as it is to let them die now.
J. D. Martin, Sociologist, Lakehead University, Ontario, Canada

Garrett Hardin has proposed an analogy, related to population growth and the supply of food in the world, that is known as the "lifeboat ethic." In this analogy, so many people crowd onto a lifeboat that it sinks and all are lost. If fewer people had been in the boat, the argument goes, those few might have reached shore safely.

Roger Revelle stated in an editorial in *Science* (1974):

> The specter, unseen by some and ignored by others, looming over the World Food Conference this week in Rome is the continuing rapid population growth of the world's poor countries. Some scientists and publicists have seriously advocated a "lifeboat ethic," saying that nations which do not *compel* human fertility control (by what means is never stated) are endangering the survival of our species—hence they should be starved out of the human race by denying them food aid. This obscene doctrine assumes that men and women will not voluntarily limit their own fertility when they have good reasons and the knowledge and means to do so.

> The sharp decline in birth rates during the past decade in a dozen developing countries belies the assumption. But one thing is clear from this experience: environmental changes can bring down birth rates only if they affect the people who have the children—the great mass of the poor who now have little hope for a better life.

Many people sent letters to the editor in reply to this editorial. J. D. Martin, a member of the sociology department at a Canadian university, wrote:

> The idea of letting people die, Revelle says, is "obscene." Well, if so, it is just as obscene to let people die in the future as it is to let them do so now.

> . . . non-Western populations will keep growing until we can't feed them, even at great cost to the quality of our soils.

> We Westerners brought it on ourselves, by saving lives through medical skill and humanitarian generosity. . . . The millions of lives saved by our medical help became the hundreds of millions of lives that are due to be lost in famines.

. . . Shall we impoverish the West in order to make the problem even worse, and in the process weaken both our land and theirs?

I think not; this is the essence of the "lifeboat ethic" which Revelle criticizes. Let too many people into a lifeboat and all will sink. The same may be true of our spaceship called Earth.

F. A. Cotton, a member of the chemistry department at Texas A & M University, echoed Martin's position:

Nobody can look without horror on the prospect, let alone the actual spectacle, of fellow human beings starving to death. It is a monstrous thing, but we live in an age of monstrosities—some still latent but imminent, unless actively forestalled—and it is literally necessary to consider not only relative degrees of monstrousness, but the fact that some monstrosities are qualitatively more ghastly than others. Overpopulation and starvation are interdependent monstrosities, but of a qualitatively different nature. I believe that the former is far more dire than the latter.

If a quarter of the people in the world starved to death next year, the human condition, in the larger sense, would not be basically or permanently changed. After a few generations, this calamity would leave no basic imprint on our collective consciousness, any more than did the deaths of one-fourth of the people in Europe in the great plague of the 14th century.

However, if the population of the world goes on increasing at the present rate for much longer, the human condition will be basically and catastrophically altered, in an irreversible way.

What do you think of Martin's argument that "we Westerners brought it [a food crisis] on ourselves" by providing medical aid and knowledge to developing countries? Should we have denied them this aid in the first place? Do you agree with Cotton that worldwide starvation can be equated with the deaths from plague in the fourteenth century? Do you agree with Revelle that the lifeboat ethic is an obscene one, or do you feel that the horrors of overpopulation call for such a drastic response?

After you have thought about these questions, you might want to read *The New York Times* editorial by Barbara Ward, reprinted in Controversy 8.2. Another way of looking at this issue is to ask: Is the nutrition problem really a population problem, as Revelle's critics contend, or is the population problem, at least in part, a nutrition problem? Could better nutrition help people in undeveloped countries solve population problems themselves? Beverly Winikoff in a 1978 *Science* article (see References), examined some of the ways these questions can be answered.

Sources: Garrett Hardin, *Atlantic Monthly*, **247** (May 1981), 60; Roger Revelle, Editorial, *Science*, **186** (November 15, 1974), 589; J. D. Martin, Letter to Editor, *Science*, **187** (March 21, 1975), 1029; F. A. Cotton, Letter to Editor, *Science*, **187** (March 21, 1975), 1030.

CONTROVERSY 8.2

Not Triage, But Investment in People

Barbara Ward, an economist and writer, wrote the following editorial in *The New York Times* (November 15, 1976):

Now that the House of Representatives has bravely passed its resolution on "the right to food"—the basic human right without which, indeed, all other rights are meaningless—it is perhaps a good moment to try to clear up one or two points of confusion that appear to have been troubling the American mind on the question of food supplies, hunger, and America's moral obligation, particularly to those who are not America's own citizens.

The United States, with Canada and marginal help from Australia, are the only producers of surplus grain. It follows that if any part of the world comes up short or approaches starvation, there is at present only one remedy and it is in Americans' hands. Either they do the emergency feeding or people starve.

It is a heavy moral responsibility. Is it one that has to be accepted?

This is where the moral confusions begin. A strong school of thought argues that it is the flood tide of babies, irresponsibly produced in Asia, Africa and Latin America, that is creating the certainty of malnutrition and risk of famine. If these countries insist on having babies, they must feed them themselves. If hard times set in, food aid from North America—if any—must go strictly to those who can prove they are reducing the baby flood. Otherwise, the responsible suffer. The poor go on increasing.

This is a distinctly Victorian replay of Malthus. He first suggested that population would go on rising to absorb all available supplies and that the poor must be left to starve if they would be incontinent. The British Poor Law was based on this principal. It has now been given a new descriptive analogy in America. The planet is compared to a battlefield. There are not enough medical skills and supplies to go round. So what must the doctors do? Obviously, concentrate on those who can hope to recover. The rest must die. This is the meaning of "triage."

Abandon the unsavable and by so doing concentrate the supplies—in the battlefield, medical skills; in the world at large, surplus food—on those who still have a chance to survive.

It is a very simple argument. It has been persuasively supported by noted business leaders, trade-unionists, academics and presumed Presidential advisers. But "triage" is, in fact, so shot through with half truths as to be almost a lie, and so irrelevant to real world issues as to be not much more than an aberration.

Take the half truths first. In the last ten years, at least one-third of the increased world demand for food has come from North Americans, Europeans, and Russians eating steadily more high-protein food. Grain is fed to animals and poultry, and eaten as steak and eggs.

In real energy terms, this is about five times more wasteful than eating grain itself. The result is an average American diet of nearly 2,000 pounds of grain a year—and epidemics of cardiac trouble—and 400 pounds for the average Indian.

It follows that for those worrying about available supplies on the "battlefield," one American equals five Indians in the claims on basic food. And this figure masks the fact that much of the North American eating—and drinking—is pure waste. For instance, the American Medical Association would like to see meat-eating cut by a third to produce a healthier nation.

The second distortion is to suggest that direct food aid is what the world is chiefly seeking from the United States. True, if there were a failed monsoon and the normal Soviet agricultural muddle next year, the need for an actual transfer of grain would have to be faced.

That is why the world food plan, worked out at Secretary of State Henry A. Kissinger's earlier prompting, asks for a modest reserve of grain to be set aside—on the old biblical plan of Joseph's "fat years" being used to prepare for the "lean."

But no conceivable American surplus could deal with the third world's food needs of the 1980's and 1990's. They can be met only by a sustained advance in food production where productivity is still so low that quadrupling and quintupling of crops is possible, provided investments begin now.

A recent Japanese study has shown that rice responds with copybook reliability to higher irrigation and improved seed. This is why the same world food plan is stressing a steady capital input of $30 billion a year in third-world farms, with perhaps $5 billion contributed by the old rich and the "oil" rich.

(What irony that this figure is barely a third of what West Germany has to spend each year to offset the health effects of overeating and over-drinking.)

To exclaim and complain about the impossibility of giving away enough American surplus grain (which could not be rice anyway), when the real issue is a sustained effort by all the nations in long-term agricultural investment, simply takes the citizens' minds off the real issue—where they can be of certain assistance—and impresses on them a nonissue that confuses them and helps nobody else.

Happily, the House's food resolution puts long-term international investment in food production firmly back into the center of the picture.

And this investment in the long run is the true answer to the stabilizing of family size . . . the whole experience of the last century is that if parents are given responsibility, enough food and safe water, they have the sense to see they do not need

endless children as insurance against calamity . . .

Go to the root of the matter—investment in people, in food, in water—and the Malthus myth will fade in the third world as it has done already in many parts of it and entirely in the so-called first and second worlds.

It may be that this positive strategy of stabilizing population by sustained, skilled and well-directed investment in food production and in clean water suggests less drama than the hair-raising images of inexorably rising tides of children eating like locusts the core out of the whole world's food supplies.

But perhaps we should be wise to prefer relevance to drama. In "triage," there is, after all, a suggestion of the battlefield. If this is how we see the world, are we absolutely certain who deserves to win—the minority guzzlers who eat 2,000 pounds of grain or the majority of despairing men of hunger who eat 400 pounds? . . .

Is this the battlefield we want? And who will "triage" whom?

being consumed and the stocks were gone. Stocks began to build again as farm productivity grew to new levels and government purchases at price-support levels continued.

In 1954, Congress passed Public Law 83-480, a temporary measure now become permanent, which aimed at reducing the growing stocks in a way that benefitted both poorer nations and American foreign policy. Under this law, the government was allowed to sell some of its surplus stocks to nations needing grain. Emergency relief could also call forth shipments from the overflowing U.S. cupboard. Famines and natural disasters were among the situations to which the United States could respond with food aid. Voluntary relief organizations and governments were the recipients who then distributed the food. Under 83-480, the U.S. participates in the World Food Program. We have been a consistent and generous supporter of this program, a UN-FAO initiative that supplies food aid for emergency situations and to nations carrying out development projects.

To restrain grain production further in order to dampen the buildup of stocks, and at the same time to

protect and even out the income of U.S. farmers, Congress has tried a variety of plans—see Box 8.1.

Farmers and Grain Reserves

Farm interests argue against a national grain reserve on the basis that for the past three decades the United States has consistently been able to export grain and still meet domestic demands. The argument that our national grain demand can surely be met is only partly true. In the presence of heavy demand for export, grain prices rise and U.S. consumers have to pay the higher price. Further, the same high grain prices that annoy the affluent citizens of this country are a deadly blow to individuals in developing nations.

In nations where perhaps 60% of the average family's budget is spent on food, a price rise for wheat or rice can make the difference between poor nutrition and malnutrition. The need for a grain reserve becomes more important as a means to dampen price increases for people in the developing nations, not simply as a source of emergency grain.

It is at times of shortage and high prices that a grain

Box 8.1 Farm Programs

The federal government attempts to stabilize farm prices and provide a floor under farmers' earnings on wheat, corn, rice, and cotton with a number of programs. Dairy farmers are covered by a separate program.

Price support loans: Loans made at Treasury interest rates with a farmer's crop put up as collateral. The per bushel loan rate is set by the Agriculture Secretary. These loans, plus interest, may be repaid after nine months. Or they may be forfeited, without interest, by transfering the collateral crops to the Agriculture Department, which may not sell them until the market rises to 105% of the loan rate.

Subsidies: Called "deficiency payments." The per bushel rate at which these are paid is the difference between the "target price" set by Congress and the average free market price in the first five months after harvest. An eligible farmer receives these automatically, whether he sells his crop, takes a loan on it, or stores it.

Reserve: Eligible farmers may designate part of their crops for the reserve, holding them on their own farms or country elevators. They receive a loan, at the loan rate, when they do. When the market reaches a "trigger price" set by Congress, these reserves may be sold and the loan, plus interest, paid off. When the

market reaches a "release price" the farmer must pay off the loan and is no longer given storage payments if he holds reserves.

Paid diversion: When the Secretary offers a "diversion" a farmer who idles a percentage of the land normally planted in these crops is paid in cash, based on the average per acre yield on the land he diverted. He may not plant another controlled crop on that land.

Payment-in-kind: Farmers who leave idle additional land are given crops from government stores in proportion to the average yield from the land idled.

Milk price supports: The support level for 100 pounds of milk is set by Congress. Any amount of milk that is not bottled as fluid milk is "manufactured" into butter, cheese or dried milk. Any of these dairy products not sold by the processors is bought by the Agriculture Department at the price support level. The department stores these products. It may not sell them in the open market but may use them in government food programs or for charity, both abroad and at home. They may also be sold in foreign markets at any price.

Source: S. King, *The New York Times*, August 8, 1983.

reserve is most needed and that farmers argue most strongly against it. The reason is that a grain reserve would dampen prices and profits in times when production levels are down in relation to demand. At present, farm interests are largely unwilling to forgo profits at such times, even in return for protection of profits when production relative to demand is high. The protection is discounted and dismissed because farmers see the near future as a time when all their products will continue to be absorbed on the world market.

When production exceeds demand, causing prices to fall, farmers want the government to buy up surplus so that prices fall no further. Thus, the ideal time, from a political point of view, for discussion of building a grain reserve is when farm prices are low due to unusually high levels of production.

In the United States, government purchases of grain for a federally held reserve will serve to protect

farmers' income in times of production excesses. What would happen if developing nations were to build grain reserves from their own harvests? Such reserves could make a very big difference in the nutritional levels of the population and could also aid the national economy of a developing country by eliminating the need for food imports, thus preventing a deficit in the balance of trade. The World Bank, through the use of loans, and the FAO are encouraging individual nations to build grain reserves, but the costs of purchase and storage are considerable.

An International Grain Reserve

Another possibility is less costly and more efficient but more difficult to achieve politically. An international grain reserve, not belonging to a particular country, could be created. Proposals for such a reserve were made in the mid-1970s. In the early 1980s, a similar international bank was started for commodities such as tin, rubber, and cocoa. In this case, the intention was to protect less developed countries, whose economies often depend on only one crop or mineral, from severe swings of the marketplace.

Resistance to an international grain reserve is great, but the benefits are large. Consider how an international agency operating a grain reserve might function. Two prices might be used to trigger the operation of the reserve. A price for grain falling below a preset low price would signal the agency to buy grain. The low-price signal to buy serves two functions. First, low prices are the time to set grain aside; holding or investment costs for the agency are kept low in this way. Low prices also mean an abundant harvest relative to demand, indicating that now is the time to buy and store grain without driving the price up significantly. Second, purchase of the excess harvest acts as a price support for grain. Farmers would find that a good portion of their excess harvest is thus absorbed without the price falling still lower. Thus, the presence of a grain reserve helps to ensure the income and prosperity of farmers in times when too much good weather works against them.

In similar fashion, when the price for grain rises above some predetermined high limit, the agency would offer its grain on the international market. The grain would be available for purchase by those poor nations for whom higher prices are a burden. The price at which the grain is sold essentially puts a lid on the international price of grain: Who would pay more than the high-signal price if grain can be purchased at that price from the international agency? True, farm profits are dampened when the agency decides to sell its grain, but the diminished profits are not the normal profits that keep a farm in business but the profits built of hardship and malnutrition for the poorest. The farmer selling grain at the grain reserve price still makes a sturdy profit.

Joseph seems to have had quite an idea.

Summary

Periodic famines, such as the mid-1980s famine in Africa, occur against a background of chronic malnutrition for over 4 million people in the world. Children suffer from kwashiorkor and marasmus, diseases caused by too little protein or too few Calories. Both are preventable by breastfeeding.

The specter of a food crisis has periodically been raised throughout history, but experts believe that the issue is really one of fair distribution rather than a shortfall in total amount of food needed. Transportation of food to areas of need, reduction of losses due to pests, a damping of food price fluctuations caused by bad weather, and changing income distribution are the main problems to be solved in order to achieve a fair distribution of food.

Issues involved in growing more food include energy requirements for fertilizers, pesticides, and machinery (a problem in poor developing nations short on energy resources); irrigation problems such as salinization, water-borne diseases, and water management; and cultivation of suboptimal land.

The Green Revolution promises great increases in food production due to high-yielding varieties of rice and wheat. But in order for the Green Revolution to work, there must be energy, water, and management inputs that may be unavailable in developing nations. Also of concern is loss of diversity in crop plants due to widespread adoption of new hybrid varieties.

The Green Revolution II promises even greater benefits from genetic engineering of plants to withstand a great variety of suboptimal farming conditions such as drought or high salt. Seed banks are important for preserving the genetic diversity on which the Green Revolution II depends. Economic and social changes are needed in many developing countries if the full benefits of agricultural technology are to be realized. Appropriate technology, or fitting technology into people's traditional way of life, is also necessary for full acceptance of the new crops and management practices. Grain reserves are a possible solution to fluctuating grain supplies and the accompanying price rises.

Questions

1. If people die from too little food every day in the poorer countries, how can we say people rarely die of starvation?
2. What special effects does poor nutrition have on children?
3. What are the Green Revolution and Green Revolution II? What problems are tied to the Green Revolution?
4. Explain how economic and political problems can be as important as introducing new technologies in terms of food production in developing countries.
5. How would grain reserves help even out fluctuations in food prices?

Further Reading

Franke, R. W., et al. *Seeds of Famine*. Montclair, N.J.: Allanheld, Osmun, 1980.

Food and population problems in Africa are perhaps the most severe in the world. Twenty-two of the poorest countries in the world are African. Population growth continues at high rates while food production languishes. Africa possesses almost half of the idle farmable land in the world, yet the prevalence of sleeping sickness and other diseases currently limit its use. This publication details why the Green Revolution seems to have bypassed Africa and what social and economic barriers appear to prevent progress in food production. It also gives some history of the region and the colonial influences that have contributed to Africa's problems.

Cowen, R. "Feeding the World's Hungry in the 1980s and '90s," *National Academy of Sciences News Report*, **36** (2) (February 1986), 4.

Scrimshaw, N. "The Politics of Starvation," *Technology Review* (August/September 1984), p. 18.

The effects of government economic and social policy on food supply in developing nations are well described in this paper.

"Harvesting the Sea," *Oceanus*, **22** (1) (Spring 1979).

This entire issue is devoted to fishery resources and food from the sea.

Intercom: The International Population News Magazine. Population Reference Bureau, Inc., 1337 Connecticut Avenue, N.W., Washington, D.C. 20036.

This publication carries a capsule on the world food situation each month, prepared by FAO. Good for up-to-the-minute information.

Popkin, B., et al. "Breast-Feeding Patterns in Low-Income Countries," *Science*, **218** (December 10, 1982), 1088.

The authors discuss the interrelationship between the food and population problems, as well as advantages and social aspects of breastfeeding in developed and developing countries.

Stutz, Bruce. "Catch as Catch Can," *Technology Review* (May/June 1984), p. 68.

The plight of U.S. fishermen, trying to get a greater share of a scarce resource, brings home the finite nature of ocean food supplies.

References

Poleman, T. J. "Quantifying the Nutrition Situation in Developing Countries," *Food Research Institute Studies*, **16** (1) (1981), 1.

Drucker, C. B. "The Price of Progress in the Philippines," *Sierra* (October–November 1978), p. 22.

Worthington, E. B. "The Greening of the Desert: What Cost to Farmers," *Civil Engineering—ASCE* (August 1978), p. 60.

Impacts of Technology on U.S. Cropland and Rangeland Productivity, U.S. Congress Office of Technology Assessment, August 1982.

Revelle, R. Editorial, *Science*, **186** (November 15, 1974), 589.

Winikoff, B., "Nutrition, Population and Health: Some Implications for Policy," *Science*, **200** (May 26, 1978), 895.

Ward, G. M., et al. "Animals as an Energy Source in Third World Agriculture," *Science*, **208** (May 9, 1980), 570.

Holden, C. "Pioneering Rural Technology in India," *Science*, **207** (January 11, 1980), 159.

Barton, K. and W. Brill, "Prospects in Plant Genetic Engineering," *Science*, **219** (February 11, 1983), 671.

Chaleff, R. S. "Isolation of Agronomically Useful Nutrients from Plant Cell Cultures," *Science*, **219** (February 11, 1983), 676.

Marn, J. L. "Plant Gene Transfer Becomes a Fertile Field," *Science*, **230** (December 6, 1985), 1148.

A GLOBAL PERSPECTIVE II

Desertification

After Man, The Desert
Drought / Causes of Desertification

Halting Desertification
Agricultural and Grazing Restrictions / Agroforestry / Need for Appropriate Technology / Comprehensive Plans

I stood on the Great Wall of China high on a hill near the border of Mongolia. . . . The slope below the Great Wall was cut with gullies, some of which were 50 feet deep. As far as the eye could see were gullies—a gashed and gutted countryside.

. . . The whole valley, once good farmland, had become a desert of sand and gravel, alternately wet and dry, always fruitless. . . . Its sole harvest now is dust, picked up by the bitter winds of winter that rip across its dry surface in this land of rainy summers and dry winters. . . .

The farmers of a past generation had cleared the forest. They had plowed the sloping land and dotted it with hamlets. . . . Each village was marked by columns of smoke rising from fires that cooked the simple fare of those sons of Genghis Kahn. Year by year the rain has washed away the soil. Now the plow comes not—only the shepherd is here with his sheep and goats, nibblers of last vestiges. These 4-footed vultures pick the bones of dead cultures in all continents. Will they do it to ours? The hamlets in my valley below the Great Wall are shriveled or gone. Only gullies remain—a wide and sickening expanse of gullies, more sickening to look upon than the ruins of fire. You can rebuild after a fire.

J. R. Smith's (1980) graphic description of a human-made desert could be matched in many parts of the world, including the United States (Figures II.1 and II.2). The Sonoran Desert in Arizona was lush grassland before overgrazing turned it into a scrub desert. In Syria, once fertile Roman farms are now bare rock. And, today in Ethiopia, as well as in the sub-Saharan countries, the worst drought in modern history has speeded up this process of desertification.

The result of the drought in countries bordering the Sahara, where there is little or no extra food to tide people over and no money to buy food from somewhere else, is famine. As many as 6 million people are starving in Ethiopia alone. International relief agencies are trying desperately to bring food to the stricken countries (Figure II.3).

In addition to relieving human suffering caused by the drought, however, we must give thought to the deeper and longer-term problem of desertification. Otherwise, these fragile, semiarid lands may be pushed into a condition where they are incapable of supporting human life, even when the rains come again.

After Man, The Desert*

Drought

Drought is by and large a natural crisis, one that occurs periodically in many parts of the world. It is a frequent visitor to the lands on the southern borders of the Sa-

---

*Old World saying, reported by R. J. Smith.

Figure II.1 Farm in Dust Bowl, Oklahoma, 1936. Severe drought in the 1930s combined with farming practices reduced much of the Midwest to desertlike conditions. (A. Rothstein)

hara. This huge desert, the largest in the world, runs from east to west across northern Africa. Geologic records indicate that similar droughts have occurred in this area at least six times since 1400 and may occur as often as three times in a century. If climatic history is a guide, this drought, too, will end and others will come. In this region climatic variability is the norm.

Causes of Desertification

Like the total rainfall, and to some extent in synchrony with it, the toll of human life and suffering has waxed and waned in the region's past. As the populations continue to grow, however, humans begin to have an environmental effect that adds to and intensifies the effects of natural drought. This interaction of drought

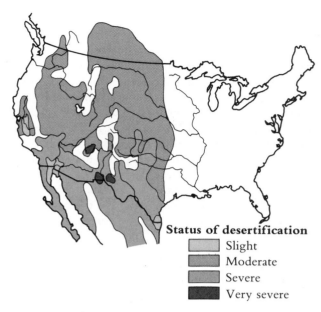

Figure II.2 Status of desertification in hot arid regions of the United States.

Status of desertification

- ☐ Slight
- ☐ Moderate
- ☐ Severe
- ☐ Very severe

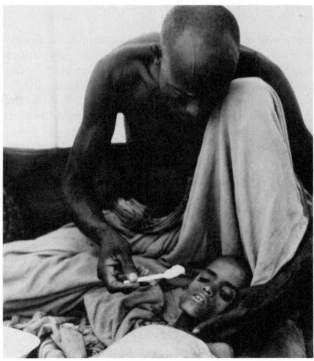

and human activities can lead to a condition known as *desertification*—a degradation of the land to the point where it will no longer sustain growth of plants or animals.

Desertification is not a problem limited to lands directly adjoining deserts or to Africa. Lands in the American southwest are undergoing desertification today, as are parts of Bolivia, Australia, and Brazil.

Overgrazing. One of the major causes of desertification is stocking of rangeland with more cattle than the range can support. The animals strip the land of vegetation, leaving it bare to wind and water erosion.

In Africa, raising cattle (as opposed to native species) entails another problem: cattle are not as resistant to drought. They require four times as much water as camels. Unlike desert animals, which respond to higher temperatures by decreased urination of a higher concentration, cattle respond by increasing volume and decreasing concentration. Apparently they cool their bodies through increased water consumption. During drought they are thus more vulnerable than native species would be.

Simplification of ecosystems. In the Sahel region, herders burn bushland in the belief that the new growth

Figure II.3 Children are often the victims of drought and famine. (Top) Small children play in the dust at a refugee camp in Jijiga, Ethiopia. (U.N. photo by John Isaac) (Bottom) A father who had lost both his wife and daughter during their 100 km journey to this camp in Bati, Ethiopia feeds his dying son. (U.N. photo by O. Monsen)

after a fire will provide tender green forage for wildlife and cattle, thus improving both hunting and grazing lands. If done regularly, however, this burning results in normal vegetation being replaced by grasslands, with few shrubs and trees, mostly pyrophytes—plants that have most of their mass underground rather than above ground. The general effects are a simplification of ecosystems, loss of nitrogen released by the burning itself, and loss of nutrient cycling previously accomplished by deep-rooted trees. Cattle graze on the grasses, reducing nutrients in the soil until the range will support only goats and sheep and finally, when desert results, only camels.

Agriculture. A second cause of desertification is intensive farming of marginally fertile arid lands. When farmers attempt to increase crop yields by eliminating traditional fallow periods, soil fertility can be exhausted. Again, the resulting bare land is subject to erosion by wind and water. Topsoil washes or blows away, leaving infertile subsoils or rock.

Firewood gathering. A third contribution to desertification occurs when increasing human populations strip areas clean of firewood. In many developing nations wood is the major fuel for cooking. In the Ethiopian highlands, increasing populations have led to the elimination of fallow periods on croplands, cultivation of steep slopes, and loss of forests to farmers and wood gatherers. The result is rocky deserts in the highlands, the extent of which have been greatly increased by the additional stress of the drought.

Political and social factors. In addition to population growth, political and social factors have influenced the spread of desertification in Africa. The lands of the western Sahel (Figure II.4) were inadvertently protected from human exploitation by raiding tribes from the thirteenth century to the early twentieth century. When colonial occupation made the area safe for herders and their cattle, agricultural populations moved northward into the Sahel, bringing their livestock. At the time the area was experiencing a relatively wet period. Thus, before the beginning of the current drought in 1968, the land appeared to support a greater number of cattle than it actually can.

The introduction of firearms in the sixteenth and seventeenth centuries led to a steady loss of wildlife in the area. Environmental consequences have been se-

vere. Birds and herbivores involved in seed dispersal were killed at the same time as habitat losses due to agriculture and human settlements decreased the number of seeds to be dispersed. In forested areas, hunting eliminated wild carnivores and made wooded areas safe for grazing cattle, sheep, and goats. These animals, grazing and browsing in the forests, destroyed the habitats of native animals. They also decreased the ground cover, which captures rainfall and thus recharges groundwater, and eliminated wild plants formerly used by rural populations in times of drought.

Cities that have expanded or new ones that have been founded have put severe stress on the surrounding area to provide food and wood for fuel and construction.

Salinization. In more developed countries, such as the United States, desertification has resulted from overgrazing of rangeland and agricultural erosion, just as it has in the less developed countries. Another serious problem in the United States, that has occurred to a lesser extent in Africa, is soil salinization. Arid lands can be irrigated to allow growth of crops requiring more water than a region provides naturally. However, evaporation of pure water from irrigation water leaves behind water concentrated in salts. If this water does not drain away, it can build up to such an extent that plant growth is stunted. Further, an impermeable crust of salt may form over the soil surface. The Senegal Delta, the Niger Delta and Lake Chad Valley, and the Tigris-Euphrates Valley join the Imperial and San Joaquin valleys in the United States as subject to this type of salinization. Thousands of acres of once fertile cropland in the United States have been sterilized in this manner.

Groundwater overuse. In the United States, desertification is also accelerated when groundwater is withdrawn faster than it can be recharged. As groundwater levels drop, native vegetation cannot survive. To save water for the needs of people in western cities such as Tucson, Arizona, many acres of once irrigated farmland have been abandoned. Without natural cover, these lands are subject to wind erosion. In the lower Santa Cruz basin of Arizona, blowing dust from abandoned or idle fields has been so severe that at times it has caused highways to be closed.

All of these human influences—overgrazing, excessive wood gathering, intensive agriculture accompanied by erosion, urbanization of arid lands,

Figure II.4 Sub-Saharan Africa. Sahel is an Arabic term meaning shore or border. It is used to indicate the lands on the edge of the Sahara Desert. It is usually meant to include Mauritania, Mali, Senegal, Niger, Chad, Burkina Faso (Upper Volta), and Gambia. The current drought has affected the countries of the Sahel as well as a number of other countries in sub-Saharan Africa. This map shows the sub-Saharan African countries affected by food shortages or inadequate food supplies as of June 1985. (Data from United Nations and from *The New York Times*, August 20, 1985.)

salinization, and groundwater overdraft—result at least in part from population pressures. Further, it is clear that the desertification to which they contribute is not a simple extension of the margins of a desert due to drought. Rather, an overuse of a relatively fragile environment appears to push the system past the point from which it can recover naturally, especially when further stressed by drought. In fact, some scientists believe desertification may prolong drought. Land stripped of vegetation reflects more sunlight back into the atmosphere than land with a plant cover. This may cause more dry air to sink toward the earth.

Halting Desertification

Is there any hope for halting desertification and recovering land already degraded? A number of suggestions have been made, and some appear to hold promise.

Agricultural and Grazing Restrictions

On relatively level land, shelterbelts can break the force of winds that scour away soil. If the trees and shrubs planted in the shelterbelt are carefully chosen, they can provide food for humans and animals, and a wildlife habitat as well.

Grazing restrictions of some sort are a necessity for arid lands and yet may be the most difficult kind of solution to impose. Often in the developing countries and on America's own Indian reservations, grasslands are exploited because people are too poor to have another choice. As populations increase, herds must increase to feed additional mouths. Rangeland is encroached upon by farming, causing remaining grasslands to be further overgrazed. The suggestion has been made that regional cooperatives based on existing local nongovernmental units (such as religious groups) might be given governance of specific land areas. This has been done successfully in Saudi Arabia and Syria where the ancient *hema* system of range reserves has been revived.

Agroforestry

Agroforestry is a mixture of herding or farming and the growing of woody plants on the same land. Such systems appear to hold great promise for arid lands.

They allow more crops to be grown on the same amount of land, provide insurance against crop failures because more than one species is harvested, improve nutrient cycling, and improve the local microclimate. As an example, a plantation of *Acacia albida* trees is recommended for those parts of the Sahel where rainfall would support it. The trees can be interplanted with a crop such as millet. Deep taproots enable the trees to reach water and nutrients unavailable to the agricultural crop, and leaf litter improves the nitrogen content of the soil. The trees are leafless during the crop season and provide shade during the hot, dry season. A harvest of *Acacia albida* pods can be used for cattle forage, while the trees also provide wood for fuel, fencing, and construction and several medicinally useful compounds. Finally, the system is usually acceptable to local farmers. This last point is one of the most important in any consideration of methods for halting desertification or reclaiming wornout lands. Unless systems are seen as logical and appealing to local populations, they will have little chance of success.

Need for Appropriate Technology

Simple transference of Western technology may result in serious problems. For example, in the 1940s irrigated rice culture was attempted in the Senegal River Valley. The project quickly fell victim to a variety of problems. Birds and rodents ate a large part of the rice grown, the soils became saline, and wind erosion caused the loss of large quantities of soil during the dry season. Because of the high costs of fertilizer and of pumping water, as well as environmental problems that seriously reduced the harvest, the project was not a financial success.

Western-based irrigation projects can also provide breeding grounds for organisms carrying schistosomiasis and malaria, and, unless carefully planned, they can disrupt local family units and other sources of local authority, leading to social disturbances such as crime, delinquency, and emotional disorders.

Comprehensive Plans

In the long view, it is clear that the Ethiopian famine is a symptom of a larger environmental disaster, desertification. In addition to the food aid that humanity requires us to give today, far-sighted people will wish to explore ways to implement plans such as the 1977 United Nations Comprehensive Plan of Action to

Combat Desertification. Such plans embody the kinds of remedies we have described. They have as their aim the development of sustainable systems of agriculture and herding in arid lands. Along with appropriate population control measures, such systems can insure that, although natural droughts may recur, a drought and famine of such magnitude need never happen again.

Questions

1. What is meant by desertification? Is it caused by drought?
2. What are the causes of desertification?
3. What are some possible solutions to the problem of desertification?

References

Smith, J. R. *Tree Crops: A Permanent Agriculture.* New York: Harper & Row, 1980.

Tolba, M. "Harvest of Dust," UNEP (Spring 1984).

Milas, S. and D. Stiles. "Recognizing a New World Crisis," UNEP, Spring 1984.

National Research Council. *Environmental Change in the West African Sahara.* Washington, D.C.: National Academy Press, 1983.

Kerr, R. "Fifteen Years of African Drought," *Science* (March 22, 1985), 1453.

Glantz, M. and R. Katz. "Drought as a Constraint in Sub-Saharan Africa," *Ambio,* **14** (6) (December 1985), 334.

Desertification of the United States. Council on Environmental Quality. Washington, D.C.: U.S. Government Printing Office, 1981.

National Park Service Photo by Richard Frear

PART THREE

Water Resource Problems

Human survival depends on a number of natural resources. Water is certainly one example; air is another, and energy resources a third. The importance of plants and animals as wildlife resources was discussed in Part One, while land and food resources were examined in Part Two and land resources will be taken up again in Part Eight.

The next three parts are concerned with the resources of water, air, and energy. We begin with water, in part because the environmental movement first took shape around efforts to protect water supplies. More than 100 years ago, people realized that water could carry disease. Because of that recognition, the profession of environmental engineering, or, as it was then called, sanitary engineering, grew up. The environmental movement, as a visible phenomenon, had begun.

Chapter 9 examines water from the ecologist's point of view. Water is a resource with unique properties, essential to all life on earth. It is a basic factor in the growth of natural communities and human civilizations. The chapter describes the various kinds of water habitats and the organisms that live there, using New York's Hudson River to illustrate several of the major kinds of water habitats.

In Chapter 10, using the Colorado River and the Ogallala Aquifer as examples, we detail how people have managed their water resources and put water to use. We also examine the deep-rooted arguments over water transfers and the use of water for development. Water is seen as a critical element for the growth and maintenance of human society.

Chapters 11–14 examine water problems from the point of view of human needs. When is water pure and when is it safe? What are the substances that contaminate water supplies, and how do we remove them? How do we handle wastewaters and what effects do they have on natural waters?

The first of these questions deals with *the water we drink*. It involves basic facts we should know about the safety and quality of our supply of drinking water.

The impurities that influence the safety of a water supply for drinking fall into three broad classes (Figure A). *Inorganic chemicals* are one class; included in it are the ions arsenate, nitrate, fluoride (at high levels), and other chemicals that can have adverse effects on our health. *Organic chemicals*, a second category, may also be dissolved in the water; some of these compounds have been linked to cancer. Finally, water may contain *microorganisms (microbes)* that cause diseases such as typhoid and cholera. Fortunately, these diseases are only a distant memory to most of us in the United States. Once widespread in this country, they still occur commonly in nations that have not yet treated their water supplies.

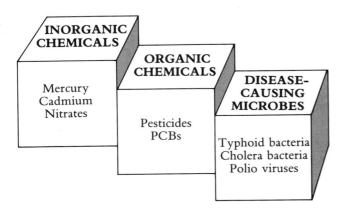

Figure A The three classes of impurities in drinking water.

In addition to being concerned with the safety of water supplies, we are concerned with other water characteristics, such as clarity, odor, and taste. Water may also be investigated for its "hardness." Hardness is caused by the presence in solution of compounds such as calcium carbonate and magnesium carbonate. Hardness decreases the effectiveness of soap and makes washing clothes, dishes, and people more difficult. In Chapter 11 we explain the possible hazards in drinking water and then look at water treatment, the methods by which we attempt to make water safe for drinking.

The effectiveness of current water treatments depends in part on the amount and kinds of wastes that we allow to contaminate water we may later wish to drink. The next two chapters in this part cover this other kind of water resource problem: that caused by *the water we waste.* Wastewaters flow from cities and industries, from mining operations, from farming and rural homes. These wastes are treated in different ways but are generally disposed of in the same way: into the nearest river or lake or into the ocean.

The problem with wastewaters is not just that they might contaminate our drinking water supplies. Waste-water can be fairly easily chlorinated so that it does not carry large quantities of disease-causing organisms, and the water we drink, even if it comes from a highly polluted source, can be treated until it is not only healthy but pleasant to drink. Thus, our insistence that wastewater be treated before it is released is not focused only on human health problems. We must also consider how the wastewater will affect the natural waters into which it flows.

If organic pollutants are not removed from waste-water, they can set up a chain reaction that robs water of the oxygen normally present. Further, certain chemicals, such as pesticides, may be directly poisonous to aquatic organisms. Fish and other aquatic creatures may not be able to live under these conditions, and other less desirable species may take over. In addition, certain inorganic elements in wastewater, such as phosphorus and nitrogen, cause excessive growths of the microscopic green water plants called algae. These "blooms," as they are called, form unpleasant scums and mats over the surface of lakes.

Water pollution control is the term given to methods of cleaning up wastewaters so that they can be released to natural waters without causing problems. What are these methods? Surprisingly, biological processes are used to purify the water.

The final chapter in this part deals with a problem of great concern to many communities today: pollution of surface and groundwaters with toxic wastes. For many years industrial and municipal wastes have been buried in landfills, abandoned wells, or mines. Only now have we become aware of how these practices can pollute groundwaters, once thought to be the safest source for drinking water.

This part closes with discussion of a water issue of truly global dimensions: the United Nations initiative to provide safe drinking water to the entire population of the world by the year 1990.

CHAPTER NINE

Lessons from Ecology: Water Habitats and Communities

The Hudson River

High in the Adirondack Mountains of New York State, the Hudson River begins as a small, clear lake, Lake Tear of the Clouds. Flowing first as a brook and then as the Opalescent River, the waters run south, joined by many other streams. At the town of Newcomb, still in the mountains, the Opalescent officially becomes the Hudson. Just above Troy, the Mohawk River joins the Hudson, which is now one of the mightiest rivers in the United States. Three hundred and fifteen miles from its origin, as it passes New York City, the Hudson meets the Atlantic Ocean in New York Bay, which is the Hudson's estuary (Figure 9.1).

The communities of organisms living in various parts of the Hudson change as conditions in the river change. Small swift streams that feed the river provide homes for different species than those in the slower moving river itself. Organisms living near the point where the river meets the ocean tides must be able to live in varying concentrations of salt.

Organisms living in the Hudson River have many problems with which to contend. For years, a number of cities have dumped raw or poorly treated sewage into the river. Industry has also been guilty of using the Hudson as a sewer. But before we deal with the problems of polluted waters, let us look at life in natural, unpolluted waters.

Figure 9.1 View of the Hudson River today from near Bear Mountain. (Courtesy of New York State Department of Commerce)

Water—A Limiting Factor for Life

Availability of Water

We pointed out in Chapter 1 that water is a necessity for all forms of life on earth. As noted in Chapter 6, in land habitats the abundance of water (a function of rainfall, humidity, and the evaporation rate) determines the kinds of communities that develop. In water environments, the availability of water changes with changes in water levels—for instance, with tides. It also changes according to differences in the salt content of water, which affect the rate at which water enters or leaves organisms. In water environments, as well as land environments, the type of community that develops depends on this availability of water.

The Watershed

Surface waters such as lakes, rivers, and oceans are highly visible features of the environment. We can easily see that they provide different habitats for living organisms than do land areas. Not as easily seen is that the two kinds of habitats, land and water, are tied together by the cycling of energy, water, and nutrients through the environment. For instance, the Hudson River is not a self-contained system. Energy comes from the sun. Organic and inorganic nutrients wash into the Hudson, by erosion and stream flow, from the river's banks and from land bordering all those streams flowing into the Hudson. Even the river water itself cycles through the hydrologic cycle—some water leaves as evaporation and some water enters the river as rainfall and stream flows. Pollutants, too, reach the river not only directly but also from the land areas surrounding it.

We can see, then, that the functioning unit is not simply the river itself but also the whole land area that drains into the river. This area is the **watershed**. In terms of understanding and maintaining the quality of natural waters, the whole watershed is the ecosystem that must be studied or managed.

Water Habitats

Water habitats are usually differentiated on the basis of salt content (saltwater versus freshwater habitats) and whether a current is present or absent (streams with swift-flowing waters versus still lakes or river pools) (Figure 9.2).

(a)

(b)

(c)

Figure 9.2 The three main types of water habitats. (a) In freshwater habitats, the water is either still, as in lakes and ponds, or moving, as in streams and rivers. (Yellowstone Lake, National Park Service photograph) (b) In marine habitats, such as oceans and seas, the water moves continuously as a result of various currents. The salt concentration is, of course, much higher than in fresh water. (U.S. Coast Guard) (c) Estuaries are partly enclosed bodies of water where salt water meets fresh water. An example is Chesapeake Bay, shown here. (Office of Tourist Development, Maryland Department of Economic and Community Development)

Freshwater Habitats

The most important physical characteristics of fresh-water habitats are turbidity, temperature, current (or lack of current), and the amount of dissolved materials, including solids (such as nitrate and phosphate salts) and gases (such as oxygen and carbon dioxide). The turbidity of water, which is a measure of its clarity, is important because turbidity affects how far sunlight can penetrate in water. Green plants can only live in the water zone into which sunlight reaches because they need sunlight for photosynthesis.

Lakes and reservoirs. Most organisms living in lakes, reservoirs, and ponds or in the quiet pools of streams are adapted to life in still waters. In the shallow water zone along the shore, light reaches all the way to the bottom. Here live rooted water plants and floating algae, as well as a variety of animal life (Figure 9.3).

Farther from shore is the area of open water. The upper layer, which light penetrates, is home to minute plants and animals called **plankton**, as well as to fish. Plankton are microscopic, drifting organisms found in lake waters as far down as light penetrates. Plankton species are, in general, unable to move against currents but drift along with water movements. Plant species, called **phytoplankton**, include many species of algae; animal species are known as **zooplankton**. Phytoplankton capture the sun's energy and turn it into organic compounds that form the basis of many of the lake's food chains. Zooplankton feed on phytoplankton and so are primary consumers in the lake ecosystem. Along with the phytoplankton in the open waters, plants in the shore zone are the producers in lakes and ponds. Although they are small compared to rooted plants, phytoplankton have an enormous rate of reproduction. Thus phytoplankton species are the more important producers in aquatic systems.

While some consumers such as frogs and snakes live along the shoreline, other consumers such as fish may range over all three zones, depending on the season and availability of food.

Living in the deeper water layer and on the bottom, where not enough light penetrates for photosynthesis to occur, are organisms that live on dead organic matter. Bacteria, fungi, small clams, and bloodworms all "reprocess" organic matter, which is then carried by currents or swimming creatures back to the other lake zones. Thus, the deep zone houses the detritus feeders in lake and pond communities.

Lakes and reservoirs often stratify, or have layers of different temperatures. Such stratification can lead to poor water quality—that is, water with little dissolved oxygen in it. This, in turn, will affect the kinds of organisms able to live in the deeper waters of a lake.

To explain how decreased quality of lake and reservoir waters comes about, we need to discuss the effects of temperature on the water in a lake reservoir. An annual cycle occurs in the temperature profile of such a body of water.

For convenience, we describe the cycle as beginning in late fall, when the weather has begun to cool. The autumn and winter winds transfer their energy to the lake or reservoir in the form of waves, which mix into the body of water. The waters are well mixed during this period. Thus, a fairly uniform quality and temperature of the water exists at all depths. Such a picture continues through the winter months.

In the spring, however, sunlight warms the waters nearest the surface, and warmer stream inflows enter the cold body of water. Above 4°C, warm water is less dense than cooler water and thus floats above it. Hence, the new warmer inflows tend to form layers near the surface of the reservoir. Similarly, surface waters that have been warmed by sunlight tend to remain near the top. The wind's force at mixing is now less effective because the warmer, more buoyant water, though pushed into the interior, rises again toward the surface.

Essentially, three layers of water form at succes-

Figure 9.3 (opposite page) Simplified lake community. In the shallow waters along lake shores grow rooted water plants such as cattails (a), water lilies (b), and muskgrass (c). These, along with algae, are the shore zone producers. Also found there are consumers such as frogs (h), pond snails (e), dragonfly nymphs (f), amphipods (d), and copepods (i). On the surface balance waterstriders (j) and backswimmers (g). In the open-water zone, algae and dinoflagellates (m, n) are producers. They are found as deep as sunlight penetrates. These producers are eaten by zooplankton such as copepods (q), cladocera (o), and rotifers (p), which are eaten by small fish such as sunfish (k) who are, in turn, eaten by larger fish, such as walleye (r) and small-mouth bass (l). The deep zones and bottom muds are inhabited by creatures that live on detritus: bacteria and fungi (s), chironomid larvae (t), clams (u) and tubifex (v).

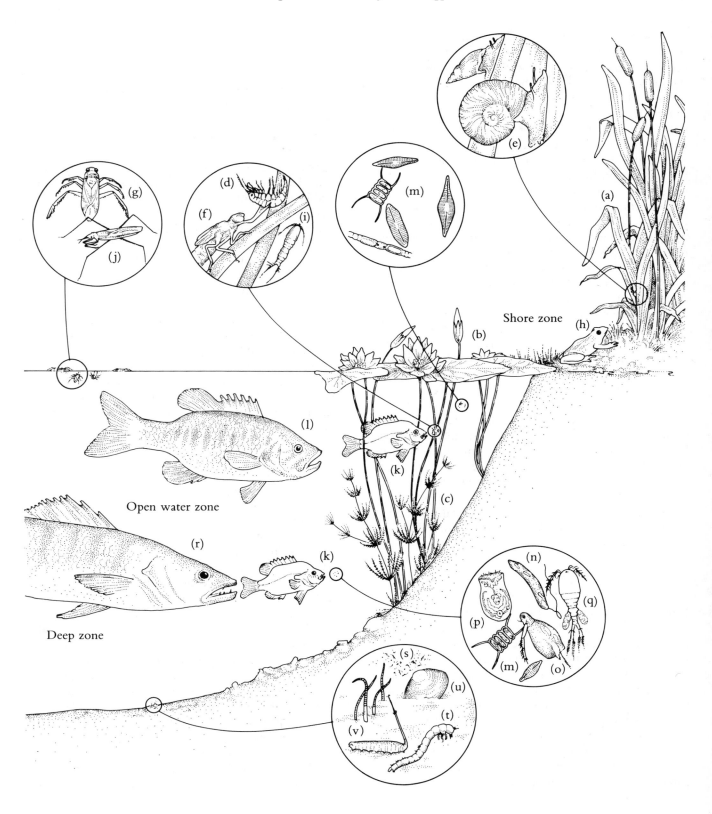

Shore zone

Open water zone

Deep zone

sively warmer temperatures. The warmest layer, called the *epilimnion*, forms at the top; it stays at the top because it has the lowest density. The coldest layer, called the *hypolimnion*, lies at the bottom; it has the highest density. Between these layers, a sharp temperature transition occurs; this middle layer is called the *thermocline*. This formation of layers of differing density and temperature is known as **stratification**.

Stratification of water into temperature layers is one of two phenomena that combine to cause a decrease in the quality of lake and reservoir water. The other phenomenon is the growth of algae in relatively clear waters. While some algae are growing, others are dying. The dead algae sink to the bottom layers; there, microorganisms in the water consume the dead algae as food. The microbes also remove oxygen as they grow and maintain themselves. Since the water in the lake has stratified and little mixing occurs between the upper and lower layers, oxygen removed in the bottom layer is not restored quickly. Thus, the bottom waters of a stratified lake or reservoir may become very low in oxygen. Because oxygen is sometimes in short supply in deep lake waters, many organisms living there have adapted to low oxygen concentrations or even no oxygen at all.

As colder weather arrives, the upper layers become cooler and cooler until they approach temperatures just less than the temperature in the bottom layers. At that point, the layering becomes unstable, and the layers "flip." This event is called the *fall overturn*, and the lake or reservoir now becomes relatively well mixed; that is, the water at all depths is of about the same temperature. The well-mixed character is sustained through the winter months.

Ponds. Ponds differ from lakes in that the shore zone is relatively large and the open-water zone is comparatively small. Often, ponds are too shallow to have a layer of water that light does not reach. Thus, photosynthesis takes place at all depths. Ponds usually have no temperature stratification because they are too shallow to prevent thermal currents from mixing the waters. Some ponds dry up during part of the year, creating particular stress on their communities. Organisms living there must have a dormant stage to survive the dry period. For instance, fairy shrimp lay eggs capable of surviving for months or longer in dry soil.

Some organisms can live both on land and in water—for example, amphibians such as frogs.

Rivers and streams. Three features of the environment in rapidly flowing waters are important to understanding the types of organisms that live there: the presence of a current; the high oxygen concentration; and the source of nutrients. A current is one of the main factors making life in a stream or river different from life in lakes and ponds. However, the difference is not found in all parts of these environments. Streams have pools or areas of quiet flow where organisms find similar habitats to those in lakes. In addition, a lake shore, where waves keep water moving, provides organisms with a habitat similar to a rapidly flowing stream or river. There are thus two types of stream or river communities: those in flowing water and those in quiet water.

A major, and very understandable, feature of organisms living in moving water is that they usually have some way of hanging on to surfaces such as rocks or stream bottoms. Some are cemented firmly to stones or other objects in the flowing waters. Others have hooks, suckers, or sticky undersides. Stream creatures also have streamlined bodies to reduce resistance to flowing water and are often flat so they can crawl under rocks to escape the pull of the current.

Because of the current, rapidly flowing streams or rivers usually have a high oxygen content. The waters, moving and tumbling over rocks, become well mixed with air and so absorb a great deal of oxygen. Organisms living in rapidly flowing water are used to these high oxygen concentrations. When pollutants that use up oxygen in water are added to streams, the clean-stream organisms cannot survive the low oxygen levels. The stream communities found in polluted and clean water are contrasted in Chapter 12.

A large part of the nutrients in streams and rivers either wash or fall into the water from the banks and surrounding watershed. Plant nutrients, such as nitrate and phosphate, and organic material, such as leaves on which detritus feeders live, enter the stream from its watershed. Stream organisms have adapted to this constant flow of fresh nutrients and also to the removal of their waste products by the current. Waters with currents thus provide an environment fundamentally different from the still waters of ponds and lakes.

Organic materials and plant nutrients are not the only materials washed into rivers and streams from

Figure 9.4 *Shad Fishermen on the Shores of the Hudson River*, by Pavel Petrovitch Svinn (1787–1839). Henry Hudson anchored his ship *Half Moon* off this point as he returned from exploring the river that was later given his name. But Indians fished for shad in the Hudson long before Hudson arrived in 1609. Fishing for some species of fish is no longer allowed in the Hudson. Unhealthy levels of the chemicals known as PCBs are now found in river fish. Despite the long passage of time and the pollution, some uses of the river have not changed much. Page 323 shows a picture of men fishing for shad on the Hudson River today. (The Metropolitan Museum of Art, Rogers Fund, 1942)

watersheds. Pesticides or industrial wastes in groundwaters may also wash in. For this reason, when river and stream pollution problems arise, we must, in examining possible solutions, take into account not only the stream itself but also the land surrounding the stream.

The Hudson as a sewer. Partly because currents usually carry wastes downstream, rivers and streams have always seemed to humans to be especially handy and inexpensive ways to dispose of wastes. For years, the Hudson River was used as a sewer for many communities along its banks. This practice resulted in a series of typhoid epidemics around 1890 in communities that not only dumped sewage but also drew drinking water from the Hudson (Chapter 11). The latest story about pollution in the Hudson River, however, goes back only to 1975, when it was discovered that as a result of industrial sewage, river fish contained dangerous levels of toxic chemicals called PCBs. PCBs are lethal in very small amounts (as little as 10 parts per billion) to many insect larvae and to some small fish. Large fish, at the top of their food chains, were found to have PCB concentrations well over the legal limit for human foods (Figure 9.4).

As a result of laws such as the Toxic Substances Control Act of 1976, the manufacture and disposal of PCBs are now controlled much more closely. For the Hudson River, however, this comes too late. (For more on PCBs, see Chapter 14.)

Estuaries

As the Hudson winds its way to the Atlantic Ocean, it reaches at last an area where its fresh waters mix with the salt waters brought by ocean tides. In the Hudson, salt and fresh waters meet in New York Bay, and some salt water moves as far up river as the city of Troy.

River mouths, salt marshes, bodies of water behind barrier islands, and coastal bays—such as New York Bay—are all estuaries (Figure 9.5). **Estuaries** are coastal bodies of water partly surrounded by land but still having an open connection with the ocean. In these areas, fresh water drains from the land and mixes with tidal currents of salt water. Estuaries, especially marshes, are often looked upon as wasteland, best dredged or filled. But this is a real misunderstanding of the role estuaries play.

Estuaries are highly productive systems and are generally more fertile than either the neighboring ocean or the fresh waters that flow into them. The reason is that nutrients are easily trapped in estuaries. The nutrients are trapped in a physical sense, first as sediments settle out from river inflows and then by the action of the tides and fresh water flow (Figure 9.6). Nutrients are also trapped in a biological sense because

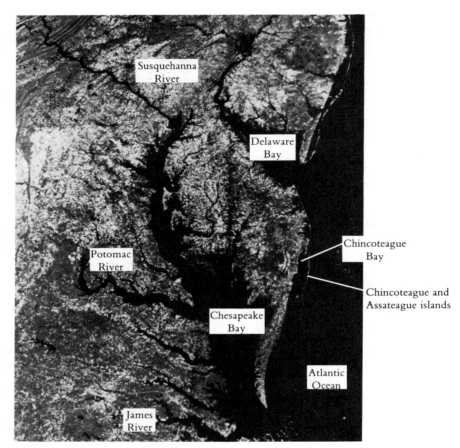

Figure 9.5 Estuaries. Chincoteague Bay is an estuary formed by the mainland on one side and the barrier islands Chincoteague and Assateague on the other. Chesapeake Bay is a huge estuary where fresh water from many large rivers—such as the James, the Susquehanna, and the Potomac—mix with salt waters from the Atlantic Ocean. (NASA)

Figure 9.6 Mixing currents in estuaries. Fresh water, which is lighter, tends to float on the heavier sea water. As one rolls over the other, mixing currents are set up that tend to recirculate nutrients. (Adapted from E. P. Odum, *Fundamentals of Ecology* (Philadelphia: Saunders, 1971], p. 354.)

they are recycled rapidly by a network of producers, consumers, and detritus feeders. Unfortunately, pollutants are also recycled and sometimes biologically magnified in estuaries, so the effect of toxic materials such as DDT can be more serious here than in a river or the ocean.

Another factor contributing to the fertility of es-

tuaries is tidal action. Tides, which cause the water in estuaries to flow back and forth, make it possible for organisms like the oyster, which feeds by filtering sea water, to sit and have their food brought to them. In the same way, their wastes are removed.

Furthermore, estuaries provide good habitats for a variety of producer organisms, from the large rooted grasses such as eel grass, turtle grass, and salt marsh grass to the tiny floating plants or bottom-dwelling algae. In fact, often more organic material is produced than can be recycled in the estuary itself. The excess nutrients flow out into the ocean and fertilize these waters. Good fishing is the result.

Estuaries serve yet another important purpose: providing a nursery for many ocean species, such as shrimp. The larvae, or immature stages, of these species find protection and food in the estuary. Some fish, such as salmon or Hudson River shad, that live in salt water but return to fresh water to breed, require estuaries as places to rest during their journey. Thus, when estuaries are unthinkingly filled in, the effects fall not

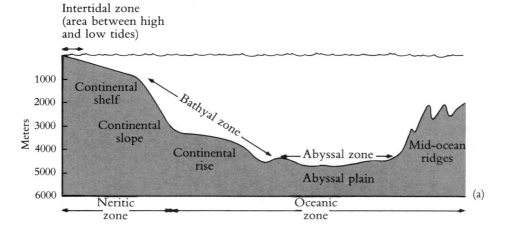

Intertidal zone
(area between high
and low tides)

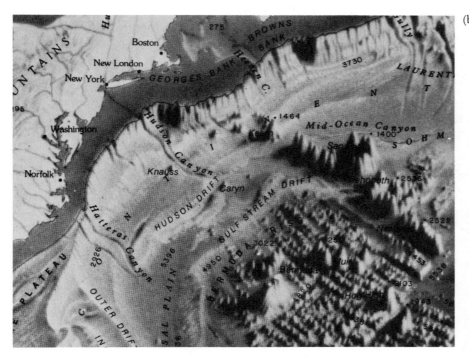

(a)

(b)

Figure 9.7 (a) A sea in cross section. The shallow water of the continental shelf is called the neritic zone, including the area where the tides cover and uncover the shore twice a day. The rest of the open sea is called the oceanic zone. The bottom of the sea along the continental slope and rise is called the bathyal zone, while along the deeper plain it is called the abyssal zone. As in fresh water, light penetrates only the top layer of water. All below this layer is in darkness. (b) Panorama of ocean bottom. Note the Hudson Canyon shown in the upper-left-hand corner of the picture. (From World Ocean Floor Panorama by Bruce C. Heezen and Marie Tharp, 1977, © Marie Tharp.)

only upon the creatures that spend their whole lives there but also upon many ocean species that use estuaries or the food produced there.

Marine Habitats

Actually, the Hudson River does not disappear into the Atlantic Ocean without a trace. Three miles southeast of Ambrose Lightship, an underwater channel begins and runs 130 miles along the ocean bottom. Geologists believe that at the end of the last ice age the shore was 150 miles farther into the ocean than it is now. They

speculate that the roaring Hudson, fed by melting ice, cut a channel through this land on its way to the sea. Hudson Canyon is a natural wonder. At some points it is 7 miles deep, deeper than the Royal Gorge of the Colorado River. But it now lies hundreds of feet below the surface of the ocean.

Besides Hudson Canyon, many interesting geologic features appear below ocean waters. Looking at the sea in profile, as if it were cut in half from top to bottom, several well-defined areas can be seen (Figure 9.7). For some distance, the ocean floor slopes gradually away from the land. This area is called the

Figure 9.8 Some of the organisms (not drawn to scale) found in communities in the region of the continental shelf. Phytoplankton species, including diatoms, dinoflagellates, and microflagellates, are the producers, which form the base of the food chains in the coastal zone. "Seaweeds," algae adapted to hold on to rocky shores as the tide washes over them, are also producers in this zone. Tiny zooplankton such as copepods, pteropods, and jellyfish (medusae), as well as the shrimplike krill, feed on the producers and are fed on by fish, squid, and sea mammals such as whales. Sea birds, large fish such as tuna and swordfish, and humans are top consumers in these food chains. Also found in this region are the larvae or immature stages of many marine species. Buried in the bottom muds or clinging to rocks are the worms, clams, snails, crabs, and bacteria that form the benthos, or bottom dwellers, which feed on detritus. (Adapted from John D. Isaacs, "The Nature of Oceanic Life," *Scientific American*, September 1969.)

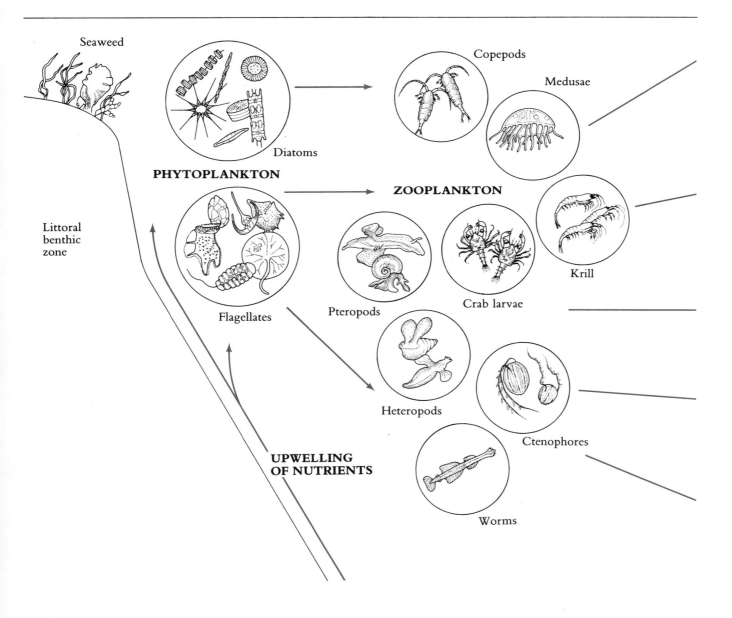

Seaweed

Copepods

Medusae

Diatoms

PHYTOPLANKTON

ZOOPLANKTON

Littoral
benthic
zone

Krill

Flagellates

Pteropods

Crab larvae

Heteropods

Ctenophores

**UPWELLING
OF NUTRIENTS**

Worms

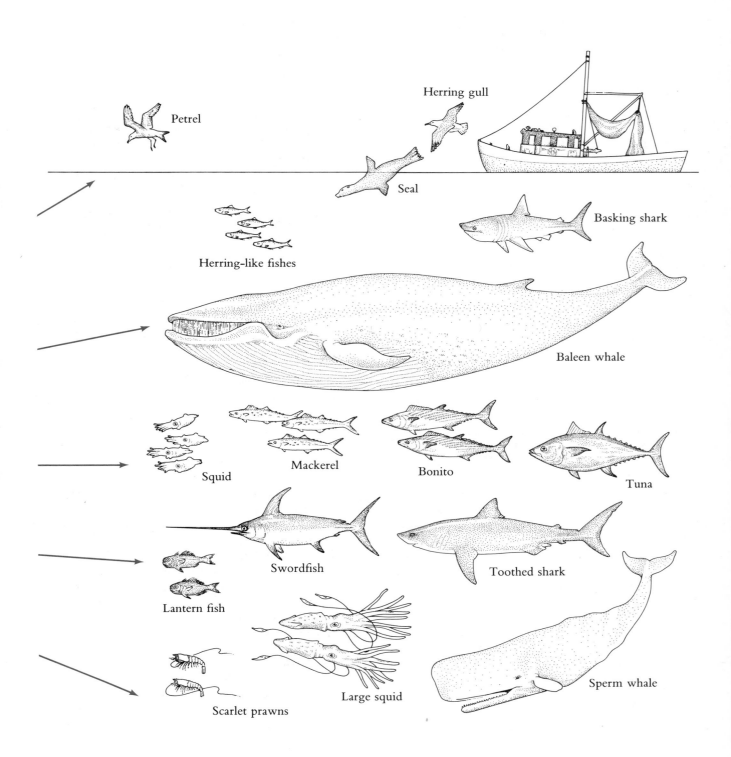

Petrel

Herring gull

Seal

Basking shark

Herring-like fishes

Baleen whale

Squid

Mackerel

Bonito

Tuna

Swordfish

Toothed shark

Lantern fish

Large squid

Scarlet prawns

Sperm whale

continental shelf. The floor then drops off sharply (continental slope) and again levels off, into the continental rise. Finally, the floor drops off once more to a level plain, the abyssal plain, 6,000–15,000 feet (2,000–5,000 m) below the surface. Toward the middle of the ocean, a series of ridges lie scattered across this abyssal plain.

Currents. Unlike lakes, where layers of water may remain still for long periods, the water in the sea moves continually. Currents are caused by a variety of forces, such as temperature differences, differences in salt content, the rotation of the earth, and winds. Because of these mixing currents, even the deep parts of the oceans have a constant supply of oxygen in the water. In some areas, along steep coastal slopes, winds continually blow the surface water away from shore, allowing cold bottom waters, rich in nutrients, to rise to the surface. This is called **upwelling**. The areas where this occurs—for instance, along the coast of Peru—are the most fertile in the seas. In general, although life is found in all areas of the sea, the major commercial fisheries are all located on or near the continental shelf. The reason is that many ocean food chains are based on the microscopic green plants, which grow best in areas of coastal upwelling.

Communities. Figure 9.8 shows some of the species found in the coastal zone. The communities that live in the intertidal zone consist of organisms specially adapted to the periodic absence of water when the tide goes out. (Some scientists are concerned that these organisms might be endangered by the development of tidal power: the use of the tides to generate electricity. We examine this more fully in Chapter 22.)

In the open ocean, species have adapted to life far from shore. In the top layer of water, where light penetrates, drifting microscopic plants and animals live. Microplankton species such as *Chloramoeba* are the main producers in the open oceans. The shrimplike krill (Figure 9.9), as well as small zooplankton species, feed on microplankton. Larger fish and sea mammals such as the baleen whale range over both the open ocean and the shore areas in search of krill and zooplankton, on which they feed. Oceanic birds such as petrels, albatrosses, and frigate birds feed on the open oceans except during breeding time, when they fly to land.

The sunlit zone in the open oceans does not support as much life per square meter as the light zone in

Figure 9.9 Shrimplike krill (*Euphausia superba*) are a vital link in ocean food chains between the microscopic producers and larger creatures such as whales. However, nations hungry for protein are now looking at the 5 cm–long krill (shown here being chased by a penguin) as a possible food source for their human populations. Krill can be made into a sort of shrimp paste. What effect massive harvesting of krill would have on ocean communities is unknown.

coastal areas does. However, the oceans cover 70% of the earth's surface, and much of this is open ocean. For this reason, the photosynthetic organisms in the open ocean are very important in world oxygen and carbon dioxide balances.

We know relatively little about communities in the deep zones of the ocean. Only recently have a number of facts come to light—quite literally, since one of the main physical characteristics of the deep sea regions is relative darkness. Since there is not enough light for photosynthesis, organisms depend on producers in the top layer of water for organic nutrients. The major portion of the organic matter reaching the deep ocean zones is probably composed of fecal pellets from zooplankton on the surface.

Diversity in deep sea communities. Much of the deep sea bottom is covered with thick layers of mud. In some areas, this appears to be formed into topo-

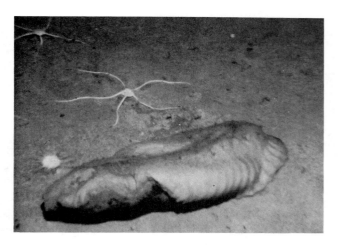

Figure 9.10 Life on the deep ocean bottom. Although a few large species are visible on the surface, most of the organisms in the deep sea are found in the mud. Here, a holothurian, or sea cucumber (*Holothuroidea*), an urchin (probably *Lytechinus*), and brittle stars (*Ophiuroidea*) share the sediment surface. (Dr. Fred Grassle, Woods Hole Oceanographic Institute)

graphic features such as ridges and cliffs. In and on the bottom muds of the plains live many species of worms, clams, and crustaceans (Figure 9.10). These organisms live in an area that maintains a stable temperature and energy supply. Although the organisms living here are small, an enormous variety of them exists. In fact, the diversity is comparable to that found in tropical rain forests and in coral reefs, two other areas noted for their great variety of species. These habitats are similar in that they are physically stable (that is, little or no changes in physical conditions take place over long time periods) and have a long evolutionary history. High diversity is characteristic of ecosystems that have not been disturbed for long periods of time.

Productivity of marine versus land ecosystems. Table 9.1 gives a comparison of the estimated productivity of the various ecosystems on earth. The estimates in the table are of gross productivity rather than net productivity; that is, the amount of produced material used by organisms themselves in respiration has not been subtracted out. Because of the loss due to respiration, as well as for a variety of other reasons,* only

---

*For example, unavoidable losses to insects and diseases, and the fact that not all parts of plants or animals are, or are considered, suitable for human consumption.

one-third or less of the gross productivity in an ecosystem is available for humans to harvest.

Table 9.1 underscores the fact that compared to most land ecosystems, the productivity of the open oceans is relatively low. Some researchers believe this is primarily due to a lack of nutrients in the open oceans. However, especially in warm oceans, it is probably also due to the fact that the primary producers are nannoplankton and picoplankton. These extremely small plankton species are used as a food source only by very specialized small planktonic animals. As we explained in Chapter 2, only about 10% of the available energy is passed from one trophic level to another; thus the addition of another level at the bottom of the trophic pyramid greatly decreases the production measured at higher levels.

At the moment, humans eat near the top of ocean food chains, thus decreasing even further the amount of productivity available to them. At one time, scientists thought that plankton might be harvested from the ocean and used as a human food source; however, productivity in the open oceans is too low to make harvesting plankton a reasonable alternative at this time. That is, too much energy would need to be invested in running the fishing boats and processing the plankton compared to the amount of food energy obtained. In addition, of course, we would have to find ways to make plankton appealing to humans as a food. It is largely because of low productivity that once widely held hopes of feeding the world from the oceans seem doomed to disappointment.

In addition, the primary producers in the ocean are important to the world's oxygen and carbon dioxide balance. Schemes to harvest primary producers in the ocean could seriously interfere with this balance.

Certain coastal marine areas are suitable for **mariculture**. Aquatic species such as clams, lobsters, and certain fish can be raised in underwater farms (see Chapter 15). However, such schemes are probably not practical in the open oceans because of the low nutrient concentrations there. These plans are also endangered by pollution of coastal waters with sewage and hazardous wastes.

In the following chapters, various factors are discussed that can disrupt the functioning of aquatic ecosystems and make waters unfit or unpleasant for human uses.

Table 9.1 Primary Productivity in Various Ecosystems

Ecosystem	Area (million square km)	Gross primary productivity (kcal/square m/year)	Total gross production (10^{16} kcal/year)
Marine ecosystems			
Coastal zone	34.0	2,000	6.8
Areas of upwelling	0.4	2,000	0.2
Open ocean	326.0	1,000	32.6
Estuaries and reefs	2.0	20,000	4.0
			43.6
Natural land ecosystems			
Deserts and arctic tundra	40.0	200	0.8
Grasslands	42.0	2,500	10.5
Dry forests	9.4	2,500	2.4
Moist temperate forests	4.9	8,000	3.9
Coniferous forests	10.0	3,000	3.0
Tropical rain forests	14.7	20,000	29.0
			49.6
Cultivated land ecosystems			
No energy inputs (no fertilizer, pesticides, etc.)	10.0	3,000	3.0
Fuel-subsidized agriculture	4.0	12,000	4.8
			7.8

Source: Adapted from E. P. Odum, *Fundamentals of Ecology*, 3rd ed. (Philadelphia: W. B. Saunders, 1971), p. 51.

Summary

To fully understand a water habitat, we must consider the land area that drains into the water—the watershed—as well as the stream, lake, or river itself. Water communities consist of photosynthetic producers (microscopic algae and more visible underwater and shore-based green plants), and the consumers that feed on them. Important factors differentiating freshwater habitats include current, turbidity, oxygen content, temperature, and dissolved solids. Marine environments differ in sunlight, currents, temperature, and salt concentration.

Productivity in water habitats is greatest in estuaries, where nutrients are trapped and protective habitats are available for the young of many species. Coastal areas and areas of upwelling are also productive and provide the basis for fisheries. Productivity in the open ocean is much smaller, but, at least in some areas, this appears to be due to the presence of extremely small plankton that form the base of the food chain.

Questions

1. Why is the watershed of a lake or river considered the important ecosystem, rather than just the river or lake itself?
2. What major factors define water habitats? Describe some of the major ways in which organisms have adapted to these habitats.
3. Why are estuaries important ecosystems?
4. What is the importance of plankton to freshwater and salt-water communities?
5. Do you think the oceans are a vast untapped resource of food to feed the hungry nations of the world? Why or why not?

CHAPTER TEN

Water Resources

How We Obtain Fresh Water

Water Sources/Surface Water Compared to Groundwater

Surface Water and Reservoirs

Uses of Reservoirs/Environmental Impact of Reservoirs

Groundwater Resources

Uses of Groundwater/Total Supplies of Groundwater

Water Resource Decisions

Extending Water Supplies Without Building More Reservoirs/The High Plains and the Ogallala Aquifer/ The Colorado River and Conflict in the Southwest/ Grand Plans

How We Obtain Fresh Water

In more and more places on the earth, we are finding that fresh water is not present in the amounts needed for growing communities. Wells are being driven deeper, water pipelines and aqueducts are being built, dams are being constructed—all with the goal of capturing more water.

In this chapter we investigate how people obtain adequate water supplies and whether the "quantity" of available water is sufficient. In the chapter that follows, we focus on the "quality" of the water we consume. By separating the two topics we do not mean to suggest that water quality and water quantity are separate issues; they are, in fact, interlocked. When engineers and planners ask whether adequate water resources are available, they are really asking: "Do we have available a sufficient quantity of water of the quality that we need?"

Water Sources

Water is drawn in two fundamental ways: from wells, tapping underground sources of water called aquifers; or from surface flows—that is, from lakes, rivers, and man-made reservoirs.

In the United States during the 30-year period from 1950 to 1980, withdrawals of groundwater and surface water (as opposed to consumption) increased by 150%, from 180 billion gallons per day to 450 billion gallons per day. Surface water withdrawals made up about 80%, groundwater about 20%. Although some of this growth could be attributed to population increase, the largest changes were due to new industrial

demands and new irrigation demands (see Figure 10.1).

Other means exist for producing potable (drinkable) water. In some wealthy technologically advanced areas, *desalination*—the desalting of sea water by such means as distillation, for example—can make even ocean water fit to drink. In water-poor regions of the

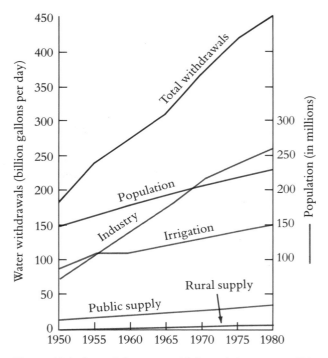

Figure 10.1 Growth in water withdrawals by sector, 1950 to 1980. (From U.S. Geological Survey, *National Water Summary 1983*, Water Supply Paper 2250.)

world without wealth, cisterns collect rainwater for human use. The extent of water production by such exotic means is negligible, however. Generally, humans rely on fresh surface water and groundwater as drinking water sources.

Surface Water Compared to Groundwater

We can examine the functions of these two water sources by comparing the human-made above-ground structures known as reservoirs with the natural below-ground water-filled spaces called aquifers.

A dam thrown across a river backs up the water flow behind its face, forming a **reservoir**. The dam releases through its gates only enough water to sustain the downstream flow, holding back high flows for gradual release later when low flows occur. Reservoirs increase the amount of water available to communities. Without a reservoir, the largest steady rate of daily water withdrawal from a flowing river can be no greater than the *lowest* daily flow rate occurring in the river during the time of operation. Any larger quantity of water could not be withdrawn steadily day after day. With a reservoir, a city or town can draw water steadily week after week, month after month, without interruption.

The above-ground reservoir is a human invention to steady the flow of fresh water through time, collecting high stream flow from one season to make water available in the low-flow season. The **aquifer**, in contrast, is a natural underground reservoir, where water temporarily resides on its route to lakes, streams, rivers, or oceans. An above-ground reservoir is a large empty space when not filled with water, but an aquifer is not necessarily empty. It may consist of free-flowing water, as in an underground cavern or stream, but it may also be water that simply fills the spaces between particles of sand and gravel. Aquifers can be huge, extending for hundreds of miles, and the volumes of water in such aquifers enormous. The volume of water stored in the Ogallala Aquifer, in the High Plains states, is probably comparable to the volume of water in Lake Huron.

Water from a surface reservoir and water from an underground aquifer are likely to be quite different in quality. In the surface reservoir, water will contain sediment picked up by the river that fills the reservoir. Some of the sediment particles settle to the bottom, depending on how long the water is detained, but particles do remain in the water drawn for human use. To remove these particles, it is common to filter reservoir water prior to use.

In surface water, organic matter from both municipal and agricultural sources is likely to be present and can cause the removal of oxygen. Algae often find the reservoir environment inviting, as it offers both the nutrients and sunlight they need for growth. Because reservoirs receive "fresh" sources of pollution, the water, if used as drinking water, needs to be fully treated. Treatment is required to remove undesirable tastes, color, and odors; to make the water clear; to make it free from hazardous chemicals; and to destroy any disease-causing organisms. However, not all reservoir water destined for human consumption is fully treated, even in the United States. Filtration is excluded on occasion for surface waters of good clarity, but the danger is that viruses and parasites may remain unseen.

Water drawn from aquifers is likely to be far clearer than water taken from surface reservoirs, especially if the aquifer has not been drawn on for long or has not been extensively depleted. Water from an aquifer is also likely to have a higher content of dissolved minerals. Groundwater will not have algae in it, since no sunlight has illuminated the water for many years. Because water reaches the aquifer by filtering through thick layers of soil, its bacterial and viral content is likely to be lower than that in surface reservoir water. However, the odor of hydrogen sulfide is common in groundwater, a result of the bacterial degradation of organic material that takes place in groundwater in the absence of oxygen.

There are exceptions to these generalizations about quality. Groundwater can become polluted by chemicals and by microorganisms. For instance, chemicals from landfills are seeping through the soil into the groundwater reservoir that provides water for the city of Tucson, Arizona. In 1964, disease-causing bacteria entered the aquifer supplying a portion of the water of the city of Los Angeles; because of inadequate chlorination, an epidemic of diarrheal illness struck the portion of the city that used the water. Once polluted, groundwater may remain polluted through many lifetimes, because aquifers are recharged (refilled) very slowly, often over hundreds of years. Groundwater supplies are thus extremely fragile resources, in that the entry of pollutants can ruin them for generations.

Because aquifers are recharged so slowly, it would seem wise to discourage any long-term uses of groundwater that occur at rates higher than the natural rate of recharge. Irrigated agriculture using groundwater much faster than its rate of recharge would be an activity to avoid. But in the High Plains, in southern Arizona, and in the San Joaquin Valley of California, irrigation with groundwater has been going on for many years. We will discuss the serious impact of these activities shortly.

Surface Water and Reservoirs

Uses of Reservoirs

Reservoirs are of two types: single-purpose and multiple-purpose. Single-purpose reservoirs provide only a single function, such as community water supply. Their operation is relatively straightforward—for instance, water-supply reservoirs release only the amount of water that is needed. Multiple-purpose reservoirs may serve a variety of purposes, including community water supply, irrigation, navigation, recreation, hydroelectric power generation, flood protection, and low-flow augmentation and environmental flow-by.

Community water supply includes water for drinking and washing, for industrial purposes, and possibly for watering lawns. Irrigation water is designated for the growing of crops; its use is often highly seasonal, with larger requirements during the hot seasons. The capability of river navigation may be maintained by steady water releases through the year. Recreation such as boating, picnicking, and the like is served by keeping a relatively constant volume of water in the reservoir so that shorelines do not fluctuate extensively. Hydroelectric power generation requires both steady releases and high water levels. Flood protection requires that the reservoir be kept as empty as possible. Low-flow augmentation and environmental flow-by imply that releases are made during times of low flow to protect the quality of the water and the species that inhabit it. Such releases dilute wastes, thereby decreasing their relative demand for oxygen from the water (see Chapter 12). Such releases may also flush out salt water from estuaries, maintaining the proper environment for the native estuarial species.

The operation of reservoirs serving these multiple functions is far more complex than that of a single-purpose reservoir because a number of these purposes are in conflict. A reservoir that serves only a water-supply function should be kept as full as possible. If the purpose of a reservoir is flood control only, it should be kept as empty as possible so the largest flood flows can be caught and then released more slowly. The operation and purposes of a reservoir can significantly influence its impact on the environment—and reservoirs have a strong impact on the environment.

Environmental Impact of Reservoirs

As flowing waters slow down, which indeed they must as they are brought to a halt by a reservoir, the sediment suspended in the turbulent waters settles out and drops to the bottom of the reservoir. Downstream from the reservoir, the clear water released into the river erodes earth from the river bank at a faster rate than the free-flowing river would have, as if to make up for the sediment lost in the reservoir. Increased erosion downstream from a reservoir is a common occurrence.

The bottom of the reservoir becomes coated with the sediments transported from upstream areas (Figure 10.2). This blanket of sediments is exposed to view periodically as the level of water in the reservoir rises and falls in response to inflows and releases. The sediments gradually build up and, unless dug out occasionally, begin to consume the storage volume of the reservoir. That is, a reservoir created to store water for water supply or to catch water for flood control can gradually lose its effectiveness unless the solids deposited in its banks are excavated from the site.

Large deposits of sediment in reservoirs may be partially prevented. Although erosion and sediment transport are natural and continuing events, farming, road development, construction of homes, and forest cutting all accelerate erosion processes by exposing fresh and unanchored soil. Careful management of the soil (see Chapter 7) can help reduce the burden of sediment carried by streams and thus help prevent the rapid deposition of sediment in reservoirs.

The unsightly mounds of sediment that become visible during times of low reservoir storage are one reason why many individuals express distate for dams. Another reason is more basic: Valued lands are lost to view and to use forever. Valued animals and plants are also lost. Not only are land species driven out but fish that inhabit the stream may also be barred. Where once they swam upstream to spawn, the reservoir wall now blocks their path.

Figure 10.2 A silted reservoir near Ithaca, New York. This reservoir has been drained because a new community water supply has been created on a nearby lake. The sediments, accumulated over some 50 years of operation, were consuming a large part of the storage volume of the reservoir. Consequently, a reasonable water level in the reservoir looked like a larger volume than was actually stored.

Figure 10.3 The long road home. This fish ladder is on the John Day Dam in Oregon. (Photo courtesy of the U.S. Army Corps of Engineers)

Reservoirs may be built so that fish are still able to pass upstream. Fish ladders consist of concrete steps down which a stream of water flows (Figure 10.3). The ladder is so constructed that fish can jump from one step up to the next, and eventually to the reservoir and upstream.

The geologic history of a region is often revealed in the layers of rock through which a river has cut its course. This history may be lost when a dam is thrown across a river. And the sheer majesty of the river's power demonstrated by the gorge may be replaced by a quiet pool of water. Each potential reservoir site has its own characteristics. It may include a mature forest or be home to a cluster of rare wildflowers. The reservoir may flood out valuable farmlands or displace families or Indian tribes from ancestral homes. Graveyards and tribal burial grounds may be forever obliterated. It is not always clear that a just compensation can be found for the people uprooted from such an area, for their loss is more than economic.

A reservoir may have still other impacts. During certain times of the year, the quality of water within the reservoir and the quality of the water released from it may be surprisingly poor. As explained in Chapter 9 during the summer and early fall, the lower layers of water in a reservoir can become very low in oxygen. This low-oxygen condition is caused by a combination of two phenomena. The first is a lack of mixing of the layers of the reservoir during the summer and early fall.

Figure 10.4 Relative withdrawals of groundwater and surface water, 1980. (From U.S. Geological Survey, *National Water Summary, 1983*, Water Supply Paper 2250.)

The second is the consumption of the dead algae by bacteria in the bottom layers of the reservoir, causing a removal of oxygen from these bottom waters. If this oxygen-poor water is released, fish and other aquatic creatures downstream from the reservoir may be harmed.

Groundwater Resources

Uses of Groundwater

Groundwater serves a more limited set of functions than surface water. In many cities, groundwater provides community water supplies. Especially in rural areas, where the cost of extending the water distribu-

tion system is high, people turn to wells to deliver their water needs. Groundwater is also used for irrigation of crops, a common practice in farming areas that are short of surface water or where the building of irrigation canals is costly. Groundwater provides the household water supply for about half the population in the United States. Surprisingly, it also supplies approximately a third of the total water used for irrigation in the United States. Figure 10.4 shows the relative withdrawals of surface water and groundwater by state in 1980. Note that groundwater withdrawals exceeded surface-water withdrawals in Arizona, Oklahoma, Kansas, and Nebraska. In a number of coastal states, surface-water withdrawals include saline water drawn for cooling steam power plants and for industrial cool-

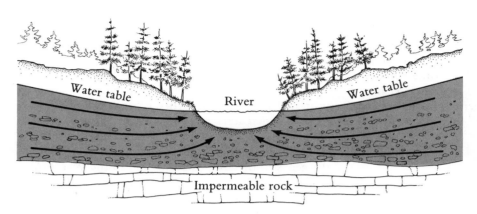

Figure 10.5 Groundwater helps to maintain stream flows. Especially in dry seasons, the relative contribution of groundwater to stream flow may be substantial.

ing. If these withdrawals are excluded in determining total surface water withdrawals, five more states—Delaware, Florida, Mississippi, Texas, and Hawaii—would join the ranks of those that withdraw groundwater at a faster rate than surface water. Even though California still doesn't make the list, note that its rate of groundwater use exceeds the rate of surface water use of any other state.

Groundwater also serves a relatively unseen and unappreciated function. Its flow contributes to and often sustains the summertime flow of streams and rivers, which themselves may be used for water supply (see Figure 10.5).

Total Supplies of Groundwater

Groundwater may seem an almost infinite resource in some places. In the conterminous United States, about 25% of the water that falls as precipitation each year becomes groundwater. Indeed, of the world's fresh (nonsaline) water, groundwater resources far exceed surface-water resources (see Table 10.1). The appearance of enormous resources is deceiving, however, because groundwater collects over hundreds and thousands of years at relatively slow rates. Rapid withdrawal of groundwater cannot be matched by rapid inflows; replacement must still occur by the same slow, steady infiltration that took place in the past. Furthermore, groundwater below 0.8 km is often too laden with salts to be useful for water supplies.

Use of groundwater provides a number of advantages to consumers. First, since groundwater may be located near its point of use, savings can be realized in piping and possibly pumping costs. Second, the firm yield can be sustained over a long period through both dry and wet seasons. (This advantage may prove to be illusory, however, if the aquifer is depleted by a consistent overdraft.) Third, in undeveloped areas, groundwater is not usually subject to bacterial, viral, or chemical contamination, although minerals and hydrogen sulfide do occur on occasion.

What is the safe yield of an aquifer? As with reservoirs, it depends on the flows that enter the aquifer. No more can be withdrawn year after year than the annual recharge rate of the aquifer—unless the user of the water is willing to see the volume in the aquifer drawn down. In a number of parts of the country, withdrawal rates are exceeding recharge rates and the water levels in the aquifers are declining. The fact is that in desert basins, rainfall only rarely recharges an aquifer. During most years, evaporation draws most water up from the surface into the atmosphere. Only during extremely wet years will enough water be present for some of it to recharge the aquifer.

Finally, in this brief description of groundwater and its uses, we should mention that groundwater is

Table 10.1 Freshwater Resources of the World

Resource	Volume (thousands of km³)
Fresh water in lakes and inland seas	125
Fresh water in streams, rivers, etc. (average)	1.25
Groundwater within 0.8 km of surface	4,200
Groundwater between 0.8 km and 4.0 km depth	4,200
Fresh water in glaciers and ice caps	29,000

Source: Adapted from H. Bouwer, *Groundwater Hydrology* (New York: McGraw-Hill, 1978).

often used in conjunction with a surface water supply. We call this blending of surface and groundwater *conjunctive use*. Because surface water supplies are more "flashy"—that is, because they exhibit greater variability than groundwater supplies—the groundwater may be used to "fill in" the shortage periods. Supplies are then "firmed" at a higher level without extensive drawdown of an aquifer.

Water Resource Decisions

Extending Water Supplies Without Building More Reservoirs

In the past, water-supply engineers would look at growing water demand and reach for new water sources to meet it. Dams, aqueducts, pipelines, and the like were the traditional means to meet demands for new water supply. Now, more than ever before, this approach is running into difficulty.

As more people express concern for the environmental alterations caused by reservoirs, new water projects are becoming less popular than they once were. Such projects are also having trouble getting off the ground because of increasing competition for water and conflicts over who owns it. Growth in water use makes the conflict fairly certain to increase as more and more water sources are allocated.

Even though fewer new water sources are available, it is often still possible to meet growing demands. One obvious means of accommodating more users with the same basic water system is to encourage people to conserve water. A higher price for water is one way to get people to look for ways to conserve. People, industry, and agriculture can all find ways to save water if given the incentive to do so.

Another way to meet growing water demands without developing new sources is by interconnection and joint use of existing water sources. We mentioned that groundwater can be used to "firm" or fill in the water availability from a reservoir. Groundwater fills up the inflow gaps, stabilizing supply at a higher level without extensive use of the groundwater resource.

The modern engineering discipline of water resources systems analysis has found methods for taking independent river basins and by means of interconnections managing them in a way that increases the system yield *above* the yield that occurs with independent op-

eration. That is, the component reservoirs of a system can reliably deliver more water when their releases are synchronized and pooled than when each is managed individually. For example, in the metropolitan area of Washington, D.C., sixteen new reservoirs had been proposed by the U.S. Army Corps of Engineers to meet the demands of the growing capital district. However, with only one of the proposed reservoirs built, a study of the water supply system showed that the needs of the area could be met to the year 2020 if releases from the existing reservoirs were coordinated properly through the seasons.

In 1981, several New Jersey communities ran short of water because of a severe drought that was fairly general throughout the northeastern United States. The state called in a consultant, who reminded officials that during World War II interconnections had been built between the major New Jersey water sources as a means of preventing possible disruption of water supplies. He also reminded them that fiercely independent communities had torn up those interconnections after the war rather than risk sharing their water. If the connections had been left intact, areas with surplus water could have helped shortage areas through the drought. The consultant, Dr. Abel Wolman, knew of the interconnections because he had been instrumental in having them built during the war, some 40 years before.

Interconnections and integrated system management are two options that can secure adequate water supplies for the future without requiring new sources and new dams. Nevertheless, across the country, many water resource projects are being undertaken, including dams for water supply and flood control, canals, hydropower installations, dredging, and water transfers. In many areas, new projects are being considered to meet emerging needs. Decisions on these projects are complex because water is so highly valued and so many people claim it. Also, water resource projects are often expensive, and who should pay for them is not always clear. In addition, water resource projects often have significant environmental effects that must be weighed.

We have selected for further discussion two areas of the country where water project decisions are currently being considered. In both cases, problems are in the offing that require resolution, and clashes are occurring over policy, economics, and environmental values.

Table 10.2   Approximate Withdrawals and Remaining Storage in Portions of the Ogallala Aquifer, 1977

State	Approximate annual withdrawal[a] (millions of acre-feet)[b]	Remaining in storage (millions of acre-feet)[b]	Years water will last with no recharge
Colorado[c]	1.0	75	75
Nebraska[d]	1.5	450	300
Kansas	2.9	750	250
Oklahoma	0.6	60	100
New Mexico	0.75	30	40
Texas	8.0	270	30

Source: Data from B. R. Beattie, "Irrigated Agriculture and the Great Plains," *Western Journal of Agricultural Economics*, **6**(2), (December 1981).

a. Withdrawal figures do not take account of natural recharge.
b. An acre-foot is the volume of water covering 1 acre to a depth of 1 foot.
c. Excludes Baca and Prowers Counties.
d. South of the Platte River.

The High Plains and the Ogallala Aquifer

In parts of Texas, Oklahoma, New Mexico, Kansas, Colorado, and Nebraska, the pumps that are supposed to draw groundwater to irrigate crops are drawing mostly air. Irrigation with groundwater is proceeding at so rapid a pace in these states that the aquifers are being drawn down faster than they are being recharged. One aquifer in particular that extends through these states, the Ogallala Aquifer, has been particularly hard hit (see Table 10.2). It is no wonder that the Ogallala has been so depleted, since about 35% of the current annual use of groundwater in the United States is now being pumped from the Ogallala. Irrigated acreage in the High Plains expanded from 3.5 million acres (1.4 million hectares) in 1950 to over 14 million acres (5.7 million hectares) in 1980.

In the area near Lubbock in Floyd County, Texas, where agriculture relies on the Ogallala, the level of the water table fell 5 feet per year during the 1970s and has fallen 200 feet (61 m) since 1940 (see Figure 10.6). (*Water table* is the term hydrologists use to describe the level of the aquifer closest to the surface.) In Grant and Finney counties, Kansas, the water table has fallen over 100 feet (31 m) since 1940. In response to these declines in the water table, wells have been drilled deeper to tap the vital water. And the water table falls still further.

A falling water table indicates that the withdrawal rate exceeds the natural rate of recharge of the aquifer. The Ogallala Aquifer that irrigates the farms of the High Plains may have declined in volume by 23% since

the mid-1950s. Not all areas that draw on the Ogallala have been so hard hit. The fall has been spotty—severe in some areas, less so in others—but indicates more widespread problems to come.

The Ogallala Aquifer covers an area the size of California; it includes southwestern Nebraska, eastern Colorado, western Kansas, western Oklahoma, northwest Texas, and eastern New Mexico (see Figure 10.7).

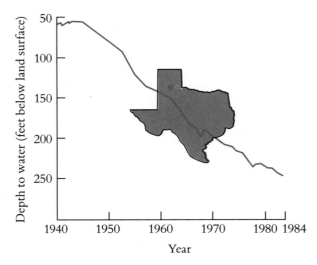

Figure 10.6 Decline of the water table in a well in the High Plains aquifer, Floyd County, Texas, 1940 to 1984. (Compiled by E. D. Gutentag from U.S. Geological Survey, *National Water Summary, 1984*.)

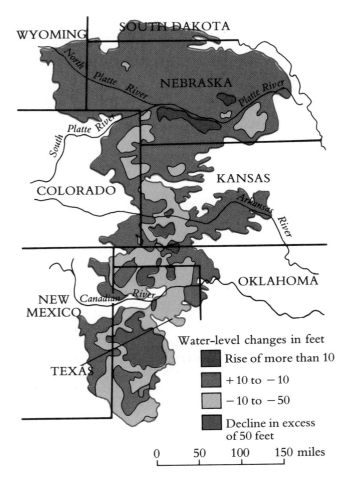

Figure 10.7 The Ogallala Aquifer in the High Plains of the United States. Changes in water level are shown from the period before development up to 1980. (From U.S. Geological Survey Yearbook, 1982)

The aquifer holds a volume of water roughly equivalent to that of Lake Huron. It underlies a region of relatively low rainfall, about 12–22 inches (30–56 cm) per year—normally not enough to support extensive irrigated agriculture. Only 0.5 to 3.0 inches of the annual rainfall in the region penetrates down to the water table and recharges the aquifer. The groundwater resource, while it does renew itself, does so on such a slow time scale that current use is essentially "mining" the water.

The size of the Ogallala Aquifer is not uniform throughout the region. In Nebraska, the water table is

not far from the surface, and the aquifer is as much as 1,000 feet (300 m) thick. Although irrigation is decreasing the water table in Nebraska, it will be a long time before falling groundwater levels threaten agriculture there (see Figure 10.8). In the irrigation area northwest of Lubbock, the aquifer is 100–300 feet (31–92 m) thick. To the south of Lubbock, the thickness falls to 25–150 feet (8–46 m).

The Ogallala Aquifer has been described by hydrologists as an "egg carton"; that is, they believe that the aquifer has large compartments with small connections rather than being a single large volume of water. This description suggests that the drawdown of the aquifer in one area (and one compartment) shouldn't have much immediate effect on the water table in an adjacent compartment. If this is the case, high use rates in Nebraska should not soon imperil groundwater supplies in Texas—which is fortunate, since Texas is threatening its own supplies without outside help.

The agricultural base of the High Plains is extensive. The region produced more than 50% of the nation's beef cattle in 1980, as well as important quantities of wheat, cotton, corn, and other grains (see Table 10.3). The value of its agricultural product was estimated at $4.2 billion in 1977, of which about $3 billion came from irrigated acreage. The impact of irrigation on yield may be seen from the fact that High Plains corn yields increased from about 50 bushels per acre in 1968 to just over 100 bushels per acre in 1977. During

Table 10.3 High Plains Agricultural Production, 1977, and Projections for 1985

Crop		Percent of national production	Percent of production from irrigated acreage
Wheat	1977	16.4	18.9
	1985	13.4	11.9
Corn	1977	13.1	93.9
	1985	13.1	96.9
Sorghum	1977	39.7	57.4
	1985	36.8	54.3
Cotton	1977	24.9	66.3
	1985	31.2	72.4

Source: High Plains Associates, *Six-State High Plains Ogallala Aquifer Regional Resources Study*, A report to the U.S. Department of Commerce, July 1982.

Figure 10.8 Center-pivot irrigation systems in Nebraska, as shown by a Skylab photo. Each circle in the photo represents the area swept out by the rotating arms of the center-pivot irrigation apparatus. Most of the increasing amount of Nebraska irrigation uses groundwater. (U.S. Geological Survey)

the same period, average U.S. yields went from 85 to 95 bushels per acre.

What will happen if groundwater pumping continues at the same rate in these areas and the aquifer continues its decline? New wells will have to be drilled and pumps lowered still further. Since the water will have to be pumped from a lower level, electrical energy costs will increase. Consequently, farmers will need higher prices for their produce to make up for the increased irrigation costs. If the farmers can't get these higher prices, they may have to go back to dry-land farming: farming without irrigation, the kind of farming in which success of the crop depends on rain—rain that doesn't always arrive in time.

Another factor, besides energy costs, that could drive the farmer back to dry-land farming is the fact that as the wells go deeper, the "sweet" water starts to decline. Water drawn at deeper depths is older water, water that has been dissolving minerals from the soil over a period of perhaps a thousand years. We call such mineral-laced waters *saline*. If the salt content is high enough, the water will not aid the growth of crops and can even harm the soil and plants. When a well starts to draw saline water, the end of groundwater irrigation is at hand.

The problem of the Ogallala is now clearly drawn and understood. In Colorado and Kansas, state legislatures have passed laws to limit the number and spacing of new wells, although enforcement of the laws is in doubt. Enforcement requires on-site inspection of farmers' well meters. Farmers can be an independent breed, and water is their lifeblood—they might be expected to resist an inspector's "intrusion."

Nevertheless, realizing that the water they began drawing in the 1950s cannot continue to be used at so rapid a rate, many farmers have taken steps to cut their use of water. One such step is to install small dams on streams on the farm property, creating ponds that can be used as a source of irrigation water. Another step is to spray-irrigate at lower water pressures, thereby using less water. Also, if irrigation is scheduled in the cooler hours, water can reach the soil and plants with less evaporation loss. The ponds created by the dams on the property can also be built in series in areas of porous soil. The water then filters through the soil and recharges the groundwater under the farm. Furrows that abut the stream may also be used for recharging groundwater.

One novel way to hold rainfall as well as irrigation water is the Texas diker. The device, developed for cotton growing, is run through the fields and forms small dams, 2.5 inches high, every 7 feet in every other furrow. Two of the small dams form a miniature reservoir and are said to be able to trap up to a 3-inch rainfall within their borders without loss.

Still another form of irrigation that could have a significant impact on water use is *trickle irrigation*. With this method, a pipe with multiple outlets is placed on the ground and extended parallel to a row of crops. Water "trickles" from the pipe into the soil near the roots of the growing plants. Water is not wasted on earth that lacks crops, and the water does not evaporate so readily because it is not sprayed into the air.

These steps to conserve irrigation water are making a difference. In 1981, a water conservation official in Lubbock estimated a one-third reduction in pumping rates over a 5-year period. However, such reduction only buys time; the water table is still falling, though at a slower rate.

An emerging technology, evaluated only experimentally thus far, is to inject compressed air into the earth to "push" water out of the unsaturated zone down to the water table level below. This water, held by capillary action in the upper unsaturated zone, ordinarily drains only slowly to the lower saturated zone. For the Texas regions of the High Plains, it is estimated that another century's worth of water could be obtained if air injection were widely utilized. At the current prices of crops, however, air injection, or *secondary recovery* as it is known, is still too expensive to use on a commercial scale.

If farmers must go back to dry-land farming, their productivity will decline. With smaller crop yields, their income and financial stability will be threatened. Prices of wheat, corn, and other food basics could rise. It has been estimated that 40% of the irrigated acres in Colorado would have to be returned to dry-land farming by 2020 if groundwater pumping continues at its present rate. That translates into about 1.3 million acres. In Texas, a 1.4-million-acre decline in irrigated cropland is projected for the period 1975 to 2000—about a 30% reduction. The situation is less serious in other states relying on the Ogallala because use began later there and the water volumes stored locally in the aquifer are greater. But it is only a matter of time before the effects of declining water supply are felt.

In an effort to determine how to sustain the agricultural productivity of the region, the federal government commissioned a $6 million study of the High Plains region. Released in 1982, the study reviewed steps at the farm level as well as regional solutions to the problem of declining water tables. Among the regional solutions considered was a system of dams, canals, pipelines, and pumping stations to bring "surplus" water hundreds of miles uphill from the Quachite, Sulphur, Sabine, Red, Arkansas, White, and Missouri rivers to the High Plains (see Figure 10.9).

Costs of the projects ranged from $6 billion to $25 billion, depending on the scale of the projects envisioned. These are the costs of moving "surplus water," but it is not clear that there is any "surplus water" that states are willing to make available (see Controversy 10.1).

The Colorado River and Conflict in the Southwest

The Colorado River is both the mighty carver of the Grand Canyon and the lowly watering trough of much of the southwestern United States. Rising in the Rocky Mountains of Colorado and Wyoming, fed by the Green and San Juan Rivers, the Colorado winds majestically south through 1,400 miles of gorge and plain to end as a mere trickle into the Gulf of California. Along the way, the waters of the Colorado are drawn off time and again to irrigate the arid plains of the southwest. On its route, deserts become verdant valleys with some of the most valued agricultural production in the world. Though its annual flow is only on the order of the Delaware River in the east, the inhabitants of much of the southwest have come to depend on the Colorado. They have thrown huge dams across it, built canals to transport its water, fought over rights to its water, allocated its flow, and gone to court over it. The Colorado, one of the most highly controlled rivers in the world, gives life to the southwest.

Of all of the regions of the United States, the arid western regions draw most heavily on their water resources (see Figure 10.10). Three regions in particular are tightly linked in terms of their water supply: California, the Lower Colorado, and the Upper Colorado. All depend to a large degree on the flow of the Colorado River.

The flow of the Colorado is committed to many users over the southwest—indeed, overcommitted—by a quirk of fate. At the time that the compact on the allocation of the river was signed in 1922, the region had just experienced a number of very wet years, years with stream flows that have rarely been seen since. Believing the future would resemble the past, the states that signed the Colorado River compact in 1922 allocated among themselves a total annual quantity of water that is no longer generally available.

To the Upper Basin, above Lee Ferry, Arizona, the agreement allocated potential water consumption of 6.7 billion gallons per day on an annual basis. The

Ogallala Aquifer
Source reservoirs
Terminal reservoir sites
Alternative water transfer routes
from reservoirs to irrigation areas

Figure 10.9 Interstate water transfer options assessed by the Corps of Engineers. (Adapted from Review Draft, Water Transfer Elements of High Plains–Ogallala Aquifer Study, January 1982, U.S. Army Corps of Engineers.)

Lower Basin was allocated another 6.7 billion gallons per day on an annual basis. In 1944, a treaty promised Mexico an additional 1.5 billion gallons per day. The total of these promises is 14.9 billion gallons per day over a year. The average daily inflow to the Colorado in both the Upper and Lower Basins in an average year is now estimated at 14.3 billion gallons per day. Thus, in about half the years the flow is expected to be less than this number; in the other half it will be greater. Even in an average year, then, the Colorado is over-committed.

Fortunately, not all users consume their full allocations at this time. The day is approaching, however, when the Lower Basin will have more users on line than water to go around.

Upper Colorado Region. In our discussion of the Colorado, we will use the water regions defined by the Water Resources Council in their 1978 National Water Assessment. The region of the Upper Colorado produces most of the flow of the entire Colorado River. This region includes portions of northern Arizona, northern New Mexico, Colorado, Utah, and Wyoming.

A number of reservoirs are located along the river and its tributaries in the Upper Colorado Region. The reservoirs both protect against floods and provide water for irrigation, for municipal use, and for industrial use. The reservoirs must also supply the water committed to the Lower Colorado Region by the 1922 Compact. Glen Canyon Dam, 17 miles upstream from

CONTROVERSY 10.1

Should Water Be Transferred to the High Plains?

. . . there is no way realistically and economically that we can sustain the kind of development we have, short of bringing in new water.
Morgan Smith, Commissioner of Agriculture of Colorado

The bottom line is that most of the High Plains area is overdeveloped and its water is overappropriated.
Keith Lebbinl, Scott City, Kansas

Is there any point in leaving the water [in the Ogallala Aquifer] underground simply to enjoy knowing that it's there?
Anonymous Hydrologist

. . . massive investment to augment the declining Ogallala is not economically efficient—not now or in the foreseeable future.
Bruce R. Beattie, Professor of Agricultural Economics, Montana State University

Now that the overappropriation of water has been recognized, the question remains as to what to do. Farmers are implementing conservation measures, but these steps only delay the day of reckoning. It is a matter of either going back to dry-land farming when the aquifer is finally exhausted (or no longer usable) or bringing in more water. The water would come via pipeline from new dams on the Arkansas and Missouri river basins. Costs of construction could range from $6 billion to $25 billion.

Are water transfers the answer? If so, who should pay for the massive projects? Or should the land go back to dry-land farming? (Remember the possible impact on crop yields, food supply, and prices of a return to dry-land farming.) Do you think groundwater should be used for irrigation. If so, under what conditions?

Sources: Morgan Smith and Keith Lebbinl, quoted in *The New York Times*, August 11, 1981, p. B4.
Hydrologist quoted in *Science 81*, **2** (June 1981), 35.
Bruce Beattie, *Western Journal of Agricultural Economics*, **6** (December 1981).

Water Resources

a Oasis in Africa

b Women walking to obtain household water in Burkina Faso

c Household water supply in Indonesia

Water problems differ greatly in the developing world versus the developed countries.

In the U.S., water problems often center around a fair division of the available water (for instance between recreational, agricultural, power supply, and municipal needs). But almost everyone in the U.S. has access to a safe, convenient drinking water supply. This is not so in developing nations. (**a**) Oasis in Africa. In too many areas people bathe in and drink the same water. (**b**) Burkina Faso. In many areas people must walk great distances for water. (**c**) Indonesia. Often their sole household water supply consists of great earthen jugs stored outside and filled by collecting rain or by carrying water.

Energy-Related Pollutants

a Cooling towers

b Oil rig in the North Sea

c Oil spill at Statue of Liberty

d RBMK reactor at Leningrad

Energy supply problems are currently problems of the developed world. As the developing nations hurry to industrialize however, they will need to face these same issues. (**a**) Cooling towers, necessary in power generation, mar the landscape with their huge bulk. (**b**) Oil rigs, such as this one in the North Sea, extract oil and natural gas from beneath the sea to fuel an industrial society, but they are responsible for continual small incidents of water pollution. (**c**) Oil spill at the Statue of Liberty. Large oil spills from offshore platforms or tankers can threaten marine life. Nuclear power carries its own set of hazards, not the least of which is the possibility of an accidental explosion such as occurred in an RBMK reactor at Chernobyl, USSR. (**d**) Here the same kind of reactor, in Leningrad, is shown.

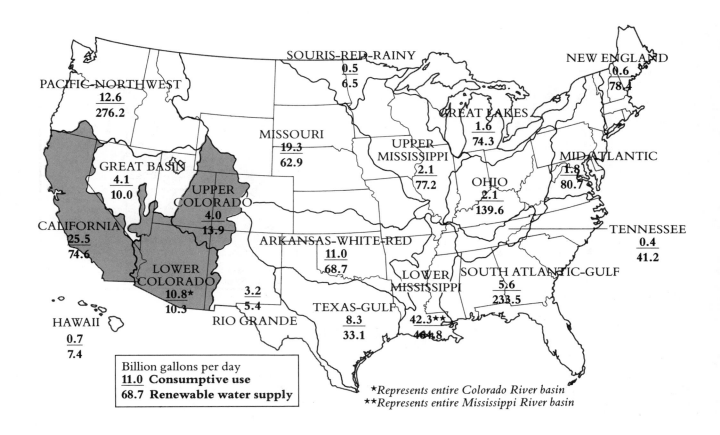

Figure 10.10 Water resources regions and consumption as a percent of renewable water supply. (From U.S. Geological Survey, *National Water Summary, 1983*, Water Supply Paper 2250.)

Lee Ferry, releases the required water to the Lower Colorado Basin.

The annual inflow to the Colorado in the upper region, before losses, averaged about 14.0 billion gallons per day from 1906 to 1975. The outflow from the region in a year when this average inflow occurs, and with the 1975 level of development in the Upper Basin, is about 10.0 billion gallons per day. Thus, some 4.0 billion gallons per day evaporate or are consumed from the Colorado in the Upper Basin, based on the 1975 level of development. Most of this consumption occurs in irrigation.

The Upper Basin has rights to a total of 6.7 billion gallons per day, or an additional consumption of 2.7 billion gallons per day over what it currently uses. By the year 2000, consumption is projected to increase by

1.1 billion gallons per day, which would decrease the flow into the Lower Basin to 8.9 billion gallons per day in an average year. At least for an "average" water year, even by the year 2000, the Upper Basin will not reach its limit of consumption and will deliver sufficient water to meet the requirements of the compact agreement.

The Lower Basin, however, is already consuming virtually all 10 billion gallons per day of the water that the Upper Basin is currently letting through. A second and potentially more immediate problem is that the virgin (undepleted) inflows into the Upper Colorado (averaged over a year) have fallen as low as 5.0 billion gallons per day (1934), even though flows have risen as high as 21.4 billion gallons per day (1917). Low-flow years are likely to happen again.

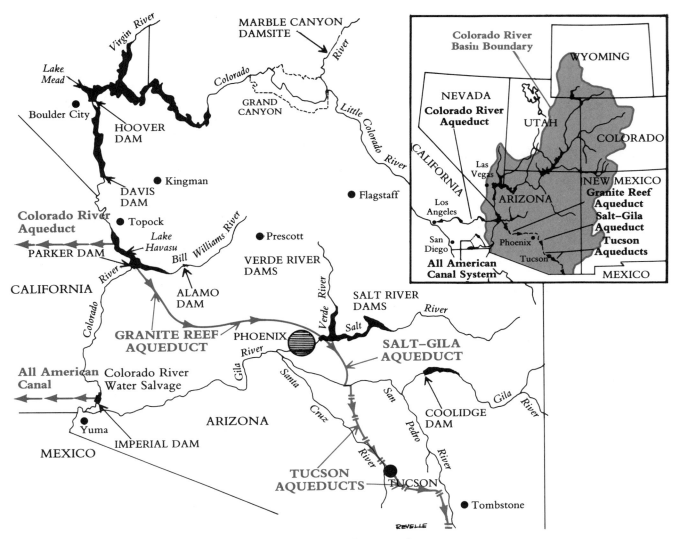

Figure 10.11 The Lower Colorado River Basin is the scene of intense conflict over the rights to water. Los Angeles, San Diego, and the farmers of the Imperial Valley have been drawing off Colorado River water for several years. Now, however, Phoenix and Tucson are claiming their share of the water, and these cities will soon have the means, through the Central Arizona Project, to physically take the water. The Central Arizona Project is made up of the Granite Reef Aqueduct, the Salt–Gila Aqueduct, and the Tucson Aqueducts. (From U.S. Department of the Interior, Bureau of Reclamation.)

The construction of dams has made possible the development of many areas in the upper region. Without the water, development would likely not have taken place. Further economic development might require more dams, and the region possesses a number of unspoiled rivers. However, these streams may qualify for designation as wild rivers and would have to remain free of dams if so designated. Thus, a conflict appears

in the offing between the preservation of wilderness areas in the region and the further development of its economic potential.

Lower Colorado Region. While the Upper Colorado Region produces most of the water, the Lower Colorado and the California regions use it most extensively (Figure 10.11). The Lower Colorado Region is

Figure 10.12 The Central Arizona Project. (Photo courtesy of the Metropolitan Water District of Southern California)

defined by the drainage area of the Colorado River south of Lee Ferry, Arizona. Lee Ferry is the historic dividing point of the Upper and Lower basins agreed on in the compact of 1922. The major state in the Lower Colorado Region is Arizona, but small portions of New Mexico, Colorado, Nevada, and eastern California are also in the region.

Beginning in the Lower Colorado Region, cities and agriculture draw off major quantities of the Colorado flow. Water for Las Vegas, Nevada is taken from Lake Mead, the reservoir formed by Hoover Dam. Next, water for Los Angeles and San Diego is drawn from Lake Havasu, the reservoir formed by Parker Dam. The Colorado River aqueduct supplies 75% of the needs of these cities. Then, irrigation water for the lush Imperial Valley of California is taken from the reservoir behind Imperial Dam. And water from the Colorado will soon be available to the desert cities of Phoenix and Tucson, Arizona (Figure 10.11). Taken together, the civil works projects on the Colorado River are truly massive; their operation literally changes the face of the land.

The Central Arizona Project, which began delivering water in 1985, will be supplying the needs of Phoenix and Tucson as well as the irrigation needs of the agricultural valley in which they lie (see Figure 10.12). For Tucson, the water is meant to replace the groundwater upon which the city has relied for years; Tucson's aquifer has been both polluted and drawn sharply down. Groundwater withdrawal in Tucson is lowering the water table 4–10 feet (1.2–3.0 m) per year; land has subsided in the nearby desert, apparently because of the withdrawal, and fissures have broken the earth near Tucson and elsewhere (see Figure 10.13).

The water supplied by the Central Arizona Project will have the possible effect, even in average years, of decreasing the quantity of water available to lower California. California's rich agricultural output may be threatened by this tightening of supplies, and California has responded with a new and controversial plan. The plan would import additional water to Southern California from the water-rich northern portion of the state. (We will discuss California's response to the Central Arizona Project in more depth in a moment.)

Recall that 10.0 billion gallons per day are currently entering the Lower Colorado Region. This quantity is more than the compact requirement of 6.7 billion gallons per day and should remain so through the end of the century. Another 0.3 billion gallons per day enter the Colorado in the Lower Colorado Region, making available a total of 10.3 billion gallons per day.

In the Lower Basin, some 8.8 billion gallons per

Figure 10.13 This fissure about 9 miles east of Mesa, Arizona is believed to be caused by a sinking of the earth's surface caused by removal of large quantities of underground water. (Photograph by U.S. Bureau of Reclamation)

day are consumed in irrigated agriculture, evaporate from reservoirs, or are exported to Nevada and California. Thus, leaving the Lower Colorado Region, the river carries barely 1.5 billion gallons per day into the Republic of Mexico; the water is also highly saline. This 1.5 billion gallons per day is the precise amount required by the 1944 treaty with Mexico.

When the Central Arizona Project begins operation, the new demand for water from the lower Colorado is expected to gradually increase to 1.07 billion gallons per day. To continue supplying the required treaty water to Mexico, some quantity of use is likely to have to give way. The probable loss is a portion of

California's water withdrawals. California has been drawing water from the Colorado at a rate of about 4.5 billion gallons per day, which is more than its court-ordered allotment of 3.9 billion gallons.

The California region. California is so blessed with mild climate and a reasonable soil that it leads the nation in the production of nearly 50 crops. Yet in the era before irrigation, cattle and sheep ranching and grain farming were the main agricultural activities. Now its leading crops are cotton, grapes, hay, citrus fruits, and tomatoes. Its coastal valleys produce most of the country's spinach, brussels sprouts, avocados, and arti-

Figure 10.14 The California Aqueduct. This aqueduct is the main means of water transfer from California's water-rich north to its water-poor but populous south. The aqueduct runs to 36 feet deep by 257 feet wide. (Courtesy of California Department of Water Resources)

chokes. In total, the 9 million acres under cultivation in the state produce a yearly crop valued at about $10 billion.

If California is blessed with climate and soil, water is its Achilles heel. Nearly 97% of those 9 million acres must be irrigated to produce their abundant yields. Water, much of it from the Colorado River, has helped convert the deserts of California into one of the most productive agricultural regions of the world. Population growth in the southern California cities of Los Angeles and San Diego has also been made possible by water imported from the Colorado River.

California has also harnessed its own rivers: Nearly 1,100 state and local reservoirs exist in California, supplemented by about 150 federal reservoirs. Ten of California's reservoirs can each store more than a million acre-feet of water. But the reservoirs are not always sufficient; in 1977, for example, Colorado River water had to carry California through a severe drought.

The locations of California's water resources and her agricultural output do not coincide. About 80% of the water needs of the state lie south of Sacramento, yet 70% of the stream flow that could be captured for use

occurs north of Sacramento. Water now flows from north to south via the California Aqueduct and the Delta-Mendota Canal (Figures 10.14 and 10.15). These canals drain water from the southern portion of the Sacramento–San Joaquin Delta down to southern California.

Thus, southern California has three sources of water: (1) local reservoirs and groundwater, (2) water from northern California, and (3) water from the Colorado. During dry years for the Colorado River, the new demand on the Colorado by the Phoenix-Tucson complex could seriously affect the ability of California to sustain its agriculture, industry, and population.

As noted earlier, of the 4.5 billion gallons per day that California has been drawing from the Colorado, only 3.9 billion are California's rightful share, according to a Supreme Court decision (*Arizona* v. *California*, 1964). Of this 3.9 billion gallons per day, the irrigation districts of southern California have rights to the first 3.4 billion under their water-supply contracts with the Secretary of the Interior. Of the 1.1 billion gallons per day being drawn by the Metropolitan Water District serving Los Angeles and San Diego, it has rights to

Figure 10.15 Aqueducts serving the Metropolitan Water District of Southern California. A vast array of dams and aqueducts are necessary to provide the municipal water needs of southern California.

only 0.5 billion. The remaining 0.6 billion gallons being drawn are at risk as the Central Arizona Project increases its draw from the Colorado River water. Obviously, California must adjust, either by using water more efficiently or by adding new water sources.

While some people see greater efficiency as the answer, others advocate a new canal from the Sacramento River in the northern portion of the Sacramento–San Joaquin Delta. The peripheral canal, as it has come to be known, would link the Sacramento River to the California Aqueduct, supplying water for ultimate transport to southern California and agricultural use. Costs would run between $3 billion and $20 billion, depending on whether you believe advocates or critics

of the canal. A 1981 statewide referendum in California said "no" to the new canal. This "no" vote may stimulate some rather painful adjustment in southern California, as growers and cities watch the faucet being tightened on water from the Colorado River.

Grand Plans

Perhaps the grandest plan ever seriously advanced for providing water for the west was the NAWAPA concept advanced by the Ralph M. Parsons Engineering Company in the early 1960s. NAWAPA stands for North American Water and Power Alliance; as the name implies, the plan was truly continental in scale.

The plan was based on drawing "surplus" water from the Fraser, Yukon, Peace, and Athabasca rivers in Alaska and Canada. New canals and tunnels, as well as rivers, would be used to distribute the water to the Canadian Plains, to the western and midwestern United States, and to northern Mexico. Water for the western United States would need to be pumped over the Rocky Mountains, but hydropower projects in the NAWAPA system would provide such power as well as additional electric generation. Water would be delivered to the Great Lakes as well via a new waterway from the Canadian Rocky Mountains to Lake Superior.

The NAWAPA system was to have over 4.3 billion acre-feet of water storage and was estimated to cost about $80 billion at the time of discussion in 1964. According to a 1967 version of the plan, total delivery of water by the system was to be 180 million acre-feet (m AF) per year, of which 136 m AF was destined for the United States, 25 m AF for Canada, and 20 m AF for Mexico. While Congress held hearings to consider the NAWAPA concept, no action was ever taken on the projects specific to the plan.

Still another grand plan was proposed by the Los Angeles Department of Water and Power in the early 1960s. The department suggested tapping the Columbia River in Washington and the Snake River in Idaho for water for the Colorado River system. The water would flow through Nevada by pipeline or canal into Lake Mead behind Hoover Dam, from which it could be distributed. The plan drew strong opposition from the governors of Washington and Idaho and was never pursued.

Such water transfers as these have never reached fruition. Yet grand plans continue to appear as new demands for water arise in the midwest and the southwest. (See Controversy 10.2.)

The water in the Great Lakes is not far from the area where energy developments are taking place. Wisconsin is reported to have been approached by coal transport companies in Montana for water from Lake Superior. The water would be used for pumping crushed coal via coal slurry pipelines to other sections of the United States.

In addition, the U.S. Army Corps of Engineers has studied the possibility of a pipeline from Lake Superior to the Missouri River Basin. The water would be used to rescue farmers who now irrigate with groundwater from the declining Ogallala Aquifer. The province of Ontario, which borders four of the Great Lakes, is particularly worried, because half of its population lives along the lakes and 60% of its hydropower is taken from the water flowing in the lakes. In 1985, the provinces of Quebec and Ontario and eight states signed a pledge not to divert Great Lakes water.

But powerful forces are at work, and another 1985 Canadian proposal has some backing. This proposal calls for damming off the James Bay from Hudson Bay in order to convert it to a freshwater lake. Water from the new lake would be piped to the Great Lakes and then sold to the United States. An engineering assessment is underway, but this area of the Arctic has a fragile ecosystem including feeding grounds of migratory birds.

Grand plans continue to be made.

Summary

Fresh water for drinking, crop irrigation, and industrial use is obtained mainly from underground aquifers and from surface waters (in natural rivers, lakes and streams, or reservoirs built by engineers). In the past, groundwater has been a less polluted water source than surface water. However, hazardous waste disposal practices as well as the use of water from deeper levels of aquifers has led to increasing chemical and salt concentrations in groundwater. Because many aquifers recharge very slowly, pollution of an aquifer can last through many lifetimes. Further, in many areas, groundwaters are being withdrawn at a rate faster than they recharge.

A reservoir may serve many purposes besides water supply, such as flood control, recreation, hydropower, and downstream pollution control. Dams built to create reservoirs may have adverse environmental impacts: loss of wildlife habitat, loss of homesites for area residents, stream erosion below the dam, and interruption of fish migration. It is often possible to provide a more secure water supply for an area by a management plan that integrates the operation of existing reservoirs rather than by building new dams and reservoirs.

Two areas of the country experiencing severe water resource problems are the High Plains (Ogallala Aquifer) and the Colorado River Basin region. Areas in the High Plains are using water from the Ogallala Aquifer at rates much faster than it is being recharged. As a consequence, underground water levels are falling, and pumping the water to the surface is becoming

CONTROVERSY 10.2

Has the West Had Enough Water Projects? And Enough Growth?

In normal years, we'll have some trouble. In drought years, we'll have a disaster.
David Kennedy, Metropolitan Water District of Southern California

The best cure for a threatening water shortage is not necessarily more water; savings in water use, or transfer of water to less consumptive, higher-yield applications . . . may offer better solutions.
Committee on Water of the National Research Council

Unless the Lower Colorado River Basin Project as planned is allowed to proceed, the nation will not keep faith with the people of the Pacific Southwest.
Craig Hosmer, former Congressman from California

The Commission believes that since the West has been won, there is no reason to provide additional interest-free money for new irrigation development in the 17 Western states.
National Water Commission

David Kennedy (quoted in *Newsweek*) says that more water is needed for Los Angeles and San Diego, but the Committee on Water argues for more intelligent use of water. Former Congressman Hosmer expresses the view that the West is "owed" more water, and the National Water Commission replies that the debt has been paid.

Together, these arguments are a miniature picture of the debate on western water policy during the past two decades. They represent the "no new water projects" philosophy and the "we need the water" philosophy.

In the west, population growth, industry, and agriculture go hand in hand with water supplies. Each new supply seems to make more demand possible. Many preservationists ask for an end to the building of water projects. They are trying not merely to protect free-flowing rivers, fish and wildlife, gorges, and na-

tive vegetation. They are also calling for an end to growth in the west. The west has grown enough, many of them say.

Has the west grown enough? Are water projects, canals, pipelines, and dams only the pets of politicians? Or do they serve and supply existing real needs? Are growing demands inevitable, and should those demands be met as long as possible? Are more water projects the best use of limited federal and state money? Where should the line be drawn, or should even the grandest plans be considered for development?

Sources:
David Kennedy, quoted in *Newsweek*, May 31, 1982, p. 33.
Committee on Water, National Research Council, *Water and Choice in the Colorado Basin*. (Washington, D.C.: National Academy of Sciences, 1968.)
National Water Commission, Press Release, June 14, 1973.

more and more costly. Conservation plans have reduced water use for irrigation, but not enough to solve the problem.

In the Colorado River region, all of the flow of the Colorado is currently being used. This system of use is being threatened by the tapping of river water by Arizona cities. It is also threatened by further development in other parts of the region using Colorado River water. Southern California, especially, which has been using Colorado water for its cities, is struggling to find replacement supplies.

Questions

1. What potential problems can arise from persistent overdraft of groundwater? Give at least two examples.
2. Describe the physical and ecological impact of a dam in place across a river.
3. How can water resources be extended or increased without building new dams?

Further Reading

Beattie, Bruce R. "Irrigated Agriculture and the Great Plains," *Western Journal of Agricultural Economics*, **6**(2) (December 1981), 287.
Takes a bit of economist's jargon to penetrate, but otherwise fun to read.

Boslough, John. "Rationing a River," *Science 81* (June 1981), p. 26.
A highly readable article with many color pictures of the Colorado.

Committee on Water of the National Research Council. *Water and Choice in the Colorado Basin*. Washington, D.C.: National Academy of Sciences, 1968.
Technical, but readable; highly iconoclastic.

Fradkin, Philip. *A River No More: The Colorado River and the West*. New York: Knopf, 1981.
The big picture of the Colorado, including historical development. May be a bit preachy in tone.

Francko, David, and Robert Wetzel. *To Quench Our Thirst*. Ann Arbor: University of Michigan Press, 1983.
A well-done general text on water, covering both quantity and quality aspects of water.

Kohrl, William. *Water and Power*. Berkeley: University of California Press, 1982.
A readable story of the political power struggle in which Los Angeles secured one portion of its water supply from the Owens Valley of California.

Shoji, Kobe. "Drip Irrigation," *Scientific American* (November 1977), p. 62.
Fairly technical but still accessible treatment of a highly efficient method of irrigation.

U.S. Geological Survey. *National Water Summary-1983* and *National Water Summary-1984*.
Remarkable compendia of water information. A bit technical, but profusely illustrated and very well written.

CHAPTER ELEVEN

Purifying Water

In this chapter, we examine how people have learned to protect drinking water supplies from contamination by bacteria, viruses, amoebae, and protozoa. We will find that one protozoan, *Giardia lamblia*, poses a new threat to the safety of our drinking water. We also discuss the methods of testing water for bacterial and viral contamination. Finally, we describe the modern processes for purifying drinking water.

Learning from Past Mistakes

Water Can Carry Disease

The notion that water can carry disease first occurred to the ancient Greeks. The physician Hippocrates, the ancient innovator of medical ethics, advised that polluted water be boiled or filtered before being consumed. Despite Hippocrates's early recommendations, only 150 years ago most people were still unaware that human diseases could be spread by water. For this reason, many communities dumped their raw sewage into the same lake or river that they, or other communities, used for drinking water.

Typhoid and Asiatic *cholera* are two diseases spread by water polluted with human wastes. These diseases attack and infect the intestinal tract of humans. Bowel discharges of infected individuals contain the pathogens (disease-producing microorganisms) that spread the diseases. If these bowel discharges enter a water source, there is a high probability that new infections will occur among those who drink the water.

In the past, typhoid was spread principally by polluted water. In the United States today, most cases of typhoid are due to contaminated food. The food was probably contaminated during its preparation by an asymptomatic carrier of typhoid (an individual who carries the pathogens but exhibits no symptoms of the disease). One famous case involved a woman who was a cook for a family on Long Island in the summer of 1906. Typhoid infections occurred among the members of this family, and an investigation showed that Mary, the cook, was a chronic carrier of typhoid. The state of New York eventually found employment for "Typhoid Mary" in a laundry to ensure that she would not endanger others.

Chronic carriers of cholera, unlike typhoid carriers, are not common. The disease is spread mostly via the water route. In the midst of an epidemic, however, individuals may temporarily carry the pathogen without showing signs of the disease. Such people are capable of transmitting the disease through contact and food handling. Cholera was entirely absent from the United States for more than 50 years. From 1973 to the present, however, we have seen sporadic cases among individuals who have eaten fish and shellfish taken from the Gulf of Mexico and its tributaries, and in 1981, a local cholera epidemic occurred on an offshore oil rig in the Gulf of Mexico.

During 1850–1900, people gradually became aware that such diseases as typhoid and cholera are associated with polluted water. Although this is common knowledge today in the technologically advanced countries, it was not common knowledge in 1854, when an epidemic of Asiatic cholera struck hundreds of residents in the London parish of St. James. The pump on Broad

Street in St. James served a large number of families and industries in the area. Its water was regarded by many as being superior to that of other wells. Nevertheless, in the 17-week epidemic of Asiatic cholera that occurred that year, over 700 of the 36,000 residents of the St. James district died. Although the death rate from cholera was over 200 per 10,000 people in St. James Parish, the two districts adjacent to St. James, Charing Cross and Hanover Square, had cholera death rates of only 9 and 33 per 10,000, respectively, during the 17-week period. Clearly, some special circumstance was responsible for the greater number of cholera deaths in St. James.

In the hope of discovering the cause of the outbreak, an inquiry committee was appointed to study conditions in the parish. The committee looked at the population density in the parish, the weather conditions, and the cleanliness of the houses. They also studied the cesspools, the sewers, and the water supply. The committee, it must be concluded, was fundamentally unaware of the cause of the epidemic. Picture, if you will, a gentleman in white wig and frock delivering the following pronouncement in the name of the committee:

> . . . a previous long-continued absence of rain . . . ; a high state of temperature both of the air and of the Thames [River] . . . ; an unusual stagnation of the lower strata of the atmosphere, highly favorable to its acquisition of impurity. . . .
>
> . . . their combined operation, either by favoring a general impurity in the air or in some other way, concurred in a decided manner, last summer and autumn [1854], to give temporary activity to the special cause of that disease.

Though still unaware of the cause, they did, however, recognize an unusual circumstance:

> But, as previously shown in the history of this local outbreak, the resulting mortality was so disproportioned to that in the rest of the metropolis, and more particularly to that in the immediately surrounding districts, that we must seek more narrowly and locally for some peculiar conditions which may help to explain this serious visitation.

One member of the inquiry committee, John Snow, began an investigation of the deaths in the area

near the Broad Street well (Figures 11.1 and 11.2). His report of the epidemic in St. James Parish revealed that 73 of the first 83 deaths in the epidemic occurred among individuals whose closest source of water was the Broad Street pump. Interviews indicated that about 90% of these 73 deaths had occurred among people known to drink from the well either constantly or occasionally. The ten additional deaths in the first stage of the epidemic occurred in households nearer to some well other than the Broad Street well; however, it was found that five of the ten households were in the habit of sending to the Broad Street pump for their water. Dr. Snow wrote in the Report of the Inquiry Committee:

> I had an interview with the Board of Guardians of St. James's Parish on the evening of Thursday, 7th of September, and represented the above circumstances to them. In consequence of what I said the handle of the pump was removed on the following day . . .

In the spring of 1855, the Reverend Mr. Whitehead, a member of the Inquiry Committee, discovered that residents of the house at No. 40 Broad Street had had an unidentified disease prior to the epidemic. He noted further that their dejecta had been disposed of in a cesspool not far from the well of the Broad Street pump. The brickwork of the cesspool was found to be defective in the sense that

> the bricks were easily lifted from their beds without the least force; so that any fluid could pass through the work . . . into the earth . . .

The main drain leading from the house was in a similar state. The cause of the epidemic had been discovered.

From the investigations at St. James, it had been learned that cholera infections could be transmitted through a water supply. It took 50 years more, however, before engineers had perfected ways of purifying water to guard citizens from disease.

Still other notable epidemics were to occur, many of them contributing to a growing awareness of the link between water supplies and disease. In the spring of 1885, the mining town of Plymouth, Pennsylvania had been drawing its water from two sources. One was a mountain stream of seemingly high quality; the other was the Susquehanna River, into which Wilkes-Barre, upstream from Plymouth, was discharging its wastes.

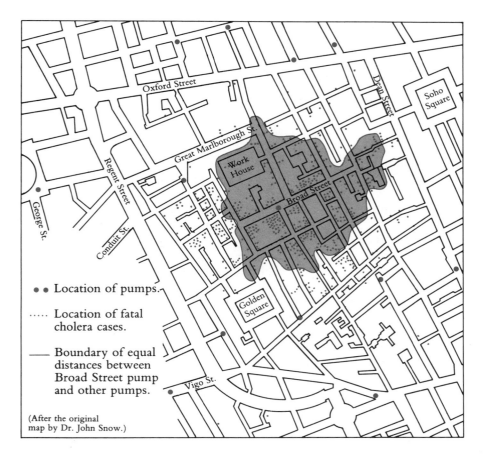

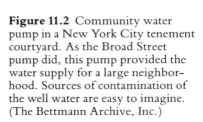

Location of pumps.

Location of fatal cholera cases.

Boundary of equal distances between Broad Street pump and other pumps.

(After the original map by Dr. John Snow.)

Figure 11.1 Asiatic cholera and the Broad Street pump, London 1854. (From Sedgwick's *Principles of Sanitary Science and Public Health*, revised and enlarged by S. Prescott and M. Horwood. Used by permission of Louise P. Horwood.)

Figure 11.2 Community water pump in a New York City tenement courtyard. As the Broad Street pump did, this pump provided the water supply for a large neighborhood. Sources of contamination of the well water are easy to imagine. (The Bettmann Archive, Inc.)

In April of that year, an epidemic of typhoid struck the town, affecting 1,100 of the town's 8,000 inhabitants; 114 people died.

Several explanations were advanced for the epidemic. Some people blamed the generally poor sanitary condition of the town for the epidemic. The sudden onset was attributed to the warm rays of the April sun falling upon the dirt in the town and thereby releasing poisonous vapors. Dr. L. H. Taylor, however, who investigated the outbreak, noted that Plymouth was no dirtier than neighboring towns and no dirtier in that year than in previous years. "Thoughtful minds," he tells us, "turned to the water supply as furnishing the true cause of the invasion."

It was the mountain stream, in fact, that carried the typhoid pathogens and brought the epidemic to the town. An inhabitant of a house located on the stream banks had contracted typhoid in late December of 1884. His illness persisted for three months, and during this period, his dejecta were thrown or deposited on snowy or frozen ground. From there, a rain or thaw had carried them down the slopes into the stream that fed the town's reservoir. An apparently pure mountain stream, trusted for its quality, had caused the epidemic.

Dr. Taylor drew two conclusions from the epidemic: first, that the dejecta of typhoid fever patients must be carefully disposed of and disinfected; second, that water companies "should be compelled to remove from the banks of their streams and reservoirs not only all probable, but all possible sources of pollution," a warning that many water managers carefully heed to the present time (see Controversy 11.1).

Many areas of the United States experienced typhoid in this era. In Massachusetts, in 1890, an epidemic was traced to the Merrimac River, which supplied water to the numerous manufacturing towns of the region. In that same year, an epidemic of typhoid occurred in New York State, and again, the path of infection followed a river, this time the Mohawk and Hudson rivers. Schenectady, Cohoes, West Troy, and Albany all experienced typhoid epidemics. Eighty years later, we find this same stretch of the Hudson River to be the route of contamination of the chemicals known as PCBs.

Filtration Is Effective

By 1872, **filtration** through sand beds was being recognized as a way to make water clear and relatively safe for drinking. In that year, Poughkeepsie, New York became the first city in the United States to install sand filters to purify its water. The filters in Poughkeepsie are still in operation over 100 years later.

The effectiveness of the sand filter in removing disease-causing organisms was demonstrated dramatically in 1892. Altona and Hamburg, sister cities on the Elbe River in Germany, shared a common boundary but did not have the same water supply. Hamburg's water was drawn from the Elbe above the two cities; Altona drew its water from a site downstream from the two cities. Thus, the supply of Altona was potentially polluted by the sewage of both cities. To produce water of sufficient clarity, Altona began to filter its water supply through beds of sand. Hamburg, however, drew its water directly from the river without any treatment.

The cholera epidemic that struck Hamburg in the fall of 1892 killed more than one resident in 100 and made nearly twice that number ill. In Altona, the cholera death rate and case rate was less than 15% of Hamburg's, and the difference was correctly attributed to the filtration of its water. The cholera cases that did occur in Altona are explained by noting that many residents of Altona worked in Hamburg. The protection provided by sand filtration was clear, and cities moved quickly in the coming years to install filters in an effort to prevent such epidemics. Hamburg was among the first to take the step.

During the early part of the twentieth century, Chicago was discharging its municipal wastes into Lake Michigan. It was also drawing its water supply from the same source via an intake pipe extending into the lake, just as it does today. Not surprisingly, typhoid fever was endemic in Chicago in this era. One epidemic, in 1885, was caused by heavy rains washing wastewater far out into the lake, out to the water intake. The death toll from cholera, typhoid, and other diseases is thought to have reached 90,000 people, an eighth of the city's total population.

The city recognized that a connection existed between its high rate of typhoid fever and its water supply, but rather than purify its water supply mechanically by filtration, Chicago chose to divert its sewage from the lake. To accomplish this, the Chicago River, which carried the city's sewage and which had emptied naturally into Lake Michigan, was diverted into a canal. The Chicago Drainage Canal, as it is called, was created to carry the river's flow and the sewage of Chicago all the way to the Mississippi River, over 350 miles distant.

CONTROVERSY 11.1

Should Water-Supply Reservoirs Be Used for Recreation?

Without strict controls over recreation, waterborne diseases become a serious problem.
American Water Works Association, 1971

No amount of properly planned recreation can generate enough contamination of reservoir water to pose any technical difficulties in treatment.
President's Council on Environmental Quality, 1975

The American Water Works Association issued the following statement on the use of water-supply reservoirs for recreation:

> The American Water Works Association supports the principle that water of highest quality be used as source of supply for public water systems. Since each water utility is responsible for its product, determination of type and extent of recreational use of impounding reservoirs shall be vested in the water utility.
>
> Water utilities of the United States and Canada provide water of the highest quality in the world. To improve further upon this record, each water utility must continue to recognize its responsibility to deliver a safe and appealing product to its consumers. A growing demand for use of reservoirs for recreational purposes, however, may make this responsibility more difficult to carry out in the future. In some areas, uncontrolled use of reservoirs already has resulted in deterioration of water quality. Without strict controls over recreation, water-borne diseases could become a serious problem. . . .

> Research on the effects of recreation on water quality should be expanded. Reservoirs used for recreation must receive close supervision. In all cases, water from reservoirs used for recreation must be treated, with the degree of treatment to be determined according to state or provincial and local utility laws, regulations, and policies. . . .
>
> Public water-supply reservoirs should not be used for recreation if other surface waters are available. Bodily contact sports such as swimming should not be allowed in public water supply reservoirs. For the purposes of this policy statement, reservoir means an impoundment reservoir subsequent to which water receives treatment before consumption. Distribution reservoirs from which water is supplied directly to the public require the strictest of controls and under no condition should be used for recreation (American Water Works Association, 1971)

The recreation referred to as permissible under certain circumstances is presumed to include boating, fishing, picnicking on shores, and so forth. The Association takes the position that water-supply reservoirs should not be used for recreation activities unless there are no other possibilities in the area. Essentially this is a weak statement. It admits that recreation on reservoirs is not so serious a threat to water quality that it must be prevented at all costs. It recognizes that such practices already exist. The only strong statement is the directive against swimming in public water-supply reservoirs. In fact, however, a number of reservoirs in the midwest and west are used for swimming.

The President's Council on Environmental Quality (1975) disagreed with the American Water Works Association. It advocated (and attempted to justify) more, not less, recreation, including swimming in larger reservoirs. The council argued that modern water treatment processes are fully adequate to safeguard water supplies.

If your community is served by a reservoir, find out whether recreational activities are permitted in the watershed area or on the reservoir. Do you think it wise to prevent any public use of the reservoir and watershed area? Can you support your opinion? If recreational uses other than swimming are permitted, what precautions would you recommend to go along with such use? That is, what design features would you incorporate in the recreational area? What activities would you restrict?

Sources:

American Water Works Association, "Recreational Use of Domestic Water Supply Reservoirs." A statement adopted by the Board of Directors on January 26, 1958, and revised January 25, 1965, and June 13, 1971.

President's Council on Environmental Quality, *Recreation on Water Supply Reservoirs*. Washington, D.C.: U.S. Government Printing Office, 1975.

This step, of course, polluted the Mississippi, and eventually Chicago was forced to treat its sewage before it could be discharged into the drainage canal.

Typhoid and cholera are not the only bacterial diseases transmitted through polluted water. Paratyphoid, dysentery, and other diarrheal diseases can also be spread by bacteria in water. The failure of the water treatment system of Detroit, Michigan led to an epidemic involving 45,000 cases of bacterial dysentery in 1926. Nor are bacteria the only microorganisms that spread disease. A form of dysentery referred to as amoebic dysentery is caused by an amoeba, a one-celled animal living in water. This disease is accompanied by a diarrhea that brings severe weakness. Chicago suffered an epidemic of amoebic dysentery caused by *Entamoeba histolytica* in 1933. Visitors to the Chicago World's Fair who stayed at two of the city's first-class hotels were the principal victims. The hotel sewer lines were found to be leaking into the water supply mains of the hotels.

Chlorination Kills Microbes

Filtration was not the only process devised to purify water. A chemical "sterilization" using chlorine and chlorine compounds was also investigated in the early part of the century. Experiments with **chlorination**

using calcium hypochlorite (also called bleaching powder) were conducted in 1896 by George Fuller in Kentucky. In the following year, sodium hypochlorite was successfully introduced to the water supply of Maidstone, England to arrest an epidemic of typhoid. The process was first used on a continuous basis in Middlekirke, Belgium in 1902. In England, the Metropolitan Water Board of London began continuous chlorination of a portion of its supply two years later, using sodium hypochlorite.

In 1908, because its supplies were polluted, the East Jersey Water Company, a private firm that furnished water to Jersey City, was ordered by the courts to make its water pure and safe. Calcium hypochlorite was introduced to the water to eliminate the bacterial contamination. The court agreed that the bacteria had indeed been destroyed and issued the following opinion:

From the proofs before me of the constant observations of the effects of this device, I am of the opinion and find that it is an effective process, which destroys in the water the germs, the presence of which is deemed to indicate danger, including the pathogenic germs. . . .

. . . The reduction and practical elimination of such germs from the water was shown to be substantially continuous. . . . I do therefore find and report that this device is . . . effective in removing from the water those dangerous germs which . . . possibly exist therein at certain times. (Johnson, 1913)

Chlorination, now sanctioned by law as a means of purifying water, was rapidly installed in many cities. Today nearly all municipal water supplies in the United States use chlorination to disinfect their water supplies, although chlorine gas has replaced calcium and sodium hypochlorite. Calcium hypochlorite still remains a valuable tool to use on a temporary basis for disinfection when disasters such as floods or earthquakes strike, or in times of war. The United States Public Health Service still tells travelers how to disinfect water by using bleach: four drops of a bleach mixed into each quart or liter of water is sufficient to disinfect water within 30 minutes.

Chlorine gas is used instead of calcium or sodium hypochlorite for a variety of reasons. It is more effective in destroying bacteria, it is easily and safely stored in large quantities, and its introduction into the water is easily regulated. We will have more to say about the specific process of chlorination when we discuss the current methods of water treatment in more detail.

Since those early times, chlorination has been properly regarded as one of the most important tools scientists have to ensure the safety of drinking water. It has prevented illnesses and deaths in untold numbers around the world. Nonetheless, some questions about its side effects on human health have been raised; we explore these questions in Chapter 14.

Viral Diseases Are Spread by Water

Viruses, too, can cause waterborne epidemics. Viruses, the simplest of living organisms, cannot reproduce themselves as bacteria can. Instead, they invade plant, animal, or bacterial cells and use the reproductive capability of these cells to produce more of their own kind. Smaller than bacteria, viruses require highly specialized equipment for detection in water samples.

Our knowledge of the potential for waterborne viral epidemics has come only recently. Scientists believe that certain polio epidemics have resulted from water supplies contaminated by sewage. Cases have also been reported in which viral respiratory diseases were spread by swimming in contaminated pools. Most proven waterborne epidemics caused by a virus,

however, have involved the hepatitis virus. The hepatitis victim suffers an inflammation of the liver. Symptoms include a feeling of weakness, nausea, and fever. The liver is unable to perform its normal function of breaking down certain pigments in the blood; these pigments build up in the body and turn the skin and the whites of the eyes a yellow color. This condition is called jaundice.

Most hepatitis epidemics have occurred in small, private water supplies or have resulted from the consumption of raw clams or oysters grown in sewage-polluted water. One large-scale epidemic, however, was recorded in New Delhi, India in 1956. In this outbreak, 50,000 people were infected with hepatitis from the city water supply, which had become contaminated with sewage. Interestingly, the water supply had been chlorinated before distribution to the public. During the hepatitis epidemic, no increase in waterborne bacterial illnesses, such as cholera or typhoid, was observed. This points up a difference between bacteria and viruses: bacteria are much more easily killed by chlorine than viruses are, although chlorination is effective against viruses if properly applied. In developing countries, where water supplies are often polluted, the presence of a group of viruses known as ECHO viruses contributes to chronic diarrheal disease.*

A New Threat to Water Supplies

Waterborne diseases in epidemic proportions have continued to occur in the United States throughout this century, although with decreasing frequency. The cities struck by these epidemics are typically small and have insufficient treatment of their water supplies. Although such epidemics are becoming infrequent in the United States, they continue to occur where water supplies are not treated, where treatment systems fail, or when water distribution systems become contaminated.

In 1980, just over 20,000 people in the United States were affected in 50 outbreaks of diseases traced to water supplies. In only 22 of the 50 outbreaks was the causative agent identified. In 30% of the cases where the causative agent was identified, the responsi-

*Other viral agents known to have caused gastroenteritis by their presence in water supplies include the Norwalk agent, the Snow Mountain agent, and the Hawaii agent.

ble agent was a little-known human parasite, *Giardia lamblia.*

Long thought to be a cause of disease only in hunters, backpackers, and others who drink untreated water, the *Giardia* protozoan is now known as the cause of many waterborne outbreaks in the United States. Symptoms of the disease, giardiasis, include diarrhea, cramps, and weight loss. The disease does not occur until 1 to 8 weeks after infection and may last two to three months. Onset of the disease is delayed while the parasite population builds up to the level that causes symptoms.

The flagellated protozoan *Giardia lamblia* has been recognized since the 1800s. It was first described by the famous Dutch microbiologist von Leeuwenhoek in the seventeenth century. The parasite has been found in about 4% of the population of the United States. Surprisingly, the same parasite can be found in beavers, coyotes, cattle, dogs, and cats. This fact explains why other warm-blooded animals can serve as a "reservoir" of infection for humans. In fact, the 1976 outbreak of giardiasis in Camas, Washington has been associated with the presence of infected beavers living upstream from the water supply.

Although *Giardia* may have been one of the responsible agents in an epidemic among U.S. residents of an apartment building in Tokyo in 1947, the first recognized outbreak among U.S. citizens occurred in 1970, when over 100 people in two U.S. tourist groups

in Leningrad became ill with the disease. Thereafter, many more cases have occurred among U.S. travelers who have consumed tap water in Leningrad. Giardiasis was also observed in U.S. travelers returning from the Spanish island of Madeira in 1976. A significant but unknown fraction of the resident population of these places is likely to have had giardiasis and to have recovered from the disease.

In the United States, more than 15,000 cases of giardiasis occurred from 1965 to 1980; the more significant outbreaks, along with causes, are listed in Table 11.1. From these outbreaks, two obvious causes stand out: the absence of any water treatment and the failure of treatment processes. Also plain to see is the evidence that chlorination alone may be inadequate to destroy the parasite. In six of the cases shown in Table 11.1, filtration did not accompany the chlorination process. Its presence might have prevented the outbreaks. The explanation for the importance of filtration and the possible failure of chlorination lies in the nature of the *Giardia* protozoan. The parasite passes out of the intestine in the form of a cyst, a resistant form of the protozoan. Chlorination, unless precisely practiced in terms of contact time and water temperature, may fail to destroy the cyst so that only filtration stands between the delivery of pure or contaminated water. The lesson is that a clear water supply, though it appears to be clean enough without filtration, may not be adequately treated by chlorination alone. The water that has only

Table 11.1 Waterborne Outbreaks of Giardiasis in the United States

Location[a]	Date	Cause	Number of cases
Aspen, Colo.	December 1965–January 1966	Well-water supply contaminated by sewage; chlorination only	123
A camp, San Juan area of Utah	September 1972	Water not treated at all	60
Meriden, N.H.	June–August 1974	Water only chlorinated	78
Rome, N.Y.	November 1974–June 1975	Water only chlorinated	4,800–5,300
Camas, Wash.	May 1976	Chlorination and filtration inadequate	600
Berlin, N.H.	April 1977	Filtration inadequate	750
Hotel, Glacier Park, Mont.	July 1977	Surface water supply not treated	55
West Sulfur Springs, Mont.	July 1977	Surface water supply only chlorinated	246
Vail, Colo.	1978	Inadequate filtration	5,000
Bradford, Pa.	1979	Water only chlorinated	2,900
Red Lodge, Mont.	1980	Water only chlorinated	—

a. Only outbreaks with more than 50 cases are listed here.

been chlorinated may appear pure due to the destruction of the coliform bacteria (the organisms that indicate the presence of sewage contamination), but the *Giardia* cyst may survive. The lesson seems plain, but numerous water supplies in the United States still do not include filtration in their treatment scheme. New York City does not filter the vast quantities of water that come from its Catskill Mountains reservoir, Boston does not filter its input from the Quabbin reservoir, and Seattle does not bother to filter its water supply.

Shen Dar Lin (1985), an expert in giardiasis from the Illinois State Water Survey, put the warning succinctly:

> Filtration of water supplies is required to provide a second barrier against the transmission of waterborne diseases. It also helps to reduce the load on the disinfection process, thereby increasing the efficiency of disinfection, which is the primary means of disease control.

Epidemic outbreaks of giardiasis have continued to occur since the last outbreak noted in Table 11.1. Nine outbreaks of giardiasis occurred between 1981 and 1983 in Colorado alone, where the parasite seems to be quite common. Whether the parasite is spreading among the populace is unknown. The possibility that the parasite might be spreading, however, should give pause to cities that only chlorinate their water supplies.

A final word of warning about giardiasis: Hikers in the backcountry can no longer drink from mountain streams without fear. The fact that beavers, muskrats, and other wild animals carry and transmit the protozoan makes suspect even these once safe sources of water. Numerous cases of giardiasis have already been reported among backpackers. We would not trust any trail method of disinfection other than boiling the water for 15–20 minutes.

How Can We Tell If Water Is Unsafe to Drink?

Coliform Bacteria Are Indicators

Three major classes of disease-causing organisms are found in water: bacteria, viruses, and protozoans. The maladies they spread range from severe diseases, such as typhoid and dysentery, to minor respiratory and skin diseases.

Examining water to isolate all the possible disease-causing organisms in it would be a difficult, costly task. To avoid such exhaustive examination, special tests have been devised. These tests detect a group of bacteria that, if present, are likely to be accompanied by disease-producing microorganisms. If this group of bacteria is absent, or if the level of these bacteria in a water sample is sufficiently small, it is doubtful that the less numerous disease-causing organisms are present.

The bacteria for which scientists test are nonpathogenic; that is, with rare exception, they do not cause disease. These bacteria are found naturally in the intestines of warm-blooded animals, including humans, and are known as **coliforms**. Although coliforms rarely cause disease, their presence in water indicates that the water may have been contaminated with raw (untreated) sewage. To drink such water is obviously unwise.

Thus, the procedure of examining water for the presence of disease-causing organisms does not involve looking for the pathogens themselves; it is sufficient to show that water has somehow been contaminated with sewage or not treated enough to kill all the coliform bacteria in it. Now let us look again at the careful wording of the court's opinion on chlorination (p. 000). We can see that the judge was referring to the coliforms when he referred to organisms "the presence of which indicate danger." The coliforms are often called *indicator organisms*.

The Environmental Protection Agency has recently drawn up drinking water regulations that set forth limits to coliform bacteria. Although the standards have actually existed for nearly half a century, they could not be enforced by the federal government unless the water was moved in interstate commerce (for example, bottled water). According to these new regulations, if more than four coliform bacteria are present in a 100 ml sample taken from water destined for human consumption, the supplier must notify the state. Further, the supplier must take action that will decrease the coliform count to less than one in each 100 ml of water.

Four coliforms per 100 ml sample is the point at which corrective action for drinking water is required. In contrast, water with more than 2,300 coliforms per 100 ml is considered unsafe to swim in. Boating is allowed in water with up to 10,000 coliforms per 100 ml (but don't fall in!). (See Table 11.2.)

In 1969, some 969 public water supplies were examined to see how pure the tap water really was in the

Table 11.2 Activities Permitted for Various Levels of Coliform Bacteria in Water

Coliform level	Activity permitted
1 coliform or fewer per 100 ml of water	Water safe for drinking
4 coliforms or more per 100 ml of water	State must be notified and corrective action taken
2,300 coliforms or fewer per 100 ml of water	Swimming is allowed
10,000 coliforms or fewer per 100 ml of water	Boating is allowed

United States. Somewhat surprisingly, 12% of the water supplies were found to exceed the permissible limits of coliform bacteria. Figure 11.3 shows that the water-supply systems in violation of coliform standards were generally in smaller communities. For instance, 79 of the 120 systems found in violation were in communities of less than 500 people, and 108 of those 120 systems served communities of 5,000 people or fewer.

The EPA found similar rates of violations in 1980. In that year, 90% of community systems met the standards for coliform bacteria, and 10% violated the standards either regularly or occasionally.

A 1982 study of rural water supply in the United States focused on private well-water supplies rather than community systems. From a random geographical sample of about 3,000 rural households, nearly 30% of the water supplies exceeded the allowable bacterial count.

Testing for Indicator Organisms

The modern method of testing for the presence and number of coliform bacteria is to filter a measured sample of water through a special filter membrane. This paper membrane is made by the Millipore Company and is called a Millipore filter. It is manufactured with pores, or holes, so small that bacteria cannot pass through, although the water can. Thus the bacteria are trapped on the paper filter when a sample of polluted water is poured through it (Figure 11.4). The filter paper is then placed on a medium containing the special nutrients that coliform bacteria need for growth. On that medium, each coliform bacterium captured on the membrane filter divides and forms a colony. These colonies are visible to the naked eye; they can be identified and counted in order to estimate the number of coliform bacteria in the water.

Determining the presence of viruses in water is not as simple as detecting the presence of bacteria. Most viruses are too small to be trapped by a filter, and they cannot be grown in a simple broth. However, just as water is not routinely examined for the disease-causing bacteria themselves, water is not commonly checked for disease-causing viruses either. Again, the presence of coliform bacteria in a water sample is taken as evidence that disease-causing viruses as well as bacteria are likely to be present in the water.

Water Treatment

Communities treat their water to remove both disease-causing organisms and harmful chemicals. In addition,

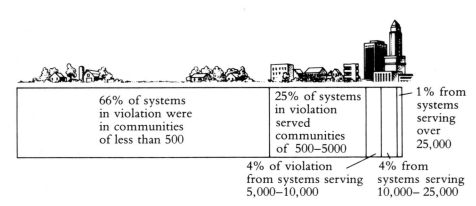

66% of systems in violation were in communities of less than 500

25% of systems in violation served communities of 500–5000

1% from systems serving over 25,000

4% of violation from systems serving 5,000–10,000

4% from systems serving 10,000–25,000

Figure 11.3 Distribution of violations of coliform standards among communities of various sizes. A study of rural households completed in 1982 showed that 29% of the homes surveyed had evidence of bacterial contamination of their water supplies, suggesting that not much change or improvement occurred in the 1970s. (Data from J. McCabe et al., "Survey of Community Water Supply Systems," *Journal of the American Water Works Association*, **62**, [1970] 670.)

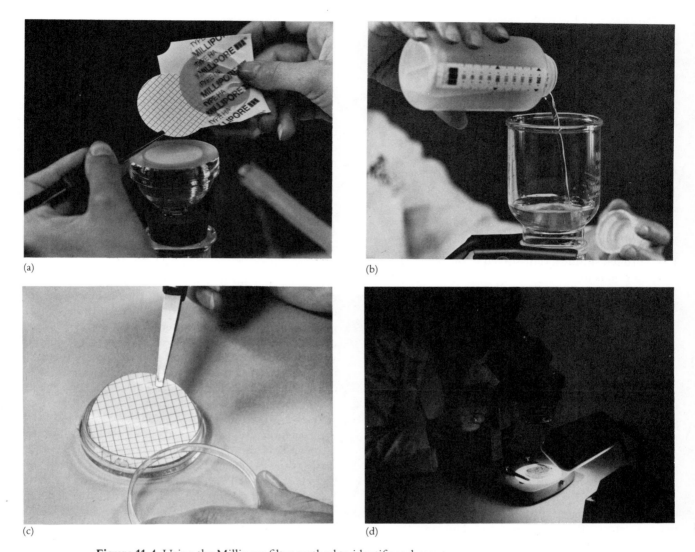

(a)   (b)   (c)   (d)

Figure 11.4 Using the Millipore filter method to identify and count coliform bacteria. (a) A sterile Millipore membrane is placed on a filter resting on the base of a funnel. (b) A glass top has been added to the funnel base, and the sample of water to be tested is poured into the funnel. The water is drawn through the funnel by applying vacuum suction. Coliform bacteria are too large to pass through the filter and are retained on its surface. (c) The membrane filter is removed from the funnel and placed in a petri dish. The petri dish contains a medium rich in nutrients that favor the growth of coliform bacteria. The dish is covered and placed in a constant-temperature "incubator" at 35°C for 24 hours. If coliforms were captured by the filtering step, colonies of the bacteria will grow up around each bacterium lodged on the filter. (d) After 24 hours of incubation, the coliform colonies are counted under a low-power microscope. In this photo, the cover of the petri dish has been removed and the bottom of the dish with the medium and colonies has been placed in the microscope field. (Photographs courtesy of the Millipore Corp., Bedford, Massachusetts)

water purification is designed to make water pleasant to drink. Not all cities purify their water in the same way because of the presence of different substances in their basic supply. We will describe a set of processes that is fairly typical among cities, and we will indicate the order in which these processes are commonly arrayed.

Removal of Taste and Odor

Tastes and odors that exist in the untreated supplies of some cities may be offensive to consumers of the drinking water and, hence, require removal. The causes of such odors and tastes may include:

1. Hydrogen sulfide gas dissolved in the water. This condition is most common in water drawn from wells.
2. Dead or decaying organic matter stemming from algae or aquatic weeds. Water from reservoirs may contain such materials.
3. Chemicals used in agriculture to control weed and insect pests and chemicals from industrial wastes.

The methods of controlling tastes and odors may be classified in two groups: (1) methods to *prevent* growth of algae and aquatic plants, and (2) methods to *remove* tastes and odors once they occur. To prevent algae and other aquatic plants from growing, chemicals may be added to the water while it is in reservoirs. Herbicides were once considered for this task but have now been rejected. The one herbicide that was thought to be unobjectionable (but never used) is now known to cause cancer. This is the compound lindane.

Copper sulfate is a commonly used substance for algae control. It may be slowly dumped from boats crisscrossing the reservoir to ensure good distribution in the water.

In the treatment plant itself, tastes and odors may be removed by aeration (exposure to the air) or by chemical treatment. Water taken from a river, from a lake, or from a community well is often aerated during treatment. The water is sprayed into the air from banks of fountains or allowed to flow over racks in a thin film. You may have seen the spraying step at a local water distribution reservoir. The aeration accomplished either by spraying or by the rack method removes dissolved gases (for example, hydrogen sulfide) from the water. With the removal of hydrogen sulfide, a substantial control of odor may be achieved. (Hydrogen

sulfide gas is what we smell when an egg becomes rotten.)

First Chlorination

After aeration, chlorine gas is added to the water to kill disease-causing organisms. This step, **chlorination**, occurs again after all other treatment steps are completed. More detail on chlorination is provided later in this chapter.

Removal of Particles and Colors

The minute suspended particles that give color to water will neither dissolve nor settle rapidly from the water; particles behaving this way are termed *colloidal*. You can conduct an experiment to illustrate the concept of a colloid. Mix a teaspoonful of rich earth (humic soil) into a gallon jug of clear water. Note how the passage of light is restricted through such water. Observe the yellow-brown color imparted to the water by the particles. These particles can be removed by a process called **coagulation**.

In the first step of coagulation, alum or ferrous sulfate is continuously added and rapidly mixed into the stream of water flowing through the plant (Figure 11.5). When the chemicals are added, a precipitate forms, called a *floc*, which consists of mineral particles that are insoluble in water. A floc particle resembles a feathery cloud in miniature. The floc then passes through a flocculation basin, where the floc is thoroughly mixed with the suspended particles.

Next, the water is pumped slowly through a **settling tank**. Here the water is detained long enough to allow most of the floc or precipitate to settle to the bottom of the tank. In the process of settling, the "blanket" of floc will capture and drag down a large share of the particles suspended in the water. The precipitate is scraped away from the bottom by a scraping bar to prevent buildup.

The water leaving the settling tank has had much of its turbidity (cloudiness) removed. In part, these early steps protect the effectiveness of the next step, filtration through beds of sand, by removing relatively large particles that could clog the filter.

In many water treatment plants, a small amount of activated carbon particles may be added along with the alum or ferrous sulfate in the first step of the coagula-

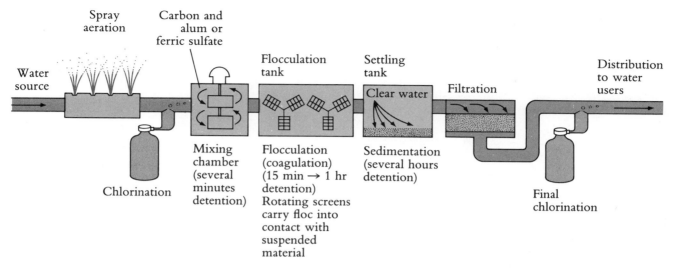

Figure 11.5 A typical water treatment plant scheme.

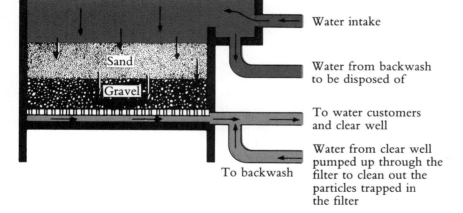

Water intake

Water from backwash
to be disposed of

To water customers
and clear well

Water from clear well
pumped up through the
filter to clean out the
particles trapped in
the filter

To backwash

Figure 11.6 A sand filter. Particles are removed when water is passed through the filter. The process eliminates most disease-causing bacteria and protozoa from a community water supply. When the filter becomes clogged with particles, water from the clear well is pumped up through the filter. This reversed flow dislodges particles and cleanses the filter for further use.

tion process. Colloidal particles bind to the carbon. When the carbon is removed in a later step, much of the colloidal material responsible for color is carried with it. Not only is color improved by this **carbon treatment**, but tastes and odors are decreased as well.

Contacting the water with activated carbon particles is a popular method of removing tastes and odors, dating to the 1930s. In that era, filters using activated carbon as the filter medium were used in Bay City, Michigan. In contrast, the recent practice has been to add a small amount of carbon in the first step of the coagulation process. Interestingly, the filtration of water through beds of granular carbon is now recommended as a part of water treatment by the Environ-

mental Protection Agency to remove low-level but important contaminants, such as pesticides and other resistant organics. Some of these compounds are now viewed as potential causes of cancer. Thus, we may see carbon filtration reintroduced as a common water treatment method.

Filtration to Remove Microbes

After the floc has settled out, the water stream passes downward through beds of sand. From top to bottom, the grains in each level of sand get larger and larger. The construction of a typical sand filter is shown in Figure 11.6.

The removal of the larger particles achieved by the coagulation step prevents the sand filter from clogging rapidly. The sand filter does cause a further removal of particles; however, the primary purpose of the filter is to capture and retain bacteria, viruses, and other organisms. Most disease-causing microorganisms are thus prevented from entering the system that distributes clean water to the city.

Periodically, the sand beds must be washed clean if they are to retain their effectiveness in filtering out microorganisms. Some of the filtered water is saved in a "clear well" for this purpose. The washing proceeds by passing the clean water up through the filter to dislodge particles retained in the spaces between the grains of sand. The dislodged particles are now suspended in the upward-flowing wash water, which is then dumped back in the river. Once the filter is clean, it is ready for continued use. The dirtier the water entering the filter, the more frequently backwashing is needed.

All of these treatment steps—aeration, chlorination, coagulation/settling, and filtration—are shown schematically in Figure 11.5.

Final Chlorination

Even though the sand filter is highly effective in removing bacteria and viruses, the removal of these microorganisms is not certain. An additional step, a second step of chlorine addition, destroys any microorganisms remaining after sand filtration. Chlorine also reacts with any ammonia that may be present in the water. Chlorine is added beyond the level required to kill all microorganisms and also beyond the amount required to react with the ammonia present in the water (Figure 11.7). This results in "free," or unreacted, chlorine in solution. One reason that chlorination is so favored as a means to disinfect public water supplies is that this "leftover" or residual chlorine remains, and a quick, simple test for it exists. When the test shows that free chlorine is present, one can be confident that any new microorganisms will be killed.

To conduct the test, add a small amount of starch plus a few crystals of potassium iodide to water from a swimming pool or to chlorinated tap water. Free chlorine oxidizes the iodide to free iodine (chlorine is simultaneously converted to the chloride ion). Free iodine in the presence of starch produces a characteristic blue-purple color, which indicates that free chlorine is available to oxidize the iodide ions. The starch-iodine test

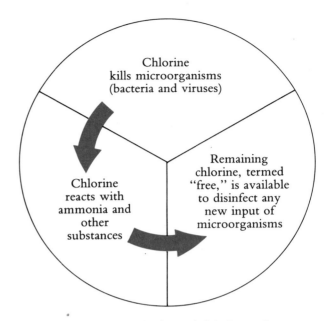

Figure 11.7 Chlorine is the favored disinfectant for water supplies. Chlorine kills microorganisms and reacts with ammonia. Any chlorine in excess of the amount needed to disinfect and react with ammonia will remain free in solution, protecting the water supply from any new sources of pollution en route to the consumer.

indicates whether free chlorine is present in sufficient concentration to destroy disease-causing bacteria. It gives a swift and reliable answer to the question: is the disinfection process operating properly—that is, is the water safe to drink? (But is this enough? See Controversy 11.2.)

Ozonation: An Alternative to Chlorination

One alternative to chlorination is disinfection with ozone. The **ozonation** process, like the chlorination process, is achieved simply by contacting water with gas. Ozone gas is a powerful oxidizing agent that destroys bacteria and viruses. In contrast to chlorination, in which chlorine may combine with hydrocarbons, ozonation does not create chlorinated hydrocarbons and may actually destroy some of the hydrocarbon compounds by oxidation. Furthermore, ozone is effective in removing color and does not produce "off" tastes and odors.

Ozonation of water supplies is practiced in a number of European cities, but chlorination is still the proc-

CONTROVERSY 11.2

Is There Such a Thing as a Zero-Risk Environment?

We are dealing with a time bomb with a 25-year fuse.
Dr. Robert H. Harris, Environmental Defense Fund

Crash programs . . . to meet "yesterday" deadlines are not understandable at all.
A. E. Gubrud, American Petroleum Institute

Several environmental groups were unhappy with how the Environmental Protection Agency set about fulfilling (or not fulfilling) the toxic pollutants portion of the 1972 Federal Water Pollution Control Act—so unhappy, in fact, that in 1975 they sued the EPA in federal court. The petition charged the EPA with failure to require that water supplies be checked for the presence of heavy metals, such as cadmium, and with failure to require proper removal of pesticides, viruses, and asbestos.

"We are dealing with a time bomb with a 25-year fuse," said Dr. Robert H. Harris, at a news conference at which the petition was discussed. He said Americans could not afford to wait to see what would happen, as they had done with cigarette smoking.

Dr. Harris noted that it often took many years for a cancer to develop. Therefore, he argues that exposure of a population to cancer-causing substances might not be reflected in an obvious rise in the cancer rate until 25 years or so after the exposure began. . . .

Dr. Harris, who is associate director of the Environmental Defense Fund's toxic chemicals program, said traces of known and suspected can-cer-causing chemicals had been found in water samples from several major cities. . . .

Dr. Harris said federal regulatory agencies had done "virtually nothing" to protect the public from cancer-producing chemicals in water supplies.

The suit was settled by an agreement between the EPA and the environmental groups, which was approved by a judge of the District Court of Appeals. The agreement required the EPA to start a huge program of experiments on 65 specified toxic pollutants. The agency also agreed to begin studies of the available methods of control and their economic effects. Many industry groups were upset by the agreement. As one industry spokesman put it:

[A] disturbing trend is emerging, ever more clearly. In the past, the scientific and technical communities have experienced difficulty enough in responding to "technology-forcing" legislation and regulation. Now, however, as a result of the Toxics Effluent Guidelines Settlement Agreement, the scientific and technical communities are being asked not merely to accelerate development of toxic pollutant controls, but to improve the whole state of toxic pollutant knowledge by several orders of magnitude. It is extremely doubtful whether all of the combined talents in government, industry, and academia could achieve that end in the time allowed by the Settlement.

The occasional acceleration of a government program in response to a clear and present environmental danger certainly is understandable. Crash programs in the absence of such danger to meet "yesterday" deadlines are not understandable at all. They arise not in response to specific, scientifically demonstrated needs, but from the belief in some quarters that *all* people can be protected at *all* times from *all* real or suspected environmental dangers—the belief, in short, that a "zero-risk" environment is attainable.

Zero-risk is not attainable in the environmental area, any more than it is attainable in any other area of human endeavor. I know of no health specialist who would maintain that there are absolute thresholds, other than zero, below which no health risks will exist for anyone.

We must continue to work hard and systematically to improve the quality of our environment and eliminate true threats to the public health. But crash programs arising out of a kind of national hypochondria not only can yield dubious and harmful results, but also are extremely wasteful. (Gubrud, 1977)

Do you feel that this is the type of question that can be settled in the courts, in the full glare of public view? Are we requiring too much of industry in this case? Can we depend on their good-faith attempts to achieve pollution control without these kinds of legal threats? Can we achieve a zero-risk environment? Should we try?

Sources:

Robert H. Harris, quoted in *The New York Times*, December 18, 1975.

A. E. Gubrud, at the Federal Water Quality Association Conference on Toxic Substances in the Water Environment, Washington, D.C., April 28, 1977.

ess most U.S. water-supply engineers choose. In the United States, ozonation is used by a number of companies that produce bottled water. They choose this method of disinfection to avoid the taste chlorination gives to water, a taste that consumers may associate with a municipal water supply. Los Angeles has recently chosen ozonation for the initial disinfection at its large new water-treatment plant. The plant provides drinking water to 80% of the city's 2.8 million residents. Chlorination is still used, however, as the final treatment step. New York City has also chosen to ozonate the portion of its water supply from its Croton Reservoirs. In 1982, about 50 water treatment plants that use ozone were in operation or under construction in the United States.

Ozone, however, does not leave any persistent free ozone residual in the water, even if added in excess of that required for reaction and disinfection. The lack of any persistent residual means that there is no quick way to determine whether all bacterial and viral contamination has been destroyed—as there is with a chlorinated water supply (see Table 11.3). The only way to see whether the water is safe is to conduct the bacterial test, which takes 24 hours (see p. 000). Furthermore, any subsequent inputs of disease-producing bacteria and viruses would not be destroyed, since no disinfectant potential remains. To protect against fresh inputs of bacteria and viruses in the distribution system, a second disinfectant, probably chlorine, would have to be used at the end of the treatment process.

Table 11.3   A Comparison of Chlorination and Ozonation

Process	Substance used	Effect	Objectionable reactions, by-products	Check on water safety
Chlorination	Chlorine gas	Kills microorganisms	Chlorinated hydrocarbons, taste and odor substances	Immediate check: safe if free chlorine is present
Ozonation	Ozone gas	Kills microorganisms	None yet identified	Must run standard 24-hour tests for coliforms

The fact that ozone does not produce a persistent residual is one drawback to its use. Another caution is that the products of its reactions with organics are still largely unidentified, although aldehydes and other simple organics have been noted. Chlorination, in terms of its effects on humans, has recently come under attack because chlorine acts on hydrocarbon compounds dissolved in water. Some of the chlorinated hydrocarbons produced from this reaction are potential causes of cancer. (See Chapter 14 for a discussion of this problem.)

Summary

It was in the middle to late 1800s that people discovered that their drinking water could carry diseases such as typhoid and cholera. Two processes, still in use today, proved effective in removing the bacteria that cause these diseases. These are filtration of water through beds of sand, and chlorination—the addition to water of chlorine at low levels. Filtration removes microorganisms; chlorination kills them. Waterborne disease-causing viruses, such as the hepatitis virus, are eliminated by these processes as well. It has recently become apparent that a protozoan, *Giardia lamblia*, poses a new threat to safe drinking water. Because it is carried by species other than humans, such as beavers, even waters that are free of human contamination must now be considered potentially unsafe. Filtration is effective in the removal of the *Giardia* cyst.

The safety of water supplies (from diseases other than giardiasis) is determined by testing the water for the presence of indicator bacteria called coliforms. These bacteria, not usually the cause of disease themselves, are normal inhabitants of the human gut and indicate the presence of fecal contamination of the water.

Water supplies are treated to remove contaminants that make the water either unpalatable or unsafe. The important steps and their functions are as follows:

1. copper sulfate addition and aeration for taste and odor removal
2. first chlorination for destroying microbes
3. coagulation/settling for removal of large particles
4. filtration for removal of disease-causing microorganisms
5. final chlorination for a parting shot at killing microorganisms

A small residual of chlorine is purposely left in the water after treatment is complete; the residual is intended to act against any new sources of contamination between treatment plant and final use; however, chlorination may also produce low levels of chlorinated hydrocarbons in the water, some of which have been found to be carcinogens.

Ozonation, the treatment of water with ozone gas, is an alternative method to disinfect water supplies. This process, widely used in Europe in place of chlorination, creates no chlorinated hydrocarbons but leaves no residual to act against new contamination.

Questions

1. Name three classes of organisms that can affect the safety of a water supply for human consumption. Give examples of each class.
2. The spread of certain diseases to epidemic levels can be described in terms of a cycle. Describe the role of water in the spread of such diseases as typhoid and cholera.
3. The engineering solution to the spread of infectious disease has been to break this cycle. List several ways in which this can be accomplished.
4. List three diseases, other than typhoid and cholera, that are spread by polluted water. Indicate the type of microorganism involved in each of the diseases you list.
5. What is the function of chlorination in supplying water for human consumption? Does this process alone ensure safety against infectious disease?
6. What is the function of filtration in supplying water for human consumption?
7. Explain what is meant by *indicator organism*. Discuss the role of coliform bacteria in monitoring water quality.
8. Where does your drinking water come from? Has your water been tested for safety recently?
9. Complete the following table, which summarizes the procedures or steps in treating water to make it safe to drink, and give the reason for these steps.

Procedure	Reason for this procedure
Spray aeration	Removes dissolved gases such as hydrogen sulfide
Addition of activated carbon	
Coagulation with alum or ferrous sulfate	
Sand filtration	
Chlorination	

10. Describe briefly the method used in testing a water sample for the presence of coliform indicator organisms. Is this method suitable for detecting virus contamination? Explain your answer.

Further Reading

Safe Drinking Water Committee. *Drinking Water and Health.* Washington, D.C.: National Academy of Sciences, 1977.
This publication is the result of the definitive and comprehensive study undertaken by the National Academy of Sciences at the request of Congress when it passed the Safe Drinking Water Act of 1974.

National Interim Primary Drinking Water Regulations. Office of Water Supply, Environmental Protection Agency, EPA 570/9-76-003, 1976.
This document, which replaces the Public Health Service Drinking Water Standards of 1962, contains the standards for all contaminants that influence the safety of water supplies. The publication contains, in addition to the standards, background material that explains the basis for the values specified in the standards. The standards provided in this document became effective in 1977 but will probably be modified in time. One new standard was issued in 1979 for trihalomethanes. Material on this standard can be found in *Federal Register*, **44** (November 29, 1979), 68, 624.

Sedgwick's Principles of Sanitary Science and Public Health. Rewritten and enlarged by S. Prescott and M. Horwood. New York: Macmillan, 1948.
This classic public health text is easy reading and especially absorbing for those interested in the historical roots of the environmental/public health movement.

Standard Methods for the Examination of Water and Waste Water, 13th ed. Prepared and published jointly by the American Public Health Association, the American Water Works Association, and the Water Pollution Control Federation, 1980. Available from the American Public Health Association, 1790 Broadway, New York, N.Y. 10019.
This book provides step-by-step descriptions of all laboratory tests used to assess the quality of water and wastewater.

Craun, G. "Outbreaks of Waterborne Disease in the United States: 1971–1978," *Journal of American Water Works Association* (July 1981), p. 360.
Outbreaks of waterborne disease continue into the present. This article reviews some recent outbreaks with explanation of the causes.

Knoppert, P., G. Oskam, and E. Vreedenburgh. "An Overview of European Water Treatment Practice," *Journal of the American Water Works Association* (November 1980), p. 597.
Focuses on differences between American and European practice, especially the reduction in chlorination, the use of ozone, and the effect on trihalomethanes (the chlorinated hydrocarbons mentioned in the chapter).

Prendville, P. "The U.S.: Rethinking Filtration and Disinfection," *Consulting Engineer* (September 1982).
Concern for amoebae and protozoans in water as well as the formation of trihalomethanes has prompted a number of cities to rethink the need for filtration and to consider alternatives to chlorine.

"Roundtable: Waterborne Giardia: It's Enough to Make You Sick," *Journal of the American Water Works Association* (February 1985), p. 14.
An informal exchange by experts on the subject of giardiasis; nontechnical.

Pluntze, J. "The Need for Filtration of Surface Water Supplies," *Journal of the American Water Works Association* (December 1984), p. 11.
"It should be employed now by most surface water systems in the United States" is the bottom line of this well-done editorial.

References

Report on the Cholera Outbreak in the Parish of St. James, Westminster, During the Autumn of 1854. Presented to the Vestry by the Cholera Inquiry Committee, July 1855. London: J. Churchill, 1855.

First Annual Report, State Board of Health and Vital Statistics of Pennsylvania, 1866.

American Water Works Association. "Recreational Use of Domestic Water Supply Reservoirs." A statement adopted by the Board of Directors on January 26, 1958, and revised January 25, 1965 and June 13, 1971.

President's Council on Environmental Quality. *Recreation on Water Supply Reservoirs.* Washington, D.C.: U.S. Government Printing Office, 1975.

Johnson, George. "The Purification of Public Water Supplies," U.S. Geological Survey Paper No. 315, 1913.

U.S. Department of Health, Education, and Welfare. *Health Information for International Travelers, 1979.* Atlanta: Public Health Service, Center for Disease Control, 1979.

Lin, Shen Dar. "*Giardia lamblia* and Water Supply," *Journal of the American Water Works Association* (February 1985), p. 40.

Gubrud, A. E. Speech, Federal Water Quality Association Conference on Toxic Substances in the Water Environment, Washington, D.C., April 28, 1977.

CHAPTER TWELVE

The Consequences of Polluting Water: Organic Wastes, Phosphates, and Nitrates

The Effects of Organic Wastes on Stream-Dissolved Oxygen

Why Organic Wastes Are Pollutants/Stream Health and Dissolved Oxygen/How to Measure Organic Wastes/Changes in Water Quality

Eutrophication

Why Phosphates and Nitrates Are Pollutants/How Can Eutrophication Be Controlled?/Helping Eutrophic Lakes Recover/Lake Erie: A Case History/Other Detergent Problems: Foaming Waters

Anatomy of a Fish Kill

CONTROVERSY:

12.1 *The Scientist and Social Responsibility: Optical Brighteners*

The Effects of Organic Wastes on Stream-Dissolved Oxygen

Why Organic Wastes Are Pollutants

Water pollution means different things to different people. To some, it means the presence of toxic chemicals; to others, it means the presence of disease-causing bacteria. To still others, pollution is seen as floating weeds and algae. To many engineers and scientists, however, water pollution is indicated by organic wastes. Such wastes can come from industry, from farming, and from cities. These organic wastes are mainly composed of carbon, hydrogen, oxygen, and nitrogen. Oxidation of the carbon, hydrogen, and nitrogen in these wastes is responsible for many of the unpleasant conditions that exist in polluted rivers and lakes.

A simple experiment. To understand what organic water pollution is, consider the following experiment. If we mix and dissolve sugar and gelatin in an open bottle of water, the resulting solution is clear. The solution does not appear to us to be "polluted." A fish should be able to live in this water (after any chlorine has been allowed to diffuse out), at least for a time. Yet the contents of the bottle become clouded in less than a week. Dissolved oxygen, which fish need to live, is used up, and a fish will die.

The organic wastes, in this case, are sugar and gelatin. Together, they cause a condition in which fish and many other aquatic creatures cannot live: a lack of oxygen. In the presence of oxygen, organic materials such as sugar and gelatin, are broken down into simpler substances by bacteria, a process known as *biochemical oxidation*. If enough species of bacteria are present, the carbon (C) in the sugar and gelatin is oxidized to carbon dioxide (CO_2). Furthermore, the hydrogen (H) in the sugar and gelatin is oxidized to water (H_2O). Last, the nitrogen (N) in the gelatin is eventually converted to nitrate ions (NO_3^-). All these reactions utilize oxygen. The net effect is to reduce the amount of oxygen dissolved in the water. (To measure the potential for oxygen removal of organic wastes, see p. 272.)

Oxygen in natural waters. When water is well mixed in the presence of air, the water dissolves a maximum amount of oxygen, called the *saturation concentration*. This is the number of milligrams of oxygen that can be dissolved in 1 liter of water before the water can hold no more. The saturation concentration achieved by thorough mixing is between 8 and 9 milligrams of oxygen per liter of water, depending on the temperature. In summer, at higher water temperatures, the saturation concentration falls as low as 8 mg/L. In winter, at colder water temperatures, the saturation concentration is nearer 9 mg/L.

If the concentration of organic material in a sample of water is sufficiently large, oxidation of the organics by bacteria and protozoa may use up all the oxygen in the sample; that is, the concentration of oxygen in water may be reduced to zero. When there is no oxygen in natural waters, we say the water is in an **anaerobic condition.** Under anaerobic or very low oxygen conditions, normal aquatic life, such as fish species, die.

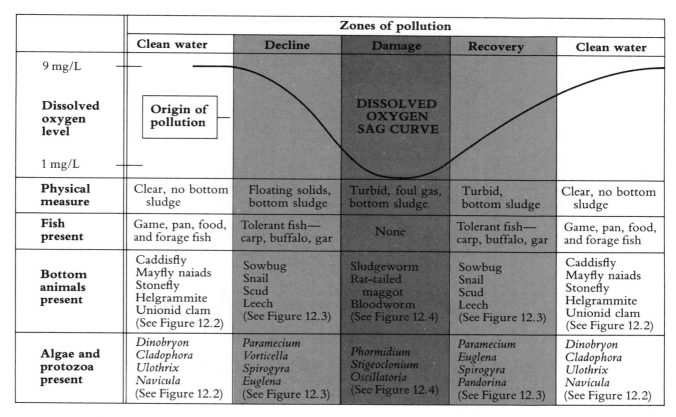

	Zones of pollution				
	Clean water	**Decline**	**Damage**	**Recovery**	**Clean water**
Dissolved oxygen level 9 mg/L — 1 mg/L	Origin of pollution		DISSOLVED OXYGEN SAG CURVE		
Physical measure	Clear, no bottom sludge	Floating solids, bottom sludge	Turbid, foul gas, bottom sludge	Turbid, bottom sludge	Clear, no bottom sludge
Fish present	Game, pan, food, and forage fish	Tolerant fish— carp, buffalo, gar	None	Tolerant fish— carp, buffalo, gar	Game, pan, food, and forage fish
Bottom animals present	Caddisfly Mayfly naiads Stonefly Helgrammite Unionid clam (See Figure 12.2)	Sowbug Snail Scud Leech (See Figure 12.3)	Sludgeworm Rat-tailed maggot Bloodworm (See Figure 12.4)	Sowbug Snail Scud Leech (See Figure 12.3)	Caddisfly Mayfly naiads Stonefly Helgrammite Unionid clam (See Figure 12.2)
Algae and protozoa present	*Dinobryon Cladophora Ulothrix Navicula* (See Figure 12.2)	*Paramecium Vorticella Spirogyra Euglena* (See Figure 12.3)	*Phormidium Stigeoclonium Oscillatoria* (See Figure 12.4)	*Paramecium Euglena Spirogyra Pandorina* (See Figure 12.3)	*Dinobryon Cladophora Ulothrix Navicula* (See Figure 12.2)

Figure 12.1 The zones in a polluted water course. (Adapted from K. Mackenthun, *Toward a Cleaner Aquatic Environment*, Washington, D.C.: U.S. Government Printing Office, 1973.)

The few species that do survive and thrive in any numbers are those especially adapted to these conditions.

In low-oxygen conditions, the normal bacterial population that uses oxygen also dies out. In its place, a population of bacteria grows that lives on sulfur instead. The sulfur atom occurs in organic wastes and is structurally similar to oxygen, except that it has one more ring of electrons. Sulfur then takes the place of oxygen in the oxidation reaction, producing, for instance, hydrogen sulfide (H_2S) instead of water (H_2O). The well-known smell of rotten eggs is the smell of hydrogen sulfide.

To summarize, oxidation of organic wastes by bacteria leads to a loss of oxygen from water. An environment in which oxygen is absent supports no fish and only a few specially adapted animal species. In this anaerobic environment, specialized bacteria produce, among other end products, the foul-smelling gas hydrogen sulfide.

Stream Health and Dissolved Oxygen

What happens at a sewage discharge. Now that we know how organic pollution can alter the dissolved oxygen concentration in a body of water, we can investigate how aquatic life responds to such pollution. When organic substances from the waste outfall of a community or industry enter a river or stream, the concentration of dissolved oxygen in the stream decreases. This is due to the oxidation of the organic material by bacteria and protozoa. Natural mixing of the stream with the air does tend to replace the removed oxygen, but not immediately. Instead, a competition arises between the oxygen-depleting forces (the oxidation of organic wastes) and the oxygen-restoring forces (the mixing of water with air). This competition produces a typical pattern of oxygen concentrations along the stream. The curve is shown in the top of Figure 12.1.

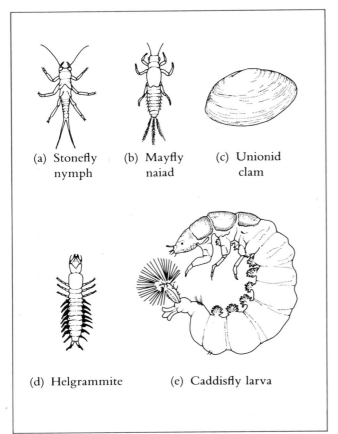

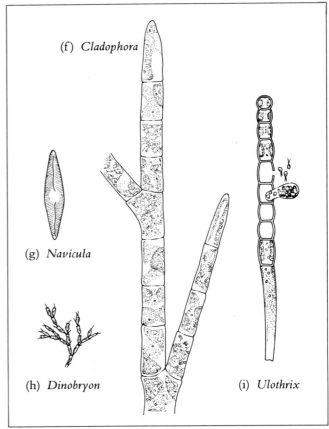

Figure 12.2 (a)–(c) Clean-water animals associated with streambed. (f)–(i) Clean-water (sensitive) algae.

The four basic sections of the stream are:

1. a clean-water zone (high level of dissolved oxygen) upstream from the pollution source;
2. a zone of decline (falling level of dissolved oxygen);
3. a zone of damage (relatively constant and low level of dissolved oxygen);
4. a zone of recovery (rising level of dissolved oxygen).

Of course, if more than one sewage outfall occurs along the stream or river, the zone of damage may extend for many miles. Such conditions may occur along rivers in major urban and industrial areas.

Clean-water zone. The clean-water zone upstream from the waste discharge may sustain fish, mayflies, clams, stoneflies, and other species (Figure 12.2). These

species require oxygen in water in order to survive. If the oxygen concentration falls, these sensitive species are among the first to disappear. Such fish as trout, bass, salmon, and minnows are among these sensitive species. While the needs of various fish species vary, biologists generally believe that dissolved oxygen levels of 5 mg/L or better are necessary to ensure the survival of fish.

Zone of decline. A zone of decline follows the introduction of organic wastes. Species that can survive at these somewhat lower levels of dissolved oxygen are referred to as *intermediately tolerant*. They are pictured in Figure 12.3. Solids from the wastewater outfall may cloud the water.

Damage zone. The damage zone, which follows the

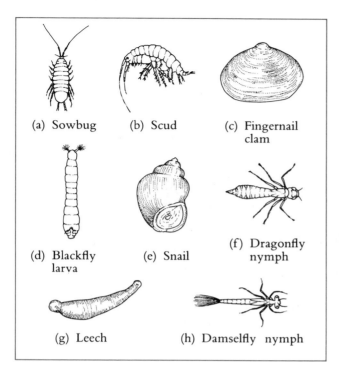

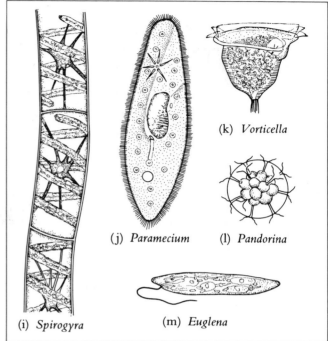

(a) Sowbug (b) Scud (c) Fingernail clam

(d) Blackfly larva (e) Snail (f) Dragonfly nymph

(g) Leech (h) Damselfly nymph

(i) *Spirogyra* (j) *Paramecium* (k) *Vorticella* (l) *Pandorina* (m) *Euglena*

Figure 12.3 Intermediately tolerant aquatic species.
(a)–(h) Intermediately tolerant animals associated with streambed.
(i)–(m) Intermediately tolerant algae and protozoa.

zone of decline, is one in which dissolved oxygen is almost gone. When dissolved oxygen has fallen to very low levels, only a few species are capable of survival, and the numerous species that characterized the clean stream have disappeared. In their place arise a group of organisms referred to as *pollution tolerant* because of their ability to survive under conditions of low dissolved oxygen. One such organism is the sludgeworm, which consumes sludge and thrives in water with as little as 0.5 mg/L of oxygen. Another is the rat-tailed maggot, a resident in the sludge of stream bottoms. The maggot is the larva of the drone fly and breathes by a long tube that reaches to the water surface. Its numbers may so increase that they cover the bottom of the streambed in a waving red sheet. Another resident of this zone is the bloodworm. This organism, like the sludgeworm, consumes sludge as its diet. These species are illustrated in Figure 12.4

In the clean-water zone, many species exist side by side, and each is moderately represented in terms of numbers of organisms. In the damage zone, only a few species survive, but they can be present in enormous numbers. If, in the damage zone, one fails to find large numbers of organisms of the pollution-tolerant species, it is likely that some chemical poison in the sewage may be preventing their increase.

Zone of recovery. A zone of recovery follows the damage zone. Here the water is likely to be clear, allowing sunlight to penetrate. The level of oxygen is on its way up to more reasonable concentrations. With the clearing of the water and the recovery of dissolved oxygen, algae may begin to grow. Their presence can result in fluctuation of the oxygen content in the water. During daylight hours, the algae produce oxygen as a by-product of photosynthesis. But at night, respiration and decomposition of algae remove oxygen from the water, yielding a low oxygen concentration. These algae-caused swings in dissolved oxygen may even deplete the stream's oxygen to such an extent that normal aquatic life, which requires oxygen, may not be reestablished. Beyond the zone of recovery, the species of the clean-water zone are found again.

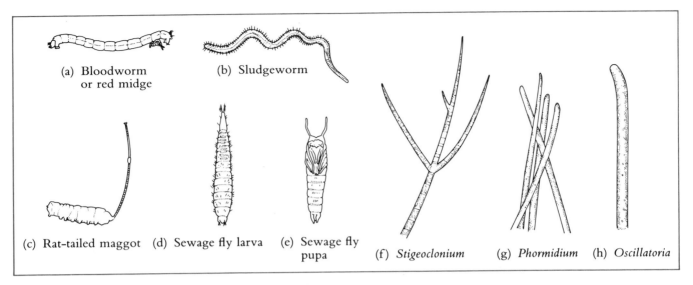

Figure 12.4 Pollution-tolerant aquatic species.
(a)–(e) Pollution-tolerant animals associated with the streambed.
(f)–(h) Pollution-tolerant algae.

How to Measure Organic Wastes

It may have occurred to you to ask how we determine the amount of organic wastes in a stream. This knowledge could help us predict the amount of oxygen that would be used up in oxidizing the organic substances. It would also help us predict what species will be present in the water.

Unfortunately, to measure the quantity of every organic substance in a waste stream is an exhausting, and perhaps impossible, job. Each substance present would first have to be identified, already an enormous task. Next, once a substance has been shown to be present, the quantity of that substance would have to be measured. Then, if we also know the oxygen consumption when bacteria oxidize one gram of each substance present, we could predict how much oxygen would be removed by oxidation of all the wastes in a liter of the polluted water.

A much easier way to measure the concentration of organic wastes in water was proposed in England near the turn of the century. The method determines not only the concentration of organics but also reveals the amount of oxygen that all the organic wastes, acting together, will eventually remove from the water. The number reported is the **biochemical oxygen de-mand (BOD)**. This is the amount of oxygen that would be consumed if all the organics in 1 L of polluted water were oxidized by bacteria and protozoa. It is reported in number of milligrams of oxygen per liter.

Suppose we are studying a sample of polluted water discharged as a waste stream from a community. The sample is reported to have a BOD of 120 mg/L. This means that bacteria and protozoa oxidizing all the organic substances in 1 L of the water would consume 120 mg of oxygen. Now suppose that 50 ml (0.05 L) of the polluted water were mixed with 950 ml of clean water. The sample has been diluted to 50 parts in 1,000 parts, or to 1/20 of the mixture. The BOD of the mixture will be (1/20) × 120 or 6 mg/L.

This dilution is similar to what happens when the flow of sewage from a community enters a watercourse that has a substantial water flow. Suppose the BOD in the discharge is the 120 mg/L we were discussing. Assume further that the community empties 1 million gallons of waste a day into the stream and that the clean stream is flowing at 19 million gallons per day past the outfall. The BOD of the mixture is diluted to 1/20 of the discharge value, or 6 mg/L. This number is the amount of oxygen that can be removed from every liter in the combined flow of stream and wastewater. Given that clean water can only possess oxygen up to about

9 mg/L, the oxygen in the mixture can be reduced to a very low level. How low it will actually go depends on many factors, but especially on the rate of oxidation by bacteria and on the natural rate of restoration of oxygen to the stream.

Thus, the BOD value tells a biologist the capability of the polluted water to deplete the oxygen resources in the stream. This is a valuable indicator of pollution, since the lack of oxygen is responsible for fish kills and also causes foul odors and populations of undesirable pollution-tolerant organisms. The BOD does not explain what organic substances are present in the water nor in what quantity they are present. Nonetheless, it provides a rapid and important insight into the maximum pollution damage the wastewater could cause.

Measuring the concentration of organic wastes. The specific procedure for determining the value of biochemical oxygen demand involves a number of steps. First, a carefully measured sample of the polluted water is diluted in a much larger, measured volume of unpolluted water. A 300 ml bottle is specially made for this test. The unpolluted water has previously been well shaken in the presence of air so that the water has absorbed as much oxygen as it can. This water is saturated with oxygen. The water may also have been "seeded" with microorganisms known to break down organic wastes in the presence of oxygen. If the BOD of polluted *river* water is to be measured, however, the needed microorganisms are likely to be present in the polluted water sample already. The mixture of polluted water and clean water is poured into the bottle, completely filling it. The bottle is then closed to the air by insertion of a glass stopper. This closure prevents any new oxygen in the air from entering the bottle. Two bottles are commonly prepared in this way from samples of the same polluted water.

One of the bottles is set aside in the dark at 20°C and taken out at the end of five full days. The darkness is necessary to prevent the growth of algae. Algae may contribute oxygen to the water as a by-product of photosynthesis. Such oxygen would interfere with the measurement of BOD and so is avoided. The other bottle is examined immediately to determine the quantity of oxygen dissolved in it (Figure 12.5).

Several methods exist to measure dissolved oxygen, including a chemical test and a test done by electrodes. The chemical test requires special solutions and procedures. The electrode test requires only the insertion of a probe (electrode) into the water. One must first ascertain, however, that the electrode is calibrated—that is, reads the appropriate value—and this requires prior testing.

Suppose the quantity of dissolved oxygen in the bottle measured immediately is 7.5 mg. Suppose further that, after five days, the quantity of oxygen in the bottle stored in the dark is 6.0 mg. No oxygen has been added to the second bottle from any other source; therefore, the difference between the initial quantity of oxygen and the quantity present after five days must be the oxygen removed by microorganisms, 1.5 mg.

Suppose a 10 ml sample of polluted water was diluted initially. Since 10 ml of this polluted water caused the removal of 1.5 mg of oxygen, 150 mg of oxygen could be removed by a 1 L sample (1000 ml) of the polluted water. The BOD of the polluted water is thus 150 mg/L.

By the end of five days, the rate of oxygen removal has become very slow. Only a little more oxygen will be removed each day after that. The five-day BOD, then, is the number commonly used to measure the oxygen-consuming ability of polluted water.

Chemical oxygen demand (COD). The BOD is an important measure of organic water pollution because it tells us how much oxygen will ultimately be removed from the water by biological oxidation of the wastes. Unfortunately, it takes five days to get such a reading, and sometimes an estimate is needed more quickly.

To meet the need for a quick measurement of organic waste concentration, water scientists have developed a chemical test that does not rely on microorganisms to do the oxidation. Instead, the test uses potassium dichromate and sulfuric acid to oxidize the organic material. Nearly all the organic matter present is oxidized by this mixture, even organic material that microorganisms could not oxidize. Thus, the test generally reports a greater concentration of oxidizable organics than the BOD test does.

Test results are reported as the *chemical oxygen demand* (COD). The units of COD are the same as BOD—namely, milligrams of oxygen per liter. As an example, suppose we had a sample of polluted water with a COD of 100 mg/L. The organics in 1 L of this water would be expected to remove a bit less than 100 mg of oxygen from a stream. The removal is less than 100 mg because the COD test oxidizes organics that microorganisms cannot.

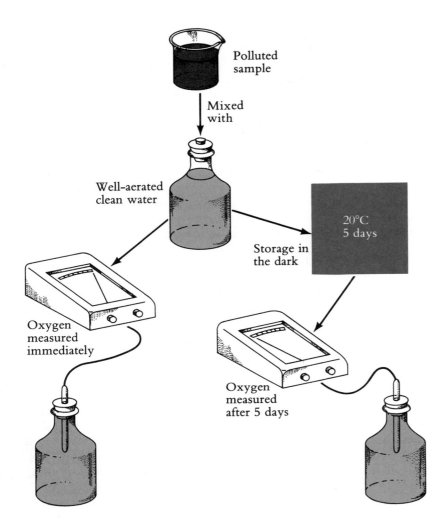

Polluted sample

Mixed with

Well-aerated clean water

20°C 5 days

Storage in the dark

Oxygen measured immediately

Oxygen measured after 5 days

Figure 12.5 Measuring the concentration of organic wastes (biochemical oxygen demand).

Changes in Water Quality

In 1972, the United States Geological Survey established the National Stream Quality Accounting Network (NASQAN). Some 504 stations monitor the water quality of the outflow from the nation's major river basins. In Figure 12.6, the changes found in dissolved oxygen between 1974 and 1981 are noted for each of the 21 water resources regions of the country. An increase in dissolved oxygen means improved water quality. On the other hand, a decrease means deteriorating water quality. Such decreases are most likely the result of more organic wastes being added to the water.

By far, the major finding has been no change. Nearly all basins reported the number of streams with increased levels of dissolved oxygen exceeded the number of streams with decreased levels of dissolved oxygen. There are some causes for concern, however. Regions reporting the number of decreases (deterioration) exceeding the number of increases (improvement) include the Arkansas White-Red, the Texas Gulf, the Rio Grande, the Upper Colorado, the Great Basin, the Pacific Northwest, California, and Hawaii.

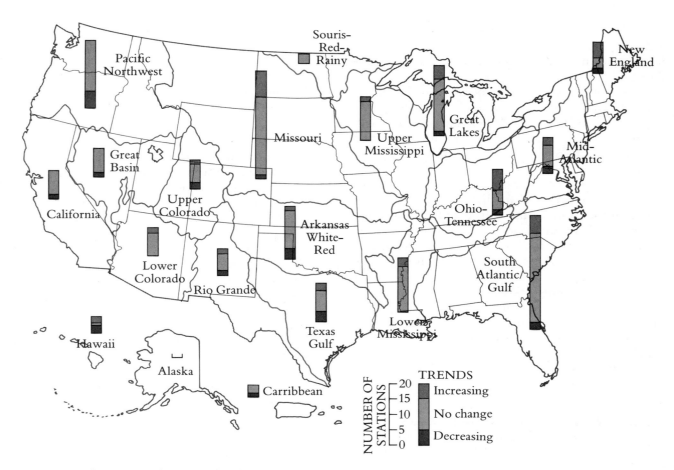

Figure 12.6 Changes in dissolved oxygen levels at NASQAN stations 1974–1981 by water resources region. (Data from U.S. Geological Survey, 1983.)

Eutrophication

Why Phosphates and Nitrates Are Pollutants

Feeding lakes. In 1974, Russell Train, director of the U.S. Environmental Protection Agency, gave warning to a meeting of the New Hampshire Lakes Region Clean Water Association: Nutrients that cause lakes to age rapidly are increasing in U.S. waters. What did he mean by this? How can a lake age?

Carbon, oxygen, hydrogen, nitrogen in nitrates, and phosphorus in the form of phosphates are some of the elements, or "foods," needed by plants for growth. Smaller amounts of many other substances, such as iron, calcium, and copper, are also required.

Lakes with large amounts of these necessary plant nutrients are called *eutrophic* (from the Greek words *eu*, meaning "well," and *trophe*, meaning "nourishment"). Relatively high concentrations of plant nutrients in eutrophic lakes allow huge amounts of aquatic plants, such as algae, to grow. This overpopulation of algae makes a lake unpleasant to swim in. Some kinds of algae, which grow in long strands, wind themselves around boat propellers, making boating impossible. The water in such a lake tends to be scummy, cloudy, or even soupy green. A rapidly growing population of algae, called a bloom, may wash onto the lake-shores in storms or high winds and die there, its decay producing bad smells.

Some of the worst effects of **eutrophication** involve changes in the dissolved oxygen content of lake waters. Sport fish have trouble living in a lake with blooms because, at night, the algae respire and use up most or all of the oxygen in the lake, leaving none for the fish.

Eutrophic lakes are clearly not the kind people enjoy using for recreation or for drinking water. Cities or industries that want to use eutrophic lake water find they must remove the algae, thus increasing costs. Almost 10,000 public lakes in the United States need treatment for the effects of eutrophication caused by excess plant nutrients.

An **oligotrophic** lake (from the Greek words *oligo*, meaning "poorly," and *trophe*, meaning "nourishment") is one with low levels of plant nutrients. Such lakes have only small populations of algae and remain sparkling clear. They are enjoyable to swim in or boat on, are delightful to look at, and should provide good drinking water.

Natural versus artificial eutrophication. Many lakes are oligotrophic when first formed. Over the centuries, however, silt, debris, and dissolved nutrients accumulate, filling in the originally deep basins and turning the waters into a nutrient-rich soup. After a few thousand years, a lake can change naturally from oligotrophic to eutrophic. This process is called *aging*.

Not all lakes age naturally. Some lakes, in the normal course of events, maintain their oligotrophic characteristics; however, any beautiful, clear lake can be made eutrophic if large amounts of plant foods are added. Most commonly, this unnatural eutrophication happens when city or industrial sewage is dumped into lakes. The sewage from a city or town is rich in the nutrients, such as nitrates and phosphates, that plants need for growth.

Limiting nutrients. Some controversy exists about which plant nutrients are most responsible for causing the premature aging of U.S. lakes, rivers, and coastal waters. In general, experts agree that phosphates and nitrates are the most likely culprits. Although there are many other necessary nutrients, these two are noteworthy because, together or singly, they are usually the **limiting factors** in natural unpolluted waters. A **lim-

iting factor** is the nutrient present in the smallest amount compared to the amount needed for growth; that is, aquatic plants such as algae will grow until all available nitrogen or phosphorus is used up (Figure 12.7). Unnatural amounts of phosphorus or nitrogen can allow more than the normal amount of algae to grow.

Nitrates and eutrophication. In most U.S. coastal waters and in lakes such as Tahoe, fed by steep mountain streams, the concentration of nitrates seems to be the factor that normally limits plant growth. Nitrates can enter natural waters from several sources. City sewage and animal feedlots (where cattle are grouped together in large pens to be "finished," or fattened for market) are two major sources, since human and animal wastes are about one-half nitrates. Fertilizer, which runs off croplands or suburban lawns during a rainstorm, also contains a large amount of nitrates.

In addition, several natural sources of nitrates exist. Volcanic eruptions and lightning can change nitrogen gas in the air into nitrates. Just as legumes found in land ecosystems can "fix" atmospheric nitrogen into a form usable by other plants, some blue-green algae found in aquatic systems can turn nitrogen from the air into nitrates. This process is called *nitrogen fixation*. Thus these algae, which are usually considered the worst nuisance plants, do not depend on an outside source of nitrate. (There is evidence that sulfate in seawater hinders nitrogen fixation. This may explain why algal growth in coastal waters is nitrate limited.) Finally, waterfowl, which feed on the shores of lakes and ponds and then fly over the water, are the source of what one wildlife expert termed "bombed in" nitrates.

Phosphates and eutrophication. In contrast to nitrogen, phosphorus in the form of phosphates is found naturally in waters in only trace amounts. Most pollutant phosphorus comes from human activities. Phosphate mines pollute certain areas of the country, such as central Florida. Fertilizer runoff contains large amounts of phosphate. Domestic sewage is high in phosphates, too, with one-half coming from human wastes and 20%–30% from detergents. Animal feedlots are sources of both nitrates and phosphates.

Although growth of algae in coastal waters and in some wilderness lakes is limited by low concentrations of nitrates in the water, algal growth in most U.S. lakes and rivers is believed to be limited by the amount of phosphate present (Figure 12.8).

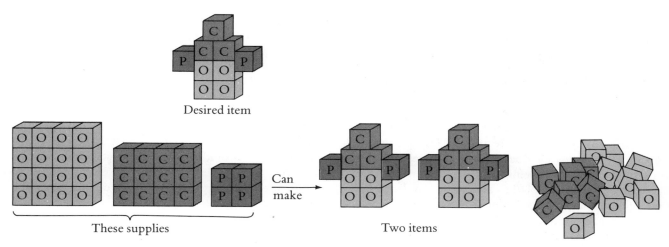

Figure 12.7 The limiting factor concept. Suppose that in order to make up an item, you needed 4 blocks of O, 2 of P, and 3 of C. If you had 16 blocks of O, 4 of P, and 12 of C, you could still make only two of the item because after that you would run out of P. In this case, P is the limiting factor. No amount of extra O or C blocks will make up for having no more P.

Figure 12.8 The role of phosphorus in lake eutrophication is dramatically illustrated in this photograph of Lake 226 in northwestern Ontario, Canada. The lake has two basins. Phosphorus, carbon, and nitrogen were added to the far basin (upper part of photo). An algal bloom covered the basin in two months. The lower basin was treated at the same time with the same amounts of nitrogen and carbon, but no phosphorus. This basin remained clear and sparkling. (Photograph courtesy of D. W. Schindler)

How Can Eutrophication Be Controlled?

Since phosphate is the nutrient that limits plant growth in most U.S. waters, and because detergents are a ma-

jor source of phosphates, it is natural to ask whether we would not be better off banning the use of phosphates in detergents in an effort to stop eutrophication.

The chemical present in the largest amount in

phosphate detergents is the phosphate builder: Dry detergents usually contain sodium tripolyphosphate (STPP) as a builder; liquid detergents may contain sodium or potassium phosphates. Builders are necessary because water often contains calcium and magnesium ions. Water that has more than 75 mg/L of calcium and magnesium (measured as calcium carbonate) is called *hard water*. The calcium and magnesium ions combine with soap to form a hard precipitate. This precipitate does not make suds and does not dissolve grease or dirt. In a similar fashion, calcium and magnesium ions can tie up the surfactant, or dirt-dissolving, molecules in detergents. Builders "complex," or tie up, these calcium and magnesium ions. Without builders, manufacturers would have to include a great deal of relatively expensive surfactant in their detergents to be sure their product performs well in hard-water areas. In addition, builders help keep dirt from reattaching itself to clothes.

A complete ban on phosphates in detergents would remove about 20% to 30% of the phosphates in sewage. In a number of areas, either a ban on or a reduction in the amount of phosphate allowed in detergents has been tried in an attempt to reverse lake eutrophication (see Table 12.1).

It is not clear, however, that a detergent phosphate ban alone will make a significant improvement in lake waters that are already eutrophic. If phosphate inputs to a eutrophic lake are not reduced below a certain level (generally by 50% or more), eutrophic growth of

Table 12.1 Legislated Detergent Phosphate Levels

State	Legal limit for phosphorus in detergents (%)	Date law went into effect
Connecticut	8.7	1972
Florida	8.7	1972
Indiana	8.7	1972
	0	1973
Maine	8.7	1972
Michigan	8.7	1972
	0	1977
Minnesota	0	1979
New York	8.7	1972
	0	1973
Vermont	0	1978
Wisconsin	0	1979, 1984[a]

Source: A. Maki et al., *Water Resources Research*, **18** (7) (1984), 893.
a. The ban expired in 1982 and was ordered again in 1984.

aquatic plants seems to continue. Removal of detergent phosphates, by itself, will not usually accomplish such a large reduction because other inputs, such as runoff from agricultural lands, are much greater sources of phosphates. For instance, phosphate inputs to the Great Lakes from runoff are now three times greater than the phosphate inputs from wastewater. If the Michigan ban on phosphate in detergents were lifted, estimates are that phosphate inputs to the Great Lakes from the Michigan area would increase by only 2%. For smaller lakes that receive wastewater from sewage treatment plants lacking the processes to remove phosphates, the contribution of detergents is probably greater. However, the detergent contribution still does not equal the contribution from nonpoint sources: runoff from agricultural fertilizers, leaking septic systems, municipal storm sewers, and cattle feedlots. (This is called nonpoint-source pollution because it is usually hard to pinpoint exactly where the phosphates or nitrates are coming from—see Chapter 13.) The total amount of nutrients from nonpoint sources can be very high, and as mentioned, in some areas it is much greater than the amount from sewage treatment facilities.

If manufacturers cannot put phosphates in detergents, what can they use instead? In some areas, water is naturally soft enough (that is, it has little or no calcium or magnesium) for soap to be used. Water can also be made soft by passing it through a water softener or an ion exchanger. In most houses, however, the water is hard enough so that some sort of detergent builder is necessary.

The choice of a chemical builder to replace phosphates must be made carefully. Just as people did not realize a few years ago that phosphates could harm the environment, we have no way of knowing whether a substitute for phosphates might not be harmful as well. If a single chemical were chosen, the possibility arises that we could be adding as much as 2 million pounds of some new material to the waters every year. (See Controversy 12.1.)

Today, phosphate-free detergents contain:

1. Carboxymethyl cellulose, an organic carbon compound that may present a problem because it may not be biodegradable.★

★"Biodegradable" is a term, along with "phosphate-free," that you often see on detergent labels. The term biodegradable is defined later in this chapter.

CONTROVERSY 12.1

The Scientist and Social Responsibility: Optical Brighteners

We must be sure that a full-sized, hungry, four-footed wolf, with teeth, is coming before we start crying out about it.
B. J. Kilbey and G. Zetterberg

I prefer not to give the product the benefit of a doubt . . .
B. Gillberg

In 1971, Bjorn Gillberg stated that he had found evidence that optical brighteners used in modern detergents caused mutations in yeast. Some evidence exists that chemicals causing mutations are also likely to cause cancer (see Chapter 26). However, scientists in another laboratory could not repeat his experiment. B. J. Kilbey and G. Zetterberg (1974) wrote:

> We do not think that our experiments indicate unequivocally that no danger exists from optical brighteners. The data are insufficient at present for this conclusion to be drawn . . . However, we must be sure that a full-sized, hungry, four-footed wolf, with teeth, is coming before we start crying out about it. For environmental biologists, this means doing all in our power to be sure that the right experiments are done, positive results are reproducible, and any artifacts of method are excluded . . . If we startle the public too many times with sensational claims that are later retracted, we run a real risk of losing our most valuable ally if and when a real crisis comes.

Gillberg (1974) replied:

> I have not made any sensational claims about brighteners; the only thing I say in my paper is

that I consider it of importance to carry on with genetic studies of brighteners against the background of my results. Research has now begun in other laboratories that should have been undertaken before the brighteners were released on the market. The benefits of a product must of course always be weighed against the risks it may create. In such a situation I prefer not to give the product the benefit of the doubt if there are some questions raised. Questions have been raised about these compounds, and I believe that it is my social responsibility to tell my fellow citizens.

Who is right? At what point should the public be told that some doubt has been cast on the safety of a product that is in common use? Must we be careful not to "cry wolf" too many times to avoid producing boredom about environmental dangers, or is it important to give people all available information as soon as it comes to light?

Sources:
B. J. Kilbey and G. Zetterberg, "Optical Brighteners," *Science*, **183** (March 1, 1974), 798.
B. Gillberg, "Optical Brighteners and Social Responsibility," *Science*, **184** (September 13, 1974), 901.

2. Carbonates and silicates, in the form of sodium carbonate. Both combine with calcium and magnesium ions to give insoluble precipitates. Large amounts of carbonates and silicates, however, can make phosphates more soluble. Vast amounts of phosphates are already captured in the muds of lake bottoms. Any substance that helps mud phosphates dissolve into lake water is not a good ingredient in a phosphate-free detergent.

3. Borates, in the form of sodium tetraborate decahydrate, or borax. Waters having more than 1 ppm of borax are toxic to most plants.

Many areas have settled for a limit on the amount of phosphate (8.7% phosphorus, which is equivalent to 50% phosphate). Phosphate-free detergents, which once accounted for 14% of the market, now have about 3%–4% of sales. The main reason for the failure of the antiphosphate campaign has been the lack of a suitable substitute.

Another way to attack the phosphate problem does exist, however. Phosphates can be removed from sewage by **tertiary treatment** of wastewater. In this method, phosphates are precipitated out of sewage before the treated water is released into lakes or rivers (see Chapter 13). About 80%–90%, or even more, of the phosphates in sewage can be removed, and the cost is not great. Further, phosphates from all sources (for example, human wastes) are removed. Unfortunately, not everyone in this country is served by a sewage treatment plant. Some have septic tanks; others have only treatment ponds. Only 1% of the people in this country are served by sewage treatment plants that have facilities for phosphate removal. This particular solution, then, is tied up in the economics and politics of building and improving sewage treatment plants.

Some communities have adopted the solution of piping municipal sewage and storm waters around threatened lakes and into some other body of water. For instance, sewage that once flowed into Lake Washington in Seattle is now routed around the lake and into Puget Sound. Although Lake Washington is slowly recovering from its gradual slide into a eutrophic state, we need only think about Puget Sound to realize that this is a temporary solution.

Recent EPA surveys show that agricultural nonpoint sources are the major source of nutrients for eutrophic lakes. Nutrients from manure and chemical fertilizers that are contained in agricultural runoff can

be controlled by good farming practices, such as strip cropping, terracing, and careful timing of fertilizer application so that fertilizer is not immediately washed away by rains.* Feedlot operators may need to install sewage treatment equipment to prevent animal wastes from enriching nearby streams.

Low-tillage or no-till farming and vegetated buffer strips along stream banks and farm fields also reduce nutrient runoff. Land management plans can reduce runoff through zoning in which the use of land near water is strictly controlled.

Helping Eutrophic Lakes Recover

Once lakes or ponds are eutrophic, simply preventing the addition of any more phosphates is often not enough to help them become clear and sparkling again. The reason is that phosphates already settled into the bottom muds can, in many cases, dissolve back into the waters. Thus, phosphate levels may remain high for years, even with no new additions. Ways of combating this problem have met with varying success. In some cases the flow of water through a lake can be increased. This helps wash nutrients and algae out of the lake. Low-phosphate water from Seattle's sewage treatment plant is used to increase the flow through Green Lake in eastern Washington.

Chemicals such as aluminum, iron, and calcium have been used to precipitate phosphate out of lake waters. The cost of this sort of treatment is about $150–$300 per hectare (depending on how polluted the lake is), and the results last about 3 years. Fly ash from power plants has also been used to precipitate phosphate. In addition, the fly ash can seal off the lake sediments, preventing phosphates from dissolving back into the water. This would provide a way for power plants to dispose of fly ash; however, the ash is known to contain heavy metals, which could be toxic to aquatic life.

Oxygen can be added directly to the oxygen-poor lower layers of eutrophic lakes. This allows fish to take up residence there again and also helps prevent phosphate release from lake muds. (Release is more likely to occur under conditions of little or no oxygen.) This method has been used in Europe for years and has been

---

*Strip cropping and terracing also reduce the damage done to farmland by erosion. These techniques are explained in Chapter 7.

(a)   (b)

Figure 12.9 The town of Vaxjo, Sweden stopped pouring sewage into overexploited Lake Trummen in 1958. (a) By 1969, the unrecovered lake contained no oxygen, no fish, and no underwater vegetation and was of little use to humans. The main problem was the rapidly increasing black muddy sediment caused by decaying plankton. Starting in 1970, the sediment was sucked out and put into settling ponds; runoff water was cleansed of phosphorus and returned to the lake. (b) This is how Lake Trummen looks now, a revitalized recreational asset. (Photographs by S. Bjork)

tried successfully in several lakes and reservoirs in the United States.

Biomanipulation refers to a group of management techniques designed to change the kinds or amounts of the various organisms living in a lake in order to control eutrophication. These techniques can include changing the pH of lake waters to favor green algae over blue-green algae, or increasing the population of lake organisms that feed on algae. How useful these techniques will be is yet to be seen. In some cases, there seems to be no alternative but to actually dredge the phosphate-rich sediments out of a lake (Figure 12.9). This is, however, the most expensive method of lake restoration.

The National Clean Lakes Group is a coalition of interested parties that has convinced Congress to allot funds to communities wishing to restore eutrophic lakes. This is basically a continuation of the Clean Lakes Program run by the Environmental Protection Agency from 1976 to 1982. In order to be eligible for a grant, the lake must be available to the public, it must be classed as fresh water, and the applicants must show that their plan for lake restoration is likely to produce long-lasting benefits. Almost 100 lakes were helped under the EPA's Clean Lakes Program between 1976 and 1982 (Figure 12.10). More information can be obtained from EPA regional offices.

Lake Erie: A Case History

The Great Lakes—Superior, Michigan, Huron, Erie, and Ontario—lie between the United States and Canada. Combined, they represent the largest reservoir of fresh water in the world; however, because they are the site of large population centers in both the United States and Canada (Duluth, Green Bay, Milwaukee, Chicago, Gary, Toledo, Cleveland, Erie, Buffalo, and Toronto, among others), the lakes are becoming more and more polluted. In fact, Lake Erie, the smallest and shallowest of the lakes, has been described as "dead." This really is not the right term to use—the lake still supports a large fish population. What has changed, however, is the kinds of fish in the lake. When the lake area was first settled in the 1700s, Lake Erie was home to large numbers of commercial and game fish. There are now no longer any blue pike, sauger, or native lake trout. Only a few lake herring, sturgeon, whitefish, and muskellunge are left, along with some walleye and northern pike. Instead, the fishery is composed primarily of yellow perch, white bass, channel catfish,

Figure 12.10 Volunteers used this truck to pull stumps out of the lake bed as part of an EPA-sponsored clean lakes project. (Photo by Harold Woodworth)

freshwater drum, carp, goldfish, and rainbow smelt.

Intense commercial fishing was responsible in part for the change in species. The introduction of new species into the lake also had unwanted effects. Sea lamprey, which feed on desirable food and game fish, found their way into Lake Erie sometime before 1921 via the Welland Canal. More serious has been the invasion of rainbow smelt, which began about 1931. The smelt became abundant, preying on the young of desirable fish such as lake trout, blue pike, and lake whitefish.

Eutrophication of the lake has also played a large part in the species change. The amount of algal growth in the lake has increased 20 times since 1919. When algae die and decay, the level of oxygen in the water plummets, especially in the cooler bottom waters. Not only does this destroy the summer habitat of many preferred fish species, but it also destroys important fishfood species, such as the burrowing mayfly. The kinds of algae in the lake have also changed. Blooms of blue-green algae are now common. In contrast to green algae, many species of blue-green algae are not eaten by fish.

In addition to providing a large portion of the nitrates and phosphates that have led to the eutrophication of Lake Erie, sewage inflows have added toxic chemicals such as mercury, PCBs, and Mirex. These chemicals accumulate in fish and have led to severe restrictions on the kinds and sizes of fish that can be eaten when caught in the Great Lakes.

In 1972, 1978, and 1983, the United States and Canada agreed on plans to clean up the Great Lakes. A central feature of the plans was building sewage treatment plants to reduce the amount of phosphate entering the lakes. Canada met the first deadline (December 31, 1975) for phosphorus reductions, but the United States, for political and economic reasons, did not.

Although the amount of phosphorus added to the lake has decreased 75% in the past 10 years, the bottom waters of Lake Erie's central basin still become oxygen-poor in the summer months. If stated goals of lowering phosphorus inputs a further 15% from 1982 levels are met, it seems likely that the lake will recover. However, this will require not only sewage treatment but also reduction of at least one-third in the amount of phosphate that runs off farmland into Lake Erie's drainage basin. Annoyed by the lack of U.S. progress, one Canadian official said the present situation is "like mixing a glass of clean water with a glass of dirty water; you end up with dirty water."

In summary, one might say that reports of the death of Lake Erie have been exaggerated. Nevertheless, if we do not make progress in reducing sewage inflows to all the Great Lakes, both because of the nutrients and because of the poisons they contain, the lakes may become effectively dead for human purposes.

Other Detergent Problems: Foaming Waters

In the early 1950s, many people noticed a strange sight. Brooks, streams, rivers, and even lakes were beginning to foam. Wherever water tumbled over stones or waterfalls, wherever winds rippled the surface, accumulations of bubbly froth were building up. The explanation for this odd circumstance was not hard to find. Huge amounts of detergents (close to 2 billion pounds per year) were being used to wash clothes. The dirty wash water ran down millions of drains and into nearby streams, lakes, and rivers. Sometimes it first passed through sewage treatment plants, and sometimes it did not. In fact, it made no difference whether it did or not. The chemical in detergents that made them foam was not removed by sewage treatment plants. The material, called alkyl benzene sulfonate (ABS), has several special properties. The property that makes it useful in washing clothes is that it helps greasy dirt dissolve in water. That is the good part. The not-so-good part is that the molecule's long, branched shape is difficult for bacteria to break down. Since treatment plants depend on the use of bacteria, and because bacteria do not have the necessary catalysts (enzymes) to destroy ABS compounds, ABS molecules began to accumulate in the early 1950s wherever wash water ran into natural waters. When the concentrations of ABS were high enough, foam formed, just as it did in washing machines and dishpans.

This problem, fortunately, could be solved. Detergent manufacturers now use surfactant molecules, which have straight, rather than branched, side chains. Such surfactants are known as linear alkyl sulfonates (LAS). They are termed **biodegradable** because bacteria are able to digest them, and so the foam has almost disappeared from our rivers, lakes, and streams.

Anatomy of a Fish Kill

In July of 1985, a huge fish kill occurred in the Trinity River near Palestine, Texas (about 65 river miles south of Dallas). The kill was characterized by oxygen levels as low as 0.8 mg/L. Dead fish floated on the water surface past a monitoring station in Anderson County at the rate of about 1,250 per hour; fish and game officials guessed that twice that number (or 2,500 per hour) were passing the station beneath the water surface. The number of dead fish in this incident was projected to exceed the count of three kills in the previous summer, which totaled 171,000 fish. Both the 1985 incident and the kills in 1984 were blamed on the organic matter from the raw (untreated) sewage entering the Trinity River from Dallas.

Why did the fish kill occur in the summer and not at other times? What were the conditions that came together to cause the massive kill at that particular time? The organic wastes were undoubtedly entering the water course throughout the year, and no comparable event had occurred in earlier seasons. Three factors can be identified as acting together to turn the large, and otherwise routine, discharge of sewage into the cause of a major fish kill. All the factors are related to the time of year the fish kill occurred—high summer.

The foremost factor was probably the low flow of water in the Trinity River upstream from the discharge. In the Northern Hemisphere stream flows decrease in the summer because of low rainfall, and the flow in the Trinity River was much diminished from its usual level. Ordinarily, the sewage discharge mixes with relatively clean water that is flowing at a relatively strong rate, and the organic wastes, with their high BOD, are diluted. Suppose that the sewage from Dallas had a BOD of 200 mg/L and was ordinarily diluted by a factor of 40 when it mixed with the flow from upsteam; the resultant flow would then have a BOD of 5 mg/L. Thus, if there were no restoration of oxygen by reaeration, 5 mg of oxygen could be removed from each liter of the mixture of sewage and clean water. But that summer the upstream flow had fallen, let us say, to half its average level. Hence, the sewage was diluted by only a factor of 20 when it mixed with the upstream flow. The resulting BOD of the mixture was 10 mg/L. Even this increased concentration of BOD, although it undoubtedly contributed significantly to lowering the level of dissolved oxygen, would not ordinarily be sufficient to deplete the stream of all its oxygen, because reaeration gradually restores oxygen to the stream. Some other factor in addition to low water flow and decreased dilution of BOD was probably involved.

Again, the time of year had a role to play. At a temperature of 60°F (16°C), which is characteristic of late spring in Texas, the dissolved oxygen level is close to 10 mg/L. The high temperatures of summer, however, often increase the temperature of river water in this portion of the country to 80°F (27°C). At this temperature the level of oxygen dissolved in the fully aereated, unpolluted upstream water is close to 8 mg/L. Thus, because of the elevated summer temperatures in 1985, the initial concentration of oxygen in the upstream water—before the sewage was mixed in—was 2 mg/L lower than its usual level. As a consequence, the BOD needed to drive oxygen levels down to zero was less.

These two factors—a higher than usual BOD level because of less diluted water coming from upstream, and a lower initial level of oxygen resources due to the high summer temperatures—could be sufficient under certain circumstances to so deplete the river of its dissolved oxygen that a fish kill would be the result. One other factor, however, is likely to have been operating: eutrophication.

Raw sewage introduced to the stream not only organic matter but also nitrogen and phosphate compounds, which you know from this chapter to be two of the key substances that promote algal blooms. Recall that algae produce oxygen by photosynthesis during the daylight hours but respire, and hence consume oxygen, through the night. Their respiration at night is not balanced by any production of oxygen, causing a temporary drain on the oxygen resources of a stream. If the stream is already marginal with respect to oxygen resources, the "algal swing" to a lower dissolved oxygen level may cause the stream to become anaerobic and therefore a hostile environment for fish and other aquatic life.

Thus three principal factors probably came together to cause the massive fish kill in the Trinity River in Texas in July of 1985: (1) a low natural flow unable to dilute the BOD of the Dallas sewage flow sufficiently, (2) a lowered level of dissolved oxygen in the clean upstream water because of the high water temperature, and (3) a further depletion of the dissolved oxygen during the nighttime hours because of the presence of algae.

Summary

Pollutants do not have to be toxic themselves in order to kill aquatic creatures or to degrade water quality. Examples of this type of pollutant are organic wastes, phosphates, and nitrates.

Organic wastes are oxidized by bacteria and other microorganisms. In the process, these organisms use up the dissolved oxygen in the water. Many desirable water species find it difficult or impossible to survive in the resultant low-oxygen conditions. The pollution potential of a sample of organic waste-containing water is measured as its BOD, or biological oxygen demand. The U.S. Geological Survey reports a general trend toward improvement in the dissolved oxygen levels in the nation's streams and rivers.

When nitrates and phosphates are added to natural waters, they can serve as nutrients for photosynthetic algae, which may then grow into huge, unsightly blooms. This is called eutrophication. At night when the algae in eutrophic waters respire and use oxygen, dissolved oxygen values can plummet, endangering other aquatic creatures.

Nitrates are the limiting nutrient in most coastal waters and steep mountain streams, while phosphates are limiting in most of the rest of U.S. waters. A major source of phosphate additions to natural waters is domestic wastewater that contains phosphates from detergents and from human and animal wastes. Phosphates from these sources can be eliminated by tertiary sewage treatment. An even larger source is the phosphate entering natural waters from nonpoint sources: fertilizer runoff, septic system drainage, and municipal runoff. Nonpoint sources can be controlled by proper land management.

Lakes that are already eutrophic may recover if additions of phosphate and nitrate are stopped. Lakes that do not recover naturally may require further treatment: dredging, oxygen addition, phosphate precipitation, or biomanipulation.

Questions

1. When organic wastes are added to natural waters, they may cause fish to die, even though the wastes themselves are not directly poisonous to fish. How do you explain this?
2. Describe how sewage changes the aquatic communities found downstream from the outfall. What are the zones that can be described in such a stream?
3. What problems are caused by too much phosphate in natural waters? Would you call phosphorus a poisonous chemical?
4. Why must we be careful about what kinds of chemicals are substituted for phosphates in detergents? Does your sewage treatment plant remove phosphates? (You could call and find out.) Does your home have soft water? What would you recommend that your family use to wash clothes? Why?

Further Reading

Mackenthun, K. *Toward a Cleaner Aquatic Environment.* U.S. Government Printing Office, Stock No. 5001-00573.
Virtually a textbook on the biology of water pollution. It is interesting, well written, well illustrated, and rarely too technical for the novice to follow. Directions for identifying algae associated with water pollution are provided.

Eutrophication of Surface Waters: Lake Tahoe's Indian Creek Reservoir. U.S. Environmental Protection Agency, Ecological Research Series, 1975.
One community's answer to the problem of lake eutrophication was to build a whole new reservoir, Indian Creek, to hold the area's wastewater so that it would not have to be dumped into Lake Tahoe. This report details some of the problems they encountered.

Pirie, N. W. "Water Weed Uses," *Water Spectrum* (Summer 1980), p. 43.
The author puts forth the idea that water weeds should be looked on as a resource rather than as a nuisance. Such an approach could control eutrophication in some cases and in others eliminate it as a problem entirely. See also the article on Lake Erie in the same issue.

Reiger, H. A. and W. L. Hartman. "Lake Erie's Fish Community: 150 Years of Cultural Stress," *Science*, **180** (June 22, 1973), 1248.
Historical perspective on the effects of civilization on the Great Lakes.

U.S. Environmental Protection Agency. *Lake and Reservoir Management.* Proceedings of 3rd Annual Conference of the North American Lake Management Society, 1984.
The latest research efforts in eutrophication, acid precipitation, and sedimentation of lakes and reservoirs.

Erichsen-Jones, John. *Fish and River Pollution.* Washington, D.C.: Butterworth & Co., Ltd., 1964.
The second chapter on fish and their oxygen requirements is of greatest interest.

Howarth, R. W. and J. Cole. "Molybdenum Availability, Nitrogen Limitation and Phytoplankton Growth in Natural Waters," *Science*, **229** (August 16, 1985), 653.
A possible explanation of why coastal waters are nitrate limited while fresh waters are phosphate limited.

Kemp, L., W. Ingram, and K. Mackenthun, eds. *Biology of Water Pollution.* Federal Water Pollution Control Administration, U.S. Department of Interior, 1967.
A collection of biologically oriented papers from journals treating stream pollution, biological waste treatment, and public health as influenced by stream pollution. The classic paper "Stream Life and the Pollution Environment" by Bartsch and Ingram is reprinted here. This article alone makes a copy worth obtaining from the library.

References

Kilby, B. J. and G. Zetterberg. "Optical Brighteners," *Science*, **183** (March 1, 1974), 798.

Gillberg, B. "Optical Brighteners and Social Responsibility," *Science*, **184** (September 13, 1974), 901.

Shapiro, J., et al. *Experiments and Experiences in Biomanipulation, Project Summary.* Washington, D.C.: U.S. Environmental Protection Agency, 1983.

CHAPTER THIRTEEN

Water Pollution Control

The Difference Between Water Treatment and Water Pollution Control

Water Pollution Control in Rural Areas and from Nonpoint Sources

Combined Sewer Systems / Primary Treatment / Secondary Treatment / Digestion / Tertiary Treatment / Chlorination / Physical-Chemical Treatment / Land Application of Wastewater / The Federal Water Pollution Control Act / The "Settlement Agreement"

Water Pollution Control in Rural Areas and from Nonpoint Sources

Control of Water Pollution in Rural Areas / Control of Water Pollution from Nonpoint Sources

Progress in Water Cleanup

CONTROVERSIES:
13.1 *Can (Should) You Save a River Without Removing the Pollutants?*
13.2 *Economics and Environmental Regulation: Pretreatment Standards*

In Koln, a town of monks and bones
And pavements fang'd with murderous stones
And rags, and hags, and hideous wenches;
I counted two and seventy stenches;
All well defined, and several stinks!
Ye nymphs that reign o'er sewers and sinks,
The river Rhine, it is well known,
Doth wash your city of Cologne;
But tell me, Nymphs! What power divine
Shall henceforth wash the river Rhine?
Samuel Taylor Coleridge

The Difference Between Water Treatment and Water Pollution Control

The term *water treatment* should be distinguished from the term *water pollution control*. These are very different activities. Water treatment and water pollution control are the two sides of the coin of water quality. Taken from a lake, river, or well, water must be made both safe and desirable to drink. **Water treatment** is the process that purifies water for the consumer, whether in the home, in business, or in factories. Safety is achieved by destroying disease-causing microorganisms. Palatability is achieved by removing tastes and odors and by making the water clear.

Water pollution control, on the other hand, is directed toward restoring the quality of the water that the consumer has used. Organic wastes and bacteria that have been added during use must be removed. Often, nitrates and phosphates need to be removed as well. The organic wastes are removed primarily to prevent depletion of oxygen in the waters that will receive the waste. The relation of organic wastes to the level of dissolved oxygen was discussed in Chapter 12. Bacteria are destroyed to prevent disease from spreading among those who later use the water for other purposes. Nitrates and phosphates are removed to avoid the eutrophication, or overenrichment, of receiving waters. The impact of nitrates and phosphates on eutrophication was also discussed in Chapter 12.

The relative positions of water treatment and water pollution control are indicated in Figure 13.1.

Water Pollution Control at Urban Point Sources

Many ways are used to combat water pollution, but the methods chosen depend on the origin of the pollution. We can distinguish two general origins of pollutants. First are those pollutants that typically arise in urban areas and enter water courses from a single pipe; we say such pollutants come from a **point source**. Point-source pollutants arise when domestic wastes and industrial wastes are collected by sewers and then carried to a treatment plant. After the wastewater has been treated, it is discharged to a waterway. We call such pollution *point source* because it occurs at a distinct and identifiable point. The concentrated runoff from large

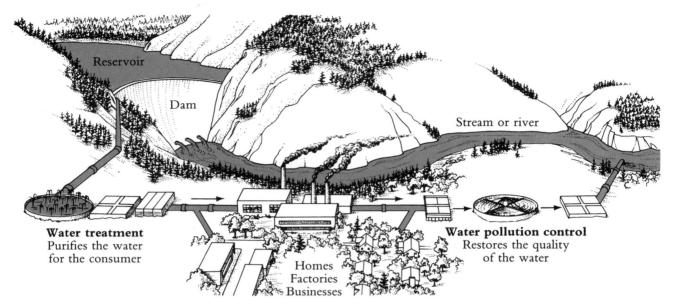

Figure 13.1 Water treatment and water pollution control. Note that water treatment precedes use of the water by the community. Water pollution control, on the other hand, follows use by the community. Water treatment "prepares" the water for use; water pollution control "repairs" it after use.

feedlot operations may be considered a point source of pollution.

In contrast are those pollutants that come from relatively rural areas and enter water bodies as runoff from the land. The runoff occurs in numerous rivulets and streams and reaches the waterway at many points. We can think of the runoff as being spread out along the waterway, and we call such runoff with its burden of pollution a **nonpoint source**.

The methods of controlling pollution from point sources and nonpoint sources are very different. To point-source pollution, we have traditionally applied the *technical fix*, which consists of chemical and biological processes that remove contaminants from the wastewater. Nonpoint-source pollution, however, is not so easy to control, for it does not necessarily lend itself to technical methods. Instead, changes in land use and in agricultural practices are needed. Changes are also needed in mining practices and in timber growing and harvesting. In this section, we discuss the technical solutions that apply to point-source pollution, and in the following section we discuss some of the changes in land management that are needed to control pollution from nonpoint sources.

Combined Sewer Systems

When wastewater enters a treatment plant (also called a water pollution control plant), it passes through a rack and a coarse screen that together prevent large objects from passing into the plant. The large objects could include such items as boots, cloth, branches, and so forth. Such objects would seem to be unexpected in municipal wastewater, but they are quite common. Their presence is a clue to a large problem that most communities in the United States continue to face: The drain system that carries storm water and the drain system that carries domestic wastewater are combined.

Runoff entering storm drains is simply water that was not absorbed into the ground. The flow of water from storms may be as much as 100 times the rate of domestic wastewater flow, or *dry-weather flow*, which averages 100 gallons per day per person. Since storm water can so greatly exceed the flow of domestic wastewater, sewage treatment plants are designed only for the domestic flow (as opposed to 101 times larger). When a storm does occur, storm water enters the sewers, and the two flows mix (see Figure 13.2). The treatment plant, however, can handle only its usual volume,

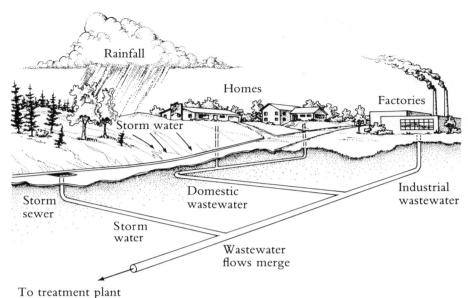

Figure 13.2 The combined sewer system of most U.S. cities. The run-off from a rainstorm enters a city's sewers and there mixes with the wastes from homes and industries. A treatment plant built for normal wastewater flow will not be able to handle the combined flow. Much of the combined flow will be diverted past the treatment plant and will enter a water body without treatment.

the dry-weather flow. The remainder must overflow into the stream or river, where it is left untreated.

Storm water carries many contaminants, including substances washed from the streets by rain. Such substances as lead (from gasoline), particles of earth, dog feces, leaf litter, and unburned fuels may be captured in the flow before it enters the sewers. Once the flow enters the sewers, it may scour up organic solids (including bacteria) deposited there by the slower moving dry-weather flow. Thus, storm water is not at all clean; it carries metals, organics, and possibly pathogenic bacteria into the watercourse.

There are three ways to alleviate this problem:

1. A storm sewer system that is separate from the wastewater sewer system can be built at a cost of billions of dollars.
2. The streets may be washed or swept on a regular basis. This action reduces the contaminants washed from streets, but it does not alter the scouring up of sediments in the sewer.
3. Large storage tanks can be inserted into the system to hold the large flows; the tanks then release these volumes gradually to the treatment plant.

Most cities have done little to deal with the problem of combined sewers because of the expense involved.

Primary Treatment

After the wastewater has passed through the rack and screen, the flow moves into the grit chamber, where large particles of earth, such as gravel, drop out of the moving stream. The flow next moves to a settling tank, also commonly called a *settler* or a *clarifier*, where the organic solids characteristic of wastewater slowly settle out. Settling is a purely physical process; no chemical or biological reactions take place. Settling removes organic solids from wastewater by slowing the velocity of the stream. As the velocity decreases, the tendency of the water to keep particles in suspension decreases, and organic particles slowly settle to the bottom of the tank. The sludge of organic solids removed by a settler may account for 35% of the organic matter in the wastewater from a typical city. This initial settling is referred to as **primary treatment**.

Secondary Treatment

The processes that follow settling are designed to remove organic materials that are dissolved (as opposed to suspended) in the wastewater. This step, the removal of dissolved organics, is referred to as **secondary treatment**.

One process, known as **activated sludge treat-**

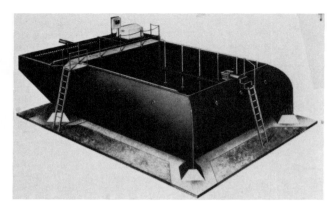

Figure 13.3 The package activated sludge plant. Often used for subdivisions, these small activated sludge plants are sold as a prefabricated steel package. The flow process is essentially the same as in the larger system, except that these small plants typically have no primary treatment, and they aerate the raw wastewater for a 24-hour period, rather than the 6–8 hours used in conventional plants.

ment, uses microorganisms in a large, well-aerated tank to break down dissolved organic substances. A steady stream of wastewater from the settler enters one end of the activated sludge tank and exits at the other end. The flow, on exit, is richer in microbes and poorer in dissolved organics (Figure 13.3). About 80%–85% of the dissolved organics in wastewater can be removed by activated sludge treatment.

The microbial population that grows steadily in the tank uses the organic substances in solution for growth and energy. Thus, the organics are removed from solution and are converted within the tank either to more microbes or to the end products of biological oxida-

tion. In the presence of oxygen, the end products include carbon dioxide and water.

The microbial solids produced in the activated sludge tank are suspended in the waste flow leaving the tank. This waste flow is passed through another settler, which removes the new solids. From the second settler, these microbial solids flow in a slurry, which is 98%–99% water, and are combined with the solids from the primary settling tank. The combined solids are then treated in a device called a digester, which will be described shortly.

A portion of the solids recovered from the flow is returned to the activated sludge tank as "seed." These solids mix with the waste stream from primary treatment and provide the initial bacterial population that will start the mixture fermenting in the tank.

Another common technology used in secondary treatment to remove dissolved organics is the **trickling filter**. Here the wastewater is distributed over and passed down through a bed of rocks in a large concrete basin. A slime of microbes grows on the surface of the rocks during long operation of the filter. The microbes in the slime remove dissolved organics for growth and energy, just as the microbes in the activated sludge process do. The slime slides off the stones and into the wastewater a little at a time. The result is a waste stream with solid materials in suspension and with much of the dissolved organics removed, just as occurred in the activated sludge tank. About 80%–85% of dissolved organics may be removed by a trickling filter, which may be as large as 200 feet (60 m) in diameter at large installations (Figure 13.4). The solids are removed in a settling tank, combined with those from the primary settling tank, and pumped to the digester.

Figure 13.4 Trickling filter. Wastewater is distributed over rocks and trickles down through the rock bed. A slime growing on the rocks removes organic material from the wastewater.

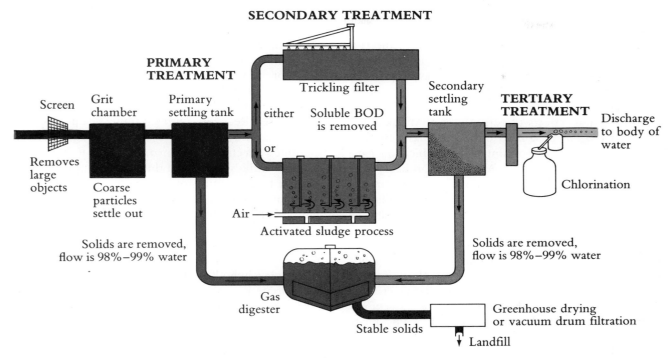

Figure 13.5 Water pollution control processes. Three distinct sets of processes are used in sequence. *Primary treatment* removes suspended solids. *Secondary treatment* removes dissolved organics. *Tertiary treatment* is designed to remove phosphorus, nitrogen, and resistant organic compounds. Digestion, not part of any of these three processes, is the means by which organic solids are degraded and stabilized.

Digestion

The process of waste digestion converts the sludge of organic solids to a stable material. The process takes place in a large, heated tank appropriately called a **digester** (see Figure 13.5). The tank is closed to the air so that oxygen does not reach the wastes. Specialized microbes, which live at high temperatures and do not require oxygen, grow in the tank. These microbes convert the wastes to stable end products, including the gases methane and hydrogen sulfide. The methane gas is often burned to provide the heat needed to keep the digester at the proper temperature. The result is a stable, nondegrading sludge, which is then dried either in greenhouses or on vacuum drums.

The disposal of sewage sludge differs across the country. For many years, Philadelphia has been dumping its sludge into the ocean, although the dumping has been done only on an interim basis lately. Many cities bury their dried sludge in landfills. Others burn it in a

special unit and bury only the ash in landfills. Some communities (such as Salem, Oregon and Madison, Wisconsin) spread the sludge on cropland, but heavy-metal contaminants in the sludge may make such use unwise. The city of Milwaukee packages its sludge as a commercial fertilizer. Milorganite, as the product is known, might be called "the sludge that made Milwaukee famous."

Primary treatment (settling) and secondary treatment (activated sludge or trickling filter) together may remove up to 90% of the organic wastes in water (Figure 13.5). Primary treatment is common in many communities in the United States. Secondary treatment is coming into much wider use.

Tertiary Treatment

Most of the organic material in wastewater is removed by primary and secondary treatment. The processes that follow secondary treatment are designed to remove plant nutrients, which are responsible for the overnourishment, or eutrophication, of lakes and rivers. The processes that follow secondary treatment are known as **tertiary treatment**. In contrast to primary and secondary treatment, tertiary treatment is quite uncommon at present.

The serious effects of eutrophication of lakes and rivers have been recognized only in the last few decades. Thus, methods to control eutrophication have not yet been widely applied. At present, probably only a few dozen such installations are operating in the United States. Nevertheless, there is promise that the processes in tertiary treatment will be added to sewage treatment plants in many areas.

The main purpose of tertiary treatment is to remove compounds containing nitrogen and phosphorus. These are the major substances contributing to the eutrophication of natural water, a process characterized by excessive algal growth. (The biological basis of eutrophication is described in Chapter 12.) Phosphate ions, found in both human wastes and detergents, are removed by one set of processes; nitrogen compounds, such as ammonia, nitrate ions, and nitrite ions, are removed by another set of processes. Tertiary treatment may also include removal by carbon adsorption of those resistant organic substances not removed by secondary treatment. All of these processes are expensive.

Phosphate removal is accomplished by chemical precipitation and settling. Chemicals such as ferrous and ferric salts, aluminum salts, or lime are added to the waste stream. When one of these chemicals is mixed well with the waste stream flowing from secondary treatment, a precipitate of solid matter forms. For example, the calcium ions from lime (calcium oxide) combine with the phosphate ions in solution to produce solid particles of calcium phosphate. These particles are removed by allowing them to settle out of the waste stream. Engineers have suggested that the addition of chemicals to "bring down" phosphates could also take place just before the primary settling tank. This would allow the collection of both organic solids and phosphate precipitates to take place at the same time. After the phosphate is precipitated and settled out, any remaining suspended matter may be removed by filtration through beds of sand, crushed coal, garnet, and gravel.

Nitrogen occurs in wastewater in the form of ammonia, nitrate ions, or nitrite ions. Unfortunately, there is no way to remove these nitrogen compounds by precipitation in the treatment plant.

Removing nitrogen that occurs as ammonia (NH_3) is important for a number of reasons. First, ammonia is harmful to fish. Second, ammonia combines with the chlorine that is added to destroy harmful bacteria and viruses. Reactions with ammonia tie up chlorine in substances known as chloramines and hence decrease chlorine's effectiveness. Third, ammonia is oxidized by bacteria in the stream, and this oxidation process may remove large amounts of oxygen from the water. The ammonium ion may be removed as ammonia gas by a physical process known as *gas stripping*. Ammonia may also be removed by biological treatment—by passing the wastewater through an aerated tank with a specialized population of microbes. These microbes convert the nitrogen in both ammonia and nitrites to nitrate ions.

Nitrate ions should be removed because they increase eutrophication, but an even more compelling reason exists for removing nitrate and nitrite ions. If high-nitrate water is used for drinking, a blood disorder in infants, methemoglobinemia, could result (see Chapter 14). Nitrate removal may be achieved by specialized populations of microbes that convert nitrate to nitrogen gas and water. Methyl alcohol must be present for this reaction, which produces carbon dioxide, water, and nitrogen gas.

A final procedure in tertiary treatment is to pass the waste stream through a tower containing particles of activated carbon. The activated carbon adsorbs dissolved organic substances that were not removed in any previous processes. The organic materials "stick" to the surfaces of the particles. This *carbon polishing* restores the wastewater to such an extent that people have considered reusing this water in public water supplies after it has been filtered and disinfected.

One of the earliest successful applications of tertiary treatment was at Lake Tahoe, Nevada in the late 1960s. The clarity of this beautiful lake was threatened by algal growth caused by phosphorus and nitrogen compounds from the community's wastewater. Scientists recognized that the usual primary and secondary treatment steps could not stop the degradation of the lake. Hence, a tertiary treatment plant was opened at Lake Tahoe in 1965. The plant removes the phosphate and the ammonia and "polishes" the water by passing it through carbon columns to remove resistant organics. The wastewater—now clear, colorless, and odorless—is never discharged into the lake. Instead, the water is pumped 27 miles to Indian Creek Reservoir, which was created expressly to receive the plant's wastewater.

Chlorination

Because the bacteria that inhabit the human intestine thrive best in a warm environment, the cold water of sewage treatment plants tends to reduce their numbers. Nonetheless, a considerable population of bacteria still exists in the waste stream even after secondary treatment. Thus, wastewater leaving American sewage treatment plants is commonly chlorinated in order to destroy disease-causing microorganisms (pathogens). While the sewage treatment plant is often merely a torturer to these bacteria, chlorination is the executioner.

In contrast to bacteria, viruses from the human intestine die away more slowly in the treatment plant. Primary treatment reduces their numbers little if at all. Secondary treatment, via the trickling filter, may reduce viruses by 40%; however, secondary treatment, via the activated sludge process, can reduce their numbers by up to 98%. The final step of chlorination reduces their numbers once again, yet even after chlorination live viruses from the human intestine can still be found in the waste stream.

The health of people swimming in or boating on the water downstream from a treatment plant is protected by destroying the pathogens in the wastewater. The process of chlorination is inexpensive compared to other methods of disinfecting wastewater. Furthermore, the effectiveness of disinfection is easily monitored by testing for free chlorine in the wastewater.

Several environmental problems do occur as a result of chlorination, however. Chlorine and the compounds it forms with ammonia are toxic to fish. Some fish can be poisoned at concentrations of chlorine as low as 0.002 mg/L (2 parts per billion). Suppose wastewater were being discharged into a river, and suppose further that the available chlorine (free and reacted with ammonia) were 2 mg/L (a fairly common level for available chlorine in such discharges). The wastewater would need to be diluted 1,000 times to decrease the concentration in the receiving body to less than 0.002 mg/L.

Another problem with chlorinating wastewater is that chlorinated hydrocarbons can form from the reaction of chlorine with hydrocarbons in the water. These

persistent and toxic substances, which have been detected in public water supplies, are suspected of being carcinogenic (cancer-causing) agents. They are discussed in greater detail in Chapter 14.

Once added, chlorine can be removed from wastewater by contact with sulfur dioxide gas, which leaves sulfate and chloride ions in solution after the process is complete. The sulfur dioxide will even strip chlorine away from the compounds it forms with ammonia, but the sulfur dioxide will not remove any chlorinated hydrocarbons that may have formed.

As we noted in Chapter 11, an alternative to chlorination of municipal wastes is ozonation. Though a powerful disinfectant, ozone disappears quickly in contaminated water and does not seem to form toxic by-products when it reacts with substances in water. Ultraviolet light is also being examined as a possible disinfectant because it would not cause the formation of undesirable by-products. Another alternative is simply not to disinfect wastewater prior to discharge. In Britain, wastewater disinfection is not an element of engineering practice. The approximate reduction of pollutants, of bacteria, and of viruses by each step in the total treatment scheme is indicated in Figure 13.6.

The processes we have discussed so far are the wastewater treatment methods used most commonly in American cities today; however, other processes and concepts are being studied or used on a small scale. These include physical-chemical treatment, land application of wastewater, and other processes (or even no treatment at all—see Controversy 13.1).

Physical-Chemical Treatment

All the secondary treatment processes we have discussed utilize microorganisms to remove dissolved organics from wastewater. Nevertheless, there are defects in this biological treatment of wastes. The most important problem is the sensitivity of microorganisms to toxic pollutants. An industry may discharge a waste that will poison the microorganisms. As the waste flows through the treatment plant, the special populations of microbes, which had been thriving in ordinary wastewater, may be partially or wholly killed. When this happens, the effectiveness of the entire plant is reduced to the levels achieved by primary treatment alone. A whole new population of microbes must develop to renew the effectiveness of secondary treatment.

In the last two decades much progress has been

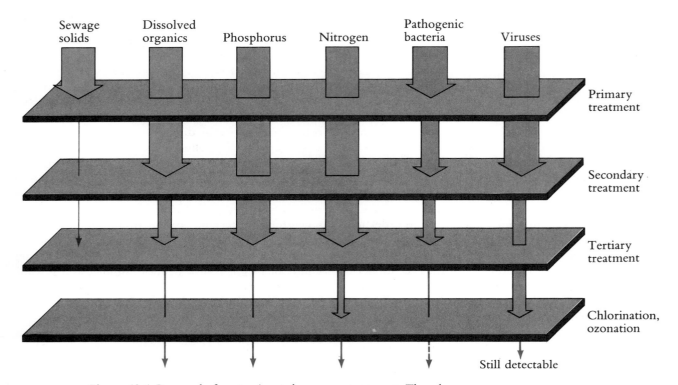

Figure 13.6 Removal of contaminants by sewage treatment. The relation between each stage of treatment and the level of removal of contaminants is shown by this diagram. If an arrow meets a treatment plane with an arrowhead and changes thickness in moving through that treatment plane, the level of the contaminant has been reduced by that stage of treatment. The relative reduction in the thickness of the arrow indicates approximately the extent of the reduction. For instance, sewage solids are nearly all removed in primary treatment. Dissolved organics, on the other hand, are not altered by primary treatment but are significantly reduced by secondary treatment. Nitrogen and phosphorus are not generally removed by primary or secondary treatment but are much reduced by tertiary treatment, and so on.

made on the *physical-chemical treatment* of wastewater, which removes the same substances as those removed by a plant with conventional primary, secondary, and tertiary processes.

Physical-chemical treatment begins as conventional treatment does, with a rack and screen. The next step, phosphorus removal, is achieved by the addition of a chemical such as lime or alum. The resulting solid particles (of calcium or aluminum phosphate), as well as the usual suspended solids, then settle out together in a settling tank, much as in the usual primary treatment step.

The wastewater is then passed through a column containing activated carbon, where dissolved organics are adsorbed onto the surface of the carbon particles. The organic removal achieved in this step matches that of the best biological treatment processes. Eventually, the carbon particles become saturated with organics; then the column is removed for cleaning and is replaced by another. Once cleaned, the first column can be placed in service again.

Physical-chemical treatment plants may soon be in service in a number of areas, including the cities of Niagara Falls, New York and Cleveland, Ohio.

CONTROVERSY 13.1

Can (Should) You Save a River Without Removing the Pollutants?

We can easily and without spending huge sums of money give back to the river the oxygen which it lacks in the summer due to pollution and a lower water flow.
Chairman of the city council, Paris, France

A major stumbling block is philosophical; aeration treats the stream instead of the pollution source.
Roy Christianson, Fox River Coordinator, Wisconsin

In addition to the methods of water pollution control discussed in this chapter, there are two quite different ways to improve the quality of rivers. These options deal with the rivers themselves and are not treatment processes in the usual sense. The first is called *low-flow augmentation* because additional water is released from an upstream reservoir to dilute the downstream concentrations of dissolved organics. The dilution is intended to add oxygen resources to the stream, thereby reducing the effect of the oxygen-demanding wastewater. For example, the wastewater from a particular community might reduce the dissolved oxygen in the average river flow during August to 3 mg/L. If the flow is augmented sufficiently, the dissolved oxygen in the increased flow may only fall to 5 mg/L.

Among environmental engineers, low-flow augmentation is humorously called "the dilution solution to pollution." It is not really pollution control, however; it is a method of water quality improvement. The same quantity of organic wastes is still entering the watercourse—only the impact is different. Until the early 1970s, whenever the Corps of Engineers proposed a reservoir project to Congress, the water quality improvement brought about by low-flow augmentation could be counted as one of the benefits, but this is

no longer the case. Environmentalists convinced Congress that low-flow augmentation was not pollution control.

In the same category as low-flow augmentation is the process known as *in-stream aeration*. This, too, is a means of improving the dissolved oxygen content of a river without removing dissolved organics. We know that dissolved organics, through the action of bacteria, remove dissolved oxygen from a stream. We also know that the flowing water of the stream captures oxygen from the air and dissolves it in the water. Why not help the natural process by bubbling in compressed air? The process is known to work, and aquatic life can be restored to the river by raising the dissolved oxygen level.

The city council of Paris, France, plans to offer the Seine River such a "face-lift." When the chairman of the city council remarked that oxygen could be restored to the river without undue cost, he did not mention that the pollution would still be present.

In-stream aeration has been in use on one U.S. waterway since 1980. The process has enabled the Chicago Metropolitan Sanitary District to discharge more soluble organic wastes (BOD) and suspended solids. Without in-stream aeration, Chicago would have to achieve a stream water quality of 4 mg/L of BOD. With in-stream aeration, the stream can now accommodate 10 mg/L of BOD. Nine other waterways have been targeted for in-stream aeration by the District. Chicago has the advantage, however, that all ten of the waterways involved have specially designated requirements for dissolved oxygen; that is, the dissolved oxygen level needed in these waterways is less than the amount required to support fish life.

In-stream aeration is also being considered for the Fox River in Wisconsin, where a sufficiently high oxygen level can be achieved for only $120,000 per year. Although the aeration equipment would be in place throughout the year, actual in-stream aeration would probably be needed for only several weeks out of every year—when flow conditions are lowest and temperature is highest. In contrast, to achieve the same water quality by advanced waste treatment (tertiary treatment) would require $10 million annually. Wisconsin officials are concerned, however, that industries will discharge more wastes or decrease treatment of existing wastes to take advantage of the greater absorbing capacity of a stream that is being aerated. In essence, the stream would be used as a treatment device.

Are in-stream aeration and low-flow augmentation as good as pollution control in restoring the quality of a river? Should these processes be considered in planning the cleanup of a river? Or should cleanup mean only the removal of wastes?

Source:
"In-Stream Aeration," *Civil Engineering—ASCE,* March 1980, p. 81.

Land Application of Wastewater

One concept of waste treatment being studied at government-supported projects is land application of wastewater, which has advantages and disadvantages. Wastewater contains phosphorus, nitrogen, and organic substances, so it is a reasonable fertilizer for crops, and, of course, it is a source of moisture. But although its use for irrigation seems attractive, there are other issues to be considered.

First of all, we know that primary treatment ought to precede any application of wastewater to the land. If it did not, the system that distributes the water could be clogged by the solids in the wastewater. Furthermore, without primary treatment a large portion of bacteria and viruses might be applied to the land. Some researchers believe that secondary treatment is not needed before applying wastewater; they assert that the land treatment will remove dissolved organics. Others believe that secondary treatment should precede any application of wastewater to the land.

Three methods of land application have evolved, each designed to accomplish a different objective. In the process referred to as *overland flow,* wastewater is applied to sloping land that has been stabilized by grasses. Gravity carries the water down the slope, across the soil, and through the grasses to a runoff collection ditch. The water does not penetrate the soil deeply, and at the base of the slope it joins other surface runoff. Because the removal processes are biological, areas where freezing temperatures regularly occur are not good candidates for overland flow.

Irrigation with wastewater is designed to fertilize crops. Irrigation could be thought of as an example of water reuse. Irrigation with wastewater is probably best suited for the arid western states. Muskegon County, Michigan and the city of Lubbock, Texas have large-scale spray-irrigation installations that apply wastewater to the land.

The third method of land application, *infiltration-percolation,* is primarily a method for recharging groundwater; a crop need not be involved. Treated wastewater is applied to the soil by pumping it into basins where the water percolates into the soil. Treatment is achieved primarily by filtration through the soil.

Nassau County on Long Island, New York plans to "polish" its sewage to drinking water quality and then recharge it to the ground instead of discharging it into the ocean, in this way maintaining the reserves of fresh groundwater. Groundwater recharging is also practiced in Israel and in the Netherlands. In the Netherlands, the dirty water of the Rhine River is recharged into the sand and then drawn out before it is used by the city of Amsterdam. In Phoenix, Arizona, the city treats the wastewater that comes from secondary treatment by recharging it into sand and gravel. This process, referred to as *soil-aquifer treatment*, recovers the recharged wastewater via wells. The water is of such good quality that it is being used for unrestricted irrigation in the Phoenix agricultural area, and it is also suitable for lakes that allow swimming. It is claimed that, after further treatment, the water is suitable for drinking.

Still another method of land treatment that has been suggested is to apply sewage to wetlands, the shallow water zones where a river or stream slowly enters a lake or bay and where special, partially submerged, shallow-water plants take root. Because the wetlands already process nutrients and organic matter, it seems natural to use this processing capability for wastewater. Water hyacinth plants seem especially well adapted to treating wastewater. Although the cost of treating wastewater by using wetlands is low compared to conventional treatment, the distance of wetlands from the wastewater source may make pumping costs prohibitively high. Artificial wetlands may be constructed, but the availability of sufficient land near an urban area may be a barrier to using such systems. Although the wild fowl and animal life that use wetlands have been mentioned in reports, it appears that little thought has been given to possible harmful effects (say of heavy metals) on these species. Sewage flows that are less than 2 million gallons per day are probably most suitable, but this corresponds to a population of only 15,000–20,000 people.

All approaches to applying wastewater to land require nearby land with no better uses (such as housing, industry, and so forth). If a suitable site is too distant, the electrical energy needed to pump the wastewater will make the project too costly. In addition, applying wastewater to land is not known to be safe in all circumstances. The Water Pollution Control Federation issued a policy statement in 1974 on land application of wastewater. It stated that "present knowledge of the effects of large-scale use of land disposal is too meager to justify widespread use for man-made wastes."

In 1976, researchers reported findings of a study on the frequency of infectious diseases in agricultural communities in Israel (kibbutzim) that spray-irrigated their crops with wastewater.* They found that the infectious disease rate was about four times higher in

these areas than in communities that did not irrigate crops with wastewater. Droplets containing bacteria and viruses were probably carried in the air by the wind, thus causing the increase in disease in the communities. Not only are populations in nearby communities at risk, agricultural workers may also be subject to infection. Most important, food crops can carry viruses, bacteria, and protozoa to the population that consumes them. An outbreak of cholera in 1970 in Jerusalem has been linked to the consumption of contaminated vegetables that had been irrigated with wastewater.

Attention must also be given to the possibility that nitrates may appear in the groundwater unless removed from the wastewater before it is applied. The presence of nitrate ions in high enough concentrations in drinking water can bring about the disease methemoglobinemia in children (see Chapter 14).

The Federal Water Pollution Control Act

Congress has attempted to control point-source water pollution with a variety of laws (see also Chapter 14), but the most basic rules are found in the Federal Water Pollution Control Act. This law was passed in 1972, further amended in 1977, and brought up for renewal again in 1985. The law is designed to go into effect in steps; some provisions will not take effect until 1989 or later. Basically, the law divides pollutants into three classes. For wastes designated as *toxic*, industries must use the best available technology (BAT) to treat their wastes before releasing them into natural waters. For pollutants listed as *conventional*, such as municipal wastes, dischargers must use the best conventional technology (BCT). Any other pollutants fall into a third class, *unconventional*. These pollutants must meet BAT standards, although waivers are possible.

If even the best available technology would not protect certain waters, stricter standards (for example, no discharge at all) must be imposed. Further, special standards called *pretreatment standards* are imposed for wastes that will pass through sewage treatment plants (see Controversy 13.2).

A special permit from EPA or from the state is required for discharge of any pollutants into navigable waters. This permit would, of course, require that the discharge meet the appropriate toxic, conventional, or unconventional treatment standards. For ocean discharges, EPA is required to set special standards.

---

*Personal communication, Hillel Shuval, Hebrew University–Hadassah Medical School, Jerusalem, Israel, 1984.

CONTROVERSY 13.2

Economics and Environmental Regulation: Pretreatment Standards

Should we not ask the EPA to show us that the benefits of
its regulations are at least as great as the costs?
R. W. Crandall

Conventional estimates of gross national product are . . .
meaningless so long as they fail to account for the cost
of . . . uncorrected levels of pollution.
R. D. DuBoff

The Environmental Protection Agency is now requiring industries that have been discharging toxic wastes
into sewer systems to set up pretreatment systems to
prevent the wastes. In this way, materials such as toxic
organic chemicals will be removed before they can
reach municipal sewage treatment plants. The measure
should help reduce toxic organics in drinking water by
stopping the discharge of these materials into those natural waters that might be used as sources of drinking
water. In addition, toxic wastes often "knock out" sewage treatment plants by killing the microbes that are a
vital part of treating sewage. An instance of this
occurred in Louisville, Kentucky, when chemicals
dumped into a sewer knocked out the city's sewage
treatment plant for 45 days, costing the city $5 million.

Of course, pretreatment is not without cost to industry. Robert W. Crandall, a Senior Fellow at the
Brookings Institution, took issue with the EPA's
regulations in a letter to *The New York Times* on
July 3, 1978:

> . . . Jorling is quoted as saying that he did not
> know how much these industrial standards would
> cost, but that it would possibly be in the "low
> billions" of dollars. This rather cavalier assess-

ment of costs is not uncommon among environmentalists who tend to view G.N.P. as a stock
which does not change with increasing regulation.
Perhaps someone should point out to Mr. Jorling
that there are not many "low billions" of dollars
in the entire industrial output of the economy. His
assurance that the costs of those latest regulations
are in the "low billions" of dollars may be likened
to the executioner's assurances to the condemned
that he will only have to fall "a few feet" with the
noose around his neck.

Recently a colleague, Edward Denison, measured the effect of environmental expenditures by
business upon economic growth and found that
by 1975 they had reduced economic growth by
more than 10 percent of its recent average. The
total cost of these regulations in 1975 was in the
"low billions" of dollars—$9.6 billion, to be ex-
act. If a single set of new regulations will cost a
few billion more, what can we expect when a
large number of new air quality standards are pro-
mulgated in the next few months? Before we proceed, should we not ask the E.P.A. to show us
that the benefits of its regulations are at least as
great as the costs?

This was answered by Richard B. DuBoff, an economist at Bryn Mawr College, on July 10, 1978, in *The New York Times*:

> Perhaps, to borrow Mr. Crandall's own language, someone should point out to him, and to Mr. Denison, that their thoroughly conventional estimates of gross national product are overstated, or meaningless, so long as they fail to account for the costs of present, uncorrected levels of pollution—of air, water and industrial inputs (the cotton "brown lung" disease but the latest case). Even a nonradical economist like Paul Samuelson warns students who use his textbook that "clearly we must adjust for any such 'bads' that escape the G.N.P. statistician whenever society is both failing to prevent pollution and failing to make power (or water or air or cotton) users pay the full costs of the damage they do." Once we make these adjustments, we see that net economic welfare grows more slowly than (conventionally measured) G.N.P.
>
> Mr. Crandall's reproof of "environmentalists who tend to view G.N.P. as a stock which does not change with increasing regulation" is another fine example of the tendency of orthodox economists to leap to policy "tradeoff" conclusions without any probing of underlying economic relationships (let alone their broader political context).

Should industry be required to pretreat wastes? Who will eventually pay for this? In what ways do we pay if industries do not pretreat wastes?

How does this controversy illustrate the way that air and water have been treated as free resources?

The "Settlement Agreement"

The Environmental Protection Agency was also supposed to decide which pollutants fall into what category. But in 1976, EPA was sued by a coalition of environmental groups on the basis that the agency was "foot-dragging" in deciding which pollutants were toxic and what discharge standards should be. Only nine pollutants had been classified as toxic at that time. A settlement agreement was finally reached in which EPA would develop standards for 65 listed pollutants on a strict time schedule to be completed no later than 1983. Nonetheless, the issuance of standards still proceeds slowly.

Water Pollution Control in Rural Areas and from Nonpoint Sources

Control of Water Pollution in Rural Areas

In rural areas and small subdivisions where there are no wastewater treatment plants, or no sewers to connect to treatment plants, households most often treat their wastes by using **septic tanks**. The septic tank is the successor to the cesspool, which was a structure resembling a well. It was lined with stones or concrete blocks without mortar. Raw, unsettled sewage flowed directly into the cesspool (Figure 13.7), so groundwater contamination was a distinct possibility.

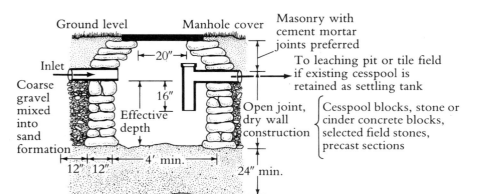

Figure 13.7 Cross-section of a cesspool. One of the first devices used to dispose of wastes, the cesspool is still a familiar concept. It is now recognized, however, that the cesspool is inadequate and dangerous in disposing of wastes. The potential contamination of groundwater is the principal objection to the use of the cesspool. (Adapted from J. Salvato, *Environmental Sanitation*, Copyright © 1958 by John Wiley & Sons, Inc., N.Y. Reprinted by permission.)

Figure 13.8 Cross-section of a typical concrete septic tank. The septic tank is the device that replaced the cesspool in rural areas. Closed to the soil by its concrete or metal sides, the device provides a modest amount of treatment to the wastewater from a home. Solids settle out in the tank and are degraded there, while a relatively clear overflow passes out to the leaching field. (Adapted from *Cleaning Up the Water*, Maine Department of Environmental Protection, 1974.)

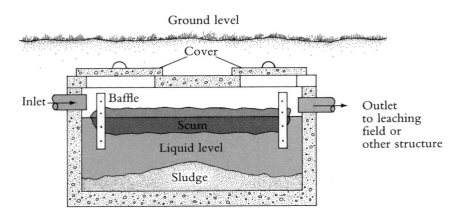

In contrast, the septic tank is a watertight tank constructed of either metal or concrete (Figure 13.8). Depending on the tank size and the size of the home it serves, the tank may hold sewage flows for anywhere from half a day to three days. During this time, solids settle to the bottom of the tank, where they are degraded by bacterial action.

The overflow from a septic tank is a liquid from which some solids have been removed. The overflow passes to a distribution box that "distributes" the water to a buried **leaching field** (Figure 13.9), which consists of perforated pipes of clay or plastic pieces. The wastewater seeps into the soil from the joints between the pieces. Microorganisms in the soil decompose the wastes. Lush grass may often be observed above the "arms" of the leaching system. The leaching system helps disperse the wastes and decreases the possibility of groundwater contamination.

If a family uses both a septic system and a well, it is good practice to keep the water supply far away (several hundred feet) and preferably uphill from the waste

disposal system. Furthermore, the soil into which a leaching field is placed should be a fairly porous one. (Porosity can be determined by a percolator test.) The soil should not have fractured or creviced rock beneath the surface, for such features could potentially channel the wastes to the groundwater. The top of the water table should be at least several feet below the pipes. Housing lots in suburbia are often required to be very large in order to handle the required leaching field for septic tank wastes.

One device used fairly widely in smaller communities and also in industry is the **waste stabilization lagoon**, or oxidation pond. The lagoon is essentially a wide shallow pond. It is typically 2–4 feet (0.7–1.2 m) deep, and raw wastewater flows directly into it. Its use depends on the availability of low-cost land and the character of the surrounding area. Odors may make its use a nuisance in populated areas.

Most ponds are designed to retain about two weeks' worth of flow from the community. Larger ponds can hold nearly two months of flow. Solids settle

Figure 13.9 Typical layout of a septic tank and leaching field. The outflow from the septic tank is distributed to a buried leaching field, consisting of clay pipes with spaces between them. The wastewater enters (leaches into) the soil through these spaces, and soil bacteria degrade the wastes further. (Adapted from *Cleaning Up the Water*, Maine Department of Environmental Protection, 1974.)

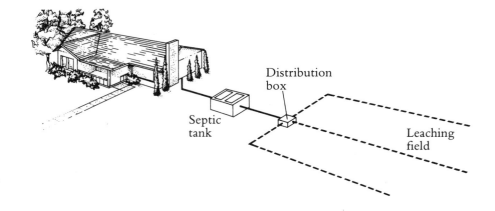

out as the wastewater flows slowly through the pond. The solids decompose in the absence of oxygen at the bottom of the pond. Near the surface, however, bacteria oxidize organic materials in the presence of oxygen. Weeds and other aquatic plants, including algae, may grow in the pond, making use of nutrients from the wastes. In addition, the algae supply oxygen to the system as a by-product of their photosynthesis. The oxygen helps keep the system operating with minimal odors. Sometimes the lagoon is aerated; that is, air is bubbled into the water, much as in the activated sludge process. The lagoon can thus act as a combination of primary and secondary treatment. In fact, primary treatment is not usually given to the waste entering a lagoon.

In 1978, the contents of a 36-acre (14-hectare) sewage lagoon in West Plains, Missouri leaked into the ground. Numerous subterranean streams carried the sewage over many miles and contaminated the groundwater in the region. Over 750 cases of diarrheal disease were associated with the leak. The incident illustrates a hazard associated with waste lagoons and septic systems: the possibility of leakage to the groundwater through channeled rock. In 1982, an internal EPA report warned that up to 90% of the waste lagoons in the United States had the potential to pollute groundwater. The report indicated that the lagoons, ranging from cattle ponds to industrial waste lagoons, numbered more than 180,000 at about 80,000 sites in the United States.

Control of Water Pollution from Nonpoint Sources

Nonpoint sources of water pollution include the many streams and rivulets that may carry pollutants off the land into major bodies of water, such as rivers or lakes. Nonpoint-source substances include organic wastes, nitrogen compounds, phosphorus compounds, and sediment or soil particles.

Dairies and livestock farms produce animal manure that, unless disposed of properly, can end up in streams and rivers. The use of feedlots, as opposed to grazing steers on pastureland, can result in a highly concentrated source of pollution. The average steer generates about fifteen times more waste than the average person; the hog twice as much. In total, farm animals may be producing ten times the waste of the U.S. population. To control pollution from feedlots

and dairy farms, waste treatment devices, such as lagoons, may be installed at the larger operations.

Growing crops can lead to water pollution because of the fertilizer applied to them. Fertilizer is likely to include compounds containing both phosphorus and nitrogen, the two elements most likely to cause eutrophication. In addition to fertilizer, decaying vegetation and soil particles from erosion may enter streams in runoff. Erosion occurs from cropland, from grazing land, from logged areas, from unreclaimed mining areas, and from construction sites. Soil particles may be suspended in the flowing water, resulting in a murky stream. On a weight basis, in fact, sediment from erosion is the number one water pollutant in our streams and rivers.

Control of pollution arising from farmland requires more careful application of fertilizers and pesticides. Contour plowing on small hills and terracing on steeper hills also help decrease runoff. (See discussion of erosion in Chapter 7.) In the last decade, many farmers have gone to *conservation tillage* and *no till farming* as labor- and fuel-saving devices. These methods involve use of a herbicide that kills weeds. The ground does not need to be turned over in the spring to eliminate weeds; hence, much less erosion occurs. These herbicides may, however, be washed into streams along with other pest-control chemicals applied to the crop. Many chemicals used to control insects and plants have been found to be cancer-causing substances. Therefore, it is important that the herbicides selected for use have as little effect on humans as possible.

Septic systems can give rise to nonpoint-source pollution. Urban runoff that does not enter storm sewers but flows from gulleys or culverts into streams is a contributor as well. The runoff from storm drains on highways carries metals such as the lead from leaded gasoline. Roads and shopping centers, with their vast impermeable areas, create conditions in which storm water runoff can occur very rapidly and, hence, carry considerable erosive power. Because of the rapid build-up of flow from highways and shopping centers, stream channels can be severely eroded thereby increasing the sediment in the flowing water.

Progress in Water Cleanup

In 1984, the U.S. Environmental Protection Agency and an association of state administrators in the water pollution area issued a joint report on cleanup progress

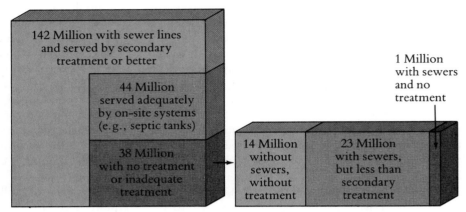

Figure 13.10 Progress in wastewater cleanup. (Data from "America's Clean Water," Association of State and Interstate Water Pollution Control Administrators, 1984.)

achieved from 1972 to 1982. During that period, a total of $56 billion was invested in water pollution control. Of this quantity, the federal government provided 64%, principally through its construction grants program; state governments provided 9%; and local governments (counties, cities, towns) provided 27%. Based on the 1982 U.S. population of 224 million people, the investment to control water pollution averaged $250 per person during the 10-year interval. What happened?

The number of people served by secondary or more advanced waste treatment went from 85 million in 1972 to 142 million in 1982, an increase of 57 million. If the investment is divided only by the people to whom secondary treatment was extended, the cost becomes $982 per person served.

While 142 million people are served by secondary treatment or better, another 44 million people appear to be served adequately by their on-site disposal systems (principally septic systems). These people, nearly all in rural areas, do not seem to need the central treatment facilities and sewers required for cities and towns. That leaves 38 million people, or 17% of the population, who have less than secondary treatment, no sewage collection, or no treatment at all (see Figure 13.10). As estimated jointly by the EPA and the states, the cost to meet the remaining needs for public wastewater systems is $118 billion. This includes not only treatment facilities but also sewer lines for an estimated 14 million people.

The new sewers and wastewater treatment plants that were extended to 57 million people between 1972 and 1982 were designed to have a significant impact on the organic wastes being discharged into the nation's streams and rivers. The oxygen-demanding organic wastes were to be cut by about 52% from the level that would have occurred in 1982 without new pollution controls.

At the same time, a 1981 report of the General Accounting Office (GAO), an arm of Congress used for policy investigation, cast serious doubt on the achievements of that decade. The GAO found an alarming rate of treatment plant failure in meeting design standards. Out of 676 plants treating 1 million gallons of sewage or more per day, the GAO selected at random 242 plants to study. When discharges of these plants were monitored for a 1-year period between 1978 and 1979, 87% were found to be violating their permissible pollutant levels for at least one month out of the year. Of these plants, 75 (31%) were found to be violating some permissible pollutant level by 50% or more for more than four consecutive months. Pollutants considered were BOD, suspended solids, and coliform levels.

These violations have been attributed to contractors' use of inferior equipment that breaks down in service. Under federal law, contractors are allowed to substitute "equal" equipment for the particular items originally specified by the design engineer. The design engineer or the city has the burden of proving that the substitute equipment is inferior to the original design. Other potential causes of failure can be identified, but the loose federal rules are thought to be a key element in the breakdown of these systems.

Summary

Water treatment purifies water for the consumer, while water pollution control restores the quality of water the consumer has used so that it can be returned to a natural body of water. The three main steps in water pollution control are (1) Primary treatment, which removes solids from wastewater by allowing them to settle out; (2) secondary treatment, which removes dissolved organics by allowing microbes to consume the organic materials; and (3) tertiary treatment, which removes nitrogen and phosphorus compounds and some resistant organic materials.

After these three steps, the water is chlorinated to kill any remaining bacteria or viruses and then relased. Alternatives to the more common treatments for wastewater include low-flow augmentation, in-stream aeration, physical-chemical treatment with activated carbon columns, land application of wastewater, and ozonation. About 63% of Americans are served by a sewage treatment plant with secondary treatment or better. Another 20% have adequate private systems such as septic tanks. About 17% have less than secondary treatment or no treatment at all.

Major problems that remain in wastewater treatment involve (1) extending sewage treatment to the unserved part of the population; (2) separating the nation's storm sewers from the domestic wastewater sewers so that wastewaters do not bypass the treatment plant during storms; (3) instituting land use practices to control the flow of wastes into natural waters from nonpoint sources; and (4) decreasing the amount of possibly carcinogenic chlorinated hydrocarbons formed during the chlorination of wastewater.

Questions

1. What is the difference between water treatment and water pollution control?
2. Complete the following chart, which lists the steps or procedures in treating sewage before it is released to natural waters and the reasons for those steps.

Procedure	What is done	Reason
Primary treatment	Settling, removes solid materials	Removes 35% of organic materials, which would otherwise use up oxygen in natural waters
Secondary treatment	Trickling filter removes: Activated sludge removes:	
Tertiary treatment or advanced waste treatment	Phosphate removal: Nitrogen removal:	
Digestion of sludge from primary and secondary treatment		

3. How is the sewage from your home treated? Does it go to a sewage treatment plant or into a septic tank or cesspool? If it goes into a septic tank or cesspool, summarize what happens in a few sentences. If it goes to a sewage treatment plant, does the plant give only primary treatment, or does it have secondary or tertiary treatment equipment, too?

Further Reading

General Sources

Environmental Pollution Control Alternatives: Municipal Wastewater. U.S. Environmental Protection Agency, Technology Transfer Seminar Publication, EPA-625/5-76-012, 1976.

A profusely illustrated, well-written document that deals with most primary, secondary, and tertiary processes for waste treatment, including European practice and emerging technologies. Since it does not go into process design, the style remains nontechnical at all times.

Fair, G., Geyer, J., and D. Okum. *Water and Wastewater Engineering*, 2 vols. New York: Wiley, 1968.

These two volumes are basic texts in environmental engineering programs and thus are very complete and explanatory. They do, however, become technical in the sense of giving mathematical treatment to many subjects.

Berger, B. and L. O. Dworsky. "Water Pollution Control." *EOS*, **58** (1) (1977), 16.

The legislative and administrative history of water pollution control efforts by the federal government are covered. The authors are knowledgeable people who have served in the government in this area. Nontechnical.

Melosi, M., ed. *Pollution and Reform in American Cities, 1870–1930*. Austin: University of Texas Press, 1984.

This book provides an historical perspective on the water sanitation history of the United States. It is a fascinating account of the political and social activities that have led to pollution control in America.

Foxen, R. "The Third Wave in Water Quality Management," *Civil Engineering-ASCE* (April 1983), p. 43.

A view of the evolution of policy on water pollution control from the 1960s to the present.

Targeted Sources

Land Treatment of Municipal Wastewater Effluents. 3 vols. U.S. Environmental Protection Agency, Technology Transfer Seminar Publication, 1976.

These three paperbacks, all with the same basic title, total about 200 pages. While there is some technical engineering material in them, on the whole they are quite readable since they were designed to educate a general audience.

Crites, R. and C. Pound. "Land Treatment of Municipal Wastewater," *Environmental Science and Technology*, **10** (June 1976), 549.

This article describes the three land application methods and reviews several examples of land application in the United States.

Bastian, R. and J. Benforado. "Waste Treatment: Doing What Comes Naturally," *Technology Review* (February–March 1983), p. 59.

Article touts land and wetland treatment of wastewater.

Reed, S., R. Bastian, and W. Jewell. "Engineers Assess Aquaculture Systems for Waste Treatment," *Civil Engineering-ASCE* (July 1981), p. 64.

Article is devoted to discussion of wetland treatment of wastewater.

Morrison, Allen. "GAO Finds Massive Failure of Wastewater Treatment Plants," *Civil Engineering-ASCE* (April 1981), p. 74.

A detailed description of the General Accounting Office study of wastewater treatment plant reliability.

References

"Nonpoint Source Pollution in the U.S.: Report to Congress," U.S. Environmental Protection Agency, Office of Water Program Operations, 1984.

"Environmental Regulations and Technology: Use and Disposal of Municipal Wastewater Sludge," U.S. Environmental Protection Agency, EPA 625/10-84-003, September 1984.

"Aquaculture Systems for Wastewater Treatment: An Engineering Assessment," U.S. Environmental Protection Agency, Office of Water Program Operations, June 1980.

"In-Stream Aeration: A Possible Alternative to Advanced Waste Treatment," *Civil Engineering-ASCE* (March 1980), p. 81.

CHAPTER FOURTEEN

Toxic Waste Pollution of Surface Water and Groundwater

The Problem of Chemicals in Drinking Water

What Is Safe Drinking Water?/Purity of Groundwater Versus Purity of Surface Water

Groundwater Contamination

Hazardous Waste Disposal/Underground Petroleum Storage Tanks/Agricultural Chemicals/Other Sources of Contamination/Cleaning Up Groundwater

Chemicals That Contaminate Drinking Water

Cadmium, Arsenic, Lead, and Nitrates/Mercury/ Drinking Water and Cardiovascular Disease/Fluoridation/ Organic Chemicals/Pesticides/PCBs/Chlorinated Organic Compounds from Drinking Water Treatment

How Safe Is U.S. Drinking Water?

Safe Drinking Water Laws/Primary Drinking Water Standards/Secondary Drinking Water Standards/U.S. Drinking Water Problems

CONTROVERSY:
14.1 ***Who Bears the Economic Burden of Pollution?***

The Problem of Chemicals in Drinking Water

What Is Safe Drinking Water?

In 1980, the town of Rockaway, New Jersey discovered the chemical trichloroethylene (TCE) in town drinking water at two to three times the recommended maximum levels. Pushed by the New Jersey Department of Environmental Protection, town officials looked around for possible solutions to the problem. Faced with drilling new wells, buying water from a nearby town, or treating their own well water to remove the TCE, officials found a fourth option. Apparently TCE was mainly seeping into one well and contaminating the whole water supply. Officials began pumping water 20 hours a day from the most contaminated well into nearby Beaver Brook. This approach seemed to solve Rockaway's problem—TCE values dropped to only trace levels. However, the New Jersey Department of Environmental Protection was not happy with the fact that the TCE-contaminated water flowed into Boonton Reservoir, Jersey City's water supply! Rockaway Mayor William Bishop declared that this wasn't a problem because TCE is volatile and thus evaporates from Beaver Brook: "We're giving the water to Boonton Reservoir, but we're giving the TCE to God."

This story illustrates the newest problem in providing people with safe drinking water: the contamination of underground sources of water with synthetic organic chemicals. Most people today are aware that, in order to be safe to drink, water must not carry harmful bacteria and viruses. But in addition, it must be free of dangerous chemicals—both inorganic chemicals, such as mercury, and organic chemicals, such as TCE.

In Chapter 11, we discussed how water is purified so that it carries no disease-causing organisms. In this chapter, we focus on the chemicals that can make water unsafe to drink. First, we examine how chemicals enter drinking water supplies in general, and second, we look at the chemicals themselves, their sources, and their effects.

Purity of Groundwater Versus Purity of Surface Water

In the past, people concerned about chemicals in drinking water focused on contamination of surface waters. Unwanted chemicals can find their way into surface waters from many sources. To a young nation with an apparently limitless supply of fresh water, it seemed reasonable to dispose of industrial waste chemicals into waterways. Mining wastes have been allowed to leach or flow unchecked into nearby rivers. Pesticides and other agricultural wastes are washed by rain into nearby rivers and lakes. Rain also washes a variety of contaminants from the air into water. Further, the salting of icy roads in winter leaves chemicals that wash into the nearest watercourse at the first thaw. Inevitably, some of these chemicals find their way into surface waters, lakes or rivers, from which drinking water is taken.

Almost half of all Americans, however, get their drinking water from groundwater, water that collects underground in the spaces between rocks or soil particles. This water used to be considered relatively pure, safe from a number of the problems that plagued surface waters. Shallow groundwaters do tend to be pure because soils and soil microbes filter out or degrade many pollutants, such as harmful bacteria or materials that make water turbid. These processes do not, however, remove most synthetic organic chemicals such as TCE. Organic contaminants are often volatile and so may eventually evaporate from surface waters, but they are trapped in groundwater. In addition, once groundwaters filter down to deeper levels, no further cleansing of pollutants takes place. Once contaminated, groundwater aquifers can remain contaminated for hundreds or even thousands of years.

In a 1982 study, the Environmental Protection Agency (EPA) found traces of synthetic organic chemicals in 17% of the randomly sampled small water systems using groundwater and in 29% of the large water systems using groundwater. Table 14.1 lists some of the more commonly found synthetic organic chemicals and shows the higher concentrations found in groundwater compared to those found in surface water.

How do these synthetic organic chemicals find their way into groundwaters? The major source is hazardous waste that has been buried in industrial and municipal waste dumps or collected in waste lagoons and ponds. In fact, of the various possible problems resulting from hazardous waste disposal, EPA ranks groundwater contamination as the most serious. When the EPA investigated 929 existing hazardous waste sites in 1982, it found groundwater contamination at 320 sites and evidence that it might exist at 326 other sites. Drinking water had been contaminated at 128 of the sites and was suspected at 213 others.

Groundwater Contamination

Hazardous Waste Disposal

Hawkins Point is a small area in Maryland bordered on three sides by water and bisected by the Baltimore Beltway. On the landward side of the beltway is a small residential community. Some two dozen families own their own homes in a peaceful, almost rural setting on this point, which juts out into the bay. On the other side of the beltway, the State of Maryland is constructing a hazardous waste dump.

Why did the state choose Hawkins Point as the place where hazardous wastes generated in the whole state of Maryland will be dumped? Can an environmentally safe dump be built there? What effect could this dump have on the people living in the Hawkins Point area? What exactly are hazardous wastes?

What hazardous wastes are. Hazardous waste is a term applied to any waste materials that could be a serious threat to human health or the environment when stored, transported, treated, or disposed of (Figure 14.1).

Toxic substances are a hazard if they leach into groundwater and enter drinking water supplies. Many such instances have been documented. Toxic chemicals from a nearby waste disposal facility have leaked into individual wells used for drinking water in small towns such as Tome, Tennessee. In other areas, the drinking water supplies of whole communities have been endangered. Accidental discharges of waste carbon tetrachloride into the Kanawha River have more than once threatened the water supply of Huntington, West Virginia.

Table 14.1 Synthetic Organic Compounds in Groundwater and Surface Water

Chemical	Location	Highest concentration found in groundwater (ppb)	Highest concentration found in surface water (ppb)
Trichloroethylene (TCE)	Pennsylvania	27,300.0	160.0
Toluene	New Jersey	6,400.0	6.1
1,1,1-Trichloroethane	Maine	5,440.0	5.1
Methylene chloride	New Jersey	3,000.0	13.0
1,2-Dichloroethylene	Massachusetts	323.0	9.8
Parathion	California	4.6	0.4

Source: Council on Environmental Quality. Reported in D. Burmaster and R. Harris, "Groundwater Contamination," *Technology Review* (July 1982), p. 57.

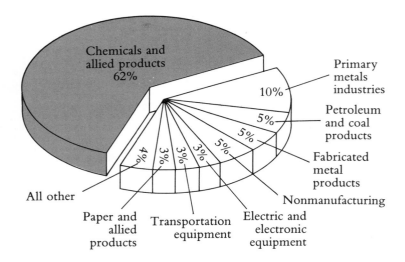

Total: 41,235,000 metric tons

Ignitable wastes present a potential fire hazard. In Elizabeth, New Jersey, a warehouse filled with stored wastes exploded with a thunderous roar in April 1980. A mushroom-shaped cloud of smoke rose over the city as 55-gallon drums of alcohol, solvents, pesticides, and mercury compounds exploded like bombs and rocketed into the air (Figure 14.2).

Corrosive wastes, such as acids, present another type of hazard. These wastes can dissolve storage containers and contaminate soils and water supplies.

A fourth category of hazardous wastes includes *reactive wastes*, those materials that can react with each other or with air or water in a dangerous fashion. In 1978, a young truck driver in Tennessee was killed while discharging wastes from his truck into an open pit. Toxic fumes were formed by liquids mixing in the pit.

Officials in Pennsylvania recently began monitoring an abandoned mine, which they learned had been used as a dump for cyanide-containing chemicals. They fear that the chemicals could react with acid mine waters and form deadly cyanide gas that could then escape to the surface.

Old dumpsites: Love Canal. There are actually two major problems with hazardous wastes today. One is what to do with the more than 34.4 million metric tons of hazardous waste generated yearly. The other is what

to do about the hundreds of abandoned dumpsites across the country that have been used for hazardous wastes in the past.

The most infamous example of an abandoned dumpsite is Love Canal near Niagara Falls, New York. At the turn of the century, William T. Love dreamed of building a model community based on the cheap power he would generate by digging a canal between the upper and lower Niagara Falls. But the project failed, leaving only a partly dug ditch. In the 1920s, the city and various industries began to use the ditch as a waste disposal dump. In 1953, Hooker Chemical Company, which then owned the dumpsite, capped it with clay soil and sold it to the School Board for $1. In the property deed was a warning that hazardous chemicals were buried in the old canal ditch.

The events that followed are still not entirely clear, but it appears that construction activities (a school and 100 homes were built encircling the canal), together with unusually heavy rains, caused the dump to fill up with rainwater and then overflow. Chemicals from this dump and three others in the area leached into groundwater. Corroding drums of waste collapsed, forming sinkholes from which chemicals evaporated into the air. Backyards, basements, swimming pools, and playing fields built over or adjacent to the old dumpsite were contaminated by a variety of noxious and hazardous chemicals. Among these chemicals were dioxins, some

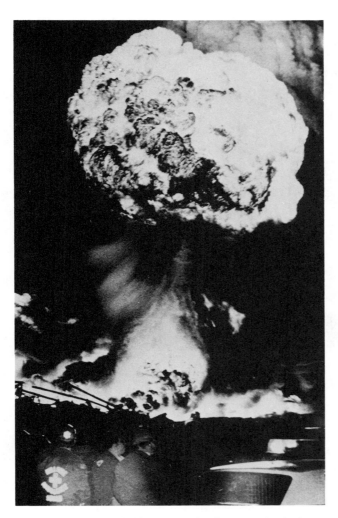

Figure 14.2 A fireball illuminates the sky over Elizabeth, New Jersey. An abandoned storage warehouse for hazardous wastes blew up in 1980 when chemicals stored there ignited spontaneously. Waste disposal officials had moved some of the chemicals out several weeks before, probably preventing a worse explosion. (Wide World Photos)

of the most toxic chemicals known. Preliminary studies seemed to show a higher than normal incidence of birth defects and chromosome damage in residents living near the canal. Over 850 families were eventually moved from the area, and their houses were purchased using federal disaster emergency funds (Figure 14.3).

It is still unclear what kind of health effects residents of the area may have suffered and how severe these effects might be. The Love Canal tragedy pointed up how poorly equipped the government was to deal with hazardous waste emergencies at former dumpsites. The mechanisms did not really exist to clean up these old sites or to protect the people living near them.

Superfund. In 1981, spurred by the Love Canal disaster, Congress passed a toxic wastes cleanup bill. Known as the "Superfund," this law initially provided $1.6 billion to be used to clean up hazardous waste sites when there was immediate danger to public health or to the environment. In 1986, after the initial appropriation had allowed no more than a good start at cleaning up dump sites, Congress appropriated another $2.5 billion per year for the next 5 years, while reauthorizing the bill. This money will be raised by a $12 tax on every $10,000 of corporate income for corporations earning over $2 million per year. Polluters, when they can be found, will be fined. Other monies will come from general revenues and taxes on crude oil, motor fuel, and chemical feedstocks. The fund does not, however, provide for the compensation of people injured by hazardous waste dumping. As directed by the bill, EPA has already identified 400 sites needing top priority cleanup.

Methods of hazardous waste disposal. There are already four dumps, two of them no longer in use, at Hawkins Point, where the State of Maryland will construct a new hazardous waste dump. Groundwater in the area is contaminated by a variety of chemical wastes and cannot be used by industry or even to water lawns. Fortunately, residents in the area have not been using well water but have city water piped into their homes. The state health department has a master plan for cleaning up the groundwater at Hawkins Point, partly in anticipation of the new hazardous waste dump that the state is locating there. In fact, some state officials claim that the area could be better off because of the new dump. Is this possible? To answer the question it is necessary to look at the state of the art of hazardous waste disposal.

Disposal of hazardous waste is governed by the Safe Drinking Water Act of 1974, the Toxic Substances Control Act of 1976, and the Resource Conservation and Recovery Act of 1976. Regulations devised under this last act require a *cradle-to-grave* monitoring system for hazardous wastes. Any company that produces more than 45 pounds (100 kg) of hazardous waste per

Figure 14.3 Love Canal home owners protest contamination of their neighborhood. Many families were forced to stay in their homes long past the time they became convinced it was unhealthy because they could not afford to move. State and federal governments eventually moved 850 families out of the Love Canal neighborhood and bought their homes; however, this was only after several years of meetings and protests. Worry about health effects, especially effects on their children, has taken a sad toll in stress-induced mental, physical, and social effects on former residents. (Citizens Clearinghouse for Hazardous Wastes)

month must fill out a manifest for any wastes leaving the plant site. The company that transports the waste must sign the manifest and give a copy to the company handling final disposal. This company must return a copy of the manifest to the company that originally generated the waste. In this way, officials at the company generating the waste can be sure that their hazardous wastes have not simply been abandoned in some remote area or otherwise disposed of improperly. Generators are responsible for alerting EPA if they don't receive notice that their waste has been properly handled.

Of course, the best disposal solution for hazardous waste would be to *recycle* it for other uses or to recover valuable materials from the waste. In some cases, this can be done. For instance, waste solvents from the electronics industry are still of high enough quality to be substituted for virgin materials used in some other industries. The EPA has been encouraging the establishment of waste clearinghouses that would match useable waste materials with potential users. About 3% of all industrial wastes could probably be reused in this way.

Wastes that cannot be reused in some way can sometimes be made much less hazardous by simple *chemical treatment*. For instance, corrosive acids or bases can be neutralized. This sometimes causes toxic heavy metals to be released from solution as solid materials

that can be filtered out. Such processes are usually more economical to carry out at large regional waste reprocessing centers. Sometimes such centers can even mix wastes from different companies in order to detoxify both. Acidic and basic wastes can be matched in this way.

Organic materials that have little or no heavy metals can be detoxified biologically. *Composting and land farming*, in which materials are spread out over a large land area so that microbes can decompose them, are examples of biological treatment of hazardous waste. Pesticides, phenols, TNT wastes, and paper mill wastes have all been detoxified by these methods. Monitoring is necessary, however, to insure that the wastes do not percolate down into groundwater before they are detoxified.

A more expensive, but possibly safer, method of disposal is *incineration* in special high-temperature incinerators. Modern incinerators are designed to destroy at least 99.99% of the organic waste material they handle. Any materials that escape up the flue are scrubbed out of the flue gas with special equipment. All of this makes incineration an expensive alternative.

Communities near incinerators have objected to them because of concerns about possible emissions. Some experts feel the answer is to incinerate on ships in midocean, far from any people. It is speculated that these ships would operate without scrubbers, since the trace amounts of hazardous wastes emitted would be a small pollution burden when mixed into the ocean. At least two such ships already operate: the *Vulcanus*, a Dutch ship, and the *Matthias II*, a German ship. The *Vulcanus* has incinerated large quantities of Agent Orange, a highly hazardous herbicide contaminated with even more toxic dioxins, for the U.S. Air Force. The EPA monitored the incineration, which took place in mid-Pacific, and found essentially no hazardous emissions.

In some areas, hazardous wastes have been pumped into *deep wells*. Controversy has arisen over the eventual fate of such materials. Proponents argue that most such wells are drilled in geologic formations that have held brine in natural isolation from fresh water for millions of years. Also, if a use is eventually found for materials in the wastes, they can be pumped up again. Critics express concern about possible leakage and poorly designed wells. They note that explosions and even earthquakes have apparently resulted from waste injection techniques.

Despite these alternatives, there are still large quantities of materials that are not recycled or reused, that have heavy metal concentrations too great to degrade or incinerate, or that are too thick to inject into wells. These wastes are commonly buried in *landfills*. A variety of factors must be considered in making a landfill "secure," that is, able to contain hazardous wastes without leakage to the environment.

The best soil for a landfill is clay because clay is less permeable than other types of soil. The clay pits in the landfill must usually be lined with another material because certain organic chemicals react with clay and cause leaks. Some liners are natural materials, such as crushed limestone or nut shells; others are synthetic membranes. The liners help contain hazardous wastes by absorbing them, or by chemically reacting with them, or by providing an impermeable barrier to leaks. A leachate system collects liquid that escapes from the pits and is monitored to detect any escaping hazardous materials. In some cases, the term *landfill* is a misnomer since, in order to keep hazardous materials above the water table, the dump may be entirely above ground, forming a huge mound 30 or more feet high (Figure 14.4).

Materials disposed of in a landfill can be further secured by solidifying them in materials such as cement, fly ash from power plants, asphalt, or organic polymers.

Most hazardous waste today is disposed of on the grounds of the company generating it. Figure 14.5 shows how the waste that is transported off a company's grounds is handled.

Siting landfills: A political matter. It is important that the monitoring of a landfill go on for a long time, at least 20–30 years and possibly forever, to be sure that the site is secure and that nothing has disturbed the cover. Disturbance of the cover was probably a major cause of leakage from the Love Canal site in upper New York State. Because of the possibility of leaks, and in spite of the technology that exists to build a modern, secure hazardous waste landfill, the public as a whole remains wary. This is understandable in light of past disposal practices and their legacy of chemical contamination of the environment. To quote Jackson B. Browning, Director of Health, Safety and Environment for Union Carbide:

> It is not terribly difficult to elicit a political consensus in support of the need for siting sound hazardous waste treatment facilities. The majority of voters will readily agree that such sites are necessary and that you should move with all deliberate speed to locate them—at the other end of the state.

Local community opposition often makes it nearly impossible to site hazardous waste landfills. In such a climate, politics rather than technology decides where a disposal site will go. A case in point is the Maryland hazardous waste dump at Hawkins Point. Although the state considered over 40 possible sites, Hawkins Point was chosen, in large measure, because of a lack of opposition. Most of the land was already owned by the state. The local political jurisdiction, Baltimore City, agreed not to protest such use of the land because the city would get, as part of the deal, a badly needed trash landfill adjacent to the hazardous waste dump. The state health department, which had to approve the site, was concerned with cleaning up the already contaminated groundwater and with seeing that the three existing chemical landfills at Hawkins Point operate in a secure manner. The only possible opponents were the residents of Hawkins Point, who were generally not affluent and who had little political influence. Although state officials felt residents would be better off after the Maryland hazardous waste landfill was sited there (due to better fire protection and stricter regulation of all the dumps in the area), residents didn't expect much im-

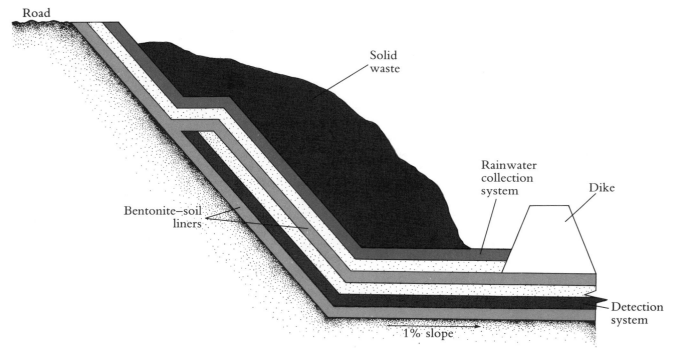

Road

Solid
waste

Rainwater
collection
system

Dike

Bentonite–soil
liners

Detection
system

1% slope

Figure 14.4 A secure landfill design. This cross-section shows an example of a secure landfill. It was built by Olin Chemical Corp. to dispose of chlor-alkalai wastes from its Charleston, Tennessee plant. The cells have a bottom liner of soil and sodium bentonite clay. A detection system is embedded in this layer to monitor any possible chemical leaks. Above this is another bentonite-soil liner, and over this is a rainwater collection system with a pump. Rainwater that seeps in is pumped out and treated. Landfills designed to handle organic liquids require a synthetic membrane liner rather than clay, since clay liners allow organic liquids to leak out eventually.

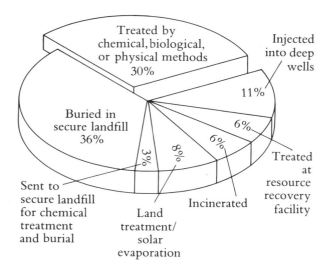

Treated by
chemical, biological,
or physical methods
30%

Injected
into deep
wells

11%

Buried in
secure landfill
36%

6%

6%

3%

8%

Treated
at
resource
recovery
facility

Sent to
secure landfill
for chemical
treatment
and burial

Land
treatment/
solar
evaporation

Incinerated

Figure 14.5 Current disposal practices. Most hazardous waste generated in the United States is disposed of or treated on the grounds of the industry that produces the waste (83%, according to *Civil Engineering—ASCE*, September 1981, p. 82). When hazardous waste is transported somewhere else for disposal, about one-third is treated to make it nonhazardous, and the remainder is incinerated, injected into deep wells, applied to land, or recycled. (Data from U.S. Environmental Protection Agency, 1980.)

provement from the siting of two additional dumps in their area. Eventually, the state offered to buy the homes of residents who wished to move out of the area.

Underground Petroleum Storage Tanks

In addition to contamination from hazardous waste sites, and of almost equal concern, is the contamination of groundwater from underground petroleum storage tanks.

In 1982, the mayor's office in Belleview, Florida began getting complaints about the drinking water. People complained that the water tasted funny and said that showering left their skin perfumed with the aroma of gasoline. Investigations discovered that two of the town's wells were contaminated with gasoline, probably from the leaking underground tanks of a nearby service station. The town now relies on trucked-in water or water piped from a local farm irrigation well, which must be boiled before it is safe to drink. This small town of 2,000 people has already spent over $30,000 in tax money on the problem and is working with a $300,000 emergency federal grant to construct a new water system.

Experts estimate that about 50,000 to 100,000 leaky tanks are buried all over the country. Most are the legacy of a gasoline-station construction boom in the 1950s and 1960s. When the stainless steel tanks were buried, no one realized they would begin to leak 20 to 30 years later. One gallon of gasoline can make 1 million gallons of water undrinkable.

Agricultural Chemicals

In agricultural areas, pesticides and herbicides percolate down into groundwater. One California study found the pesticide DBCP (banned in 1977 because it is a suspected carcinogen) in 359 community water supplies and in 30 school wells at levels of 1 part per billion (ppb) or greater. In another study, one out of seven wells in the California Central Valley area tested positive for DBCP.

Other Sources of Contamination

In addition to these synthetic organic contaminants, inorganic chemicals leach into groundwater from road salt used to de-ice roads in the winter. In several states,

salt water has intruded into groundwater resources because large quantities of groundwater have been pumped out for agricultural and domestic use. A further problem, in such states as Ohio and Illinois, is chemical contamination of groundwater with brines that are pumped into the ground during oil production from oil wells.

In other areas, septic tanks pose a threat. Even the chemicals used to clean septic tanks are a hazard because the major solvents in them are organic chemicals such as trichloroethane.

Cleaning Up Groundwater

An example of the possible extent and cost of groundwater contamination is found at Colorado's Rocky Mountain Arsenal. Waste dumps on the property, which were administered by the U.S. Army Chemical Corps and by Shell Chemical Company, have leaked waste synthetic chemicals into groundwater, contaminating 30 square miles of the aquifer over which the arsenal is sited. Numerous wells in the area have been closed. The Army has spent over $200 million to clean up the aquifer and to prevent further contamination. It is expected that $1.8 billion more will be needed to finish the job.

It first became apparent that groundwater beneath Rocky Mountain Arsenal was contaminated when, in the 1980s, nearby crops irrigated with groundwater were killed. Two attempts to solve the problem failed. Wastes were pumped from unlined surface ponds into asphalt lined basins; then, when the basins were full, the remaining wastes were injected deep into geologic strata underneath Denver. A series of earthquakes occurred during this injection, however. Some experts believed the waste injection was causing the earthquakes, so deep-well injection was stopped. Meanwhile, the asphalt lining on the new ponds was eaten away by solvents in the waste materials. In 1975, in the face of continued pollution of groundwater, a new strategy was devised. An underground barrier wall was built at one end of the property to prevent the flow of contaminated groundwater. Now, the water that collects behind the wall is pumped out, treated with granulated activated carbon to remove contaminants, and added back on the other side of the wall.

The Rocky Mountain Arsenal case illustrates three of the possible actions when groundwater is contaminated:

1. containment of the source of contamination with an underground barrier;
2. pumping out already contaminated water; and
3. treatment of the groundwater to remove contaminants, after which it is returned to the ground.

A fourth possibility is to pump the contaminated water out and dispose of it elsewhere, as Rockaway, New Jersey did. This is more a displacement of the problem than a solution, however. A fifth option involves pumping water (not necessarily contaminated water) from one place to another in order to charge the underground flow of water and so contain a pollution source or keep it away from wells that are in use.

The Rocky Mountain Arsenal case illustrates one other feature of cleaning up groundwater: the high cost involved. In fact, the most common solution to groundwater contamination involving a drinking water source is to drill a new well somewhere else. This is due, in part, to the high cost and technical difficulty of cleaning up contaminated groundwater. It is also a result of other considerations, however. A new well or a hookup to a noncontaminated source can often be arranged much more quickly than the contaminated groundwater can be cleansed. When an upset and angry public needs to be supplied with safe water, the deciding factor may be time, not cost. In the same way, if certain standards need to be met within a few years, an expensive barrier wall may be built rather than a less expensive pumping system that could take ten times as long to achieve the same result. One economic fact is clear, however. It is cheaper to prevent groundwater pollution than to clean it up afterward.

At the present time, groundwater protection is viewed as a state responsibility. A 1984 GAO study found that all states have some sort of program for groundwater protection. The extent of the existing programs varies widely, however. Between 1977 and 1983, Ohio authorities estimated they spent $35,000 on groundwater protection efforts, while California spent $5,160,000, and Texas spent $22,969,700. Provisions in the 1985 reauthorization of the Safe Drinking Water Act require each state to formulate a plan to protect groundwater drinking water sources. A major part of such plans will involve mapping aquifers and patterns of land use. This will both identify problems and permit protection of aquifers by zoning regulation. Other federal laws that pertain to protection of groundwater are the Solid Waste Disposal Act, the Resource Con-servation and Recovery Act, the Toxic Substances Control Act, the Clean Water Act, and the Comprehensive Environmental Response Compensation and Liability Act (Superfund).

Chemicals That Contaminate Drinking Water

Cadmium, Arsenic, Lead, and Nitrates

Cadmium, arsenic, lead, and nitrates are examples of inorganic chemicals that contaminate water supplies.

The natural occurrence of cadmium in water in more than minute amounts is almost unknown. In the past, detectable levels were usually the result of contamination from mining or industrial wastes. Cadmium is used in the manufacture of such products as paints, alloys, light bulbs, pesticides, and nuclear reactor parts. Wastes from electroplating plants have contaminated groundwater in the United States. Experts now worry that cadmium-containing products burned at dumps release cadmium into the air and that those buried in landfills are contaminating groundwaters.

The substitution of coal for petroleum in electric power generation is expected to increase airborne cadmium, arsenic, and lead concentrations. From the air, these pollutants will be washed into natural waters by rain. In addition to the increased air and water concentrations, there is evidence that cadmium and lead in sewage sludge or phosphate-based fertilizers may be increasing the amount of cadmium and lead in foods. Smokers seem to face an additional risk from the cadmium, arsenic, and lead found in tobacco products. (The topic of chemicals in cigarette smoke is covered more thoroughly in Chapter 28.)

The main effect of cadmium in the body appears to be in the kidneys, where it accumulates. In addition to kidney disease, cadmium may contribute to high blood pressure. Some scientists feel that cadmium levels found in the kidneys of the general population in the United States already approach half the level known to be toxic, and there is concern that increasing environmental levels will endanger health.

Cadmium may be removed by water-softening treatments used at drinking-water plants. On the other hand, drinking water that is relatively acidic and high in oxygen content may corrode pipes easily and pick up cadmium after leaving the treatment plant. The

EPA has set a limit of 0.010 mg/L for cadmium in drinking water.

Most of the arsenic used in this country (80%) is used in agricultural pesticides and defoliants. Because arsenic is also found in smoke from burning coal, surface waters near coal-fired plants or in agricultural areas may be subject to arsenic pollution. In some areas, arsenic from rock contaminates groundwaters to such an extent that they are not suitable as drinking water sources.

Lead from leaded gasoline is a common air pollutant that is washed down by rains into surface waters. The lead problem is covered in Chapter 21.

Iron and manganese may also contaminate waters that serve as community water sources. In most cases, high levels of these metals are a result of acid mine drainage (see Chapter 16).

In the past, nitrate contamination of drinking water was occasionally a problem in rural wells. Poorly built wells can be contaminated by nitrates from nearby septic systems or by fertilizers that have back-siphoned into the wells during spraying or diluting of the fertilizers.

Water containing more than 10 mg/L of nitrate is considered unsafe for drinking, mainly because it is likely to be poisonous to infants. Why are only infants affected? Apparently, in some infants the stomach does not yet produce enough acid to prevent the growth of the bacteria that convert nitrates (NO_3^-) to the highly poisonous nitrites (NO_2^-). Poisoned infants develop a disease called methemoglobinemia, in which their red blood cells cannot carry oxygen. Bacteria that convert nitrates into nitrites cannot grow in the stomachs of older children and adults who produce the proper amount of acid. There is some evidence that vitamin C, from orange or tomato juice, can prevent nitrite poisoning in infants, perhaps by combining with the nitrites.

Recently, scientists have decided that nitrates in drinking water may be a problem to adults and older children as well because of the possibility of the formation of N-nitroso compounds in the stomach, from nitrates. It is not known whether this actually occurs; however, increasing concentrations of nitrates in shallow groundwater, primarily due to the use of nitrate fertilizers and human wastes, makes such a possibility one of serious concern.

In parts of Israel, for instance, some wells are increasing in nitrate concentration by 2 mg/L or more per year. Plans for recharging groundwater supplies with treated wastewater, which is high in nitrates, will increase this problem. In eastern England, nitrates in soil above drinking water sources are percolating slowly down, causing some experts to refer to them as a "time bomb" that will cause serious problems in the future. In the United States, there exists clear evidence of increasing nitrate pollution of surface waters and shallow groundwaters. Concern is growing that deeper groundwaters could be similarly affected as nitrates move slowly downward. The drinking water standard for nitrates is currently set at 10 mg/L in order to protect infants from methemoglobinemia.

Mercury

In 1953, people living in the region around Minamata Bay in Japan began to suffer from a mysterious nervous disease. Symptoms included a narrowing of their field of vision and a lack of coordination. One unusual aspect of the epidemic was that animals and birds seemed affected as well as people. Because of this, public health officials were led to suspect that an environmental poison was causing the epidemic. The symptoms were, in fact, those of mercury poisoning.

Mercury has long been known to be a poison. The expression "mad as a hatter" originated in the times when many workers in the felt hat industry, who were exposed to high mercury levels in their work, suffered from mental problems. In mild cases, mercury causes symptoms that mimic mental and emotional disorders: insomnia, inability to accept criticism, fear, headache, depression, and generally exaggerated emotional responses.

In Minamata, some 120 people were affected, and 46 died before investigators realized that people and animals were being poisoned by eating mercury-contaminated fish and shellfish from the bay. The origin of the mercury was a plastics factory located on a stream leading into Minamata Bay. Although mercury is toxic to fish as well as to people, mercury levels in the water were not high enough to prevent fish and shellfish from living there. Two natural processes changed the factory's waste product from a trace chemical in the environment to an epidemic-causing pollutant.

Transformation and biological magnification. First, a *transformation* of the mercury was taking place. There are several forms of mercury. Elemental mercury (used in thermometers) and inorganic salts of mercury (for instance, mercuric chloride) are excreted relatively quickly from the body. Much more poisonous are the alkyl mercury compounds, such as methyl mercury and ethyl mercury. These compounds are excreted very slowly from the body—perhaps only 1% of the total amount present is removed each day. Although a great deal of the mercury that finds its way into natural waters is in the form of inorganic mercury, the mercury found in fish is almost always the more toxic methyl mercury. Studies have shown that bacteria in the bottom muds of lakes and rivers, in the slime on the fish themselves, and in the mucus in fish's stomachs are capable of transforming inorganic mercury to methyl mercury. Some of the mercury released into Minamata Bay was in the form of methyl mercury, but a good deal more methyl mercury was apparently formed by bacteria.

Second, *biological magnification* (see Chapter 2) increased mercury concentrations in fish and shellfish to levels many times greater than those found in the bay water. Fish and shellfish in the bay concentrated the methyl mercury to levels that were toxic to the humans who ate the seafood.

The possibility of these two processes, transformation of a substance in the environment and magnification by living organisms, must always be taken into account when a decision is made about the hazard of a particular chemical.

Mercury in water. Mercury enters natural waters from many sources. The English River in Ontario, Canada, has been contaminated with mercury discharged from a chlorine–caustic soda plant. Mercury levels in fish caught in the river have been reported to be as high as in fish from Minamata Bay. Furthermore, cats fed on fish from the river have shown symptoms of mercury poisoning (see Controversy 14.1).

In the United States, scientists have estimated that chlorine–caustic soda plants released ¼ to ½ pound (100–200 g) of mercury for each ton of caustic soda (sodium hydroxide) produced until the early 1970s. Today, strict U.S. laws prohibit the discharge of mercury by industry; however, in areas where mercury was formerly discharged—for instance, near pulp and paper mills or chlorine–caustic soda plants—mercury in the bottom mud still contaminates the water and the organisms living there (Table 14.2). Fishing restrictions have been set in many states because fish are magnifying the mercury dumped into the water many years ago.

One of the largest discharges of mercury in the United States seems to have been in Tennessee at the Oak Ridge Y-12 plant, which made weapons components. In 1983, private investigations by an Oak Ridge employee indicated that there might be serious mercury contamination of plants and fish in the area around the facility. It was eventually found that some 2.5 million pounds of elemental mercury had been discharged into the environment, most of it probably leaking slowly from deep fissures and cracks in the ground under the plant. About 500,000 pounds was discharged directly into East Fork Poplar Creek. Fish caught close to the plant contain twice the legal mercury limit of 1 part per million (ppm).

Downstream, fish have higher than normal levels of mercury, although the level is still below the legal limit. Investigations have shown that most of the mercury spilled into East Fork Poplar Creek was washed downstream and buried in the sediment of two large reservoirs, Watts Bar and Chickamauga Lake. How much of a hazard this presents is not yet clear (see Controversy 14.1). Natural cleansing of mercury-contaminated lakes occurs when mercury-containing sediments are sealed off by a layer of clean sediments. As long as the sediments remain undisturbed, clay sediments can effectively isolate mercury in a lake bottom for a period of 10 years or more. On the other hand, where mercury has settled into crevices in rocky beds, swiftly flowing rivers and streams may release mercury to water and fish over much longer periods of time.

Mercury in air. Mercury is not only a water pollutant but also an air pollutant from sources such as coal-fired power plants and mercury-ore refining plants. The fossil fuels (coal and oil) contain mercury that is released

Table 14.2 Mercury Found in Organisms Living Upstream and Downstream from a Paper Mill

Organism	Living in stream above the paper mill (parts per million)	Living in stream below the paper mill (parts per million)
Sowbug	65	1,900
Burrowing alderfly	49	5,500
Caddis fly	—	1,700

Source: A. G. Johnels, et al., "Pike and Some Other Aquatic Organisms in Sweden as Indicators of Mercury Contamination in the Environment," *Oikos,* **18** (2) (1967), 323.

CONTROVERSY 14.1

Who Bears the Economic Burden of Pollution?

[Oak Ridge is] a relatively affluent city . . . populated by scientists and engineers who have other life pursuits than sports fishing.
Officials at Oak Ridge Tennessee Plant

This town is structured around the needs of the elite and no one else.
Jimmy Fuzzell II, resident, Oak Ridge, Tennessee

[Mercury pollution of the reservation waters was] the last nail in the coffin.
**Resident of Grassy Narrows Reservation, Ontario, Canada
(quoted in Shkilnyk, 1985)**

Environmental improvement is sometimes viewed as a concern of relatively wealthy people who don't have to worry about more basic necessities, such as jobs, food, or doctor bills. That this can be a real misreading of where the pollution burden often falls is illustrated by two cases of mercury pollution: the pollution of the Wabigoon–English River system in northwestern Ontario, and the Oak Ridge Y-12 plant incident in Tennessee.

A chlorine–caustic soda plant in Dryden, Ontario discharged 3.3 g of mercury per metric ton of chlorine produced through the 1970s. Between 1962 and 1970, 9,000–11,000 kg of mercury were discharged into the Wabigoon River on which the plant is located.

Fish caught in this river system show mercury levels comparable to those found in Minamata Bay. Further, two cats fed on fish from the river were found to have symptoms of Minamata disease, or mercury poisoning. Concern has been voiced because two Ojibway reservations are located nearby—the Grassy Narrows and the White Dog Reserves on the lower English River, which is part of the Wabigoon River system. The Ojibway Indians eat a diet high in fish, and government studies have found that many of them have excessive mercury levels in their blood. In addition, a Japanese group that came to study the Indians found signs of Minamata-like mercury poisoning, such as visual problems and lack of coordination.

Mercury pollution of the food supply was not the only problem the Ojibway Indians had to contend with. In 1963 the Canadian government had begun moving residents from their reservation to a new site on a logging road. The intention was to provide electricity, medical care, and schooling for them, which could be more easily accomplished at the new site. For the Indians, however, separation from their ancestral lands and from their hunting-fishing culture was devastating. Alcoholism became a problem for many who were unable to handle the loss of their former way of life and who were unable to fit into the new lifestyle offered them. Still, sociologists who originally investigated believed the reaction of the Grassy Narrows

Indians was extreme compared to similar cases they had studied among other North American Indian groups. It was then discovered that the Indians might well be suffering from mercury poisoning. The effect, both of the poison and of experiencing this further loss of control of their environment, seems to have been an almost total disintegration of the social fabric of the community.

In another case of mercury contamination, officials at the Oak Ridge complex in Tennessee were informed that fish caught in the creek near the Y-12 plant had high mercury levels. The officials responded that, since most people in the town were scientists and engineers, they would not be spending a lot of time fishing and eating mercury-contaminated fish. True or not, there was indeed a group of people in town who were not affluent engineers and scientists, who did commonly fish the creek, and who swam in it as well. This group,

the black community of Scarboro, consisted of 1,500 people who lived along the East Fork Poplar Creek. High unemployment in the community had encouraged a fair amount of fishing by Scarboro residents before the mercury pollution was discovered.

In these two cases of mercury contamination, what factors led to the unequal burden of pollution, which fell on the poor rather than on the wealthy members of society? Can you think of another case in which this unequal burden is common? Can you think of instances in which the more affluent members of society are at higher risk than the less affluent? Is there a difference between the two types of cases?

Source:
A. Shkilnyk, *A Poison Stronger Than Love* (New Haven, Conn.: Yale University Press, 1985).

---

to the air when they are burned. It is estimated that some 5,000 tons of mercury may be added to the air each year from the burning of fossil fuels.

Mercury in industrial and consumer products. About 6 million pounds of mercury is used each year in the United States in electrical devices, thermometers, fungicides, dental fillings, drugs, and paints. Although three-quarters of this mercury could be recycled, at least half is not recycled; that is, it finds its way into the environment and eventually into natural waters.

Mercury could be called a *permanent pollutant* in that, once released to the environment, it appears to be cycled from the air to water, to organisms living in the water, to human food supplies, and perhaps to humans themselves in seemingly endless cycles. Many years go by before environmental mercury at the bottom of lakes or oceans can be covered with such thick layers of mud that it becomes harmless (Figure 14.6).

Because of its toxicity and its tendency to accumulate in living organisms, the standard for mercury in drinking water is set at 0.002 mg/L. Fish with more than 1 ppm mercury are considered unsafe for human consumption.

Drinking Water and Cardiovascular Disease

Concern about the amount of sodium in drinking water is a relatively recent development. In Japan in 1957, scientists first noted that those middle-aged men who seemed to be at greatest risk of dying from cardiovascular disease (heart attacks and strokes) lived in communities with soft water rather than in communities with hard water. Since then, this effect has been confirmed in several other countries. It is still not clear whether hard water has a protective effect, whether soft water has a harmful effect, or whether some combination of both effects is the case. Studies of the minerals in hard and soft water have been interesting but inconclusive.

There is some evidence that either calcium or magnesium in hard water may have a protective effect. On the other hand, high concentrations of sodium are present naturally in some soft water, and the main methods of water softening (as well as some water treatment processes) add sodium to water. In areas with waters naturally high in sodium, the sodium values may reach 300–1,700 mg/L. Water softening in exceptionally hard water areas may result in water with sodium levels of

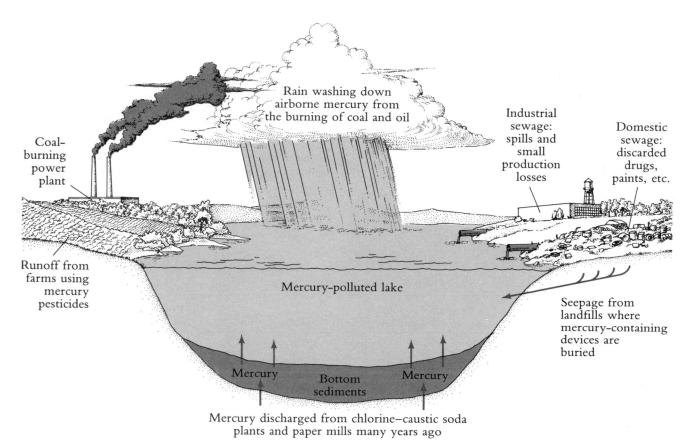

Coal-
burning
power
plant

Rain washing down
airborne mercury from
the burning of coal and oil

Industrial
sewage:
spills and
small
production
losses

Domestic
sewage:
discarded
drugs,
paints, etc.

Runoff from
farms using
mercury
pesticides

Mercury-polluted lake

Seepage from
landfills where
mercury-containing
devices are
buried

Mercury   Bottom   Mercury
              sediments

Mercury discharged from chlorine–caustic soda
plants and paper mills many years ago

Figure 14.6 Mercury pollution of a lake. In this imaginary lake, mercury enters the waters from industrial and domestic sewage and in seepage from landfills where mercury batteries, switches, and other devices are buried. Rainwater washes mercury-containing pesticides into the lake and also washes down mercury in the air from the burning of fossil fuels. Furthermore, mercury washed into the lake many years ago has settled into the bottom muds, where it is slowly being converted to toxic methyl mercury by bacteria and is then entering food chains. Unfortunately, although this particular lake is imaginary, the processes shown are not. Examples of all of them can be found in many areas of the country. In developing countries, too, mercury pollution is becoming a problem. In 1980, Djakarta Bay, in Indonesia, was reported to be seriously polluted with mercury, and 85% of the seafood eaten in Djakarta comes from this bay.

200–300 mg/L. Medical evidence strongly points to excessive sodium intake as an aggravator, if not a cause, of at least certain types of high blood pressure. Only about 10% of a person's intake of sodium is from water, however; 90% is from food. This is one reason for the difficulty in determining whether sodium in water has a significant effect on health.

In addition, soft water is more likely than hard water to leach harmful materials, such as cadmium and lead, from water pipes. Both cadmium and lead have been shown to raise blood pressure. Finally, other factors, such as smoking and exercise habits, also affect the incidence of cardiovascular disease, making it difficult to sort out the effects of drinking water.

Although there are no standards for sodium in drinking water, in 1982 the EPA began to require that public water systems test for sodium and report their results. The EPA also suggests drinking water sodium levels of 20 mg/L as a goal. This level would allow people on a strict sodium-restricted diet (500 mg per day) to drink the water.

Utility companies can switch treatment processes in order to decrease sodium addition; for instance, a softening process involving hydrogen ion exchange adds no sodium to water. In addition, plumbing in households with home water softeners can be changed so that cold-water taps carry only unsoftened water.

Fluoridation

Table 14.3 summarizes another drinking water issue— fluoridation. Fluoride is a trace chemical added to water to help prevent tooth decay. Although the American Dental Association is in favor of adding fluoride to drinking water supplies, many people still feel that fluoride may be harmful. We may, in fact, be reaching optimum levels of fluoridation in the United States because fluoridated water contributes fluoride to canned drinks and other food items. (See the Further Reading section for more on this.)

Table 14.3 Percentage of State Populations Using Fluoridated Water, 1975

State or district	Percent of population using fluoridated water	State or district	Percent of population using fluoridated water
Alabama	31.5	Missouri	42.0
Alaska	42.7	Montana	26.7
Arizona	31.2	Nebraska	45.8
Arkansas	38.2	Nevada	3.0
California	22.0	New Hampshire	13.3
Colorado	83.8	New Jersey	21.4
Connecticut	79.4	New Mexico	63.8
Delaware	39.5	New York	66.1
District of Columbia	98.4	North Carolina	45.4
Florida	35.8	North Dakota	50.8
Georgia	41.4	Ohio	41.4
Hawaii	6.4	Oklahoma	63.3
Idaho	33.5	Oregon	10.7
Illinois	86.2	Pennsylvania	46.2
Indiana	61.2	Rhode Island	66.3
Iowa	61.9	South Carolina	52.8
Kansas	51.5	South Dakota	61.5
Kentucky	51.0	Tennessee	66.9
Louisiana	23.2	Texas	58.7
Maine	40.6	Utah	2.4
Maryland	68.1	Vermont	37.0
Massachusetts	21.6	Virginia	51.2
Michigan	76.1	Washington	39.8
Minnesota	71.5	West Virginia	50.7
Mississippi	24.7	Wisconsin	62.2
		Wyoming	21.3

Source: Morbidity and Mortality Weekly Report, U.S. Department of Health, Education, and Welfare, 1977.

Organic Chemicals

Organic chemical contaminants are found in both surface and groundwater sources for drinking water. Reasons for groundwater becoming increasingly contaminated with organic chemicals were detailed earlier in this chapter. Taking surface and groundwater sources together, there are three main sources of organic chemical contamination: (1) agricultural use of pesticides, (2) chemical waste disposal practices, and (3) the treatment process used in water purification and water pollution control. Table 14.4 lists some of the more than 700 organic chemicals identified in drinking water so far. They were chosen for the table because all have the potential to cause cancer; however, it is not yet clear how much of a health risk they present separately or combined. (See Chapter 26 for more on the hazards of carcinogens in the environment.)

Pesticides

Pesticides, which are discussed in Chapter 7, are chemicals designed to kill insects and other pests. These chemicals are widely used. In fact, they have become part of our daily lives. A study by the U.S. Environmental Protection Agency showed that DDT, a pesticide once very popular in this country, can be found at levels of a few hundred parts per trillion in many U.S. waters, including drinking water supplies. Other studies have shown that Americans have detectable levels of DDT and other pesticides in their fatty tissues.

How do pesticides find their way into the water we drink? There are several ways. Pesticides run off farmlands during rains, either dissolved in the rainwater or attached to soil particles. Rainfall itself may be contaminated with pesticides that have evaporated or that remain in the air after the spraying of crops or woodlands. Industrial plants manufacturing pesticides may accidentally spill or intentionally discharge pesticides into nearby rivers and lakes.

The situation with regard to pesticides in surface water appears to be improving. A study published in 1984 by the U.S. Geological Survey, in which 22 pesticides were monitored, reported that fewer than 10% of almost 3,000 water samples showed detectable levels; however, this is balanced by increasing reports of groundwater contamination by pesticides.

Table 14.4 Known or Suspected Organic Chemical Carcinogens Found in Drinking Water

Human carcinogens	Animal carcinogens
Known	Dieldrin
Vinyl chloride	Kepone
Suspected	Heptachlor
Benzene	Hexachlorobenzene
Benzo(α)pyrene	Chlordane
	DDT
	Lindane (γ-BHC)
	Ethylene dibromide
	Benzenehexachloride
	PCBs
	Ethylenethiourea
	Chloroform
	Acrylonitrile
	1,2-Dichloroethane
	Tetrachloroethylene
	PCNB
	Carbon tetrachloride
	Diphenylhydrazine
	Aldrin
	Triclorophenol

PCBs

PCBs are a good example of organic chemicals that are found in drinking water as a result of waste disposal practices. The abbreviation PCB actually refers to a whole family of similar chemicals. Because they do not burn readily, they have been widely used to transfer heat from one material to another.

PCBs are used in electrical devices, such as transformers and capacitors, where a spark could ignite a flammable material. PCBs have also been used in a wide variety of other ways: as solvents for paint and ink, in pesticide sprays, and in plastics. In fact, so much PCB has been manufactured and used, we now find small amounts of contaminants in water all over the world (Figure 14.7). Like mercury, PCBs are concentrated by fish and other water dwellers (Table 14.5). Even people retain PCBs. About 1% of the U.S. population has more than 3 ppm PCBs in their fatty tissues, mainly as a result of PCBs in food. The rest of us have about 0.5 ppm.

Although the effects of large amounts of PCBs are

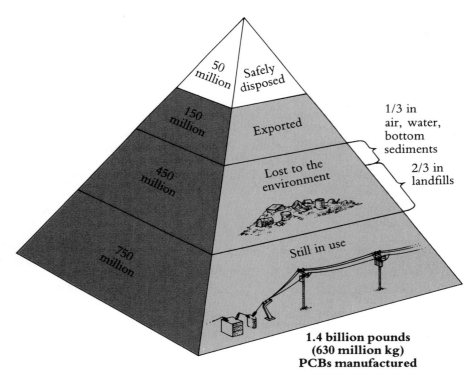

50 million — Safely disposed

150 million — Exported

450 million — Lost to the environment

1/3 in air, water, bottom sediments

2/3 in landfills

750 million — Still in use

1.4 billion pounds (630 million kg) PCBs manufactured

Figure 14.7 The fate of PCBs. Of the 1.4 billion pounds (630 million kg) of PCBs manufactured since 1929, some 450 million pounds (203 million kg) are now in the environment. About 750 million pounds (350 million kg) are still in use. Some of what is in use can be expected to enter the environment. Since PCBs are broken down very slowly, environmental levels of PCBs will probably decrease very slowly.

Table 14.5 Concentration of PCBs in Cayuga Lake Trout as a Function of Age

Age of fish (years)	Mean PCB concentration (ppm)
1	1.3
2	2.1
3	1.8
4	3.2
6	7.0
7	6.3
8	11.7
12	15.5

Source: C. A. Bache et al., "Atmospheric Carbon Dioxide: Its Role in Maintaining Phytoplankton Standing Crops," *Science*, **177** (September 29, 1972), 1192.

known, the effects of small amounts ingested over a long period of time are not as clear. People exposed to PCBs at work have developed nerve, skin, and liver ailments. In the laboratory, PCBs have been found to cause birth defects and cancers in animals (possibly at levels as low as 100–300 ppm in their diet).

PCBs in the environment. By 1970, enough concern had been expressed about the contamination of the environment with PCBs that the only U.S. manufacturer, Monsanto Chemical Co., announced it would sell PCBs only for use in closed systems—that is, systems in which no PCBs would be released to the environment. Allowable uses included the manufacture of devices such as capacitors and transformers but did not include PCBs in paints, inks, or plastics.

Soon after these restrictions went into effect, studies showed PCB concentrations in some rivers to be dropping. For instance, in rivers flowing into Green Bay, Wisconsin, levels dropped from 0.15–0.45 parts per billion to almost nothing. Many people heaved a sigh of relief—unfortunately, prematurely. In August 1975, the New York State Commissioner of Environmental Conservation issued a warning about eating salmon and bass from the Hudson River or from Lake Ontario because the fish had been found to contain well over the 5 ppm of PCBs then considered safe. In fact, federal researchers sampling fish for toxic materials notified the commissioner that striped bass and salmon taken from the Hudson River and Lake Ontario had as much as 37 ppm of PCBs.

Figure 14.8 Hudson River fishermen. Closing a river to commercial fishing can mean the loss of thousands or even millions of dollars to the industries depending on those fishing grounds. It also means a loss of livelihood to the fishermen involved. (Edward Hausner/NYT Pictures)

General Electric had been discharging PCBs into the Hudson for almost 20 years, possibly as much as 48 pounds (22 kg) per day.

Meanwhile, the EPA announced that at least twelve other plants in the United States were also dumping PCBs into waterways or municipal sewage systems. Surveys showed that PCBs could be found in the drinking water of cities all over the country.

As a result of hearings held by the New York State Department of Environmental Conservation, General Electric was judged guilty of violating state water-quality laws. The chairman of the hearings called the situation a regulatory agency failure as well as corporate abuse because General Electric made no secret of what it was doing.

Since 1976, the Hudson has been at least partly closed to commercial fishing (Figure 14.8). In 1984, the EPA lowered the allowable level of PCBs in fish to 2 ppm, forcing New York State officials to extend fishing bans to other waters in and around the state. Fishing restrictions are also in effect in other parts of the country due to PCB contamination. Fishing has been restricted in Lake Michigan since 1971 and in New Bedford Harbor, Massachusetts since 1979.

In 1982, New York State began work on a dredging plan to remove the worst PCB contaminated muds from several "hot spots" in the Hudson. The project has been held up, however, by lack of a suitable spot to dispose of the muds. In fact, some environmentalists are beginning to wonder whether dredging and disposal problems might cause worse environmental contamination.

Further study showed more rivers, including the St. Lawrence, to be contaminated with PCBs. A wide variety of fish from these waterways contained over the allowable limit of 5 ppm. Fish from the Hudson River, however, still had the distinction of containing higher PCB levels than fish from anywhere else in the country. Bass tagged in the Hudson River have been caught off the Massachusetts coast. In fact, the Hudson River is a major spawning ground for the entire East Coast from Massachusetts to Delaware. Thus, pollution of the Hudson has widespread effects.

Where were the PCBs coming from? A check of manufacturing plants on the river showed that two General Electric plants, both making capacitors, had permits from the federal government to discharge PCBs into the river. Further investigation showed that

PCB control—still a problem. The United States Congress passed an amendment to the 1976 Toxic Substances Control Act that banned the manufacture of PCBs in the United States after January 1979. The EPA followed through by forbidding any discharge of PCBs into U.S. waters. In addition, rules were made to phase out the use of PCBs, even in totally closed systems.

How much of a problem remains then? Although PCBs are no longer allowed in new manufacturing systems, some 750 million pounds (350 million kg) of PCBs are still in use in electrical equipment, mostly in capacitors and transformers used by electric utility companies (Figure 14.9). These devices may leak while in service, they will contaminate natural waters if improperly disposed of, and capacitors can explode, scattering PCB-containing oils into the environment.

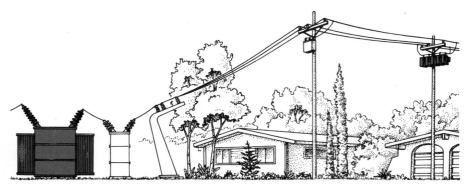

Figure 14.9 PCB-Containing transformers and capacitors. Although PCBs can no longer be used in the manufacture of electrical equipment, many old PCB-containing transformers and capacitors (shown in color) are still in service. In many cases, utility companies do not know whether older equipment contains PCBs or not. (Most capacitors made before 1977, however, are believed to contain over 500 ppm PCBs.) Leaks from this equipment, as well as improper disposal of old transformers and capacitors, are responsible for continuing pollution of the environment with PCBs.

A continuing series of contamination incidents is being reported by the press. In some cases, soil and water or whole buildings are reported contaminated by exploding capacitors or abandoned transformers. In other cases, PCBs find their way into human food supplies. (The problem of accidental contamination of food by PCBs and other materials is discussed in Chapter 28.)

The Department of Agriculture is attempting to prevent further accidents by requiring meat, poultry, and egg-processing plants to replace all PCB-containing equipment. In 1985, the EPA ordered removal of all equipment containing PCBs from public buildings by 1990.

Chlorinated Organic Compounds from Drinking Water Treatment

The third major concern about organic compounds in drinking water involves the chlorination process in water purification and water pollution control. The chloroform and carbon tetrachloride listed in Table 14.4 may of course come from industrial wastes, but they are more likely to have resulted from the chlorination of drinking water to kill disease-causing microorganisms or sewage.

In 1967, the Environmental Protection Agency began to study why drinking water drawn by the city of New Orleans from the lower Mississippi River smelled and tasted "fishy" or "oily." The source of the problem seemed fairly clear. A number of petroleum, chemical, and coal-tar product industries discharge wastes into the Mississippi. Thus, it was no surprise that the 1972 EPA report on the problem stated that chemicals in the wastes from these industries were responsible for the tastes and odors in New Orleans drinking water.

The report also noted, however, that chlorination of drinking water might be adding to the problem because chlorine, in the form used to disinfect drinking water, can react with natural or pollution-related organic materials to form chlorinated hydrocarbons. Some of these chlorinated hydrocarbons are known to cause cancer.

Since that report was issued, the chlorination aspect of the problem has caused the greatest controversy. Chlorinated hydrocarbons and other organic compounds have now been found in many water supplies. In one survey of 129 municipal drinking water plants that chlorinate water, chlorinated organic compounds at levels of a few parts per billion were found in the finished water of almost every plant. Chemicals found included chloroform and carbon tetrachloride, both known carcinogens. In the experiments that showed these chemicals to be carcinogenic, however, the doses of chloroform and carbon tetrachloride were much higher than humans could get by drinking New Orleans water. The problems involved in deciding whether the low levels in water are hazardous are explained in Chapter 26.

It is unlikely that we will stop chlorinating drinking water. The immediate risk of incurring epidemics of waterborne bacterial diseases such as typhoid fever and viral diseases such as hepatitis or polio is too great. Instead, the focus has been on removing organic compounds from drinking water.

Not only has attention been given to removing organic compounds, but water treatment methods are being changed as well in an effort to reduce the formation of chlorinated hydrocarbons.

The EPA is now requiring many water suppliers to reduce the level in drinking water of a group of chemicals called trihalomethanes, a group that includes chloroform. Trihalomethanes are one class of compound that may be formed when chlorine reacts with hydrocarbons dissolved in water. Reduction of trihalomethane levels can be accomplished in several ways:

1. The source of raw water can be changed to a less polluted one, decreasing the amount of organic material able to be chlorinated to trihalomethanes.
2. Adjustments can be made in the water treatment scheme, such as changing the point at which chlorine is added.
3. Other methods of disinfection, such as the use of ozone, can be tried.

How Safe Is U.S. Drinking Water?

Safe Drinking Water Laws

In December 1974, President Nixon signed into law the Safe Drinking Water Act. Several unsuccessful attempts had been made in other years to pass such a law, but it required some hint of an environmental disaster to propel the law through Congress. In this case, the hint of disaster was the publicized discovery of a number of possibly carcinogenic substances in the drinking water of New Orleans.

The Safe Drinking Water Act directs the Environmental Protection Agency to establish regulations ensuring the purity and safety of United States drinking water. Although each state is expected to be responsible for setting and enforcing its own standards, the federal government can legally step in if the states' regulations are not at least as stringent as federal regulations, or if the state does not enforce its regulations. This is a major change. In previous years, only suggested guidelines were set by the U.S. Public Health Service. The federal government had no power to regulate any water supply except that used by interstate carriers such as buses or trains. Fifty out of 57 states and territories have approved drinking water programs and so have been granted full control of their drinking water safety, or *primacy*, by EPA.

Primary Drinking Water Standards

National interim primary drinking water standards were issued in 1975. They have been amended several times since then. At present they cover the substances listed in Table 14.6. Sodium monitoring is also mandatory; however, the suggested limit, 20 mg/L, is not enforced by the federal government.

Water suppliers must examine their water distribution system for components such as lead-soldered pipes or asbestos cement pipes, which could allow hazardous materials, such as lead, cadmium, or asbestos, to be leached into finished water after leaving the plant. The water itself must also be checked for its corrosive action (mainly related to its acidity or alkalinity) to determine how likely it is to leach hazardous materials out of the distribution pipes, but at present, there are no standards for corrosivity of drinking water.

Secondary Drinking Water Standards

There are also secondary drinking water standards for a variety of materials (Table 14.7). These are recommended standards, not mandatory ones. In general, there is no significant health hazard if these secondary standards are exceeded, but their presence makes water unpalatable or causes problems in household uses such as clothes washing.

The EPA has published *suggested no adverse response level* (SNARL) documents for a variety of other organic compounds. Even though these are not legal standards, some states and communities have adopted them as levels above which they will close wells.

The standards set by the EPA are interim standards; they will be revised as new information becomes available. The EPA has also begun research into areas where information is needed.

U.S. Drinking Water Problems

Is the Safe Drinking Water Act insuring safe water for all Americans? An ongoing study by the National Research Council reported in 1980 that 30 to 40 states still had serious drinking water problems (Figure 14.10). Problems included bacterial contamination of water supplies, chemical contamination by toxic waste chemicals, and salt-water intrusion into groundwater drinking water sources. Nearly every state east of the Mississippi reported problems, as did several nonindustrial western states.

Table 14.6 National Interim Primary Drinking Water Standards

Contaminant	Maximum allowable level	Contaminant	Maximum allowable level
Inorganic chemicals		*Inorganic chemicals*	
Arsenic	0.05 mg/L	*Chlorophenoxyl pesticides*	
Barium	1.00 mg/L	2,4-D	0.10 mg/L
Cadmium	0.010 mg/L	2,4,5-TD	0.01 mg/L
Chromium	0.05 mg/L	Total trihalomethanes	0.10 mg/L[a]
Lead	0.05 mg/L	(the sum of bromodichloromethane, dibromochloromethane, bromoform, and chloroform)	
Mercury	0.002 mg/L		
Nitrate (as N)	10.00 mg/L		
Selenium	0.01 mg/L		
Silver	0.05 mg/L	*Coliform bacteria*	less than 1/100 ml/L
Fluorides (depending on temperature)	1.4–2.4 mg/L	*Turbidity*	1 turbidity unit[b]
Organic chemicals		*Radioactivity*	
Chlorinated hydrocarbon pesticides		Radium 226 and 228	5 pCi/L[c]
Endrin	0.0002 mg/L	Gross alpha-particle activity	15 pCi/L[c]
Lindane	0.004 mg/L	Beta-particle and photon radioactivity	4 millirem/year[d]
Methoxychlor	0.1 mg/L		
Toxaphene	0.05 mg/L		

a. Applies only to community drinking water systems serving more than 10,000 people.

b. For an explanation of this term, see Standard Methods for the Examination of Water and Waste Water, American Public Health Association, American Water Works Association and the Water Pollution Control Federation, 13th ed. Available from the American Public Health Association, 1790 Broadway, New York, NY 10019.

c. pCi/L = picocurie per liter, a standard unit used to measure radioactivity.

d. A millirem is a measure of the dose of radioactivity received by a human, weighted to account for the biological damage that a particular form of radioactivity can do to various organs (see Chapter 27).

Table 14.7 National Secondary Drinking Water Regulations

Contaminant	Recommended standards
Chloride	250 mg/L
Color	15 color units
Copper	1 mg/L
Corrosivity	noncorrosive
Foaming agents	0.5 mg/L
Iron	0.3 mg/L
Manganese	0.05 mg/L
Odor	Threshold odor number = 3
pH	6.5 to 8.5
Sulfate	250 mg/L
Total dissolved solids	500 mg/L
Zinc	5 mg/L

In 1981 there were 32 reported outbreaks of disease due to contaminated drinking water in which 4,430 people became ill. Most of these cases were due to bacterial contamination of water supplies. It is now clear that the main problems occur in small water systems (those serving fewer than 10,000 people each). Larger systems consistently deliver water of higher quality than do smaller ones, mainly because they can afford better treatment processes and more highly trained personnel.

In a 1982 study of rural households, most of which draw their water from individual wells, almost one-third were found to have bacterial contamination of their drinking water. Surprisingly, high levels of cadmium, mercury, lead, and selenium were also found (Table 14.8).

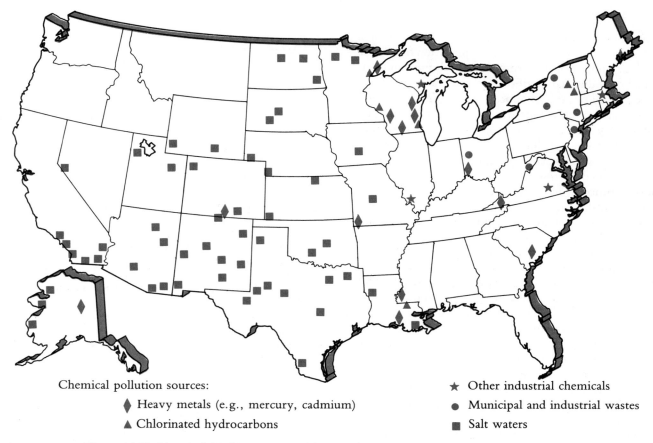

Chemical pollution sources:

♦ Heavy metals (e.g., mercury, cadmium)

▲ Chlorinated hydrocarbons

★ Other industrial chemicals

● Municipal and industrial wastes

■ Salt waters

Figure 14.10 Chemical drinking water problems in the United States. (From U.S. Water Resources Council and the Conservation Fund.)

Table 14.8 Problems in Rural Drinking Water Supplies

Contaminant	Percentage of households with water exceeding standard or recommended level
Coliforms	28.9
Nitrates	2.7
Sodium	14.2
Lead	9.2
Selenium	13.7
Cadmium	15.9
Mercury	24.1
Silver	4.7

Source: National Statistical Assessment of Rural Water Quality, U.S. Environmental Protection Agency, June 1984.

Problems do still exist, then. More technical help is needed for small systems. In addition, groundwater drinking water sources need to be identified and protected from contamination by hazardous chemicals.

In 1986, Congress reauthorized the Safe Drinking Water Act, adding 2 new major provisions. One requires the EPA to decide within 3 years whether and how to regulate an additional 60 drinking water contaminants, many of which are synthetic organic chemicals. EPA has been considering regulations on these chemicals, but Congress and many environmental groups feel the agency has not moved quickly enough. A second provision requires states to devise plans to protect groundwater drinking water sources.

Summary

Groundwater drinking water sources, on which 50% of Americans rely, are in danger of pollution from toxic chemicals buried in hazardous waste dumps and municipal or industrial landfills. Groundwater pollution, especially by organic chemicals, is the major problem resulting from hazardous waste disposal in the United States today. The category of hazardous waste includes toxic substances, ignitable wastes, corrosive wastes, and reactive wastes. Such wastes can sometimes be recycled, treated chemically, composted, or incinerated. Most commonly, however, they are buried in landfills. Secure landfills can be designed but must be monitored for long periods of time, possibly forever, to be sure there is no leakage of hazardous chemicals.

A serious problem exists with the hundreds of older dumpsites across the United States that are currently leaking waste materials into groundwater. Superfund is a bill designed to provide funds for the cleanup of these old dumpsites.

An equally serious problem is groundwater contamination by petroleum from underground storage tanks. These were buried in the 1950s and are now starting to corrode and leak. In certain sections of the country, groundwater contamination by agricultural pesticides, road salt, or oil drilling brines is a serious problem.

The high cost of cleaning up contaminated groundwater sources and the slow, natural renewal of such sources makes prevention of pollution a necessity. Most states are currently engaged in mapping aquifers and devising land-use plans to protect them.

Toxic inorganic chemicals that can contaminate drinking water include cadmium, arsenic, lead, nitrates, and mercury. Concern is also expressed over inorganic chemicals that may be involved in the cause or prevention of cardiovascular disease: sodium, calcium, and magnesium. Hazardous organic chemicals appear in drinking water from three main sources: use of agricultural pesticides, chemical waste disposal of such materials as PCBs, and the treatment of water with chlorine. This last causes the formation of possibly carcinogenic chlorinated hydrocarbons.

Drinking water safety in the United States is governed by national interim primary standards, which specify maximum allowable levels of ten inorganic chemicals, seven organic chemicals, coliform bacteria, turbidity, and radioactivity. Studies suggest that drinking water problems do still exist in the United States, especially in rural areas.

Questions

1. Contrast groundwater and surface water as sources for drinking water. What problems are each subject to?
2. What are hazardous wastes?
3. What methods are available for the disposal of hazardous wastes?
4. How would you feel if you discovered that your state planned to site a hazardous waste dump on vacant property near your home? What questions would you ask of state authorities? What safeguards would you feel were absolutely necessary to protect yourself and your family?
5. Explain why the mercury concentration found in the waters of Minamata Bay was much lower than that found in the fish and shellfish of the bay. What does this imply for people living in that region?
6. Suppose you are a public health official in charge of deciding whether to allow a manufacturer to sell a new chemical that is supposed to kill nuisance plants in lakes. List three questions you would want answered before you decide whether the chemical seemed safe to add to the water environment.
7. Why is there a recommended limit for sodium in drinking water?
8. PCBs have been found in water all over the country, but only in trace amounts. Nowhere have sampling studies found enough PCBs to be immediately poisonous to humans. Give three reasons why PCBs at these low levels are still a serious problem.
9. Should General Electric be legally responsible for cleaning PCBs from the Hudson River in addition to paying fines for exceeding allowed discharge levels? In a broader sense, should industries be responsible for cleaning up environmental pollution caused by illegal discharges? by accidents? by employee negligence? by "acts of God"? Oil companies are legally obligated to clean up oil spills that occur during drilling off the coast, whatever the cause of the spill. Is this an acceptable burden on a particular industry? Are there other ways to achieve the cleanup?
10. How can chlorination of drinking water contribute to the amount of toxic organic chemicals found in drinking water?
11. Not all trace minerals are undesirable. Some give water the flavor to which we are accustomed. Try the taste of distilled water, which has none of these minerals: catch the steam from a teakettle on a piece of aluminum foil (be careful—steam burns are painful) and let the drops run into a glass. Compare the taste to that of water straight from the tap.

Further Reading

Evans, R. J., J. D. Bails, and F. M. D'Itu. "Mercury Levels in Muscle Tissues of Preserved Museum Fish," *Environmental Science and Technology*, **6**(10) (1972), 901.
One of the unusual aspects of environmental mercury contamination is that fish near the top of salt-water food chains, such as tuna or swordfish, seem to have dangerous levels of mercury even in areas where no human sources of that element are found. This paper gives an introduction to the literature on the subject of mercury in food fish.

Giam, C. S., et al. "Phthalate Ester Plasticizers: A New Class of Marine Pollutant," *Science*, **199** (January 27, 1978), 419.
Phthalates are chemicals used in the manufacture of polyvinyl chloride plastics. Because of the widespread use of these plastics, they have become an environmental pollutant. Is this a problem? The evidence is not in yet.

Borelli, Peter. "To Dredge or Not to Dredge," *Amicus Journal* (Spring 1985), p. 18.
A well-written summary of the Hudson River's PCB problem.

Drinking Water and Human Health. Chicago: American Medical Association, 1984. Order from AMA, P.O. Box 10946, Chicago, Ill. 60610.
A valuable compendium of drinking water papers on all aspects, from federal and local regulations to the possible risks of cancer or cardiovascular disease from substances in drinking water.

Shkilnyk, A. *A Poison Stronger Than Love*. New Haven, Conn.: Yale University Press, 1985.
A regional development specialist looks at the social disruption of an Indian community and finds environmental mercury pollution.

Maugh, Thomas H. "Biological Markers for Chemical Exposure," *Science*, **215** (February 5, 1982), 643.
Summarizes books and conferences on the problems of monitoring toxic substance escape from dumpsites and of determining the effects on neighboring populations.

Zeighami, Elain. "Drinking Water and Cardiovascular Disease," *Oak Ridge National Laboratory Review*, **4** (1984).
A good discussion of how substances in drinking water may contribute to or protect people from heart attacks and strokes.

"Love Canal: A Boyhood is Poisoned," *The New York Times* (June 9, 1980), p. B1.

"A Tangle of Science and Politics Lies Behind Study at Love Canal," *The New York Times* (May 27, 1980), p. A1.

"Love Canal Residents Under Stress," *Science*, **208** (June 13, 1980), 1242.
The effects of living near a leaking chemical waste site can be as damaging psychologically as physically. These articles give some idea of the stress Love Canal residents lived with.

Bloom, G. F. "The Hidden Liability of Hazardous-Waste Cleanup," *Technology Review* (February/March 1986).
The economics of hazardous waste disposal and who eventually pays for it are considered in this article.

References

Maugh, Thomas H. "Burial Is Last Resort for Hazardous Wastes," *Science*, **204** (June 22, 1979), 1295.

"Research Needs for Evaluation of Health Effects of Toxic Chemical Waste Dumps." Proceedings of a Symposium, Environmental Health Perspectives, December 1982.

"Environmental Monitoring at Love Canal," vol. 1. U.S. Environmental Protection Agency, May 1982.

"Groundwater Pollution: Environmental and Legal Problems." American Association for the Advancement of Science, 1984.

"Federal and State Efforts to Protect Groundwater." Report to the Chairman, Subcommittee on Commerce, Transportation and Tourism, Committee on Energy and Commerce, House of Representatives, 1984.

"Groundwater Contamination by Toxic Substances: A Digest of Reports." Committee on Environment and Public Works, U.S. Senate, November 1983.

"Mercury Pollution in the Wabigoon–English River System." Summary of Technical Report, Canada-Ontario Steering Committee, 1983.

Madison, R. J. and J. O. Brunett. "Overview of the Occurrence of Nitrate in Groundwater of the United States." National Water Summary, U.S. Geological Survey, 1984.

Calabrese, E. "Inorganics in Drinking Water and Cardiovascular Disease," *Water Research Quarterly*, **3** (4) (1985).

Gilliam, R. J. "Pesticides in Rivers of the United States." National Water Summary, U.S. Geologic Survey, 1984.

"National Statistical Assessment of Rural Water Conditions." U.S. Environmental Protection Agency, June 1984.

Weber, W. J. "Organic Contamination: Whistling Past the Graveyard," *Journal WPCF*, **58** (January 1986), 12.

A GLOBAL PERSPECTIVE III

Safe Drinking Water for the Whole World

A Burden of Suffering

Waterborne and Sanitation-Related Diseases

*Diseases Spread by Contaminated Drinking Water/
Diseases Spread by Contact with Contaminated Water/
Diseases Spread by Lack of Washing*

The Drinking Water and Sanitation Decade

Motivation/Training Local People/Progress and Problems

A Burden of Suffering

U.S. drinking water problems pale before the obstacles to obtaining safe drinking water in developing countries. Fully half of the children in the world who die each year die from diseases carried by water, and 98% of these children are in developing countries. In fact, the lack of a safe water supply and of a safe way to dispose of human wastes is responsible for waterborne diseases that account for 80% of the death and disease in the world today. The suffering and disability caused by the lack of safe drinking water are almost incalculable, as is the economic burden of people too ill to work or care for themselves. Figure III.1 illustrates how much of the world still lacks access to safe drinking water.

What importance does the World Health Organization (WHO) place on water supply and sanitation for the developing world? Dr. Halfdan Mahler, Director General of WHO, put it this way, "The number of water outlets per thousand inhabitants is, in my opinion, a better health indicator than the number of hospital beds per thousand inhabitants." In response to the problems of a lack of clean water, the United Nations and WHO began a campaign in 1981 to bring safe drinking water to all people. The campaign is called the International Drinking Water and Sanitation Decade.

Waterborne and Sanitation-Related Diseases

What are the diseases caused by lack of clean water that afflict people in the developing world and that the U.N. wishes to bring under control? They include diseases spread by drinking contaminated water, by contact with contaminated water, and by lack of washing.

Diseases Spread by Contaminated Drinking Water

Cholera, typhoid, amoebic dysentery, viral diarrhea, and viral hepatitis are diseases spread through contaminated drinking water (Chapter 11). These diseases have disappeared almost completely from the industrialized Western world. In the developing world, however, such diseases still strike with discouraging frequency among the poorest people, especially among the children of the poor. The burden of poverty and malnourishment is bad enough, but the burden of ill health to the 500 million to 1 billion children under the age of 5 who experience diarrheal diseases each year is crushing. It is estimated that 5 million children die each year of diarrheal diseases. That is 10 deaths every minute of every year. These diseases are the result of consuming water infected with the disease-causing microorganisms. The water is infected because it has somehow received human wastes, either directly or indirectly.

Another disease that results from drinking contaminated water is dracontiasis, often referred to as Guinea worm disease because of the meter-long worm that causes the disease. The larvae of this worm enter the water from feces of infected individuals. The larvae then infect a common algae that enters the human body in contaminated drinking water. In the body the larvae mature to adult worms, which migrate to just below the skin; there they produce blisters, especially of the

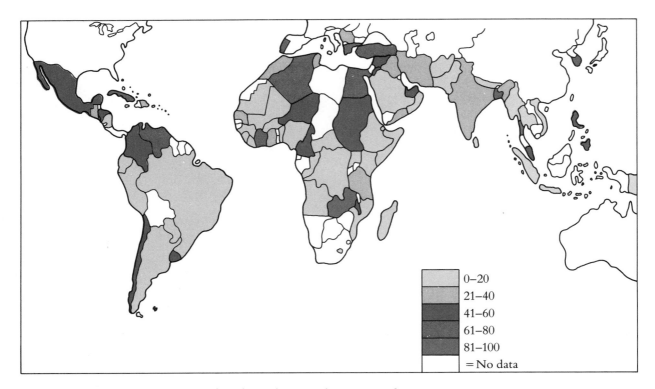

Figure III.1 Percent of rural population with access to safe water in 1980. These figures represent optimistic WHO estimates; some countries' figures are probably lower than shown. (From Cairncross and Feachem, *Environmental Health Engineering in the Tropics* [New York: John Wiley & Sons, 1983].)

feet. Between 10 and 48 million cases of this potentially crippling disease are thought to occur annually. The disease is most common in the Indian subcontinent, Africa, the Middle East, and northeastern South America.

Diseases Spread by Contact with Contaminated Water

It is thought that over 200 million people, 5% of the world's population, are infected with schistosomiasis (bilharzia). The disease organism is a schistosome, a parasitic worm that lives in the veins of infected individuals, causing liver and urinary disease. Eggs of the worm enter the water mainly through the urine but also through the feces of infected individuals. The eggs hatch to larvae, which then reside for several weeks in freshwater snails. Leaving the snails, the larvae enter the human body by penetrating the skin of an individ-

ual who comes in contact with the infected water. Such people may be bathing, washing, or simply wading in the water. The larvae and snails may contaminate irrigation water in fields and infect people who come in contact with such water.

The cycle of schistosomiasis can be interrupted by proper disposal of urine and feces and by providing clean water for washing, bathing, and drinking. Chemicals may also be used to kill the snails that carry the larvae.

Another disease spread by contact with contaminated water is leptospirosis, a disease characterized by fever, chills, headache, vomiting, and muscular aches. The disease organism is deposited in water through the urine of wild or domestic animals (cattle, dogs, and swine). People who bathe in contaminated water, workers in rice fields and sugar fields, and farmers in general are at greatest risk of contracting the disease in this way. Presumably the organism enters through wounded skin.

Legend:
- 0–20
- 21–40
- 41–60
- 61–80
- 81–100
- = No data

Diseases Spread by Lack of Washing

Ascariasis, another worm disease, causes digestive upsets, abdominal pain, and vomiting. Complications can lead to death. The disease is not spread by drinking water. Nonetheless, hands soiled with human feces will transmit the ascariasis eggs to foods. Sanitary disposal of excreta and washing hands before eating would block transmission of the disease.

Although ascariasis is uncommon in the United States, salmonellosis, a bacteria-caused form of food poisoning, is frequently reported. This disease may be spread through naturally contaminated food (such as chicken), or through foods handled by people with hands soiled by feces. Here again, hand washing before meal preparation is an effective tool for control. Thorough cooking of animal foodstuffs, especially chicken, that may carry the organism, is essential.

From simple eye diseases (conjunctivitis), to food poisoning, to typhoid fever, the range of diseases that can occur from lack of washing is enormous. The simple act of washing the body and washing clothes can help to prevent this host of diseases.

Scabies, caused by a mite that burrows under the skin, is characterized by lesions of the skin with intense itching. Yaws is a bacterial disease also characterized by lesions of the skin, but the bacterium is more destructive of tissue and causes bone lesions as well. Yaws is thought to be spread by contact with the "weeping" lesions of infected people; the disease may last for years. The transmission of trachoma, an eye inflammation that can lead to blindness, can often be broken by personal washing. Similarly, the transmission route of typhus, an often fatal disease marked by fever and chills, can often be blocked by frequent bathing and washing clothes.

The Drinking Water and Sanitation Decade

In an attempt to alleviate this tragic burden of disease and death, the United Nations instituted its International Drinking Water and Sanitation Decade campaign. The goal is to bring safe drinking water and sanitary waste disposal to all of the world's people by 1990.

The cost, of course, is likely to be tremendous. Estimates have risen from $140 billion to $175 billion and continue to increase. Over the time period involved, however, such a sum might be found. Of probably greater concern is fostering the motivation to achieve safe drinking water and sewage disposal in developing countries themselves.

Motivation

Motivation is a serious problem at all levels. On the governmental level, water supply programs must compete with industrialization schemes that promise much quicker returns. Yet it is clear that good health is necessary for people to be productive and that safe drinking water is necessary to good health. Even on an individual level, motivation to secure safe drinking water may be lacking. For many people in developing countries today, the notion that human wastes can contaminate water and cause disease is a new one:

> In a village in West Nile, Uganda, I was asked by the schoolmaster and school committee to look at the water supply. They were taking water from a seepage—you couldn't call it a spring—on the side of a hill. Just above it were several houses with no latrines. I said I thought there must be contamination, that it was likely to increase, and that they could have an epidemic of diarrhea. One woman did not agree with me. She held up a glass bottle of water and said in Lugbara: "Look, this water is clear, it flows, and it tastes all right. Show me where the disease is in this water." A long discussion followed. It was a crowded meeting, and most of the people there were unconvinced by anything I said. (Scotney, 1984)

The idea that wastes can contaminate water has not long been a part of Western knowledge either—see Figure III.2.

It is a simple matter to say "Provide clean water and educate people to use it," but it is a formidable task to carry out this program. There are problems not only in finding the appropriate technology but in convincing people to use a distant hand pump. The nearness of the pump is a major factor in whether safe water or polluted water will be chosen for use.

We know that groundwater will in general be safer to use than streams, rivers, or springs, but we need to warn people of the hazard of having privies nearby that could contaminate well water. We know, too, that a

Figure III.2 For lack of technology and lack of an understanding of the relationship between health and sanitation, America's rivers and streams were simultaneously used as sources for drinking water and the disposal of wastes during much of the nineteenth century. Household wastes often contaminated the groundwater that seeped into the rivers and streams that served as sources of drinking water. (The Bettmann Archive, Inc.)

water supply should be covered to prevent entrance of infectious material.

People, especially children, often need to be convinced to use latrines rather than leaving their feces on the soil. Privacy needs to be provided in the latrines or people will reject their use. The simple act of hand-washing after defecation may prevent untold disease, but people must be made aware of the connection between sanitation and lack of disease.

Training Local People

Besides motivation, a second need is for trained people to insure that the water system continues to work after it is installed. Many a new well has silted up or its pump

has broken down, leading the well to be abandoned because no one knew how to fix it. If hand pumps are provided, someone should be trained to repair them and provision should be made for obtaining the parts that may be necessary. Manufacture of pumps in the country of use may be needed.

These problems are intertwined with the issues of appropriate technology and respect for peoples' traditional ways of doing things. That is to say, if the people who are to use a well are involved in its design, construction, and location, they are more likely to use it and to help maintain it.

A final need, money, is of course important, but it is not at the moment the primary stumbling block to water projects. The World Bank, the U.S. Agency for International Development, and regional development banks all have money to lend for well-planned projects.

Progress and Problems

How well is the U.N. initiative doing? According to reports, the greatest progress is still in the area of planning. However, the needs and aims of the decade have become more clearly defined. It is becoming increasingly clear that education and involvement of the peoples who will benefit from the new water systems will insure their successful and continued operation. It may well take more than a decade, then, to achieve the stated goal, but the goal remains a worthy one: that of safe drinking water for every human on this planet (Figure III.3).

Questions

1. Make a chart of diseases, by name, that are spread through lack of clean water. Note which diseases enter the body in drinking water, which through lack of sanitation, which by contact with polluted water, and so on. Indicate how the diseases are spread where possible.

Further Reading

Dallaire, G. "U.N. Launches International Water Decade; U.S. Role Uncertain," *Civil Engineering—ASCE* (March 1981), p. 59.

Morrison, A. "In Third World Villages, A Simple Hand-pump Saves Lives," *Civil Engineering—ASCE* (October 1983), p. 68.

Figure III.3 A new tap in Kobobati, Ethiopia. (Photo: Seitz/UNICEF)

Benenson, A. *Control of Communicable Diseases in Man*, 11th ed. Washington, D.C.: American Public Health Association, 1970.

Wolman, A. "Reaching the Goals of the International Drinking Water Decade," *Journal of the American Water Works Association*, **77** (1985), p. 12.

All these items are relatively nontechnical and easy to read. The last article is written by Abel Wolman, the individual who suggested at the founding meetings of the World Health Organization in the 1940s that one of the organization's goals must be safe drinking water and environmental sanitation.

Reference

Scotney, N. "Water and the Community," *World Health Forum*, **5** (1984), 234.

(Department of Energy)

PART FOUR

Conventional Sources of Energy: Resources and Issues

In the fall of 1973, the United States came abruptly to the end of an era. This era had been marked by the growth of cities, sprawling across the landscape; by record levels of auto ownership and use; and by the spread of interstate highways, slicing across the nation. Transit and train travel dropped dramatically during this period. It was a time in which it was fashionable to erect glass buildings with windows that did not open. Insulation was largely ignored. Cars were generally large with high horsepower ratings and low efficiencies. It was, in short, an era of cheap energy.

The sudden increase in oil prices that occurred in 1974 after the Arab oil embargo was a clear notice to the United States that the era of cheap energy was coming to an end. Inflation, combined with a decreased energy resource base in the United States, contributed to the death of the era. A more fundamental reason, however, was the realization on the part of a cartel of oil producers that the industrial world was feeding at their trough.

It was and is true that Western Europe, the United States, and Japan consume far more energy than they themselves produce from their own resources. The industries of these nations require oil in such vast quantities that the entire industrial structure of the Western world could be threatened if oil supplies were slowed or interrupted. The nations that formed the oil cartel realized how vulnerable the Western world had become, and the cartel set new prices for oil. These prices were high enough to make the industrial world suddenly aware that it had unwittingly become dependent on a group of supplier nations capable of draining off their money in huge gulps.

The oil embargo in 1973–1974, followed by the sixfold price increase in the spring of 1974, was the first signal. In 1973, the United States was paying only $2 per barrel for foreign oil. By 1981, the United States was paying an average of $37 per barrel for its imported oil. The price of oil dropped by almost a third in 1982, however, as worldwide recession cut oil use drastically.

The signals from the oil cartel cannot be ignored. The United States and the rest of the industrialized world must learn, and are learning, to conserve energy and need to discover and utilize new energy sources.

With the embargo and following price rise, the Western world entered a new era, "the era of energy conflict," in which energy users are squeezed by energy producers. Such conflict logically precedes times of scarcity. Since our energy resources such as coal, oil, natural gas, and uranium are limited, energy use will eventually encounter the barrier of limited resources, ushering in the era of scarcity. Although this has not happened yet, we now have warnings of scarcity to come, warnings communicated by the higher prices we are paying for energy of all kinds. What has happened can be seen as fortunate in the sense that we have received an early warning of problems still only glimmering on the horizon.

The embargo and meteoric price rise were the results of a reduced energy resource base in the industrial world and an enormous energy base in certain less developed countries. The reality of the global resource situation has now brought us squarely up against the issue of energy resources, and it has happened a great deal sooner than most people would have predicted.

The reason the cartel could raise the price of oil so high so quickly was that the cost of finding and producing new oil in the industrial world was much higher than the price charged by the oil-producing nations. There was room for a price increase. Furthermore, in the short term, the industrial nations had no option but to pay the price, since new oil could not be discovered and produced quickly enough to take the place of oil from the cartel.

In a way, the drastic price rise illustrates a classic argument of resource economists. These economists have said that resources are never actually used up but simply become so costly that substitute resources begin to be found. Resource economists envision that as more of a resource is consumed, additional quantities become more difficult for suppliers to find. To continue to sell the resource at a profit when it is increasingly costly to find and produce, the supplier must charge a higher price. As the price of the resource is pushed higher by continued use and gradual depletion, users become interested in seeking substitutes that are less costly.

Substitutes can even replace the original resource and become more widely used than the resource they displace. Whale oil, shale oil, and coal oil are examples of energy resources from earlier years that were eventually displaced by a cheaper substitute. These oils were used to light lamps in colonial days and into the middle 1800s. It was in the last half of the nineteenth century that kerosene, a liquid derived from petroleum, replaced these oils because of its lower cost.

The harder it is to obtain a resource, the higher its price. This notion is so basic that the very statement of reserves depends crucially on the price of the resource in the marketplace. Oil in the North Sea, which cost $5–$6 per barrel to produce in the middle 1970s, was

not even counted as a reserve in the early 1970s, when the world price of oil was $2 per barrel.

The main point to be learned from events taking place from 1974 to the present is that the energy crisis is the result of the geographic and political separation of producers and suppliers. The crisis is not yet one of global demand butting up against global supply. Such a crisis may one day come to pass, but the likelihood of its happening is now far less, for we have had a clear warning notice nailed on our collective door.

Now as we approach the 1990s, decisions are being made that will influence our long-term security and way of life. These decisions, such as whether to proceed with synthetic fuel production, can be made on the basis of momentary conditions of plenty. Or they can be made with the realization that our conventional energy resources are limited, especially with regard to oil, and that much of our supplies are controlled by other nations. Now, while oil prices are momentarily low, is the time to further tax gasoline. Now, when we don't need synthetic fuel by tomorrow, is the time to make ready for the day when we do. The decisions we make in this decade could free us from energy insecurity or haunt us for decades to come.

In Part Four, we discuss conventional ways of generating power and the conventional energy resources regularly used. In Chapter 15, we describe the fundamental limit in the efficiency of converting heat energy to work or to electricity. This limit, imposed by the second law of thermodynamics, explains the source of one of our basic inefficiencies in energy use. Ways to improve the efficiency of electric generation are also noted. An example of the first law of thermodynamics, "energy can neither be created nor destroyed," is also found in this chapter. This example, thermal pollution, results from disposing, into natural waters, of the heat that cannot be turned into electricity.

With this background in energy use, we go on, in Chapter 16, to an investigation of conventional sources of energy. Coal, the troublesome, dirty fuel in which the United States is so rich, is explored first. Its abundance and uses are contrasted to the host of environmental problems it creates. From the impact of strip mining on the land to the impact of acid rain on forests, there are numerous points at which coal can degrade the environment.

Oil and gas, two fuels cleaner than coal but in shorter supply, are discussed in Chapter 17. More than most other resources, oil and gas are surrounded by economic, political, and conservation issues that must be aired. What are our current sources of these fuels, and what are the patterns by which we consume them? Strong clues on how we should conserve oil and gas will emerge from the answers to these questions. New sources of oil and natural gas are also being sought. For instance, oil-bearing shale exists in abundance in the Rocky Mountains, but the environmental consequences of developing this resource are very great.

In Chapter 18, we discuss uranium resources, which can serve to power nuclear electric plants. Nuclear energy, however, poses basic questions to society about safety, risk, and the spread of nuclear weapons. It is a resource that, if fully developed, will demand the most rigid control and security. The "nuclear option" raises grave issues, to say the least. One issue, of global concern, is the possibility of a "nuclear winter" if nuclear power resources are used to wage war instead of to generate electric power. This issue is covered in Global Perspective IV.

Our energy use also produces global water pollution problems. While the transport and production of oil pollute the oceans, the disposal of waste oil provides an unnoticed but large source of the oil that reaches our waters. This second issue of global interest is also covered in Global Perspective IV.

In the next major part of the book, Part Five, the effects of these conventional energy sources on the air we breathe are discussed in greater detail. We explore how the pattern of consumption of various fossil fuels is at the root of most air pollution difficulties. We seek new sources of fossil fuel to extend our resource base, but we can also seek new sources of energy not linked to fossil fuels—energy from tides, winds, the sun, and the earth's heat. These new sources, which we refer to as natural sources, not only extend our resource base but also have far less impact on the environment. Many air and water pollution problems are simply swept aside with the use of these natural sources. In Part Six we explore these alternatives to conventional energy resources. Do they provide a solution to our energy crisis, or do they themselves have undesirable environmental effects? Finally, we look at methods of energy conservation, which at least offer a solution guaranteed to do no harm to the environment.

CHAPTER FIFTEEN

How We Use Energy Resources

Energy Conversion and the Second Law of Thermodynamics

Historical Overview

We commonly utilize energy resources in three ways. First, we may release heat energy by burning fossil fuels and use this energy directly as heat for homes, schools, factories, and shops. Second, we may convert this heat energy into work, using refined oil to drive machinery and to power automobiles, trucks, trains, planes, and ships. Last, we may convert the heat energy from burning fossil fuels or from the fission of uranium into electricity and then use this transformed energy either for heat or to do work. Falling water is also commonly used to generate electricity. In effect, electricity acts as a middleman between energy resources and final use (Figure 15.1). Just as the presence of a middleman in a market creates higher prices, the use of energy in the form of electricity creates a markup as well—in this case, an energy loss, or penalty.

The practice of converting resources into electrical energy arose for many reasons. In some cases, it was not possible to make effective use of certain energy sources except by converting them to electricity. Before the era of electricity, the energy from falling water (hydropower) could be used only to drive machinery at the waterfall itself. Textile mills, grain mills, and lumber mills were operated at waterfalls in early manufacturing history. Hydropower had no other use until the advent of its use to generate electricity, which then made it possible to operate machines at sites distant from the falling water. Likewise, uranium fission is not generally used to fulfill energy requirements except in the form of electric energy. As with hydropower, electricity from uranium fission can be used not only to drive machinery but also to produce heat for homes (an inefficient use), for hot water, and for other uses.

The fossil fuels, in contrast to falling water, were first recognized as sources of light and heat, not as sources of energy to power machinery. Wood and coal, and often dried peat, were burned for heat in homes and commercial buildings, and coal was used to provide the heat needed to manufacture iron. *Carbon oil*, from coal, was used in lamps. Not until the invention of the steam engine in the 1700s was the potential of fossil fuels for driving machinery discovered. By the early decades of the 1800s, locomotives in the United States were burning coal to generate the steam for power. And by the early decades of the twentieth century, coal was burned to provide steam to generate electricity, although the production of electricity in this manner was inefficient at the time.

Thus, fossil fuels began to assume a new role in the industrial world. Not only could they provide heat and light but, through the use of engines and electricity, they became sources of power for industrial machinery and transport.

These historical events color our present attitudes. The use of coal and other fossil fuels to generate electric power for machines is appealing; since electric power can be transmitted from place to place, machinery can be located far from the site of electricity generation. Electric heating of homes is also appealing; no fuel is consumed in the home, and no ashes or smoke are produced there. But there is a penalty for using energy

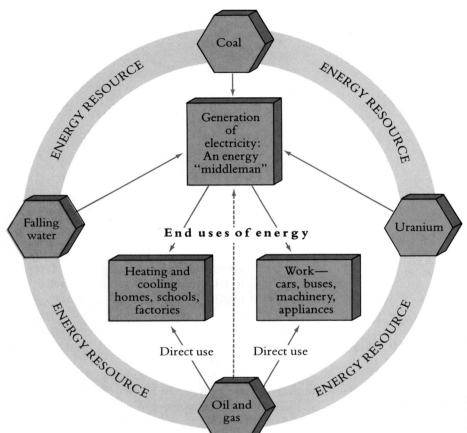

Figure 15.1 Common routes for the use of energy resources. Electricity is an energy middleman that takes the basic energy resource, processes it, and transfers it to the final user. Like any middleman, it demands that a toll be paid. Coal, uranium, and falling water are all processed into electricity before their primary use. Oil and natural gas, however, are used directly to produce heat and work.

in the form of electricity, a penalty far beyond the extensive pollution at the point of electricity generation. This penalty is an enormous loss of efficiency—an enormous waste of the heat energy in fossil fuels.

The Second Law of Thermodynamics Demonstrated

According to the second law of thermodynamics, there is an inherent limit to the amount of heat energy that can be converted into work and hence into electricity. The reverse is not true; that is, other forms of energy can be converted completely into heat. This limit to the efficiency of converting heat into work was not an obvious concept to the early designers of engines that used heat for power. An engineering text tells us that:

> . . . the low efficiencies encountered in the conversion of . . . heat to work were at first attributed to incorrect design of the mechanism

employed, or to practical difficulties in operation. Even after careful improvement, however, low efficiencies prevailed. When extensive efforts at improvement proved fruitless, it was finally inferred that there existed some underlying natural limit to the conversion of heat to work. (Weber and Meissner, 1957)

In fact, most of the possible design improvements to increase the efficiency of generating electric power from steam have now been made. The modern coal-fired power plant, which produces steam to turn a turbine for generating electricity, has reached 40% efficiency; that is, 40% of the heat in the coal that is burned is converted into electric energy. The efficiency of oil-fired power plants is close to this range as well (Figure 15.2). Conventional approaches to electric power generation cannot be expected to increase this efficiency by very much. Nuclear power plants also produce steam to turn a turbine, and these plants have attained effi-

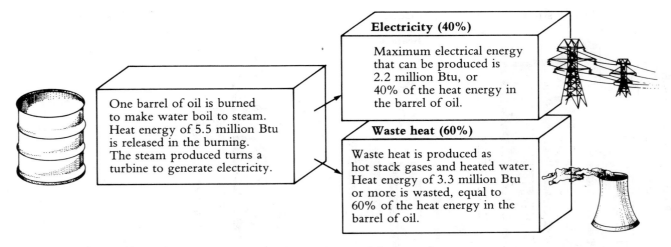

Figure 15.2 The inefficiency of generating electricity: the example of oil.

ciencies of only 30%–32%; that is, only 30%–32% of the heat of fission is converted into electrical energy. Attempts to increase this conversion percentage have yet to succeed.

The generation of electricity, then—although a great convenience—wastes much of the heat energy of the fossil fuels: coal, oil, and natural gas. It is especially wasteful when the electric energy is only converted back into heat at its point of use. Because so much of the energy we use is in the form of electric energy, we need to discuss more completely the sources and uses of electric power as well as the impact upon the environment that accompanies this use.

Electric Power

A flick of the finger and we have

light,
sound,
visual communication,
heat,
hot water,
refrigeration, and
air conditioning

all from electric energy.

A flick of the finger and we also have

river valleys buried in sediment,
sulfur oxides, acid rain, acid mine drainage,
oil pollution,

particulate matter,
nitrogen oxides and smog,
strip mining,
thermal pollution,
radioactive wastes, and
plutonium for bombs

all from electric energy.

Growth in Use

Electric energy is one of the most easily used forms of energy. One electric line enters a house, and the energy to light the home, cook and bake, warm water, and run machines and appliances is all at our fingertips. The use of electricity for household tasks is steadily increasing. Electrical appliances are becoming more widely available, as Figure 15.3 shows. Moreover, because electricity is such a convenient form of energy, it is expected to capture more and more of the energy market. According to the Energy Information Administration, the quantity of electric energy used in the United States will increase by about 45% from 1985 to 1995, even though total energy use will increase by only about 19% in that same period (Figure 15.4).

It should be noted again that the quantity of energy *consumed* for electric generation at power plants is not, in fact, the quantity of electrical energy *generated* at power plants. Nor is it the quantity consumed in end uses such as heating water. It is, instead, the far larger quantity of energy that was released in the burning of

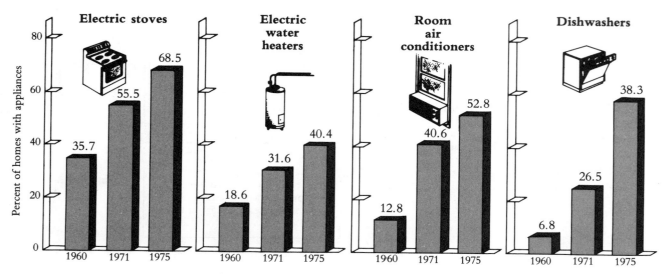

Figure 15.3 Growth in percentage of homes with certain electric appliances. (Note that the number of homes increased by about 25% from 1960 to 1971.) (From "Energy and the Environment: Electric Power," The President's Council on Environmental Quality, August 1973; and *EPRI Journal*, December 1977.)

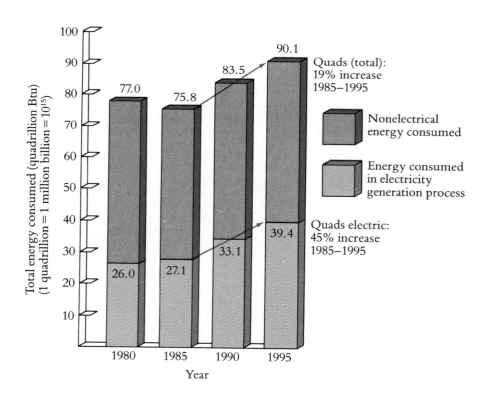

coal or fissioning of uranium at the power plant to produce the electrical energy that is ultimately used in application or lost in the process of transmission. The quantity of electrical energy that can actually be put to use is about 30% of the energy consumed for purposes of electricity generation. This is a result of the inherent inefficiency of electricity generation, a consequence of the second law of thermodynamics.

Generation of Electricity

To understand how fuels are used to generate electricity, we need to examine the operation of a typical thermal electric plant (Figure 15.5). By *thermal electric plant*, we mean a power plant that utilizes heat to produce steam, which turns a turbine. A coal-fired plant is a thermal electric plant; so is a plant that burns oil or gas and a plant that utilizes the heat from nuclear fission. The principle of operation is the same in all three.

Heat is produced by combustion (or fission). That heat is used to boil water for steam. The steam, at a high temperature and under high pressure, is used to turn a turbine. The turbine rotates an armature in a magnetic field, inducing the flow of electric current and thus generating electricity. As the steam leaves the turbine, its pressure and temperature are reduced. This "spent" steam is converted back to water by a condenser, through which cooling water flows.

In the process of condensing the steam, the cooling water is itself heated. Once it has been heated, the cooling water may either be returned to the body of water from which it was taken or passed through towers to

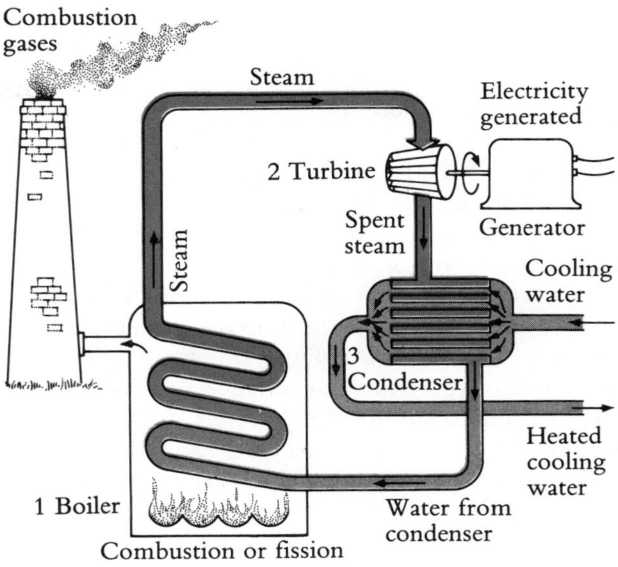

Figure 15.5 Elements of a typical thermal electric power plant.

be cooled and used again in the condenser. The water produced from the condensed steam reenters the boiler to be reconverted to steam and recirculated to the turbine.

A coal-powered station loses 60 units of heat to the environment for every 40 units converted to electricity; it works at 40% efficiency. In other words, 1.5 units of heat are wasted per unit of energy produced. A nuclear plant operating at 30% efficiency wastes 70 units of heat for every 30 units converted to electrical energy;

Figure 15.4 (opposite page) Projected growth in total energy consumption and growth in energy consumed in electricity generation in the United States. Projections for 1990 and 1995 assume intermediate economic growth. (The energy consumption numbers are given in quadrillions of Btu, the British thermal unit of heat energy that will raise the temperature of 1 pound of water by 1°F. A quadrillion Btu, or quad as it has come to be known, is equal to the heat energy in about 180 million barrels of oil, or in about 40 million tons of coal. In 1980, in the United States, one quad could have powered 12 million cars for a year or heated 20 million well-insulated homes for a year.)

To understand the term *energy consumed in electricity generation*, examine the data for 1980, when 26 quads of energy were consumed in the generation process. This quantity is not the amount of electric energy actually available for end uses—heating hot water, operating machinery, lighting, drying, making toast, and so forth. It is, instead, the amount of energy required to produce the electric energy used in end uses. It is a quantity that is about two and one-third (2⅓) times larger than the amount of electrical energy actually utilized. It is the energy in the coal, oil, and gas that was burned at the power plant plus the total energy produced by the fission process at nuclear power plants. Of the 26 quads consumed in the generation process, only 30%, or 7.8 quads, was actually utilized. (Data from Energy Information Administration.)

thus, 2.33 units of heat are wasted per unit of energy produced. A nuclear plant loses to the environment about 55% more energy than the coal plant per unit of electric energy produced. The problem is compounded for a nuclear plant because all its waste heat is transferred to the condenser cooling water. A coal-fired power plant, in contrast, directs only 75% of its waste heat to the cooling water. The remainder of the heat from the coal-fired plant goes up the smokestack.

The combination of power plant types from which we draw electricity has been constantly changing for a variety of reasons. During the late 1960s, coal-fired power plants were being converted to residual oil in an effort to reduce sulfur dioxide levels in urban areas. With the onset of the 1973 oil embargo and the massive increase in oil prices, coal plants with better pollution control, or those burning lower-sulfur coal, began to replace the oil-fired plants. Because of more certain coal availability, a program to convert to coal was carried out in an effort to reduce oil imports. From the late 1960s to the late 1970s, new nuclear plants were being added quickly to the stock of generating stations. But as resistance to nuclear power grew, the number of orders for new nuclear stations dropped to only several plants per year in the late 1970s, and to none in the 1980s. Thus, the combination of plant types, as shown for 1985 in Figure 15.6, is actually a snapshot of an evolving system. By the time you read this, gas use in electricity generation will have fallen because natural gas will have been diverted for use in heating homes. Coal use will have increased because much of the new generating capacity will be in coal-fired units. Nuclear use will have grown because the nuclear units already under construction will have come "on line" during

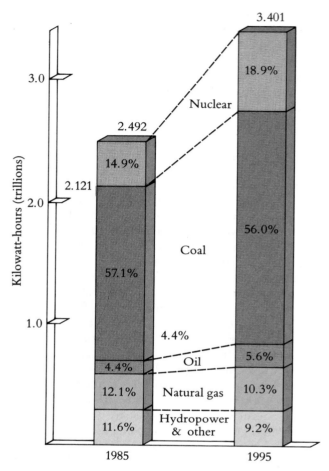

Figure 15.6 Fuels for electric generation in the United States. (Numbers may not total due to independent rounding.) (Data from Energy Information Administration.)

Box 15.1   Beware the Meter

Now, there is not the least reason to believe that electricity exerts any better moral influence than gas, and we know that when a number of reasonably Christian men form themselves into a gas company they immediately become pirates of the most merciless and extortionate character.

Why should we look for better things from the electric light companies? They expect to have us at their mercy, and they will be as merciless as the gas men.

We shall have electric meters in our cellars that will be as mendacious and unprincipled as the gas meters, and the moment we refuse to pay for ten thousand feet of electricity which we have not used, our lights will be cut off and we shall be left to candles and kerosene.

(Excerpted from an editorial that appeared in *The New York Times,* June 11, 1881)

this period. Hydropower will have lost a portion of its market share because only a few new large hydro-projects are possible.

That electric generation is not a clean source of power is vividly shown in Table 15.1. Here we can see that emissions from electric generation account for a high proportion of the annual emissions of three major air pollutants. The impact of electric power generation on air pollution is detailed in Part Five.

Thermal Pollution

The generation of electricity produces waste heat that must be disposed of. When this waste heat is released into the environment, it can have harmful effects. For this reason it is classed as a pollutant: **thermal pollution**.

If you will look back to Figure 15.5, you can see cooling water entering the condenser. This cooling water passes through the condenser pipes, and spent steam condenses around the pipes. The cooling water picks up the heat from the spent steam as it allows the

Table 15.1 Sources of Selected Air Pollutants

Source	Particles (million tons per year)	Sulfur oxides (million tons per year)	Nitrogen oxides (million tons per year)
Electric generation	4.5[a]	16.0	7.1
Residential and commercial combustion	0.6	2.9	1.1
Industrial combustion	2.0	3.2	2.7
Other industrial processes[b]	14.4	7.5	0.2
Transportation	0.8	1.1	11.2
Miscellaneous	12.8	0.4	2.4
Total	35.1	31.1	24.7

Source: Energy/Environment Fact Book (March 1978), USEPA and USDOE EPA-600/9-77-041.

a. If not under control, this amount could be multiplied by a factor of 10.
b. Grinding, spraying, demolition, construction, mining, cement plants, grain handling, etc.

spent steam to condense to water. The heat in the cooling water must then be disposed of in some way.

There are two basic types of cooling arrangements in electricity generating plants. In one type, cool water is drawn from a nearby body of water, is used to condense steam, and then is discharged in a heated state back into the body of water. This is called *once through* or *open-cycle cooling*. In the second type, *closed-cycle cooling*, the heat in the cooling water is released to the atmosphere through a cooling tower. Thermal pollution of water results from open-cycle cooling.

Biological Effects of Thermal Pollution

Off Turkey Point on Biscayne Bay, Florida in the early 1970s, biologists measured an area of almost 75 acres (30 hectares) that was barren of aquatic life. Surrounding this, another 100 acres was only sparsely populated. Yet Biscayne Bay is known to yield over 600,000 pounds per year of marketable seafoods such as spiny lobsters and stone crabs. Sport fish abound in the bay and a valuable bait-shrimp fishery is located there. Many fish and shellfish use the bay as breeding grounds, and a variety of wading birds feed in the shallows. What caused the apparent ecological disaster at Turkey Point, in an area ordinarily teeming with aquatic life? The barren area was centered around an effluent canal from Florida Power and Light's Turkey Point Power Plant. At that time, the plant took cool water from the bay and returned heated water through the outflow. It was this heated water that caused a biological desert in the midst of the bay.

Temperature is an important factor in the well-being of all organisms. For each species, there is a particular temperature range that supports life. Some organisms have adapted to living in the hot springs at Yellowstone National Park, where temperatures may reach 70°C. There are even fish in arctic waters that survive being frozen in ice. But for any particular species, the temperature range necessary for survival is relatively narrow; in some cases, very narrow indeed. For instance, some organisms that build coral reefs in the Caribbean can withstand no more than a few degrees change in temperature.

Warm-blooded animals, such as humans, have evolved a variety of mechanisms for keeping their body temperatures in the proper range: digestion of food produces heat, and sweating increases heat loss when the body is too hot. Most aquatic creatures, however, are not able to maintain a particular body temperature. These creatures must stay the same temperature as the water in which they live. Those that cannot move, such as adult oysters or rooted plants, are at the mercy of water temperatures. Beyond certain limits, they simply cannot survive. Those that can swim, such as fish, move about to find suitable temperatures. This is called behavioral regulation of temperature.

Furthermore, if given a choice, such creatures as fish, which can move about, spend most of their time in areas where the temperature is at a level they seem to "prefer." Thus, in the winter fish may congregate in warm-water outflows from power plants, not because they can't live in the naturally cold waters but because they prefer their environment to be somewhat warmer. If the power plant must shut down during the winter for repairs or refueling, a fish kill can result. The fish will not be able to withstand the normal, cold water temperatures because they have become used to the warmer effluent temperature.

Survival and well-being of organisms. Thermal fish kills are relatively rare events. Moreover, although fish kills are dramatic evidence of the effects of thermal pollution, less obvious effects can be even more serious. Temperature can affect the ability of organisms to reproduce without actually killing them directly. For instance, trout require cool summer temperatures for the formation of eggs and sperm. Although adult fish might be able to survive in warm summer water, they will not reproduce. For certain insects, hatching is triggered by increasing temperature. If waters are artificially warmed, the hatching temperature may be reached earlier in the year than is normal. At the Hunterston Generating Station in Ayrshire, Scotland, a species of intertidal sand-dwelling copepod hatches earlier than usual due to a heated effluent. It then dies off in greater than usual numbers because not enough food is available early in the year for the larvae.

Fish that are not killed by high temperatures still may be unable to catch food. Even more subtle changes have been shown to occur that cause thermally shocked fish to be "picked out" by predators. Heat-stressed fish may also be more susceptible to disease. In the long run, these sorts of effects can be just as devastating to the population as a direct thermal kill.

The effect of excess heat on ecosystems. Temperature can affect the whole community structure of

Figure 15.7 Wood piling riddled by shipworms. Warm-water effluents from a New Jersey Central Power and Light company nuclear plant on Oyster Creek appears to be allowing the spread of shipworms into Oyster Creek. The worms require warmer temperatures than are normally found in Oyster Creek. The larvae of shipworms eat their way into wooden structures like ship hulls or dock pilings, growing from pinhead size to more than 2 feet (0.6 m) long and .5 inch (1.25 cm) in diameter. (Carl Gosset/NYT Pictures)

an aquatic environment. For instance, different species of freshwater algae compete for light, space, and nutrients. Temperature changes can alter the competitive position of different species, even though the changes are not severe enough to be lethal (Figure 15.7).

Overall, the effect of heat is to simplify aquatic communities; that is, fewer species are found, although there may be many individuals of each species. One study found that fewer than half as many species were found at 31°C than at 26°C. Another 24% disappeared at a temperature of 34°C. Such a simple ecosystem may be less stable than the original more complex one.

Many natural waters are subject to seasonal variations in temperature. This variation allows different species to be dominant at different times, and therefore a greater number of species are able to compete for space and nourishment in a given area. A partial explanation for the effect of thermal pollution on communities may be that thermal additions can even out this temperature fluctuation. Thus, fewer species are able to coexist in a given area.

A large number of thermal-effect studies have been carried out with respect to the location of power plants. The studies seem to show that obvious harmful effects on ecosystems are more likely to occur from power plants located on naturally warm waters because organisms living in warm regions are often already near their upper thermal limits. The additional heat from the power plant pushes the organisms over their thermal limits.

Better Ways to Dispose of Waste Heat

Using waste heat. The thermal pollution problem can be approached in two ways. Ideally, the waste heat, which is, after all, a form of energy, could be used, rather than simply dumped into nearby bodies of water (Figure 15.8). Some of the possible beneficial uses of waste heat from power plants include (Fields, 1971):

1. water desalinization
2. agricultural uses—irrigation, frost protection
3. waste disposal—desalinization or demineralization of sewage water, sterilization and drying of sewage

Figure 15.8 Peach blossoms protected by ice formed when warm water was sprayed on them. Orchards protected this way yielded full crops in years when nearby unprotected crops were damaged by frost. (Research done under auspices of Corvallis Environmental Research Laboratory.)

4. sterilization of drinking water, using heat instead of chemicals
5. refrigeration, using gas-absorption refrigeration
6. climate control—district heating and cooling, greenhouse heating, melting arctic ice and snow into irrigation water for arid regions
7. heating intake waters at power plants, preventing fouling of pipes
8. transportation—keeping shipping lanes and harbors ice-free
9. rerefining of waste oil
10. aquaculture—luring fish for catching, growing tropical species farther north
11. power from new energy technologies—thermoelectric elements, and so forth
12. wildlife protection, such as warm-water ponds for water fowl
13. airport safety—defogging and deicing runways
14. mining—hot water and steam for hydraulic mining techniques
15. space heating

If none of these suggested uses are possible, technical solutions such as closed-cycle cooling can be used.

Closed-cycle cooling. The problems that result from thermal pollution are all caused by open-cycle cooling, the use of water to receive waste heat. Technologies have been developed, however, that can transfer much of the waste heat from cooling water to the atmosphere. These technologies are called *closed-cycle cooling*, and they utilize cooling ponds, cooling canals, or cooling towers. Closed-cycle cooling provides two important benefits to the environment. First, if heat is transferred from the cooling water to the atmosphere, there is no need to discharge heated water to a lake or a river. In this way, damage to aquatic life is avoided. Second, not only is thermal pollution avoided, but the amount of cooling water withdrawn from a lake or a river can be cut to a few percent of the amount normally withdrawn for open-cycle cooling; the water that has been made cool can be reused to condense steam again and again. Thus, withdrawal of water from the main body can be cut dramatically, allowing the water to be used in other ways (see Figure 15.9).

Cooling towers (Figure 15.10), cooling ponds, and cooling canals (Figure 15.11) make use of the concept that when water evaporates, a great deal of heat energy is absorbed by the evaporating water molecules, allow-

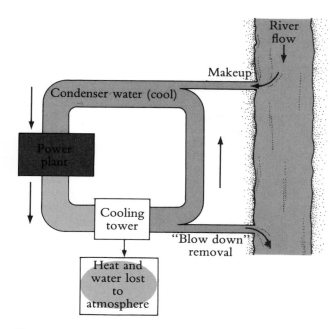

Figure 15.9 Closed-cycle cooling. Since the cooling water circulates continuously through the plant in closed-cycle cooling, it would seem that no further water need be withdrawn or discharged after an initial quantity of cooling water is taken into the plant. Unfortunately, this is not the case; the cooling tower loses some water to the atmosphere in the process of evaporation, and another small fraction of the water is drawn off and discharged. Water loss is made up by drawing new (makeup) water into the cooling system.

Figure 15.10 A typical evaporative cooling tower. Note its enormous size relative to its surroundings. Such structures can be seen for miles. This is a picture of the Trojan Nuclear Power Plant on the Columbia River near Prescott, Washington. (EPA Documerica)

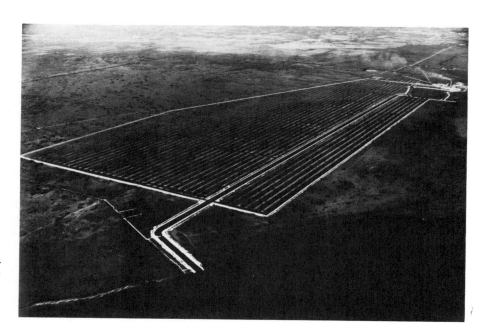

Figure 15.11 The Turkey Point Power Plant no longer discharges water into Biscayne Bay. Instead, a 6,000 acre (2,400 ha) canal system cools condenser water. In this photograph of the canal system, the power plant is in the upper right. (Photograph courtesy of Florida Power and Light Company)

ing them to shift from the liquid to the vapor state. In evaporative cooling, evaporation of a small amount of cooling water withdraws a large amount of heat from the water that remains behind; this is the source of the cooling effect. The temperature drop of water cooled by a cooling tower is on the order of 14°C.

The use of the evaporative cooling principle is widespread. Bedouins keep their waterskins wet on the outside; the evaporation cools the water within the skin. In the American Southwest, evaporative coolers are often used in place of air conditioners to cool homes.

Where land is too expensive for a cooling pond, towers are virtually the only cooling option available. **Cooling towers** are more complicated than ponds and come in several types. The *natural-draft evaporative cooling tower* is a massive concrete structure (Figure 15.10). The natural-draft tower is referred to as a "wet" cooling tower because a portion (up to 2.5%) of the recirculating water is lost (evaporated) to the atmosphere. Addition of a wet natural-draft cooling tower to an electric plant may add about 5% to the cost of constructing the plant and about 1% to the residential electric rate.

Another type of wet cooling tower utilizes fans to move air from outside the tower into and through the structure. This *mechanical-draft tower* requires electrical energy to operate, so that its operating cost is higher than that for the natural-draft tower. In addition, the fans may be noisy and often need repair. Nonetheless, since the towers are much smaller than natural-draft towers, their construction costs are much lower.

For several reasons, cooling towers are not perfect controls for thermal pollution. The towers are enormous both in size and in visual impact. For a wet cooling tower, a height and base diameter of 400 feet (121 m) is not uncommon. Such a structure is the equivalent of a 40-story building and may dominate the landscape for many miles.

Wet towers may also present other problems to the area in which they are located. To illustrate, a pair of wet natural-draft towers on the Monongahela River at Fort Martin, West Virginia are used to cool water from two 540-megawatt coal-fired power plants. The towers decrease the temperature of the condenser water from 45°C to 32°C. Of the 250,000 gallons per minute cooled by each tower, about 6,000 gallons per minute evaporate. That quantity of evaporation is approximately equivalent to a room 10 feet (3 m) on a side with a ceiling height of 10 feet (3 m), and this quantity of water

evaporates every minute. The towers must be taller than the depth of the river valley, which is 300 feet (92 m) at this site. If they were not, mists could settle in the valley, enshrouding the area in a permanent fog.

Closed-cycle cooling via towers or ponds has become the main way in which new steam electric plants dispose of waste heat. The use of these cooling methods is required on new plants (with some exceptions) because of the almost total ban on heated discharges imposed in 1974 by the Environmental Protection Agency. Many of the plants built in the 1960s and early 1970s also utilize closed-cycle cooling. An EPA ruling forced this action on many of the larger plants built between 1970 and 1974. The need to conserve water led still other plants in water-short areas to adopt closed-cycle cooling.

Although new steam electric stations built on inland waters are likely to employ closed-cycle cooling methods, not all new plants will require it. Power plants on the cold ocean or on cold estuaries may escape the requirement. Power companies can win exemptions for their plants if they can demonstrate that the heated discharges will not alter the ecological balance in the receiving waters; that is, they must show that the shellfish, fish, and wildlife native to the area will be able to survive and reproduce. Such a demonstration is most likely to succeed on cold natural waters, such as the oceans and bays of the Atlantic Northeast and the Pacific Northwest.

Energy Today

In the chapters that follow, we describe how we produce and apply our principal energy resources: coal, oil and gas, and nuclear energy. To understand the necessity of these particular resources, we need to explore how we currently employ them. That is, we need to know how they fit into our current energy picture and how (from a technological point of view) society has structured itself to put them to use. This information will help us determine how we can conserve energy and fuels.

Figure 15.12 shows how we consumed energy in 1985 in the various sectors of the economy: transportation, industry, and residential/commercial activities. In general, the pattern of consumption has not changed drastically over the past quarter century. Roughly the same proportions have been used in each economic category from 1950 to the present; however, actual energy

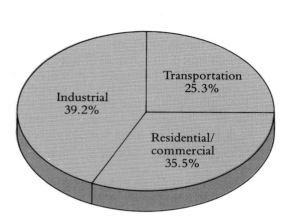

Figure 15.12 U.S. energy consumption by economic sector, 1985. (Data from Energy Information Administration.)

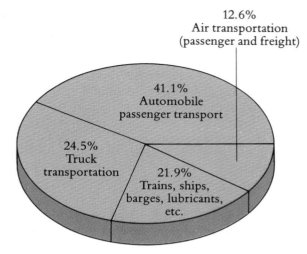

Figure 15.13 How 19.2 quads of transport energy was consumed in the United States in 1985. (Data from Energy Information Administration.)

consumption increased by more than 100% between 1950 and 1985. A population increase of about 50% accounts for a portion of this increase in consumption, but the average annual energy consumption per person increased about 55%, from 210 million Btu in 1950 to 330 million Btu in 1985. (Btu is the British thermal unit of heat energy that will raise the temperature of one pound of water by 1°F.)

Transportation accounts for about 25% of the energy we consume each year. Nearly all of this transport energy in the United States is consumed as some form of petroleum. In 1985, about 41% of this energy was burned in automobiles, 25% in transportation of goods by truck, and 13% in airplane operations. Railroads, buses, pipelines, ships and barges, military air and ground vehicles, and other minor consumers used the remainder of U.S. transport energy (Figure 15.13). Although nearly all of the transportation energy in the United States comes from oil, in Europe this is not the case. Trains still play an important role in Europe, and most rail systems in western Europe have been electrified, with the electricity being generated from coal and nuclear energy.

We can break down the household category of energy use as well. In 1985, about 61% of the energy consumed in the household category was used to keep people either warm or cool (Figure 15.14). Heating used 57% of all energy in this category, air conditioning 4%, and water heating 17%.* In Figure 15.14, the space heating category and the water heating category are further broken down by fuel type to show the contributions of the various fossil fuels. As can be seen, natural gas furnishes more than half of the energy in both categories.

As the energy uses in various sections expand, so must our use of fuels. Figure 15.15 displays U.S. energy growth from 1965 to 1985, and it projects the growth to 1995. From this chart, we can see that petroleum and natural gas together will have fallen from 74% of total energy use in 1965 down to 61% of energy use by 1995. Coal, nuclear power, and hydropower in aggregate will have grown from 26% of energy use in 1965 to 39% in 1995. These three energy resources are used in electric power generation, so this increase reflects a steady increase in our reliance on electrical energy. (Should the private sector profit from this growth? See Controversy 15.1.) In Figure 15.15, note

---

*The figures are for 1985, but a hotter summer or colder winter could change the numbers, especially as air conditioning spreads to more of the population.

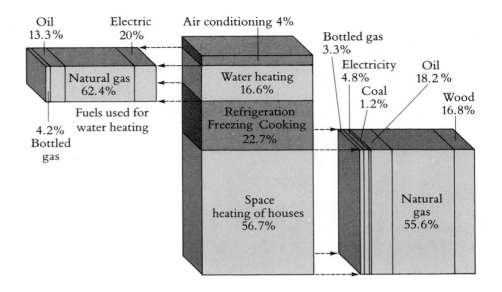

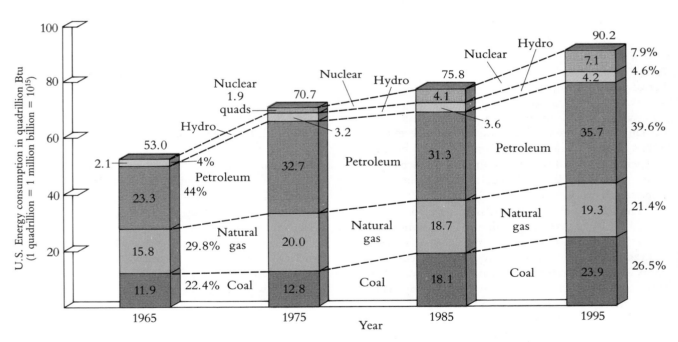

Figure 15.15 Pattern of growth in United States energy consumption by fuel type.

that the actual quantity of petroleum and natural gas that is used remains almost steady from 1975 to 1995, suggesting that these fuels are retaining their original markets in home heating and transportation.

Prior to the 1950s, coal had been used extensively to heat homes, but it lost ground quickly to natural gas and oil. Coal was dirty and produced large quantities of ash, which the homeowner added to the trash. Coal also required a large storage area in the home (the coal bin), and the coal furnace needed frequent shovel feeding. Natural-gas heating systems are compact, are automatic, and need no storage area in the home. Oil

CONTROVERSY 15.1

Should Electric Companies Be Publicly or Privately Owned?

[The] impetus toward municipal ownership now is the higher rates of the private utilities and the feeling that local ownership would be more responsive to public needs.
Alex Radin, general manager, American Public Power Association

Higher electric bills are simply a reflection of the higher cost of everything a utility must buy to provide service.
W. Donham Crawford, president, Edison Electric Institute

About 80% of the power generated in the United States comes from investor-owned utilities; however, a number of electric companies are not investor-owned but are owned by local governments. These publicly owned companies generally have lower electric rates than investor-owned utilities and are more receptive to public opinion. The electric rates are lower because a municipally owned utility does not have to pay dividends to stockholders, pay federal taxes on its profits, or pay local property taxes.

The interest on the bonds issued by an investor-owned utility is taxed as income for the owner of the bond. In contrast, the interest on the bonds issued by state and local governments is not liable for federal taxation. Because interest on the municipal bond is not subject to taxes, purchasers are willing to accept a lower interest rate. Thus, the municipal utility is able to raise money for new construction at a lower cost than the investor-owned utility. As a consequence of this and of the lack of taxation on its property or income, the municipally owned utility is generally able to offer lower electric rates to its customers.

However, since a municipal utility escapes local taxes, the local government must obtain its required revenues by increasing taxes on its remaining property. Because of the nonpayment of local taxes by the utility the lower electric rates may, in part, be an illusory sav-

ing for customers. Lower electric rates due to tax-free bonding power and the lack of stockholder dividends are very real, though.

In addition, public utilities should also be more willing to save the public money by using peak-load pricing—that is, charging higher prices for electricity generated at the hours of peak use. The resulting lower peak demands mean less investment required for new capacity and thus lower electric rates. Privately owned utilities, in contrast, like to expand their capacity and investment because they are allowed to earn a rate of return on their investment.

Finally, utilities owned by local governments are expected to be more accessible to public opinion on such issues as location of facilities and whether to use nuclear power. On the negative side, one must face the issue of efficiency—whether public enterprises are as efficient as private ones. It is this issue that brings opposition to public power.

Do you think power companies should be publicly owned? If yes, why? If not, why not? Do you think a publicly owned power company will be as efficient as an investor-owned company? more responsive?

Source:
Quotes from *Wall Street Journal*, October 23, 1974.

heat, although requiring a tank for storage, is also compact and automatic. These fuels, which were inexpensive when they began to become popular, displaced coal in most space heating applications and still hold most of this market today.

Since the oil embargo of 1973–1974, which caused sharp increases in oil prices as well as shortages and price increases in natural gas, the nation has begun to reevaluate the importance of its coal resource. It is the fossil fuel in which we are richest; perhaps 20% of the world's supply of coal lies within our borders. All plans to fill the growing U.S. demand for energy focus upon coal as a central element. Our coal, however, is a mixed blessing. It has serious effects upon the land and upon our health. Because coal is now seen as so important to the nation's economic health, and because it also has a large negative impact, we need to explore fully the consequences of our plans for its use. As we embark on a new path, we should be aware, to the largest extent possible, of the path's condition—it may be a rocky road indeed. These issues are further discussed in Chapter 16.

Energy in the Future

Fusion?

In the midst of recurring energy shortages that foreshadow an era of scarcity to come, the question occurs again and again, "What about fusion?" Fusion is not only a specific atomic event that scientists hope to harness to produce electric power; to many, it has become the name for the "dream of limitless energy." Is fusion one piece of the answer to energy-supply problems of the future? How soon could it contribute to our basic needs for energy? We need to examine fusion in order to approach answers to these questions.

While **nuclear fission** refers to the *splitting apart* of the nuclei of large "heavy" atoms with a release of energy, **nuclear fusion** refers to the *combining* of the nuclei of certain "light" atoms, such as two atoms of deuterium, an isotope of hydrogen. This combination, which takes place constantly in the sun and other stars, releases vast quantities of heat energy. The hope is that the fusion reactor of the future will capture fusion heat and convert it to electric power.

The fuel for fusion reactors is deuterium, an isotope of hydrogen with one proton and one neutron in its nucleus. (Ordinary hydrogen has one proton and no neutrons in its nucleus.) Deuterium is so abundant in the surface waters of the earth, and the fusion reaction is so energetic, that if the energy from deuterium reactors were fully extracted, only 8 pounds of water (about a gallon) could supply the energy equivalent of 300 gallons of gasoline. Unfortunately, it is not clear that fusion can ever be exploited because we seem unable to "house" the fusion reaction. Thus, cost estimates for fusion actually have no basis since we are not at a stage where we can say that fusion will work.

Containment is needed for a fusion reaction because fusion takes place only at the temperature of a star: 100,000,000°C. Since no substance is known to remain solid at this temperature, scientists have devised several novel concepts to "hold" the fusion reaction. These approaches include the famous *tokamak* concept in which the fusion reaction is held in place by powerful magnetic fields. Another concept, which does not use magnetic confinement, is laser fusion, in which converging laser beams focus on a nearly microscopic pellet of "fuel." The heat from the laser beam produces the temperatures needed for fusion. Many nations are now conducting fusion experiments.

The environmental safety of the future fusion reactor is an often-mentioned advantage for fusion. In fact, there may be a significant quantity of radioactive waste to dispose of. Most fusion reactions now being considered produce high-speed neutrons that bombard the walls of the reaction vessel, making the wall radioactive and thus, a waste to be disposed of.

When active research on fusion began in the late 1950s, the first fusion power was estimated to be 20 years away. Today, more than a third of a century later, scientists are still saying, "20 years away." It makes sense to pursue dreams; dreams and ideas are the forerunners of reality. It makes sense to pursue fusion, but we should understand that it cannot contribute to our needs "here and now."

Fuel Cells

The *fuel cell* is the name given to the concept of producing electricity from the chemical reaction of hydrogen and oxygen. The aerospace program pushed this concept far enough along that fuel-cell power plants are now being built, though the first plants are small. At the site of electric generation, fuel cells produce very little in the way of air pollution, thermal pollution, or noise pollution. Hence, as urban neighbors, these modular units are expected to be more welcome than the smoky, sooty, coal-fired plant or the nuclear power plant with its built-in hazard potential.

Whereas the oxygen needed for a fuel cell can be supplied from the air, hydrogen gas must be obtained by chemical processing. Hydrogen may be obtained by processing a liquid hydrocarbon fuel such as naptha or by processing natural gas or petroleum gas. Two first-generation fuel-cell plants, each with about 5% of the power capacity of large modern coal or nuclear plants, have been installed, one in New York City and the other in Tokyo, to demonstrate the feasibility of fuel cells. An inverter is needed on these plants to convert the direct current electricity produced by the fuel cell to the alternating form that we use throughout the country. The Tokyo plant began successfully producing electric power in 1983. The plant in New York City was delayed by regulations and component failures, and by the time it was ready to generate power in 1984, the electrolyte had migrated away from the electrodes, preventing any power production. United Technologies, builder of the New York plant, and the Consolidated Edison Electric Company, its owner, have asked Congress for money to replace failed components at the plant. Though the New York plant never produced power, its concept has been proven in the Tokyo plant.

United Technologies was expected to offer utilities an 11-megawatt commercial fuel cell by 1986, and Westinghouse will have a fuel cell product for sale in a few years. In Japan, Fuji and Mitsubishi will have fuel cells available in about the same time frame. Here is a new technology ready to take hold.

In a second-generation concept of the fuel cell, two fuels combine with oxygen: hydrogen and carbon monoxide. Both of these fuels are produced in the initial stage of coal gasification (see Chapter 17). At present, this cell is the subject of research; full-scale cells are at least a decade away. The fuel cell's advantage of almost pollution-free power generation would disappear if the coal gasifier were nearby, as it is a generous producer of air pollutants.

The hydrogen-using fuel cell is one example of how hydrogen may be used in the future. Hydrogen is not a new energy resource, however. Instead, it is a new way to convey energy. Hydrogen might be produced from the electrolysis of sea water. The energy for electrolysis might be obtained by generating electrical energy from the wind; then, the wind would be the primary energy source. Hydrogen might also be produced from sea water using electricity from con-

ventional power plants; in that case, the fuel that fired the power plants would be the base energy resource.

In addition to being used in fuel cells, hydrogen is being considered as a fuel for the internal combustion engines of automobiles. It is also being considered to power turbine engines, such as those in jet planes, and since a similar turbine engine is used to produce extra electrical power at times of peak demand, hydrogen might be considered for those turbines as well. Such future uses of hydrogen are not at all certain. Some experts see such uses in their crystal balls; others do not.

New Developments in Conventional Electricity Generation

Although improvements in the efficiency of conventional electricity generation have come almost to a halt, novel and unconventional routes of improvement remain open. The thrust of most of this research is to put processes together to capture the waste heat from electric generation. We will discuss three promising ideas. The first, called *district heating*, uses spent steam for heating. The second, known as *combined cycle*, puts together the turbine engine and the conventional boiler/turbine system. The last, *magnetohydrodynamics*, is designed to extract energy from the hot exhaust gas of a coal-fired electric plant.

District heating. One option for making electricity generation more efficient is to use waste heat—the heat from combustion or fission not converted to electrical energy. The spent steam from the power plant can be transported via underground pipes to homes, offices, and factories to provide space heating during winter months. District heating, as this process is called in Europe, has been achieved in Finland and elsewhere, but its prospects are diminished by the large quantity of pipes that must be placed to deliver the heat to customers. To decrease heat losses, the concept also requires small power stations whose customers for the heating service are located nearby. Since smaller plants are less economical than larger ones, and since most Americans do not like to live near power plants, today's prospects for district heating are not good.

Combined cycle. Describing the combined cycle method requires a brief description of the turbine engine, a device used to power jet aircraft. Improved turbine engines have been adapted to power generation and have become common equipment in electricity-generating stations. The turbine engine is turned by the combustion gases from the burning of kerosene or natural gas, instead of by steam. Only about 25% of the heat energy produced is converted to electric power, and the fuel is expensive compared to coal. The equipment, however, is relatively inexpensive compared to the boiler/turbine system of conventional plants. Electric utilities buy these turbine generators and hold them on standby until power loads reach levels beyond the capacity of their base-load equipment. At those times, the turbine engines are pressed into service. For these peak loads, a slightly more expensive fuel does not drive costs too high.

The combined cycle idea is really quite simple. The turbine engine and the boiler/turbine system are put together (Figure 15.16). The burning gases in the turbine engine generate power from the rotation of the engine shaft. The very hot gases discharged from the turbine engine are then used to boil water to steam. The steam, in turn, is used in the ordinary way to turn another turbine for additional electricity generation.

Magnetohydrodynamics (MHD). The last major extension of modern generating plants, magnetohydrodynamics, makes use of the hot gases from combustion. The combustion gases are seeded with potassium, which ionizes into charged particles in the high-temperature flow. The hot gases with ionized particles are directed through a channel surrounded by coils to induce a magnetic field. The movement of charged particles through the magnetic field sets up a current flow that is drawn off by electrodes spaced along the channel. Upon exiting the channel, the hot gases are used to produce steam, which is then used to turn a turbine in the ordinary way.

The combined MHD-conventional system could reach efficiencies of up to 60%; that is, 60% of the heat in coal would be converted into electrical energy. If such plants are built, they will be far less polluting than the 40%-efficient conventional plants. For every kilowatt-hour of electrical energy produced, the 40%-efficient plant will consume 50% more fuel than the 60%-efficient plant. That means 50% more thermal pollution, particles, sulfur oxides, and nitrogen oxides come from the conventional plant.

One of the problems still to be overcome is obtaining materials that will not corrode in the high

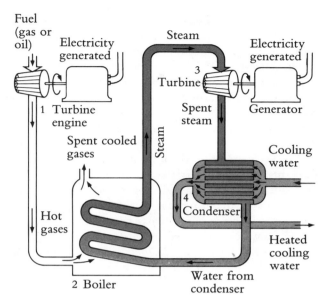

Figure 15.16 Combined cycle concept for electric power generation. This concept puts together the conventional power generation scheme with "front-end" generation using a turbine engine. The hot gases from the turbine engine are used to boil water to steam for conventional electric generation.

temperatures (2000°C) of the MHD equipment. No demonstration generators have yet been built and operated, although a good-sized developmental unit has been built in the Soviet Union. The MHD research in the United States seems to be well funded at the experimental level, but whether MHD power generation will be economical is still not known.

Summary

Electricity acts as a middleman between energy resources, such as coal or falling water, and the final use of that energy. This middleman allows energy to be transported easily or to be used in a more convenient form. The cost of using electricity as a middleman is a loss of efficiency as the energy is transformed into work, and work into electricity. The second law of thermodynamics states that there is a limit to the amount of heat energy that can be transformed into work. Modern coal-fired plants are 40% efficient, nuclear plants are 30%–32% efficient.

In a thermal power plant, heat from combustion or fission boils water to steam. The steam then turns a turbine and generates electricity. The spent steam is condensed back to water in a condenser and then circulated to the boiler again. Coal will increasingly be used to generate electricity in the future because of public resistance to new nuclear power plants and because natural gas is being diverted to heat homes. Nevertheless, the amount of electricity generated by nuclear power will increase for a time because nuclear plants have already been started and will be coming on line.

Electric generation from fossil fuels releases many pollutants, including particles, sulfur oxides, and nitrogen oxides. Both fossil fuels and nuclear power produce waste heat that can cause thermal (heat) pollution.

In open-cycle cooling, waste heat is disposed of in a body of water; in closed-cycle cooling, the heat is released to the atmosphere. Open-cycle cooling leads to thermal pollution of water, in which heat leads to the destruction of aquatic organisms such as fish by exceeding their ability to withstand changes in temperature. Thermal pollution can also affect an organism's reproduction, food-gathering ability, disease resistance, and avoidance of predators. Higher temperatures may change the structure of aquatic communities, tending toward the simplification of communities. Power plants located on naturally warm waters are most likely to do harm.

Waste heat from power plants could be put to a number of beneficial uses, such as home heating and agricultural frost protection. Closed-cycle cooling uses a cooling tower, pond, or canal to evaporate waste heat into the atmosphere. Cooling towers tend to be large structures, and the evaporating water may cause mist in nearby areas. Plants built on inland waters will probably employ closed-cycle cooling in the future, but those on the ocean or cold estuaries may still use open-cycle cooling.

Although U.S. energy consumption increased by more than 100% between 1950 and the present, the relative use in different sectors remains approximately 40% industrial, 35% residential/commercial, and 25% transport. Energy use is expected to increase by an additional 45% over the next 10 years. Transportation energy is mostly used in the form of petroleum. Petroleum provides about 40% of total U.S. energy use, natural gas and coal account for about 25% each, and hydropower and nuclear power account for about 5% each. By 1990, coal, nuclear power, and hydropower are expected to increase their share of energy use in the United States, and petroleum and natural gas will ex-

perience a decrease in their share of energy use. The actual amount of energy used in the form of oil and gas will stay about the same but will have a smaller share of the growing total of energy use.

One possible future source of energy is fusion, the combining of the nuclei of certain light atoms. However, serious difficulties in the design of a "vessel" to contain the fusion reaction make it hard to predict if, or when, fusion power will become a reality. Similarly, the fuel cell, in which hydrogen and oxygen react, is a power source that seems to be some years away, at least on a commercial scale. Three techniques to improve the efficiency of conventional electric generation include district heating, combined cycle, and magneto-hydrodynamics (MHD).

Questions

1. From what you have read about electric power, do you think that the use of electric autos will help solve the pollution problems of the United States? Explain your position as fully as possible. Will it help one kind of geographical area and hurt another?
2. According to the U.S. Energy Information Administration, electric energy consumed in *end uses* for 1980, for 1985, and for projected years (all in quadrillion Btu) is:

1980	7.80
1985	7.92
1990	9.68
1995	11.50

   Yet Figure 15.4 shows higher numbers (again in quadrillion Btu):

1980	26.0
1985	17.1
1990	33.1
1995	39.4

   What is the explanation for these different figures on electrical energy consumption?
3. Explain how thermal pollution illustrates both the first and second laws of thermodynamics.

Further Reading

Phillips, Owen. *The Last Chance Energy Book.* Baltimore: Johns Hopkins University Press, 1979.
    This short, 140-page book is an eminently readable, en-

tertaining, and thoughtful treatment of our energy dilemma. Its discussion of alternative energy sources is well done, as is the treatment of the theory of resource exploitation.

Energy: The Next Twenty Years. Report of a Study Group sponsored by the Ford Foundation and administered by Resources for the Future. Cambridge, Mass.: Ballinger Publishing Co., 1979.

Oil, coal, nuclear power, solar power, projections, economics—all are here, thoroughly documented and comprehensive; one of the most complete studies of the 1970s.

Hayes, E. T. "Energy Resources Available to the United States," *Science*, **203** (January 19, 1979), 233.

Although this work is a nontechnical treatment of conventional sources, the reader will find very little discussion of natural sources of power.

Kash, Don, et al. *Our Energy Future*. Norman: University of Oklahoma Press, 1976. (Originally published as *Energy Alternatives: A Report to the President's Council on Environmental Quality*, 1975.)

Slightly dated, but still a comprehensive source of information on most conventional energy alternatives, resources, and uses.

Electric Power Research Institute. "MHD: Direct Channel from Heat to Electricity," *EPRI Journal* (April 1980), p. 21.

Bush, R. M., et al. "Potential Effects of Thermal Discharges on Aquatic Systems," *Environmental Science and Technology*, **8** (6) (1974), 561.

The effects of heat on different species of organisms are compared in this paper, which is intended to help in the choice of suitable sites for new power plants.

Crawshaw, L., et al. "The Evolutionary Development of Vertebrate Thermoregulation," *American Scientist*, **69** (September/October 1981).

A fascinating discussion of temperature regulation in organisms.

"The Dry Look in Cooling Towers," *EPRI Journal* (December 1985), p. 34.

A possible advance in controlling thermal pollution may come from the use of ammonia, rather than water, in cooling towers, as this paper explains.

"Keep Biofouling at Bay," *EPRI Journal* (July/August 1984), p. 23.

Not only waste heat but also chemicals may be discharged by power plants in cooling water. This paper looks at current problems and regulations.

References

Weber, H. C., and H. Meissner. *Thermodynamics for Chemical Engineers*, 2nd ed. New York: Wiley, 1957.

Fields, S. R. "Morphological Analysis of Beneficial Uses of Waste Heat from Power Plants," *U.S.A.E.C.* HEDL-TME 71-97 (1971), p. 8.

Rose, D. J., and M. Feirtag. "The Prospect for Fusion," *Technology Review* (December 1976), pp. 21–43.

Fickett, A. P. "Fuel-Cell Power Plants," *Scientific American*, **219** (6) (December 1978), 70–76.

Waldrop, M. "Compact Fusion: Small Is Beautiful," *Science*, **219** (January 14, 1983), 154.

Kintner, E. "Casting Fusion Adrift," *Technology Review* (May/June 1982), p. 64.

Lidsky, L. "The Trouble with Fusion," *Technology Review* (October 1983), p. 32.

CHAPTER SIXTEEN

Coal: A Mixed Blessing

oal took us out of the era in which only human muscle, wind, and water power were available to manufacture goods. Coal brought us into the era of machinery powered by combustion. Watt's steam engine, powered by coal, made it possible for us to have machinery that was not linked to rivers. Coal was also the fossil fuel that first changed our methods of transportation—coal powered the steam locomotive.

Although the major use of coal in the United States today is as a fuel for steam electric power plants, it was once an important fuel for home heating and for transportation. After World War II, homeowners began to switch from coal to oil and to the newly available natural gas. Railroads also converted from coal-fired locomotives to the more dependable and cleaner diesel engine.

There are four kinds of coal: peat, lignite, bituminous, and anthracite. All are derived from natural processes acting on wood, and all are used as fuels. These basic forms reflect differing levels of carbon content.

Peat, the least developed form of coal, is still highly woody. It is found in bogs where wood has been immersed in water and has decayed in the absence of oxygen. Historically, peat has been used only for heating.

Lignite, or brown coal, has a slightly greater heat and carbon content than peat, but it still has less than the higher forms of coal. Most lignite resources of the country can be mined from the surface (strip mined).

Bituminous coal is characterized by a higher heat and carbon content than lignite. Subbituminous coal is a grade of coal whose heat content is just below that of bituminous coal but above that of lignite. Of the four types of coal, we rely most heavily on bituminous coal because of its relatively high carbon content and its abundance in the United States.

Anthracite coal has the highest carbon content of all coals. However, its current production and use in the United States is small because its reserve base is so low.

Coal is not a pure substance. In addition to carbon, it contains inorganic material that remains after coal has been burned; we know the material simply as ash. Sulfur also occurs in coal, sometimes as iron sulfide and sometimes combined with organic compounds. This sulfur, when burned, is responsible for much of the acid rain that falls on the northeastern United States. Arsenic is also present in coal, as are radioactive elements. In fact, coal is the dirtiest of the fossil fuels.

Dirty though it is, it is a magnificent source of heat energy. Burning 1 pound (0.454 kg) of bituminous coal releases 13,000 Btu, or 13,700 kilojoules (kj) of heat energy. Furthermore, coal is the most plentiful fossil fuel we possess in the United States.

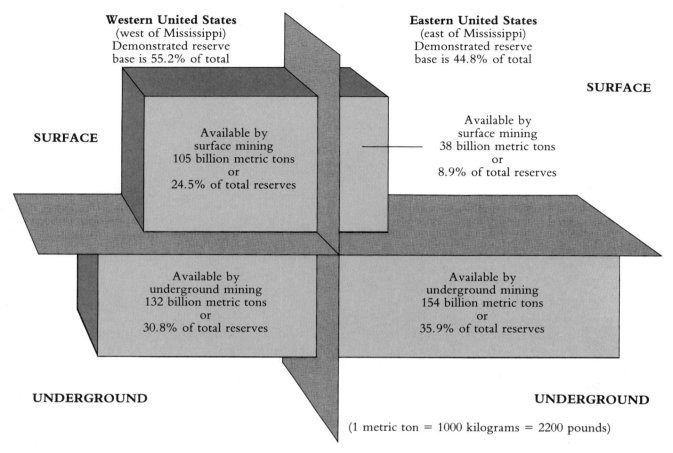

Western United States
(west of Mississippi)
Demonstrated reserve
base is 55.2% of total

Eastern United States
(east of Mississippi)
Demonstrated reserve
base is 44.8% of total

SURFACE

SURFACE

Available by
surface mining
105 billion metric tons
or
24.5% of total reserves

Available by
surface mining
38 billion metric tons
or
8.9% of total reserves

Available by
underground mining
132 billion metric tons
or
30.8% of total reserves

Available by
underground mining
154 billion metric tons
or
35.9% of total reserves

UNDERGROUND

UNDERGROUND

(1 metric ton = 1000 kilograms = 2200 pounds)

Figure 16.1 Demonstrated reserve base of coal in the
United States as of January 1, 1980 (430 billion metric tons).
Note that numbers may not total due to independent round-
ing. (Data from Energy Information Administration.)

Coal and Its Uses

How Much Coal Is There?

As we have seen, coal has a range of carbon content,
and this determines its heating value. Peat has the
lowest heating value, followed by lignite, then by
subbituminous and bituminous coal, and finally, by
anthracite, which has the highest heating value. To give
an accurate accounting of our coal reserves, we need
to provide reserve estimates for each major kind of
coal. We also need to know how much of the reserves
can be surface mined and how much must be mined
underground.

Of the 430 billion metric tons* of discovered coal

---

*One metric ton = 1,000 kilograms = 2,200 pounds = 1.10 En-
glish ton.

reserves in the United States, 55% occurs in the western states (see Figure 16.1). Of these western coal
reserves, 44% can be obtained by surface-mining
methods, while the remaining 56% requires underground mining. Eastern coal, in contrast, is concentrated as underground reserves, which constitute about
80% of total eastern coal. The total coal reserves in the
United States amount to about 20% of the world's coal
reserves.

These reserves are discovered resources; that is, we
are reasonably confident these resources exist, but they
are not fully recoverable by current mining technology. The coal obtained from underground mining is
only about 50% of the actual coal in the deposits, since
supporting structures, or pillars, must be left in the
mine. In contrast, surface mining recovers 80%–95%

of a coal deposit, leaving only minor amounts of residue. Even if we apply these recovery factors, American coal reserves are sufficient for several centuries at the present rate of use.

Coal reserves must also be described by rank (type), since heating values per pound differ. About 92% of the coal reserves in the United States are of either bituminous or subbituminous grade. Lignite is the other major coal resource, at about 6% of the total. Virtually all of this low-heating-value coal is mineable at the surface. Anthracite coal reserves, all of which require underground mining, make up only about 2% of U.S. coal reserves. Because anthracite constitutes only a tiny portion of the coal market, we will not consider it further here.

The heating value of bituminous coal is 13,000 Btu per pound (30,200 kj/kg), and the value of subbituminous coal is 10,000 Btu per pound (23,200 kj/kg). Lignite provides only about 7,000 Btu per pound (16,250 kj/kg). Using the heating values of the different coals, we can restate U.S. reserves of coal in Btu as well as in tons (Table 16.1). On the basis of tons, bituminous coal makes up 51% of our coal base, but on the basis of energy, it makes up nearly 60%.

The coals of various ranks are not evenly distributed in the United States. Lignite and subbituminous coal, for instance, are found in large quantities only west of the Mississippi River. In contrast, 85% of the bituminous coal reserves are concentrated in the eastern portion of the United States; only about 15% of this resource lies west of the Mississippi (Figure 16.2).

Certain states have much more coal than others. In the east, coal resources are concentrated in Ohio, Illinois, Pennsylvania, West Virginia, and Kentucky. In the west, North Dakota, Montana, Wyoming, and Colorado have the richest coal resources (Figure 16.3).

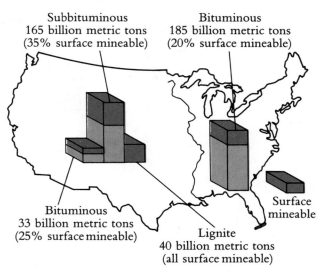

Figure 16.2 Distribution of grades of coal as of January 1, 1980. Note that numbers in figure may not add to numbers in Table 16.1 due to independent rounding. (Data from Energy Information Administration.)

About 45% of the land area of North Dakota and 41% of Wyoming contain substrata of coal-bearing rock. About 35% of Montana and 28% of Colorado sit atop coal resources. Taken together, Montana and Wyoming account for 40% of the tonnage of U.S. coal reserves.

Nearly 85% of the coal that is less than 1% sulfur by weight is found west of the Mississippi. Thus, to obtain low-sulfur coal in the eastern United States, where a larger percentage of the coal is used, the east must bring its coal from the western United States. These imports may come in by train or barge, or if the newest technology takes hold, by coal-slurry pipeline; however, the lower heating values of most western coals require greater quantities to be shipped.

How Is Coal Being Used?

The use of coal is growing, after a period in which it had fallen steadily (see Figure 16.4). The long decline in coal consumption came about as the railroads switched from coal to diesel fuel and as homeowners switched from coal to gas and oil. The present upward trend is the result of the growing demand for electricity, a demand that is increasingly being filled by coal-fired electric power generating stations.

With discovered reserves of about 400 billion metric tons, coal appears to be a secure energy source for the next several centuries. In 1985, about 817 million

Table 16.1 Coal Reserves in the United States, as of January 1, 1980

	Billions of metric tons	Quadrillion Btu (Quads)[a]	Quadrillion kilojoules	Percent of coal resource, based on energy
Bituminous	217	6206	6547	59.4
Subbituminous	165	3630	3830	34.7
Lignite	40	616	650	5.9

a. 1 quad = 1 quadrillion Btu = 1,000 trillion Btu.

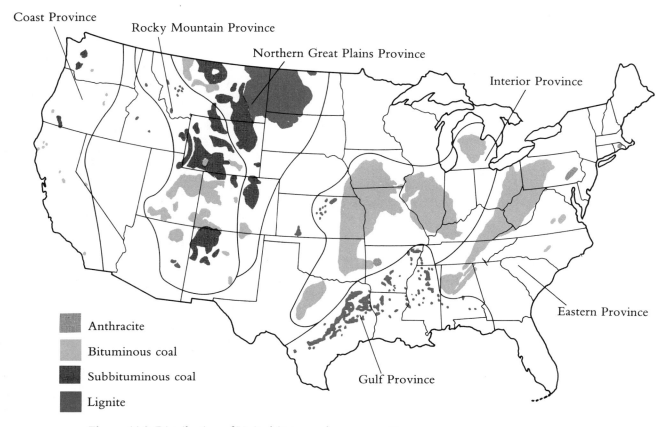

Figure 16.3 Distribution of United States coal resources. (From Bureau of Land Management, 1974.)

metric tons of coal were mined in the United States. Of that tonnage, about 55% was obtained by surface-mining methods. The dramatic growth in surface mining is shown in Figure 16.4. In 1985, more coal was surface mined in the United States than was produced by all methods in 1960.

Coal is used most in the electric utility industry. In 1985, coal consumed in the generation of electricity amounted to about 84% of the annual coal consumption in the United States. Other uses of coal include coking in the iron and steel industry (5.5%) and general industrial use (9%). (See Figure 16.5.) A small amount of coal is still used in home heating, although this use has fallen dramatically since the end of World War II. Exports of coal were 71 million metric tons in 1983. Foreign countries that purchase our coal generally use it in their iron and steel industries. The consumption of coal in 1995 is projected to expand by 30%, with vir-

tually all of the increased use going into electricity generation.

Environmental and Social Impact of Coal

With so much coal available, it appears to make good sense to get the maximum use from this fuel. It can replace such costly or scarce fuels as oil and gas in electricity-generating stations. Also, it can be converted from solid form to a liquid fuel, or from solid form to a gaseous fuel, although there are costs linked to such conversion.

The National Coal Association, an industry organization, calls coal "America's ace in the hole." Although such organizations urge us to let the mining industry go in and get it, coal does not rise from the earth like magic. Both the natural environment

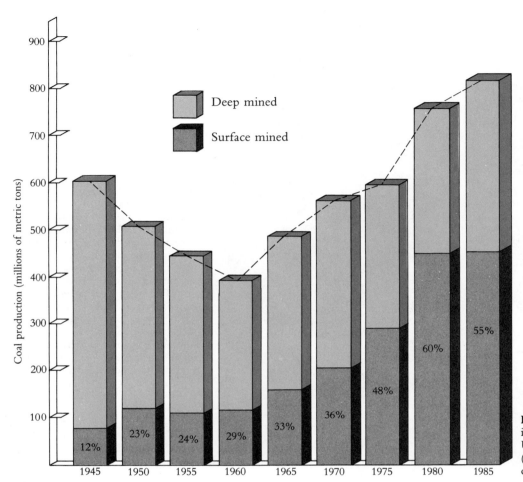

Figure 16.4 Coal production in millions of metric tons the United States 1945–1985. (Data from U.S. Department of Energy.)

humans may be sacrificed along the way. Moreover, at its point of use, the burning of coal pollutes the air, injures vegetation, and affects human health.

Surface (Strip) Mining

How coal is removed. Whether a modern coal company decides on surface or underground methods of mining a coal seam depends on the depth of the seam below the surface of the earth. The more layers of soil and rock (called overburden) that must be removed, the greater the cost to mine the seam from the surface. The decision also depends on the thickness of the seam, since the amount of coal recovered will influence profits.

Much modern surface mining (also referred to as strip mining) depends on giant equipment (Figure 16.6). To justify bringing this costly equipment to a site, a large deposit of coal must be present. Thus, a small number of huge surface mines account for a large percentage of the coal produced each year in the United States. In 1982, the 13 surface mines operating in Campbell County, Wyoming supplied nearly 9% of the nation's annual coal production. Operating at full capacity, these mines could have supplied about 16% of 1982 production, or 126 million metric tons.

There are two kinds of strip mining; their difference stems from the different land forms on which they operate. **Contour mining** is used in hilly terrain where coal seams lie beneath hills. A power shovel cuts

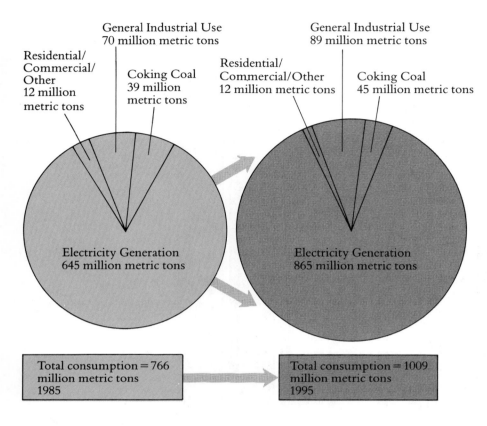

Residential/
Commercial/
Other
12 million
metric tons

General Industrial Use
70 million metric tons

Coking Coal
39 million
metric tons

General Industrial Use
89 million metric tons

Residential/
Commercial/Other
12 million metric tons

Coking Coal
45 million metric tons

Electricity Generation
645 million metric tons

Electricity Generation
865 million metric tons

Total consumption = 766
million metric tons
1985

Total consumption = 1009
million metric tons
1995

Figure 16.5 Consumption of coal in United States by sector, 1985 and 1995. 1985 total = 766 million metric tons, not including exports. 1995 projected total = 1,009 million metric tons, not including exports. Note that coal use in electricity generation is projected to expand by nearly 35%. Exports of coal in 1983 (not shown) were 71 million metric tons (1985 exports not available). Exports of coal projected for 1995 (not shown) are 96 million metric tons. Totals may not equal sum of components due to independent rounding. (Data and projection from Energy Information Administration.)

a groove into a hillside, exposing the coal seam. The overburden may be discarded to lower levels of the hill, but reclamation is more difficult if this is done (Figure 16.7). When the overburden becomes too thick to remove, giant auger drills, up to 7 feet in diameter, may bore into the coal seam to bring out more coal.

When no attempt is made to restore the area, a *highwall* remains (Figure 16.8). This vertical cut into the side of the hill may have vegetation at the top, followed by a layer of soil and rock. Then comes a dense black stripe—the coal left behind when further removal of the overburden became too costly. Such a stretch of highwall may run for miles. Over 20,000 miles (32,000 km) of highwalls ring the hills of the Appalachian region, scars from a not-too-distant era when profit came before environment.

Area mining, in contrast to contour mining, operates on flat terrain (Figure 16.9). Area mining lays up the soil in rows, exposing the coal in a long path. When the coal is removed from that open path, the overburden on the coal next to the path is excavated and

dumped into the mined area, exposing another long path of coal. The process is repeated for row after row, leaving hills of rubble resembling miniature mountain ranges (Figure 16.10).

In larger surface operations, a power shovel may be utilized along with a dragline. The power shovel scoops overburden and loads coal into trucks. The dragline is used primarily to move overburden from place to place. The bucket of the largest dragline can carry up to 200 cubic yards (168 m³) of earth. This is equivalent to a room 27 feet (8.1 m) wide in both directions with an 8-foot (2.4 m) ceiling. The weight of earth in the bucket of the dragline may run to 200 tons (182 metric tons).

Reclamation of surface-mined land. Even into the 1970s, strip miners were allowed to remove coal and simply depart. The result was a surrealist picture of huge mounds of rubble, sometimes arranged in rows, sometimes strewn haphazardly over the landscape. The land from which the coal was torn was often left use-

(a)

(b)

Figure 16.6 The equipment used in strip mining is massive. (a) This is the bucket of a dragline machine at the Medicine Bow Coal Mine in Hanna, Wyoming. The bucket holds 78 cubic yards, a volume equal to that of a 16-foot square room with an 8-foot ceiling. (b) This dragline operating at a surface mine at Marrisa, Illinois is so large (20 stories tall) that it had to be assembled at the mining site. (Photos courtesy of U.S. Department of Energy)

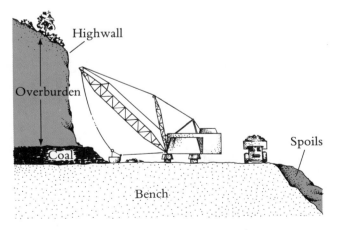

Figure 16.7 Contour mining using a power shovel. Disposal of the spoils down the hill slope makes later reclamation more difficult. (Adapted from D. Kash et al., *Energy Alternatives*, U.S. Environmental Protection Agency, EPA 430/9-73-011, 1973.)

less and without vegetation. Black-striped highwalls ringed the tops of hills, cutting off any plant and animal life from the nearby surroundings.

When land is left unreclaimed, soil particles can be washed off the barren slopes of the hills and can fill streambeds. Because the stream channels are decreased, the flows from large rainstorms cannot be carried in the channel and the stream will flood over its banks more frequently. Furthermore, the barren hillsides with loose earth are unstable. A soaking rain can increase the weight of a soil mass and decrease its hold to the stable earth; the result is landslides.

Eventually, vegetation will return to many of these areas, but the process is slow. As we will see later in this chapter, the rubble left on the land is acidic because of the presence of the impurities in coal; this acid condition hinders the regrowth of grasses and trees. In addition, acid rivulets run off these lands and enter nearby streams. Acid waters do not typically support

Figure 16.8 A highwall is left after a vertical cut has been made into a hillside to expose a coal seam for mining. The dense black stripe is coal that was too costly to recover. Highwalls, which may run for miles, isolate the hills from many animal species and so destroy valuable wildlife habitat. Leaving highwalls is now illegal, but the hills of Appalachia are scarred from pre-1977 operations.

aquatic life, and humans cannot drink such stream water.

Reclamation can prevent such scenes of devastation. The concept of **reclamation** is simple: restore the original contour of the land—if possible, with topsoil on top—and reestablish vegetation to "anchor" the soil.

Where area mining is utilized on gently sloping or flat areas, the topsoil is saved so that it can be replaced on top of the regraded rubble. Once the topsoil has been replaced, fertilizer and grass seed are applied. With adequate rainfall and 6 inches (15 cm) or more of topsoil, the vegetation has a good chance of becoming established, preventing rapid runoff of water and checking the erosion process.

Where contour mining has occurred, special methods have been designed to repair the damage. In one reclamation procedure, *full backfilling*, the spoils are pushed back up the bank and fully cover the highwall (Figure 16.11). The restoration matches the surface of the filled area to the hill surface at the top of the highwall so that water from runoff will not fall rapidly over a sharp slope. To prevent erosion gullies from forming, terraces may also be cut around the restored hillside to catch and slow the water that runs off the slope.

Backfilling is an effort to repair damage after it has been done. In contrast, the *modified block cut* is designed to minimize damage during the process of mining (Fig-

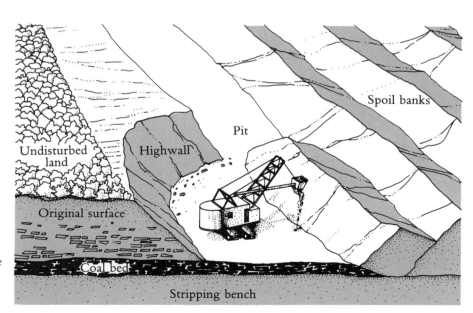

Figure 16.9 Area mining, one of the methods of surface mining for coal. It is used when the land above the coal seam is flat or gently rolling.

Figure 16.10 Area strip mining left unreclaimed. This land near Nucla, Colorado was area mined for coal by the Peabody Coal Co. Rapid erosion of soil occurs from such areas, clogging streams and making them subject to flooding. Acid conditions in the rubble and soil hinder the regrowth of trees or other vegetation. The land is unfit for human or wildlife habitation. (EPA-Documerica, Bill Gillette)

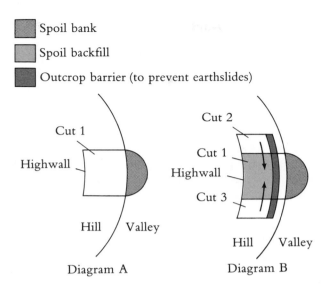

Figure 16.12 The modified block cut is one of the best methods to reclaim contour-mined land and appears to cost no more than ordinary mining. In Diagram A, an initial cut is made into the hill. Diagram B shows the second and third cuts being made; the spoils from these cuts are deposited in the void left by the first cut. The spoils from later cuts are deposited in the new voids created by cuts 2 and 3. (Adapted from D. Kash et al., *Energy Alternatives*, U.S. Environmental Protection Agency, EPA 430/9-73-011, 1973.)

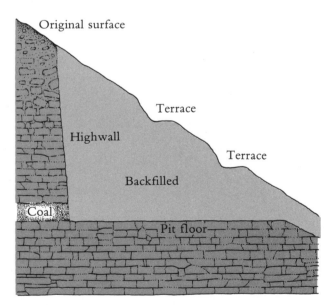

Figure 16.11 Full backfilling of contour-mined land. This procedure is used to reclaim hill slopes that have been contour mined. The terraces are used to channel runoff gradually down the hill. (Adapted from D. Kash et al., *Energy Alternatives*, U.S. Environmental Protection Agency, EPA 430/9-73-011, 1973.)

ure 16.12). In the first step, or cut, overburden is excavated from just above the coal seam and is dumped downhill. The second cut and subsequent cuts are smaller in size and are taken at the same elevation as the first, but to either side of the initial cut. The overburden removed from the second and third cuts is dumped into the adjacent space made by the first cut. More cuts are taken to either side of these initial cuts, and the spoils are discarded into the spaces made by the second and third cuts, and so on.

These are only a few of the many methods that have been used to restore contour-mined land. The modified block cut is reported to have been successful on slopes of 20° and above. In addition, the modified block cut appears to be no more expensive than ordinary contour mining, suggesting that it may become an important procedure in the future.

Replanting a strip-mined area. The best revegetation cannot soon bring back forested areas. Forest wildlife cannot survive in grassland; their food supply and nesting places do not exist. Thus, the ecological balance is unavoidably altered by strip mining, even with the best of reclamation.

The success of reclamation appears to be dependent on rainfall. In regions where rain is sparse, it may happen that the best of methods are inadequate to restore vegetation to the land. The concern here is mainly with the vast coal fields of the west, where huge strip-mining operations are underway and increasing in scope; the consequences of such operations in low-rainfall areas are difficult to predict.

A committee of the National Academy of Sciences (1974) studied the issue of western surface mining from the standpoint of the potential for reclamation of the lands. The panel indicated that areas with less than 10 inches of annual rainfall were probably not good candidates for reclamation because full restoration of native vegetation might not be possible within a reasonable time scale.

The committee felt that sandy soil supporting only desert shrubs probably cannot be restored to its original form because no topsoil is available to cover the earth after surface mining. Because so few seeds would be present in the sterile, exposed soil, the growth of new plants would be extremely slow. The natural succession of plant species, in which one diverse community of plants replaces another, would be long delayed in this difficult environment, so that 100 years or more might pass before the native vegetation returned.

Figure 16.13 summarizes the views of the committee on the probabilities of successful restoration of four vegetative zones encountered in the western coal states. While the recommendations were made for coal-mined land, they apply equally well to any mining activity.

Surface mining law established. In 1977, after years of bitter debate and disagreement, Congress finally passed a law establishing partial control over strip mining. The Surface Mining Control and Reclamation Act forbids leaving highwalls after mining. It also forbids dumping overburden on slopes of over 20°. The legislation demands that strip-mined lands be returned to approximately original contours. Prime farmlands, according to the legislation, are not to be mined unless they can be returned to their original productivity. Furthermore, a tax is collected on coal currently being produced. This tax is to be used to pay for restoration of land that was marred in the era before reclamation. While enforcement of these standards is left to the states, the Department of the Interior may step in when the states fail to act.

Although reclamation costs could run from $1,000 to $5,000 per acre ($2,500 to $12,500 per hectare), depending on the slopes involved, it will still be profitable to surface mine coal. Further, strip-mined coal will still be cheaper, in general, than deep-mined coal. Also, a 35¢ tax is now collected on each ton of coal mined, and about $250 million is available each year for use in reclaiming abandoned mines.

Surface mining: Is the issue settled? During the 1970s, federal laws to control strip mining were proposed again and again, only to be defeated either by Congress or by presidential veto. One group in Congress favored a complete ban on all strip mining of all minerals. On the opposite side, the coal companies and other mining companies argued for no regulation. Disagreements were fierce as each side had its story to tell. Strip mining was ravaging counties in Kentucky, Virginia, and West Virginia, where irresponsible operators would simply devour a coal seam and depart, leaving rubble and destruction in their wake. At the same time, responsible companies were restoring the land they had mined. In general, however, state laws on surface mining were weak★ and the conscience of the operator was all that people had to rely on. In such cases, pangs of conscience meant lower profits for operators.

Now, as strip miners operate in the west under a new law, it seems wise to review the local impact of strip mining as it took place in the 1970s. Perhaps the most eloquent spokesman against strip mining was the Honorable Kenneth Heckler, representative to Congress from West Virginia. He gave a speech in the House of Representatives titled "The Hills of Appalachia Are Bleeding." The following statement is from that speech:

> The human suffering of those who live near strip mining sites is pitiful. The blasting and the bulldozers have frequently sent boulders onto the property and even into the homes of those on the fringes of strip mining. I have called at homes where embarrassed owners have almost cried because they cannot draw me a glass of water, for their water supply has turned black or brackish. The entire Appalachian area is honey-combed with moonscapes. Foundations of homes have

---

★A notable exception was Pennsylvania, where a tough law was enforced by a colorful and tough administrator, William Guckert.

(a) *Ponderosa pine and mountain brush*

". . . where precipitation is favorable for plant growth and where rather deep, fertile soils have developed, [such sites] do not present a revegetation problem; it must be remembered, however, that many years are required to grow mature pines."

(b) *Mixed grass plains*

". . . a rather high probability for satisfactory rehabilitation . . . predicting such results assumes the best technology will be applied, including the addition of topsoil . . ."

(c) *Sagebrush foothills*

". . . a delicate reconstruction process in the handling of substrata and top soil is demanded for favorable rehabilitation . . . in some places disturbed ground may have to be repeatedly reseeded . . ."

(d) *Desert*

"Disturbing such areas for the surface mining of coal amounts to sacrificing such values permanently for economic reward."

Figure 16.13 Reclaiming surface-mined lands in the West. (Quotes are from National Academy of Sciences, *Rehabilitation Potential of Western Coal Lands*, Cambridge, Mass.: Ballinger, 1974.) (Photos: [a] Kathleen Sullivan; [b,c] U.S. Bureau of Land Management; [d] National Park Service, Richard Frear)

been destroyed. The giant mounds of earth which are pushed over the hillsides are unstable. They swell with the rains, get heavy, crack and slide. They erode away, washing rock and soil down the hillsides into the streams below . . . Do you want more areas of the country to become instant Appalachias? Do you want to allow these strip-and-run exploitation artists to rip up your land the way they have ripped up ours? (Heckler, 1973)

Surface mining is now increasing dramatically in the west. To illustrate, 101 million metric tons of coal (17% of total production) were produced west of the Mississippi in 1975. By 1985, production had increased to 288 million metric tons, and by 1995, it is expected to reach 445 million metric tons, or 40% of total production. In 1975, a portion of Death Valley National Monument was being surface mined for coal by the Tenneco Company. Borax and talc are still being mined there. Montana, Wyoming, North Dakota, and other western states are being surface mined extensively.

The law that now governs surface mining is the Surface Mining Control and Reclamation Act of 1977, described earlier. In terms of western mining, there are several flaws in that legislation. When a state fails to enforce the law, allowing illegal and destructive strip mining, much damage can be done before the federal government discovers the operation and is able to take action. The act dictates that prime farmlands cannot be strip mined unless their productivity can be restored, but who is to judge what has not yet been determined even by scientists? In addition, there is no prohibition on surface mining the desert, where we know that restoration of the land is extremely unlikely. Finally, it appears that an administration unsympathetic to the law, and to the environmental values it is meant to preserve, has the power to limit enforcement of the law by simply cutting the budget and the staff of the enforcement agency until it is powerless.

In 1984, the Committee on Government Operations of the U.S. House of Representatives issued House Report 98-1146 on the subject of enforcement of the Surface Mining Act of 1977. The Committee concluded that the Department of the Interior had "failed miserably to efficiently and effectively carry out its responsibilities for assessing and collecting civil penalties and for implementing other enforcement provisions of the Surface Mining Control and Reclamation Act." The Committee accused the Department of the

Interior of "either a lack of full understanding, or a conscious disregard, of the requirements of the statute." The Committee made specific recommendations for the department to undertake to improve its performance. In 1985, the Committee investigated to see how enforcement had changed. In HR 99-206, titled "Office of Surface Mining: Beyond Reclamation?" the Committee once again castigated the Department of the Interior. That report noted "a virtual breakdown of the civil penalty collection system" and "a failure to comply with Federal Court orders requiring implementation of the Act."

Thus, the issue of coal surface mining is not yet laid to rest. As the 1980s proceed, you may wish to observe the effect of vastly increased surface mining on the states of the western plains and ask your Senator, Representative, and President why the surface mining law is not being enforced.

Underground Mining

How coal is removed. In an underground coal mine, men and machines cut the mineral from the coal face and load it into cars or onto conveyors for removal to the surface. The **room-and-pillar** method of mining is the traditional means of extracting the coal (Figure 16.14). From a central passage, parallel rooms or tunnels up to 20 feet (6 m) wide may penetrate the coal seam for 200–300 feet (60–90 m). Another set of parallel tunnels driven perpendicular to the first results in a tic-tac-toe pattern of rooms. The pillars of coal between the rooms, along with large timbers and roof bolts, hold the earth up. In room-and-pillar mining, about 45%–50% of the coal in the seam is actually brought out. Nearly three-quarters of the cutting and loading is now done by machine, while the remaining one-quarter is still accomplished by the more labor-intensive operations involving either hand cutting or hand loading.

Of growing interest is the highly efficient method imported from Europe known as **longwall mining**. Only in the last decade has this procedure been used in the United States. Longwall mining removes the coal without disturbing the layers of earth above the coal seam and without creating permanent deep mines (Figure 16.15). Movable jacks hold the roof of the mine in place while mining is going on, but when the coal has been removed, the jacks are moved forward to the new mining room. When the jacks have been moved, the

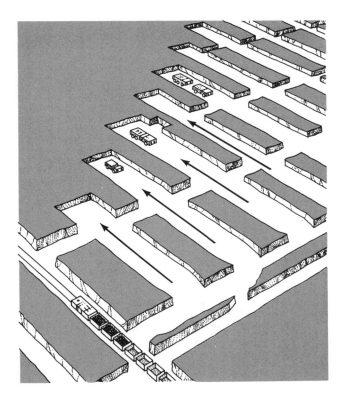

Figure 16.14 Looking down on a room-and-pillar mining operation. Arrows indicate the direction of further mining. The tunnels may be up to 20 feet (6 m) wide.

earth collapses above the mined space since no columns support it. The surface also subsides (sinks) as the jacks are gradually moved forward, but otherwise the surface remains untouched. In addition, most of the collapsed area has been tightly sealed off from water or oxygen. No tunnels into the interior of the mine are left behind, so water has no easy route to escape to the surface, and oxygen has no easy route to enter. Thus the formation of acid mine drainage is largely prevented.

Coal waste piles or refuse banks. Once coal has been extracted from a mine, it may be sent directly to customers, such as the electric utilities. Two-thirds or more of current bituminous coal production moves directly to users in this way. The remainder is sent to a preparation plant, where it is crushed, washed, and graded by size.

The wastes from the preparation plant consist of low-grade coal, shale, slate, and coal dust. These wastes are stored in huge hills known as **refuse banks** or "gob" piles. Water contaminated with coal dust may be produced from the washing process. This water may be settled in ponds to remove the coal particles, but in some cases the water is discharged directly into

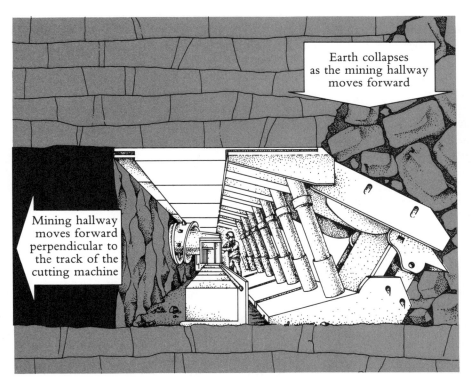

Earth collapses as the mining hallway moves forward

Mining hallway moves forward perpendicular to the track of the cutting machine

Figure 16.15 Longwall mining. The cutting machine proceeds along the coal seam, enlarging the width of the hallway. As the hallway is enlarged, the roof supports are moved forward, closing the room and allowing the earth behind the supports to collapse. No underground tunnels are left when the longwall mining operation has been completed. (Adapted from *Coal*, Utah Power and Light, 1982.)

Box 16.1 West Virginia Flood Toll at 60 with Hundreds Lost

Bodies Found in River 24 Miles Away After Water Behind Coal-Waste Pile Rushes Down Over 14 Communities

MAN, W. VA., FEB. 27—Up on Buffalo Creek, where people used to live, only the bulldozers and the helicopters belonged today. The people were gone, scattered by yesterday morning's flash flood that was unleashed by the collapse of a coal waste pile accumulated over the last 15 years. . . .

The tragedy occurred around 8 A.M. yesterday

when a huge pile of waste material from the operations of the Buffalo Mining Company gave way on a mountain outside Man. The wall of water, estimated by some residents as ranging from 20 to 50 feet high at the beginning, rushed downhill through 14 mining communities, wrecking most of the houses in its path and displacing as many as 5,000 persons. . . .

Source: *The New York Times*, February 28, 1972.

---

streams. One practice is to dam the streams with the refuse bank, thus creating a settling pond behind the dam. In West Virginia in 1972, such a coal waste pile gave way, resulting in a devastating flood (see Box 16.1). A vice president of the Pittston Coal Co., the owner of Buffalo Mining, indicated that the company considered the flood to be "an act of God."

Refuse banks that are not used to dam streams may also be a source of acid water when rainfall drains through them. One infamous coal pile in Aberfam, Wales collapsed in the 1960s, burying a school and all inside. These mounds of coal debris may often catch fire, producing air pollution, and can smolder for years. One bank in the northern Appalachian region was discovered to be burning during a 1963 study; it had been burning since 1884, despite efforts to control it. Controlling the fire in a refuse bank may require actually burying the mound under an earth cover. Some 800 refuse banks of varying composition are known to exist in the anthracite region of Pennsylvania alone.

Subsidence. The removal of coal from underground mines leaves another legacy: **subsidence**, the sinking, or settling, of the ground surface. The practice of removing some of the supporting pillars for their coal at the end of a mining operation increases the risk of subsidence. As the roof supports in a mine give way, the mine fills with overburden; sink holes and cracks may appear at the surface, and tremors may shake the area.

Though only a nuisance when it occurs in farmland or forest, subsidence is a serious problem when human dwellings are nearby. Roads and foundations may crack, buckle, and crumble as the earth settles into

firmer position below. Railroad tracks may be bent and lose their support, gas mains may break and threaten explosion, and sewer lines may crack. In some areas of Pennsylvania, homes have had to be abandoned for safety.

Acid mine drainage. Another effect of underground coal mining, and of surface mining as well, is **acid mine drainage**. The process that creates acid mine drainage is much the same for underground or surface mines. Water may enter an underground mine directly through the mine shaft, or groundwater may seep into the mine through the layers of soil. Sometimes the earth above the mine has subsided. In this common situation, water may easily penetrate the mine through the sink hole that has been formed. It is even possible that entire surface streams may go underground through such sink holes. Iron pyrites from the coal remaining in the mine dissolve in these waters.

Iron pyrite is known chemically as ferrous sulfide, a compound of iron and sulfur. It is also well known as the shiny golden mineral "fool's gold." When ferrous sulfide dissolves in water, oxygen in the water begins to oxidize both sulfide and ferrous ions. The oxidation, in which bacteria play a role, converts the sulfide ion to sulfate ion and the ferrous ion to ferric ion. Oxygen from the water is used up in the conversion, resulting in such low levels of oxygen that the normal aquatic life of the stream is threatened. The effects of low levels of dissolved oxygen include the destruction of the fish population, the normal insect population, and much of the normal microscopic plant and animal life. These effects are further discussed in Chapter 12.

Other reactions make matters worse. The ferric

ion combines with hydroxyl ions in the water to form the yellow-brown precipitate ferric hydroxide, the substance called *yellow boy*. Yellow boy may sink and coat the bottom of the stream. Creatures that feed on the stream bottom, such as freshwater crayfish, will thus be deprived of their food source. Visitors to the coal-mining areas of Appalachia often remark on the yellow-orange color of the streambeds in the region.

Although yellow boy coats the streambed and the low levels of dissolved oxygen harm aquatic life, the worst consequence is the acidity of the water. The precipitation of ferric hydroxide leaves an excess of hydrogen ions in solution. The water is now rich in both hydrogen ions and sulfate ions. This situation is equivalent to adding sulfuric acid to water. The highly acidic water with low levels of dissolved oxygen and a stream bottom coated with yellow boy create an environment unfavorable to the survival of aquatic species.

Acid mine drainage arises from strip-mined areas in the following way. An unrestored strip-mined area has numerous ditches covering the landscape. Rain leaves pools of surface water in these ditches, and this water comes in contact with coal still in the soil. The surface water from these ditches follows a winding route to a stream, coming in contact with more coal en route. The surface water leaches iron pyrites from that coal, and this substance gives rise to the acid waters, and to yellow boy.

Acid mine drainage has historically been a regional problem. Pennsylvania, West Virginia, and neighboring coal states have had the greatest share of the problem. An estimate in the 1970s by the Environmental Protection Agency placed the total length of streams affected by acid mine drainage at more than 11,000 miles, mostly east of the Mississippi River; however, the surface mining of coal in the western states was only beginning in this period. Because so much of the nation's coal now originates in the west, the regional aspect of the problem of acid mine drainage is disappearing. Its control is a problem of national scope.

Solutions to the problem of acid mine drainage.
Even though coal-mining operations will undoubtedly continue and expand, a number of ways exist to prevent or reduce acid mine drainage. The methods are applied in different situations and are not equally successful.

Where an area is to be deep mined, it is helpful if

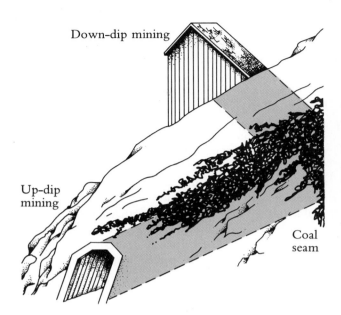

Figure 16.16 Two methods of reaching a coal seam. In down-dip mining, surface water may enter through the downward-sloping shaft or via groundwater flow. It may also condense on the walls of the mine because of humid conditions there. Such waters are likely to be safely trapped in an abandoned mine. In contrast, water that enters the up-dip mine via groundwater flow and through condensation flows out naturally down through the mine shaft itself. When such a mine is abandoned, it is extremely difficult to seal the entrance.

the mining company uses *down-dip*, as opposed to *up-dip*, recovery (Figure 16.16). The shaft of the down-dip mine slopes downward to the coal seam; the shaft of the up-dip mine angles upward to the coal seam. The pooled waters in a down-dip mine have difficulty escaping to surface waters.

Unfortunately, traditional methods of coal mining favor up-dip mining. In years gone by, mine shafts would be carved into the hillside at an upward angle so that coal could be loaded onto carts at the coal face and allowed to roll down to the outside of the mine. Water that accumulated in the mines from groundwater seepage, or that "wept" off the walls due to condensation, would simply run out of the mine by gravity. Abandoned mines are responsible for a high portion of the acid mine drainage that occurs.

How, then, can we control the pollution from long-abandoned mines? You might think one way would be to seal the entrance to the mine to prevent the escape of acid drainage. However, when the mine shaft rises upward at an angle from the entry, water accumulates behind the seal, and the tremendous pressure of a column of water can cause the seal to leak or give way. In contrast, where the mine shaft is level or where it falls away at an angle from the entrance, mine sealing has been effective.

Another option to control acid drainage from abandoned mines is to prevent surface water from entering the mines. This can be done by filling sink holes and by rerouting gulleys that may sometimes flow above an abandoned mine. If the stream channel can be routed around the mine site, surface water will be unable to seep into the mine. Another method of controlling acid mine drainage is to treat acid waters chemically with crushed limestone to neutralize the acidity. This last technique can be applied to acid drainage from both active and abandoned mines.

Surface mining compared to underground mining. Thus we see that the impact of coal mining on the environment is enormous. Where coal is mined underground, human safety is threatened by mine explosions and collapse. Underground miners are subject to black lung disease, a crippling shortness of breath. Acid drainage from underground mines pollutes streams and rivers.

Where coal is surface mined without adequate reclamation, unchecked erosion chokes streams with particles of earth. Mountains of rubble degrade the landscape, and highwalls ring the scalped hillsides.

Why is surface mining capturing such a large share of the coal-mining market? It is a matter mainly of economics: The price of surface-mined coal at the mine is roughly about half the cost of coal from underground mines. The low price of surface-mined coal is due to the technology of surface mining, which has eliminated a great deal of human labor. A surface "miner" can haul each day about twice the tonnage hauled by an underground miner. In addition, the special safety precautions to avoid cave-ins and explosions are not needed in surface mining, nor are the health precautions to restrict particles in the air. (See Controversy 16.1.)

The price of surface-mined coal has also been low because, until 1977, very few states required the reclamation of land that was scarred by the surface-mining process. The Federal Surface Mining Control and Reclamation Act of 1977 changed the rules for reclamation, requiring restoration of disturbed land. The difference in price due to a lack of adequate environmental protection in surface mining has now di-

minished somewhat, but the gap in price remains. Wherever transport costs do not alter the price structure, surface-mined coal will still be cheaper than coal mined from underground mines.

Transportation of Coal

Coal does not reach its destination without having an impact on the environment. Coal may move from mine to point of use in several ways: by train, by barge, by slurry pipeline, as synthetic natural gas, or as electricity. In recent years, the *unit train*, a train devoted exclusively to moving the huge quantities of coal needed by coal-fired electric plants, has come into use. Coal may also be moved by barge on a portion of the trip from mine to plant. The most recent development in coal transportation is the *coal-slurry pipeline*. In this system, a mixture of water and crushed coal is pumped from mine to point of use. Finally, the energy in coal may be transported by converting coal to a synthetic natural gas (SNG) or by converting it directly into electricity at a coal-fired generating station.

Train transportation of coal seems a harmless enough activity at first glance; after all, the trains move on existing lines along rail networks long established and accepted. In fact, the vastly increased movement of coal will have an impact. The coal car is an open car, so dust from coal is scattered along the right of way. Where trains pass through communities, the dust will affect people as it settles from the air. A 1,000-megawatt coal-fired electric plant has a coal requirement of about 12,000 tons (10,900 metric tons) per day. This is the quantity delivered daily by a single unit train of about 100 cars.

As an example of the impact of coal trains, the town of Littleton, Colorado regularly experiences the disruption of passing coal trains. In 1981, Littleton, a town of 28,000, was split daily by the passage of 35 trains, each carrying coal eastward out of Wyoming, and the number of trains was projected to reach 80 or more by 1990. Not surprisingly, people were considering leaving Littleton.

Coal trains may be moving enormous distances. Contracts have already been completed for the sale of Montana coal to a power plant in Paducah, Kentucky, on the Ohio River, and to Detroit, Michigan, on Lake Huron. Coal from Wyoming has been sold to power companies in Illinois and Indiana, where coal production is already large. These sales were probably based on the low sulfur content of western coal. A portion of

CONTROVERSY 16.1

Strip Mining: What Are the Implications of Expansion for Jobs? for the Environment? for Human Health?

[S]uperscale western mining means . . . between twenty-five thousand and forty thousand jobs will be lost in the East.

Arnold Miller

Black lung and silicosis are rare diseases among strip miners. Very few men are injured by rockfalls in surface mines. In an underground mine, if something goes wrong, there is nowhere to go; the miner is surrounded by rock . . .

David L. Kuck

As surface mining expands its share of the coal market, the number of mining injuries and fatalities declines. From the viewpoint of worker health, it appears that strip mining holds an edge over deep mining. Also, strip-mined coal costs less to produce than deep-mined coal. Yet Arnold Miller (1973), former president of the United Mine Workers, sounded like an environmentalist when he commented on strip mining:

> Fifty years ago they promised to develop Appalachia; and they left it a wreckage. Now they promise to develop the Great Plains. They will leave it a ruins.

As environmentalists, we hear Miller warning one section of the country of what the coal industry did to another, warning of the destruction of the land and of vegetation. In counterpart to Miller's concern for the environment, David Kuck (1974), in a letter to *Science*, asked:

> Which resources are we most concerned about conserving—human lives, the terrain, the vegetation, or the mineral?

Miller, however, has another motive for his warning, a motive he does not hide:

A headlong commitment to superscale Western mining means that over the next five years between twenty-five thousand and forty thousand jobs will be lost in the East. Of course, that concerns us as a union of miners. It concerns us also because we have lived through an unending depression in Appalachia . . . Finally it concerns us because you cannot turn underground coal production on and off like a light switch. If we arrive at a rational fuels policy (finally) and decide to strengthen our emphasis on Eastern mining, the mines will not be there, and neither will the miners.

Do you agree with Miller now that you understand his motivation? Defend your position. Do you think coal mining in the east should be fostered? If you do, can you think of ways to keep coal mining in the east alive?

Sources: Arnold Miller, *Center Magazine* (November/December 1973).
David Kuck, *Science*, **183** (January 11, 1974), 4120.

these long-distance trips may be undertaken by barge. Coal trains reaching the Mississippi River may transfer their loads to barges, which complete the trip to market. Coal trains may also unload at Lake Superior ports, leaving barges to finish the journey to the midwestern cities on the Great Lakes.

The coal-slurry pipeline is viewed as a serious competitor to the train and barge as a means of coal shipment. First attempted successfully near the turn of the century, coal-slurry pipelines are now attracting favorable attention from coal companies. Although a coal-slurry pipeline was delivering coal to London in 1914, not until 1957 was a 100-mile (160 km) pipeline for coal established in Ohio, because of the state's high rail rates for coal. Extending from Cadiz, Ohio to Cleveland, Ohio, the 10-inch (25 cm) pipeline was capable of moving 1.3 million tons (1.2 million metric tons) of coal each year. The pipeline closed in 1963 because local rail rates had finally become competitive.

The pipeline had been well publicized, however, and its success encouraged another installation in Arizona. There, the Peabody Coal Co. was to supply a power plant in the southern tip of Nevada, some 270 miles (432 km) from the mine. Because of the rugged terrain, direct rail haul was considered unfeasible. A roundabout routing by railroad, avoiding the mountains, would have been too costly. Completed in 1970 by the Black Mesa Pipeline Co., the pipeline now carries a river of water and finely ground coal through the deserts and mountains of the rugged southwest. The mixture is about 50% water and 50% coal. With the Black Mesa experience in mind, companies have been planning more coal-slurry pipelines.

Unfortunately, surface water is scarce in the west; its most important uses there are for irrigation, community water supply, and recreation. (See Controversy 16.2.) Water destined for these purposes should not be diverted for coal-slurry pipelines. As a consequence, the Black Mesa pipeline, as well as the pipeline proposed from Wyoming to Arkansas, draws water from deep wells rather than consume precious surface water (see Table 16.2).

Even in the face of water shortages, the slurry pipeline remains attractive because of its comparatively low cost. When the coal must move long distances and a railroad bed does not yet exist, rail haul will not compete. Neither will rail/barge shipment, nor the transmission of electricity.

Health and Safety of the Coal Miner

Mine Safety

Coal mining, more than any other process of fuel resource extraction, has an impact on human life itself. We often read and hear of mine disasters in which rescue operations have failed, of cave-ins or coal mine explosions. Several hundred coal-mining deaths occur each year. It is a grim, grimy, and dangerous business. Although statistics tell us that mines are getting safer, coal mining is still one of the most hazardous jobs in modern times. Some dangers are immediate—a loading car may break loose and careen down a slope, striking a miner. Other dangers are delayed: The sooty air, laced with coal particles, may implant in a miner's lungs and eventually destroy his ability to breathe.

About 60% of the injuries in underground coal mining are related to roof falls and face falls (landslides in miniature). About half of the deaths in underground coal mining stem from roof falls alone. Hauling coal underground is also extremely hazardous, and accidents are often fatal. About 15%–20% of underground accidents occur in hauling operations. Explosions and fires account for another 15%–20%. Small mines, with 50 or fewer employees, experience a fatality rate three times that of larger mines.

Explosions in coal mines result from the ignition of methane gas (though coal dust itself can explode). Pockets of methane commonly occur in underground coal deposits. If these pockets go undetected and gas

Table 16.2   Comparison of Water Consumed in the West for Different Uses of Coal[a]

Coal uses[b]	Thousands of acre-feet consumed per year[c]
Rail transport	less than 1,000
Coal-slurry pipeline	9,200
Coal gasification	28,000
On-site electric generation	44,000

Source: U.S. Geological Survey, Open File Report 77-698, August 1977.

a. The figures indicate acre-feet of water consumed for each 12.5 million tons of coal that are either transported by rail or slurry, or consumed in coal gasification or in on-site electric generation.

b. Data are for Yampa River Basin.

c. One acre-foot = 3,050 cubic meters.

CONTROVERSY 16.2

What Should Be Done with the Coal in the West?

Assuming coal is to be mined, and there is every indication that it will continue to be, the options are as follows:

1. ship it east and west to population centers via train or coal-slurry pipeline;
2. burn the coal in generating stations in the west and transmit the electricity to the population centers;
3. gasify the coal, removing sulfur, and send the gas via pipeline to the population centers.

If the coal is shipped to population centers, it will be burned there. Even if scrubbers are installed, the total pollution burden on individuals living in the cities will be increased. Not only will pollution levels rise, but the number of individuals exposed will increase as well. The pollution impact of burning the coal is thus more serious; however, even when electric utilities build their plants in the remote areas of western states, they are being forced by federal government regulation to install such devices as scrubbers and precipitators.

On the other hand, a task force, not connected with the government, composed of industrial executives and environmentalists, recommended that new coal-fired power plants be built in the regions where the power will be consumed. That is, the people who get the electricity are the ones who should have to live with the environmental effects of generating it. The issue is whether to keep clean areas clean or to allow clean areas to be polluted in order to prevent polluted areas from being degraded still further.

There are other issues as well. The panel of the National Academy of Sciences that studied strip mining in the west referred to the huge water demands that would be placed on the arid west if energy development takes place in that region:

If water resources are to be committed to large electrical generation and coal gasification plants, huge supplies of water will be necessary. Based on current technologies and projected rates of consumption many power plants are planned for an operating life of less than fifty years. What will be done with the proposed aqueducts, dams, and water when the coal economy no longer needs them? Does society wish to commit water resources to an ephemeral industry at a particular place?

Water demands aside, it has also been pointed out that if coal is to be gasified (made into a synthetic natural gas) and sulfur removed in the process, there is no reason why gasification should not take place in the east. Not only could the coal be gasified in the east, where water is more abundant, but eastern coal could be used rather than western coal. Thus, water is not taken from a place where it is scarce, and neither gas nor coal need be transported long distances. Unfortunately, coal gasification plants also produce air pollution, again placing the pollution burden on the east.

These questions of policy are not easily settled; the debate will continue. Where do you stand on the issue of where energy development should go and why? Do you think the plants should go in remote areas of the west? Should they have pollution control equipment? Why did the task force feel that people who live in population centers should bear the pollution burden?

Source: National Research Council. *Surface Mining: Soil, Coal and Society*. Washington, D.C.: National Academy Press, 1981.

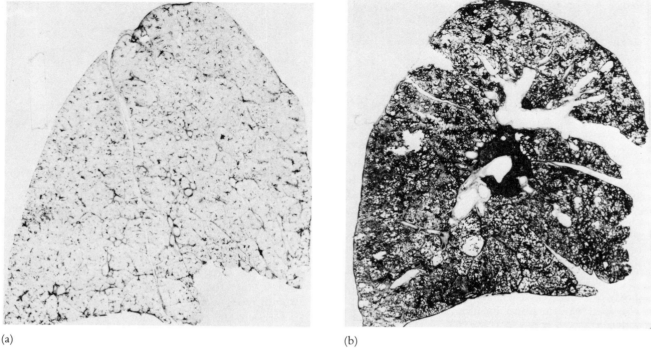

(a)   (b)

Figure 16.17 A comparison of the lungs of a nonminer and a coal
miner. (a) A section from the lung of a man who died at age 86 and was
never a coal miner. (b) A section from the lung of a man who died at
age 78 and was an underground coal miner for 36 years. Although
the man did not smoke, he suffered from severe emphysema, a lung
disease in which breathing is badly impaired. (Photos courtesy
of Dr. Frank Green, Chief, Pathology Section, Appalachian
Laboratory for Occupational Safety and Health)

builds up in the mine, a simple spark can detonate an
explosion. While the explosion itself can injure and kill,
the rock falls that result may be even more hazardous,
sometimes sealing miners inside a portion of a mine.
Explosions from methane gas can be prevented by adequate ventilation of the mine, but to be sure the mine
air is safe, methane detection equipment must be used.

The high accident rate in the coal industry leads to
a great number of disability cases. Unable to work,
injured former miners may also require extensive medical attention. Not only do accidents contribute to disability, but working conditions do as well. Performing
heavy labor in a stooped position and crawling on
hands and knees both lead to inflammations of the
joints—conditions known as *beat hand* and *beat knee*.

Black Lung Disease

Accidents, fatalities, crippling deformities—and finally, we come to **black lung**: a wheezing and shortness of breath, an inability to climb stairs or perform
labor, a disease so disabling that some victims are only
able to sit in chairs, their lungs destroyed by the particles they inhaled. The public and the coal miners call
this condition black lung disease because the lungs of
miners who die from it are black (Figure 16.17). The
physician knows the disease as *coal workers' pneumo-
coniosis* (pronounced new-mō-cō-ni-ō'-sis) and abbre-
viates the name as CWP.

The disease is caused by the inhalation of dust composed of minute particles of carbon and rock. The car-

bon particles account for the black color of the lungs. Fibers grow within the lungs around the sites where particles have deposited. These fibers can, if they grow extensively within the lung, destroy its normal elasticity. The elasticity of the lung, its ability "to bounce back," is what makes breathing easy for most of us. A miner's lung, overgrown with fibers due to coal dust, lacks the ability to bounce back. For him, breathing is torture. There appears to be no treatment for black lung, only for the bacterial infections that are side effects of the disease; that is, the symptoms of black lung cannot be relieved, and the disease cannot be cured.

The risk of black lung disease appears to increase with the quantity of coal dust inhaled. Two factors influence this quantity significantly. First, the greater the number of years spent underground, the greater the quantity of dust inhaled. The specific mining job—whether dusty or relatively clean—also influences the risk of contracting the disease. Cutting-machine operators show a larger risk compared to workers stationed outside the mine. To bring the number of new cases down, the levels of dust in the mine must be brought under control. Although a federal standard for dust in coal mines has been set, the extent of compliance with the standard is not clear because of the absence of checks on company data.

How many individuals are afflicted with black lung disease? The question requires answers on a number of levels. Although there were about 400,000 working coal miners in 1947, mechanization of the mines had reduced the number of miners to 220,000 by 1982. However, in the 1920s there were about 700,000 active coal miners. This means that many men have retired from the mines in the past 30 years or so. We need to know this fact to understand why over 250,000 claims of black lung disease have been approved by the Social Security Administration. These claims were submitted either by miners or their widows, and more claims are continually being approved. As of 1978, more than $1 billion per year was being paid in benefits to coal miners or their widows. These are tax dollars, not dollars from the coal industry, where the miners contracted their disease.

A bill to amend the 1969 Black Lung Law was enacted by Congress in 1978. Although the original law called for the coal industry to pay black lung compensation, the coal industry successfully fought this law in court, and it was never implemented. The 1978 Black Lung Benefits Reform Act established a Black Lung Disability Trust Fund, which comes from a tax on each ton of coal produced. Money from this fund would be used to pay any newly established claims of black lung disease.

Impact of the Coal-Fired Electric Plant

Coal, as we pointed out earlier, is used mainly in the generation of electricity. The burning of coal produces carbon dioxide but only a small amount of carbon monoxide. Carbon dioxide is of concern because of its hypothesized effect on the earth's climate. (See Global Perspective V for further discussion of carbon dioxide and climate.) The three most important pollutants in the stack gases of coal-fired power plants are sulfur oxides, particulate matter, and nitrogen oxides.

Sulfur oxides injure plants, materials, and people (see Chapter 20). Sulfur dioxide was present in the air in great quantities during the air pollution disasters in Donora, Pennsylvania, in the Meuse Valley in Belgium, in the infamous "London Fog," and in episodes in New York City. A single 1,000-megawatt coal-fired electric plant burns 4–5 million tons of coal a year. If the coal is 2.5% sulfur (not an unusual level), the plant will discharge 200,000–250,000 *tons* of sulfur dioxide per year. As the nation uses more and more coal, the sulfur dioxide problem will multiply unless something is done. That "something" consists of two steps, which may be used in combination: ridding the coal of its sulfur by cleansing it, and "scrubbing" the stack gases (see Chapter 20).

Particles, or particulate matter, are "the partners in crime" of sulfur oxides. Particles are known to aggravate human respiratory problems. Particles discharged into the atmosphere when coal is burned at power plants are known as fly ash. If the emissions of fly ash were not controlled by electrostatic precipitation, the annual production of particles from a 1,000-megawatt plant could be as high as 200,000–250,000 tons. Instead, because particles are so well controlled, only about 10% or less actually reaches the atmosphere.

Nitrogen oxides are also produced when coal is burned at electric power plants. Although nitrogen oxides have direct effects on our health, they are noted primarily for their interaction with hydrocarbons to produce ozone, which aggravates diseases of the respiratory tract. Control of nitrogen oxides appears to be a very difficult technical problem.

Coal also contains arsenic in small quantities. Arsenic, a carcinogen, is released in fly ash or particulate matter when coal is burned. An estimated 1,000–3,000 tons of arsenic are released annually from coal burning. If an electric plant has a scrubber or an electrostatic precipitator, 70%–90% of the arsenic is removed.

Selenium is also present in small amounts in the fly ash discharged by coal-fired power plants. Much of this selenium seems to appear in the smallest fly ash particles, which often elude capture by electrostatic precipitators. High doses of selenium in grasses ingested by cattle would be injurious.

The last category of emissions from coal-fired power plants is unexpected: radioactive elements and their decay products, which are found naturally in fly ash. The quantity of radiation from a coal-fired electric plant is very low, less than 1% of natural background radiation, about the quantity of radiation from a nuclear plant operating under the stringent standards set in the mid-1970s (McBride et al., 1978).

Summary

The four basic types of coal are (from low to high in carbon content and, thus, in heating value) peat, lignite, bituminous, and anthracite. Two major impurities in coal are sulfur and ash. Of the total tonnage of U.S. coal, 55% of the reserves are in the west, mainly as lignite and subbituminous, which are often low in sulfur. About half of western coal must be mined underground; the other half is surface mineable.

In the east, 80% of the coal (mainly bituminous, high-sulfur coal) is in deposits that require underground mining. Most U.S. coal (84%) is used for electricity generation. Coal resources in the United States appear sufficient for the next several centuries, even with the projected 30% increase in use between 1985 and 1995.

Surface mining of coal is cheaper than deep mining. The two basic types of strip mining are known as contour mining and area mining. Both types, unless the land is properly reclaimed, can lead to useless wasteland that is unable to support vegetation. Soil sterility, erosion, acid drainage, and isolation of areas by highwalls are all problems that occur when strip-mined land is left unreclaimed. The most important factor in reclamation is replacement of topsoil over the exposed strip-mined earth. However, where rainfall is sparse, as in semiarid or desert areas, reclamation that allows a return of the former vegetative cover may not be possible. The 1977 Surface Mining Control and Reclamation Act requires reclamation of strip-mined lands and forbids highwalls or dumping of overburden on steep slopes. A tax on coal is also now collected to help repair previously strip-mined lands. In the period from 1980 to 1988, news reports indicated that the Department of the Interior did not appear to be enforcing the Act as vigorously as environmentalists would prefer.

Two methods of underground, or deep, mining are room-and-pillar mining and longwall mining. Deep mining is much more hazardous to miners than surface mining. Underground miners are exposed to explosions and cave-ins. They may be crippled as a result of working conditions, and they breathe coal dust that eventually causes black lung disease. Other environmental problems from coal mining include the formation of refuse banks, which can catch fire or which can lead to flash floods if they collapse; water pollution from coal washing; and subsidence of land over deep mines that collapse. Transportation issues include the air pollution problem of coal dust blowing from open train cars and the use of scarce water in western coal-slurry pipelines.

Acid mine drainage occurs from both surface and deep mines as water dissolves iron pyrite (ferrous sulfide). This compound reacts in water to form sulfuric acid and yellow boy (ferric hydroxide). Dissolved oxygen is removed from the water as a consequence of the reactions. This low-oxygen, high-acid water, runs in streambeds coated with ferric hydroxide and causes aquatic life to be destroyed. Some mines can be sealed to prevent acid mine drainage; in other cases, surface water can be diverted to prevent it from entering mines. Acid water can also be treated with crushed limestone to neutralize the acidity.

When coal is burned, it produces many air pollutants, including carbon dioxide, sulfur oxides, particulates, nitrogen oxides, arsenic, selenium, and radioactive materials.

Questions

1. What are the four types of coal, ranked by heating value? Which is most abundant in the United States?

2. Why are lignite and subbituminous coal shipped from western states to the midwest and farther east? Give two reasons.

3. Where are the majority of surface-mineable coal reserves located: in the eastern United States or in the western United States? Where are the reserves of low-sulfur coal concentrated: in the eastern U.S. or the western U.S.?

4. What single use consumes two-thirds of all coal mined each year in the United States?

5. Why is surface-mined coal gradually replacing coal from deep mines? You may wish to give several reasons, but one is most important.

6. What are the two basic kinds of strip mining, and how does a mine operator decide between them?

7. What is a highwall, and how does it come about?

8. When strip-mined land is not reclaimed, that is, when it is left scarred after mining, a number of problems occur. List and briefly explain three of the problems.

9. Strip mining is on the increase in the west. What are the two vegetative areas with the least probability of being successfully reclaimed?

10. The Surface Mining Law passed in 1977 calls for reclamation of lands previously stripped. Where will the funds come from to do this?

11. What is acid mine drainage?

12. What environmental problems does acid mine drainage cause?

13. Why could you say that acid mine drainage is one of the environmental costs of generating electric power?

14. How does coal dust affect a miner's health? Why do the miner's lungs lose their elasticity?

Further Reading

Ackenhail, Alfred. "Pennsylvania Erases Its Mining Scars," *Civil Engineering Magazine* (October 1970), p. 54.
    A short but extremely well-written article that covers causes, effects, and simple control approaches.

Lackey, James. "Aquatic Life in Waters Polluted by Acid Mine Waste," *Public Health Reports*, **54** (1939), 740–746.
    A classic article identifying the nature of the problem of acid waters. It is nicely written and well illustrated.

Coal—Bridge to the Future, Report of the World Coal Study. Cambridge, Mass.: Ballinger, 1980.
    Probably the most comprehensive study of coal on a worldwide basis.

Perry, H. "Coal in the United States: A Status Report," *Science* (October 28, 1983), p. 377.
    Harry Perry is a reknowned expert on coal, and this article is complete and authoritative.

National Research Council. *Surface Mining: Soil, Coal and Society*. Washington, D.C.: National Academy Press, 1981.
    This clearly written volume by a committee of the National Academy of Sciences updates and expands *Rehabilitation Potential of Western Coal Lands*, a previous effort of the National Academy of Sciences.

Surface Mining and Our Environment. Washington, D.C.: U.S. Department of Interior, 1968.
    Vivid color pictures of strip mining and of the resulting erosion and acid mine drainage.

The American Coal Miner. The President's Commission on Coal, 1980.
    A report on community conditions and living conditions in the coal fields. Many photographs. Covers housing, health, safety, black lung disease, company towns. The meaning of coal as a resource is expanded by this document.

U.S. Environmental Protection Agency. "Energy from the West: Summary Report," EPA 600/9-79-027 (August 1979).
    This report explores in some detail the environmental, economic, and social issues associated with western energy development.

References

Kash, D., et al. *Our Energy Future*. Norman: University of Oklahoma Press, 1976.

"Demonstrated Reserve Base of Coal in the United States on January 1, 1980," Report of the Energy Information Administration, U.S. Department of Energy, DOE-EIA-0280(80).

McBride, J., R. Moore, J. Witherspoon, and R. Blanco. "Radiological Impacts of Airborne Effluents of Coal and Nuclear Plants," *Science*, **202** (December 8, 1978), 1045.

National Academy of Sciences. *Rehabilitation Potential of Western Coal Lands*. Cambridge, Mass.: Ballinger, 1974.

Miller, Arnold. *Center Magazine*. (November–December 1973).

Kuck, David L. *Science*, **183** (January 11, 1974), 4120.

Heckler, Kenneth. *Regulation of Surface Mining*, Part I. Hearings before Subcommittees of the Committee on Interior and Insular Affairs, House of Representatives, Serial 93-11, 1973.

CHAPTER SEVENTEEN

Oil and Natural Gas

A Brief History of Oil in the United States

In the decade just preceding the Civil War, a black oily liquid was being collected from salt wells near Pittsburgh, Pennsylvania. The producers originally bottled the liquid as a medicine, but some of the liquid was processed by a small "refinery." The refinery could produce five barrels of "carbon oil" each day for use in oil lamps. This type of carbon oil competed with whale and coal oil for the lamp oil market, eventually capturing the lighting market.

From these modest beginnings, a little over a century ago, the oil industry has grown to be the largest industry in the world, dominated by giant multinational companies. The principal instrument, or rather vehicle, propelling this growth has been, of course, the automobile. This turn-of-the-century invention has accounted for a steadily growing demand for oil through most of this century.

Until the 1950s, the United States produced nearly all the oil it consumed. In that decade, the United States found itself importing cheap foreign oil in larger and larger quantities. Even into the early 1970s, oil on the world market was only about $2 per barrel, and by the late 1970s the proportion of U.S. demands met by imports was approaching 40%.

The 1958 discovery of oil on the North Slope of the Brooks Range in Alaska brought a dozen oil companies exploring for the riches buried in that frozen land. The nation decided to develop Alaskan oil as a means of establishing greater independence from for-

eign sources. Alaskan oil extended our proved reserves by nearly 25%, or 10 billion barrels. In the mid-1970s, the Trans-Alaska pipeline was cut across the Brooks Range to the Alaskan port of Valdez for shipment of oil via tanker to West Coast ports. When the first Alaskan oil finally arrived, the West Coast did not need it (as predicted by those who advocated a pipeline through Canada to the midwest). A new pipeline to carry Alaskan oil inland may yet be built.

In 1978, drilling began offshore in the eastern United States, but no commercial finds of oil or gas were reported. Large oil fields were found off the coast of eastern Canada. Offshore finds and production continue in the Gulf of Mexico and off the coast of California.

Finding and Producing Oil and Natural Gas

We link oil and natural gas together not only because they are often found together geologically but because these two preferred fossil fuels are both in short supply in this country and require imports to fill demands. We also link these two fuels because conservation efforts can help make them last longer.

Liquid and gaseous hydrocarbons, known collectively as **petroleum**, are found in underground reservoirs within sedimentary rock; both liquids and gases often occur together. Whether they are present in any particular sedimentary rock is always unknown until drilling is undertaken. This uncertainty makes predictions of recoverable oil in the world a subject of sharp controversy.

Because oil reservoirs are deep underground, precise knowledge about the limits of a field and the volume of oil in place are not available. A large number of wells, many of them dry holes, would have to be drilled at considerable expense in order to define the field size precisely. Quite naturally, petroleum engineers tend to estimate volumes of oil on the low side so that their recommendation for the development of a field is sure to have a positive payoff.

The quantity of oil that flows out of the well by natural pressure alone is referred to as **primary recovery** and averages about 20% of the oil in place. Since about 1960, petroleum engineers have developed methods to force up more of the petroleum. These **secondary recovery** techniques involve either injection of water beneath the oil or injection of gas above the reservoir to place increased pressure on the oil in place. Secondary methods can increase the yield of oil reservoirs to 50% or 60% of the oil in place. Newer *tertiary* methods, such as injection of carbon dioxide, nitrogen, or steam, may push recovery to the 90% level.

What Are Oil and Natural Gas?

Scientists believe that petroleum was derived from the algal **plankton** in the ocean. In much the same fashion that coal was derived from woody land plants, buried plankton are broken down by bacteria in the absence of oxygen. Chemical and physical processes, operating over many thousands of years, further convert the organic material into petroleum compounds.

Oil consists mainly of liquid hydrocarbons, whose only elements are carbon and hydrogen. About 90%–95% of oil by weight is usually hydrogen and carbon, about 80% or more being carbon alone. Sulfur and oxygen may each account for up to 5% of the oil's weight. Oil with a sulfur content of less than 1% is referred to as *sweet crude oil*. High-sulfur oil, because of the odor of hydrogen sulfide, is called *sour crude oil*.

Oil is distilled, or *refined*, in a distillation column where trays are used to separate the components (Figure 17.1). Petroleum gases exit the top of the tower; liquids collect at various levels in the tower. The components that boil first condense at the topmost tray. Those that boil last, the tars and pitches, collect at the bottom of the column. On the trays between are the

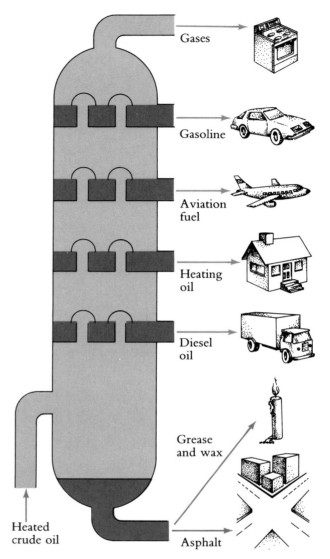

Figure 17.1 The refining of crude oil into its components. The components collect at levels that reflect how easily they boil. Gasoline and dissolved gases boil out first, followed by aviation fuel and heating oil. The most difficult components to boil, the tars and pitches, collect at the bottom of the refining tower. (Adapted from *The Open University, The Earth's Physical Resources, Block 2, Energy Resources*, The Open University Press [Walton Hall, Milton Keynes, United Kingdom, 1973].)

various components, separated in an order reflecting the ease with which they boil.

Natural gas, in contrast to oil, may be as little as 65% carbon by weight; its hydrogen content is variable. Although sulfur content is usually low, nitrogen levels may be high, up to 15%. The same kinds of sediments and geological conditions that give rise to oil also give rise to natural gas. It is not unusual, therefore, to find oil and gas together. Often the gas is found dissolved in the liquid hydrocarbons. In this situation, the gas is separated from the oil when it is brought to the surface. Gas is also often found trapped above the liquid petroleum; the industry calls this *gas cap* gas. Gas is most often found alone, but about 25% of natural gas is found in the search for oil.

Oil and Gas Resources

Oil Resources of the United States

If we were to ask the question, "How much oil is there in the United States?" the answer would require at least two parts. First, the proved oil reserves in the United States were estimated at 27.3 billion barrels as of January 1, 1985. The term **proved oil reserves** means that this oil is known to remain in portions of the oil fields where production drilling has already begun. Furthermore, we are assured that this oil is profitable to recover. These proved reserve quantities, however, have been falling in the United States since the early 1960s because new discoveries have failed to keep pace with our rate of production.

The second part of the answer would be: There is actually more oil than this, but for one reason or another we cannot count this oil at the present time. On the basis of experience, we can estimate small portions of oil resources as quite likely to be recoverable when new but proven methods of recovery are utilized. These are called **indicated reserves**. Other larger portions of resources, **inferred reserves**, can be estimated on the basis of drilling to date; that is, we already have limited evidence of their existence. Finally, much larger quantities, for which we have much less assurance, are **undiscovered resources**. The latest available published estimates of these quantities for the United States are displayed in Figure 17.2. Included in this figure are not only undiscovered resources but also indicated, inferred, and proved reserves, as well as oil extracted so far in the United States.

The quantities of undiscovered oil may be estimated on the basis of geological evidence, by the use of statistical methods, by extrapolation of past trends, or by combinations of these. Unfortunately, these procedures can lead to very different estimates of the oil remaining to be discovered. Some of the most recent, widely cited studies provide estimates of undiscovered liquid hydrocarbons ranging from 72 billion to 200 billion barrels (*Energy: The Next Twenty Years*, 1979). The lowest estimate (72 billion barrels) of unproven U.S.

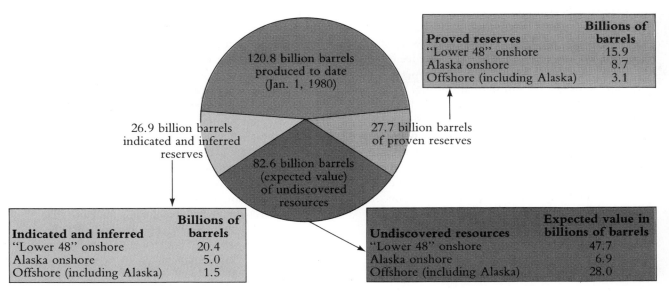

Figure 17.2 Oil resources of the United States by region and category. Data as of January 1, 1980. (Data from U.S. Geological Survey Circular 860, 1981.)

oil reserves was provided by M. King Hubbert of the U.S. Geological Survey (See Box 17.1).

There is one further type of oil resource: oil not presently profitable to produce. Some is already discovered; the remainder is yet to be discovered. We mentioned earlier that North Sea oil became profitable after the price of oil jumped from $2 per barrel to $12 per barrel. In this regard, we do not yet count the 600 billion barrels of oil that can be derived from oil shale in any of the above categories because it is not yet prof-

Box 17.1 A Fish Story

M. King Hubbert, who provided the lowest recent estimate of U.S. oil reserves remaining to be discovered, is one of the most renowned energy analysts in the world. In a chapter of a Congressional Committee Print, *Project Interdependence*, Hubbert explains the philosophical basis of his calculations in an amusing way:

The problem of estimating the ultimate amount of oil or gas that will be produced in any given region has been aptly likened to that of estimating the abundance of fish in a lake by the success in fishing. Although the fish in the lake are not seen by the fisherman, if he catches a fish by almost every cast the inference is justified that the lake is teeming with fish. If, after such a lake has been heavily fished by vacationers for a number of seasons, the same angler is able to catch only a fish or two per day, he is justified in the inference that the lake has been about fished out.

Similarly, in a virgin petroleum-bearing region, if most of the initial exploratory wells succeed in discovering oil fields, the inference is justified that the given region is rich in petroleum resources. When, at a more mature stage of exploratory drilling, it is found that a steadily decreasing fraction of the exploratory wells drilled succeed in finding oil, the inference is unavoidable that most of the oil in the region has already been discovered, or that the lake is about fished out.

However, in the case of the lake, if fishing is discontinued for a few years, the lake will become restocked, whereas in the oil-field analogy this cannot happen because oil fields do not breed. There was only a fixed and finite number of oil accumulations in the sediments of the region initially and every time one of these is discovered, the number remaining is reduced by one. It is inevitable that as the remaining fields become fewer in number, and probably also deeper and smaller in size, the difficulty of making a discovery must accordingly increase. Conversely, this record of fewer discoveries with increased exploratory drilling affords one of our more reliable means of estimating how far we have progressed toward the ultimate discoveries likely to be made in the region.

To many of us, a few fish per day sounds like a great deal, so Hubbert must have some very special fishing spots. Of course, Hubbert is right about how we might respond to a lake that once yielded fish in plenty and now provides only a few per day. Suppose, however, that we had a cabin on the lake and a canoe with which to explore. How quickly would you as a fisherman abandon your lake? Would you take other actions first in terms of tackle and bait? What actions? How about in terms of exploration? What advanced methods might you think of to locate the fish? Might you consider catching fish that were less desirable than those in your earlier catches? In the context of oil, a less desirable fish might be oil shale. Can you think of others? Perhaps a different way of cooking would make them better to eat. You begin to understand now how the oil companies have responded to their fewer successes on the U.S. mainland.

Hubbert's story gives us another insight about oil. Because oil exploration is likened to fishing, we are drawn to a comparison of fish to oil. Fish is a renewable resource; unless we remove nearly all the fish, the fish will breed and repopulate the lake. Oil is a non-renewable resource; each removal is final and irrevocable. A much better energy resource on which to rely is one that is renewable, such as solar energy.

Source: Project Interdependence, Chapter 19, "World Oil and Natural Gas Reserves and Resources." Washington, D.C.: U.S. Government Printing Office, 1977.

itable to produce. The heavy oil available from tar sands in Venezuela is another example of oil not yet profitable to extract and refine.

In the early 1980s, a remarkable search for oil began in the Arctic Ocean, north of Alaska and Canada. The international oil companies invested tremendous sums to find and produce oil in one of the most inhospitable environments on earth, where ice breaks up and re-forms annually, where subzero temperatures and winds gnaw at workers, and where waves threaten structures built to find and withdraw the oil.

However, from 1982 through 1985, the price of oil was falling as the OPEC (Organization of Petroleum Exporting Countries) cartel felt the pinch of a worldwide recession, as competition increased from other oil-producing nations, and as conservation measures took hold. The falling price of oil and the huge investments required have now slowed the search for oil in the Arctic offshore—indeed, they have slowed the search for oil throughout the world.

United States and World Use of Oil

In 1984, world oil production was about 53 million barrels per day. Of this total, the United States used 15.7, or 30%. The United States previously consumed an even greater proportion of world oil production, but other nations have increased their consumption dramatically, thus decreasing the U.S. share.

In 1979, U.S. consumption was 18.51 million barrels of oil per day. In 1981, in response to rapid increases in the world price of oil and a severe recession in the United States, this figure fell to 16 million barrels. By 1984, conservation measures had taken hold, and oil use had declined to 15.7 million barrels, even though the price of oil was still falling from its 1979 high.

In 1984 gasolines accounted for 43% of consumption of refined petroleum products; middle distillates, which are used for home heating oil, diesel fuel, and kerosene (jet fuel), made up 18% of consumption; and residual fuel oil, which is used for heating commercial and industrial buildings and as a fuel in a decreasing number of steam electric plants, accounted for 9% of oil consumption (Figure 17.3). The total of all other uses, including asphalt, plastics, chemical manufacture, and so forth, gives us the remaining 30%.

Although the United States used 15.7 million barrels of oil per day in 1984, it produced only 10.3 million barrels per day. The remainder, or 5.4 million barrels

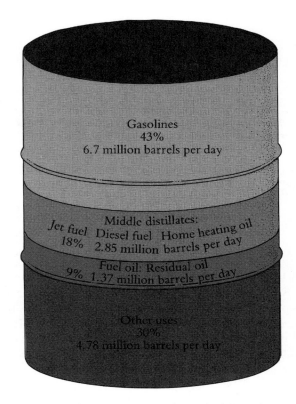

Figure 17.3 Oil consumption in the United States by use category in 1984. (Source: Energy Information Administration.)

per day, had to be imported. Such levels of imports (between 30% and 40% of total use) have been common since the early 1970s. The contribution of imports and U.S. production in 1984 are illustrated in Figure 17.4.

The pattern of consumption and crude oil importation for the United States changed markedly from 1978 to 1983. Consumption and imports both fell—consumption by about 20% and imports by about 30%. At the same time, the source of U.S. imports also changed significantly. In 1978, over 40% of our oil imports were of Middle East origin. Only about 12% of oil imports came from the Americas. By 1983, imports of Middle East oil had fallen to about 12% of our total oil imports, and oil from countries in North and South America accounted for 35% of the total oil imports by the United States.

Proved reserves of oil are not distributed equally throughout the world. Figure 17.5 shows the inequity. As we can see, the Middle East has by far the greatest

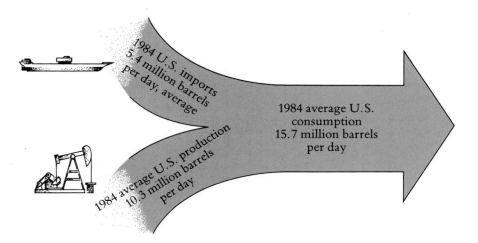

Figure 17.4 Production, imports, and consumption of oil in the United States in 1984 (figures are rounded). (Source: Energy Information Administration.)

proved reserves in the world. Nonetheless, discoveries and production in other parts of the world have been increasing since the price hikes of the 1970s. The dominance of the Middle East in the world oil market has thus been reduced, to a degree. However, the depletion of other nations' resources is likely to lead eventually to a renewed dependence on the vast oil resources of the Middle East; that is, such renewed dependence is likely unless such actions as conservation and alternate energy development are taken in the present. Also shown in Figure 17.5 are estimates of the undiscovered resources by world region.

The future statistics on U.S. oil production and oil importation are uncertain for a number of reasons. If imported oil is taxed, if OPEC raises its prices steeply once again, or if major oil discoveries occur offshore in the United States, domestic production could climb. A shale-oil industry is still a distinct possibility, though now more distant.

If the economy is not healthy and many people are out of work, the use of the automobile will diminish as

people try to save money. This was seen in the period following the 1973–1974 embargo, when the country went through an economic slump because of inflation in the price of oil and other goods. A decrease in automobile use, and hence in the use of gasoline, was seen again in 1982 when a severe recession gripped the nation. Any combination of these various events could occur, which explains the enormous uncertainty in predicting the U.S. oil future.

Natural Gas Resources, Production, and Use

We can describe natural gas resources in the United States in the same way we described oil resources. Proved reserves are in place and profitable to produce. There are indicated and inferred reserves for which we have some degree of evidence, and finally, undiscovered resources are yet to be found. Table 17.1 provides estimates of natural gas resources in these various cate-

Table 17.1 Natural Gas Resources of the United States[a]

Location	Production to date	Proved reserves	Indicated and inferred	Undiscovered (expected value)
Lower 48 states	519.3 (14.69)	123.3 (3.49)	132.1 (3.74)	390.2 (11.04)
Alaska onshore	1.2 (0.03)	30.0 (0.85)	4.4 (0.12)	36.5 (1.03)
Offshore (including Alaska)	57.5 (1.63)	38.2 (1.08)	41.0 (1.16)	167.0 (4.73)
Total	578.0 (16.4)	191.5 (5.42)	177.5 (5.02)	593.7 (16.80)

Source: U.S. Geological Survey Circular 860, 1981.
a. In trillions of cubic feet (trillions of cubic meters in parentheses).

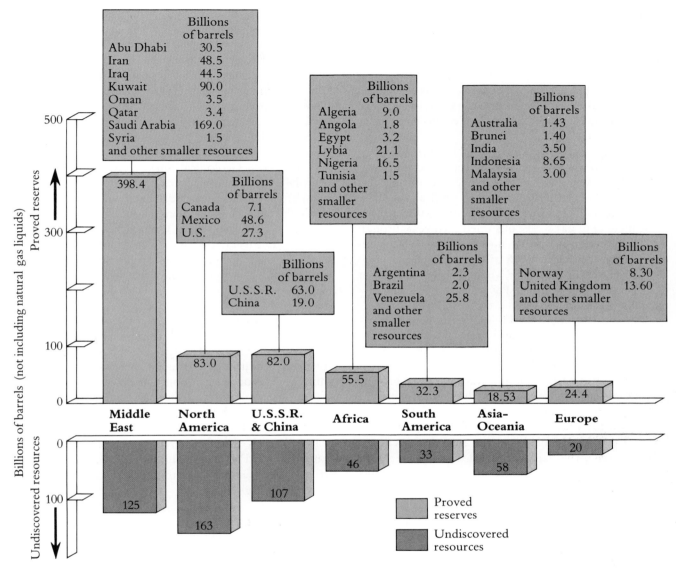

Figure 17.5 Distribution of world crude oil, 1985. (Reserves are from *Oil and Gas Journal* and are as of January 1, 1985. Undiscovered resources are from Open File Report 85-248 of the U.S. Geological Survey.)

gories and indicates the potential in Alaska and offshore for natural gas.

Proved reserves of natural gas were falling through most of the 1970s. The natural gas industry explained this phenomenon as being the result of government control of the price of natural gas sold in interstate commerce. The industry argued that the regulated

price did not allow them enough profit to justify looking for new sources of gas. Finally, in 1978, the industry convinced Congress that an end to control was necessary. In that year, Congress passed, and the President approved, a bill that would phase out all price controls on natural gas by 1985. Since then, the price of natural gas has risen, as all predicted. The average price to the consumer jumped from $2.56 per 1,000 cubic feet in 1978, to $3.68 in 1980, to $5.17 in 1982, and reached $6.12 in 1984. The gas being purchased by most consumers was a mixture of old gas (controlled in price) and new gas (higher priced), and its price rose as new gas replaced old. Since all controls on natural gas ended in 1985, the price of natural gas is expected to continue its rise in the second half of this decade (see Controversy 17.1).

Although natural gas reserves fell during most of the 1970s, from 1978 to 1983 they remained fairly stable, within 5% of 200 trillion cubic feet. Consumption of natural gas was fairly stable also until about 1980. Between 20 and 22 trillion cubic feet (0.51–0.62 trillion cubic meters) of gas were consumed each year through the 1970s, but a steady decline began in 1980, with consumption falling to 17.65 trillion cubic feet in 1983. Of this annual consumption, about 1 trillion cubic feet (0.028 trillion cubic meters) were imported annually from Canada by pipeline.

In the early 1980s, the United States began to import natural gas by tanker, principally from Algeria. In order to save space during shipment, the gas was made liquid by cooling it and putting it under pressure. At the coastal terminals, the liquid was converted back to gas and injected into pipelines that already carried U.S.-produced natural gas.

Although liquefied natural gas (LNG) was still being received in small shipments at U.S. coastal terminals in 1983, two factors limited the imports. First, U.S. gas supplies stopped declining as a result of allowing higher prices for U.S.-produced gas. Second, Algeria had been asking exorbitant prices for its liquefied natural gas.

Other Sources of Gas

Geopressured methane. If there ever was a phantom resource, it is the natural gas known to be dissolved in deeply buried caverns of salt water. Trapped subterranean waters are at high temperatures and under enormous pressure. In this environment, natural gas may be formed from the breakdown of oil deposits and should be found dissolved in the salt water.

If the salt water caverns themselves can be tapped, a very large resource of natural gas may await us. The theory is that if the pressure on the reservoirs can be released, perhaps by allowing some of the water to escape as steam, the gas will come out of solution forc-

ibly and can be captured at a well. Controversy abounds over whether this gas can, in practice, be recovered, but experts agree it is there. A Chevron well driven 21,000 feet (6,300 meters) into the earth in Louisiana is said to have produced geopressured gas in 1977, but the company asserts it was an ordinary gas deposit. There are potential environmental problems if we use this resource, such as disposal of hot brine and concern over subsidence above the cavern. The resource may, however, be worth thinking about.

The U.S. Geological Survey estimates that there may be 24,000 trillion cubic feet (680 trillion cubic meters) of gas in the geopressured zone of the continental United States. This estimate should be compared to current annual use in the United States of about 20 trillion cubic feet (0.57 trillion cubic meters) of gas; this is a 1,200-year supply at the current rate of use. How much of the geopressured gas could be recovered is not known.

Methane associated with coal. There are other ways to produce a methane gas of the quality presently produced from drilling. One such method is to remove the methane produced in coal deposits before opening the mine. This procedure not only produces gas but makes the mine safer to work. In fact, the Equitable Gas Company of Pittsburgh has been quietly producing natural gas from a coal field since 1949. Still another source of natural gas is the shale created in the geologic era known as the Devonian. There may be as much as 500 trillion cubic feet of gas in these deposits in the eastern and central United States.

Methane in tight sands. One unusual plan for gas production from wells is quietly slipping into oblivion. Known as the Plowshare Program,★ the undertaking began in the 1960s under the leadership of the Atomic Energy Commission (now part of the Department of Energy). Underground nuclear explosions were to be used to free gas contained in "tight" formations by fracturing the rock walls between small deposits.

Although three experiments were conducted in New Mexico and Colorado, the program came to an end because residents of the states where the atomic explosions were to be set off were concerned about the

---

★A hopeful name derived from the biblical phrase ". . . They shall beat their swords into plowshares, . . ." (Isaiah, 2: 4).

CONTROVERSY 17.1

Social Effects of Natural Gas Pricing

Let the Bastards Freeze in the Dark!
A popular bumper sticker in oil- and gas-rich states

"I stayed in bed to keep warm and to keep from being sick,"
she said. "My brother used to sit by the oven."
An unidentified elderly woman in Queens, New York,
describing how her brother had gotten frostbite and gangrene

The conflict over natural gas pricing probes deeply into one of the controversies of our free-enterprise society. On the one hand, we have a society that believes in free enterprise as a practical and efficient means of providing the goods and services people demand. On the other hand, we have chosen to provide social security and medical care to the elderly. We attempt to be at once a society that supports a free market system and that cares for those who have the least.

The issue of natural gas pricing illustrates the conflict between these objectives. The notion of a free marketplace involves people who need a product and producers who can provide it. When many producers can provide a product, let us say bread, we assume the producers compete; that is, we assume they lower their prices to similar levels so that they can continue to capture a portion of the demand. Unfortunately, perfectly free markets do not exist for every product.

A free market for the sale of natural gas, which moves in interstate commerce, has not existed for some years. The market has been controlled by the government because it was thought to be open to abuse. The rationale was that monopolies could occur when a single pipeline supplied an urban area. A single gas-supply firm could charge very high prices for natural gas because it would be the only supplier and because gas is essential. To prevent excessive prices in this monopoly situation, the government controlled the price of natural gas.

Government control of the price of gas sold interstate, however, appears to have been too tight. The low price of this gas led many millions of families in urban areas of the eastern United States and the midwest to heat their homes with natural gas. Meanwhile, gas sold within the state in which it was produced became a valued good. In addition, its price was not regulated. Gas was used in producer states for fertilizer manufacture and chemical manufacture and for electric power generation, as well as for heating homes and hot water. The price of gas sold within the state in which it was produced soared to four and five times the regulated price for interstate gas. With such a market in the producing states, it was no wonder that producers did not want to sell gas on the interstate market.

Supplies to the east and midwest did not keep pace with demand; in the late 1970s, new customers for natural gas had to be turned away by the local utilities. Some industries had their supplies of natural gas diverted to homes and schools as severe winter weather set in. The producers told us that the government should not regulate prices. Mobil Oil Company put it this way in a *New York Times* ad:

What is needed is decontrol of prices of new supplies of natural gas. . . . This approach would give producers the incentive to step up their already active search for gas . . .

In another *Times* advertisement, Mobil added,

The government's focus on low prices to the consumer has ignored his need for secure and adequate supplies.

In 1978, Congress ordered the regulation of natural gas prices to end by 1985, with a gradual phasing out of price controls. It would be a mistake, however, to suppose that the decision of 1978 will no longer be debated. When Congress originally ordered the regulation of natural gas in 1954, the gas industry lobbied and buttonholed Congress for deregulation the next year, the year following that, the year after, and so on, until 1978, almost a quarter of a century later. And the vote to end regulation in 1978 was very close. In 1985, we had our first taste of a "free" market in natural gas.

Some senators and representatives, as well as ordinary citizens, do not trust the oil and gas industry and can be expected to work to reestablish price regulation to prevent skyrocketing natural gas prices. In a sense, the issue of regulation boils down to whether or not you trust the oil and gas industry.

It is clear that both sides have something of value in their argument. Low interstate prices have caused shortages as new supplies have been sold within their state of origin, but low prices have also protected the consumer. An unregulated natural gas industry has the potential to exploit the consumer. Furthermore, if prices rise sharply, reserves may come up, but the poorer people among us will be less able to afford heat for their homes. Do you have suggestions for what we should do? Do you favor the government's removing or restoring controls on natural gas? Why or why not?

Sources:
Mobil advertisements, *The New York Times*, January 18, 1976; January 26, 1977; December 18, 1976.

---

effects. In addition, the gas itself had a low level of radioactivity.

Methane from organic wastes. Another way to obtain gas is to capture the methane produced by sanitary landfills, the areas where trash and garbage are buried. The decomposition of the buried garbage in the absence of oxygen is known to produce methane, sometimes contaminated with the foul-smelling gas, hydrogen sulfide. The methane level, however, is good and the gas can be burned.

In addition, methane has been produced for some years in sewage treatment plants in the device known as the anaerobic digester. The methane is burned in the plant to provide the heat needed for the digester and elsewhere in the plant.

Methane may also be produced from animal wastes that may build up on cattle feedlots. The decomposition of these wastes in the absence of oxygen produces methane. Plans to use gas from cattle manure have been formulated by the Peoples Gas Company, a midwestern firm.

Synthetic Fuels

The term *synthetic fuels* refers to liquid and gaseous fuels derived from coal or shale oil (or other sources), rather than from naturally occurring petroleum and natural gas. The name does not indicate a precise class of compounds, only fuels derived from other than their usual source.

Synthetic fuels have been around for a long time. Oil derived from shale rock and oil from coal may be the synthetic fuels of today, but a century ago, these fuels, along with whale oil, were the only liquid fuels available. In the first half of the nineteenth century, before we learned to exploit petroleum, the lamps of North America were lit by whale oil, coal oil, and shale oil.

Synthetic gas is not a new idea either, although the quantity and quality of the product now being considered is quite different from earlier versions. Gas was being manufactured from coal in the early 1800s on a wide scale, first in Britain and then in the United States. The product of the local "gasworks" was known as *town gas* or *illuminating gas*. This low-heating-value gas was used for community lighting, for heating homes, and for cooking. By the end of World War II, natural gas had displaced town gas in the United States. In Scotland, town gas is only now giving way to natural gas from the North Sea.

Synthetic fuels derived from coal, tar sands, or shale oil have the potential to multiply fuel supplies in North America by a factor of 10 or more.

Synthetic Fuels from Coal: Gasification and Liquefaction

Coal gasification processes are of two types, based on the quality of the product. One process produces a gas of relatively low heating value, consisting primarily of carbon monoxide and hydrogen; the process is referred to as *low-Btu gasification*. The second process produces a gas with a heating value nearly that of natural gas, and the process is known as *high-Btu gasification*. The gas from this second process may be referred to as synthetic natural gas (SNG), and it consists primarily of methane.

This high-Btu gas is a reasonable substitute for the natural gas transported by pipeline to many sections of the country. The low-Btu gas, however, is likely to be used in industries only at the place where it is produced, or it may be used to generate electric power near the site of production.

Although other processes are being studied, the only commercially proven process for coal gasification is the Lurgi process, which originated in Germany before World War II. In all processes, the undesirable constituents of coal are replaced with hydrogen, eventually producing methane in the high-Btu gas.

In the 1980s, a coal gasification plant was built in the United States as one part of an effort to develop synthetic fuels. The Great Plains coal gasification plant produced 137.5 million cubic feet of high-Btu gas per day in 1984 and 1985. It consumed about 1 ton of coal for each 10,000 cubic feet of gas produced, or about 4.7 million tons of coal per year.

A number of such plants were being built or were on the drawing boards in the late 1970s, and all used the Lurgi process, a process currently used in Sasol, South Africa. The stories of these plants, as well as all other synthetic fuel plants, can be told together.

In the early 1980s, a dozen commercial-scale demonstrations of synthetic fuel production were planned around the country. With an aggregate value of $23 billion, shared between industry and government, these projects would have provided the technology for coal gasification, coal liquefaction, and oil extraction from shale. These projects would not necessarily pay for themselves if oil prices dropped, but they were viewed as vehicles to create technologies that would protect American security if oil supplies were cut off, as in 1973 and 1974. They required government subsidy both to build and to operate, but Congress and three presidents (Ford, Carter, and Reagan) affirmed their importance, so industry forged ahead. But by 1985, Congress and President Reagan had left the synthetic fuel industry in a shambles. One by one, the projects threw in the towel as government support was withdrawn, as oil prices slipped downward, and as the promise of financial reward diminished.

In July 1985, one of the few commercial-scale projects remaining, the successful Great Plains coal gasification plant, was put on notice that its support would end in the spring of 1986, when the Department of Energy would cease to subsidize the cost of the gas it was selling. The project had been built within its $2 billion budget, and the U.S. Department of Energy had been well aware of its need to aid the company in its production phase. The federal government tossed away its $1.5 billion investment, and the industry partners lost $543 million. The executives of the industries—Tenneco, American Natural Resources, Transco Energy, Mid Con, and Pacific Lighting—were bitter about the government pullout. It will be understandable if their interest in new long-term projects with government cosponsorship is low. There are efforts in Congress, however, to rescue the Great Plains plant; one bill would have the plant produce a liquid fuel and would require the Department of Defense to purchase 10,000 barrels of jet fuel per day from the plant.

Synthetic Fuels from the Tar Sands of Athabasca

In contrast to the picture in the United States, where synthetic fuel developments are now largely confined to the laboratory scale, the picture in Canada is different. There, a different resource is being determinedly developed, cautiously but steadily. In the northern portion of Alberta, Canada is a region with deposits of a sand that is coated black with pitch, or tar; these deposits are being exploited to produce oil. The petroleum deposits in the Athabasca region, instead of being trapped in an underground reservoir, have migrated up into porous sands near the surface of the earth. The tar, a hydrocarbon known as *bitumen*, may account for up to 16% of the weight of the sandstone. Three tons of a rich tar sand, one that is 14% or more bitumen by weight, is sufficient to produce two barrels of hydrocarbon liquid. Geologists estimate there may be 300 billion barrels of oil in the tar sand deposits, about ten times the proved oil reserves in the United States.

Efforts in Alberta to extract the oil have had a difficult history. The oil has proven costly to produce, and mechanical breakdowns have plagued the operations. In 1985, however, a general optimism accompanied oil company efforts to free bitumen from the sandy deposits. A Sun Oil project, known as Suncor, was begun in 1963 and was producing 45,000 barrels of oil per day by 1985. The project was scheduled to expand by 6,500 barrels per day by 1990. The Syncrude project, begun in 1978 by a consortium of oil companies, had a production capacity of 130,000 barrels per day in 1985 and was investing $1.5 billion to raise capacity by 10%. Imperial Oil (an Exxon subsidiary) has established a project at Cold Lake, Alberta, which was scheduled to produce 75,000 barrels per day by the end of 1986 and 94,000 barrels per day in 1988. Oil from the tar sands has not always been such a heady business. In 1982, the Alsands project lost half its initial ownership when five oil companies abandoned their $13 billion venture. One of the remaining partners, Canada's state-owned oil company Petro-Canada, was back in 1985 with a $3 billion plan to produce oil from the tar sands; a feasibility study of the project was begun that year. Other smaller projects from Amoco Canada and Royal Dutch Shell were also underway by 1985.

Tar sands are found in the United States, but most deposits are small in comparison to Canada's. By far the largest U.S. deposits are on federal land in Utah; 19–29 million barrels of liquid hydrocarbons might be produced from these deposits. Venezuela has large deposits of tar sands, which it was working vigorously to exploit until the price of crude oil began to fall in 1982.

Synthetic Fuels from Oil Shale in the Rockies

Another source of liquid hydrocarbons is the shale rock that contains the organic material known as *kerogen*. A

liquid that is much like oil can be extracted from the kerogen by distillation. The organic matter locked in the shale is the result of geologic processes acting on the ancient sediments accumulated in inland lakes. The world's largest deposits of this oil-bearing rock are apparently in Wyoming and Colorado, where the equivalent of 600 billion barrels of oil may be present. Compared to ordinary U.S. proved reserves (about 30 billion barrels of oil), this resource is enormous.

Why have we been so slow to develop this resource? Clearly, it could solve any current oil shortage. There are several reasons for the long delay. First, removing the oil requires the mining of rock. Between .5 and 2 barrels of oil can be extracted from each ton of rock, leaving 1,700 pounds of waste rock to dispose of. If we were to produce 1 billion barrels of oil from this shale, or about one-sixth of our annual oil consumption, we would produce about 900 million tons of waste rock. This waste rock would have a huge impact on the land, and the mining operation would be massive; caution in pursuing this resource is thus appropriate.

The Institute of Ecology provided a review of the government's Environmental Impact Statement on oil-shale leasing. That report indicated other problems besides disposal of the waste rock, itself a considerable obstacle. Oil-shale processing is expected to cause the release of mercury, cadmium, and lead into the air. Water used in the mining operation would be laced with dissolved salts. This brine must be disposed of in a way that will not contaminate groundwater, and revegetation of mined areas would probably be difficult because of the saline content of the waste rock.

Production of one barrel of shale oil requires 2–6 barrels of water, mining and retorting 1–2 tons of oil shale, and 100 square feet of land.

Production of one barrel of shale oil produces ⅓ pound of airborne dust, 2½ pounds of polluting gases, 2–5 gallons of contaminated water, and 1–1½ tons of spent shale.

The second reason we're not producing oil from shale has to do with cost. Estimates of the cost to produce a barrel of oil from shale are uncertain. In 1977, some said shale oil could be produced for as little as $8 per barrel; others thought the cost of production (not the sale price) might be as much as $28 per barrel. More recently (1980), the Office of Technology Assessment (OTA), an arm of Congress, estimated that shale oil would have to sell for $48 per barrel to give the manufacturer a 12% return on investment and, hence, a reasonable profit.

In 1984, the Synthetic Fuels Corporation was hoping shale oil could be marketed at $69 per barrel, but the price of oil on the world market was $28 per barrel. Even if the government were to subsidize 50% of the manufacturer's construction cost, the price to earn a fair return could only be reduced to $34 per barrel, according to OTA. With such uncertainty, energy companies have been reluctant to continue investing in the production of oil from shale.

One shale-oil project remains in Colorado. The Parachute Creek shale-oil plant, owned by Union Oil (Unocal), was completed in 1983 but has not begun regular operations. The plant continues to experience operating difficulties; shale oil and the remains of shale rock gum up its pipes and shafts. The plant has price guarantees for its oil that were agreed to by the now-abolished Synthetic Fuels Corporation. The price guarantee agreement runs to the year 2002 or up to 10 years after the first commercial production of oil. As a consequence, Unocal is likely to continue its determined effort to build a technically successful shale-oil plant.

The Economic and Political Power of a Cartel

The OPEC oil cartel has had enormous influence on the quality of life in the United States. It could be argued that the cartel's pricing policy will have long-term benefits in stimulating our use of renewable energy resources. It could also be argued that high prices will eventually lead us to develop shale-oil processes, which could have adverse effects on the states in which the shale is located. In the short run, OPEC pricing policy will have marked effects on our ability as a nation to afford improvements to our environment. When oil prices increase, inflation increases, and citizens ask whether we can afford pollution control. To understand the impact of OPEC on our lives, we must take a closer look at the organization that controls 60% of the world's oil.

To explain the power of the cartel, we must first discuss the notion of a monopoly. A single-firm monopoly is an enterprise that is alone in its field. No other firm exists that supplies the same product or service to customers.

Any time one company produces a large portion of the total demand for a product, the other firms that supply the product are in a happy but risky situation. If the giant firm has the financial resources, or if its production costs are low in comparison to competitors, it can decrease its price to a level at which competitors cannot make a profit. If the competitors are driven from the scene, the giant becomes a monopolist. On the other hand, if the giant firm does not care to increase its share of the market for some reason, competitors are in an excellent financial position. They can share the monopoly power of the giant firm by selling at an inflated price.

In the international oil market today, the giant firm is represented by OPEC. The competitors are the international oil companies, many of which are based in the United States. Even though OPEC consists of a number of countries, they have still been able to dominate the marketplace as though they were a single firm; that is, they set the price at which they sell their oil. Such an organization, which attempts to dominate the international market, is referred to as a *cartel*. In order to ensure that each member of the cartel is able to obtain the price set by OPEC, each country participating in OPEC agrees to produce so many million barrels of oil per day and no more.

Since OPEC both sets a price and allocates production to the member nations, each member is, in effect, agreeing to a particular revenue from oil sales. The stability of the cartel is anchored in the agreement that each member will continue to accept its stated share of production so that it can obtain the price set by the organization. The cartel—any cartel for that matter—may not be stable if one of its members needs increased revenues or if new suppliers appear on the scene.

In our brief discussion of monopolies, we mentioned that the other producers in a near-monopoly situation would be in a happy but risky situation. The international oil market illustrates this point well. Just before the 1973–1974 embargo, oil produced within the United States was selling on the U.S. market for $3.50 per barrel. Just after the oil embargo of 1973–1974, OPEC began to demonstrate its strength; it set the price of its oil at $11 per barrel. The OPEC price eventually rose to $34 per barrel but fell back to $26 in 1985.

When OPEC was able to maintain high prices for the oil, producers were in the happy situation of those who operate in the shadow of a monopolistic firm. The price of their oil rose to the giant firm's price. U.S. and international oil companies were able to reap a handsome benefit from the cartel's control of the market.

The situation of the producers was risky, however. In their search for greater profits, the firms invested heavily in exploration, in production, and in research

Figure 17.6 Legs of the production platform nearing completion, destined for the Brent Field in the North Sea. This structure will be almost entirely underwater once in place. The investment to produce this oil is evident. (A Shell photograph)

for methods of enhanced recovery. To see the origin of the risk, we use the example of the North Sea, where the pace of exploration and development quickened during OPEC's rise to power in the 1970s. In the middle 1970s, an offshore oil platform, built to withstand the lashing winds and towering waves of the North Sea, cost about $200 million. This investment had to be made before the first drop of oil could be produced from a field (see Figure 17.6). It was estimated that North Sea oil cost on the order of $5–$6 per barrel to bring in; production costs were even more. As long as the cartel keeps its price high, however, the investment is well spent. It is of interest that the oil in the North Sea was not even counted as a reserve in the early 1970s. The world price for oil was then about $2 per barrel, so that North Sea oil was not profitable to produce. If the cartel were to decide to lower its price to the $8–$10 range per barrel, the North Sea investments

could again no longer be profitable. With prices above $10 per barrel, most North Sea oil continues to flow.

Could the cartel lower its prices so drastically? The plentiful oil in the Middle East costs less than $.50 per barrel to produce; transporting oil via tanker to market costs only about $1 per barrel. OPEC has awesome power to cut prices and destroy the competition's investments. This is the risky situation to which we referred earlier. Synthetic fuels derived from oil shale and coal faced similar risks and lost when oil prices fell in the mid-1980s.

The falling prices of oil in 1986 were the result of too much oil chasing a shrinking market. North Sea oil from Britain and Norway and oil from Mexico, Indonesia, and elsewhere was cutting away at OPEC's market. Meanwhile, conservation was decreasing the demand for oil.

In a desperate move, OPEC agreed to disagree and to let each member get what price it could for its oil. Prices fell as low as $10 per barrel on the spot market in April 1986. OPEC hoped that the resulting revenue losses of Britain, Mexico, and Norway would force these countries to agree to limit their production. In effect, OPEC hoped to expand its membership. Because oil from the British sector of the North Sea is owned by independent oil companies, such agreement was not forthcoming. Prices gradually rose back to $18 per barrel by June of 1986. Oil prices are likely to fluctuate for the foreseeable future but will probably remain in the $12–$18 range—unless political events cause supply interruptions.

Summary

Petroleum is a term for liquid hydrocarbons, which are formed by physical and chemical processes acting on algal plankton that were buried in ocean sediments. Sweet crude oil is 90%–95% hydrogen and carbon plus less than 1% sulfur. Sour crude oil may have up to 5% sulfur. Crude oil is separated, or refined, into less complex mixtures for specific uses; examples are gasoline, heating oil, and residual fuel oil.

Natural gas generally contains less carbon than oil does, but it has more hydrogen and nitrogen; it contains little or no sulfur.

Worldwide, the largest amount of crude oil appears to be located in the Middle East (almost 400 billion barrels of proved reserves). The United States has 27.3 billion barrels of proved reserves—reserves we are confident exist and can be profitably extracted. There are, in addition, an estimated 26.9 billion barrels of indicated and inferred oil reserves and an estimated 82:6 billion barrels of undiscovered resources in the United States. Shale oil constitutes another oil resource but is currently too expensive to produce. Nearly 600 billion barrels of shale oil could be produced from American oil-shale rock, most of which is in Colorado.

The United States used an average of about 15.7 million barrels of oil per day in 1984, down from 18.5 million barrels per day in 1979. Until the 1950s, the United States produced almost all the oil it used, but by the late 1970s, imports accounted for about 40% of demand—a level of importation that has continued to the present. The United States now accounts for about 30% of all oil consumed worldwide, but its share has been declining as other nations have increased their use of oil. In 1978, over 40% of U.S. imports were from the Middle East, but by 1983, only 12% were; imports from North and South America make up the difference.

The largest portion of oil consumed is used for gasoline (43%), with lesser amounts used for home heating, diesel and jet fuel, and residual fuel oil. The manufacture of asphalt, plastics, chemicals, and a variety of other smaller uses accounts for 30% of oil consumption.

In 1983, U.S. consumption of natural gas was 17.65 trillion cubic feet (0.029 trillion cubic meters). The United States has some 200 trillion cubic feet of proved reserves of natural gas, about 175 trillion cubic feet of indicated and inferred reserves, and perhaps 600 trillion cubic feet of undiscovered resources.

The price of natural gas was controlled until 1978, when controls began to be phased out. This decontrol of natural gas prices has led to an almost threefold increase in price but has also slowed the fall in proved reserves. Other possible sources for natural gas are geopressured methane, methane in coal fields, methane in tight sands, and methane generated by the decomposition of organic wastes.

Liquid and gaseous hydrocarbons can also be synthesized from sources such as coal (by liquefaction or gasification), heavy oil in tar sands, and shale rock. Except for efforts in Canada, most projects to recover synthetic fuel from these sources have come to a halt due to lack of government funding. Oil shale represents a possibly huge resource of fuel, but the environmental problems that accompany its production (waste rock, water pollution, airborne dust and gases) have not been satisfactorily resolved.

OPEC is a cartel of oil-producing countries formed to control the pricing and production of oil for the world market. Currently other producers, such as U.S. oil companies, benefit from the high prices charged by the oil cartel. However, because Middle Eastern oil costs only about $1.50 per barrel to produce and ship, compared to $5 or more per barrel for such sources of oil as the North Sea, the cartel has tremendous economic power. It has the potential to cut its price to a level that would drive many oil producers out of business.

Questions

1. What are the two basic elements in oil? What other elements or substances may be present in the oil?
2. Why is crude oil distilled or refined? What are some of the substances into which it is separated?
3. In terms of oil, what are proved reserves? How do these differ from undiscovered resources? To what extent do experts, agencies, and companies agree on the oil remaining to be discovered?
4. What portion of the oil that we use is consumed in gasoline combustion? What are some other uses of oil? What portion of the oil used worldwide is consumed in the United States? Why is this portion falling?
5. What part of our oil consumption was imported in 1984? What portion of this was from the Middle East?
6. Why was the price of natural gas controlled by federal law in the first place? What were the effects of price control as seen by the gas and oil companies?
7. What three sources of synthetic oil exist in North America? How will the price of OPEC oil affect whether they will be recovered?
8. What are some of the untapped resources of natural gas or synthetic natural gas? What two factors are generally responsible for our not using these resources in quantity?

Further Reading

General

International Petroleum Encyclopedia.
      See latest yearly edition for the facts, as the industry sees them, on oil production, reserves, and locations around the world. Many tables and charts.

Kash, D., et al. *Our Energy Future.* Norman: University of Oklahoma Press, 1976.
      As thorough a description of energy technology and re-

sources as can be found. Language is largely nontechnical and new words applying to the technologies are defined as they are used.

Metz, W. "Mexico: The Premier Oil Discovery in the Western Hemisphere," *Science*, **202** (December 22, 1978), 1262.

"The Natural Gas Shortage," *Business Week* (September 27, 1976), p. 66.
A very readable and relatively unbiased article describing the background to the natural gas shortage.

Hefner, Robert. "The NGPA Is Working Well," *Oil and Gas Journal* (December 28, 1981), p. 224.
This article indicates the impact of the gradual decontrol of natural gas prices on availability of gas.

Cook, James. "The Great Oil Swindle," *Forbes* (March 15, 1982).
The message is: The energy crisis isn't over.

The Geopolitics of Oil. Prepared for Senate Committee on Energy and Natural Resources (November 1980).
This book is summarized in an article in *Science*, **210** (December 19, 1980), 1324.

Seltz-Petrash, A. "Billion Dollar Oil Project to Provide Energy Security," *Civil Engineering* (August 1980), p. 41.
This two-part article discusses both the construction and political aspects of the Strategic Petroleum Reserve.

Oil Shale/Tar Sands/Coal Gasification and Liquefaction

Fletcher, K., and M. Baldwin, eds. *A Scientific and Policy Review of the Final Environmental Impact Statement for the Prototype Oil Shale Leasing Program of the Department of the Interior*. The Institute of Ecology, 1973.
Understandable and worth reading for the environmentalist viewpoint.

Maugh, T. "Tar Sands: A New Fuels Industry Takes Shape," *Science*, **199** (February 17, 1978), 756.
Oriented toward the science, technology, and policy aspects of tar sands exploitation.

"Oil from Shale Is Still a Distant Hope," *Business Week* (April 23, 1979), p. 126.
Nontechnical, oriented toward problems of oil companies interested in developing the shale oil resource.

"Oil Shale and the Environment." Office of Research and Development, U.S. EPA, EPA 600/9-77-033, October 1977. Available from National Technical Information Service, Springfield, Virginia.
Recommended: nontechnical and profusely illustrated.

"Uncertainty, Cost/Price Squeeze Hit Fledgling Synfuels Industry," *Oil and Gas Journal* (May 24, 1982), p. 21.
Discusses the drop in oil prices and its effect on synfuel projects.

References

Energy: The Next Twenty Years. Report by a study group sponsored by the Ford Foundation and administered by Resources for the Future. Cambridge, Mass.: Ballinger, 1979.

Project Interdependence, Chapter XIX, "World Oil and Natural Gas Reserves and Resources." Washington, D.C.: U.S. Government Printing Office, 1977.

"An Assessment of Oil Shale Technology," Office of Technology Assessment, Congress of the United States, 1980.

"Gulf Successful in Gasifying Steeply Dipping Coal Beds," *Oil and Gas Journal* (December 28, 1981), p. 71.

CHAPTER EIGHTEEN

Nuclear Power

Light–Water Nuclear Reactors

How They Operate/Radioactive Wastes and Radiation Releases from Light-Water Reactors/Safety of Light-Water Reactors/Chernobyl: The Worst-Case Disaster Occurs

Problems in the Nuclear Fuel Cycle

From the Mine to the Power Plant/Fuel Reprocessing and Spent Fuel Rod Storage/Mixed-Oxide Fuel and Plutonium Transport/International Safeguards/Storage or Disposal of Wastes from Fuel Reprocessing

Liquid–Metal Fast Breeder Reactors

Economic and Social Aspects of Nuclear Power

CONTROVERSIES:
18.1 *Is Nuclear Energy an Inevitable Result of Technical Progress?*
18.2 *Does Reprocessing Produce Weapons-Grade Plutonium?*
18.3 *Does the Public Have the Right to Know (How Easy It Is to Steal Nuclear Material and Build an Atomic Weapon)?*

It began under a cloud—a mushroom cloud rising over a Japanese city. The nuclear electric power industry has never been able to shake off the reputation acquired by that beginning.

The scientists who created the atom bomb also envisioned that the atom could be used peacefully as a means of generating electric power. Under the "Atoms for Peace" program of President Eisenhower, private corporations were given the privilege of owning nuclear reactors. Industries then began to join in projects with the Atomic Energy Commission to develop electric plants powered by nuclear energy.

The first experimental production of electricity from a nuclear reactor occurred in 1956, when a boiling-water reactor began operating at Argonne National Laboratory in Illinois. In the following year, a pressurized-water reactor at Shippingport, Pennsylvania began delivering 60 megawatts of electrical power. The size of new plants increased quickly as operating experience grew. By 1963, several nuclear plants were delivering 200 megawatts of electric power, and commitments had been made for the larger Oyster Creek plant in New Jersey and the Nine Mile Point plant in New York. These new plants were to each deliver up to 600 megawatts of power. The nuclear industry was poised for growth.

A tide of commitments to nuclear plants began in 1965. Seven orders for plants were placed that year, 20 the following year, and 30 in 1967. The rate then began to fluctuate but continued strong into the 1970s. By mid-1974, almost 240 nuclear plants had been ordered, and most new plants on order were to provide 1,000 megawatts or more of electrical power. The new technology had taken hold with remarkable speed.

However, between 1974 and 1978, only 13 new orders were placed for nuclear reactors to be sited in the United States, and but two were ordered in 1978. Not one order for a nuclear reactor to generate power in the United States has been placed since that time. And in 1978, 13 reactor orders were cancelled; then 8 were cancelled in 1979, 16 in 1980, 6 in 1981, 9 in 1982, 6 in 1983, and 8 in 1984. In all, 66 reactor orders were cancelled between 1978 and 1984. The tide had begun to recede.

The course of nuclear power development was slowed at first by debate in the scientific community and in Congress about the possible levels of radiation released routinely from the plants. Scientists have also argued over the safety of the containment structure in terms of its ability to prevent the accidental release of radioactive material. Debate continues on whether plutonium ought to be used as a replacement nuclear fuel for uranium as a means to extend fuel supplies, because plutonium might be stolen and used for making atomic bombs. Several states have had referendums on the further development of nuclear electric power in their jurisdictions.

Is it the cloud at the beginning that has alone limited the acceptance of nuclear electric power? That is, is it only a problem of public misunderstanding of the actual safety of nuclear power? Or is there more substance to the debate than simply poor public relations?

Understanding the controversy that swirls around

nuclear power requires a discussion of the complete nuclear power system as it has evolved to the present time. Only with such a discussion can we knowledgeably examine the positions of the opponents and proponents of nuclear power and judge their validity. In no other way can the reader formulate sound opinions.

Light-Water Nuclear Reactors

How They Operate

To understand the operation of nuclear power plants, it is helpful first to examine the nuclear reaction known as **fission**. Fission occurs when a neutron strikes the nucleus of **uranium-235** or certain other heavy atoms. The reaction consists of the breaking apart (fissioning) of the atom into two relatively large fragments, accompanied by the release of a large amount of heat energy and **gamma rays**. The large fragments, or fission products, are themselves atoms, each consisting of a portion of the electrons and a portion of the nucleus of the "parent" atom. The fragments are often radioactive and decay in gradual steps to stable atoms, releasing radiation at each step.

Not only are fission products, energy, and gamma rays released, but neutrons are also ejected in the fission reaction. When one of these neutrons strikes another uranium-235 atom, it can cause the uranium-235 atom to fission in turn. Still more fragments, heat energy, and neutrons are released, and the fission reaction is thereby continued or sustained. The sustained fission reaction is the basis for the design of both the atomic bomb and the nuclear power plant.

Two fundamental reactor types are commonly used to capture the heat energy from the fission process: the **boiling-water reactor (BWR)** and the **pressurized-water reactor (PWR)**. Their technology is relatively proven and reliable, and they promise to be the workhorses of nuclear electric generation for the next several decades. Other reactors are being developed and investigated, but their contributions to electrical energy supply are likely to be small in the near future.

The boiling-water reactor and the pressurized-water reactor are actually very similar. In both types, uranium-235 undergoes controlled fission in a **core**, generating heat energy. In the interior of the core, long cylindrical metal rods containing the nuclear fuel are arranged vertically (Figures 18.1 and 18.2). Fission

Figure 18.1 Inspection of fuel rods. Westinghouse assembles nuclear fuel rods at a fabrication plant at Columbia, South Carolina. In this photo, the rods are undergoing final visual inspection. Each rod contains uranium dioxide pellets, the "fuel" for the nuclear reactor. At this stage, the rods are not yet emitting dangerous radiation. (Westinghouse Photo by Jack Merhaut)

takes place within the fuel rods and provides the energy to heat water that is flowing through the core, around the rods. The core of the nuclear reactor is thus the "boiler" of the nuclear power plant. Because neutrons and gamma rays are produced by the fission process, the core must be heavily shielded to prevent the exposure of workers to these dangerous particles and rays. Layers of concrete and iron or steel are used to encase the core.

The nuclear fuel in the rods is in the form of pellets of uranium oxide. About 3% of the uranium in the fuel is in the form of uranium-235, which is the isotope that undergoes fission. The remainder of the uranium in the fuel rod is uranium-238, an isotope that cannot undergo fission. As the reactor operates over a period of time, the amount of uranium-235 within the rod decreases and fission products build up. The fission process gradually slows down because the fission products capture some of the neutrons that sustain the chain reaction. After about three years, the rod is no longer

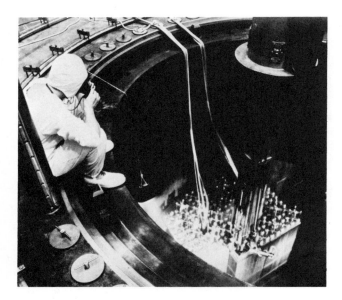

Figure 18.2 Fuel-loading operation. A technician monitors the fuel-loading operation at Unit No. 1 of the Calvert Cliffs Nuclear Power Plant, owned by the Baltimore Gas & Electric Company. Each fuel bundle—of a total 217 bundles—is being positioned by an automatic fuel-handling machine, shown extending down to the upper core level. (Combustion Engineering, Inc.)

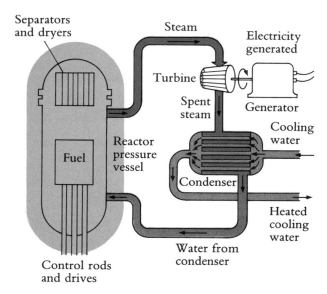

Figure 18.3 Boiling-water reactor (BWR) power plant. Water is boiled to steam in the reactor pressure vessel, just as in the boiler of a coal-fired electric plant. The steam turns a turbine to generate electric power and then is condensed to a liquid prior to return to the "boiler."

an efficient source of heat because of a low rate of fission; it is then replaced by a fresh fuel rod. The spent fuel rod is now relatively low in uranium-235 and rich in fission products.

This, in general, is how the core operates, both for the BWR and the PWR, but the two reactor types differ in how the heat energy is transferred to the circulating water. The boiling water reactor functions in precisely the same way as the boiler in a fossil-fuel power plant—that is, the water flowing through the core is converted directly to steam, and the steam is used to turn a turbine (Figure 18.3).

The pressurized-water reactor, in contrast, does not convert the water in the core to steam. Instead, the reactor maintains enormous pressure on the water to prevent it from boiling. All the heat energy from fission goes into raising the temperature of the water, which is cycled from the core to a steam generator, where its heat is transferred to water flowing in a secondary loop. The water from the core returns from the steam generator back to the core to be heated again. This water, circulating between the core and the steam

generator, flows in what is called the primary loop. The water in the secondary loop boils to steam as it leaves the steam generator under pressure. This steam in the secondary loop is used to turn a turbine and generate electricity (Figure 18.4).

Thus, in the BWR steam is generated in the core itself and is used directly to turn a turbine. In contrast, in the PWR steam to turn the turbine is produced in a steam generator and flows only in the secondary loop.

In both reactors, the steam that finally leaves the turbine lacks the energy to generate more electrical power, though it is still quite hot. This steam is condensed to liquid water in a condenser. Here, heat from this "spent" steam is transferred to a flow of cold water, called *cooling water*. The water from the condenser circulates back to be converted to steam once again. In the PWR, the conversion to steam takes place in the steam generator. In the BWR, it occurs in the core itself.

The cooling water that condenses the spent steam is warmed as a result. The effects of cooling water on aquatic life are discussed in the section on thermal pollution in Chapter 15.

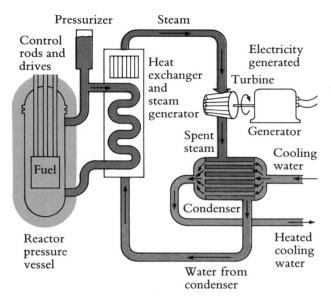

Figure 18.4 Pressurized-water reactor (PWR) power plant. In contrast to the BWR, where steam is produced in the reactor core, only liquid water flows through the primary loop of the PWR. The water, at a very high temperature and pressure, boils water to steam in the secondary loop. The operation of the secondary loop then is the same as that of the primary loop of the BWR.

Radioactive Wastes and Radiation Releases from Light-Water Reactors

Just as fossil-fuel power plants emit pollutants, so also do nuclear power plants. Whereas emissions from fossil-fuel power plants include the commonly known chemical air pollutants, emissions from nuclear power plants are radioactive elements. These radioactive elements stem almost entirely from the fission process.

The fissioning of uranium-235 produces the radioactive forms of the elements krypton, xenon, cesium, strontium, and iodine, among others. Although the fission process is supposed to be confined to the interior of the fuel rod, these fission products may escape into the water in the core through defects in the rod's metal shell. Fission products, then, appear in the water in the primary loop. The water that circulates through the core contains not only fission products but also metal that has corroded from the shell of the fuel rod. This metal becomes radioactive when struck by neutrons from the fission process. The treatment of the water to remove various radioactive elements produces gaseous, liquid, and solid wastes.

Gaseous wastes removed from the circulating water include radioactive forms of the gases krypton, xenon, and nitrogen. The plant discharges its gaseous wastes through tall stacks, which allow the substances to be widely dispersed. The total amount of radioactivity routinely discharged from the stacks of a PWR is far less than that from a BWR, an amount already vanishingly small.

If the water that circulates in the primary loop should leak from a pipe or from a seal or joint, it must be considered a radioactive waste and must be cautiously handled and disposed of. Although *liquid wastes* are, for all practical purposes, water, radioactive substances are dissolved or suspended in the water. When the liquid wastes are treated to remove these substances, new solid, liquid, and gaseous wastes are created.

Solid radioactive wastes, which are relatively low-level wastes, are encased in concrete within 55-gallon metal drums and shipped to commercially operated, federally licensed, special sites for burial. Although the sites are commercially operated, the grounds are supervised by the states. Six such sites have been in operation, located in Illinois, Kentucky, Nevada, New York, South Carolina, and Washington. Political pressures have caused the Illinois, New York, and Kentucky sites to close, seemingly permanently. The other sites have been closed periodically as well, largely because of political pressures. Six federally operated (as opposed to state-operated) sites accept low-level wastes from government-owned facilities. These are in Washington, Nevada, Idaho, New Mexico, South Carolina, and Tennessee.

How well do nuclear plants prevent routine radioactive releases to the environment? When we consider the potential for such releases, the answer is "Exceptionally well." Most plants are meeting the rules set by the federal government on releases, and in the last two decades these rules have become very stringent. The rules call for routine releases of less than 5% of the natural background radiation. It should be added that these guidelines were not always so strict and that criticism by scientists helped to bring about new and more strict guidelines.

In addition to the spent fuel, one other type of waste arises from nuclear power plants: the plants themselves. No major plant has yet been fully "decommissioned," although the decommissioning of the Shippingport, Pennsylvania plant and of Three Mile

Island Unit 2 has begun. The Shippingport plant is being dismantled and the reactor taken by barge down the Mississippi through the Gulf of Mexico and Panama Canal and up the Pacific Coast to be finally buried on government land in Hanford, Washington. Decommissioning of the plant should be completed by 1988.

In general, if a plant is to be dismantled and hauled away, the cost could run to $100 million for the modern 1,000-megawatt plant. This is about 5%–10% of its initial cost, and it is a cost not yet taken into account by utilities. The alternative to decommissioning is to seal and guard the site and reactor for 100 years or more.

Safety of Light-Water Reactors

Routine releases of radioactivity are negligible. That is, on a day-to-day basis, a resident near a nuclear power plant is exposed to less radiation from the plant than from natural background levels. Day-to-day emissions are not the target of critics, however; it is the potential release of radiation if a serious accident or sabotage occurred within the plant. One particular form of accident is of more concern than all others.

Recall that water circulates through the reactor core to withdraw the heat produced by the fission process. If the circulating water were suddenly lost, the heat in the core would accumulate rapidly, and the contents of the core could melt. In the jargon of the nuclear industry, this is called a *loss-of-coolant accident*. This kind of accident might occur if a steam or water line carrying water into or out of the core should break. The high pressures in the line would probably cause most of the water or steam to be rapidly ejected from the core through the break. The temperature in the core would rise rapidly, melting the fuel rods. This extreme event is known as a *meltdown*. Radioactive xenon and krypton gases would be released from the melted rods and would enter the building through the break in the pipe. The molten mass of fuel and fuel rod shells would fall to the floor of the reactor vessel, perhaps melting through the floor and falling into the containment room that surrounds the reactor core. Although more radioactive gases would be released into the containment area, the containment room is designed so that the leakage of radioactivity to the outside will be very small. A water spray system is poised to condense the steam in the containment room and reduce the temperature there.

The first line of defense against a meltdown, however, is at the reactor core. The reactor core is also equipped with a spray system. The *emergency core cooling system* is designed to inject water automatically into the core in the event of a break in the coolant line; its purpose is to cool the reactor core and prevent melting. The ability of the emergency core cooling system to flood the core with water was finally tested in 1979 at a test reactor, more than 20 years after the first commercial nuclear power plants went on line. The initial test did prove successful, but it was run at less than full-power operation.

A loss-of-coolant accident is regarded as the most serious accident that might occur at a light-water reactor. The severity of the accident, in terms of its impact on the population, would depend on how effectively the emergency core cooling system functioned. If a core meltdown were not prevented by the emergency spray, the impact would then depend on how effectively the containment room prevented radioactivity from entering the environment. And if radiation escaped, the impact would depend on the wind speed and direction and on the presence of a populated area downwind from the plant.

Because reactor safety is so prominent an issue in the nuclear power debate, the Atomic Energy Commission (now the Nuclear Regulatory Commission) directed that an extensive study (WASH-1400) of the problem be undertaken. In the late 1970s, Professor Norman Rasmussen of MIT conducted the study, which cost $4 million, involved more than 100 people, and took three years to complete. The study concluded that the odds against the worst-case accident occurring were astronomically large. The worst-case accident projected an estimate of about 3,000 early deaths and $14 billion in property damage due to contamination. Cancers occurring later as a result of the event might number 1,500 per year. The odds against such an event, however, were estimated at 10 million to 1, given that 100 reactors are in operation.

The probability of a meltdown is not so small, however. With 100 light-water reactors in operation (see Figure 18.5), the probability of a meltdown is 1 in 100 per year. Thus, although the worst-case accident is estimated as only a remote possibility, the meltdown alone is not such an unlikely event. The Rasmussen study attempted to evaluate the probability that a meltdown will turn into a serious, life-threatening accident and concluded that the safety features engineered

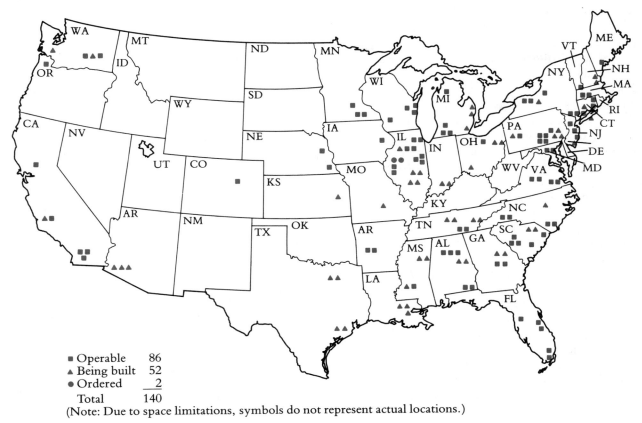

■ Operable   86
▲ Being built 52
● Ordered   2
Total   140

(Note: Due to space limitations, symbols do not represent actual locations.)

Figure 18.5 Status of nuclear power reactors in the United States as of January 1, 1984. (From U.S. Department of Energy.)

into the plant are likely to prevent such serious consequences.

The Rasmussen study received scathing criticism from a panel of 21 scientists, economists, and political scientists assembled by the Ford Foundation for a study of nuclear power. The panel felt that the study seriously underestimated the uncertainties of reactor safety. Flaws in methodology, they asserted, could make some of the probability estimates of accidents low by a factor of 500 (Nuclear Energy Policy Study Group, 1977).

Another panel, created in 1977 by the Nuclear Regulatory Commission itself, also reviewed the Rasmussen report and found it lacking in a number of respects. In particular, the estimates of risk set forth in the report were criticized as being subject to much larger errors than the report implied, errors that could not be reduced because of a fundamental lack of data. That is, whereas the report provided estimates of the risk of various accidents, the panel felt the risks could not be stated nearly so precisely.

Until 1979, the issue of nuclear safety lay buried in the jumble of questions that had piled up about nuclear power. People were concerned not only about nuclear safety but also about thermal pollution, radioactive waste disposal, fuel reprocessing, the spread of atomic weapons, and the transport of nuclear material. In 1979, however, at a pressurized-water nuclear reactor near Harrisburg, Pennsylvania, an accident occurred of a character not thought possible. Nuclear safety suddenly came bubbling to the top of the nation's and the world's nuclear concerns.

The reactor incident at the Three Mile Island (TMI) Nuclear Power Station in Middletown, Pennsylvania, began at 4 A.M. on March 28, 1979, with the failure of two pumps in the secondary cooling water loop of Reactor 2. This was the first in a series of events that eventually resulted in a partial meltdown of the reactor core and the most serious accident at any commercial nuclear reactor in the United States. In the incident, the metal rods and some of the fuel began to melt. Although no one died as a direct result of the TMI accident, it could have easily escalated into an event far worse—as happened in Chernobyl in 1986.

The accident at Three Mile Island resembled only slightly any of the many scenarios in the Rasmussen report. In that report, the accidents of interest were the major loss-of-coolant accidents, resulting from large equipment failures. But the "accident" at Three Mile Island was really a collection of small events—small equipment failures and human errors—and it nearly led to a complete core meltdown. Releases of significant and potentially lethal amounts of radioactive material were fortunately avoided.

The role of human operators at Three Mile Island was terribly important; their actions made the accident far worse than it could have been. Poor control room design and poor information display contributed to delayed and inadequate response. During the first 8 minutes of the accident, the operators were subjected to 100 different alarms of bells, buzzers, and lights.

Further, the response of local, state, and federal authorities revealed a most disturbing lack of preparation for such accidents. The public was confused by contradictory information from many sources. No usable evacuation plan existed, although many people left the area voluntarily, and children and pregnant women were advised to leave.

Cleanup operations at the damaged plant were expected to be finished in 1988 at a final cost of about $1 billion. Whether the reactor will ever operate again is in doubt. A second, undamaged reactor unit at Three Mile Island was allowed to restart in October 1985, despite a Pennsylvania referendum that voted against reopening.

Interestingly, in 1985 an equipment failure similar to the one that began the incident at TMI occurred at the Davis-Besse Nuclear Plant in Oak Harbor, Chicago. In contrast to the event at Three Mile Island, new instrumentation and quick operator response prevented a meltdown.

Chernobyl: The Worst-Case Disaster Occurs

On April 28, 1986, technicians engaged in routine monitoring at a nuclear power plant in Sweden began to note the presence of unexpected levels of radioactivity on personnel and in the atmosphere. Levels of radioactivity in the atmosphere rose to ten times the natural background and continued high for several days. The Swedish government became involved. Noting the pattern of wind circulation and the origin of the winds, the Swedes focused on the Soviet Union as the source of the radioactivity and demanded an explanation. As more European countries began to pick up radiation in the atmosphere, the explanation finally arrived. The world's most terrible nuclear power disaster had occurred in the Ukraine. Although the worst was past, the disaster was actually still in progress.

Ultimately, the world learned that the event had begun early in the morning on April 26. A violent explosion had followed a series of experiments at the 1,000-megawatt Reactor IV at the multiunit Chernobyl Power Station near Kiev on the Black Sea; a seemingly unquenchable fire followed the explosion. The explosion had blown the roof off the unit and had ejected vast quantities of radioactive materials high into the atmosphere.

Among the radioactive elements released were cesium-137 and iodine-131. Iodine-131 is a substance that can enter the human diet (by falling on grasses that milk-producing cows may eat) and that concentrates in the human thyroid gland. Up to 50% of the radioactivity in the emissions may have been from iodine-131. Some estimates placed radio-iodine and radio-cesium emissions at half of the quantities of these elements then in the core.

Estimates of total radiation from fallout have varied, but one source compared it to the total of radiation from all atmospheric weapons tests. Another source suggested that the radiation released to the atmosphere rivaled that from about 300 atomic bombs. Soviet officials later estimated that 1%–3% of the reactor's 180 tons of fuel were ejected into the atmosphere, but some Western experts suggested that as much as 10% of the fuel had entered the atmosphere.

Evacuation of almost 50,000 people began on April 27 and was completed by May 8. Firefighters fought the blaze initially, but water and chemicals could not be used because of the possibility of further explosions. By mid-May, nearly 300 firefighters and power station personnel had been hospitalized for radiation exposure. By late June, 26 people had died as a result of the accident, although only 2 had died in the initial explosion. A number of those hospitalized had been given bone marrow transplants in an effort to save their lives.

Those hospitalized for radiation exposure will generally not live out their expected life spans.

Once it was realized that the fire could not be fought by conventional means, helicopters began dropping sand, boron, dolomite, and clay onto the open reactor. These neutron-absorbing materials slowed the fissioning of uranium-235 and decreased the production of heat and radioactivity. Gradually the fire diminished. Even as the fire was being fought, sandhogs began tunneling beneath the reactor, hollowing out a space in which to pour a several-foot layer of concrete—as part of a plan to entomb the reactor and isolate it from the groundwater and the nearby river. The Soviet press called the concrete structure being built around and under the reactor a "sarcophagus." Much of the agricultural land around the reactor remains contaminated and will remain closed for years until radioactivity levels subside.

What went wrong at Chernobyl? Was the seriousness of the accident due to the special design of Soviet nuclear power reactors? What are the implications for nuclear reactors in the West?

The first reports reaching the West suggested that Unit IV lacked the concrete and steel containment structure that surrounds the reactor in the light-water reactors used so extensively in the United States and Western Europe. But it became apparent shortly that the reactor *was* surrounded by a containment. The reactor also appears to have had much of the modern safety equipment that U.S. reactors have. The reactor itself, however, was of a special and uncommon design favored by the Soviets, known as the RBMK-1000.

The RBMK, like the light-water reactors, has nuclear fuel encased in thin rods. These thin rods are in a larger tube through which water flows to take away the heat being generated by the nuclear reaction. The larger tube is embedded in a graphite (carbon) column; many columns are placed together. Graphite absorbs neutrons from the fission reaction and prevents the reaction from speeding up. Graphite is not used in the Western-style light-water reactors; the large volume of water that flows past the fuel assemblies in the light-water reactor serves the same function as the graphite. The long-burning fire in Unit IV was burning graphite.

Analysts think that a rapid-fire sequence of events outran the safety systems of the reactor, producing a power (heat) surge that could not be cooled fast enough. The RBMK design is known to be susceptible to power surges. Some of the water-carrying tubes are

thought to have ruptured, leading to a further heat buildup and the melting of the fuel tubes. The exceedingly hot fuel came in contact with water, producing steam. The steam reacted with the graphite, producing hydrogen, and an explosion followed, blowing the roof off the reactor building.

The common view of many U.S. nuclear experts is that American reactors are not susceptible to the power surges that the RBMK design is. However, experts disagree on the ability of the American-style containment structure to withstand a steam explosion. A panel convened by the Nuclear Regulatory Commission concluded, without conducting experiments, that such a breaching of the containment was almost impossible; other nuclear scientists challenged that view. (See Controversy 18.1.)

Problems in the Nuclear Fuel Cycle

From the Mine to the Power Plant

A system for nuclear electric generation contains more than individual nuclear power plants. It encompasses a cycle that begins with the mining of uranium and proceeds through use of uranium at the plant to the recycling and/or disposal of the contents of spent fuel rods. Serious safety questions arise at a number of points along the way. Figure 18.6 illustrates the operations in the nuclear fuel cycle, along with the transportation of products from one operation to another.

The first operation in the cycle is the actual mining and milling of uranium ore. This step is followed by the extraction of an oxide of uranium called "yellowcake." In the past, at older mining operations, the tailings from the processing of uranium ore were simply dumped on the ground (see Chapter 27); they are, however, a potential hazard.

Both uranium-235 and uranium-238 are found in the extracted uranium oxide. However, only about 0.7% of the uranium is present as uranium-235, which is the isotope necessary for fission. Because the natural concentration of uranium-235 is so low, a step is needed to increase the concentration. This step is called **enrichment**, and the original process for enrichment is known as *gaseous diffusion*. In the gaseous diffusion process, the level of uranium-235 is increased to 3% of the total uranium present, although it can be increased further if, perhaps, a bomb were being made. The Department of Energy is spending $100 million a year on

CONTROVERSY 18.1

Is Nuclear Energy an Inevitable Result of Technical Progress?

The anti-nuclear movement . . . seeks to undermine the scientific and technological revolution which has created the modern world.

Alex A. Vardamis

If man is to survive on this earth, . . . it will be through his wisdom in choosing between those innovations that he can control and those that he cannot.

George Dryfoos

The excerpted letters that follow appeared in *The New York Times* on successive July days in 1978, shortly after demonstrations at the nuclear plant site at Seabrook, New Hampshire. They raise issues and questions that are worth exploring.

To the Editor:

The recent anti-nuclear-energy demonstration in Seabrook, N.H., should be placed in perspective. The animating rationale of the protest movement is based upon a rejection of progress and technology. . . .

The anti-nuclear movement is doomed to failure, for it seeks to undermine the scientific and technological revolution which has created the modern world . . . The Seabrook demonstrators . . . are temporary and inconsequential impedimenta to the tide of history and are destined to become a forgotten footnote to the constructive advancements of science in this century . . .

Alex A. Vardamis, July 6, 1978

To the Editor:

Questioned as to the possibility of another blackout in New York this summer, Con Ed Chairman Charles Luce sensibly pointed out that in a complex system of machines, operated by fallible human beings, there can be no 100 percent guarantee.

His words should be carved in stone for the benefit of those who see nuclear power as the answer to our future energy needs . . .

The catalogue of hazards ranges from an admitted inability to guarantee that storage of nuclear wastes will not fatally pollute our land, water and atmosphere, to the very real possibility that proliferation of weapons-grade fuel will make possible the production of explosive devices . . .

If man is to survive on this earth, it will not be through his genius for technical innovation. It will be through his wisdom in choosing between those innovations that he can control and those that he cannot.

George Dryfoos, July 7, 1978

Who was Ned Ludd? And who were the Luddites? Is Vardamis suggesting that antinuclear demonstrators are modern-day Luddites? Is the comparison a valid or invalid one? Defend your position. Has the United States ever rejected an otherwise useful technology, that others have gone on to develop, simply for environmental reasons?

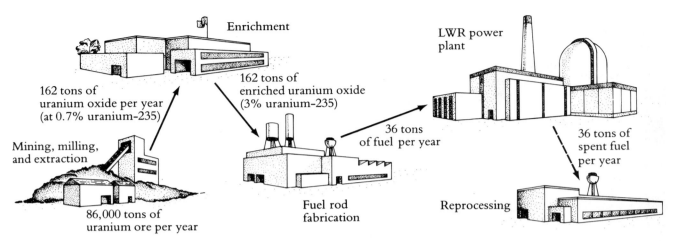

Figure 18.6 The nuclear fuel cycle. The numbers on the diagram are approximate annual requirements for one 1,000-megawatt (electrical) pressurized-water reactor.

research on a new enrichment process, laser enrichment of uranium. Two gaseous diffusion plants currently operate in the United States, one in Paducah, Kentucky, the other in Portsmouth, Ohio. (The original plant in Oak Ridge, Tennessee was mothballed in 1985 because of an oversupply of enriched fuel on the world market, primarily from France and Germany. The possibility of resuming operations at Oak Ridge was left open if demand for enriched uranium fuel picked up sufficiently.)

The enriched uranium from the gaseous diffusion plant is next placed in fuel rods at a fuel fabrication plant. The fuel rods are shipped, in turn, to the power plant where, as we discussed earlier, they produce heat in the core of the reactor. When fission products have built up in the rods to a level that slows the chain reaction, the spent rods are removed from the core. Such spent rods were originally intended to go to a fuel reprocessing plant, where remaining fuel resources could be recovered. At the moment, however, in the United States the rods are simply piling up. The reason for this buildup will be discussed shortly.

Fuel Reprocessing and Spent Fuel Rod Storage

The most serious environmental problems in the fuel cycle begin after the spent fuel rods have been removed from the core of the power plant. Because quantities of reusable fuel are still in the rods, the advocates of nuclear power have viewed the step of fuel reprocessing

as a crucial one. However, hazardous radioactive elements, the fission products, are also in the rods.

Of the reusable fuels in the rods, there is first the uranium-235 that has not fissioned and can still be used if it can be recovered. Another substance present in the spent rod that can also be used as a nuclear fuel is *plutonium*. Plutonium is not present in the rod initially but is formed, or "bred," from uranium-238 during the process of fission. Recall that uranium-238 does not participate in the fission process, even though it is the most abundant uranium isotope in the rod. However, during the fission process, a small but significant portion of the uranium-238 is converted by nuclear bombardment to plutonium. The fuel reprocessing plant is designed to separate the useful uranium-235 and the plutonium from the hazardous and unwanted fission products (Figure 18.7).

Through most of the 1970s and 1980s, no reprocessing plant was operating in the United States. The plant at West Valley, New York had been in operation for only a few years when it closed for repairs at the beginning of the decade; it never reopened. The General Electric reprocessing plant built at Morris Plains, Illinois never opened its doors because of design problems. And the Allied General Nuclear Services plant at Barnwell, South Carolina was unable to meet its licensing requirements, so it, too, never opened.

Storage at the power plant. Although a fuel rod is not particularly hazardous when it enters the core, a spent fuel rod contains the products of uranium fission.

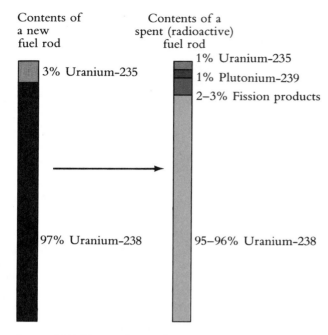

Contents of
a new
fuel rod

3% Uranium-235

97% Uranium-238

Contents of a
spent (radioactive)
fuel rod

1% Uranium-235
1% Plutonium-239
2–3% Fission products

95–96% Uranium-238

Figure 18.7 New and spent fuel rods. Plutonium-239 is created when neutrons are captured in the nucleus of uranium-238. This capture process creates uranium-239, which decays in two steps to plutonium-239.

We might refer to the intensely radioactive fission products from the "burning" of uranium as the "ashes" of the nuclear fuel cycle. Unlike the ashes from coal, though, these substances emit lethal quantities of radiation and remain at high temperatures because of the radiation. The rods must therefore be unloaded from the reactor core through an underwater canal to minimize exposure of plant personnel to radiation. Each year about one-third of the rods in the core are removed and replaced; this amounts to the removal of 36 tons of spent fuel annually for the typical 1,000-megawatt light-water reactor.

The spent rods are removed through the underwater canal to storage in a basin also filled with water. In the first few months of storage in the basin, much of the fast initial radioactive decay within the rods takes place. Most of the dangerous iodine-131 disappears via the decay process in this interval. The rods can then be placed with greater safety into heavily shielded canisters for transport to a fuel reprocessing plant (Figure 18.8).

Because no fuel reprocessing plants have been operating in the United States for almost two decades,

spent fuel rods have been accumulating at power plants for some time. During the Carter Administration, the plan to deal with spent fuel rods relied on **away-from-reactor (AFC) storage** as the means of relieving the space squeeze on nuclear power plants. The concept was to transport the spent rods to several centrally located sites around the country, where they would be stored intact in swimming-pool-like vaults. There they would remain until the government could decide what to do next in the way of final disposal. With the arrival of the Reagan Administration in 1981, the plan for away-from-reactor storage was abandoned. As of 1985, no decision had been made nor plan formulated as to what to do with the spent fuel rods accumulating at the operating nuclear power plants. However, the site at Clinch River, Tennessee where the Department of Energy had been building a breeder reactor, has been considered as a storage location for the accumulating spent fuel rods.

By the late 1970s, a number of nuclear plants had already begun reracking the spent fuel rods in their storage basins to make more room, in order to proceed with refueling on schedule. There is, however, a limit to the number of spent fuel rods that can be stored at power plants. The Nuclear Regulatory Commission

Figure 18.8 Crash test of a spent fuel cask. A rocket-propelled rail car carrying a 74-ton spent fuel cask smashes into a concrete wall at a speed of 81.4 mph. The impact demolished the front of the rail car, turning it into a mass of twisted metal, but the cask was essentially undamaged. Such crashes are meant to duplicate the worst possible accident conditions in order to predict how well containers will survive. (Sandia Laboratories)

has predicted that 27 of the 73 nuclear plants that were operating in 1982 will have filled their storage pools by 1990. Recognizing that the capacity to reprocess spent fuel is far in the future, if ever, utilities are now building their new plants with very large water-filled basins; these basins can store as much as 15 years' accumulation of spent fuel rods. In fact, we may soon find shipments of spent fuel rods moving on major highways as utilities transport the rods from plants that are short of storage capacity to newer plants with remaining capacity for fuel rod storage.

These methods of dealing with spent fuel rods may be termed the "throw away cycle" in the sense that it appears that the rods will never be opened. They will simply be stored until the levels of radiation and radioactive elements have decreased enough to allow the rods to be disposed of as ordinary, nonhazardous wastes.

At the fuel reprocessing plant (West Valley Days).

In 1966, the first fuel reprocessing plant in the United States began operation at West Valley, New York, a town just south of Buffalo. Privately owned by Nuclear Fuel Services, the plant processed spent fuel rods from light-water nuclear power plants, separating the unwanted fission products from the reusable fuels, uranium and plutonium. When the plant closed in 1972 to modernize and expand, it had already processed 600 metric tons of spent fuel.

In 1976, Nuclear Fuel Services announced that it could not reopen its West Valley plant. It offered as reason for its action the additional expense of meeting new earthquake-protection requirements. In fact, the motive for the decision to remain closed may have been the presence of a carbon-steel tank holding about 600,000 gallons (2.25 million liters) of high-level radioactive wastes. A tank of similar material in Hanford, Washington had eventually leaked its radioactive wastes because its steel corroded.

No one was certain how to prevent the tank at West Valley from leaking. An estimate in the mid-1970s for the cost to transfer the wastes safely to a stainless-steel tank was $600 million. In contrast to the tank that leaked at Hanford, the West Valley tank does have double walls. Nevertheless, eventual corrosion and leakage were still seen as certain.

In 1980, Congress directed the U.S. Department of Energy to "clean up" the West Valley plant—specifically, to remove the wastes from the tank and encase

them in glass. The leak of high-level wastes "waiting to happen" at West Valley, followed by federal action to prevent a serious situation, was one result of the plant's six years of operation. Other activities at West Valley do not bolster our confidence in the nuclear industry's ability to handle nuclear wastes. Surveys of the environment around the plant found radiation doses to people to be very small, but the control of radiation received by employees inside the plant was less successful. The company did not build into the plant the capability for remote repair of broken equipment, apparently as a means to save money in plant construction. Employees repaired broken equipment by hand, even though the broken machines were probably contaminated by radioactivity. This resulted in high doses of radiation to the regular employees of the plant. The "dirty jobs," those with the worst exposures, were given to unskilled labor hired off the streets of Buffalo. The unskilled workers were sent in to do a job, were allowed to continue until reaching a maximum level of radiation exposure, and were then taken off the job. For their daring or ignorance, they received a half day's pay for as little as a few minutes of work (see Figure 18.9).

As mentioned earlier, two other reprocessing plants have been built. The General Electric plant at Morris Plains, Illinois never worked and never opened.

Figure 18.9 Remote repair and maintenance of equipment. The West Valley reprocessing plant was built without such facilities and consequently had many "dirty" jobs that had to be done by human hands.

The Allied General Nuclear Services★ plant at Barnwell, South Carolina was never licensed to operate by the Department of Energy, although it was substantially complete. Supplemental federal funding required for the Barnwell plant ceased in late 1983, and the companies that built Barnwell have filed suit against the federal government to recover their investments. Although the facility could conceivably be brought back to a state of readiness, most movable heavy equipment was sold off to recover a portion of costs.

The closing at West Valley, the failure at Morris Plains, and the delays at Barnwell all contributed to an uncertainty in the utility industry about the future of nuclear power. In the late 1970s, it was apparent that the electrical industry's enthusiasm for nuclear power had waned. The number of orders for new reactors to be sited in the United States fell to zero in 1979, and no reactors for the United States have been ordered since. This halt in orders meant a declining rate of new nuclear units going into service in the mid-1980s.

Arguments for and against reprocessing. Three reasons have been put forward for the reprocessing of spent fuel. First, some experts have asserted that there "may be" a reduction of costs for nuclear fuel if the uranium and plutonium remaining in the spent fuel are purified and recycled. Although the reprocessing of the fissionable uranium-235 and plutonium can decrease the need for mining and enriching new fuel, there is controversy over how much savings this makes possible. Those who favor reprocessing estimate up to a 20% savings in the cost of electrical power if spent fuel is reprocessed. Other scientists estimate savings in electrical costs of only 1%–2% if reprocessing of spent fuel is carried out.†

The second reason cited for fuel reprocessing in the United States is to decrease dependence on foreign sources of uranium. If foreign suppliers have a large share of a nation's market, they could raise prices or could put an embargo into effect in order to influence American foreign policy. France and Germany, two industrial nations that have decided to reprocess nuclear fuel, have little uranium resources of their own. They also have relatively low amounts of other energy re-

sources. Reprocessing makes these nations less dependent on uranium fuel imports.

The third reason given for reprocessing is to reduce the difficulty of disposing of radioactive wastes. This reason, however, has been vigorously debated, since if no reprocessing takes place, the high-level wastes remain neatly packaged inside the fuel rods.

Probably one of the reasons for the halt to reprocessing in the United States was the need to ship plutonium by road or rail from the reprocessing plant to the plant that manufactured the fuel rods. If plutonium, the raw material for atomic bombs, were hijacked, the risks would be great. If spent fuel were not reprocessed, there would be no plutonium recovery or plutonium shipment.

Concerned that plutonium could be hijacked for nuclear weapons, U.S. diplomats have asked Britain, France, and Germany to reconsider their decisions to reprocess nuclear fuel. Yet Britain is building a reprocessing plant at Windscale in Cumbria, with a capacity twice that of Barnwell. The British public is aroused, however, because of a nuclear accident at an earlier plant at Windscale in 1972. The British people have also protested a decision to use the plant at Windscale to reprocess the spent nuclear fuel of Japan. The nation's energy authority was accused of making Britain the world's nuclear "dustbin" (a British term for trash can). A number of delays in reprocessing have occurred as the British ponder their decision.

In Marcoule, France, an existing plant offers a reprocessing capacity of 1,000 metric tons (MTU) per year. At La Hague, France, an 800-MTU plant has been built that is scheduled to be expanded several times over. In 1985, Germany announced plans for a reprocessing plant to be located at Wackersdorf, near the border with Czechoslovakia.

In addition, France and Germany are selling their reprocessing technology. France has contracted to sell reprocessing technology to Pakistan; West Germany has offered to sell reprocessing technology to Brazil. France and Germany have agreed under pressure not to export reprocessing technology in the future, but their agreements with Pakistan and Brazil may still be honored.

PUREX process. Why did the United States oppose these sales and work against the ownership of reprocessing plants by other countries? The answer lies in what reprocessing—specifically, the PUREX (pluto-

★A partnership of General Atomic and Allied Chemical.

†See *Nuclear Power, Issues and Choices* (Cambridge, Mass.: Ballinger Publishing, 1977).

Figure 18.10 A crane prepares to place a spent fuel cask inside a vertical decontamination structure. Note the receiving pool in the foreground, within which the washed-down cask will later be opened so that fuel bundles can be removed. (Photograph courtesy of Nuclear Fuel Services, Inc.)

nium and uranium extraction) process—does. It produces plutonium that can be used in bombs.

The PUREX process begins with the mechanical chopping of fuel rods into fragments; the fuel is then dissolved in nitric acid (Figure 18.10). Uranium and plutonium are extracted from the solution by mixing it with a solvent that selectively dissolves these elements. The reprocessing plant produces three separate products: (1) uranium, a portion of which is uranium-235; (2) plutonium; and (3) wastes (the fission products). The amount of plutonium that could be recovered annually from the spent fuel of a 1,000-megawatt light-water reactor is about 550 pounds (about 250 kg), or enough to produce about 15 atom bombs per year.

The leverage that the United States used under the Carter Administration to induce Europe and Japan to withhold nuclear technology from nonnuclear nations was the fact that the United States is a principal supplier of enriched uranium to the European nations and Japan. That supply, Europe and Japan were told, could be interrupted if exports went forward. Hints have surfaced that the Reagan Administration may no longer care to deter the export of nuclear technology to nonnuclear nations. This could be a signal that U.S. suppliers may also be allowed to compete with their European counterparts in overseas sale of nuclear technology. The impact of such policies on the spread of nuclear weapons is predictable.★

CIVEX process. In the PUREX process, uranium and plutonium are recovered separately. Nonetheless, it is possible to modify the PUREX process so that uranium and plutonium are recovered together as a mixture. The mixture of uranium and plutonium would be mainly uranium-238, a substance that cannot be used for bombs. Only about 1% would be uranium-235 and 1% would be plutonium. Since this material could not be used for bomb building, it would be much less valuable to would-be hijackers. The uranium-plutonium mixture can be used in fuel rods if it is first mixed with uranium extra-enriched in uranium-235. The final mixture would then be about 3% fissionable material. The modification of the PUREX process is called "coprocessing" and lately has been called CIVEX (for civilian extraction).

The CIVEX process would, in addition, purposely leave some small quantity of fission products in the recovered uranium-plutonium mixture. This radioactivity from the fission products would make the mixture physically unapproachable for humans. Only remote handling would be possible. This mixture of uranium and plutonium, made dangerously radioactive by the fission products, will seem much less attractive to would-be hijackers when it moves from the reprocessing plant to the fuel fabrication plant. Advocates of CIVEX claim that a nation given only this reprocessing technology would not be able to produce the pure plutonium needed for bombs. The separation of plutonium from the mixture would, however, require only one additional chemical step. Coprocessing would also continue to produce fission products (high-level wastes) from reprocessing, just as the PUREX process does. (See Controversy 18.2.)

★"Reagan Changes Course on Non-Proliferation," *Science* (June 25, 1982), p. 1388.

CONTROVERSY 18.2

Does Reprocessing Produce Weapons-Grade Plutonium?

Conflicting views exist on the potential for plutonium recovered from spent fuel rods to be used in bombs. W. P. Bebbington (1976) tells us:

> It is not generally appreciated that during the long exposure of fuel in a power reactor, there is an accumulation of plutonium isotopes other than plutonium-239, particularly plutonium-240, which makes it much more difficult to assemble a supercritical mass of plutonium without an efficient premature explosion. Weapons-grade plutonium is made with much shorter exposure in the reactor.

In contrast, an anonymous official in Washington is quoted in *Science* (April 1, 1977) as saying, "For proliferation, you can't find anything worse than PUREX. It was developed for bombs."

A member of the Nuclear Regulatory Commission also addressed the issue of the suitability for bombs of plutonium recovered from the spent fuel of power reactors.

> There is an old notion, recently revived in certain quarters, that so-called "reactor-grade" plutonium is not suitable to the manufacture of nuclear weapons . . .
>
> The obvious intention here is to create the impression there is nothing to fear from separated plutonium from commercial power plants. This is not true. . . .
>
> The fact is that reactor-grade plutonium may be used for nuclear warheads at all levels of technical sophistication . . . we now know that even simple designs, albeit with some uncertainties in yield, can serve as effective, highly powerful

weapons—reliably in the kilotron range. (Gilinisky, 1977)

In support of this latter view, we note that of the seven nations that had tested nuclear weapons by 1983, six had obtained the materials for their bombs from reprocessing. It should be pointed out that the material reprocessed was not necessarily produced in nuclear power reactors.

There is controversy also on the fate of plutonium produced from civilian reactors in Britain; some of this plutonium has been provided to the U.S. Department of Energy. It has been alleged that a portion of this plutonium has or will be used in weapons (*Science*, April 27, 1984).

From the above excerpts, it is clear that experts disagree on the usefulness of reactor plutonium for atomic weapons. What would you tell people if you were asked whether plutonium recovered from nuclear power plant rods is useful for bombs? Given that the experts disagree, suppose the issue of plutonium recovery for nuclear power generation were on a national election ballot as a yes or no question—how would you vote and why? Why might you like more information on who these "experts" are in order to make your decision? Does where these statements were published influence your opinion?

Sources:
W. P. Bebbington, "The Reprocessing of Nuclear Fuels," *Scientific American*, **235** (6) (December 1976), 30.
Victor Gilinisky, "Plutonium, Proliferation and Policy," *Technology Review*, 79(4) (February 1977).
"Congress, DOE Battle Over British Plutonium," *Science*, **224** (April 27, 1984), 365.

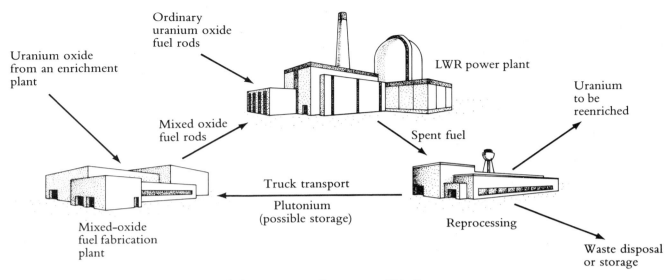

Figure 18.11 Uranium and plutonium recycling system. This diagram shows the nuclear fuel cycle as it was envisioned to operate in the early days of nuclear power in the United States. At present, no recycling system exists here, although such systems do exist in Britain and France.

Mixed-Oxide Fuel and Plutonium Transport

The conventional reprocessing of spent fuel (PUREX) produces separate streams of uranium, plutonium, and waste material, the last consisting principally of fission products. The uranium is transported elsewhere to be enriched in uranium-235. The plutonium can be transported to a special fuel fabrication plant. There it is mixed with uranium that has been enriched in uranium-235; the mixture is then encased in fuel rods (Figure 18.11). These *mixed-oxide* fuel rods can be used in today's light-water reactors. It is called mixed-oxide fuel because both uranium and plutonium are present in the oxide form.

If spent fuel is reprocessed, should plutonium be recovered for use as a supplemental fuel in light-water reactors? One issue is most important to us here: transportation of the plutonium. Plutonium is likely to be moved by truck along ordinary highways from the reprocessing plant to a plant where the mixed-oxide fuel will be made. Although the truck's contents will be clearly labeled "radioactive" to warn the public, there will occasionally be collisions, overturns, flat tires, and so on.

The standards that the specially designed packing must meet are strict, so that the contents of the shipment cannot escape should an accident occur. Furthermore, the shielding that prevents radiation exposure must stay intact so that no radiation escapes. In addition, the component packages of the shipment cannot come in close enough contact to one another to allow a chain reaction (and explosion) to occur.

The older design package for plutonium transport was built up with three layers of steel and insulation, all surrounding about 10 pounds (4.5 kg) of plutonium. Improved designs call for even more insulation and may allow transport of up to 15 pounds (6.4 kg) of plutonium in a package. At 40 drums per truck, about 600 pounds (0.26 metric tons) of plutonium can be moved in a single shipment. If plutonium were being extracted from the spent fuel of 100 operating plants, each of 1,000-megawatt capacity, 100 such shipments of plutonium could be expected annually.

We are concerned about the movement of plutonium on the highways for two reasons. First, it is a dangerously poisonous substance; inhaling the most minute quantities can cause cancer. Second, forceful redirection of the shipment (hijacking) could occur. The plutonium could be used by terrorists for either atomic bombs or dispersal weapons. (See Controversy 18.3.)

CONTROVERSY 18.3

Does the Public Have the Right to Know (How Easy It Is to Steal Nuclear Material and Build an Atomic Weapon)?

> The choice [on nuclear safeguards] is ours to make as a nation, and we believe it should be made on broad economic and social grounds after full public discussion.
>
> **M. Willrich and T. Taylor**

> As seasoned professionals know only too well, truly constructive contributions to an effective, balanced safeguards system . . . require competent, dedicated and sustained hard work—nearly always at low profile.
>
> **G. R. Keepin**

A book published in 1974 by Ballinger Publishing Company set off a controversy in the nuclear community. The book, *Nuclear Theft: Risks and Safeguards*, was written by Mason Willrich and Theodore Taylor. Willrich is a professor of law who has been involved with the U.S. Arms Control and Disarmament Agency. Taylor is a nuclear physicist, now a corporation executive, who has been involved with the design of atomic weapons.

The second chapter of *Nuclear Theft* contained a rather general discussion of bomb building, with extensive quotations from an encyclopedia entry by J. S. Foster, a nuclear weapons expert. The authors chose to omit material, though the references were not classified, on the specifics of bomb building. Willrich and Taylor conclude that:

> Under conceivable circumstances, a few persons, possibly even one person working alone who possessed about ten kilograms of plutonium oxide and a substantial amount of chemical high

explosive (the implication is that the material will be squeezed by the explosive to decrease the needed critical mass) could, within several weeks, design and build a crude fission bomb . . . This could be done using materials and equipment that could be purchased at a hardware store and from commercial suppliers of scientific equipment for student laboratories.

In their opening chapter, Willrich and Taylor explain their reasons for writing the book and for the level of detail they provide:

> [H]ow much does the public need to know about these matters? This question haunts us, and we believe it merits discussion before proceeding further.
>
> To us, the most compelling argument against informing the public about the risks of nuclear theft is that such an effort might inspire warped or evil minds. . . . However, a large amount of in-

formation in much greater detail than we present here is already in the public domain. . . .

. . . when security risks are inherent in a long-term activity, which is clearly the case with nuclear theft, the public in a democratic society has a right to know, and those with knowledge have a duty to inform. . . .

. . . The years just ahead provide the last chance to develop long-term safeguards that will deal effectively with the risks of nuclear theft. Once the material flows in the nuclear power industry are as enormous as expected a few years from now, it will be too late. (pp. 2–4)

Willrich and Taylor closed their introductory chapter with an appeal to readers to arrive at their own conclusions on risks and safeguards, saying that the choice belongs to the people and (by implication) not to the technical experts.

A number of nuclear professionals, however, felt that Willrich and Taylor had provided aid to would-be terrorists by collecting so much information in one source on ways to steal and use nuclear materials. G. R. Keepin (1974), who reviewed the book in a magazine read by nuclear professionals, disagreed with Willrich and Taylor. His position is that the public would be better off if the book had not been printed, that the nuclear industry was moving along satisfactorily with safeguard methods:

Without elucidation, this reviewer would simply record here the opinion that it is both unseemly and counterproductive for a former professional in the weapons field to speculate—however hypothetically—on how a would-be diverter might proceed with design and fabrication of an illicit atomic bomb. Genuine concern with safeguarding . . . nuclear materials could surely

take more constructive forms than indulging in what could turn out to be self-fulfilling prophecy.

The dictates of reason and prudence would appear to reject Taylor's assertion that "it seems necessary to be quite specific" in order to make the risks of nuclear terrorism credible, and to convince the public of the gravity and urgency of the nuclear materials diversion problem. . . .

The important over-all point to be made here is that—notwithstanding certain glaring shortcomings and sins of the past—the AEC and much of the nuclear industry are in fact making great strides toward effective, stringent control of the nuclear materials which are the lifeblood of that industry. (p. 76)

Does the public have the right to know that a technology is unsafe if publishing this information could increase the risks? Would you rather leave the management of nuclear material to the experts, as Keepin suggests, or did Willrich and Taylor make the right decision? Is the discussion here in itself a disservice because it brings the controversy to light?

Here are some references to assist your understanding of this controversy: Gillette, R. "Nuclear Safeguards: Holes in the Fence," *Science*, **182** (December 14, 1973), 1112. Lapp, R. E. "The Ultimate Blackmail," *The New York Times Magazine* (February 4, 1973), p. 13. McPhee, J. "The Curve of Binding Energy," *New Yorker* (December 3, 10, and 17, 1973). McPhee, J. *The Curve of Binding Energy: A Journey into the Awesome and Alarming World of Theodore B. Taylor.* New York: Farrar, Straus, & Giroux, 1974.

Sources:
M. Willrich and T. Taylor, *Nuclear Theft: Risks and Safeguards* (Cambridge, Mass.: Ballinger Publishing, 1974), pp. 2–4.
G. R. Keepin, in *Nuclear News*, **17** (12) (September 1974), 76.

Perhaps the most effective way to reduce the possibility of hijacking is to place the reprocessing plant and mixed-oxide fuel fabrication plant at the same site. The plutonium produced at the reprocessing plant would need to be moved only to an adjacent building to be combined into fuel rods. Nuclear power plants might be located at this same site as well.

International Safeguards

The term *safeguards* refers to the procedures and methods of preventing both the misuse of nuclear materials and the spread of nuclear weapons. A nation building a nuclear power plant may be subject to international treaties that force it to accept safeguards.

When the United States first began to export nuclear technology, it insisted on treaty guarantees from the buying nations that neither the material nor the technology would be used for atomic weapons. The United States reserved the right to inspect all records and reports from the buyers, as well as the nuclear power plants themselves. Over a thousand inspections were conducted by the United States through 1974, but an international agency has now picked up this function. This agency is the International Atomic Energy Agency (IAEA). The IAEA's purpose is to "ensure that special fissionable and other materials, . . . equipment, . . . and information made available by the Agency . . . are not used in such a way as to further any military purpose."

The IAEA does not itself protect fissionable material; this function is left to the nation that owns the material. Nor can the IAEA prevent theft or sabotage, even if it is aware of them. All "police-type" responsibilities are reserved to the owning nation. If a prohibited act is detected, the IAEA may notify the supplier nation so that it can end its cooperation agreement with the violating country, but that is virtually all the agency can do.

The nuclear Non-Proliferation Treaty (NPT), which gave the IAEA its power to monitor and inspect, is designed to prevent new nations from obtaining nuclear weapons or from obtaining nuclear explosives for "peaceful uses." However, a party to the treaty can legally make all the preparations needed to manufacture a nuclear weapon as long as it does not actually assemble the warhead. A summary of the treaty, which was established in 1970, is given in Table 18.1.

Since 1970, more than 100 nations have signed the treaty, and most have approved it internally as well. Nevertheless, many nations have not signed, among them Argentina, Brazil, Chile, France, India, Israel, Pakistan, South Africa, and Spain. China agreed to IAEA guidelines in 1983.

Of the nations that have not signed the NPT, India was able to explode an atomic bomb in 1974, violating a treaty with Canada, which was supplying it with nuclear technology. By diverting nuclear material from a power plant to the manufacture of explosives, India was able to manufacture a nuclear explosive, illustrating the hazard of giving nuclear technology to a nation that has not signed and ratified the NPT. Sales of full nuclear technology, including reprocessing, have also been made to Brazil (by West Germany) and to Paki-

Table 18.1 The Non-Proliferation Treaty

Article I prohibits the transfer of nuclear weapons or other nuclear explosive devices (including devices for peaceful nuclear explosions) to any state. . . . Nuclear-weapon states are also forbidden to assist non-nuclear-weapon states to acquire nuclear weapons or explosive devices.

Article II prohibits non-nuclear-weapon signatories from manufacturing or otherwise acquiring nuclear weapons or devices, including peaceful nuclear explosives.

Article III obligates the non-nuclear-weapon parties to accept international safeguards . . . to ensure that there is no diversion of nuclear material to the manufacture of nuclear explosives.

Source: Stockholm International Peace Research Institute, "Preventing Nuclear Weapons Proliferation." Stockholm, Sweden, January 1975.

stan (by France). Although these nations deny intentions to obtain nuclear weapons, neither has signed the Non-Proliferation Treaty and hence cannot be inspected by IAEA. (In 1985, a citizen of Pakistan was arrested in the United States for attempting to ship an A-bomb trigger to Pakistan.) There is a great deal of insecurity in providing such nations with nuclear technology.

Besides the Non-Proliferation Treaty, a number of European nations belong to the European Atomic Energy Community, known as EURATOM.* In the 1950s this organization created a safeguards system including inspection and audit. The IAEA accepts the safeguard findings of EURATOM, but the IAEA reserves the right to conduct independent studies. The IAEA cannot, however, investigate procedures in France, since France has not signed the NPT.

Unfortunately from the safeguards standpoint, it appears from recently published information that a reprocessing plant could be secretly built in as little as 4 to 6 months. Once in production, the plant could process spent fuel rods from light-water reactors and produce enough plutonium for half a dozen bombs per month. The time for the discovery of violations and the application of sanctions against the offending nation is terribly short.

*France, Belgium, Luxembourg, West Germany, the Netherlands, and Italy were initial members. The United Kingdom, Ireland, and Denmark joined later.

Storage or Disposal of Wastes from Fuel Reprocessing

The safe storage of high-level wastes from a fuel reprocessing plant is an issue of great importance to the nuclear industry. The plans developed for storing these wastes are drawn from experience in nuclear-weapons manufacture, which produces liquid wastes similar to those from fuel reprocessing plants.

At the Hanford site in Washington, at the Savannah River site in South Carolina, and at the National Reactor Testing Station in Idaho, experience has accumulated in handling these high-level liquid wastes. At Hanford, a concrete tank with an exterior carbon-steel liner was used initially. At Savannah River, a carbon-steel tank was placed inside a concrete tank, and steel was used to line the outside. Leaks have occurred from both these tanks because of corrosion of the steel. Exposure of the public has been negligible, but now the underground storage sites of the tanks must be controlled for several hundreds of years.

The corroding or rusting evidently took place at points of stress in the metal tanks. New tanks are now being heat-treated to relieve such stresses. Although no leaks have yet occurred from tanks so treated, a long time passed before the leaks were discovered in the original types of tanks. The tanks used at the Idaho site, on the other hand, are of stainless steel, and leaks from these tanks have not been reported.

The era of liquid wastes from reprocessing is ending, however. At the Idaho site, as well as at Hanford, methods to solidify the liquid wastes were developed. Solidification makes it much more difficult for these wastes to contaminate the environment. Concrete and clay are viewed as useful substances into which the wastes can be physically mixed and fixed in solid form. Fixing the wastes into glass may provide even greater safety because of the continuous character of the solid form of glass. Studies of this method are under way. New regulations for high-level wastes, if reprocessing should ever resume in the United States, call for the wastes to be solidified before shipment from the fuel reprocessing plant.

A number of proposals have been studied for safe storage alternatives or for disposal of high-level wastes. One of the most unusual suggestions is to rocket the wastes into deep space, but the costs to dispose of many tons of high-level wastes in this way are very high. Furthermore, one cannot rule out an accident at the launching pad or a failure to exit the earth's gravity due to a failure of the boosting rockets.

Other unusual suggestions have been put forward, all involving burial in the deep earth. One concept would see the wastes placed in the geologic trench in the ocean where the ocean's tectonic plates are gradually sliding beneath those of the continent. Another idea is to dispose of the wastes in a deep shaft drilled into the earth's core or in a cavity carved out by a nuclear explosion. In none of these processes can the wastes be retrieved at some future time. Nor can the results of these plans be carefully evaluated with present knowledge.

As early as 1955, however, nuclear scientists and engineers were considering the possibility of storing high-level wastes in deep geologic formations, such as salt beds. Natural underground salt beds exist in a number of places in the United States and apparently remain stable in both their physical and chemical characters for thousands of years. The judgment that salt beds are a safe place to store nuclear wastes has been reaffirmed many times, but recent criticism suggests that the heat generated by the wastes could alter the salt formations.

In 1972, the Atomic Energy Commission selected a site near Lyons, Kansas, for Project Salt Vault. The project at Lyons was designed to show that storage in salt beds would work, but the project was abandoned when it was discovered that salt mining (by dissolving salt in water) within the formation was being conducted not far away. The unknown effect of mining on the stability of the formation was the key factor in ending the project at Lyons. The idea of storing high-level wastes in bedded salt, however, was not abandoned because of this setback. A waste isolation pilot plant (WIPP) is being constructed in New Mexico to test the idea further.

Liquid-Metal Fast Breeder Reactors

So far we have discussed the light-water reactors that utilize the fission of uranium-235 as a source of heat to generate electric power. But surprisingly, light-water reactors have been viewed by nuclear advocates as a short-term source of energy. These people see light-water reactors as a means to supplement and replace fossil fuels until the time when more secure sources of

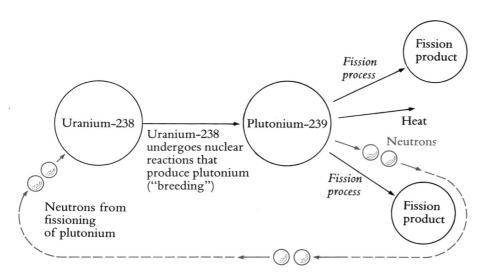

Figure 18.12 Nuclear reactions in the breeder reactor. Plutonium is "bred" by bombarding uranium-238 with neutrons. These neutrons are furnished by the fission of plutonium-239. The fission process produces heat, which eventually is used to make steam and releases the neutrons needed to breed more plutonium-239 from uranium-238.

nuclear power, such as the breeder reactor, become available.

The **liquid-metal fast breeder reactor (LMFBR)** is viewed as the successor to the light-water reactor. It is under development in the U.S.S.R., Britain, Japan, France, the United States, and possibly other countries as well. Although the first commercial breeder reactor was built in Scotland, the United Kingdom is now cooperating with France, whose Phoenix breeder began delivering electric power in 1973.

The light-water reactor uses uranium-235 for fission, but uranium-235 is not as plentiful in nature as uranium-238, the dominant component of uranium ore. The breeder reactor utilizes the abundant uranium-238 to produce power. The idea of using uranium-238, even though it does not naturally undergo fission, is ingenious.

In our discussion of light-water reactors, we pointed out that the fission of uranium-235 "breeds" plutonium-239 from uranium-238. Like uranium-235, plutonium-239 undergoes fission when properly struck by a neutron. In much the same fashion as uranium-235, plutonium breaks apart into two fragments, with a release of neutrons and energy. The neutrons released sustain the fission of more plutonium-239, producing heat. The neutrons also "breed" more plutonium from uranium-238 atoms. The process, illustrated in Figure 18.12, both produces heat from fission and "breeds" more fuel to undergo fission. This explains the word *breeder* in the name of the reactor.

Instead of using water to withdraw heat energy

from the core, this reactor uses a liquid metal (pure melted sodium) to withdraw the heat of plutonium fission from the reactor core. Liquid sodium has some advantages. Sodium melts at 210°F (99°C) but does not boil until it reaches 1640°F (893°C). Because its boiling point is so distant from its melting point, it is unlikely to "flash" to a vapor if an accidental rupture of a line occurs. The loss-of-coolant accident, which is a concern in light-water reactors, is therefore thought to be much less probable in a breeder reactor. In addition, the sodium in the core will not slow down the high-energy neutrons released in the fission process, which are essential for breeding more plutonium.

The design of the LMFBR, shown in Figure 18.13, calls for a primary loop containing liquid sodium, which circulates through the core of the reactor. This sodium, which will become radioactive as it circulates through the core, picks up the heat produced by the fission process. A second loop also contains circulating liquid sodium. This sodium picks up the heat from the first loop in a heat exchanger. A third loop contains water that is boiled to steam by the heat from the second loop. The steam turns the turbine, is condensed, and returns to the boiler.

The fuel rods in the LMFBR contain a mixture of plutonium oxide and uranium oxide. In time, as fission proceeds, fission products build up in the rods and "poison" the reaction by capturing a portion of the neutrons, just as occurs in light-water reactors. When neutron capture slows heat production too greatly, the rods are withdrawn and sent to a reprocessing facility.

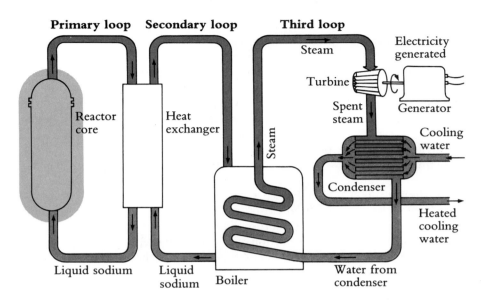

Primary loop **Secondary loop** **Third loop**

Steam

Electricity generated

Turbine

Spent steam

Generator

Reactor core

Heat exchanger

Steam

Cooling water

Condenser

Heated cooling water

Liquid sodium

Liquid sodium

Boiler

Water from condenser

Figure 18.13 Schematic diagram of the liquid-metal fast breeder reactor (LMFBR).

There, plutonium and uranium are separated from the fission products. The plutonium content will have increased during the breeding cycle in the reactor. The fission products are wastes and require disposal. New mixed-oxide fuel rods are fabricated from the plutonium and from uranium.

The familiar problems of reprocessing, plutonium movement, and disposal or storage of fission-product wastes are all present in the breeder reactor concept. Furthermore, the quantities of plutonium produced are enormous, even in comparison to the quantities potentially produced by light-water reactors. A core meltdown after a loss of the sodium coolant would pose an even greater problem than a meltdown in a light-water reactor, because the core contains so much more of the highly toxic plutonium. Thus, although the breeder reactor is an idea of cunning ingenuity for extending fuel supplies, the problems and issues that surround its development must be critically examined. The United States, although it has abandoned work on a demonstration breeder plant at Clinch River, Tennessee, still conducts $300 million worth of breeder reactor research each year.

Economic and Social Aspects of Nuclear Power

The terror and destruction of Hiroshima and Nagasaki weighed on the consciences of the nuclear scientists who built the first atomic bombs. Many of these sci-

entists saw in the awesome new technology hope for supplying electrical power, for turning the power of the atom to peaceful uses; these scientists pushed their dream. It was a dream that would in part release them from the responsibility they bore for the birth of atomic power.

These scientists then did not know or foresee all that would eventually be known about radiation effects. Nor could they foresee all the issues involved in the safe disposal of radioactive wastes. But they had confidence that all the problems in the nuclear fuel cycle could be resolved, and so they persevered. They were very successful at pushing their dream, and they were very successful at resolving many of the problems of the nuclear fuel cycle. They were the architects of the current age of nuclear electrical power. To some, however, the dream of the nuclear scientists was ill considered, was too hastily pushed, and would allow too many nations to possess nuclear weapons.

Three decades have now elapsed since President Eisenhower opened nuclear technology to private use. It is logical to ask, "What is the future of nuclear electric power in the United States and in other parts of the world?" On the one hand, more and more nuclear plants are coming on line in the United States, and the fraction of our electrical power supplied by nuclear energy is steadily increasing (see Chapter 15). On the other hand, the nuclear industry—the collection of companies that build the generators and the major components that are assembled at the power plant sites—is in trouble. There have been no orders for nuclear reactors to be constructed in the United States since 1978. Only orders for reactors in overseas countries have kept the U.S. nuclear industry afloat. What are the barriers that have stalled and halted, at least for the moment, the spread of nuclear technology in the United States? What is the impact of the sinking status of the U.S. nuclear industry?

The factors that have slowed construction and halted new orders for nuclear reactors are multiple and interrelated. Difficulties in obtaining full licensing for the operation of new plants have so delayed these plants coming on line that their costs have soared. Of course, the licensing procedures were designed to ensure safe operation of the plants. From about $1 billion for a 1,000-megawatt nuclear plant in the mid-1970s, the cost rose to on the order of $1.5 billion by 1985. Construction costs, driven up by delays and by high interest rates, are one aspect of added costs. The price of uranium fuel has also risen; it quadrupled in the late 1970s from $10 per pound to $40 per pound when an international cartel of nations and companies successfully imitated the tactics of OPEC, the oil cartel; it has remained high since that time. The economics that pointed utilities toward nuclear energy and away from fossil fuel have thus been confounded.

The delays in licensing are part of an effort to ensure the safety of each nuclear plant and to ensure that the impacts of each plant on its local environment are acceptable. The fact that nuclear plants are less efficient than coal-fired plants means that more heat is discharged to the cooling water and that thermal pollution effects must be more closely examined. Safety has become an even greater concern since the reactor incident at Three Mile Island. In large part, the concern with safety must reflect the point of view of the public at large and not simply be the opinion of the engineers, scientists, and others who regulate the nuclear industry. Thus, the slowing of the industry growth represents the effect of public opinion. In addition, the search for disposal sites for radioactive wastes has been slowed by hostile receptions in many states. The concern is that storage will not be as safe as promised.

The nuclear industry's orders for new plants have ground to a halt, and cancellations of previous orders continue. Yet plants that have cleared the hurdles are still being built. Nuclear engineers are needed to staff them. What of the training of new nuclear engineers in the United States? With declining enrollments, future staffing of plants may have to draw on foreign-trained nuclear engineers.

We are witnessing a very unusual event in the history of any nation. We are gradually rejecting nuclear electrical power. The current plants and plants in the building stages will be with us for a long time, but the trend is clear. For the moment, coal appears to be the route chosen for the next generation of power plants; other choices are not on the horizon.

Overseas, the trends are not so clear; smaller nations want nuclear power, want to be part of the nuclear club. France, Germany, and Britain, though public opposition exists, are building the entire fuel cycle, including reprocessing. Nuclear electrical energy will fill a large portion of their energy needs. The United States is either being left behind or is choosing another route to the future.

Summary

Nuclear electric power dates only from 1956, when the first experimental reactor produced power at Argonne National Laboratory. The industry experienced rapid growth until 1974, when a decline in orders set in. Since 1978, no new reactors have been ordered for power production in the United States and many reactor orders have been cancelled.

The principle of nuclear electric power is the sustained fission (breaking apart) of uranium-235, in fuel rods located in the reactor core. This fission produces heat, radioactive wastes, gamma rays, and neutrons. Water is boiled to steam by the heat from the fission process. The steam turns a turbine, generating electricity. Two reactor types are in wide use in the United States: boiling-water reactors (BWR) and pressurized-water reactors (PWR).

The waste products from the fissioning of uranium-235 include the radioactive forms of krypton, cesium, strontium, and iodine. Remarkably little radioactive material escapes to the environment during normal operations. Eventually the plant itself wears out and, because it is radioactive, must either be disposed of or guarded for a century. This problem has not yet been faced. The safety concerns for nuclear plants center on the possibility of a loss-of-coolant accident, in which water is lost from the core (such as through a pipe break), causing the fuel rods to heat up uncontrollably, and on the misuse of plutonium, which forms in the rods during fission.

The architects of nuclear power originally intended that the fuel in spent rods would be reprocessed to recover the fissionable plutonium, which can also be used as a fuel, as well as to recover the uranium mixture that can be reenriched in uranium-235. The idea was to avoid as much further mining or importation of uranium as possible. Unfortunately, the fissionable plutonium is suitable for building atom bombs. Production and shipment of plutonium from reprocessing creates the risk of hijacking or diversion of the plutonium for use in nuclear weapons. The amount of plutonium that could be produced each year from the spent fuel of one 1,000-megawatt nuclear plant would be sufficient to produce about 15 atom bombs. Although Britain and France are reprocessing spent fuel, the United States no longer does so, and spent fuel rods are piling up in huge water pools at the electrical plants.

The International Atomic Energy Agency has the power to monitor and inspect the nuclear activities of nations that have signed the nuclear Non-Proliferation Treaty (NPT) as a means of preventing the production of nuclear weapons. The IAEA has no power to halt such activities but instead notifies the nation that supplied the technology to the offending nation. France, India, Pakistan, Brazil, Israel, and several other nations have not signed the NPT and cannot be inspected.

Storage concepts for the wastes from fuel reprocessing have evolved to the point that encasement of the radioactive wastes in glass is now considered optimum. Storage of the glass-encased wastes in deep salt mines is being explored.

A concept that would make use of the abundant uranium-238 is the liquid-metal fast breeder reactor (LMFBR), which relies on plutonium as its principal fuel. The plutonium is continually created in the core of the reactor by neutron bombardment of uranium-238. Liquid sodium is used to carry off heat from the core. France has built an LMFBR, but the United States has cancelled plans for one, though research on the concept is still being conducted.

The United States appears to be abandoning nuclear power as a permanent means of power generation. Yet other nations are increasing their reliance on nuclear energy steadily and plan to continue to do so. The citizens of the United States appear to have decided that nuclear energy is not worth the multiple risks associated with it.

Questions

1. What are the two basic types of light-water reactors?
2. Describe the basic differences between the two reactor types.
3. As the reactor operates, what eventually happens to the level of uranium-235 in the fuel rods? to the level of fission products?
4. What is the function of the heat exchanger, as shown in Figures 18.3 and 18.4, in the electric generation process?
5. What accounts for the presence of fission products in the water that circulates through the core of a nuclear reactor? Why should we be concerned about the ultimate disposal of water in which these fission products are dissolved?
6. What is done with the solid radioactive wastes generated by nuclear power plants?
7. Explain what is meant by core melt, or meltdown. What would cause a meltdown? What safeguards are designed into nuclear power plants to prevent this from happening?
8. What is meant by fuel reprocessing? What substances are separated out? Which of these substances can be used again?
9. Discuss the pros and cons of reprocessing as an alternative to indefinite underground storage of spent fuel rods.
10. Why did France and Germany proceed with reprocessing in the 1970s?
11. Why did the U.S. work against the sale of reprocessing technology to Brazil and Pakistan?
12. What rights does the nuclear Non-Proliferation Treaty give to the International Atomic Energy Agency (IAEA)?
13. Does the IAEA have the right to seize fissionable material to prevent manufacture of nuclear weapons?
14. List and discuss briefly at least three methods that have been proposed for the storage and/or disposal of high-level radioactive wastes.

Further Reading

General References

Bupp, I., and J. Derian. *Light Water: How the Nuclear Dream Dissolved.* New York: Basic Books, 1978.

Weinberg, A. "Social Institutions and Nuclear Energy," *Science,* **177** (July 7, 1972), 27–34.

Weinberg, A. "Salvaging the Atomic Age," *The Wilson Quarterly* (Summer 1979), pp. 88–112.

"The Future of Nuclear Power, a Global View," *Civil Engineering-ASCE* (August 1981), p. 74.

Weaver, K. "The Promise and Peril of Nuclear Energy," *National Geographic* (April 1979), p. 459.

Lester, R. "Is the Nuclear Industry Worth Saving?" *Technology Review* (October 1982), p. 39.

Hileman, B. "Trends in Nuclear Power," *Environmental Science and Technology*, **16** (7) (1982), 373A.

"A Meltdown for Nuclear Power," *Business Week* (January 30, 1984), p. 18.

Kaku, M., and J. Trainer (eds.). *Nuclear Power: Both Sides.* New York: Norton, 1983.

Nuclear Plant Reactor Safety

"NRC Panel Renders Mixed Verdict on Rasmussen Reactor Safety Study," *Science*, **201** (September 29, 1978), 1196.

Shapely, D. "Reactor Safety: Independence of Rasmussen Study Doubted," *Science*, **197** (July 1, 1977), 29–31.

"The Accident at Three Mile Island (The Need for Change: The Legacy of Three Mile Island)." Report of the President's Commission, J. Kemeny, Chairman, October 1979.

Marshal, E. "Ultra Safe Reactors, Anyone?" *Science*, **219** (January 21, 1983), 265.

Weinberg, A., and I. Spiewak. "Inherently Safe Reactors and a Second Nuclear Era," *Science*, **224** (June 29, 1984), 1398.

Mackenzie, J. "Finessing the Risks of Nuclear Power," *Technology Review* (February/March 1984), p. 34.

Reprocessing, Wastes, and Nuclear Weapons Spread

Lester, R. K., and D. J. Rose, "The Nuclear Wastes at West Valley, New York," *Technology Review* (May 1977), pp. 20–29.

Spector, L. *The New Nuclear Nations.* New York: Random House, Vintage Books, 1985.

Spector, L. "Nuclear Proliferation: The Pace Quickens," *Bulletin of the Atomic Scientists*, **41** (1) (January 1985), 11.

Gilinisky, V. "A Common-Sense Approach to Nuclear Waste," *Technology Review* (January 14, 1985), p. 14.

Gilinisky, V. "Plutonium, Proliferation and Policy," *Technology Review* (February 1977), p. 58.

Dunn, L. *Controlling the Bomb: Nuclear Proliferation in the 1980s—A Twentieth Century Fund Report.* New Haven: Yale University Press, 1982.

Bernstein, C. "A Salt Solution for Nuclear Wastes," *Civil Engineering/ASCE* (February 1986), p. 3.

References

Nuclear Energy Policy Study Group, S. Keeny, Jr., Chairman. *Nuclear Power, Issues and Choices.* Cambridge, Mass.: Ballinger Publishing, 1977.

"Risk Assessment Review Group Report to the Nuclear Regulatory Commission." Prepared for the U.S. Nuclear Regulatory Commission, MUREG/CR-0400, September 1978.

Marshall, E. "Lessons of Chernobyl," *Science* **233** (September 26, 1986), 1375–1376.

Schwarzschild, B. "Cause and Impact of Chernobyl Accident Still Hazy," *Physics Today* (July 1986), pp. 17–23.

Levi, B. "Soviets Assess Cause of Chernobyl Accident," *Physics Today* (December 1986), pp. 17–20.

Norman, C. "Hazy Picture of Chernobyl Emerging," *Science* (June 13, 1986), pp. 1331–1333.

Wilson, R. "A Visit to Chernobyl," *Science*, **236** (June 26, 1987), 1636.

Keepin, G. R., "Nuclear Materials Safeguards: A Professional Speaks Out," *Nuclear News* (September 1974), p. 76.

Willrich, M., and T. Taylor, *Nuclear Theft: Risks and Safeguards*, Cambridge, Mass.: Ballinger Publishing, 1974.

A GLOBAL PERSPECTIVE IV

Oil Pollution and Nuclear Winter

Oil Pollution of the World's Waters
*Where the Oil Comes From/Biological Effects of Oil/
Lessening Oil Pollution and Its Effects/Economic and
Social Effects of Mixing Oil and Water*

Nuclear Winter
*Atmospheric Dust and Climate/Scenario for Nuclear
Winter*

he production and use of energy can have a number of detrimental effects on the global environment: pollution of the oceans with oil; acid rain and the buildup of carbon dioxide from the burning of fossil fuels; and the destruction of ozone in the atmosphere by jet planes. Still another global environmental insult may stem from power production. The spread of nuclear weapons may accompany the widening use of nuclear electrical power among the smaller nations. A nuclear war begun by these nations or by an exchange between the United States and the Soviet Union could lead to excessive ash particles in the atmosphere and the onset of a "nuclear winter." In the next three Global Perspectives we examine global aspects of pollution from energy production and use. The first issue discussed, oil pollution, is basically the study of small, continuous insults to the world's oceans, resulting, so far, in local degradation. The second issue is the event called nuclear winter. Here we are concerned with the possibility of environmental disaster on a truly global scale—the cessation of all plant growth on earth as a consequence of the misuse of atomic energy.

Acid rain, carbon dioxide buildup, and the ozone problem are examined in later Global Perspectives.

Oil Pollution of the World's Waters

"Oil pollution." The words bring to mind pictures of wrecked tankers grinding over submerged rocks or of geysers of flaming oil shooting from well blowouts. Yet, historically, such dramatic happenings have accounted for only a small portion of the 2–5 million tons of oil added each year to the world's waters.

Most of the oil spilled into natural waters has been the result not of accidents but of normal operations. Even in 1979, the worst year on record for accidents, twice as much used motor and industrial oil entered the oceans as was spilled in oil-well or tanker accidents (Figure IV.1).

Where the Oil Comes From

Used motor and industrial oil. In most years, half of the oil entering natural waters is used motor and industrial oil from sewer outflows and from runoff. Although anyone found dumping oil into a waterway

in the United States is subject to heavy fines, there are still few convenient depots for the collection of such wastes as used automobile oil. Thus, used oil most often finds its way into sewers or garbage dumps. From there it runs into nearby waterways.

Waste oil can be collected, refined, and then used again, although this is a complicated process. Many substances are added to oil that allow it to be used longer at higher temperatures and at increased bearing loads. Some examples of materials added to oil include compounds of barium, calcium, zinc, magnesium, phosphorus, and chlorine. Occasionally even vegetable and animal fats are added. Along with these additives, used oil picks up contaminants from the air and the machine itself.

However, new processes have been developed that appear to remove all contaminants and produce an oil that works as well as virgin oil. One of the main problems involves collecting waste oil so it can be reused. In some areas, citizen groups have organized waste oil collection campaigns. This involves setting up depots where car owners can drop off used oil after they change the oil in their cars; it also involves educating people about the need to dispose of oil properly. The Federal Energy Administration has prepared a "Waste Oil Recycling Kit" to show citizen groups how to begin recycling centers for used oil in their communities.

Normal shipping operations. All ships collect both oil and water in their bilges. The simplest method of disposing of this oily water is to pump it into the ocean. An international treaty, written in 1973 by the United National Intergovernmental Maritime Consultive Organization (IMCO), forbids the discharge of oily wastes near shores and limits the amount that can be dumped on the high seas. However, several important shipping nations did not sign the treaty. In areas where dumping is prohibited by the treaty, illegal dumping still goes on, especially at night when it is hard to detect.

A further contribution of oil pollution is made by oil tankers. Many tankers pump sea water into empty oil tanks as ballast and use sea water to wash out tanks between cargoes. When such a tanker nears an oil-loading port, it has tanks full of oily water to dispose of. For many years, the water was simply pumped back into the ocean. Newer ships are fitted with separate ballast and oil tanks or with slop tanks where oil can be recovered from ballast water. Further, some

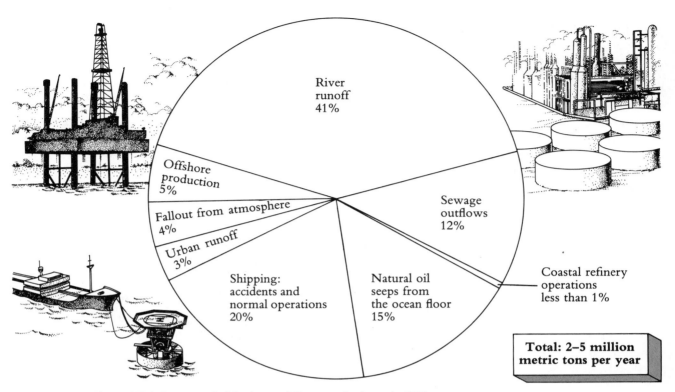

River
runoff
41%

Offshore
production
5%

Fallout from atmosphere
4%

Urban runoff
3%

Shipping:
accidents and
normal operations
20%

Natural oil
seeps from
the ocean floor
15%

Sewage
outflows
12%

Coastal refinery
operations
less than 1%

**Total: 2–5 million
metric tons per year**

Figure IV.1 Sources of oil in the world's waters in the early 1980s.
Most of the oil was used motor and industrial oil carried in runoff and
sewage inflows. Of the portion from shipping operations, the largest
part was due to normal procedures. In most years, only a small portion
came from accidents. (Adapted from J. W. Farrington, in *Petroleum in
the Marine Environment*, L. Petrakis and F. Weiss, eds. [Washington,
D.C.: American Chemical Society, 1980]; and from *Oil in the Sea: Inputs, Fates and Effects*, [National Academy Press, 1985].)

tankers now wash out tanks with oil rather than with
water. As of 1986, ships serving U.S. ports must be
fitted with equipment to reduce oil pollution from normal operations.

Tanker accidents and offshore blowouts. Although tanker accidents and offshore oil-well blowouts
normally account for only a small percentage of the
total amount of oil pollution, they result in most of the
visible damage. The reason is that a grounded tanker
or an offshore blowout releases an enormous quantity
of oil at one time.

Because it is cheaper to ship oil in large quantities,
the size of the tankers traveling the oceans is increasing
rapidly. The average-size supertanker now in use is as
long as three football fields and carries about 250,000
tons. Damage from the wreck of the 233,690 ton

Amoco Cadiz in 1978 included total loss of the oyster
beds along the Brittany coast, as well as effects on the
fishing and resort industries. Total costs eventually exceeded $30 million.

The blowout of the Mexican oil well Ixtoc in the
Gulf of Mexico in 1979 released more than 3 million
gallons (about 10,000 tons) of oil before it was brought
under control nine months later. Oil and dead sea birds
washed up on the beaches of barrier islands in South
Texas, more than 3,600 miles (6,000 km) away. Oil
from a 1968 well blowout off the coast of California in
the Santa Barbara channel caused the death of at least
3,600 birds.

Such accidents have left a legacy of mistrust about
the safety of drilling for oil in marine waters. Yet the
stakes, in terms of energy resources, are high. Some
13.6 billion barrels of oil are believed to lie under-

ground in Britain's North Sea fields, while 12 billion barrels of oil are probably yet to be found offshore under the U.S. continental shelf. Governments want to accelerate the development of these resources, both to decrease reliance on foreign oil and to increase government revenues from the royalty payments oil companies make on the oil they obtain.

Oil-well blowouts are not the only problem connected with development of oil resources in the marine environment. During normal well-drilling operations, drilling muds and cuttings are released and chronic oil spills occur. The drilling muds contain a variety of chemical additives, including heavy metals as well as diesel oil. Later, while oil is pumped from the well, water comes along with it. This oil-and-water mixture is usually separated on the oil rig. The water, which still contains an average of 25 ppm of oil, is discharged back into the sea. In addition, transporting the oil to shore involves small spills due to the loading and unloading of barges or to pipeline leaks.

The extent of the problem of both accidental and intentional oil spills is brought home by Figure IV.2, which shows the locations of visible oil slicks in the early 1980s. Most of these slicks were found along the main tanker routes between the Persian Gulf, Europe, and Japan and between the United States and Europe.

Biological Effects of Oil

Oil and birds. Birds suffer during an oil spill because the oil soaks into their feathers, ruining the waterproofing and insulating qualities. The birds can no longer keep warm or float. Estimates of the number of birds killed during an oil spill are often low because many birds simply sink out of sight. As birds try to preen away the oil, they swallow it and are blinded and poisoned.

Oil also contaminates or destroys sea birds' natural foods. Diving birds are especially affected, as they dive through the oil slick again and again in search of food (Figure IV.3).

Detergents and marine life. When the tanker *Torrey Canyon* went aground in 1967, spilling 30 million gallons (about 100,000 tons) of oil, it was the first major oil spill ever to occur. Not much was known about the eventual effects of oil on marine life. The main concern

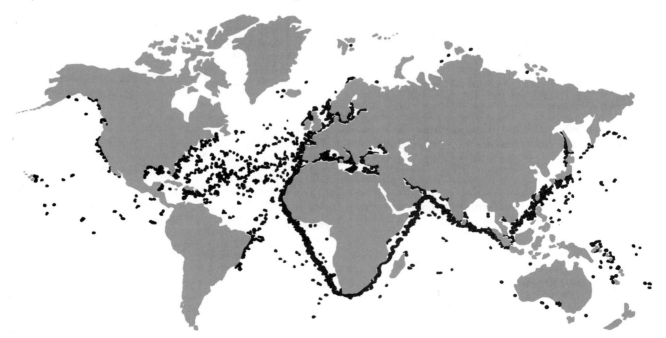

Figure IV.2 Locations of visible oil slicks, early 1980s. Remote monitoring techniques allow the location of oil spills and discharges in the world's waters. (From International Oceanographic Commission, UNESCO.)

Figure IV.3 Rescue workers attempt to clean oil from birds. Methods for washing oil off the birds also remove much of the natural waterproofing from their feathers. Thus, the birds must be kept in captivity until they molt and grow a new set of feathers. It is extremely difficult to keep and feed large numbers of wild birds for any length of time. As a result, bird salvage rates have averaged only a disappointing 1%–15%. (New York Zoological Society Photo)

of almost everyone involved was that the oil would foul the vacation resort beaches along the coasts of England and France. Since the summer holiday season was about to begin, local hotel keepers and the government directed almost all their efforts toward getting rid of the oil as quickly as possible. Because oil and water do not mix well, detergents were used to break up the oil into small droplets and enable it to mix into the water and wash away. Although this method was fairly successful in cleaning up the oily beaches, it proved disastrous to sea creatures.

We now know that detergents make oil more toxic to marine life. The oil-detergent mixture sticks to wet surfaces, such as fish gills, where oil alone would not stick. The detergent enables oil to penetrate deeply into the sand and kill even those creatures who might otherwise burrow to safety. Further, those early detergents used low-boiling petroleum hydrocarbons as solvents—exactly those fractions of crude oil most toxic

to marine life. Because of the enormous amount of detergent used (2 tons per ton of oil), a massive kill of marine life resulted. Except for a few anemones, everything within a quarter of a mile from shore and in waters up to 7 fathoms deep perished.

Newer detergent formulas no longer contain toxic solvents. Nevertheless, they are still hazardous to marine creatures. The Environmental Protection Agency limits their use in the United States to oil fires or other cases where human life or the major portion of a bird population would be endangered by an oil spill.

Effects on ecosystems. In addition to its effects on individual aquatic organisms, oil affects whole ecosystems. There is a salt marsh in Southampton, England where an oil refinery discharges, each day, 1,500 gallons (5,800 L) of water contaminated with very low levels of oil (10–20 ppm). This chronic pollution has killed off over 90 acres (36 hectares) of marsh grasses in the area around the refinery. With the grasses gone, the sandy soil began to erode so that the affected area is now lower than its surroundings. Birds and other aquatic creatures that once found food in the area have been forced to move elsewhere. Thus, very small amounts of oil can, over a long period of time, have serious effects on an aquatic community.

In areas where many oil spills occur, such as harbors or the Main Pass oil field in the Gulf of Mexico, changes in the *kinds* of organisms are becoming apparent. For example, one study showed that organisms growing in bottom sediments in Timbalier Bay in the Gulf of Mexico were mainly two hardy species known to take over in polluted areas. The Gulf of Mexico has been contaminated with oil for so many years that it is no longer possible to find an unaffected area to see what the natural communities were like.

In the North Sea, on the other hand, drilling operations began in 1973, and studies have been carried out since then. The studies have found gradual increases in the amount of oil in sediments around the drill sites. In addition, definite decreases have been seen in the number of species found and in the total number of organisms found. The area in which these decreases are seen becomes larger as time goes on.

Both oil and oil tars contain some cancer-causing substances. Several studies of shellfish grown in polluted waters have shown that they have an abnormally high number of growths similar to human cancer tumors. Oil, which is concentrated by shellfish such as oysters and clams, may be at least a partial cause of these tumors.

Biological recovery after an oil spill. After oil is spilled, recovery time would certainly include the length of time it takes for all traces of oil to disappear. However, it also includes the time necessary for the polluted area to be repopulated with the kinds and sizes of organisms that lived there before the spill. If a spill does not kill all the resident organisms, those left begin to repopulate the area, once poisonous parts of the oil have disappeared. Organisms from other areas also begin to move in, either by swimming and floating (e.g., larvae) or by creeping in from nearby colonies (sea grasses). Competition among species and predation begin to establish a balance between the different groups. But how long does oil persist in the environment after it has been spilled?

When the coastal barge *Florida* ran aground in Buzzards Bay at West Falmouth, Massachusetts in 1969, there was one positive aspect. Scientists at nearby Woods Hole Oceanographic Institute were well equipped to investigate closely the effects of the spill. They discovered that even though oil may have disappeared from the surface of the water, its effects can still be far-reaching and serious. In Buzzards Bay, heavy seas and onshore winds ensured that the spill of home heating oil was well mixed with the water. In a few

days, no oil could be seen floating on the surface. Five years later, however, oil could still be found in the sediments. Where the oil was found, there were fewer organisms than in neighboring regions. The oiled sediments had also eroded away to some degree. This seemed to allow the oil to spread further. The area in which oil was found was much larger than the original spill. Certain of the sea creatures studied, such as fiddler crabs, were still absorbing oil into their bodies. This oil caused them to behave strangely and prevented the population of crabs from increasing to the levels that existed before the spill. Scientists therefore warn that although oil may have only short-term effects in some areas, in others, such as marshes, the oil becomes mixed into the sediments and lasts for many years (Figure IV.4).

Lessening Oil Pollution and Its Effects

Moderating the effects of offshore drilling. One result of offshore drilling accidents has been an improvement in the safety devices designed to prevent blowouts. Producing oil wells have safety valves designed to close off the wells if storms or earthquakes destroy the platform. Experts believe that these safety devices will work about 96% of the time.

Still, because safety devices are not 100% effective and because there is always room for human error, a

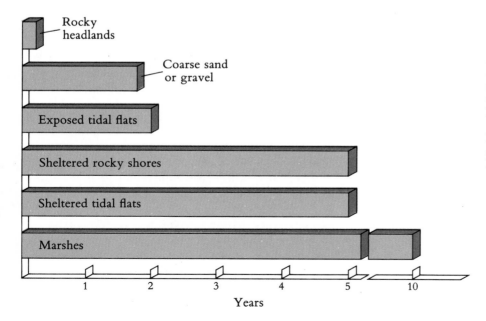

Figure IV.4 Persistence time for oil. On exposed rocky headlands, oil may disappear in days or weeks, but it can still be found 5 years later (as asphalt) in sheltered rocky shores. In sheltered tidal flats or marshes, the oil is still found 5–10 years or more later. (Data collected after the *Amoco Cadiz* oil spill, from E. R. Gundelach et al., in *1981 Oil Spill Conference* [Washington, D.C.: American Petroleum Institute, 1981], p. 525.

high probability exists that some spills and blowouts will accompany any offshore drilling. According to a report by the Council on Environmental Quality, Alaskan areas are unusually hazardous with respect to offshore drilling. An earthquake of a magnitude of at least 7 on the Richter scale can be expected in the Gulf of Alaska every 3–5 years. An earthquake in 1964 wiped out the original town of Valdez. (The new town is across the harbor from the terminal end of the Alaska pipeline.)

Environmentalists are rightly concerned about possible effects of offshore drilling on the fishing industries. Bristol Bay, one of the areas slated for development, has the richest fishing grounds in Alaska. The Outer Continental Shelf Bill, passed in 1978, includes strict safety procedures and also specifies that oil companies must set up a compensation fund for fishermen and landowners hurt by offshore spills.

Improving tanker safety. The U.S. Department of Commerce estimates that as many as 85% of shipping accidents are due to human error. Part of the problem may be due to the existence of "flags of convenience." Certain countries, notably Panama, Liberia, Singapore, and Cyprus, have much lower standards for the ships and crews to which they grant registration. Because it is cheaper to register in such countries (both because standards are lower and because taxes are lower than in countries like the United States), a large proportion of the world's shipping fleet is registered in them. The Law of the Sea Treaty deals with this problem by confirming the right of any state to set standards for vessels entering its ports. (However, the treaty is yet to be signed by all the countries concerned and may founder on the issue of rights to seabed mining of minerals.) Safety features such as double hulls and bottoms on tankers, special tanker lanes, and traffic controls should certainly be considered to prevent disastrous spills. Size limitation should also be explored for tankers serving the United States.

Cleaning up oil spills. If an oil spill does occur, what can be done about it? The oil slick must first be kept from spreading, with barriers such as floating booms. The oil can then be soaked up with adsorbent materials or skimmed off the surface of the water. Various devices are available to skim oil from water (Figure IV.5). However, booms and skimmers perform well only in calm waters, on small to moderate spills. Ocean cur-

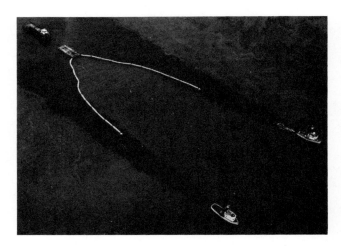

Figure IV.5 Fishing boats tow an oil skimmer through an oil slick to demonstrate how the booms (V-shaped arms) funnel oil into the skimmer (at the point of the V), where it is skimmed off the surface of the water and collected. Note the darker area of clean water following in the wake of the skimmer. Such systems do not work in choppy waters, where the oil slops over the top of the booms. (Photo courtesy of Clean Seas Inc., Santa Barbara, California)

rents can carry oil under the booms, and in rough seas oil does not form a slick at all but rather droplets that mix into the ocean. In a storm, little can be done to prevent spilled oil from coming ashore on beaches. When oil does reach beaches, it is still best cleaned up by broadcasting straw and raking it up again (Figure IV.6). According to a report by the Maine Department of Environmental Protection (1976):

> Once the volume spilled exceeds the capabilities of present state of the art technology, man can then only realistically chip away at the oil with available recovery machines, walk behind the spill on the beaches physically recovering oil with rather primitive tools, attend to claims for various oil spill related losses, and measure the effects by counting nature's victims as they wash up on the beach.

Economic and Social Effects of Mixing Oil and Water

People living in coastal areas are understandably uneasy about offshore drilling. While a country as a whole may benefit from the oil, the people living in the coastal

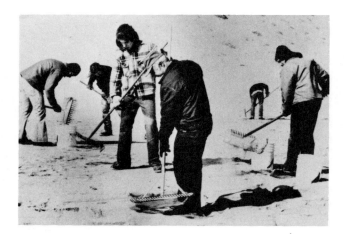

Figure IV.6 On Nauset Beach, Cape Cod, Massachusetts, workers use rakes to clean up tar balls (which are what is left of oil spills after the low-boiling fractions have evaporated). Thirty miles of beach were littered with the tar balls, which ranged in size from raisins to footballs and washed ashore for a period of several weeks. The cleanup cost over $100,000. (NYT Pictures)

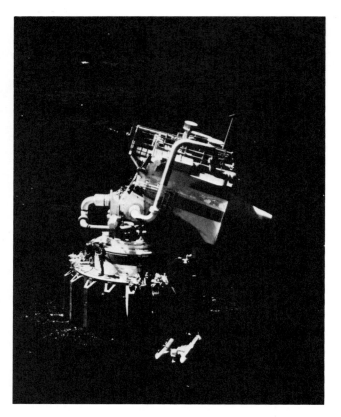

Figure IV.7 A single-point mooring (SPM) facility can load tankers of any size in any weather condition. Crude oil is fed from land through a submarine line to this structure, standing in deep water several miles offshore. (An Exxon photo)

states bear all the environmental costs of producing the oil. These include possible oil spills from drilling and transporting the oil, as well as water and air pollution from onshore development of oil-refining facilities. Localized overcrowding may occur as workers stream in to develop the oil fields. In some economically depressed areas, people may welcome the possibility of oil-related jobs. However, unless there is careful planning by local officials and agreements are drawn up between the oil companies and the towns, few local people may be hired. In such a case, the native population will suffer the inflation, housing shortages, overcrowding of schools, roads, and recreational facilities, and higher taxes that accompany an oil boom, without sharing in any of the money it brings (see Controversy IV.1).

The development of deep-water ports presents some of the same economic and social problems onshore as drilling for oil does on the continental shelf. A deep-water port is a complex of buoys and a platform. It is located in deep water, some distance from the coast. Such distance is necessary to provide deep enough water to float the supertankers bringing the oil (Figure IV.7). Currently, U.S. ports are served by tankers in the 50,000-ton range because harbors are not deep enough to accommodate supertankers. From an

environmental point of view, one benefit of the deep-water ports is that they can be located out of the mainstream of normal harbor traffic, thus reducing the possibility of tanker collisions. However, oil spills during loading and unloading, as well as pipeline leaks, pose a threat to ocean communities. In addition, dredging ship channels and other construction activities produce material that can cause disposal problems.

One of the best sites for a superport is the Palau Islands in Micronesia. The sociological effects of building a port there, however, are expected to be enormous (see Controversy IV.2).

Nuclear Winter

In the short run, the threat of nuclear war depends on the willingness of the superpowers to find peaceful

CONTROVERSY IV.1

Are National Interests More Important than Local Concerns?

The Interior Department now appears willing to accept the fact that the people of California want a program . . . that balances the need for oil with the need to protect important fisheries, tourist areas and other environmentally sensitive and economically important areas.
Lisa Speer, Natural Resources Defense Council

This is a classic example of the broad public interest versus narrower local interests, and in this case local concerns do not always reflect national or even regional needs.
James Watt, former Secretary of the Interior

Offshore oil drilling in several areas has been vigorously opposed by many environmental groups and sometimes by the states themselves. Clearly, such development could cause oil pollution due to catastrophic accidents or even just normal operations. For instance, oil is believed to lie beneath the sediments along two-thirds of the northern California coastline. Wildlife also abounds along the coast, including all of the nation's endangered southern sea otters.

Coastal areas directly opposite proposed drilling sites fear the social and economic upheavals a large influx of workers brings. This was a major concern when Britain started producing North Sea oil. Britain's Orkney Islands were a remote, self-contained, rural area before the huge influx of oil and construction workers arrived, increasing the demand for schools, housing, and other municipal services. Similar upheavals occurred during construction of the Alaskan pipeline. But such concerns are not limited to offshore drilling proposals. Communities and environmental interest

groups also protest the siting of nuclear power plants, hazardous waste dumps, and oil refineries. James Watt (see above quote) appears to feel that national interests must take priority over local concerns.

Do you agree that national priorities (such as a need for oil to reduce dependence on foreign sources or to increase federal revenues) come before the interests of local residents? What efforts can the federal government make to reduce harmful impacts on local residents? Do you feel that the federal government should provide compensation for unavoidable harmful effects? Would this change your opinion about the first question?

What about potential harm to local wildlife or plant populations? Should local feelings about this be taken into account?

Sources:
Lisa Speer, quoted in *The New York Times* (June 3, 1981), p. A20.
James Watt, quoted in *The New York Times*, (July 17, 1985), p. A14.

means to resolve continuing conflicts as well as means to defuse crises that may occur. In the long run, however, the importance of the superpowers to the equation for controlling conflicts is likely to decrease, and the actions of the less powerful nations become more crucial. The transition is the result of the almost inevitable spread of nuclear weapons. This spread of nuclear arms is directly related to the spread of nuclear technology, especially the components of the fuel cycle used to support nuclear electric power. The technologies for enrichment and for reprocessing are the key components that could make weapons-grade fissionable material such as uranium-235 and plutonium available to the smaller nations.

Nuclear war is an abhorrent concept to most individuals because of the death and suffering that would result from the first-order consequences of nuclear explosions (blast, fires, radioactivity). Nonetheless, it had long been thought that the results would at least be localized to the nations that were combatants in the war. The hypothesis of a "nuclear winter" has diminished even that faint hope. A nuclear winter spreads the miserable consequences to the entire globe.

Atmospheric Dust and Climate

In the summer of 1783, Benjamin Franklin was serving as U.S. Ambassador to France. He observed during that unusually cool summer a dry fog over France, a fog that the sun could not seem to evaporate. An early autumn followed the cool summer, and a severe winter followed the early autumn. Franklin conjectured that the dry fog and cool weather were the result of an eruption of the Laki Volcano in Iceland reported early that year.

Some 30 years later, Europe experienced another cool spell. Three more volcanic eruptions had occurred between 1812 and 1817: Sonfriere on St. Vincent Island in 1812, Mayon in the Philippines in 1814, and Tambora in Indonesia in 1815. The massive explosion of Tambora is thought to have thrown 37–100 cubic miles of ash, dust, and rock into the earth's atmosphere. The year after the Tambora eruption, 1816, became known as the "year without a summer" in the northeastern United States: Temperatures averaged 2°–3°C less than normal. In New York's Madison County, frosts occurred in every month of the year. Crops failed in the cold weather not only in the United States but across Europe.

Figure IV.8 Huge quantities of ash and dust were ejected into the atmosphere by the eruption of Krakatoa in Indonesia in August 1883. By the completion of the eruption, the island Anak Krakatoa, on which the volcano had stood, had completely disappeared into the ocean. Severe winters followed in the 1880s as two more volcanoes erupted, one in 1886, the other in 1888. The link between volcanic ash and dust and climate is the basis for the nuclear winter hypothesis. (Used by permission of the Smithsonian Institution Press, from *Krakatau, 1883: The Volcanic Eruption and Its Effects*, by Tom Simkin and Richard S. Fiske, Figure 2, p. 16. © Smithsonian Institution, Washington, D.C., 1983.)

In 1883, Krakatoa, near Java, exploded with a ferocity unknown in recorded history before or since; the eruption was reported to be heard in the Indian Ocean over 3,000 miles (4,800 km) away (Figure IV.8)! Within a few months, the ashes that had entered the stratosphere circled the globe, principally in the mid-latitudes. And three years later, in 1886, they were just disappearing when the ashes and dust from the Tarawera Volcano in New Zealand were rocketed into the stratosphere. In 1888, Bandai-San, a volcano

CONTROVERSY IV.2

Save Palau: For What? For Whom?
(Economic and Social Effects of Deep-Water Ports)

Two hundred islands in a chain almost 100 miles (161 km) long make up the Palau Island district of the U.S. Pacific Trust Territory.* A few are inhabited; some are only a few yards across. But all are scattered among the coral reefs located in deep, clear, blue Pacific waters. The location of the islands and the depth of the surrounding waters make the area an ideal port for supertankers carrying oil from the Middle East to Japan. In Palau, the oil could be loaded off the supertankers and into small tankers, which are able to enter Japan's relatively shallow harbors. However, the dredging and blasting needed for construction of the superport would destroy parts of the coral reef and the very delicately balanced community of marine life it supports. Further, the construction workers, oil-port personnel, and all of the accompanying hustle and bustle might overwhelm the predominantly farming and fishing culture of the native islanders. The following quotes are from an article by Andrew Malcolm in *The New York Times*, February 7, 1977:

> An outside project of this magnitude is too big for this little place. It would control our politics, our economy, our institutions, our lives. It is far beyond us. And after centuries under foreigners, we want to be free, man, free.
>
> **Moses Uludong, a leader of the Save Palau Organization**

---

*In January 1980, the islands became the Republic of Belau.

Speaking as an individual, the superport would be an environmental disaster of the worst order. The people would trade a few years of money for their whole fragile environment and culture. The potential for environmental disaster here is much greater than even on the Alaskan pipeline.
Mr. Owen, Palau Conservation Officer

I can't accept all this so-called environmental concern. The world has always had pollution and overcome it. These people want to keep Palau a human zoo so they can come and swim and take pretty pictures and then go home to their own lives while the people here starve without work. Save Palau, they say. For what? For whom?
Roman Tmetuchl, Businessman

It is never easy to draw conclusions when environmental protection appears to be pitted against the human desire to upgrade living standards. What environmental damage do you think the superport might do in Palau? What sort of disruption might the construction crews cause? (Consider, for instance, what services they will need, where they will live, what they will do for recreation.) How likely is it that native Palau Islanders will either be hired by or provide services to the construction and operating crews of the port? Would you side with Mr. Owen or Mr. Tmetuchl? If you were an islander, with whom would you side?

in Japan, ejected still more fine particles into the atmosphere.

The 1880s were reported as very cold years across the United States. In 1888, a severe snowstorm swept the eastern coast of the United States from Washington, D.C. to Maine. Drifts of 40 to 50 feet of snow were reported in New York State and New England, with snowfall averaging about 4 feet over most of the area. Houses were buried in some communities, as were trains en route. People caught outdoors were buried alive in the swiftly arriving storm. At sea, the storm was known as the "Great White Hurricane" and in New York City, which was utterly paralyzed by 15- to 20-foot drifts, the storm became a legend as the "The Blizzard of '88."

In the United States, William Humphreys, a Weather Bureau scientist, and in Britain, Professor H. H. Lamb, pieced together these volcanic events and the reports of frigid weather. They concluded that the shrouds of dust circling the earth from these eruptions were reflecting sunlight away from the earth and causing it to cool. As the dust settled from the stratosphere and rained out of the troposphere, more normal weather returned.

Scenario for Nuclear Winter

The volcanic dust hypothesis is at the root of the prediction that a severe winter of snow and ice would quickly follow any significant exchange of nuclear weapons. Correlations of the climatic effects of eruptions and volcanic material ejected are not feasible because of a shortage of data. Nonetheless, it is conjectured that the ash and dust resulting from nuclear blasts, although different than volcanic material, would likely enter the stratosphere, block sunlight from reaching the earth, and cause the earth to cool quickly. It is further thought that urban fires caused by nuclear blasts could become *firestorms*—tornadolike winds that pull a fire into a central vertex cone—that could destroy cities and eject large quantities of ash into the stratosphere. (Such firestorms were seen in the German cities of Hamburg and Dresden in the heavy bombing in the months just before the end of World War II.) Wildfires in forests and grasslands would likewise be expected to produce large quantities of smoke.

The nuclear winter hypothesis has been supported by a mathematical model of dust and smoke formation, spreading, and fallout in the atmosphere. There are enormous uncertainties in the assumptions and data used in the model, but these uncertainties can be partially bypassed by examining the model results for ranges of possible values of the basic parameters and inputs. For example, the size of an exchange in a nuclear war might conceivably range from 1,000 megatons to 10,000 megatons. Even the lower-level exchange could bring on a nuclear winter, although it is possible that the higher level could result in only minor climatic effects. Air bursts and the use of low-yield nuclear weapons would produce less nuclear dust than high-yield weapons exploding at the land surface. On the other hand, air-bursting nuclear weapons would be expected to set more fires burning in both urban and forested areas than would weapons exploding on the land surface.

Fires in urban areas would be expected to raise more smoke into the stratosphere than fires in forested areas because the concentration of combustible materials in cities would lead to more intense fires and possibly to firestorms. The fine smoke particles from these urban fires could reach the stratosphere and reside there for several years before settling out into the troposphere and then washing out in rain. These smoke particles would block the rays from the sun, thereby preventing the rays from warming the earth.

The presence of particles of whatever sort would cause the stratosphere to warm; at the same time the surface of the earth would cool by radiating heat into the upper air. The surface temperature would decrease because the earth would be radiating energy away and because it will have lost most of its source of heat, sunlight. Infrared energy radiated to the earth from the upper atmosphere would simply not be enough to counter the cooling trend.

A nuclear winter would be characterized by a darkened and overcast sky lasting for many weeks on end. Land temperatures would drop by 40°C over a period of months. Ocean temperatures might fall only several degrees centigrade during this time because of the high heat capacity of water. The relatively warm and moist air over the ocean would come in contact with colder air over the land along the coast, creating storms with extraordinary amounts of snow.

It is not thought that an ice age would be triggered by a nuclear exchange. The earth would eventually recover to normal temperatures, but the nuclear winter could last for several years and extend across the globe, ending only when most dust and smoke have settled out of the stratosphere. The failure of crops and destruction of feed animals could lead to an end of the human era. And the destruction of most vegetation could bring an end to most higher forms of animal life.

Questions

1. What is the major source of oil pollution in the world's waters? Why are tanker accidents and oil-well blowouts significant, even though they contribute only a small percentage of all the oil spilled in water?
2. Suppose your class decided to start a waste oil recycling project in your community. How would you begin? How would you explain the need for such a project to citizens in your community?
3. What are the major effects of spilled oil on the water environment?
4. What do volcanic eruptions and the possibility of nuclear winter have in common?
5. Describe the major environmental effects of a nuclear winter.

Further Reading

Wilson, S., and K. Hayden. "Where Oil and Water Mix," *National Geographic* (February 1981), p. 145.
An exploration of the wildlife refuges along the Texas coast, where the protection of whooping cranes and oil production are in a fragile balance.

"The Language of Oil." New York: Mobil Oil Corporation, 1974.
This small booklet clearly defines 100 commonly used terms relating to oil production and oil pollution. It is a handy reference to have while reading on the subject of oil pollution. For copies, write to Mobil Oil Corporation, 150 E 42nd Street, New York, NY 10017.

Howarth, Robert. "Fish Versus Fuel: A Slippery Quandary," *Technology Review* (January 1981), p. 61.
A clear discussion of the long- and short-term effects of spilled oil on the marine ecosystem.

Dolensek, E. P., and J. Bell. *Help! A Step by Step Manual for the Care and Treatment of Oil-Damaged Birds.* New York: New York Zoological Society, 1978.
A clear, comprehensive manual intended for volunteers who want to help save birds injured by oil pollution. Any coastal environmental group might like to have copies on hand. Available from: Publications Dept., New York Zoological Society, Bronx, NY 10460.

Kerr, R. A. "Oil in the Ocean: Circumstances Control Its Impact," *Science*, **198** (December 16, 1977), 1134.
This article provides a good summary of several studies done to determine the long-term effects of spilled oil.

Ehrlich, Paul, Carl Sagan, Donald Kennedy, and Walter Roberts. *The Cold and the Dark: The World After Nuclear War.* New York: Norton, 1984.
Describes the physical and biological consequences of nuclear war. The authors each contribute sections of the book. See also the careful review of the book by Joseph Smith in *Bulletin of the Atomic Scientists*, January 1985, pp. 49–51.

Turco, R., O. Toon, T. Ackerman, J. Pollack, and C. Sagan. "Nuclear Winter: Global Consequences of Multiple Nuclear Explosions," *Science*, **222** (December 23, 1983), 1283–1300.
Describes many of the assumptions of the nuclear winter model as well as various war scenarios.

Francis, Peter. *Volcanoes.* New York: Penguin Books, 1974.
Eyewitness accounts of eruptions and compelling information on climate effects of volcanic eruptions. This fact-filled book is surprisingly fun to read.

U.S. Department of Commerce, National Oceanic and Atmospheric Administration. *American Weather Stories.* Washington, D.C.: U.S. Government Printing Office, 1976.
Description of the year without a summer and the "Blizzard of '88" and the relation of volcanism to weather.

"NRC Panel Envisions Potential Nuclear Winter," *Science*, **226** (December 21, 1984), 1403.
A well-done news account of the report by a special committee of the National Research Council, the research arm of the National Academy of Engineering and National Academy of Sciences.

References

Sanders, H., et al. "Long-Term Effects of the Barge *Florida* Spill." EPA PB 81 144–792 (January 1981).

Oil in the Sea: Inputs, Fates and Effects. Washington, D.C.: National Academy Press, 1985.

Petrakis, S. L., and F. T. Weiss, eds. *Petroleum in the Marine Environment.* Washington, D.C.: American Chemical Society, 1980.

Brinkman, D. W. "Used Oil: Resource or Pollutant?" *Technology Review* (July 1985), p. 48.

Maine Department of Environmental Protection. *Statistical Report—Oil Spills, 1976.*

PART FIVE

Air Pollution

hen we burn fossil fuels—coal, oil, and gas—we furnish heat for homes, energy for mobility, and power for the production of goods. Fossil fuels have become our servants, and our skill at capturing their energy has transformed civilization. This mastery of energy has brought abundance and comfort. But energy has turned out to be a dangerous servant.

Our misuse of fossil fuels has resulted in air pollution episodes so severe that people have died. Numerous pollutants foul the air we breathe. Solid particles such as lead, asbestos, and soot are dispersed in the air. Liquid droplets of hydrocarbons and sulfuric acid float in the atmosphere. Gases such as carbon monoxide, nitrogen oxides, and sulfur dioxide are dissolved in the atmosphere that surrounds us. The contaminants in the air have direct biological effects on humans. Breathing may be impaired; diseases of the heart and lung are complicated and made worse. In addition, the landscape is influenced because vegetation is also sensitive to air pollutants. Construction materials such as mortar and metal are corroded by certain pollutants, and

the fibers of natural fabrics are damaged by other pollutants.

Near Los Angeles, the most sprawling city in the country and the most dependent upon the automobile, smog is destroying the pine trees in the hills that surround the city. Almost 200 miles away, about the distance between New York and Boston, the desert holly, a plant that grows in the isolation of Death Valley, is threatened by the creeping smog from Los Angeles. The smog results from reactions between hydrocarbons and nitrogen oxides—products of automobile exhausts.

In the northeastern United States, as well as in Scandinavia, acid rain is turning inland lakes increasingly acid, destroying fish life and other aquatic species. The acidity is caused by the solution in rain of sulfur oxides, pollutants produced primarily from the burning of coal and residual oil, two sulfur-containing fuels. Oxides of nitrogen, principally from combustion occurring at fossil-fuel power plants and from the combustion of gasoline in automobiles, also contribute.

Part Five begins with a chapter on air pollution ecology (Chapter 19). The concepts in this chapter are essential to understanding how the air pollutants produced when energy is generated affect climate and weather as well as the natural cycles of materials such as carbon, sulfur, and nitrogen. Chapter 20 discusses sulfur oxides and particulate matter. Chapter 21 focuses on pollutants produced by the automobile: photochemical air pollution, which stems from the reactions of hydrocarbons and the oxides of nitrogen, and carbon monoxide.

Our discussion of air pollution concludes with two issues of global or international dimensions. The first is acid rain and forest decline. Acid rain is now observed in both North America and Europe as well as in the industrialized countries of the Far East. Serious forest decline has been linked tightly with acid rain. The second global issue is the greenhouse effect and the role of carbon dioxide in warming the surface of the earth, a truly global phenomenon. We explore, as well, the consequences of that warming in terms of sea level rise.

CHAPTER NINETEEN

The Ecology of Air Pollution

The Donora Story, October 1948

During the last week of October 1948, a heavy smog settled down over the area surrounding Donora, Pennsylvania. Weathermen described it as a temperature inversion. . . . Smogs of short duration are not unusual and except for discomfort due to irritation and nuisance of the dirt and poor visibility, no unusual significance is attached to such occurrences.

This particular smog encompassed the Donora area on the morning of Wednesday, October 27. It was even then of sufficient density to evoke comments by the residents. It was reported that streamers of carbon appeared to hang motionless in the air and that visibility was so poor that even natives of the area became lost.

The smog continued through Thursday, but still no more attention was attracted than that of conversational comment.

On Friday, however, a marked increase in illness began to take place in the area. By Friday evening the physicians' telephone exchange was flooded with calls for medical aid, and the doctors were making calls unceasingly to care for their patients. Many persons were sent to nearby hospitals, and the Donora Fire Department, the local chapter of the American Red Cross, and other organizations were asked to help with the many ill persons.

There was, nevertheless, no general alarm about the smog's effects even then. On Friday evening the annual Donora Halloween parade was well attended, and on Saturday afternoon a football game between Donora and Monongahela high schools was played on the gridiron of Donora High School before a large crowd.

The first death during the smog had already occurred, however, early Saturday morning—at 2 A.M., to be precise. More followed in quick succession during the day and by nightfall word of these deaths was racing through the town. By 11:30 that night 17 persons were dead. Two more were to follow on Sunday, and still another who fell ill during the smog was to die a week later on November 8.

On Sunday afternoon rain came to clear away the smog. But hundreds were still ill, and the rest of the residents were still stunned by the number of deaths that had taken place during the preceding 36 hours. That night the town council held a meeting to consider action, and followed with another on Monday night. By this time emergency aid was on its way to do whatever possible for the stricken town . . . ("Air Pollution in Donora," 1949)

What are *inversions*, those weather conditions that can combine with pollution to cause such havoc to human health? In this chapter we explore that subject and other aspects of the atmosphere and climate on earth. But before looking at air pollution, we need to discuss the characteristics of clean air.

The Earth's Atmosphere

Components

Although we usually speak of the material we breathe simply as "air," it is actually a mixture of substances. Clean, dry air is mostly (99%) nitrogen gas and oxygen gas. However, clean air contains several other substances that, although present in only small quantities, are still very important. **Carbon dioxide**, which has a strong effect on the earth's atmospheric temperature, is one example (Table 19.1), and **ozone** gas is another. Water vapor is also an important component of the atmosphere. The amount of water vapor varies from 0% by volume in dry air to about 4% in humid air. Dust particles from both human and natural sources (for example, volcanoes) may also be a significant component of air, even though present in a relatively small percentage.

Origin of the Oxygen in Air

The oxygen in air is essential for respiration by plants and animals. Ozone, which is discussed more fully in Global Perspective VII, provides protection from the sun's ultraviolet rays, which can be harmful to life. However, neither oxygen nor ozone has always been part of the earth's atmosphere.

Scientists believe that 4.5–5 billion years ago, the earth's atmosphere was similar to the mixture of gases released when volcanoes erupt—primarily water vapor, carbon dioxide, and nitrogen. During the period when the earth was cooling and becoming more solid, heavy rains washed out most of the carbon dioxide.

Table 19.1   Components of Clean Dry Air[a]

Element	Percent by volume
Nitrogen (N_2)	78.08
Oxygen (O_2)	20.94
Argon (Ar)	0.93
Carbon dioxide (CO_2)	0.03
Ozone (O_3)	less than 0.00005

a. Small amounts (less than 0.002%) of neon, helium, methane, krypton, and hydrogen are also present. Nondry air, of course, also has water vapor in it.

The oxygen now in the atmosphere came from a completely different source: green plants.

At some point, more than 3 billion years ago, simple cells, which were nourished by chemicals floating in the water in which they lived, evolved into organisms capable of photosynthesis. While using the sun's energy and carbon dioxide to make their own organic compounds, they also produced oxygen. This oxygen began to build up in the atmosphere, and a new era began (Figure 19.1). Part of the oxygen (O_2) was changed by sunlight into ozone (O_3).

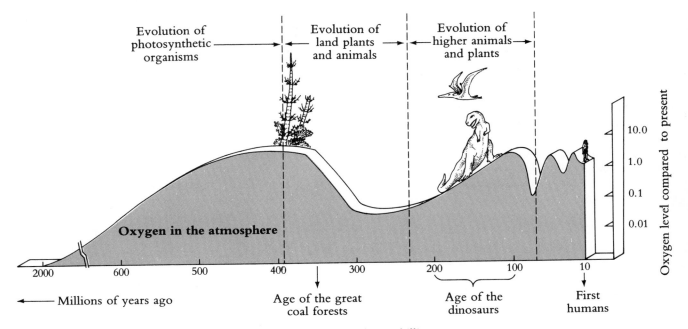

Figure 19.1 Origin of oxygen in the atmosphere. About 2 billion years ago, the level of free oxygen in the earth's atmosphere began to rise. After a protective layer of ozone was formed from part of the oxygen, land plants and animals evolved. Oxygen levels have varied a great deal as levels of production and use have changed. (From E. P. Odum, *Fundamentals of Ecology*, 3rd ed. [Philadelphia: Saunders, 1971].)

Distribution of the Atmosphere

Like the term *air*, the word *atmosphere* is often used loosely to describe the whole envelope of gases surrounding the earth. However, the components of the atmosphere are not distributed uniformly throughout this envelope. Scientists who study the atmosphere recognize several zones at different heights above the earth, depending on their temperature characteristics (Figure 19.2).

The layer closest to the earth's surface is called the *troposphere*. In this layer, 5–10 miles (9–16 km) high, most of what we call weather occurs. All rainfall, almost all clouds, and most violent storms occur in this layer. The temperature in the troposphere usually decreases as we go up. Above the troposphere is the *stratosphere*. Temperatures in this layer first remain constant and then begin to increase with height. Most of the ozone is concentrated in the stratosphere and this is responsible for the temperature increase; that is, ozone absorbs ultraviolet light, which then causes the stratosphere to become heated. Even higher, above 30 miles (50 km), is the *mesosphere*, a zone in which temperatures again fall. Finally, there is the *thermosphere*, which begins about 50 miles (80 km) above the earth's surface. There is no well-defined upper boundary to this layer. In the thermosphere, temperatures again rise with height.

Most of the gases are present in about the same percentages in the troposphere, stratosphere, and mesosphere. However, atmospheric pressure decreases with height. That is to say, the air becomes less and less dense going upward; 90% of the atmosphere is concentrated within 16 km of the earth's surface. You may already be familiar with this fact if you have ever traveled into high mountain areas. Difficulty in breathing results from oxygen concentrations being lower at higher altitudes.

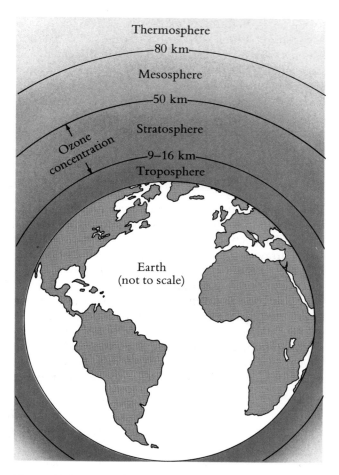

Figure 19.2 Atmospheric layers. Diagram of the earth's atmosphere, divided into layers on the basis of temperature. Heights given are approximate because they vary from place to place around the globe.

Climate on Earth

Weather Versus Climate

The term *climate* was introduced in Chapter 6 in relation to soil formation. Here we will contrast two closely related terms, *climate* and *weather*. Both are composed of the same elements: temperature, precipitation, wind, humidity, air pressure, and cloudiness. However, **weather** describes the day-to-day changes in these elements, while **climate** is a long-term description of these factors for a given region. Climate is not simply average weather, but also includes the kind of extremes of heat, cold, wind, and so forth that occur in the region.

Human Influences on Climate

A number of human activities influence world climates in several ways.

Influences on temperature. Carbon dioxide (CO_2) in the atmosphere acts to absorb radiation from the earth that would otherwise escape into space. By absorbing this radiation and reemitting it, carbon dioxide keeps the earth's atmosphere warmer than it would be otherwise. Carbon dioxide in the atmosphere has been increasing since the late 1860s, paralleling the growth of industry. However, world temperatures increased only until the 1940s, when they began to drop. This could be the result of a natural cooling trend, which would be even more severe without the CO_2 effect. The addition of carbon dioxide to the atmosphere from the burning of fossil fuels is discussed in Global Perspective V.

The cooling trend just mentioned could also be partly due to the effects of particles in the atmosphere. Volcanoes are natural sources of airborne particles. When volcanoes erupt, they may spew huge quantities of dust high into the stratosphere, where it can remain for several years. These particles can reflect solar radiation away from the earth, causing a cooling of the atmosphere. In fact, notable cooling trends have been experienced after volcanic eruptions. Not all scientists agree that particles from human sources (discussed in Chapter 20) contribute to a cooling trend, however; some believe that such particles may have a net effect of warming the atmosphere. (See the references at the end of this chapter for a more complete discussion of the many factors involved.)

Chlorofluorocarbons may also have effects on cli-

mate. These compounds, once used extensively as propellants in spray cans, appear to destroy ozone in the stratosphere (see Global Perspective VI) and seem to absorb infrared radiation. This latter effect could cause atmospheric warming. What will the final outcome be? We don't know.

Whatever the outcome, it seems fairly obvious that humans are influencing the earth's temperature through climatic processes that are not well understood. We are concerned about these possible changes because, by a sort of domino effect, humans feel economic, social, and possibly even political effects from changes in climate. For instance, Figure 19.3 shows how a volcanic eruption in 1815, by cooling and worsening weather conditions, probably caused the price of flour in London to double 18 months later. Think of the social and political events that might follow this sort of climatic event today.

Changes in climate affect agriculture because of the possibility of crop failures: temperatures may drop too low, rainfall may be too much or too little, or severe storms may arise. Wildlife is also affected. For example, the survival of fish eggs or of young hares depends on certain temperatures. Furthermore, effects of climate on one species may, in turn, affect other organisms in food webs. An interesting example of this sort of delayed effect is pointed out in Figure 19.4.

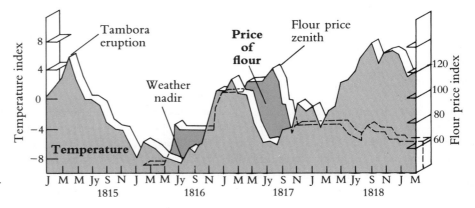

Figure 19.3 Effects of Tambora eruption on temperature and possible effect on the asking price for a sack of flour in the London commodities market. The temperature indices are 5-month running sums of departures from the long-term mean for the corresponding month, using data from historical reconstruction of the central English climate. Flour prices are from *Duffy's Farmers Journal*. Newspapers of the time describe the effect of inclement weather conditions on harvesting operations and crop growth following the eruption. We expect a lag in the effect of weather on crop prices, because most countries have about a one-year carryover of grain in storage. (Data from K. E. Watt, *Principles of Environmental Science* [New York: McGraw-Hill, 1973].)

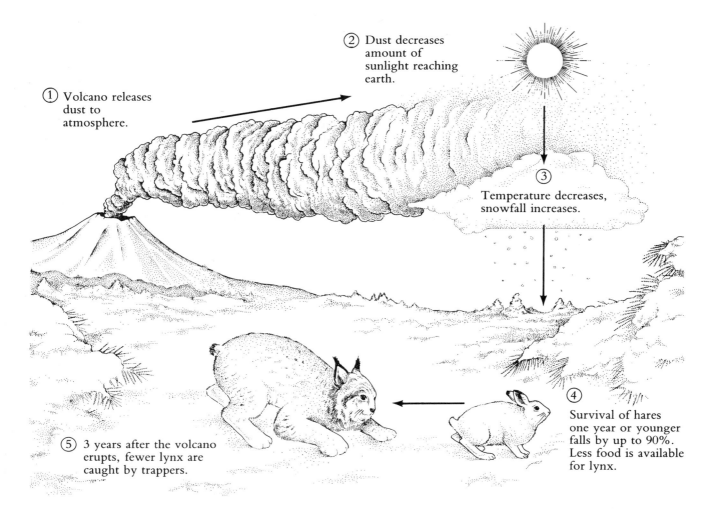

① Volcano releases dust to atmosphere.

② Dust decreases amount of sunlight reaching earth.

③ Temperature decreases, snowfall increases.

④ Survival of hares one year or younger falls by up to 90%. Less food is available for lynx.

⑤ 3 years after the volcano erupts, fewer lynx are caught by trappers.

Figure 19.4 The volcano–temperature–snow–hare–lynx–trapper effect. Note that the effect of the volcano was not seen in the lynx population until 3 years after the eruption. In a similar way, the effects of human-caused climate changes may not be seen immediately. (Data from K. E. Watt, *Principles of Environmental Science* [New York: McGraw-Hill, 1973].)

Cities and climate changes. Infrared photos taken by satellites show another characteristic change in climate due to human activities: the urban *heat island*. That is to say, cities are warmer than the surrounding countryside. This is partly due to changes in the surface of the land when countryside is changed into city. The concrete and asphalt areas of the cities absorb and hold heat better than soil and vegetation, and they release the heat more slowly at night. Another factor is waste heat from air conditioning and space heating. The pollution over cities may also contribute, since carbon dioxide, water vapor, and particulates absorb and re-radiate some of the long-wave radiation from the city. This radiation would otherwise escape into space. Cities also appear to increase the amount of rainfall over the city itself and even downwind of the city.

Acid rain. Another way that human activities are affecting climate involves a change in the quality of rainfall. The production of sulfur oxides (Chapter 20) and nitrogen oxides (Chapter 21) is causing rainfall to become more acid in many parts of the world. This acid rain then interferes with the growth and development of plants and animals. (See Global Perspective V.)

Gaia. The evolution of the earth's atmosphere to its present oxygen-rich composition is the subject of a book by Lovelock (1979), who calls both his concept and his book *Gaia*.

The argument of Gaia is that the earth's atmosphere is in a homeostatic state. That is, if events were to occur that pushed the atmosphere away from its present composition, righting mechanisms would correct the composition back to its present status. Those mechanisms include the natural processes of air, earth, sea, plants, and animals. The mechanisms would operate on a global scale but in geologic time (millions of years) rather than within the human time frame. Gaia is an exciting concept, but it may also be a dangerous one; scientists may argue that insults to the earth's atmosphere will be corrected naturally, but they may not mention that those corrections will take millions of years. The Gaian hypothesis, beautiful though it is, has the potential to be abused.

Air Pollution Episodes: The Awakening

Our recognition, on a national or international scale, of the permanent threat of air pollution comes long after serious events in the past momentarily focused our attention on these hazards. These events serve today as constant reminders of the consequences of doing nothing. As such, they are the touchstones of the clean air movement.

The earliest recorded incident of air pollution in the United States was in 1948 in Donora, Pennsylvania (described at the beginning of this chapter).

Four years later, in December 1952, an air-pollution episode gripped the city of London for five successive days. More than 4,000 deaths occurred in that interval as a result of the pollutants that hung in the air. These deaths, attributed to the London fog, were in excess of those normally expected to occur during that time of year. The elderly and those with chronic respiratory diseases were often the victims and suffered most. In the years following, episodes occurred again in London, but fortunately the tragic extent of the 1952 fog was never repeated.

Episodes have occurred in New York City, and alerts have been announced in other major cities in recent years. The science of statistics and the monitoring of air pollution have provided new information. When

our senses have failed to tell us of the decline in the quality of our air, chemical sensors have detected pollutants. When the effects on health have not been visible to the untrained eye, statisticians have shown deterioration of health at times of high pollution levels.

These episodes are not the result of sudden and vast discharges of pollution from many sources. That is, we cannot contend that they are caused by certain individuals or industries that have taken a momentary irresponsible action. In fact, industries have probably been operating in a business-as-usual fashion. The pollutants present during an episode are typically only a portion of the normal output from domestic, industrial, and transportation activities.

Instead, an air-pollution episode represents a massive accumulation, a gathering together, of these ordinary pollutant discharges on a huge scale. The massive accumulation is the result of weather conditions—specifically, inversions that hinder the natural mixing of the atmosphere and prevent the natural dispersion and scattering of pollutants.

Inversions

Inversions clearly illustrate how human activities can interact with the natural phenomena of weather and climate to cause serious ecological or health disturbances.

As noted in Chapter 2, we depend on the second law of thermodynamics to help dispose of air pollutants. That is, pollutants emitted from smokestacks tend naturally to disperse and become diluted in air to nonharmful levels. Winds help speed this mixing, as do upward air currents, which carry pollutants aloft to mix into the upper atmosphere. A condition can arise, however, in which the air layers are very stable. The pollutants, instead of mixing into the upper atmospheric layers, are restricted to the layer of air close to the ground. There the pollutants accumulate to unhealthy levels. An **inversion** is an unusual weather condition in which air temperatures in the troposphere do not decrease with height. Rather, colder air sits beneath a warmer layer of air. In this situation, the colder air, which is heavier, does not rise up through the lighter air layer above it. Pollutants accumulate below the "lid" of warmer air.

Inversions commonly form during the cold, clear nights of fall. On a clear day in the fall, the sun's rays heat the surface of the earth, which, in turn, heats the layer of air next to the surface. However, earth is a better radiator of heat than the atmosphere. Thus, during the night, the earth radiates heat into space. As the earth cools, it chills the layer of air next to the surface. By morning, a temperature inversion exists: Cold air lies close to the earth, while the air above is still fairly warm. Once the sun rises, the earth's surface is warmed again, and so is the layer of air next to the surface. The inversion then disappears as the day advances. These so-called surface inversions tend to be shallow except where there are valleys. Here, the cold air drains off the uplands during the night, forming deeper inversions that may be slow to disappear.

A more lasting inversion may occur as a result of the sinking of a high-pressure air mass (anticyclone). As such air sinks to lower altitudes, it becomes compressed and its temperature rises. This phenomenon often sandwiches a lower layer of cool air between the warm subsiding air and the earth's surface, giving rise to an inversion aloft. Sometimes both kinds of inversions can occur together.

Once the inversion is formed, pollutants begin to accumulate in the cooler lower layer. Autos carrying people to work and trucks carrying goods discharge carbon monoxide, nitrogen oxide, and hydrocarbons to the atmosphere. Industrial activities discharge these contaminants plus sulfur oxides and particulate matter. Concentrations of pollutants in the cold lower air are increased even further. As the sun begins to warm the

lower air, vertical thermal currents are set in motion, rising until their temperature has reached that of the surrounding air. However, if the cold layer is thick and extends far above the surface of the earth, the currents may be unable to escape into the upper air. The concentrations of contaminants increase through the day because little pollution escapes. Sometimes, these conditions are repeated for several days, and the level of contaminants becomes increasingly dangerous. Strong winds are needed to break up the cool lower layer. Sometimes they arrive in time; sometimes, as in Donora, Pennsylvania in October 1948, they do not.

Because the emissions leading to and continuing during an episode are occurring at no more than a normal rate, if an episode happens once in a particular city, the potential exists for an episode to happen again, when the weather conditions that trap pollutants recur. The potential for an episode will decrease only if some positive action is taken to reduce the output of pollutants.

Sadly, the weather conditions that now threaten to bring on air-pollution episodes were regarded in the past as among the finest weather of the year (Figure 19.5). In rural areas, the fine clear days of fall that follow cool cloudless nights are still enjoyable. In polluted urban areas, however, such weather is a danger signal. Extended over many days, such weather may lead to the inversions that act to trap air pollutants in the lower layers of the air. Weather conditions such as these sub-

Figure 19.5 The sunny days of fall are beautiful to behold in rural areas. In urban areas, however, these same days can create air-pollution episodes. If the air over a city remains windless for too many days, the pollutants from autos, industries, homes, and shops may become dangerously concentrated, to the point at which human health can be harmed. (USDA–Fish and Wildlife Services; photo by Wilbrecht)

jected the town of Donora to a deadly concentration of pollutants more than three decades ago. Only the control or removal of air-pollution sources can prevent episodes from happening again.

What are the air pollutants that so injure our health? Where do they come from? How do they harm us? How can we control them? In the chapters that follow we shall describe the various air pollutants, their origins, their effects, and the methods to decrease their discharges.

Summary

Air is a mixture of gases, primarily nitrogen, oxygen, carbon dioxide, and argon. Smaller amounts of other important gases, such as ozone, are also present, as is water vapor. The source of all the oxygen is photosynthesis.

The earth's atmosphere is composed of several layers in which air becomes progressively thinner, or less dense: the troposphere, the stratosphere, the mesosphere, and finally, the thermosphere.

The word *weather* is used to describe day-to-day changes in temperature, precipitation, wind, humidity, air pressure, and cloudiness. *Climate* is a long-term description of these factors and their extremes for a particular region of the earth. Humans have influenced climate by increasing the carbon dioxide and particulate content of the atmosphere, by releasing substances that cause acid rain, by creating heat islands around cities, and by the release of chlorofluorocarbons.

Inversions are tropospheric conditions in which the normal decrease in temperature with height is reversed. A warm layer atop a cold one traps pollutants that would ordinarily escape by rising into the atmosphere. The realization that such inversion episodes could be deadly when combined with the release of air pollutants was slow to come. However, this phenomenon is now well documented.

Questions

1. What are the major components and the important minor components of the earth's atmosphere?
2. What is an inversion? How can inversions increase the effects of air-pollution emissions? Outline the process that results in the formation of an air-pollution inversion. Begin with a description of the "normal" temperature stratification of the earth's atmosphere.
3. Briefly describe human influences on climate. What do you think the final effect will be?

References

"Air Pollution in Donora, PA." Public Health Bulletin No. 306, U.S. Public Health Service, 1949.

Lutgens, F. K., and E. J. Tarbuck. *The Atmosphere*, 2nd. ed. Englewood Cliffs, N.J.: Prentice-Hall, 1982.

Kerr, R. A. "Climate Control: How Large a Role for Orbital Variations?" *Science*, **201** (July 14, 1978), 144.

Levin, H. L. *The Earth Through Time*. Philadelphia: Saunders, 1978.

Watt, K. E. *Principles of Environmental Science*. New York: McGraw-Hill, 1973.

Lovelock, J. E. *Gaia*. Oxford and New York: Oxford University Press, 1979.

CHAPTER TWENTY

Sulfur Oxides and Particulate Air Pollutants

Sulfur Oxides

Sulfur Comes from Coal and Oil/Oxidation of Sulfur to a Corrosive Mist/Effects on Materials, Plants, and People/ Ways to Reduce Sulfur Dioxide Emissions/Progress in Control of Sulfur Oxides

Particulate Matter

Particles Stem from Combustion and Industrial Activities/ Physical and Biological Effects of Particles/Health Effects of Particles and Sulfur Oxides/Particulate Matter in the Air and Cancer/Ways to Control Particle Emissions/Progress in Control of Particulate Matter: An Uphill Battle

Pollutant Standards Index and Summary of Standards

CONTROVERSY:
20.1 *Costs of Air Pollution Control*

Sulfur Oxides

Sulfur Comes from Coal and Oil

Sulfur compounds in the air derive mainly from the burning of sulfur-rich fuels, such as coal and heating oils. These **sulfur oxide** compounds have polluted the air in many sections of the country and, when dispersed by tall stacks, are the principal cause of acid rain. Nonetheless, the fuels from which the sulfur comes are needed to produce heat, generate electricity, and power machinery.

Not all fuels contain significant quantities of sulfur. Some coals may be only 0.5% sulfur, while others may be as much as 6.0% sulfur. Coal is used extensively in steel manufacture, but it is used primarily for generating steam to produce electricity. While the average sulfur content of the coal used in electric power generation is on the order of 2.5%, the coal used in making steel must be much lower in sulfur. In fact, long-term contracts with the steel industry tie up a large proportion of the low-sulfur coal in the United States.

For every million metric tons of coal burned in electric power generation, about 25,000 metric tons of sulfur are released. Of course, the sulfur is not released in the elemental form but mainly as sulfur dioxide gas. In 1985, about 645 million metric tons of coal were burned for the sole purpose of generating steam to produce electricity. How many tons of sulfur were released?

Sulfur is also present in crude (raw or unrefined) petroleum, but probably less than 1% of the petroleum is sulfur. Refining coaxes much of the sulfur out of such common petroleum products as gasoline and kerosene. The waste sulfur compounds are *flamed* at the refinery. As you drive by a refinery, you may see the waste sulfur being burned to sulfur oxides at the top of a tall metal stack. Gasoline and kerosene, then, are responsible for only a small portion of the sulfur compounds in the atmosphere. In the United States, probably less than 5% of the yearly release of sulfur compounds comes from the burning of gasoline. Home heating oil is also relatively low in sulfur, having an average sulfur content of about 0.25%.

In the refining process, much of the sulfur is shifted to residual oil, one of the most dense of the refined products. Anywhere from 0.5%–5.0% of residual oil may be sulfur, although special refining steps may be used to further reduce the sulfur content. Alternatively, low-sulfur crude oil may be used at the outset so that a low-sulfur residual oil is produced. Residual oil is used to heat apartments and institutions such as schools and hospitals. As recently as the early 1970s, residual oil was being widely used to produce steam from electric generation; utilities chose residual oil then because of its relatively low cost and its low-sulfur content as compared to coal.

Natural gas, in contrast to coal and oil, is almost free of sulfur. From this standpoint, it is an environmentally sound fuel. In 1983, burning coal and oil in electric generating plants and for heat produced 16.8 million tons of sulfur, or 87% of the total sulfur oxides (20.8 million tons) emitted that year (see Figure 20.1).

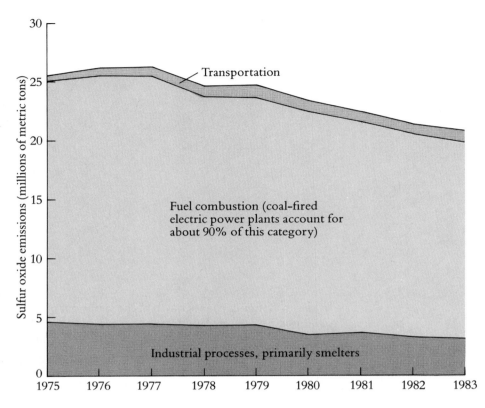

Figure 20.1 Sulfur oxides emission trends. Even with wider application of control technology, increased coal burning keeps emissions high. "Fuel combustion" refers to the burning of coal and oil for energy in electric power plants and factories, as well as the burning of these fuels in commercial and industrial buildings and residences for space heating. Such stationary sources are distinguished from mobile sources such as cars, trucks, trains, and planes. "Industrial processes" refers to the production of sulfur oxides from such activities as the smelting of ores. The sulfur referred to here comes from the ore in these cases and not from the burning of the fuel. (Data from U.S. EPA.)

Sulfur compounds also enter the air from the smelting and refining industries, as well as from other sources.

Oxidation of Sulfur to a Corrosive Mist

When coal or oil is burned, the sulfur in the fuels is oxidized. Two compounds are formed: sulfur dioxide and sulfur trioxide (see Figure 20.2). Less than 3% of the sulfur will oxidize to the trioxide form during the initial burning. The remainder is sulfur dioxide, the primary form in which sulfur enters the atmosphere. Sulfur dioxide in the air is oxidized gradually by oxygen in the air to trioxide. Any sulfur trioxide formed will immediately react with water vapor to yield sulfuric acid (H_2SO_4), which is present in the air as a fine mist of liquid droplets. This highly corrosive mist eats away many materials, including such building materials as marble and mortar.

Sulfur dioxide produced in combustion also reacts with water vapor in the atmosphere to produce sulfurous acid (H_2SO_3), a weak acid that also reacts gradually with oxygen to produce sulfuric acid. Thus, sulfuric acid is produced by two routes. When air is dry, the route that includes sulfur trioxide is the major route;

when humidity is high, the route that includes sulfurous acid is the major one.

Oxides of calcium and iron are also formed in the process of combustion. These oxides enter the air in great quantities when coal is burned, but particles from the burning of oil are far less numerous. The oxides often react with sulfuric acid to yield calcium and iron sulfate in particle form. That is, calcium oxide may react with sulfuric acid to produce calcium sulfate and water. Iron oxides may react in a similar fashion. The reactions are summarized in Figure 20.2.

Such sulfate particles plus the sulfuric acid droplets may account for 5%–20% of the particulate matter found in urban air. It is thought that much of the sulfur dioxide is converted to the sulfate and sulfuric acid forms within a few days from the time of its emission. During this time, winds can transport the pollutants hundreds of miles.

Effects on Materials, Plants, and People

Many materials may be attacked by the mists of sulfuric acid. In addition to carbonate building materials such as marble and mortar, metals such as steel, copper, and

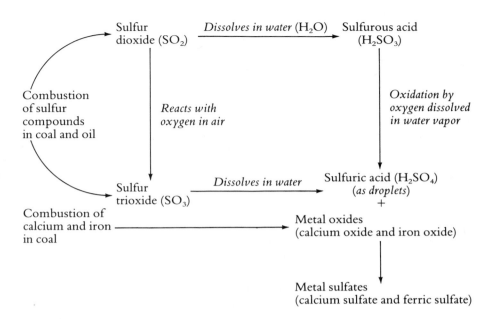

Figure 20.2 The reactions of sulfur in air.

aluminum are also corroded, and common cloth fabrics are damaged as well.

High concentrations of sulfur dioxide and its derivatives may cause acute injury to plants. Where sulfur dioxide concentrations reach nearly 3,000 $\mu g/m^3$, leaves, needles, and other plant tissues may look bleached. Eventually, leaves or needles may take on a scorched, reddish-brown appearance, and the damaged foliage falls from the plant. Smelters that refine copper ores and lead ores containing sulfur may pollute extensive areas with their gaseous emissions. For example, a near absence of vegetation may be observed in an amazing scene of devastation around the International Nickel Company smelting operation at Sudbury, Ontario.

Even where sulfur dioxide levels average only on the order of 100 $\mu g/m^3$, such as may occur in cities, plants may have a yellow discoloration. Fruit trees such as apple and pear and forest trees such as ponderosa pine and tamarack (larch) are susceptible to damage from sulfur oxides. The cotton plant is also susceptible, as are alfalfa and barley.

Elevated levels of sulfur oxide have been linked to human illnesses and even to deaths. In air-pollution episodes in New York, Osaka, and London, investigators have noted an increase in the normal death rate following periods of high concentrations of sulfur oxides. Effects on human health of sulfur oxides and of particulates are difficult to distinguish because the two kinds of pollutants tend to interact. The effects of the two together are discussed later in this chapter.

Respiratory illnesses such as bronchitis are seen to increase with sulfur oxide levels. One study found that the illness rate increased in an area where the average annual concentration was only 100 $\mu g/m^3$. Not long ago such levels were observed in a number of areas in the United States. Only Pittsburgh and Salt Lake City come near such levels now.

To understand how severe sulfur oxide air pollution has been in the past and where we stand today in its control, we need to describe the air-quality standards for these substances. The daily standard for sulfur oxides set in 1971, based on the maintenance of human health, is 365 $\mu g/m^3$ (an average over 24 hours), and this level is not to be exceeded more than once per year. An air pollution *alert* is to be called when the 24-hour average exceeds 800 $\mu g/m^3$. The annual standard, the average over one year, is 80 $\mu g/m^3$.

Rural areas have background concentrations of about 0.5 $\mu g/m^3$ but urban areas have levels of 50 to 100 times greater. The people of Chicago, for example, lived at an incredible average concentration of 470 $\mu g/m^3$ in 1964. Levels in Chicago have decreased considerably since the air-quality standards were issued in 1971.

During the famous London fog of 1952, the concentration of sulfur oxides reached 4,000 $\mu g/m^3$. In a 1962 episode in New York, the average concentration during one day was 2,500 $\mu g/m^3$. Such concentrations are still occurring in some places in the world. In Ankara, Turkey in January 1982, sulfur dioxide levels reached 2,835 $\mu g/m^3$ during a period of several days.

All industries and businesses, except hospitals and bakeries, were ordered closed in an attempt to reduce emissions. Ankara's air pollution problem is chronic because it uses so much high-sulfur lignite coal and high-sulfur fuel oil.

Sulfur oxides have clearly been shown to aggravate and cause respiratory problems by increasing resistance to air flow in the pulmonary tract. That alone is reason enough to control sulfur oxide emissions. Sulfur dioxide may, however, have an additional uncounted effect. This gas has been shown to be a **cocarcinogen** in rats. In the presence of benzo(α)pyrene, sulfur dioxide increases the frequency of cancer occurrence above the natural rate. At the moment, we can say that sulfur dioxide has not been proven to be either a **carcinogen** or a cocarcinogen in humans, but it is suspect. (Carcinogens are further discussed later in this chapter.)

Finally, sulfur dioxide is the principal factor in producing acid rainfall, which is turning lakes acid and which is thought to be responsible for widespread forest decline in the United States, Canada, and the European countries. (See Global Perspective V.)

Ways to Reduce Sulfur Dioxide Emissions

Cleaning up coal itself. In addition to the possibility of using expensive low-sulfur coal, emissions of sulfur dioxide from power plants can be reduced by cleansing the coal of its sulfur prior to use. Two forms of sulfur occur in coal: inorganic and organic. *Inorganic sulfur* is the sulfur present as pyrites; these are the metal sulfides, such as iron sulfide (also called iron pyrite). *Organic sulfur* is sulfur that is chemically bonded to the carbon in the coal. Special washing steps are sufficient to remove inorganic sulfur from coal. The organic sulfur, in contrast, requires chemical treatment for removal. Sulfur as pyrites (metal sulfides) may range from 30% to 70% of the total sulfur present in coal, but on average, equal quantities of organic and inorganic sulfur are present.

To cleanse coal of the inorganic (mineral) sulfur, the coal is first crushed to expose the mineral veins. The ground coal is then mixed with water in a large tank. Pyrite has a higher density than coal and hence sinks faster; the cleaned coal is skimmed from the top of the tank. As much as 500–1,000 tons of coal can be processed per hour. Washing is most effective on coals with higher proportions of pyrites.

Chemical cleaning of coal is a general name for a variety of processes, all of which remove the organic sulfur that is bound to the carbon in coal. Several dozen processes are competing for research and development funds in this area. Some of the chemical methods remove both the pyrite sulfur and a portion of the organic sulfur.

One promising cleaning method developed by the Batelle Memorial Institute mixes finely ground coal into water containing sodium and calcium hydroxide. Treated under pressure and high temperature, the coal is cleansed of most pyrites and of half the organic sulfur; these substances remain in the liquid phase. The coal is washed, dried, and then ready for use. The method has the potential to make the high-sulfur coal from the eastern United States an acceptable fuel in coal-fired electric plants. The coal could potentially be used without the need to *scrub* sulfur dioxide from the stack gases, but a commercial-scale application is needed to assess the cost.

One process that appears to remove up to 60% of the organic sulfur is known as **solvent refining of coal (SRC)**. Ground-up coal is mixed with the organic solvent anthracene; nearly 95% of the carbon dissolves in the anthracene. Exposed to high pressures and temperatures, the solution is then treated with hydrogen. The coal can be recovered either as a solid or as a liquid. The solid material, black and brittle, has less than 1% ash, though the original coal may have had anywhere from 8% to 20% ash. Sulfur is reduced to less than 1% as well, and the heating value is increased by 30% on a per-ton basis. Solvent refining of coal, if it turns out to be commercially feasible, has bright prospects. With sulfur content lowered sufficiently and ash nearly eliminated, it may be possible not only to avoid scrubbing the stack gases but also to avoid electrostatic precipitation for particle removal.

Scrubbers. A number of methods are being and have been developed to scrub sulfur dioxide from the gas that exits the smokestack. Unfortunately, the term *scrub* is not scientifically descriptive; it simply means that the stack gases are cleansed of sulfur dioxide. The operations of the various scrubbers are based on chemical reactions with the sulfur dioxide in the stack gas of coal-fired power plants. The chemicals formed in these reactions may be waste products or marketable items.

Of the more than 50 control concepts, the single most thoroughly tested and reliable process is the lime-limestone wet scrubber. About 85% of the scrubbers in place in 1982 were of the lime-limestone type. In this device, the flue gas is passed through a slurry mixture of limestone and lime in water. Sulfur dioxide is absorbed into the slurry and reacts to form calcium sulfite and calcium sulfate (gypsum). The flue gas is not only cleansed of about 80% of its sulfur dioxide, but it is also cleansed of 99% of the fly ash it carried. The flue gas has been cooled so much, however, that it must be reheated to make it buoyant enough to rise up the chimney for discharge to the atmosphere.

The resulting sludge must be disposed of in some way. Because it is a gloppy mass, it is not by itself a good landfill material. However, it appears that a hardener such as fly ash can be added to react with the slurry, producing a claylike substance.

Because sludge disposal is seen as a serious problem, scrubbers that produce a marketable product instead of sludge appear very attractive. One such scrubber produces sulfur of high purity; another produces a dilute sulfuric acid. The sulfuric acid is not economical to transfer across large distances, but the high-purity sulfur, which is used in such products as pharmaceuticals, industrial chemicals, and fertilizers, is quite compact.

Another concept combines both coal cleaning and scrubbing. Cleaning mineral sulfur from coal is relatively inexpensive but is limited to the fraction of sulfur contained in pyrite (metal sulfide) form. If this fraction of the sulfur is removed, and if the stack gas produced by burning the cleaned coal is then scrubbed, the resulting level of sulfur dioxide may meet emission standards at a reasonable cost.

During most of the 1970s, portions of the utility industry kept up a steady criticism of scrubbers. Financed principally by the American Electric Power Company, the largest electric power company in the country, a massive ($3.6 million) advertising campaign derided the scrubbers, saying they were unreliable and expensive and that the amounts of sludge were enormous. However, as experience with scrubbers has grown, their reliability has proven to be high. The mountains of sludge predicted have not materialized either. A typical 1,000-megawatt coal-fired plant is expected to require 400–700 acres for the disposal of both fly ash and scrubber sludge over a lifetime of 30 years. The EPA estimated a total acreage requirement for scrubber sludge and fly ash of 18,000 acres by 1985 and of 63,000 acres by the year 2000.

In the 1970s, the opponents of scrubbers advocated control of sulfur oxides by the use of very tall stacks that would disperse the gas more widely. When levels at the ground became hazardous, the boiler could burn a low-sulfur fuel until the episode was past. Alternatively, it was suggested that power generation be cut back during such times. The opponents of scrubbers also suggested that the industry should wait to see how well coal cleaning turns out before installing scrubbers.

Now scrubbers are an accepted technology, although acceptance is still grudging. By 1982, some 225 scrubbers (or flue gas desulfurization units) were either in place, being built, or on order. In all, 100,000 megawatts of coal-fired electric power capacity would ultimately be covered by these units, about 35% of all coal-fired capacity. This is the rough equivalent of 100 full-sized modern coal-fired plants. Further, as new generating plants are planned, the number of units will increase still further.

Sorbent injection is still another technology that appears to be effective in removing sulfur oxides. Though not yet a commercially proven process, its simplicity and low cost make it appealing. In one concept, dry, powdered, reactive minerals are sprayed into the flue gas. Sulfur dioxide reacts with the dry sorbent chemicals to produce a dry sulfur compound, which is removed in a baghouse as particulate matter with the fly ash. (See page 463 for a description of how particulate matter is removed from stack gases using a baghouse.) Another similar process sprays droplets of lime-water slurry into the flue gas. Reaction produces calcium sulfite and sulfate, which are dried in the hot gases and then removed as particles. Both of these processes are attractive because of the reduced volume of product.

Fluidized bed. Coal washing (physical cleaning) and scrubbers are existing technologies that reduce sulfur dioxide fumes. In contrast, the fluidized bed combustion system is an experimental design for removing sulfur dioxide. The fluidized bed system burns coal as a conventional boiler does, but the coal is mixed with granular limestone and is layered on metal plates. Further, the air for combustion comes from below, passing through holes in the metal plates up past the coal. The air flow is so strong that the particles of coal and limestone are lifted, or floated (fluidized, as the name implies), above their bed on the plate.

The limestone reacts with the sulfur dioxide from the burning coal to form fine particles of calcium sul-

fate, which are carried off in the stack gases and removed with the fly ash, probably by an electrostatic
precipitator. The temperature of the burning coal in the
fluidized bed is lower than the temperature of burning
coal in a boiler; as a consequence, fewer nitrogen oxides
are formed. Thus both sulfur dioxide and nitrogen oxides are lower for the fluidized bed combustion system.
Commercial-scale demonstration projects were begun
in 1984 in Kentucky, Minnesota, and Colorado. These
three plants should be operating by the end of the decade, and their success could bring us a new and effective
control technology.

Progress in Control of Sulfur Oxides

What progress has been made since 1970, and where do
problems remain? In 1977, the sulfur dioxide standard
was violated in parts of Pittsburgh on more than 20%
of the days. During the same year in parts of Salt Lake
City, the standard was violated on more than 50% of
the days, and the alert level was exceeded on nearly
10% of the days.

By 1983, only Pittsburgh was not meeting the annual average standard and the daily standard for sulfur
oxides. All other metropolitan areas were meeting both
standards. Success appeared to be at hand. Yet progress
in reducing emissions of sulfur oxides from 1975 to
1983 was not at all substantial. By 1983, national emissions had been reduced by only 20% from their 1975
level (Figure 20.1). Now emissions have the potential
to increase as more coal-fired power plants are brought
on line.

At the same time that emissions were slowly declining, the air quality in urban areas was increasing
fairly steadily. What was happening?

The geography of sulfur oxide emissions was simply being shifted (see Controversy 20.1). Air pollution
control requirements forced polluters in urban areas to
change fuels or otherwise control emissions. New
power plants were forced to be sited at more remote
locations. The atmospheric burden of sulfur was simply being more widely spread. In addition, tall stacks
were being used to discharge sulfur oxides high up
in the atmosphere, thereby decreasing ground-level
concentrations.

Thus, the total atmospheric burden of sulfur oxides was being decreased, but only a little. In essence,
we were taking sulfur oxides and putting them "somewhere else," which has turned out to be mountain lakes

and evergreen forests. From ground-level pollution we
created acid rain.

Particulate Matter

Particles Stem from Combustion and Industrial Activities

Particulate matter is another serious air pollutant. Unlike other pollutants we have discussed, however, particles are not a single chemical type. Instead, numerous
solid and liquid compounds are dispersed in the air
from many sources. Transportation, fuel combustion,
industrial processes, and solid waste disposal all contribute generously to the atmospheric burden of particles. Although we mention them briefly here, two
types of particles with special properties and effects are
discussed more completely elsewhere: Lead is treated
in Chapter 21; asbestos is discussed in Chapter 27.

The burning of coal produces solid particles in the
air—not only ash particles (calcium silicates) and carbon particles, but also metal oxides such as calcium and
ferric oxide. The metal oxide particles may react with
the mist of sulfuric acid droplets. The reaction produces still other particles, the metal sulfates. The sulfuric acid droplets themselves are particles derived
from the reaction of sulfur trioxide with water vapor.
Both the acid droplets and the sulfate particles, then,
are derived in a large part from coal burning (see Figure 20.2).

The quantity of particles derived from coal burning is enormous. Fortunately, however, a large proportion of the particles are removed from the stack gases.
In 1983 in the United States, about 2 million metric
tons of particles were released to the atmosphere from
coal and oil combustion, most of it from coal-burning
electric plants. Probably seven times that quantity was
generated, but most of it was removed. The 2 million
metric tons from fuel combustion, however, made up
approximately 30% of the 6.9 million metric tons of
particles reaching the air from all sources in 1983. Only
a small amount of ash from fuel combustion comes
from the burning of oil. Whereas 80–120 pounds of ash
are emitted per 1,000 pounds of coal, only about 2
pounds of particles stem from burning the same
amount of oil.

The burning of gasoline and diesel fuel produces
liquid droplets in the air. Liquid hydrocarbons (compounds of carbon and hydrogen) and liquid derivatives

CONTROVERSY 20.1

Costs of Air Pollution Control

The good and sweeping intentions of many
environmentalists are now an obstacle blocking those less-
fortunate Americans who desire economic justice.
Bayard Rustin

. . . using that resource [air] as an inducement to promote
development is as short-sighted as would be the reckless use
of any other resource.
Cubia L. Clayton

Faced with the high costs of pollution control, it is
natural that one of an industrial executive's first impul-
ses may be to relocate where the standards are less
strict. This kind of reaction is, of course, viewed as a
serious problem by labor and local political leaders,
who are understandably concerned with a loss of jobs
and with a lack of growth in their areas.

Bayard Rustin (1975), a labor leader, wrote:

> Many in the environmental lobby, in a hysteri-
> cal effort to reverse the effects of the industrial
> revolution in a few years, have supported sweep-
> ing and uninformed legislation which has had un-
> necessarily harmful economic consequences. The
> Clean Air Act is but one example. It has resulted
> in the closing down of hundreds of plants, from
> drop forges to specialty organic chemical plants.
> The environmental damage of the plants forced to
> close had been minimal and could have been easily
> corrected. The economic damage of the legisla-
> tion has been severe.
>
> The poor black in the ghetto or the unem-
> ployed white worker would not find very much
> with which to identify in [Senator Stuart] Udall's
> semi-revivalist call that we must no longer "over-

indulge ourselves" or seek "to satisfy unlimited
greed and desire for luxury." They would agree
that "we must change our way of life," but they
would take this to mean rather more luxury than
they have been accustomed to in the past. The
good and sweeping intentions of many environmentalists are now an obstacle blocking those
less-fortunate Americans who desire economic
justice.

But this is not the whole story. An official of the
New Mexico Environmental Improvement Agency
has noted that:

> . . . environmental quality is too often viewed as
> the tool with which to bargain.
>
> The difficulty, apart from environmental deg-
> radation, is that this approach fails to consider the
> reality of air quality as a natural resource which is
> as depletable as any other. Whether one agrees
> philosophically with national air-quality stan-
> dards, they do exist, and their existence means the
> end of the age-old concept of an unlimited air re-
> source. Hence using that resource as an induce-
> ment to promote development is as short-sighted

as would be the reckless use of any other resource.

The imposition of a uniform, nationally designated Class II ceiling . . . actually ensures more development than would otherwise occur. This is because (i) those states that desire to use air quality as an inducement to development will not be allowed to develop at the expense of neighbors who are interested in maintaining as much of a quality environment as possible, and (ii) a tighter ceiling than that imposed by national standards will help impress on everyone that air is a depletable resource and that new industry must be required to utilize the best control technology in developing new energy supplies.

The question of available technology is the crux of the problem. Existing industry faced with the problems of the retrofit of control devices is finding the job difficult and expensive. In many cases, the result has been an unwillingness to accept the fact that the job of control is even possible. But a difficult job is not the same as an impossible one, particularly in the case of new industry where controls can be made an integral part of plant design.

The end result is that, rather than having an air shed used up by three or four inadequately controlled industries, more industry can be accommodated. (Clayton, 1976)

Who is right? How would you balance air cleanup and jobs? Can you think of ways the poor and unemployed could benefit from air pollution control?

Sources:
Bayard Rustin, quoted in *The New York Times*, August 11, 1975.
Cubia L. Clayton, *Science*, **193** (September 10, 1976), 953.

of hydrocarbons come from the incomplete combustion of gasoline and diesel fuel. Still another kind of particle results from the photochemical reactions of nitrogen oxides and hydrocarbons in the air. These photochemical, or sunlight-stimulated, reactions produce liquid organic substances that are scattered as tiny droplets in the air. The term **smog** has been coined to describe the resulting foglike condition. Photochemical smog is particularly evident in cities such as Los Angeles, where automobile use is excessive. In the early 1970s, up to 40% of the particles in the Los Angeles air stemmed from the use of automobiles. Automobile exhaust also emits particles of lead, which derive from the lead compounds put into gasoline to increase its octane.

Surface mining of coal and of other substances produces large quantities of particles. The refining of ores and the manufacture of metals are among the industrial processes releasing particles to the air. The grinding and spraying that accompany construction are also sources. Taken together, these industrial activities may account for more particle emissions than occur from coal combustion.

In the recent past, asbestos escaped to the atmosphere from new building construction, where it was being sprayed into place as insulation. Although asbestos use has now come to an end, this substance continues to escape into the atmosphere during the demolition of older buildings. Finally, the incineration of solid wastes may, in some cities, be a significant source of particles when incinerators are centrally located.

Physical and Biological Effects of Particles

Particles may settle on surfaces, leading to a dirty gray appearance. In addition to soiling, particles may cause corrosion by acting as the centers from which corrosion spreads. The annual repair of these surfaces carries a significant cost.

Sulfuric acid droplets may damage plant tissue. Compounds from photochemical reactions produce a burn on the leaves of many vegetables, such as beets, celery, lettuce, and peppers. Inert, unreactive particles may simply soil plant surfaces.

Particles have a subtle effect on weather. They may act as nuclei upon which water vapor condenses. Prolonged periods of fog may result from high levels of particles in the air. Increased rainfall has also been seen in areas where particle levels have risen. Furthermore, particles may be reflecting solar energy away from the earth. As such, they may have been partly responsible for the small but noticeable cooling in the Northern Hemisphere during the last quarter century. The recent eruptions of Mount St. Helens and of El Chichon in

Mexico introduced such quantities of particles to the global atmosphere that emissions due to human activity were dwarfed. These volcanic emissions, however, do not contribute to the particle levels in our cities except on rare occasions.

Health Effects of Particles and Sulfur Oxides

The deadly duo. We discuss sulfur oxides and particles together for two reasons. First, high levels of both these pollutants have been measured during some of the worst air-pollution episodes recorded. By worst, we mean those episodes with the largest numbers of deaths and illnesses. Second, sulfur oxides and particles in the air are known to potentiate one another. That is, when both substances are present at high levels, the effects on health are worse than when only one of the substances is present.

Why have sulfur oxides and particles appeared together in high concentrations? The answer is that they have had a common source: the burning of coal. Not only is coal burned for electric power generation, but before World War II, it was used widely to heat homes in the United States.

Sulfur oxides and particles aggravate respiratory disease. When we first recognized that contaminants fouling the air could cause death, sulfur oxides and particles were on the scene. In the Meuse Valley of Belgium in 1930, in the tragedy at Donora, Pennsylvania in 1948, and in the London fog of 1952, a killer struck. Today, there is no longer any doubt that sulfur oxide pollution in the presence of particles was the killer.

For a time, we did not understand why the problem in these pollution episodes was always sulfur oxides *plus* particles. We have long known that particles act as a focus upon which water vapor condenses. We now understand that sulfur dioxide quickly dissolves in the water droplets, producing a highly acid, highly corrosive mist. It is this sulfurous acid mist that is responsible for so much damage to life and health.

In the London fog of 1952, more than 4,000 deaths were attributed to the excessive pollution levels. The number of individuals who became ill was obviously far greater. Sulfur dioxide levels may have reached 4,000 $\mu g/m^3$ of air. The death toll was heaviest among the elderly, a population group already afflicted with lung and heart diseases. This appears to be the way sulfur oxides and particles act, by worsening the al-

ready existing respiratory and cardiac diseases. Although many deaths during and after the fog were due to such respiratory diseases as pneumonia and bronchitis, investigators were able to blame pollution levels because more deaths occurred than had been expected for that time of year. The 4,000 deaths attributed to the fog were those deaths *in addition to* the number expected for that time of year.

Since those early episodes, scientists have accumulated masses of data, some indicating health influences even at very low levels of air pollutants. In New York City, for instance, when sulfur dioxide levels (averaged over 24 hours) exceeded 500 $\mu g/m^3$ of air, a small increase in deaths was noted. The daily standard set for sulfur dioxide is 365 $\mu g/m^3$ (averaged over 24 hours), so the margin of safety between the standard and the level at which excess deaths have occurred is not very large.

At annual average concentrations for each pollutant of 100 $\mu g/m^3$ or greater, higher rates of bronchitis among adults and respiratory infections among children have been observed. The standard set for the annual average concentration of sulfur dioxide is 80 $\mu g/m^3$ and of particles, 75 $\mu g/m^3$; again, there is not a wide margin of safety. In the period 1960–1965, many major cities exceeded these standards. In 1983, some 44 urban and metropolitan areas were still violating the annual particle standard (see Figure 20.3). Only Pittsburgh and Salt Lake City were violating the annual sulfur oxide standard.

In London, since the devastating fog, particle levels have been cut from an annual average of 300 to 50 $\mu g/m^3$ and sulfur dioxide has been cut from an annual average of 300 to 200 $\mu g/m^3$. Investigators in London now find much less disease that is related to air pollution.

Particles. This is not to say that particles alone do not have an effect on human health. The frequency of respiratory infections such as colds and bronchitis is seen to increase as particle levels increase. Furthermore, at high concentrations of particles, deaths in excess of the number expected for the time of year have been seen to increase.

The daily health-based standard for particle concentrations has been set at 260 $\mu g/m^3$ (averaged over a 24-hour period), and this level is not to be exceeded more than once per year. At a particle concentration of 375 $\mu g/m^3$, an air pollution alert is to be declared, at

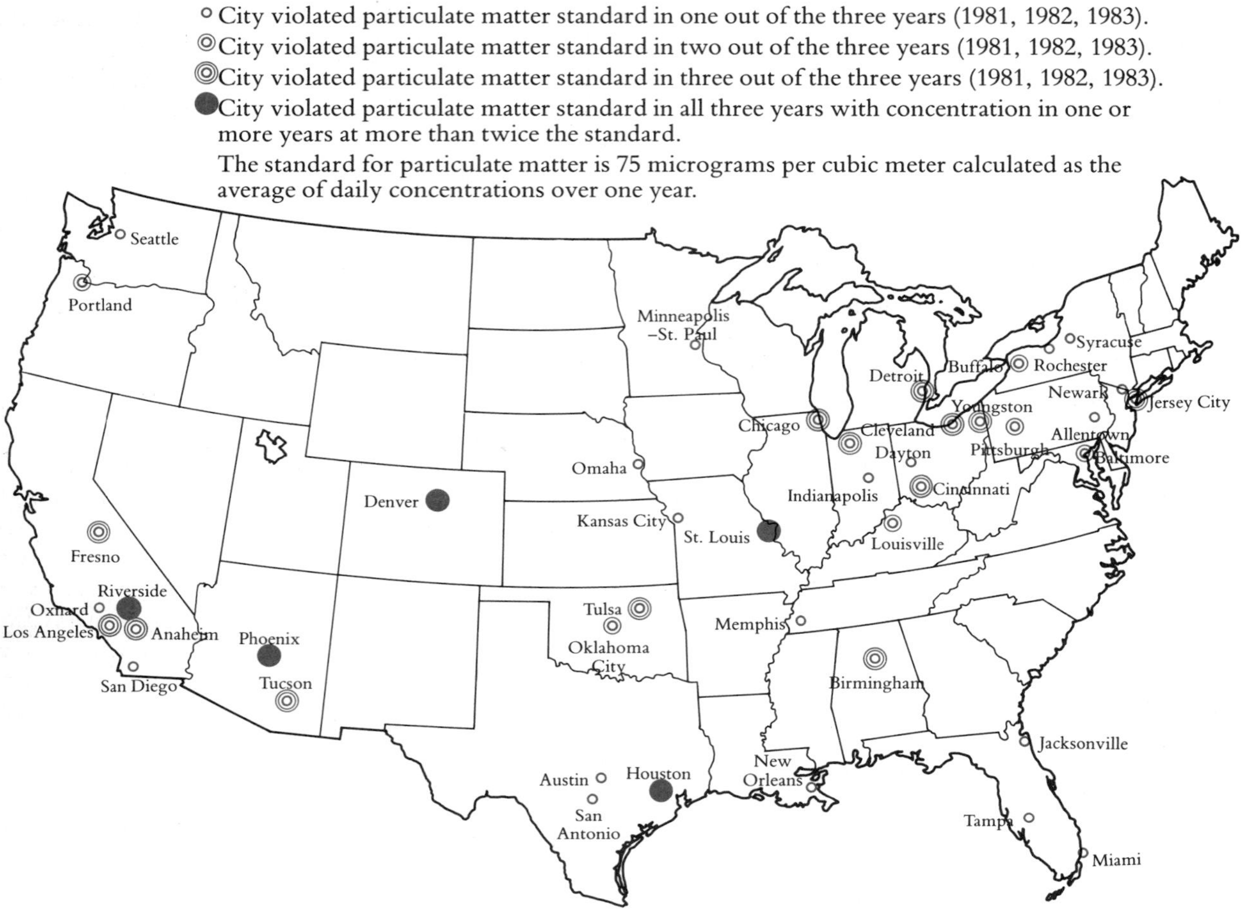

○ City violated particulate matter standard in one out of the three years (1981, 1982, 1983).
◎ City violated particulate matter standard in two out of the three years (1981, 1982, 1983).
◎ City violated particulate matter standard in three out of the three years (1981, 1982, 1983).
● City violated particulate matter standard in all three years with concentration in one or more years at more than twice the standard.

The standard for particulate matter is 75 micrograms per cubic meter calculated as the average of daily concentrations over one year.

Figure 20.3 Cities not meeting particulate matter standards in 1981, 1982, and 1983. Many (44) cities were not achieving the health-based standard for particulate matter in the period from 1981 to 1983. Cities in California and Arizona had high particle levels due especially to the presence of droplets of liquid associated with photochemical air pollution. In most other areas, particle levels are mainly due to industrial processes and power plant emissions.

which time industries might be requested to curtail or postpone their activities.

Sulfate particles. The sulfates we mentioned earlier have been implicated in human health problems. These sulfates are among the smallest particles in urban air and so have easy entry to the lungs. Because sulfate particles are small, they remain in the air longer than larger particles and are subject to long-distance transport over hundreds of miles. The sulfates include the sulfuric acid droplets as well as ammonium, calcium, magnesium, and iron sulfates, among others. These particles are apparently formed by the chemical reactions between emitted substances in the air, rather than being themselves emitted from a combustion source. The sulfates are implicated not only in human health problems, but also in acid rain. Much of the sulfate in the air undoubtedly stems from power plant emissions.

No air-quality standard has yet been stated for sulfates. Nonetheless, a 1975 position paper from the U.S. EPA suggested that health effects begin to show up at sulfate levels of 10 μg/m^3 (averaged over a 24-hour

Air Pollution

a   Air pollution in Egypt

b   Statue in Italy

c   Adirondack lake

d   Czechoslovakian forest devastated by acid rain

An energy-related environmental problem that both developed and developing nations already share is air pollution. As industrialization and the use of automobiles grow so does air pollution. (a) An inversion captures pollutants in Cairo, Egypt. (b) Pollutants in the air can soil buildings and statues, such as this beautiful fountain in Italy. (c) Air pollution also contributes to acid rain. This picturesque lake in the Adirondack mountains appears untouched by pollution, but acid rain has already raised the acid content of its waters to the point that fish can no longer breed there. The spruce bordering the lake show browning on the tips of their needles — the first sign of acid rain damage. If the situation continues, this landscape may one day resemble the following one in Czechoslovakia. (d) Both conifers and hardwoods once grew on this hillside in the Erzgebirge Range, but now all are dead, destroyed by an acid rain fed by the sulfur emissions of coal-burning power plants. The soil here has become highly acidic to a depth of 10 centimeters. Liming on a massive scale is being attempted in this area in a desperate effort to restore the soil's fertility.

Natural Sources
of Power

a Old windmill

b Wind farm

c House with solar collectors

d Solar energy

Natural sources of power offer the hope of relief from energy-related pollution to both developed nations, such as the U.S., and to developing nations. (**a**) Farmers and other rural residents have used windmills to draw water and grind grain for centuries. (**b**) New "wind farms" promise to generate electricity without the air and water pollution that accompanies conventional methods of generation. (**c**) The technology for heating household water has been available in the U.S. since the late 1800s. This house has solar panels on the roof. (**d**) In developing nations today, solar energy offers limitless and potentially pollution free power. Here a traveler in an arid region uses solar power to boil water.

day). Data on sulfates have not been as extensive as on other pollutants, but measurements in August 1977 showed almost all of the northeastern United States having 10 or more days with sulfate levels at or above $10 \ \mu g/m^3$.

High sulfate concentrations occur in areas where sulfur dioxide emissions are high, but they are higher still at times when ozone, another air pollutant, is present in high concentrations in the air. The relation exists because ozone hastens the oxidation of sulfur dioxide to the trioxide form and hence fosters the formation of sulfuric acid and metal sulfates.

Rainfall appears to bring sulfate levels down rapidly. Sulfate concentrations do not appear to peak in a geographic sense but tend to be spread widely across regions. The spreading is the result of the small size of sulfate particles. Winds can transport sulfate particles farther than larger particles before they are deposited.

Particulate Matter in the Air and Cancer

Although particles are known to play a role in stimulating respiratory diseases, particles in the air have also been implicated as a cause of cancer. A **carcinogen** is a substance that has been shown in the laboratory to produce cancer among experimental animals. Alternatively, the evidence that it produces cancer may come from the observation of occupational groups inadvertently exposed to the substance. An example is the increased risk of lung cancer among asbestos workers.

A number of carcinogenic substances have been found in polluted city air, but no direct evidence exists that these compounds cause cancer in humans at the concentrations observed. Nevertheless, the substances are "suspicious" because, in a number of countries, investigators have found more cases of lung cancer among urban dwellers than among rural inhabitants. Remember, however, that individuals who live in cities may have industrial jobs and different habits, such as smoking and so on. Since it is now clear that smoking is a principal cause of lung cancer, scientists have had to eliminate or correct for smoking in their studies. Even after this is done, the conclusion remains that urban dwellers suffer more lung cancer than rural residents.

In addition to these surveys within countries, studies have been made on British migrants to the United States. Of the study participants, those born in Britain who came to reside in the United States had two-thirds the frequency of lung cancer as the permanent residents of Britain. However, the incidence of their lung cancer still exceeded that of the general population of the United States. It bears mentioning that up to the present time, British lifestyles have differed from our own. Coal has been the main fuel for heating homes and buildings and for generating electric power, as well as for the production of *town gas*, a gas with properties resembling natural gas but manufactured from coal. Furthermore, the number of autos per 1,000 people in Britain has been about half that in the United States.

The airborne substances under greatest suspicion as potential carcinogens are polycyclic aromatic hydrocarbons, an example of which is benzopyrene. Known to cause cancer in experimental animals, polycyclic aromatic hydrocarbons have been found free or associated with particles of soot in urban air in the United States and Britain.

Ways to Control Particle Emissions

Numerous devices are available to reduce particle emissions. These include the settling chamber, the after-burner to ignite and burn particles, and the **electrostatic precipitator**. First used in the collection of fly ash in 1923, the precipitator is now in use at nearly 1,000 plants in the United States. The operation of the precipitator consists, first, of attaching electric charges to fly ash particles. As the flue gas moves between the plates of the device, ions or electrons formed by a high-voltage discharge bombard the particles of fly ash and provide them with charges. The charged particles are attracted to and deposited on the grounded metal plates, or collection electrodes. The plates must be cleaned periodically by vibrating them or rapping them with a mechanical device (Figure 20.4). The devices can attain efficiencies of particle collection up to 99% and can operate with little maintenance while drawing little electric power.

An alternative to the electrostatic precipitator is the **baghouse**, which is composed of fabric bags to capture the fly ash. (See Figure 20.5.) The fabric filters have a long history of use in industries such as grain mills, asbestos factories, and cement plants. Electric utilities have been purchasing these units on a trial basis in the last few years in order to evaluate them for use in power plants. The principle of the baghouse is simple; it is a very large vacuum cleaner. Air is drawn up through the bags, whose fine weave material traps particles. Only clean air exits.

Despite impressive efficiency (over 99%), utilities

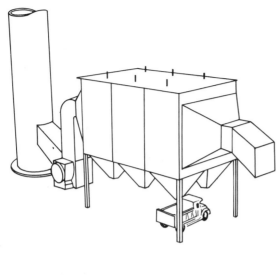

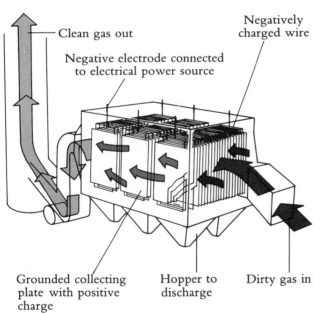

Figure 20.4 Electrostatic precipitator. Dirty air flows between negatively charged wires and grounded metal collecting plates. The particles in the flowing air become charged and are then attracted to the plates, which hold the accumulated dust until it is periodically knocked into hoppers. The clean air is then pumped out through the stack. (Adapted from *Controlling Air Pollution*, American Lung Association, 1974.)

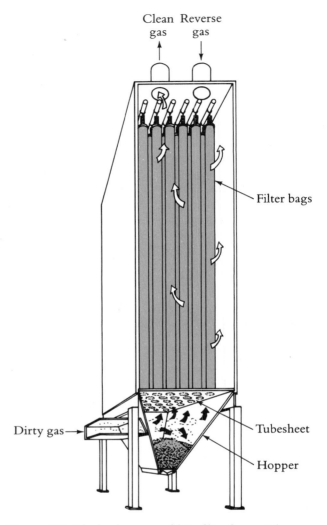

Figure 20.5 The baghouse, or fabric filter for particle removal. This unit, some 35 feet (11 m) tall, may contain several hundred filter bags. The bags are 12 inches in diameter. Some 8 to 12 units may be used in a large-scale application, such as at a power plant. Baghouses are a promising technology for both particle and sulfur oxide removal. (Drawing courtesy of The Electric Power Research Institute.)

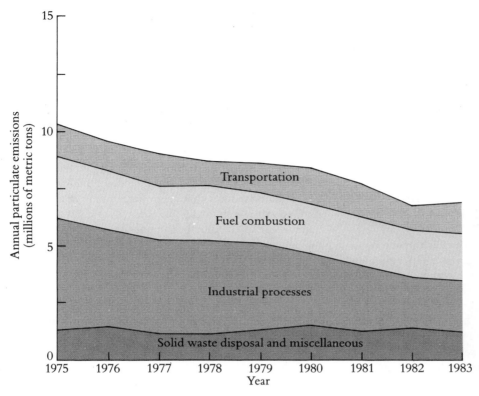

Figure 20.6 Trend in particle emissions. Note that particle emissions do not seem to be declining. Improvements in particulate air quality will be unlikely unless the trend in emissions turns down again. The fuel combustion category is electric power plants. In the period from 1975 to 1983, emissions in this category declined by only 23%. In the same interval, emissions from industrial processes decreased by 54%.

have only recently begun to use this technology to any significant degree, citing lack of experience with this device. The first baghouse unit was installed in 1973. By 1983, however, a number of utilities had begun to recognize the potential of the baghouse for increasing removal efficiencies. In that year, more than 100 baghouse units were either in operation, under construction, or in design, representing about 21,000 megawatts of electrical capacity, the equivalent of 21 coal-fired 1,000-megawatt power plants. Utilities have become convinced that the baghouse rivals the electrostatic precipitator in efficiency, reliability, and cost. In addition, the baghouse can be used to capture sulfur oxides. Sorbent injection can be used to convert gaseous sulfur dioxide to particles, and these particles can be removed effectively by the baghouse. The removal of particles and sulfur oxides in a single step could mean a large cost savings.

Particulate control methods have been widely applied and have had a substantial impact on air quality. Although particulate emissions were steadily declining during the 1970s, the recent trend is agonizingly slow (Figure 20.6).

Progress in Control of Particulate Matter: An Uphill Battle

Progress in controlling particle emissions appears to have stalled; the trend in emissions was turning flat in the mid-1980s (Figure 20.6). In the 20-year period from 1960 to 1979, the average concentration of particles in the air at a number of monitoring sites had fallen by 32%. By the end of the 1970s, however, improvements in air quality as measured by particle levels seemed to come only slowly. In 1980, about 21% of the population still lived in areas where the annual standard (75 μg/m^3) was exceeded.

According to the National Commission on Air Quality, particulate levels actually increased in 16 major areas of the nation between 1975 and 1979. These areas included Buffalo, Baltimore, Seattle, Houston, St. Louis, Denver, Kansas City (Missouri), and Portland

(Oregon). In addition, five areas of the nation exceeded the daily particle standards on more than 140 out of 365 days in 1979: Phoenix (285 days), Los Angeles basin (202 days), Las Vegas (175 days), Tucson (156 days), and Spokane (146 days).

In its 1981 report, the National Commission on Air Quality indicated that many urban areas would be unable to achieve the standard for particles in the air by 1982. Figure 20.3 shows 44 cities and metropolitan areas around the country that failed to meet the particle standards in at least one year during 1981, 1982, and 1983. Five cities (Houston, St. Louis, Denver, Phoenix, and Riverside, California) had annual particle levels greater than twice the standard in at least one of those three years.

Not only is there much to accomplish in the control of particles, but a new concern is emerging as well—namely, that the air-quality standard for particles, as presently stated, is inadequate to protect people's health. The present standard is stated as a maximum weight per cubic meter. The removal of large particles from the air might achieve the standard for some cities, but many small particles could still remain. Larger particles are less likely to reach the lungs because of respiratory defense mechanisms, so their removal is not as important in the first place. On the other hand, fine particles—those measuring less than 2 or 3 microns—are more likely to reach the lungs and have a serious effect, yet they add little to the mass of pollutants in the air. A new air-quality standard for fine particles in the air is needed.

Pollutant Standards Index and Summary of Standards

The Pollutant Standards Index is a means by which the general public is informed of air quality. A radio or television announcer may say that the state health department has declared air quality to be moderate, good, or perhaps unhealthful. *Moderate* air quality means that one or more of the pollutants exceed 50% of the air-quality standard but that no pollutant exceeds 100% of its air-quality standard. *Good* air quality, on the other hand, means that none of the pollutants exceed 50% of the air-quality standard. There is one exception to measuring each pollutant against its standard: Ozone is measured differently, and its benchmarks are indicated in Figure 20.7.

When one or more pollutants exceed 100% of the air-quality standard, but none exceed the respective alert level, the air is regarded as *unhealthful*. During intervals when the air is unhealthful, persons with existing heart or lung disease should reduce exertion and outdoor activity. During such a period, even healthy people show evidence of pollutant irritation.

An *alert* is declared when one or more of the five major pollutants exceed the value listed in Figure 20.8 in the row labeled "Alert," but when none exceed the

Table 20.1 Primary (Public Health) Standards for Air Quality (As of January 1983)

Pollutant	Time period of standard[a]	Maximum permissible concentration	Equivalent nonmetric concentration[c]
Suspended particles	annual geometric mean[b]	75 µg/m³	
	maximum 24-hour concentration	260 µg/m³	
Sulfur oxides	average annual concentration	80 µg/m³	.03 ppm
	maximum 24-hour concentration	365 µg/m³	.14 ppm
Carbon monoxide	maximum 8-hour concentration	10 ml/m³	9.0 ppm
	maximum 1-hour concentration	40 ml/m³	35.0 ppm
Oxidants/ozone	maximum 1-hour concentration	240 µg/m³	0.12 ppm
Nitrogen dioxide	average annual concentration	100 µg/m³	0.053 ppm
Hydrocarbons	maximum 3-hour concentration	160 µg/m³	

a. All standards based on concentrations over 24 hours or less are not to be exceeded more than once per year.

b. The geometric mean is the antilog of x, where x is the sum of the logarithms of the data divided by the number of data points.

c. ppm = parts per million.

Index value	Air quality classification	Pollutant levels					Air quality regarded as
		Particulate matter (24-hour), $\mu g/m^3$	Sulfur dioxide (24-hour), $\mu g/m^3$	Carbon monoxide (8-hour), $\mu g/m^3$	Ozone (1-hour), $\mu g/m^3$	Nitrogen dioxide (1-hour), $\mu g/m^3$	
500 — Significant Harm		1000	2620	57.5	1200	3750	
400 — Emergency		875	2100	46.0	1000	3000	Hazardous
300 — Warning		625	1600	34.0	800	2260	
							Very unhealthful
200 — Alert		375	800	17.0	400	1130	
							Unhealthful
100 — NAAQS★		250	365	10.0	235	No short-term standard for nitrogen dioxide	
							Moderate
50 — 50% of NAAQS		75[a]	80[a]	5.0	180		
							Good
0		0	0	0	0		

★Annual primary NAAQS (National Ambient Air Quality Standard).

Figure 20.7 Pollutant standards index and air-quality categories. An *alert, warning,* or *emergency* is declared when any one of the pollutants exceeds the appropriate value but none exceed the value listed in the next, more serious category. (Data from Pilot National Environmental Profile, 1977, U.S. Environmental Protection Agency, October 1980.)

values in the row labeled "Warning." During an alert, the air quality is said to be *very unhealthful.* Health scientists advise that those with heart and lung disease *and* those who are elderly should stay indoors and reduce exertion.

A *warning* is declared if the concentration of one or more of the major pollutants exceeds the appropriate value in the row labeled "Warning," but none exceed the values in the row labeled "Emergency." Now, not only those cautioned during the alert, but *all* people should reduce outdoor activity.

An *emergency* occurs when the concentration of one

or more of the pollutants exceeds the appropriate value in the row labeled "Emergency," but none yet exceed the values in the row labeled "Significant Harm." In addition to the previous advice, *all* people should now remain indoors and avoid physical exertion.

Finally, when one or more pollutants exceed the level associated with significant harm, air quality is at its lowest ebb. Vehicle use may be curtailed by closing all but the most vital government offices and by closing banks, schools, food stores, and the like. The standards for the major air pollutants are summarized in Table 20.1.

Box 20.1 Trends in Air Quality—Challenge in Display

We've all been warned: there are "lies, damn lies, and statistics." We present two sets of *statistics* on nationwide sulfur dioxide concentrations and one on particles. Figure B.1 appeared in a widely distributed bulletin from the U.S. Environmental Protection Agency. The bulletin, "Trends in the Quality of the Nation's Air," was intended for the general public. Figure B.2 appeared in a draft technical report from the U.S. Environmental Protection Agency; this report was intended for scientists and engineers; public viewing was unlikely.

In Figure B.1, we have added two solid black lines to indicate the same time period used in Figure B.2. Figure B.1 shows dramatic progress between 1964 and 1971. Note, though, that the bottom portion of the graph (shown with dotted lines) was missing as it originally appeared. A quick glance at Figure B.1 gives the impression that levels have been driven nearly to zero, when that is not at all the case. The impression of dramatic progress is achieved by compressing the time scale in which the decreasing trend is shown. The impression of dramatic progress is also heightened by expanding the scale of the vertical axis on which the change is shown. The time axis is 25 years in Figure B.1 but only 9 years in Figure B.2. Figure B.2 shows only steady, undramatic progress, and the bottom portion of the graph is shown.

Furthermore, Figure B.2 shows the range of data

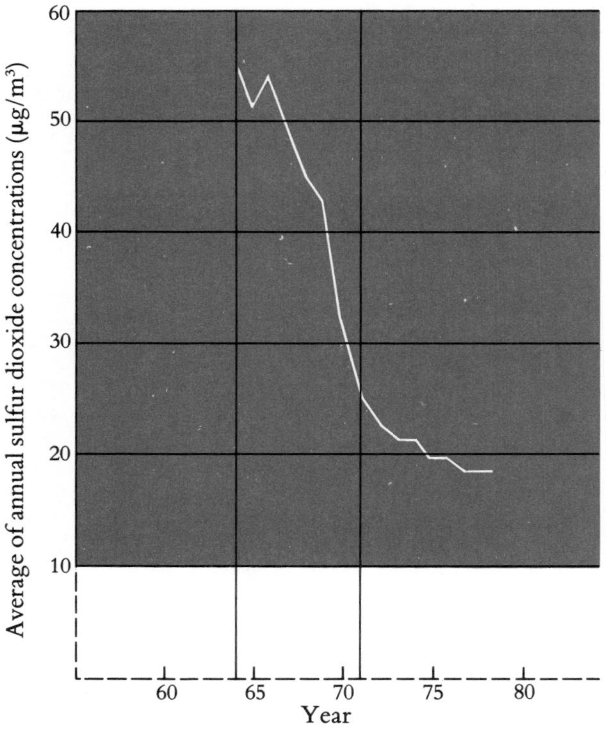

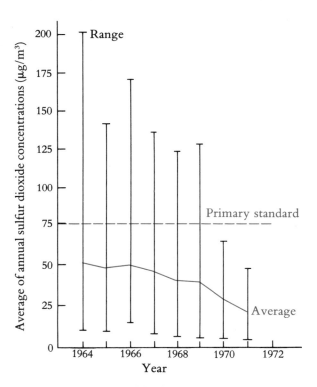

Figure B.1 Average of annual sulfur dioxide concentrations over many monitoring stations. This graph was intended for a wide audience, primarily of lay people. (Adapted from "Trends in the Quality of the Nation's Air," U.S. Environmental Protection Agency, October 1980.)

Figure B.2 Average sulfur dioxide concentrations are shown for the same 32 monitoring stations as in Figure B.1. This graph was intended primarily for an audience of scientists and engineers. (From Draft Technical Report, U.S. Environmental Protection Agency.)

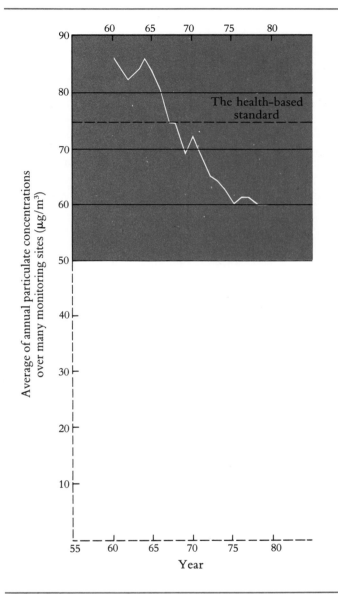

used to calculate the average. Both the highest and lowest numbers that entered the calculation of the average are shown. Some very low numbers, representing very good air quality, were used to calculate the trend. Neither the highest numbers, representing the worst air quality nor the lowest numbers are shown in Figure B.1. By not showing the worst air quality, an impression that all is well is created.

The same bulletin reported particles in the air in a similar way; the graph that appeared in "Trends in the Quality of the Nation's Air" is reproduced here as Figure B.3. We have added dotted lines to show the portion of the graph that was missing in the bulletin. We have also added the health-based standard to the graph. With the bottom portion of the graph missing, a quick look at Figure B.3 suggests there is little left to accomplish in controlling particles in the atmosphere. How wrong that conclusion is can be seen by looking back at Figure 20.3.

Figure B.3 Particle levels in the atmosphere have been declining since about 1960. (Adapted from: "Trends in the Quality of the Nation's Air," U.S. Environmental Protection Agency, October 1980.)

Summary

Sulfur oxides in the air are derived mainly from the burning of coal for electric power generation; sulfur is a contaminant in coal. The sulfur oxides react with water vapor in the air to form sulfurous and sulfuric acid in droplet form. In turn, sulfuric acid reacts with iron and calcium oxides in the air to form iron and calcium sulfates. The metal oxides in the air are also the result of burning coal. These droplets and particles can account for 5%–20% of the particulate matter in urban air. The acids and particles, which are also acidic, dissolve marble and mortar, corrode metals, damage fabrics, and injure plants. They also harm human health and contribute to acid rain.

Inorganic sulfur in coals, such as the sulfur in ferrous sulfide, can be removed by crushing the coal and then washing it with water. Various chemicals can be used to remove portions of the organic sulfur that is bound chemically to the carbon in coal. In addition to being able to remove some of the sulfur from the coal itself, the oxides of sulfur can also be removed from the stack gases by scrubbers or by sorbent injectors. Sulfur emissions may also be reduced by new coal-burning technologies, such as the fluidized bed process.

Although air quality in terms of sulfur oxides has improved considerably in cities over the past 15 years, the 1983 total of sulfur oxide emissions had decreased only about 20% since 1975. Many of the sources of sulfur oxides were simply shifted to more rural locations during this period.

Unlike sulfur oxides, particulates have diverse origins. Large amounts come from burning coal, but equally large amounts derive from industrial activities, such as the mining and refining of ores, the manufacture of metals, and the construction of buildings. Transportation contributes particles such as lead, hydrocarbon droplets, and photochemical smog. Incineration of solid wastes and demolition of asbestos-containing buildings also release particles to the air.

Particles soil and corrode materials, and they harm plant tissue. They affect weather by increasing fog, and they may affect climate by reflecting sunlight away from the earth. Particulates and sulfur oxides act together to adversely affect human health. Particles form the nucleus for the formation of water droplets, and sulfuric acid quickly dissolves in those droplets to produce a corrosive mist. Sulfate particles (iron and calcium sulfate) may also have adverse effects on human health. In addition, particulates may contribute to the carcinogenic potential of urban air, although no proof for this observed association is yet at hand.

Particle emissions from power plants can be decreased 99% by using an electrostatic precipitator or a baghouse. Particulate concentrations have decreased by about one-third over the past 20 years, but 21% of the U.S. population still lives in areas where the particulate standard is exceeded. Further, the standard based on weight may still allow fine particle concentrations that damage human health.

The Pollutant Standards Index classifies air quality as good, moderate, or unhealthful, depending on the amount that various pollutants exceed air-quality standards. Higher pollutant levels lead to announcements of successively more serious conditions: that is, an alert, a warning, an emergency, and a condition of significant harm.

Questions

1. Where do sulfur oxides come from? Where do particles come from? What industrial activity using what fuel produces the most sulfur oxides?
2. Why does the gas sulfur dioxide cause mist to be acid?
3. What kinds of particles are produced by an auto?
4. Why do some individuals in the electric power industry say that tall smokestacks are the correct response to air pollution from sulfur oxides? What does the Environmental Protection Agency suggest instead? What kinds of pollution will be prevented if sulfur dioxide is removed from stack gases?
5. Why are particles regarded as an "accomplice" of sulfur dioxide?
6. What is the meaning of *excess deaths* and how does this concept fit into a study of the health effects of air pollution?

Further Reading

U.S. Environmental Protection Agency. *Air Pollution Engineering Manual*, 2nd ed., Washington, D.C.: U.S. Government Printing Office, 1978.
  Although this is a highly detailed technical work, a number of portions are general descriptions that can be understood by the layperson; there are even discussions of the history of the technology. Whether all the details given here are of interest, however, depends on the intent and training of the reader.
Air Quality Criteria for Particulate Matter, U.S. Environmental Protection Agency, AP-49, 1969.
Control Techniques for Particulate Matter, U.S. Environmental Protection Agency, AP-51, 1969.
Air Quality Criteria for Sulfur Oxides, U.S. Environmental Protection Agency, AP-50, 1969.
Control Techniques for Sulfur Oxide Air Pollutants, U.S. Environmental Protection Agency, AP-52, 1969.
  While technical, these documents are very clearly written and very complete.
Goldsmith, B., and J. Mahoney. "Implications of the 1977 Clean Air Act Amendments for Stationary Sources," *Environmental Science and Technology*, **12,** No. 2 (February 1978), 144–149.
  A discussion of the Prevention of Significant Deterioration (PSD) section of the Clean Air Act Amendments of 1977.
Waldbott, G. *Health Effects of Environmental Pollutants*, 2nd ed. St. Louis: C. V. Mosby, 1978.
  This well-illustrated book (136 pages of illustrations) has a wealth of information and documentation on all major air pollutants. It is a technical work, but clearly written.

The book could serve as an excellent reference for the health aspects of air pollutants.

To Breathe Clean Air. Report of the National Commission on Air Quality, March 1981, (Available from the Superintendent of Documents, Washington, D.C., 20402).

A thorough compilation of regulations and recommendations. An excellent source of data.

Sulfur Emission: Control Technology and Waste Management, U.S. Environmental Protection Agency Decision Series, EPA 600/9-9-79-019, May 1979.

Description of the scrubbers and their impacts.

Hileman, B. "Particulate Matter: The Inhalable Variety," *Environmental Science and Technology,* **15,** No. 9 (September 1981), 983.

The new standards proposed by the U.S. EPA for particles with diameters less than 10 microns are discussed here. The conclusion is that EPA is leaving no "margin of safety" as dictated by the Clean Air Act.

References

Rustin, Bayard. Quoted in *The New York Times,* August 11, 1975.

Clayton, Cubia L., *Science,* **193** (September 10, 1976), 953.

Controlling Air Pollution. American Lung Association, 1974.

Patterson, Walt. "A New Way to Burn," *Science,* **83** (April 1983), 67.

CHAPTER TWENTY-ONE

Pollutants from the Automobile: Carbon Monoxide, Nitrogen Oxides, and Photochemical Pollutants

Carbon Monoxide

Motor Vehicles and Smoking: The Major Sources / What Auto Emission Controls Have Brought / The Effects of Carbon Monoxide on Human Health

Photochemical Air Pollution

What Is Photochemical Pollution? / Effects of Nitrogen Dioxide on Human Health / Effects of Photochemical Oxidants on Human Health and on Plants / Progress in Controlling Photochemical Air Pollution / Controlling Pollutants from Automobiles: The Record Examined / Lead Compounds in the Air

In this chapter, we focus on two broad classes of air pollutants: carbon monoxide and the photochemical air pollutants. Carbon monoxide is clearly motor-vehicle generated. In 1983, 70% of the annual emissions of carbon monoxide were from highway vehicles. In most major cities, 90% or more of the carbon monoxide burden is due to motor vehicles.

Photochemical air pollutants are the result of reactions that occur among primary pollutants, which come largely from motor vehicles. The primary pollutants are nitrogen oxides and hydrocarbons. Motor vehicles are responsible for 45% of the yearly emissions of nitrogen oxides and 36% of the yearly emissions of hydrocarbons. With a record like this, it is no wonder that so many air-pollution control efforts have focused on the automobile.

Carbon Monoxide

Motor Vehicles and Smoking: The Major Sources

When carbon is not oxidized completely, the colorless and odorless gas **carbon monoxide** results. Carbon monoxide surrounds the city dweller in concentrations greater than any other air pollutant. Because the gas is colorless and odorless, however, we cannot detect it with our senses. It cannot be seen; it cannot be smelled; it is present nevertheless.

The greatest source of carbon monoxide in our cities is the motor vehicle. Over 120 million vehicles are on the road in the United States, discharging carbon monoxide into the atmosphere. In most cities, over 90% of the carbon monoxide in the air comes from the incomplete combustion of carbon in motor fuels. The reaction is:

$$C \quad + \quad \tfrac{1}{2}O_2 \quad \longrightarrow \quad CO$$

carbon in motor fuel   oxygen   carbon monoxide

Whereas incomplete combustion of carbon produces carbon monoxide, complete combustion of carbon produces carbon dioxide (CO_2) as the end product.

There is another source of carbon monoxide to which people are exposed, but only smokers and their neighbors have this special privilege. We can compare the individual who smokes moderately and lives in a clean environment with an individual who does not smoke and lives in a highly polluted environment. The smoker daily absorbs twice as much carbon monoxide as his nonsmoking counterpart in the highly polluted environment.

The carbon monoxide in air or in tobacco smoke is inhaled and then absorbed into the bloodstream, where it competes with oxygen for the hemoglobin molecule. Hemoglobin, a complex protein present in blood, transports oxygen from the lungs to the cells and carries carbon dioxide from the cells back to the lungs. Carbon monoxide, however, attaches more strongly to hemoglobin than oxygen does. The more carbon monoxide present in air, the more hemoglobin is "tied up," and the less oxygen can reach the cells. An analogy is that carbon monoxide fills up the bus, leaving little room for the important passengers, the oxygen molecules. For this reason, at high enough concentrations, carbon monoxide is a deadly poison.

In workplaces such as tunnels and loading platforms, carbon monoxide may reach concentrations of 70 mg/m^3. After 8–12 hours at such a concentration, a worker may lose the service of 10% of his hemoglobin. If an individual were exposed to an average of 16 mg/m^3 for an 8-hour period, about 3% of his hemoglobin would be unavailable for oxygen transport. This level is not uncommon on city streets. At this point, the impact of smoking on the availability of hemoglobin can be seen. Even the light smoker ties up 3% of his hemoglobin *by smoking alone.* This is equivalent to being in a room with a carbon monoxide concentration of 16 mg/m^3. The moderate smoker achieves nearly a 6% loss of useful hemoglobin.

The standard for carbon monoxide in the atmosphere is 10 mg/m^3, averaged over 8 hours. This level is not to be exceeded more than once a year. We can see that smoking alone produces a personal environment worse than that required by the air-quality standards. In fact, a light smoker has twice the amount of hemoglobin tied up as a nonsmoker breathing air in which the carbon monoxide level is approaching the air-quality standard.

In 1977 a Los Angeles suburban resident or a resident of Phoenix, Arizona would have experienced about 50 days per year on which carbon monoxide levels exceeded the standard. A resident of Portland, Oregon would have experienced nearly 60 days in which the standard was exceeded; a Seattle, Washington resident, nearly 90 days. In 1976, levels above the standard were exhibited in Los Angeles on 120 days. New York City, specifically Manhattan, was even worse. The automobile, our marvelous instrument of mobility, was the principal cause.

What Auto Emission Controls Have Brought

The initial control of carbon monoxide emissions was achieved by increasing the ratio of air to fuel in the gasoline engine. Additional air is provided to burn the gasoline more completely so that less carbon monoxide is in the exhaust. Some vehicles also utilized *air injection.* Air was mixed with the exhaust stream for a final combustion of the carbon monoxide to carbon dioxide.

$$CO + \tfrac{1}{2}O_2 \longrightarrow CO_2$$

The typical motor vehicle of the mid-1960s ex-

hausted an average of 73 g of carbon monoxide in every mile of travel. A standard of 23 g of carbon monoxide per mile was set for 1971 vehicles; new passenger cars met the standard easily. By 1981, carbon monoxide emissions of only 3.4 g per mile were being achieved by new autos.

To achieve this standard, the exhaust gases are mixed with a stream of air in the presence of a catalyst. Further oxidation of the remaining carbon monoxide occurs in this **catalytic converter**. The catalyst system appears, for the present, to be the chosen method of reducing carbon monoxide emissions. Nevertheless, auto manufacturers are probably studying the potential of the stratified-charge, dual-carburetor engine. This system has been used in the Japanese Honda and has achieved the required reduction in carbon monoxide.

To assess more precisely the progress in carbon monoxide control, we need to examine the trend in total carbon monoxide emissions. We also need to compare levels of the gas against the atmospheric standard that has been set for it in a number of areas around the country. The standard is 10 mg/m^3 averaged over 8 hours, not to be exceeded more than once per year.

The annual nationwide emissions of carbon monoxide have been falling gradually since 1976 as new model cars with the catalytic converter have gradually replaced older, less efficient vehicles (Figure 21.1). Total emissions from transportation activities declined from 64.3 million metric tons per year in 1976 to 47.7 million metric tons per year in 1983, a drop of 25%.

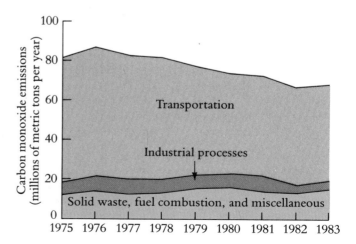

Figure 21.1 Trend in annual emissions of carbon monoxide, 1975–1983. (Data from U.S. Environmental Protection Agency, 1985.)

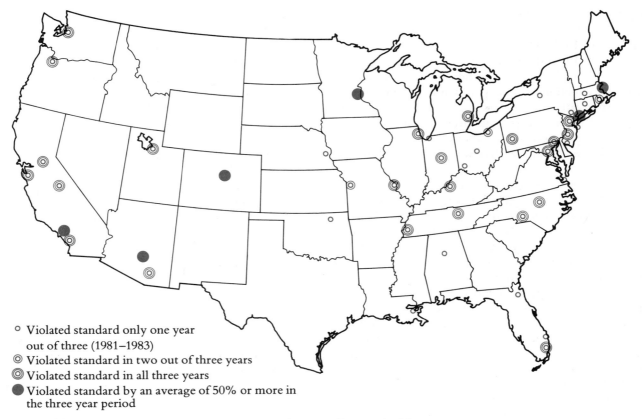

○ Violated standard only one year
  out of three (1981–1983)
◎ Violated standard in two out of three years
◎ Violated standard in all three years
● Violated standard by an average of 50% or more in
  the three year period

Figure 21.2 Status of compliance with the air-quality standard for carbon monoxide, 1981–1983. (Data from U.S. Environmental Protection Agency, 1985.)

This decrease appears relatively small given the tight level of control on carbon monoxide emissions from new cars (from 23 g per mile in 1971 to 3.4 g per mile in 1981, an 85% reduction). One reason for such a small decrease is the number of vehicle miles driven—a quantity that steadily increases each year because more vehicles are on the road. Another reason is probably a lack of maintenance of the emission control systems. Maintenance becomes an important issue because the effectiveness of the catalytic converter declines over time.

In 1981, 1982, and 1983, the carbon monoxide air-quality standard was being violated in 50 metropolitan areas around the country (Figure 21.2). Of these areas, 6 exceeded the standard by an average of more than

50% in these three years. Those cities were Boston, Denver, Los Angeles, Minneapolis, New York City, and Phoenix. The remainder violated the standard in at least one of the three years, and some violated the standard in all three years. Those cities failing to achieve the standard in all three years, as well as those with an average violation greater than 50%, are unlikely to achieve the air-quality standard soon. Urban areas with growing populations and growing use of automobiles may even get worse. Examples are Portland (Oregon), Norfolk-Virginia Beach, and Raleigh-Durham.

Although we have focused on problem areas, there has been progress in carbon monoxide control. In 1975, the mean number of times the carbon monoxide standard was exceeded at 174 monitoring sites around the country was forty-one per year. That is, at the typical or average site, forty-one violations would have been expected in 1975. This number has been steadily decreasing. In 1983, the mean number of violations was down to eight. It is likely, though, that the cities with excessive violations of the standard experience more frequent violations than those nearer to achieving the standards. Nevertheless, the decrease in the mean number of violations does indicate that many people and places are living with fewer episodes of poor-quality air.

Compulsory inspection of emission controls on autos to detect equipment failure appears to be needed. Because of the choice to use a catalytic converter to control emissions, and because the converter eventually fails in service, it is necessary that emission control be inspected and repaired. Otherwise, we cannot hope to continue progress toward achieving the air-quality standards. A change to a more reliable technology, such as to the stratified-charge engine, might decrease the need for annual inspections. Even with inspection and repair, the battle to improve air quality is uphill since total vehicle mileage continues to grow steadily. The relentless growth in vehicle mileage could be countered, however, by building and operating superior public transportation systems—systems that would attract and hold a wide ridership.

The Effects of Carbon Monoxide on Human Health

Carbon monoxide disrupts oxygen transport in the body. For more than a decade, scientists have suspected that the concentrations of carbon monoxide found in our cities are harmful. Nevertheless, it has only been in the last few years that the needed data have been obtained. We now know that carbon monoxide in the air is a serious health hazard.

In an atmosphere rich in carbon monoxide, death results from asphyxiation. This is another way of saying that the body tissues become starved for oxygen. At lower concentrations, other more subtle effects are noted.

To understand the danger of small concentrations of carbon monoxide, we need to review the process that supplies oxygen to the tissues of the body. Oxygen is brought into the lungs with each breath. At the alveoli, the small sacs at the end of the treelike branches

of the lungs, the oxygen gas is transferred to the bloodstream. In the blood, the oxygen attaches to hemoglobin, a complex protein molecule carried in red blood cells (erythrocytes). The red cells transport the oxygenated hemoglobin through the arteries of the body and finally to the capillaries, the narrowest tubes of the arterial circulatory system. Here, oxygen is transferred across the walls of the capillaries to the cells.

Carbon dioxide, one of the waste products of cell activity, flows in the opposite direction, from the cell into the bloodstream. Some of the carbon dioxide takes the place of the oxygen that was attached to hemoglobin, and some of the gas dissolves in the blood fluid as bicarbonate ions. The blood, now rich in carbon dioxide, returns via the veins to the lungs. There, carbon dioxide diffuses from the blood into the alveoli, while oxygen from the air in the sacs moves into the blood. The carbon dioxide is then exhaled in the breath.

This normal pattern of transportation is disturbed when carbon monoxide is present in the air we breathe. Even minute quantities can disrupt oxygen transport, because carbon monoxide is about 200 times more attractive to hemoglobin than is oxygen. The carbon monoxide attaches tightly to the hemoglobin, depriving oxygen of its carrier to the cells. The greater the amount of carbon monoxide present in the air, the more hemoglobin is "tied up" and unavailable for oxygen transport. When hemoglobin binds carbon monoxide to itself, it is called *carboxyhemoglobin*. In contrast, when hemoglobin has oxygen attached to it, it is called *oxyhemoglobin*.

Table 21.1 shows how even small quantities of carbon monoxide gas in the air produce high levels of carboxyhemoglobin in the blood. Note that the table lists carboxyhemoglobin percentages after 8–10 hours of breathing contaminated air. This level is termed the *equilibrium value*. A longer exposure at that concentration will not increase the percentage of carboxyhemoglobin any further. Also note that even when the air is free of carbon monoxide, some small amount of hemoglobin is tied up. This carbon monoxide stems from natural body processes.

Effects of low carbon monoxide levels. The effects of low levels of carbon monoxide on health have been surmised from experimental data rather than from real-world observations. The use of experimental data is necessary because when outdoor carbon monoxide concentrations are high, the concentrations of other

Table 21.1 Carbon Monoxide and Hemoglobin

Level of carbon monoxide (mg/m³)	Estimated percentage of hemoglobin tied up as carboxyhemoglobin after 8–10 hours[a]
0	0.4
5	1.0
10	1.6 (1972 air-quality standard and resulting carboxyhemoglobin)
20	2.9
30	4.1
40	5.4
45	6.0 (level for a moderate smoker)
50	6.6
60	7.8

a. This is the percentage of hemoglobin to which carbon monoxide is attached. This hemoglobin is not available to carry oxygen to the cells.

pollutants in the air are typically also high, and the effects cannot be separated.

Individuals with increased levels of carboxyhemoglobin are subject to two important effects. One is a decreased ability to perceive one's personal environment. This perception has been measured by a number of tests. For instance, individuals have been asked to report on sound signals. At levels of carboxyhemoglobin in the range of 2%–5% of total hemoglobin, the signals were often missed. The ability to tell which of two tones was longer in duration decreased when carboxyhemoglobin was in the range of 2.5%–4%. The processes of the mind are also interfered with. Simple tests, such as adding columns of numbers, took longer to complete as carboxyhemoglobin levels increased. The ability to distinguish a light becoming brighter also decreased. Tests of brightness perception showed fewer correct answers even when carboxyhemoglobin levels were as low as 3%.

In experimental situations that produced levels of 10% unusable hemoglobin, the skills needed to drive a car were impaired; responses to brake lights and to the speed of the auto ahead were poorer. The possible influence on safety is obvious. Carbon monoxide levels may rise to 60 mg/m³ or more in freeway traffic, producing carboxyhemoglobin levels close to those at which skills are impaired. One scientist found higher

carboxyhemoglobin levels in individuals involved in auto accidents, but we are still unable to conclude that carbon monoxide is a cause of auto accidents.

The past decade has brought us much new information on carbon monoxide's relation to heart attacks. Earlier, medical scientists suspected that carbon monoxide might be a cause of heart attacks because they found that more heart attacks occurred during periods of high carbon monoxide concentrations. But because other pollutants were also at high concentrations at the same time, no firm conclusions could be drawn. Now the suspicion is being substantiated by new data from individuals with angina pectoris.

Angina pectoris is a chronic form of heart disease characterized by chest pain; it is less severe than an acute heart attack, which is life-threatening when it occurs. Individuals who suffer from angina pain have been tested for their susceptibility to carbon monoxide (*Air Quality Criteria for Carbon Monoxide*, 1970). They were first asked to breathe air with carbon monoxide at levels sufficient to raise their carboxyhemoglobin to 3%; then they were asked to exercise. At this blood level of carboxyhemoglobin, when the subjects were stressed, the onset of angina pain arrived sooner than expected under normal conditions; furthermore, the pain continued longer than expected. Data reported in 1981 now indicate that even a 2% level of carboxyhemoglobin will aggravate angina.

Angina is only one form of heart disease. In total, 35% of the annual deaths in the United States are attributed to some form of heart disease. Carbon monoxide is known to decrease the supply of oxygen to the tissues. One tissue that is frail if deprived of oxygen is the myocardium (the muscle tissue of the heart). Experiments with angina patients tend to support the contention that carbon monoxide may be an agent in causing heart attacks. (We do not say that carbon monoxide causes heart disease itself, which is most frequently defined as a narrowing of coronary blood vessels.)

Another observation may support the idea that carbon monoxide is one factor in bringing on heart attacks. Smoke inhaled from a cigarette may contain as much as 4% carbon monoxide. Heavy smokers may have carboxyhemoglobin levels as high as 10%–15%. We know from statistical studies that when people quit smoking, their risk of heart attack is quickly reduced. It appears that a causative agent has been removed from their personal environment. That causative agent may be carbon monoxide, but other substances are also absent when they stop smoking, so we are still uncertain of the relationship.

Photochemical Air Pollution

What Is Photochemical Pollution?

A photochemical reaction requires light energy. Certain pollutants in the atmosphere—nitrogen oxides and hydrocarbons—undergo photochemical reactions. These reactions produce new pollutants, including ozone, aldehydes, and exotic organic compounds. The new pollutants are referred to, in sum, as **photochemical air pollution** because they arise from photochemical reactions.

Thus, the emission sources responsible for photochemical air pollution are the sources of nitrogen oxides and of hydrocarbons. By the same reasoning, the control of photochemical air pollution requires controlling emissions of nitrogen oxides and hydrocarbons.

Nitrogen oxides and hydrocarbons: By-products of combustion. While nitrogen oxide is produced naturally from such occurrences as forest fires, the activities of human society are responsible for the high concentrations that occur in cities and near industries. During the high-temperature combustion of fossil fuels, two types of reactions occur that produce nitrogen oxides. In the first type, oxygen from the air and nitrogen *from the fuel* react to produce nitrogen oxides. Coal may typically have about 1% nitrogen content. Oil and gas may have 0.2%–0.3% nitrogen content; it is this nitrogen that is likely to be oxidized.

In the second type of reaction, oxygen from the air reacts with nitrogen *from the air* to produce nitrogen oxides. Thus, a fuel with no nitrogen in it can produce nitrogen oxides during combustion. Nitrogen oxides are produced whether the fuel is natural gas, coal, gasoline, or home heating oil. Approximately 95% of the annual emissions of nitrogen oxides stem from the combustion of fossil fuels. About 40% of the total emissions are from motor vehicle operation and other transportation modes. Another 30% result from the burning of natural gas, oil, and coal in electric power plants. Industrial combustion of fossil fuels produces an additional 20% of the total. The manufacture of explosives and nitric acid, two noncombustion sources, also result in emissions of nitrogen oxides (see Figure

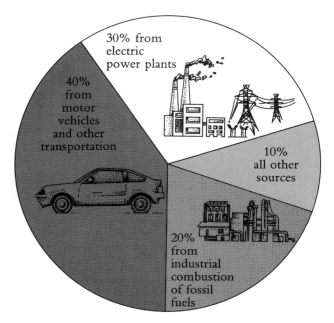

Figure 21.3 Sources of nitrogen oxide emissions. (Data from *To Breathe Clean Air*, National Commission on Air Quality, 1981.)

21.3). Of the three basic fossil fuels, the burning of natural gas in all uses accounts for about 20% of all nitrogen oxide emissions; the burning of coal, about 25%; and the burning of oil, about 47%.

Hydrocarbons are released from many sources. Methane gas is present naturally in the atmosphere due to emissions from coal fields, gas, and petroleum fields. Methane emissions also stem from fires, and emanations of methane arise from swamps as well. (The will-o-the-wisp seen in swamps is actually burning methane gas.) Methane, however, does not typically react in the atmosphere. Thus, even though methane background levels may reach a milligram per cubic meter, it does not concern us as an air pollutant since it appears to be harmless at these concentrations. Methane is, however, a *greenhouse gas*. That is, just as carbon dioxide does, it reflects skyward-radiating heat back to the earth. In the 1980s, methane's concentration in the stratosphere was found to be increasing. It is thought that increasing levels of methane in the stratosphere may accelerate the predicted warming trend from carbon dioxide buildup.

The hydrocarbons that do concern us are mainly by-products of the activities of civilization. Currently, one-third of the annual emissions of hydrocarbons stem from motor vehicle operation. In the cars of the

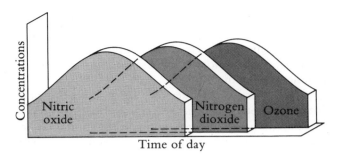

Figure 21.4 Levels of nitric oxide, nitrogen dioxide, and ozone during and after the morning "rush hour."

1960s, hydrocarbons evaporated from the carburetor and the fuel tank. They also escaped around the piston during compression. Most important, however, incomplete combustion left unburned hydrocarbons as droplets and gases in the exhaust stream. Although these sources of hydrocarbon emissions have now come under some control on new vehicles, significant quantities of hydrocarbons still enter the atmosphere from transportation activities. Smaller sources of hydrocarbons are petroleum refining and petroleum transfer operations.

A complex of reactions. Levels of photochemical air pollution closely follow the pattern of automobile use. During morning and evening rush hours, peak emissions of nitrogen oxides and hydrocarbons occur. These are the substances that react to produce the photochemical air pollutants.

Nitrogen and oxygen unite during the high-temperature combustion of fuel in an automobile engine to produce the gas nitric oxide, which enters the atmosphere. In several hours' time, the level of nitric oxide in the air decreases substantially. During the period of that decline, the level of nitrogen dioxide rises to a peak. Later, as the nitrogen dioxide level declines, the concentration of a third gas, ozone, increases. Ozone levels, too, then decrease (Figure 21.4).

The reactions that produce high concentrations of nitrogen dioxide and ozone are not perfectly understood. Nevertheless, certain key relationships have been identified. The presence of hydrocarbons and the photochemical properties of nitrogen dioxide make these complex reactions possible. These reactions produce a typical pattern of pollutant levels over a period of time. Nitric oxide rises during the period of high emissions from motor vehicles. Nitrogen dioxide rises

and nitric oxide falls due to the reaction of hydrocarbons with nitric oxide. After nitric oxide levels have fallen, ozone levels rise due to the photodissociation of nitrogen dioxide.

Still other pollutants result as consequences of these reactions. Hydrocarbons react with nitrogen dioxide to produce peroxyacyl nitrate (PAN) compounds. Ozone reacts with hydrocarbons to produce aldehydes. Eventually, winds disperse all of these contaminants. Ozone, nitrogen dioxide, PAN compounds, and aldehydes are called photochemical air pollutants because they arise from sunlight-stimulated chemical reactions. (See Figure 21.5.)

The term **oxidant** is applied to compounds capable of oxidizing substances that oxygen in the air cannot oxidize. Nitrogen dioxide, PAN compounds, ozone, and aldehydes are all oxidants. A concentration of oxidants is reported in terms of the weight of all such substances per cubic meter (m^3) of air. The air-quality

Figure 21.5 The buildings in New York City fade into a haze of pollution. In the city itself, respiratory problems are likely to be increasing. (National Archives, Documerica Collection)

standard for photochemical air pollutants was originally given in terms of an allowable oxidant concentration, 240 μg/m³. An oxidant concentration did not indicate how much of any specific compound was present, although ozone is usually the largest component.

Unfortunately, measurement of all the oxidants in polluted air is difficult. As a consequence, only ozone is actually measured, and the air-quality standard for photochemical oxidants (240 μg/m³) has been converted during the Reagan administration to the standard for ozone alone. Ozone accounts for only about 75% of the oxidants in the air, so 240 μg/m³ of ozone is equivalent to an even greater concentration of oxidants. Hence, a standard that counts ozone as all oxidants is a weak standard.

Effects of Nitrogen Dioxide on Human Health

As we mentioned earlier, the combustion of fossil fuels in engines and furnaces produces such intense heat that pollutants are formed from two natural substances in the air: nitrogen and oxygen. These pollutants are the nitrogen oxides. They are formed at the extremely high temperatures that exist in internal combustion engines, such as that of the automobile. The combination also takes place in boilers such as those in the coal furnaces of electric power plants. About 90% of the nitrogen oxides are produced in the form of nitric oxide (one atom of nitrogen plus one atom of oxygen). The remaining 10% are in the form of nitrogen dioxide (one atom of nitrogen and two atoms of oxygen).

Most of our information on health effects concerns nitrogen dioxide. Initially, nitrogen dioxide composes only 10% of nitrogen oxide emissions; however, a complex series of chemical reactions in the air converts much of the nitric oxide to the more hazardous nitrogen dioxide form.

Nitrogen dioxide is a foul-smelling gas. Even at concentrations as low as 230 μg/m³, about one-third of a group of volunteers were able to detect its presence. The ability to detect the gas disappeared, however, after only 10 minutes of exposure, although people reported a dryness and roughness of the throat. Even these irritations vanished after prolonged exposure to levels 15 times the odor threshold discussed here.

Not only does nitrogen dioxide affect the sense of smell, it decreases night vision, the ability to adapt one's eyes to see in the dark. This phenomenon has been observed at levels as low as 0.14 mg/m³. The visual and olfactory responses to nitrogen dioxide may be called sensory effects. More important are the effects of nitrogen dioxide on disease and on the functioning of the human body.

Two effects of nitrogen dioxide on body functions have been noted. One is the greater effort required in breathing; medical investigators refer to this as increased airway resistance. The response has been observed in normal healthy individuals at gas concentrations as low as 0.056 mg/m³. Individuals with chronic lung diseases experienced breathing difficulty at only 0.038 mg/m³.

In addition, measurements by a team of Czech scientists indicate that nitrogen dioxide gas can attach to hemoglobin, just as carbon monoxide does, thus keeping hemoglobin from carrying oxygen to the tissues. It is already accepted that nitrites in food can attach to hemoglobin, forming methemoglobin.

A number of studies have noted increased respiratory disease in areas polluted by nitrogen dioxide, but the presence of other pollutants at relatively high levels makes the results less useful, and firm conclusions on the causative agent become impossible. On the other hand, the study by Shy and others of over 4,000 school children and parents in several portions of Chattanooga, Tennessee does point toward nitrogen dioxide alone as causing increased respiratory disease (Waldbott, 1978). Although we say *causing*, a more accurate statement would be that nitrogen dioxide made people more susceptible to the *pathogens* that cause respiratory disease. Nitrogen dioxide from a nearby TNT plant was by far the most concentrated of air pollutants in the study area. Researchers observed more colds, bronchitis, croup, and pneumonia among the population group exposed to these higher nitrogen dioxide levels than among a less exposed but otherwise similar population nearby.

Investigators have also sought to link nitrogen dioxide to increased mortality (death) in addition to increased morbidity (disease) rates. Statistical analyses indicate that areas with higher nitrogen dioxide levels do have a greater number of deaths from heart disease and cancer. In the two notable studies, though, the presence of other pollutants and the limited number of sampling stations in each study area made it difficult to draw firm conclusions.

Individuals with chronic (continuing) respiratory diseases, such as emphysema and asthma, and individuals with heart disease may be more sensitive to direct effects of nitrogen dioxide. We do not yet know whether this is true; what we do know, however, is that nitrogen dioxide is associated with increased cases of short-term respiratory disease. Individuals with chronic heart and respiratory disease are more likely to develop complications from these short infections—dangerous complications, such as pneumonia. On the order of 10%–15% of the population in the United States are thought to have some form of chronic respiratory disease.

This line of reasoning leads us to conclude that the standard for nitrogen dioxide should be set at a level that protects the population from increased respiratory infections. (See Controversy 21.1.) The standard set for nitrogen dioxide is an average annual concentration that should not be exceeded. Its value is 100 $\mu g/m^3$. The allowable concentration is set at one-tenth the level known to decrease resistance to respiratory disease. This is a fairly common margin of safety employed in setting standards. On the other hand, one of the statistical investigations—not fully accepted for reasons just cited—linked an increase in the incidence of cancer to levels of nitrogen dioxide that were near the annual air-quality standard.

At present, no short-term standard (for instance, an average daily concentration) has been set, although a value was briefly proposed in 1971 and then abandoned. In the past, peak (instantaneous) levels of nitrogen dioxide in excess of 400 $\mu g/m^3$ have been commonly observed in major cities.

Effects of Photochemical Oxidants on Human Health and on Plants

The standard for oxidant concentration in the air, as pointed out earlier, is 240 $\mu g/m^3$ averaged over 1 hour. This concentration is not to be exceeded more than once each year. The new standard is for ozone alone, rather than for all oxidants as the previous standard was, even though other oxidants can account for as much as one-fourth of the total oxidants present in the air. The new standard is also 20% higher than the level at which alerts were previously announced.

In 1979, the standard was revised to this level from 160 $\mu g/m^3$, a level set in the early 1970s. During 1974–1975, the old standard was violated on 38% of the days at one or more of the stations in Los Angeles and in Philadelphia. Denver and Washington reported such

violations on 27% of the days in 1974–1975, and Cincinnati and Houston noted violations on 23% of the days in the same interval. Thus, the relaxation of the standard from 160 to 240 $\mu g/m^3$ gives the appearance of significantly better air quality because of less frequent violations.

The new standard appears to be a compromise between air quality and the expense of controlling pollution. At the level of the new standard, performance of a high school cross-country team was observed to deteriorate. Asthmatics have been noted to have an increased frequency of attacks in the range of 300–500 $\mu g/m^3$, the lower level a mere 25% above the current standard. The fraction of the population afflicted with bronchial asthma has been estimated at 3%–5%. Individuals with chronic lung disease have also been found to be affected by photochemical oxidants. Eye irritation is a significant effect of photochemical air pollution. At hourly averages as low as 200 $\mu g/m^3$, such eye irritation could be expected to begin. The standard, or rather the air quality goal to be attained, would not prevent eye irritation. An unanswered question remains. "Is eye irritation a health effect?" Does it indicate that other effects are beginning?

A number of studies of laboratory animals were also used in setting the old standard. In the presence of oxidants in the air, mice became more susceptible to bacterial infections; red blood cells and the cells of heart muscles became misshapen. A three-month exposure to ozone levels at about two-thirds of the old standard produced a decrease in the weight of rats and disrupted the physiological chemistry of the animals. Hamsters were found to have damage to the chromosomes of white blood cells on short-term, low-level exposures.

Long before effects of oxidants on health were documented, investigators found damage to plants. Tiny dark spots called *stipples* were discovered on the leaves of plants in the 1940s. The condition was first observed in Los Angeles, but has since been seen across most of the United States. The spots are the result of elevated ozone levels. Such levels result when excessive automobile use leads to high concentrations of hydrocarbons and nitrogen oxides.

Laboratory studies show that many common vegetables are damaged by ozone and also by PAN compounds. Ozone also affects citrus trees by causing the fruit to ripen and fall at a smaller size. Many plants show damage at oxidant concentrations of 80–160 $\mu g/m^3$, levels that are below the new standard. In 1982,

Figure 21.6 The automobile is responsible for much of the air pollution in cities. It is the source of most of the carbon monoxide, and the auto generously contributes nitrogen oxides and hydrocarbons to urban air. Nitrogen oxides and hydrocarbons are the substances out of which photochemical smog is formed. (EPA Documerica)

the Office of Technology Assessment released figures on the value of crop damages from air pollution. The report studied the damages from only one air pollutant, ozone. In addition, the study limited its analysis to just four crops: soybeans, corn, wheat, and peanuts. The conservative estimate of damages to these four crops from ozone was between $1.9 billion and $4.5 billion per year, as reflected in lower crop yields.

Progress in Controlling Photochemical Air Pollution

The control of oxidants is accomplished by controlling hydrocarbons and nitrogen oxides. Because the automobile is largely responsible for these pollutants, efforts are under way to decrease the amounts from this source (Figure 21.6).

The hydrocarbons that blow past the pistons in an automobile engine are now being recycled to the combustion chamber. Those that evaporate from the carburetor and fuel tank are also recycled. Emissions of

hydrocarbons from the exhaust system are decreased by mixing more air with the gasoline when it is burned. Beginning with the 1975 models, many new cars were equipped with a catalyst system. The catalyst system oxidizes unburned hydrocarbons in the exhaust to carbon dioxide and water vapor. Unleaded gasoline must be used by vehicles with the catalyst system to prevent "poisoning" the catalyst and making it ineffective.

Another catalyst system is used to control emissions of nitrogen oxides. This device reduces the nitrogen oxides back to molecular nitrogen.

Power plants that burn coal, gas, or oil are another large source of nitrogen oxides. A shift to nuclear electric power would reduce nitrogen oxides from this source, but there are many problems with nuclear power. (See Chapter 18.) Decreasing the air intake in gas and oil boilers may also decrease emissions of nitrogen oxides from these sources. In general, changes in the combustion process in industrial burners seem to offer a great deal of promise for controlling nitrogen oxides.

What has been the impact of the control measures applied so far? Recall that photochemical air pollution results from the reactions of nitrogen oxides and hydrocarbons. Progress is measured, then, not only in terms of ozone and oxidant levels but also in terms of emissions and levels of nitrogen oxides and hydrocarbons. In 1981, the National Commission on Air Quality concluded that additional controls on hydrocarbons and nitrogen oxides would be required for a number of areas to meet the ozone standard.

On a nationwide basis, nitrogen oxide emissions remained nearly stable between 1975 and 1983. The lack of any decrease in emissions, even though new cars have nitrogen oxide emission controls, is due to several factors. First, total vehicle mileage increased by 25% in this interval. Second, emission controls came late in this period. Third, coal-fired electric power has increased steadily (see Figure 21.7).

Emissions of hydrocarbons from 1975 to 1983 have remained relatively steady. Although there was some reduction in total hydrocarbon emissions from transportation sources, the 25% increase in annual vehicle miles traveled prevented emission controls on autos from having a more significant effect (see Figure 21.8).

Since emissions of nitrogen oxides are not decreasing appreciably and since emissions of hydrocarbons are also fairly constant, it is not surprising that many

CONTROVERSY 21.1

Should Society Protect the Unhealthy from Air Pollution?

Would it not be better to close the EPA and buy each person sensitive to carbon monoxide a condominium in Key West?

Paul MacAvoy, professor of economics, Yale University

Should the entire populace assume the burden of preventing aggravation of a disease in a relatively small group of people who unfortunately live in large cities?

R. Jeffrey Smith

In setting standards . . . we must be concerned with the health effects on the most vulnerable in our population rather than upon the healthy groups.

Edmund Muskie, U.S. Senator from Maine

For over a decade, we have sought to protect particularly sensitive citizens, such as children, the aged and asthmatics, from polluted air. I don't think the American people would stand for abandoning these sensitive populations by misguided use of cost/benefit analyses . . . It is useless to pretend that some kind of utilitarian calculus can give us the answers to what are essentially moral and political questions.

Henry Waxman, U.S. House of Representatives

As originally written, the Clean Air Act required the Environmental Protection Agency to set air-quality standards at levels that would protect the public health "with an adequate margin of safety." The *public health* meant not only the health of the general population but the health of those most susceptible to damage from air pollutants. Those with chronic respiratory disease, such as emphysema and asthma, and those with heart conditions, such as angina, are among the people most susceptible to ill effects from air pollutants. Children, because their rate of breathing is more rapid than that of adults, in some cases show greater susceptibility to air pollutants. The original Clean Air Act assumed that protection of the most sensitive segment of the population was a reasonable proposition, but this assumption has been challenged by industry and economists, who claim some substances are too expensive to control. The National Commission on Air Quality does not agree with industry about costs, but notes that: "The costs of meeting primary air-quality standards are best taken into account in determining what control programs should be implemented in specific areas of the country, not in establishing a national air quality standard to protect public health" (*To Breathe Clean Air*, 1981).

Paul MacAvoy (1981) suggests that carbon monoxide control is so costly that the solution is to relocate people sensitive to carbon monoxide. It is now clear that carbon monoxide can aggravate angina pectoris, an extremely common form of heart disease, at levels in the air not far above the air-quality standard. MacAvoy suggests that it costs too much to protect people with angina. His solution is a restatement of the sentiment of conservative economists: "Let them vote with their feet," which translated means "If they can't handle poor air quality from a health standpoint, they should move."

There are two issues here in suggesting that people move if they don't like air quality. The first is the cost of moving. Do all people have the money to move? Are there social costs as well as economic costs involved in moving? What are the social costs? Are the social costs higher to the elderly? The second issue is the choice of where to move. An example of the risks in choosing a destination occurred in a family we know. Because of his heart condition, the father moved to an area noted for clean air. But in 15 years Tucson, Arizona went from being one of the cleanest cities to being one of the most polluted cities in the nation. If pollution is not controlled everywhere—if the most susceptible are not protected everywhere—there may well be no safe place to move.

Does anyone in your family have angina? Do you think they need to be protected by safe air quality? Would they consider moving? If they would consider moving, is their destination safe both today and in the future?

Two other common diseases that make people susceptible to poor air quality are emphysema and asthma. Sulfur dioxide has a particular impact on people with these lung conditions. Asthma is a disease that often begins in childhood. What bearing does this have on MacAvoy's suggestion that afflicted people move? Do you think we should protect the most vulnerable in our population?

Sources:
Paul MacAvoy, quoted in *Science*, **212** (June 12, 1981), 1251.
R. Jeffrey Smith, quoted in *Science*, **212** (June 12, 1981), 1251.
Edmund Muskie, "Air Pollution, 1970." Part I, Hearings Before the Subcommittee on Air and Water Pollution of the Committee on Public Works, U.S. Senate, March 16, 17, 18, 1970, p. 74.
Henry Waxman, Joint Hearing Before the U.S. Senate Committee on Environment and Public Works and the U.S. House of Representatives Committee on Energy and Commerce, March 2, 1981, p. 59.
To Breathe Clean Air, Report of the National Commission on Air Quality, 1981, p. 8.

sections of the country are having difficulty meeting the ozone standards. They would have even greater difficulty if the allowable level had not been increased in 1979 from 160 to 240 $\mu g/m^3$. That 50% increase in allowable level, by itself, may have removed 10–20 cities from the list of cities not meeting the standard. Figure 21.9 indicates areas of the country that did not meet the ozone standard in 1982 but that were expected to meet it by 1987. Figure 21.9 also shows the areas that were unlikely to meet the standard even by 1987.

Ozone has been a problem in southern California for many years and continues to blight that area because of an extremely high rate of automobile use. The eastern United States is not far behind, however; severe ozone pollution now afflicts the northeast corridor stretching from Boston down to Washington, D.C. In 1979, about 30% of the population in this densely inhabited area was exposed to 10 or more days in which the ozone standard was exceeded. In fact, the entire northeastern United States from Washington, D.C. north to the Canadian border, bounded by the Atlantic Ocean on the east and Ohio on the west, has poor air quality in terms of ozone.

As bad as the northeastern United States is, southern California is still worse. In 1979, the ozone standard was exceeded in the Los Angeles basin on 194 days, in San Diego on 66 days, and in the Ventura-Oxnard area on 73 days. Houston, Texas logged 48 days above the standard in that same year; St. Louis County logged 50 days. The sunbelt cities of Reno, Las Vegas, Phoenix, Tucson, and Albuquerque likewise exhibit poor air quality in terms of ozone. In total, as the 1980s opened, nearly 144 million people lived in areas that were violating the ozone standard. With emissions of nitrogen

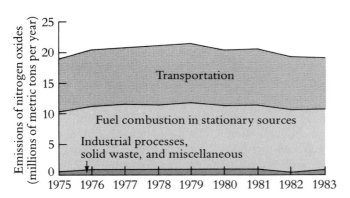

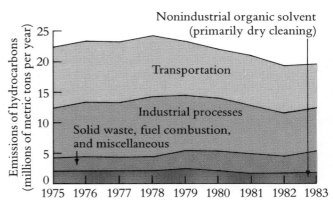

Figure 21.7 Trend in nationwide emissions of nitrogen oxides. Nitrogen oxides from "fuel combustion in stationary sources" refers to the burning of coal, oil, and gas for energy in electric power plants and factories, as well as the burning of these fuels in commercial, industrial, and residential buildings for space heating. "Industrial processes" refers primarily to the manufacture of nitric acid. (Data from U.S. Environmental Protection Agency, April 1985.)

Figure 21.8 Trend in nationwide hydrocarbon emissions. (Data from U.S. Environmental Protection Agency, 1985.)

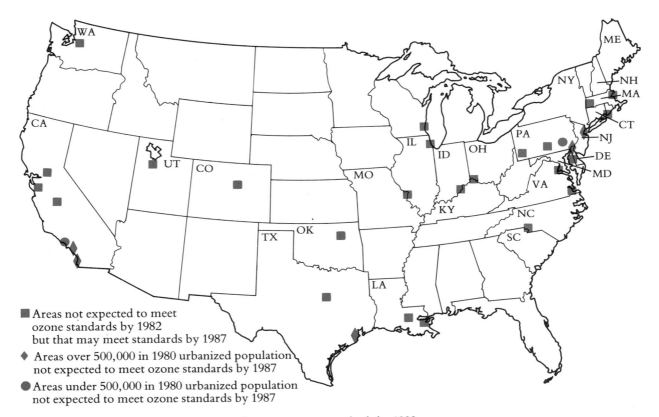

■ Areas not expected to meet
   ozone standards by 1982
   but that may meet standards by 1987

◆ Areas over 500,000 in 1980 urbanized population
   not expected to meet ozone standards by 1987

● Areas under 500,000 in 1980 urbanized population
   not expected to meet ozone standards by 1987

Figure 21.9 Areas not expected to meet ozone standards by 1982 or by 1987. (Data from *To Breathe Clean Air*, National Commission on Air Quality, 1981)

oxides and hydrocarbons fairly steady, progress toward achieving the ozone standard can be expected to be agonizingly slow.

Controlling Pollutants from Automobiles: The Record Examined

The record of success in controlling pollutants from automobiles has been positive overall, largely as a result of the willingness of Congress to become involved on a regular basis, instead of delegating responsibility for regulation to a commission. With control of auto emissions, a new concept in environmental regulation began—namely, *technology forcing*. (See Controversy 21.2.)

In the Clean Air Act of 1970, Congress directed a 90% reduction in automobile emissions of carbon monoxide, nitrogen oxides, and hydrocarbons. The reduction was to be measured from the emission levels then occurring in 1970 model automobiles. When the auto industry was unable to meet the emission standards, Congress and the administrator of the Environmental Protection Agency were forced to grant delays. To a large degree, however, only delays were granted; to a considerable extent, the basic goals remained firm. Only the standard for nitrogen oxides emissions was eventually loosened.

Table 21.2 shows the time pattern of emission control that resulted from the Clean Air Act of 1970. Except for nitrogen oxides, the original goals set for 1975 and 1976 were finally met in 1981. The law stated that a 90% reduction was to be achieved from 1970 emissions levels. The figures in column 3 of Table 21.2 do not correspond to a 90% reduction from column 2 because the testing procedure has changed several times since the original goal was set. If the 1970 emissions had been determined by the current method of measurement, the figures in column 3 would corre-

spond to a 90% reduction from the 1970 emissions so measured.

Why has it taken so long to bring auto emissions almost under control? Was the means of regulation wrong? Should Congress not have set emission standards for automobiles but left the task to the Environmental Protection Agency? Should Congress have paid more attention to the costs of achieving the emission standards? In fact, Congress commissioned a study on costs and feasibility by the National Academy of Sciences as part of the Clean Air Act of 1970. The Academy reported in 1972, two years after the act, that emission control was feasible and that costs were reasonable. As Mills and White (1978) described it:

> Congress had earlier [before 1970] simply authorized the administrative bureaucracy to set standards: now Congress itself was setting standards—and in the process showing contempt and vindictiveness for both the automakers and the federal agencies.

Congress might have been more trusting had not the automakers delayed implementing then-feasible technology for exhaust control in California in the 1960s. Congress was suspicious that the auto companies were simply foot-dragging again.

To this day the auto industry lobbies for the relaxation of auto emission standards, and it will probably continue to do so until it decides to take credit for the cleanup.

A different problem confronts us today, however: How do we make sure that pollution-control devices on autos are adequately maintained and are removing the quantities of pollutants they are supposed to remove? Vehicle inspection to be sure the devices are working has been strongly resisted by motorists. Tampering with pollution control devices has been found to occur on 12%–19% of the cars sold with these

Table 21.2 Auto Emissions and Emission Standards in Grams per Mile

Emissions	(1) Typical auto of the mid-1960s	(2) 1970 standards	(3) Originally required for 1975	(4) Originally required for 1976	(5) Achieved in 1980	(6) Achieved in 1981
Hydrocarbons	11	2.2	0.41	—	0.41	0.41
Carbon monoxide	80	23	3.40	—	7.0	3.4
Nitrogen oxides	4	—	—	0.40	2.0	1.0

CONTROVERSY 21.2

Air Pollution Control: A "Policy Beyond Capability"?

We think this is a necessary and reasonable standard to impose upon the [auto] industry.
Senator Edmund Muskie

This bill would write into legislation concrete requirements that can be impossible . . .
Senator Griffin

The 1970 Clean Air Act, called the "toughest air pollution law," was designed to promote the air quality we would like to have, not the quality of air we were sure we had the technology to achieve. Ever since then, scientists, industrial leaders, and politicians have been arguing over whether this was a good idea. Is it reasonable to base a law on what has been called "policy beyond capability"—that is, what you hope to achieve rather than what you are fairly sure can be accomplished?

In the Senate floor debate on the bill, the following exchange occurred:

Mr. Griffin: Did the committee have any hearings in this session on this problem as to the state of the art—on the likelihood or possibility that this goal can be reached by 1975?

Mr. Muskie: Yes, we had testimony jointly before the Commerce Committee and before our committee from the automobile companies on the state of the art. With respect to this specific deadline, no.

Mr. Griffin: On this particular bill?

Mr. Muskie: No.

Mr. Griffin: No hearings?

Mr. Muskie: The deadline is based not, I repeat, on economic and technological feasibility, but on considerations of public health. We think, on the basis of the exposure we have had to this problem, that this is a necessary and reasonable standard to impose upon the industry. If the industry cannot meet it, they can come back. . . .

Mr. Griffin: . . . without adequate expertise, without the kind of scientific knowledge that is needed—without the hearings that are necessary and expected, this bill would write into legislation concrete requirements that can be impossible— and that will literally force an industry out of existence. . . . (*Congressional Record*, September 21, 1970, pp. 5160–5195)

What do you think? Does setting standards that we are not positive we can meet act as a spur to the development of pollution control? Are we asking for trouble with this type of policy? What about the air we have to breathe? How much do you think Americans are willing to give up, if that is necessary, to assure themselves of clean air? You may wish to compare this instance of "technology forcing" with the automobile fuel efficiency requirements sometimes being enforced by the U.S. Department of Transportation (see Chapter 25 and the section in this chapter, "Controlling Pollutants from Automobiles: The Record Examined.")

devices. (No wonder vehicle inspections have been resisted.) The enemy is no longer the auto companies; in the words of comic strip character Pogo Possum, "We have met the enemy and he is us."

Lead Compounds in the Air

Lead is found in water and food in trace amounts. It is also present in layers of old house paint. And lead is in the air because gasoline contains lead additives to make it burn more smoothly. These souces of lead, taken together, put urban children at great risk of developing lead poisoning.

Lead is a cumulative poison; that is, it gradually builds up in the human system because of a slow rate of natural removal. Its presence in the atmosphere adds to the burden of lead we carry in our body. Lead slows the rate at which the bone marrow produces red blood cells. Lead also blocks the body's manufacture of hemoglobin.

In addition to breathing lead in the air, small children in poorer and older urban areas are exposed to another and more dangerous source of lead: peeling paint. Lead has long been used as a pigment in house paints, but since World War II, titanium dioxide has been gradually replacing it. Since the end of 1973, lead has been banned from most paints in the United States. Nevertheless, houses from the prewar era remain. Today many of these houses have fallen into disrepair, and the paint is crumbling from their walls. Mostly poor people inhabit these structures, and their children may eat the poisonous paint chips.

Eating paint chips is tragically common. Small children of all social and economic classes are often observed eating nonfood items. It is the child's way of sampling the environment. In this case, the sample is deadly indeed. A 1 g paint chip may contain up to 50,000 μg of lead. One estimate suggests that up to 2% of the children living in decaying prewar dwellings may have levels of lead in their blood exceeding the toxic limit.

The dangers of lead poisoning from eating paint chips increase if there is lead in the atmosphere. Where lead concentrations in the air are highest, people already carry a burden of lead in their blood and tissues because the body absorbs and retains a portion of the inhaled lead. City residents, then, are closer to the threshold levels of lead at which the symptoms of poisoning appear. Because children have a threshold level about half that of adults, they are far more susceptible than adults to lead poisoning. For many years, lead in the air has been pushing lead tissue levels in children too close to the poisoning threshold.

City air has been fouled with lead particles from the combustion of leaded gasoline. The act of breathing moves these particles into the lungs. Lead has been measured in the air of cities at monthly averages up to 5 μg/m^3. At this concentration, a 1-year-old inhales and absorbs almost as much lead from the air as he is absorbing from food and drink. The child's exposure to this toxic metal, then, can be nearly doubled because of the presence of lead in gasoline. The inhalation of street dust, which has been found to have high levels of lead, is still another route by which children are exposed to lead.

Lead poisoning in a child is first seen as a variety of symptoms, including appetite loss, problems in discipline, and a lack of interest in play. The disease progresses to constipation, vomiting, seizures, and finally, coma.

Although lead poisoning may lead to death, in milder cases it causes mental retardation. Even levels below the poison threshold appear to cause subtle inadequacies in learning. Lead has been shown to cause cancer in rats, suggesting that a similar effect in humans might be found; however, the fact is little mentioned because the immediate health effects of lead are so great.

Lead has been added to gasoline for many years as either lead tetraethyl or lead tetramethyl. These compounds are used to improve the antiknock quality of gasoline; they eliminate the "ping" during engine operation. These substances are not absolutely necessary, however. A slightly higher-priced refining process produces a fuel with the required antiknock properties, so that lead need not be present in gasoline. In the past, lead levels have ranged from 2.5 to 4.23 g of lead per gallon of gasoline. About 75% of the lead in gasoline is released to the atmosphere in the exhaust stream of cars. In 1975, about 147,000 metric tons of lead were added to the atmosphere from gasoline combustion.

The hazard to health from acute and chronic lead poisoning is one of the reasons for the 1974 decision to eventually make gasoline lead-free. The other reason for that decision is that lead poisons the catalytic converter, the device that further burns the hydrocarbons and carbon monoxide remaining in exhaust gases.

In 1977, the Environmental Protection Agency es-

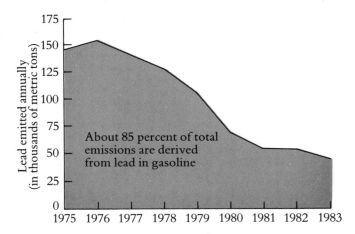

Figure 21.10 Trend in lead emissions into the atmosphere. Lead emissions have been declining since 1975 as older cars that require leaded gasoline have been replaced by newer vehicles. In 1985 and 1986, the EPA decreased the allowable lead content of gasoline from 0.5 g per gallon to 0.1 g per gallon in order to prevent misfueling of cars with catalytic converters.

tablished a standard for lead in the atmosphere of 1.5 $\mu g/m^3$ averaged over one month. For the past decade, the number of older cars that need leaded gasoline has been declining. As a consequence, sales of leaded gasoline, and hence, lead emission into the atmosphere, have been declining (see Figure 21.10).

Unfortunately, the slightly lower price of leaded gasoline tempted consumers. In 1985, it was estimated that, nationwide, 16% of the cars equipped with catalytic converters had been illegally fueled with leaded gasoline. To prevent illegal misfueling and damage to the converter, the EPA finally acted decisively on leaded gasoline in 1985. On July 1 of that year, the lead limit was reduced to 0.5 g of lead per gallon of fuel. As of January 1, 1986, the maximum allowable lead level was further reduced to 0.1 g per gallon. This level, automotive engineers agree, should be adequate to protect older engines.

This reduction of lead in gasoline will not cure the lead poisoning of children, but it will remove one element of risk. Lead poisoning remains a public health problem of major importance. To guard against new cases, parents and children must be informed of the risk involved in eating paint chips. Detection programs are necessary to find those children already near the toxic limit. Where cases are found, the surface from which the paint peeled should be covered up, or the paint should be completely removed.

Summary

Carbon monoxide as an air pollutant results when the carbon from gasoline, and to a lesser extent from other fuels, is only partly oxidized. In most cities, autos are responsible for 90% of the carbon monoxide emissions. Carbon monoxide is also produced in cigarette smoke. When the gas is inhaled, no matter the source, it ties up hemoglobin in the bloodstream, decreasing the amount of oxygen that can be carried to the cells. A light smoker has twice the hemoglobin out of service than a nonsmoker who breathes air at the air-quality standard.

Since 1971 controls on automobiles, especially use of the catalytic converter, have been decreasing carbon monoxide emissions per vehicle mile. Unfortunately, the increasing number of vehicles on the road has raised annual vehicle miles substantially, partly undoing the technical progress in control. Thus, an 85% reduction in carbon monoxide emissions per vehicle mile yielded only a 25% reduction in annual emissions from 1976 to 1981. In the period from 1981 to 1983, 50 major metropolitan areas were still violating the carbon monoxide air-quality standards. Boston, Denver, Los Angeles, Minneapolis, New York City, and Phoenix were the worst offenders. In growing areas such as Portland, Oregon, where vehicle traffic is getting worse, air quality in terms of carbon monoxide is getting worse as well.

At low levels of carbon monoxide in the air (resulting in 2%–5% of hemoglobin out of service), processes such as the perception of light brightness and the addition of columns of numbers are found to decrease. With only 3% of their hemoglobin tied up, individuals with angina experienced more rapid onset of heart pain when they exercised.

Another class of air pollutants, the oxides of nitrogen, stem from combustion of fossil fuels. About 40% of annual emissions are from gasoline combustion, and 30% are from fuel combustion at electric power plants. Hydrocarbons, another air pollutant, stem principally from automobiles (35%) and from industrial processes (35%). The reactions of nitrogen oxides and hydrocarbons in the presence of sunlight produce the photochemical air pollutants, of which the most prominent is ozone. Ozone is called an *oxidant* because of its strong oxidizing power.

Although most of the nitrogen oxides (90%) are produced as nitric oxide (one N atom and one O atom), an additional oxidation step quickly yields nitrogen dioxide (NO_2). Nitrogen dioxide in the atmosphere at levels above the air-quality standard increases the number of cases of respiratory disease, such as colds, bronchitis, croup, and pneumonia, probably by making people more susceptible to the pathogens that cause the disease. People with chronic (continuing) respiratory diseases, such as emphysema and asthma, are more likely to have serious complications (for example, pneumonia) because of exposure to nitrogen dioxide.

Photochemical oxidants in the air have been found to cause asthmatics to have more frequent attacks—at levels only 25% above the current air-quality standard. Asthmatics constitute 3%–5% of the population. Ozone damage to both commercial and ornamental plants is extensive (in the billions of dollars); damage would continue even if ozone levels were brought to the current ozone standards.

Modifications of automobile engines and exhausts have been utilized to bring nitrogen oxides, hydrocarbons, and photochemical oxidants under control. The catalyst system oxidizes unburned hydrocarbons to carbon dioxide and water vapor and converts nitrogen oxides back to nitrogen gas, a relatively unreactive form. Changes in industrial combustion, such as decreased air intake, are also being used to decrease nitrogen oxide emissions. Even with these changes, increases in coal combustion at electric plants and increased vehicle miles for autos have kept annual emissions of nitrogen oxides and hydrocarbons at nearly the same high levels. As a consequence, cities are having trouble meeting the ozone standards. Photochemically produced ozone has become a problem in the east as well as in California, its birthplace. In the early 1980s, 144 million people lived in areas that were violating the ozone standard.

The quest for clean air brought us the catalytic converter, but that device is poisoned by the lead in gasoline. As a consequence, lead, a long-time additive in gasoline, is finally being removed. Lead compounds in the air from the combustion of leaded gasoline have been increasing the burden of lead in people's bloodstreams. Children are at special risk of elevated lead levels. Prior to 1973, lead had been a major component of paint. Although it is now banned from paint, children living in decaying, inner city neighborhoods are often exposed to peeling paint, which may cause lead poisoning when eaten. Lead in the air has increased lead levels in these children, making poisoning more likely.

Although both lead sources have ended, peeling paint remains, and the problem of childhood lead poisoning continues.

Questions

1. What are the chemical categories of air pollutants produced by the automobile? List three categories. Which categories are known to affect our health?
2. Why is carbon monoxide a health hazard?
3. What measures have been taken to reduce carbon monoxide emissions by motor vehicles? Are these measures working?
4. Nitrogen dioxide decreases the ability of the eyes to adapt to seeing in the dark. Can you think of a common activity where personal safety could be threatened by the decreased ability to adapt to darkness after brightness? Explain. Is the activity likely to be undertaken in air with elevated levels of nitrogen dioxide? Why?
5. What are the names of the photochemical air pollutants? Why are they called *photochemical* air pollutants? How are they controlled?
6. The air-quality standards for photochemical pollutants were set in such a way as to protect the health of a portion of the population that was particularly susceptible to the substances. Who are these people?
7. List five ways in which urban children are exposed to lead that may enter their systems.
8. What is pica (you will need to look this up), and how is it related to the occurrence of lead poisoning?
9. Why was the lead level in gasoline reduced in 1985 and 1986 to just 0.1 g per gallon?

Further Reading

Controlling Air Pollution. American Lung Association, 1974. A well-done short book on methods to control air pollution from stationary and mobile sources. Drawings and text are excellent. Writing level is for the layperson.

Air Quality Criteria for Hydrocarbons. U.S. Environmental Protection Agency, AP-64, 1970.

Air Quality Criteria for Nitrogen Oxides. U.S. Environmental Protection Agency, AP-84, 1971.

Air Quality Criteria for Photochemical Oxidants. U.S. Environmental Protection Agency, AP-68, 1970.

Control Techniques for Carbon Monoxide, Nitrogen Oxide, and Hydrocarbon Emissions from Mobile Sources. U.S. Environmental Protection Agency, AP-66, 1970.
   Although technical, these documents are very clearly written and very complete.

References

Air Quality Criteria for Carbon Monoxide. U.S. Environmental Protection Agency AP-62, 1970.

Waldbott, George. *Health Effects of Environmental Pollutants.* 2nd ed. St. Louis: C. V. Mosby Co., 1978.

To Breathe Clean Air. Report of the National Commission on Air Quality, 1981, p. 8.

MacAvoy, Paul. Quoted in *Science,* **212** (June 12, 1981), 1251.

Congressional Record, September 21, 1970, pp. 5160–5195.

Mills, E. S., and L. F. White. "Auto Emissions: Why Regulation Hasn't Worked." In *Approaches to Controlling Air Pollution,* A. F. Friedlander, ed. Cambridge, Mass.: MIT Press, 1978.

A GLOBAL PERSPECTIVE V

Global Air Pollutants

Acid Rain and Forest Decline

*Sources of Acid Rain/Biological Impacts of Acid Rain/
Other Effects of Acid Rain/What Makes Some Lakes and
Soils Sensitive to Acid Rain?/Forest Decline/Should
Action Be Taken Now?/What Can Be Done About Acid
Rain and Forest Decline?*

Carbon Dioxide and Global Warming

*The Carbon Cycle/Effects of Carbon Dioxide Buildup/
Other Greenhouse Gases*

he effects of most air pollutants are, by and large, felt locally in the areas relatively near where they were emitted. Twenty to thirty miles might be the typical range in which the effect of an air pollutant is felt. One air pollutant category, however, seems to have a longer reach. Tall stacks on coal-fired power plants and at smelters often disperse sulfur oxides across great distances, sometimes across international boundaries. Reaching the earth as particles or in an acidic rainfall, these pollutants play havoc with the acidity of natural waters and with forest growth.

Another pollutant, carbon dioxide, a normal constituent of the earth's atmosphere, knows no boundaries. Produced as a result of fossil fuel combustion and the burning of wood, carbon dioxide levels in the air are steadily building up uniformly around the world. Though not directly harmful to plants or animals even at current elevated levels in the air, carbon dioxide is, nonetheless, acting as a barrier to the escape of heat from the earth's surface. A warming of the earth could have awesome consequences for future generations.

In this Global Perspective, we discuss in more detail the international and global aspects of these two pollutants, sulfur oxides and carbon dioxide.

Acid Rain and Forest Decline

Sources of Acid Rain

First in the Scandinavian countries, then in the northeastern United States and southeastern Canada, then in northern Europe, Taiwan, and Japan, scientists have been finding that rainwater—water thought of as among the purest available in nature—has become highly acidic. This **acid rain** is the result of the presence of both sulfur oxides and nitrogen oxides in the atmosphere. Sulfur oxides, as we have pointed out, are the result of burning the fossil fuels that contain sulfur: Coal is the prime source of sulfur among the fuels, oil is second, and natural gas is a distant third. Although nitrogen oxides also result from the burning of fossil fuels, the main source of the nitrogen is likely to be the air itself, since most fossil fuels are relatively low in nitrogen. The high-temperature combustion of all three fossil fuels results in significant emissions of nitrogen oxides.

Oxides of sulfur originate largely from coal-fired power plants and smelters. Oxides of nitrogen also stem to a very large degree from power plants, but automobiles are another significant source (see Chapter 21). Power plant emissions are often from tall stacks; hence, sulfur oxides and nitrogen oxides from these sources are likely to get caught up in long-distance transport. Ground-level emissions, on the other hand, are less likely to be captured by high-altitude winds and transported the long distances so characteristic of acid rain.

Sulfur dioxide and nitric oxide undergo chemical reactions in the atmosphere. The sulfur dioxide is oxidized to sulfur trioxide, which then dissolves in water droplets to form sulfuric acid. Nitric oxide is oxidized to nitrogen dioxide, which dissolves in water droplets to form nitric acid. These two acids, as well as salts of these acids, are responsible for acid rain. The more of these acids that are present in the atmosphere, the more acidic the rainwater becomes.

The two acids do not contribute equally to acid rain. In the northeastern United States, about 15%–30% of the acidity is due to nitrates (nitric acid); the remainder is attributed to sulfates (sulfuric acid). In California, in contrast, nitrates are the most common component of acid rain. Measurements of acid rain in Pasadena showed 57% of the acidity due to nitric acid and the remainder (43%) due to sulfuric acid. The large portion of acidity due to nitric acid is the result of the high tonnage of nitrogen oxide emissions from automobiles in southern California.

Acid particles are also deposited dry from the atmosphere onto vegetation and soil. These are typically the sulfates of calcium and magnesium (see Figure 20.1, where the chemical reactions of sulfur are diagrammed). This *dry deposition* component of acid deposition is increasingly thought to be an important mechanism by which acidity is reaching the earth.

The measure of acidity of water is the number of hydrogen ions per liter of water. Water molecules (H_2O) are normally dissociated into hydrogen ions (H^+) and hydroxyl ions (OH^-). In a sample of pure water, we would expect to find about 0.0000001 (one ten-millionth) of the water molecules dissociated into hydrogen ions and hydroxyl ions; in pure water the two ions are present in approximately equal numbers. A solution with equal concentrations of hydrogen and hydroxyl ions is called *neutral*. It is neither acidic (having more hydrogen ions) nor basic (having fewer hydrogen ions).

Rainwater is not pure; it comes in contact with and dissolves carbon dioxide, a natural component of the atmosphere. The solution of carbon dioxide in water produces carbonic acid, a weak acid. The concentration of hydrogen ions relative to the number of water molecules in unpolluted rainwater containing dissolved carbon dioxide would be about 0.000001, or one hydrogen ion per million molecules of water. If you count the zeroes after the decimal you will see that the hydrogen ion concentration has increased by a factor of 10 due to the solution of carbon dioxide. This level of hydrogen ions is presumed to be the natural condition for rainwater.

The rain falling in New England, on the other hand, has a ratio of about 0.0001 hydrogen ions per molecule of water (1 hydrogen ion per 10,000 water

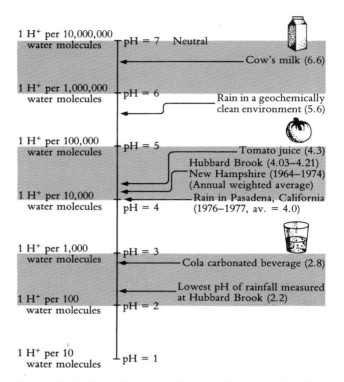

Figure V.1 Relation between pH and acidity, including the pH levels of some common substances and of rain in different environments. The lower the pH, the more acidic the substance.

molecules). This is about a 100-fold increase above the expected concentration of hydrogen ions in rainwater. New England has acid rain because the air is downwind from the major industrial centers of the northeast, where sulfur-bearing fossil fuels are burned in enormous quantities.

Acidity is not commonly measured by the ratio of hydrogen ions to water molecules. Instead, it is measured by the negative logarithm of the hydrogen ion concentration, which is called the **pH**. Thus, $-\log_{10}$ (.0000001) = 7, and a pH of 7 indicates water that is neither acid nor basic, but neutral. The rain in New England, with about 0.0001 hydrogen ions per molecule of water, has a pH of $-\log_{10}$ (.0001), or 4. In general, the lower the value of pH, the more acidic the water (see Figure V.1). A pH of 4 for rainwater used to be very unusual. Now pH values even less than 4 are commonly observed in some areas.

In Figure V.2, the average pH of rainfall in the eastern United States is displayed. From 1955–1956 to 1972–1973, the region with an average pH of 4.5 and

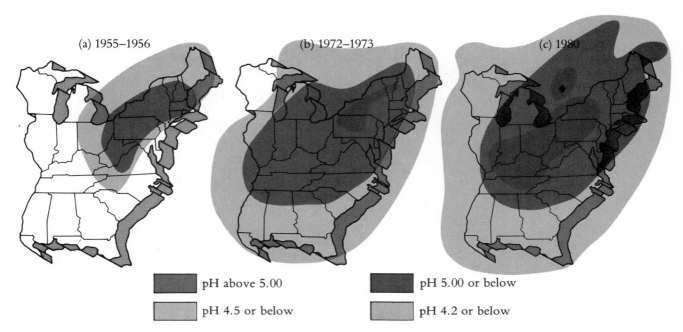

(a) 1955–1956   (b) 1972–1973   (c) 1980

pH above 5.00   pH 5.00 or below

pH 4.5 or below   pH 4.2 or below

Figure V.2 The weighted annual average of pH of precipitation in the eastern United States in 1955–1956, 1972–1973, and 1980. (From *Nitrates. An Environmental Assessment*, National Academy of Sciences, 1976 and Ministry of the Interior, Canada.)

below is seen to have spread from nine eastern states to include portions of all the states east of the Mississippi except Florida. In addition, a region with pH below 4.2 has appeared over New York, Pennsylvania, and Vermont in the 1972–1973 map.

In the most recent map, 1980, the region of pH 4.2 and below has spread to a wider area. The 1980 map also provides convincing evidence of the connection between sulfur oxides and acid rain. The small teardrop area with pH less than 4.2 surrounds Sudbury, Ontario. At about the position of the diamond in the teardrop is the 380 m stack from the smelting operations of the International Nickel Company. The operation at Sudbury emits more sulfur oxides than any other single source in North America. Prevailing winds are oriented, on average, in the same direction as the long axis of the teardrop. The lakes downwind from the stack are acidifying. The cause of that circle of acid rain, and the causes of acid rain in the United States, are not in doubt.

To further make the connection, Figures V.3 and V.4 indicate the major state and provincial sources of sulfur dioxide and of nitrogen oxides. Note how emissions of both pollutants cluster in the states where the pH is at its lowest level.

Europe suffers from the same plague. Extensive coal burning, particularly in the United Kingdom and central Europe, appears to be causing acid rain even more severe than in the United States and Canada (see Figure V.5).

Biological Impacts of Acid Rain

Of all the biological impacts of acid rain, the most obvious is the reduction and even elimination of fish populations in lakes that have been made acidic. In Sweden, Norway, and eastern North America, commercial and sport fishing have suffered as fish populations have declined or disappeared.

In the numerous lakes of the Adirondack Mountains at high elevations (610 m and above), acid waters are now the norm. In 1975, 51% of these lakes had pH values less than 5 and 90% of these low-pH lakes had no fish. In total, 45% of the high-altitude lakes were without fish populations; only 4% of these lakes were without fish in the 1930s. It has also been found that one-third of the more than 2,000 lakes in southern Norway are devoid of fish, a change that has taken place since the 1940s.

It is clear that acid waters have something to do

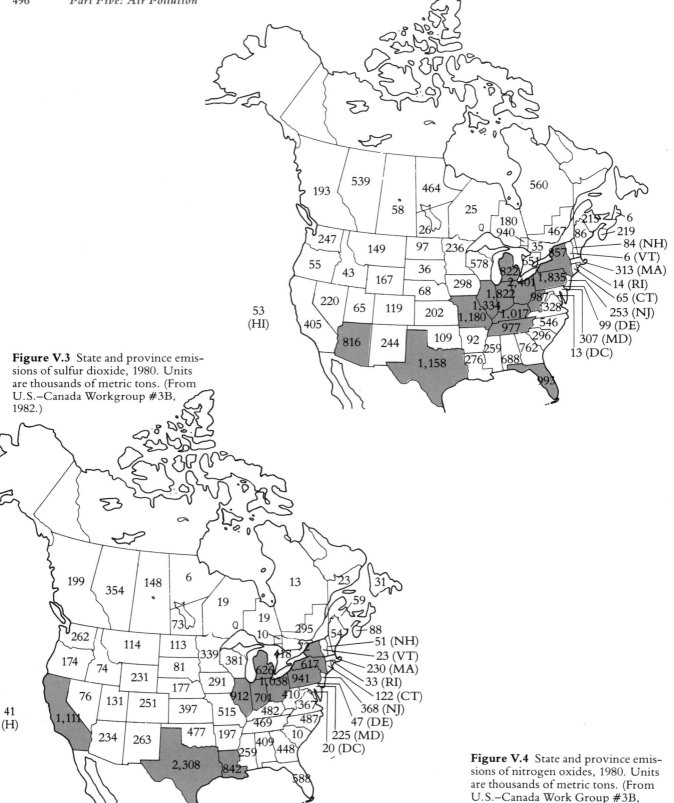

Figure V.3 State and province emissions of sulfur dioxide, 1980. Units are thousands of metric tons. (From U.S.–Canada Workgroup #3B, 1982.)

Figure V.4 State and province emissions of nitrogen oxides, 1980. Units are thousands of metric tons. (From U.S.–Canada Work Group #3B, 1982.)

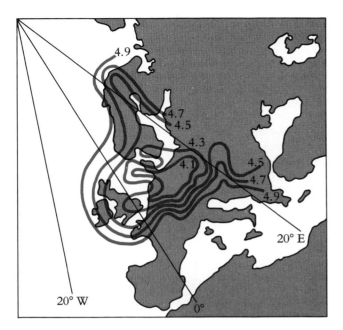

Figure V.5 Mean pH values for Europe based on available data from 1978 to 1982. Acid rain afflicts Europe as well as North America. The burning of high-sulfur coal is the principal cause. (From UNECE, 1984.)

with declining fish populations, but it is generally not the case that mature fish are killed in massive numbers by the acidity in these waters. Instead it appears that acid waters are preventing fish from reproducing. Female fish may not be able to spawn (release their eggs) in acid waters, and if they are able to do so, eggs and larvae may die at an increased rate. Thus, it is common in lakes that are in the process of becoming more acidic to see no young fish, only mature adults.

Many areas where fish populations have declined due to acid rain experience cold winters, with the accumulation of snow. When the snow pack melts, a rush of acid water may enter the lake, turning the lake sharply acid. The melting of the snow pack and the sharp increase in acidity coincide roughly with the spawning activity of fish. Thus, it appears that spawning is subjected to the maximum acid conditions that may occur through the year. As fish populations decline, it would be expected that those species that feed on fish, such as the bald eagle, the loon, and the osprey, mink, and otter, would also see some decline.

Scientists are predicting that populations of frogs, toads, and salamanders may all be reduced by acid rain. Many of these species breed in the temporary pools of water found during spring rains; the pools are likely to be more acid than the lakes because they are supplied

only by the acid rainwater. Again, these species are likely to decline quietly rather than in massive numbers; new young individuals will simply not survive to join the population.

Acid rain is thought to have an effect on plants as well as on animals. Changes in plant species where acid rain falls are difficult to detect; long periods of time are thought to be necessary to pick up changes such as forest growth. To fill the gap, experiments were performed in the laboratory in which plants were irrigated by water that had been made acid to the same extent as rainwater in the northeastern United States. The water was sprayed on pine trees and tomato plants to mimic the way rain arrives. Pine needles grew to only half their normal length in these experiments, and fewer tomatoes were produced than normally. The experiments have been prophetic, as you will see when you reach the heading "Forest Decline."

Other Effects of Acid Rain

At the moment, scientists have not detected any direct impact of acid rain on human health, but there are clearly possibilities for such effects. All the possibilities revolve around the increased ability of acid water to dissolve or otherwise act on minerals. Mercury in natural waters may be converted to monomethyl mercury under acid conditions. Fish will accumulate monomethyl mercury in their flesh. High mercury concentrations in fish have been found in areas where acid rain is falling and lakes are acidifying. Mercury is a known human poison, which has contaminated fish in the past under other circumstances (see Chapter 14).

If reservoirs used for drinking water were to become acidic, toxic metals from the watershed may dissolve in water being used for human consumption. In addition, acidic drinking water could dissolve lead from household plumbing systems. Such a case has apparently already occurred in New York State. There, the highly acid water from the Hinckley Reservoir near Amsterdam, New York, leached lead from household plumbing, causing concentrations of lead above the drinking water standard. The water had stood in pipes overnight; once this water was run out of the system, lead concentrations returned to their typically low levels.

Acid rain also damages mortar and stone by chemically reacting with the calcium and magnesium in them. Irreplaceable statuary is particularly at risk from such damage. In addition, iron products and other metals are very susceptible to corrosion from acid rain. Swedish investigators found a high correlation between acid rain and the corrosion rate of steel.

What Makes Some Lakes and Soils Sensitive to Acid Rain?

Some lakes seem not to be subject to becoming acidic; these lakes or their watersheds seem to have the capability of neutralizing acid additions. Other lakes, based on their present conditions, are obviously sensitive to acid rain. Vulnerable lakes typically are fed by a watershed in which the bedrock contains igneous rock (such as granite) or metamorphic rock (such as gneiss). These rock types are resistant to solution; water flowing over them dissolves little in the way of minerals and hence the lakes into which they flow have very "soft" waters.

The minerals in sedimentary rock, on the other hand, are more easily dissolved by water flowing over them. Lakes whose watersheds are composed of sedimentary bedrock tend to have "hard" water, water rich in dissolved minerals. If limestone bedrock (a sedimentary rock composed of calcium carbonate) is present in the watershed, runoff tends to be a "hard" water. Lakes whose watersheds are composed of sedimentary bedrock tend to resist becoming acidic because the carbonate minerals neutralize acidity.

The Adirondack lakes have watersheds whose bedrock is largely granite. The waters in the lakes are typically soft, low in dissolved minerals. Acid rain makes a big impact on these lakes. Regions of North America whose lakes are thought to be sensitive to the impact of acid rain are shown in Figure V.6.

Figure V.6 Areas containing lakes sensitive to acid precipitation. (From *To Breathe Clean Air*, National Commission on Air Quality, March 1981.)

The composition of the bedrock is not the only factor in determining whether lakes are sensitive to acid additions. The soil in the region also plays a role. Watershed soils that are rich in soluble minerals, such as calcium or magnesium carbonate, tend to protect lakes because these minerals neutralize acidity. As an example, although granite bedrock is common in areas of Maine, the lakes in these areas do not seem to have acidified, apparently because of the neutralization of acid rain in the lime-bearing soil—this in the face of rain with an average pH of 4.3. Some lakes in Florida, in contrast, have acidified despite having a watershed in which there was sedimentary bedrock. These lakes apparently did not get much groundwater inflow and the surface flows drained over soil in which little calcium carbonate was present.

Not only are lakes and their populations sensitive to acid rain, but soil and land-based ecosystems are sensitive as well. Soils have differing sensitivity to acid rain, just as lakes do. Again, soils that are derived from sedimentary materials, such as carbonate minerals, or that are rich in organic material tend to neutralize acid rain. Those soils that stem from rocks such as granite and gneiss, which resist solution, tend to be easily acidified. About 70%–80% of the land in the eastern United States possesses soils sensitive to acidification. Acid rain may leach the already small mineral content and plant nutrients from these soils, decreasing their productivity. That such productivity effects have occurred has been hard to demonstrate except under laboratory conditions. Such possibilities, however, should be enough, coupled with the other known effects of acid rain, to stimulate action.

Acid rain has no respect for boundaries of states or nations. Britain and northern Europe export acid rain to Sweden and Norway. Emissions in the United States contribute to acid rain in Canada, and Canada donates emissions that produce acid rain in the United States. Canada and the United States negotiated for much of the early 1980s on how each nation would handle its sulfur oxide emissions and which nation bore the most responsibility for acid rain. The discussions were not always pleasant. It was clear that the United States was the greater producer of sulfur oxides and that prevailing winds blow toward Canada from the eastern United States. Nonetheless, the Reagan Administration contended that it could not be determined how much of each nation's production crossed the boundary and that more research was needed on the cause of acid rain.

Forest Decline

The term *forest decline* carries two meanings. It can refer simply to a slowing in the growth of trees, as reflected by a decrease in the thickness of tree rings. The technical description for this slowing in growth is "a decline in forest productivity." An individual looking at a forest that has experienced such a decline in productivity would be unable to recognize what has occurred. Biologists, however, can detect the drop in productivity. The other meaning for *forest decline* is actual damage to trees, even tree death. Such a decline is not hard to observe.

Arthur Johnson at the University of Pennsylvania has studied the thickness of tree rings of both red spruce and pitch pine in order to detect whether a decline in forest productivity has occurred. His red spruce cores were drawn from a number of stands in the northeastern United States; his pitch pine cores from southern New Jersey. Thirty to forty percent of the cores he studied revealed a menacing pattern. Beginning in the 1960s, the thickness of the rings—the extent of new growth—narrowed substantially, and in many trees the thickness did not recover.

In Tennessee, scientists have found a similar decline not only in red spruce and pitch pine but also in short-leaf pine, hemlock, and fraser fir. They have discovered, too, that the decline extends to deciduous trees—hickory and yellow birch. In Germany's Black Forest, damage and decline extend from spruce to oaks, beeches, mountain ash, and sycamore. A plant disease that afflicts so many species of trees, deciduous and coniferous, is unheard of. Nor can drought fully explain what is happening to so many trees. Most scientists agree that air pollution and acid rain are the cause, but the evidence is circumstantial. That is, air pollutants and acid rain are present at the scene, but other factors, drought in particular, may be operating.

If a decline in forest productivity were all that had been observed, people might be less concerned; but decline has turned to damage and damage to death. In Germany's Black Forest, one out of every five trees shows moderate damage already; two of every three trees show symptoms of damage, and, diseased trees extend to fully two-thirds of the area of the forest. In 1984 it was estimated that three-quarters of the fir trees in Germany are in some way affected (see Figure V.7). In Germany the phenomenon is called *waldsterben* (literally "forest dying" or "forest perishing"), and it has been observed on over 50% of Germany's forested

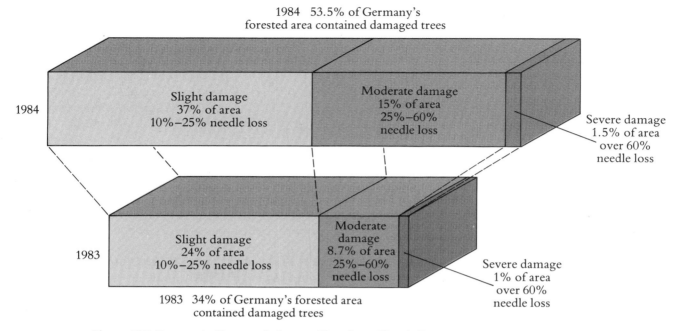

1984   53.5% of Germany's
forested area contained damaged trees

1984

Slight damage
37% of area
10%–25% needle loss

Moderate damage
15% of area
25%–60%
needle loss

Severe damage
1.5% of area
over 60%
needle loss

1983

Slight damage
24% of area
10%–25% needle loss

Moderate
damage
8.7% of area
25%–60%
needle loss

Severe damage
1% of area
over 60%
needle loss

1983   34% of Germany's forested area
contained damaged trees

Figure V.7 Damage in Germany's forests. (Data from Electric Power
Research Institute.)

area, some 6.25 million acres (2.5 million hectares). Germany's forests in Bavaria are also severely affected. In terms of the red spruce, the species that seems most sensitive to damage, a yellowing and early loss of needles is seen. The dropping of needles begins at the top of the tree (the crown) and proceeds down. And it begins at the outermost tips of branches and moves inward toward the trunk. Roots may deteriorate, boughs sag. Insects and fungus growths invade the bark of the tree. High levels of toxic metals, including aluminum, cadmium, copper, and manganese, may be found in the layer between the rings and the bark.

Anecdotal descriptions of Czechoslovakia, where a high-sulfur brown coal is burned extensively, suggest a scene of devastation, the destruction of whole forests. A half-million-acre (200,000 hectare) area of damaged trees has been reported. One-fifth of the area is said to contain dead trees. Poland, where the brown coal is also used, is reputed to be suffering even more extensive damage; damaged trees are found on 1.25 million acres (500,000 hectares) of forested land area. Forest death has also been discovered in Austria, Switzerland, Sweden, East Germany, the Netherlands, Romania,

the United Kingdom, and Yugoslavia. In southwest Sweden, every tenth tree is damaged or dying. The spruce trees in the Scania section of Sweden hold their needles only a single year (healthy adult trees normally hold their needles for 3 years). The beech trees in Scania, as in the afflicted areas of West Germany, are losing their leaves in August; the leaves fall while they are still green. The bark of the beeches is swollen and cracked; craters the size of the palm of a hand pit the trunks of the trees.

Tree damage has now been found in the United States: in the Adirondack Mountains of New York, the Green Mountains of Vermont, the White Mountains of New Hampshire, and the Great Smoky Mountains of eastern North Carolina. Red spruce in these forests have exhibited dramatic dieback, but the extent has not been estimated as it has in Germany and other European countries.

In all these instances, there are certain features that seem to occur again and again. First, acid deposition afflicts all the regions. Long-distance transport has brought sulfur oxides and nitrogen oxides to the areas, resulting in acid rainfall and dry deposition of metal sulfates. Second, in a number of situations, the forests are at relatively high elevations and stand in the cloud cover for a substantial portion of the year. This is true of the Black Forest and of Vermont's Green Mountains, where clouds may surround the trees for a quarter of the year. Cloud moisture may have a pH as low as 3.5, even more acid than the rain, suggesting that the stress of an acidic environment could be a likely mechanism for tree damage. Third, droughts preceded the intervals in which tree-ring decline began to be noted. If drought were the only culprit, however, recovery in growth would be expected after weather patterns improved. In one study, recovery in tree-ring growth did occur in about 20% of the tree cores examined, but ring growth never recovered in 40% of the samples.

The fourth common characteristic is the status of the soils. Soils in these high-altitude forests are subject to acidification because they are thin and contain relatively little organic material. Essential minerals such as magnesium and calcium may be leached from the soil by the hydrogen ions and lost in runoff, hence becoming unavailable to the growing tree. On the other hand, the acidity of rainfall may cause aluminum, an element toxic to plants that is normally chemically bound in mineral form and unavailable, to be "mobilized" or made chemically available in the soil. It may then enter the tree roots, where it may disrupt normal root function. Fifth, chemical analysis of leaves and needles of declining trees shows about a 10% higher sulfur content than that found in the foliage of healthy trees, at least in the United States.

Sixth, considering mountain forests, especially in the southern United States and in Europe, relatively high ozone concentrations have been observed in the air—higher than those observed at lower elevations, and even higher than in some cities. Ozone has long been known to damage evergreens, but high ozone concentrations on mountain slope environments is an unusual finding. A possible explanation is the presence of nitrogen oxides in the air. We know (see Chapter 21) that nitrogen dioxide reacts in a complex way with hydrocarbons in the presence of sunlight to produce ozone. But where are the hydrocarbons coming from? They may come from the trees themselves. The Great Smoky Mountains are so named because of the haze of *terpenes*, produced by the trees, that hangs over the forests. Terpenes are highly reactive hydrocarbons pro-

duced by the trees of the coniferous forest. The terpenes may be reacting with nitrogen dioxide in the presence of sunlight to produce the ozone.

These are the factors: acid rain, high elevations and cloud cover, drought preceding damage, soil acidification and mineral change, sulfur in needles and leaves, and ozone in the atmosphere. Together they characterize what seems to be becoming a Northern Hemisphere ecological disaster.

Should Action Be Taken Now?

Four benchmark scientific reports point to the need for action on acid rain. In 1981, a report of a panel of the National Research Council (the research arm of the National Academy of Sciences and the National Academy of Engineering) concluded that a 50% reduction in acidity of rainfall was needed to protect forests and lakes. The report did not say how to accomplish the 50% reduction in acidity—only that such a reduction was needed.

A second panel, appointed by President Reagan's science adviser, was chaired by William Nierenberg, director of the Scripps Institution of Oceanography. The official news release of the study's conclusions was made available in 1983, although the actual report was delayed until 1984. The panel recommended "reductions from present levels of emissions of sulfur compounds beginning with those steps that are most cost effective in reducing total (acid) deposition." The panel further concluded that "actions have to be taken despite incomplete knowledge." The report warned: "If we take the conservative point of view, that we must wait until the scientific knowledge is definitive, the accumulated deposition and damage to the environment may reach the point of irreversibility."

In 1983 another committee of the National Research Council that had also been appointed to study acid rain made its report. The committee, chaired by Jack Calvert of the National Center for Atmospheric Research, concluded that the extent of acid rain in eastern North America "is roughly proportional to sulfur emissions from power plants, industries, and smelters." The link of acid rain to sulfur emissions was thus firmly supported by the committee. The report made clear that a 50% reduction of sulfur oxide emissions would be expected to produce a 50% reduction in acid rain.

In 1984, a fourth report was released, this one from the Office of Technology Assessment (OTA), an arm of Congress. This report estimated that 3,000 lakes and 23,000 miles of streams (about 20% of lakes and stream miles sensitive to acid rain) are at risk of being acidified or are already acidic. In terms of threats to human health, OTA estimated that as many as 50,000 extra deaths occur annually as a result of sulfates and other particulates in the air. Patients with cardiac and respiratory problems are affected the most. The OTA projected that 10 million tons of sulfur oxides (about 50% of the annual emissions) could be removed at a cost of $3 to $4 billion, increasing electric rates by only about 2%–3%. The report indicated that this level of removal is likely to be able to protect all but the most sensitive aquatic areas.

What Can Be Done About Acid Rain and Forest Decline?

Ministering to the patients. Limestone has been applied to the acidified lakes in Sweden and to those in the Adirondack Mountains of New York chiefly on an experimental basis. The limestone, which is mainly calcium carbonate, may be applied in a crushed form or a water slurry. Liming decreases the acidity of a lake and provides some resistance (a buffering capability) to further acid inputs. If the lake or pond flushes quickly—that is, if its inflow and outflow rates are high—the liming will have to be repeated every 1 or 2 years. Because most of critically acidified lakes in the Adirondacks cannot be reached by road, liming them will require application by helicopter or airplane, increasing the expense. Electric utility spokespeople and coal industry spokespeople are recommending this alternative.

Limestone may be applied to the declining forests as well. Liming slows the rate at which soil acidifies and raises the pH toward neutral. It is expected, however, that the growth rate in a limed soil will not be optimum. In Germany's Black Forest, foresters have fertilized the soil with a combination of magnesium sulfate (at 700 pounds per acre) and limestone (at 2,000 pounds per acre). Trees not too far injured have recovered under this prescription. Foresters guess that this process might buy the trees 5 to 10 years while the air is being cleaned up.

The fact that trees recover under this treatment suggests that explanations for forest decline are not far off the mark. Magnesium is apparently being leached from the soil, since restoring it helps, and soil acidity is also a causative factor, since decreasing soil acidity also helps.

Somehow, these seem like heroic measures to keep dying patients alive. However, the patients will fully recover only if they are given clean air to breathe.

Coping with the cause. The technical means of cleaning up the air and controlling acid rain and forest decline are clear; they are the same means that should be used to control sulfur oxide and nitrogen oxide emissions. At power plants, this means scrubbers to remove sulfur oxides and combustion modifications to decrease the formation of nitrogen oxides.

In 1982 Norway, Finland, and Sweden suggested a 30% reduction in sulfur discharges from 1980 levels by 1993. Eight nations have now pledged to achieve such a reduction. They are the original three plus Denmark, West Germany, Switzerland, Austria, and Canada. Great Britain, France, and East Germany have declined to make this pledge. Canada has, in fact, set a goal of a 50% reduction in sulfur emissions, proceeding without the cooperation of the United States. Although a goal has been established, the Canadian government has yet to establish deadlines for the reduction or to set penalties for noncompliance. In 1985, on the other hand, the Canadian province of Ontario ordered its major sulfur polluters to clean up their discharges to achieve a one-third reduction from 1980 levels by 1994. Four corporate dischargers account for 80% of Ontario's sulfur dioxide emissions.

Although automobiles are only a minor source of sulfur emissions, they are a major source of nitrogen oxides. To control nitrogen oxides from autos, catalytic converters and engine modifications are used. (Control of automobile emissions is discussed further in Chapter 21.) The United States has applied these controls, but vehicle miles keep climbing, preventing substantial progress at the emissions standards that have been set.

In contrast, no European nation now imposes emission controls on automobiles, although the auto manufacturers obviously have the technology—they attach it to cars exported to the United States. In 1984, in response to the shocking news of forest decline, Germany tried to impose unleaded fuel and catalytic converters on the nation by 1986. The auto lobby beat back this threat. Now the law calls for new large cars to have emission controls by 1988 and new small cars by 1992. Older cars—uncontrolled—will remain on the highway until they wear out.

Carbon Dioxide and Global Warming

The Carbon Cycle

Burning of fossil fuels produces two oxides of carbon. One, carbon dioxide, is not poisonous, although it could change the earth's climate. The other, carbon monoxide, has known harmful effects on humans.

As we have said before (see Chapter 1), plants use carbon dioxide, and animals and plants produce it. Plants use carbon dioxide and produce oxygen during **photosynthesis**. Most plant and animal species consume oxygen in respiration. They also produce carbon dioxide as a waste product of that respiration. Hence, carbon dioxide is naturally present in the atmosphere in reasonable quantities. A normal sample of air contains about 0.03% carbon dioxide by weight.

Removal of carbon dioxide. Carbon dioxide is not only consumed in photosynthesis but also dissolved in the oceans. When carbon dioxide dissolves in the ocean, carbonic acid is produced at a *very low concentration*. The carbonic acid dissociates in part to bicarbonate and carbonate ions. These ions combine with calcium and magnesium (from the natural weathering of rocks), which have been carried into the ocean. The reaction of the calcium ions with carbonate ions produces calcium carbonate. We know this relatively insoluble substance as limestone. Magnesium and calcium may jointly react with the carbonate ion to produce dolomite. These precipitation reactions remove carbonate from the water, making room for more carbon dioxide to dissolve.

Thus, there are two natural mechanisms for carbon dioxide removal from the atmosphere: (1) solution in the ocean, followed by precipitation; and (2) the utilization of carbon dioxide by green plants in photosynthesis (see Figure V.8).

Additions of carbon dioxide. Carbon dioxide is now being produced in massive quantities from the burning of fossil fuels. In 1985, the United States alone was annually burning about 650 million metric tons of coal in electric power plants. The nation also consumed a comparable tonnage of oil that year. The rest of the world consumes about twice the quantity of petroleum products that the United States does, so the worldwide consumption of fossil fuels is enormous. This global increase in combustion is sharply forcing up the rate at which carbon dioxide enters the atmosphere.

In addition to the amount of carbon dioxide produced annually by green plants, fossil fuel combustion now adds another 5%–7% each year. About half this additional quantity is accumulating. That is, it is not being consumed in photosynthesis, nor is it being dissolved in the ocean; it is remaining in the atmosphere (see Figure V.9).

The annual amount of carbon dioxide from combustion is expected to grow continuously. One projection of the use of coal for electric power generation in the United States shows consumption doubling between 1980 and 2000.

The extent of the increase in fossil fuel combustion will depend on energy conservation measures that may be taken. It will also depend on the extent to which nuclear and solar energy replace coal, oil, and natural gas. Total fossil fuel combustion might be expected to increase by 3%–4% annually through the year 2000. On that basis, the amount of carbon dioxide in the atmosphere could potentially increase by as much as 24% from today's level by the year 2000. Another factor that could hasten the increase in carbon dioxide levels is the destruction of tropical rainforests for agriculture, as is happening in countries like Brazil. The release of carbon dioxide occurs when the forests are destroyed, because trees hold 10 to 20 times more carbon dioxide per unit of land area than do most agriculture crops.

Effects of a Carbon Dioxide Buildup

The greenhouse effect. For about three-quarters of a century, scientists have been aware of the buildup of carbon dioxide. It is now being carefully watched. The concern is that carbon dioxide will trap heat in the earth's atmosphere. The effect of carbon dioxide has been compared to the effect of the glass panes of greenhouses. The panes let sunlight into the greenhouse, and the greenhouse is warmed by the solar radiation. At

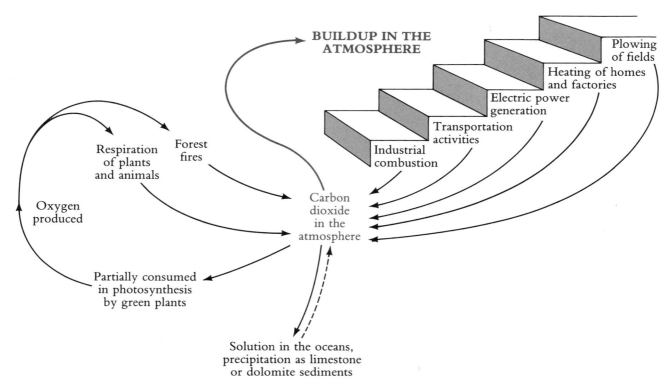

Figure V.8 The carbon dioxide cycle as modified by human activity. If human activities are excluded, carbon dioxide is only added to the atmosphere by the respiration of plants and animals and by forest fires. It is removed from the atmosphere by photosynthesis and by solution in the oceans. The removal and restoration rates were in rough balance before humans began burning fossil fuels. The combustion of fossil fuels has upset the balance and carbon dioxide is now accumulating in the atmosphere.

night, heat radiates away from the building. However, the glass panes decrease the rate of heat radiation out of the structure.

Carbon dioxide is expected to act in a similar way. Solar energy would continue to reach the earth without being affected by the gas, and the earth would be warmed. But the radiation of heat away from the earth would be slowed by the carbon dioxide. The earth would then heat up (Figure V.10).

From the turn of the century until World War II, a small warming trend was noted. Since about 1940, however, the trend has reversed, and a modest cooling has occurred that scientists feel is the definite trend of the earth's temperature at this time. Thus, we have not

yet seen the global increase in temperature that has been predicted. From appearances alone, the theory does not appear to have stood up.

Other factors may be operating, however. In the first place, the global cooling trend is regarded by some as being part of totally natural changes in the earth's climate and temperature, perhaps due to the gradual and cyclic changes in the earth's orbit. The trend may also be related to extensive volcanic activity ejecting small particles high up into the earth's atmosphere.

In 1980, Mount St. Helens in the state of Washington spewed forth enormous quantities of volcanic dust. In 1982, El-Chichón, a volcano in Mexico, rocketed even more massive amounts of dust into the strato-

Figure V.9 Atmospheric concentration of carbon dioxide at Mauna Loa Observatory, Hawaii. Combustion of fossil fuels is producing carbon dioxide faster than green plants can use it and faster than the ocean can dissolve it. As a consequence, levels of carbon dioxide in the atmosphere are building up and will build up faster as combustion rates increase. Scientists suggest that carbon dioxide may prevent the earth from radiating heat back to space and so cause a warming trend. Most scientists now agree, barring unforeseen volcanic activities, that gradual warming and gradual sea level rise will probably occur within the next decade. (Data from "Characterization of Information Requirements for Studies of CO_2 Effects," U.S. Department of Energy, Carbon Dioxide Research Division, DOE/ER-0236, December 1985. Data were obtained by C. D. Keeling.)

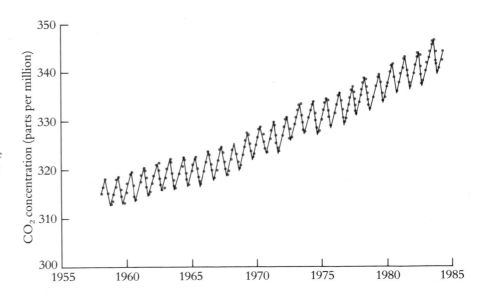

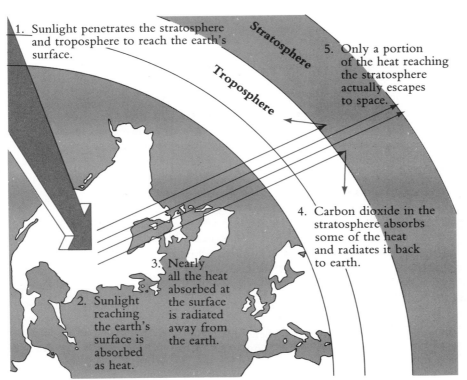

1. Sunlight penetrates the stratosphere and troposphere to reach the earth's surface.

Stratosphere

Troposphere

5. Only a portion of the heat reaching the stratosphere actually escapes to space.

4. Carbon dioxide in the stratosphere absorbs some of the heat and radiates it back to earth.

3. Nearly all the heat absorbed at the surface is radiated away from the earth.

2. Sunlight reaching the earth's surface is absorbed as heat.

Figure V.10 The greenhouse effect of high levels of carbon dioxide on the heat balance of the earth.

sphere. A volcano in Indonesia was also releasing volcanic dust into the atmosphere during this period. Such particles, suspended in the atmosphere, reflect sunlight away from the earth, robbing the earth of the heat that the sun's radiation would have provided. Years with extremely cold winters and cold summers followed the massive eruption of Mt. Tambora in Indonesia in 1815. This volcanic activity then and the release of particles to the upper atmosphere may temporarily be producing a countereffect to the greenhouse effect.

Particles not only stem from volcanic activity but are also discharged in abundance by our industrial society. Combustion produces not just carbon dioxide but also atmospheric particles in tremendous quantities. Observations show that the fraction of sunlight reaching the earth has been decreasing. Presumably, this has been due to particles suspended in the atmosphere. Scientists who have studied the carbon dioxide question suggest that the cooling trend may, in fact, even be fortunate, for it affords us temporary protection from the earth's beginning to heat up.

Projections of world temperature and sea level. Many investigators now agree on the ultimate effect of carbon dioxide in the atmosphere. A National Academy of Sciences committee of experts (1979) has estimated the impact of a doubling of atmospheric carbon dioxide levels. They see a global average surface warming of 3°C, with a 50% error to either side, as the result of such a doubling of carbon dioxide levels from 400 ppm to 600 ppm. From 1880 to 1980, the carbon dioxide level went from 280–300 ppm up to 335–340 ppm, a 10%–12% increase.

We have already experienced a sea level rise of 4 to 6 inches over the last hundred years. This rise occurred in the same time interval in which the earth was undergoing a mean global warming of 1°F (0.4°C). Because water expands as its temperature is raised, the sea level rise has generally been attributed to a combination of a thermal expansion of the upper layers of the ocean and glacial melting.

Estimates of the rise in sea level in the next 100 years due to thermal expansion of sea water and the melting of the ice sheets range from as little as 15 inches (0.38 m) to as much as 7 feet (2.11 m). These projections cannot be refined further with available knowledge, but a sea level change of only 1 foot (0.30 m)

would be expected to cause 60–250 feet (20–80 m) of beach on the open coast to erode away. The waves from ocean storms would then reach farther inland, threatening manmade structures. The marshes that accompany coastal estuaries would drown, and in cases where new marshes did not form farther inland, important wildlife habitats would be destroyed. This is the low end of projections for effects of carbon dioxide increase.

A conceivable figure for the percentage increase in the carbon dioxide level by the year 2000 is 25%. Such an increase could produce a change in the average worldwide temperature of about 1°C. This temperature change might be even larger in the polar regions, because the haze of particles that surrounds industrial civilization is less concentrated in those areas. Furthermore, the melting of ice in the polar regions would reduce the local reflectivity and increase the absorbance of light and heat there. Thus, the polar regions are expected to react to changing carbon dioxide levels more quickly than other areas of the earth.

One pair of scientists estimate that 50,000 km^3 of ice has melted since 1940, raising the global mean sea level an average of 3 mm per year during that period (5.4 inches over the entire interval since 1940). Mean global temperature, however, has decreased by 0.2°C during this interval. Such a quantity of polar ice melting requires the input of a vast amount of heat. Thus the ice sheets absorbing heat and melting could be delaying the onset of surface warming.

If the average antarctic temperature were to increase 5°C, enough ice could melt in Antarctica to raise the sea level by 15–25 feet worldwide. That would put New Orleans and large portions of Florida under water. Drought-prone regions in North America and Asia could also develop as global climate shifts.

Corrective mechanisms, faint hope. To be fair, some scientists, lead by S. G. Idso, see these predictions as grossly overstated. Their physical measurements produce results in conflict with the predictions from the global models of carbon dioxide and temperature. Built-in corrective mechanisms may also take effect to soften the heating trend. One likely possibility is that a warming trend would increase evaporation from the oceans. This, in turn, would increase the cloud cover and decrease the quantity of sunlight reaching the

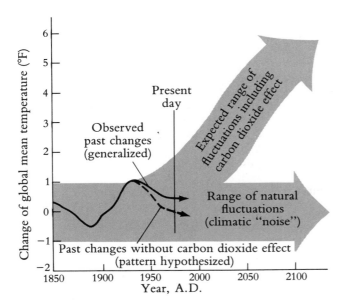

Figure V.11 Time trends of global mean temperature with and without the carbon dioxide effect. By about the year 2000, scientists expect to see the trend in mean global temperature begin to deviate significantly from the pattern it would normally have been expected to follow. (From J. Mitchell, "Workshop on the Global Effects of Carbon Dioxide from Fossil Fuels." U.S. Department of Energy, May 1979, p. 98.)

earth. Unfortunately, water vapor, like carbon dioxide, is a greenhouse gas, and its increased concentration in the atmosphere would trap more heat. The sunlight-blocking effect of increasing cloud cover, then, would be countered by an even slower heat loss from the earth. The net effect would be an acceleration in the temperature increase.

Apparently, the argument of built-in corrective mechanisms cut little ice with the National Academy of Sciences. Its expert committee asserted, "We have tried but have been unable to find any overlooked physical effect that could reduce the . . . global warming to negligible proportions."

A prediction. When will the carbon dioxide effect be seen as a clear increase in temperature, distinct from other trends? Scientist J. M. Mitchell (1979) suggests that by the turn of the century the upward trend of surface temperature will become apparent (see Figure V.11). He also notes that "by the time the CO_2 disturbance will have run its full course, more than a millennium into the future, temperature levels in the global atmosphere may possibly exceed the highest levels attained in the past million years of the earth's history" (p. 98).

Other Greenhouse Gases

In the mid–1980s, new evidence was coming to light about increased levels of other greenhouse gases. Both methane and chlorofluorocarbons were found to be accumulating in the atmosphere. The reasons that methane is entering the atmosphere in increasing quantities are uncertain, but the gas is produced by cattle when they digest cellulose material. Methane is also produced underwater in flooded rice paddies. Worldwide cattle and rice production have both been increasing rapidly to sustain an increasing population in the past several decades. Chlorofluorocarbons are used as solvents, refrigerants, and spray-can propellants; these substances are far more effective than carbon dioxide at blocking the escape of heat energy from the earth. The increases in both these gases give rise to deeper concern about global warming. Indeed, estimates of the rate of global warming may now prove to have been far too conservative.

Questions

1. What two gases are primarily responsible for acid rain? What are the principal sources of these gases?
2. What and where is the largest sulfur oxide source in North America?
3. Name two ways acid rain can affect human health.
4. Would you expect lakes in the Rocky Mountains to be as sensitive to acid rain as those in the Adirondacks? Why or why not?
5. What evidence, relative to forest decline, can be gained from studying tree rings?
6. What tree species seems to be hardest hit by decline? Name some other species being affected.
7. Why have tree pests such as insects or fungi been ruled out as the culprit causing forest decline?
8. What is *waldsterben*?

9. Name the six factors that seem to be operating in most cases of forest decline.

10. What can be done at the site to restore an acidified lake? A declining forest?

11. Describe the carbon cycle.

12. What is limestone and what role does it play in removing carbon dioxide from the atmosphere?

13. Name an additional way in which carbon dioxide is removed from the atmosphere.

14. Explain why carbon dioxide's effect in the stratosphere is likened to the role of a greenhouse.

15. What two factors could be operating that may have prevented a global warming trend since World War II?

16. Explain why increased cloud cover, caused by increased evaporation, is not expected to prevent a global warming.

17. Name two other greenhouse gases building up in the earth's atmosphere.

Further Reading

Acid Rain

All articles are readable and nontechnical.

Luoma, J. "Troubled Skies, Troubled Waters," *Audubon*, November 1980.

"How Many More Lakes Have to Die," *Canada Today*, **12** (2) (February 1981). Available from Canadian Embassy, 1771 N Street, N.W., Room 300, Washington, D.C., 20036.

Likens, G. "The Not So Gentle Rain." In *Yearbook of Science and the Future*, Encyclopaedia Britannica, 1981, pp. 212–227.

Gorham, E. "What to Do About Acid Rain," *Technology Review* (October 1982), p. 59.

Likens, G. "Acid Precipitation," *Chemical and Engineering News*, **54** (November 22, 1976), 29.

Cowling, E. "Acid Precipitation in Historical Perspective," *Environmental Science and Technology*, **16** (2) (1982), 110A.

Burton, P. "Acid Rain: The Water That Kills," *National Parks* (July/August 1982), p. 9.

"Acid Rain." United States Environmental Protection Agency, Office of Research and Development, EPA-600/9-79-036, July 1980.

La Bastille, A. "Acid Rain: How Great a Menace?" *National Geographic* (November 1981), p. 657.

Office of Technology Assessment. "Acid Rain and Transported Air Pollutants: Implications for Public Policy." Washington, D.C.: Congress of the United States, 1984.

Maugh, T. "Acid Rains Effects on People Assessed," *Science*, **226** (December 21, 1984), 1408.

Johnson, A., and T. Siccama. "Acid Deposition and Forest Decline," *Environmental Science and Technology*, **17** (7) (1983), 294A.

Hileman, B. "Forest Decline from Air Pollution," *Environmental Science and Technology*, **18** (1) (1984), 8A.

Kiester, E. "A Deathly Spell Is Hovering Above the Black Forest," *Smithsonian*, **16** (8) (November 1985), 211.

Postel, S. "Air Pollution, Acid Rain and the Future of Forests." World Watch Institute Paper 58, March 1984.

Stumm, W., L. Sigg, and J. Schnoor. "Aquatic Chemistry of Acid Deposition," *Environmental Science and Technology*, **21** (1) (1987), 8.

Ruben, E., et al. "Controlling Acid Deposition: The Role of FGD," *Environmental Science and Technology*, **20** (10) (1986), 960.

Carbon Dioxide

Barth, M., and J. Titus (eds.). *Greenhouse Effect and Sea Level Rise: A Challenge for This Generation*. New York: Van Nostrand Reinhold, 1984.

A collection of high-quality technical papers.

ReVelle, R. "Probable Future Changes in Sea Level Resulting from Increased Atmospheric Carbon Dioxide." In *Changing Climate*. Washington, D.C.: National Academy Press, 1983.

An assessment by the scientist who first verified the increasing levels of carbon dioxide in the atmosphere.

ReVelle, R. "Carbon Dioxide and World Climate," *Scientific American*, **247** (2) (August 1982), 35.

An expert's analysis of current scientific views on carbon dioxide's effect on climate, plus a historical perspective.

Kerr, R. "Doubling of Atmospheric Methane Supported," *Science*, **226** (November 23, 1984), 954.

A readable news report on the increase in another greenhouse (heat-trapping) gas.

Kerr, R. "Trace Gases Could Double Climate Warming," *Science*, **220** (June 24, 1983), 1364.

A readable report on the family of greenhouse gases.

Rose, D., M. Miller, and C. Agnew. "Reducing the Problem of Global Warming," *Technology Review* (May-June 1984), p. 49.

A description of policy options for control of carbon dioxide. Emphasis is on energy conservation.

Woodwell, G. "The Carbon Dioxide Question," *Scientific American*, **238** (1) (January 1978), 34–43.

A clear exposition by a research scientist in the field of the global carbon cycle.

References

National Academy of Sciences. "Carbon Dioxide and Climate: A Scientific Assessment." Washington, D.C.: National Academy Press, 1979.

Mitchell, J. "Workshop on the Global Effects of Carbon Dioxide from Fossil Fuels." U.S. Department of Energy, May 1979.

(U.S. Department of Energy)

Natural Sources of Power and Energy Conservation

e found in Parts Four and Five that many pollutants stem from the use of energy and the production of electric power. The list of such pollutants is surprisingly long and the quantities of pollution remarkably large. It is logical to ask, do the methods of energy use and power generation have to be accompanied by degradation of the environment? While it is true that human activities have some unavoidable effect on the environment, technologies do exist whose environmental impacts are far less than others. It is practically impossible for us to have no effect on the environment in which we live, because living itself consumes and utilizes energy. (This point was explained in terms of thermodynamic laws in Chapter 2.)

Though human life must have some impact, there are natural balances, or righting mechanisms, that tend to keep environments and the natural communities living in them in an equilibrium, where changes come only slowly. Nonetheless, in many cases, human activities overbalance these mechanisms, causing rapid changes with which neither humans nor wildlife tend to cope well. Conventional power generation that produces quantities of air and water pollutants is one such activity. Direct combustion and use of fossil fuels is another.

In this sixth part we shall look at natural sources of power: hydropower, tides, wind, geothermal energy, and the sun. These methods of power generation seem to promise softer impacts on the environment than combustion of fossil fuels or the fissioning of uranium. In addition, most of these energy sources are renewable, in the sense that nature makes them available virtually forever.

It is remarkable to think that only two centuries ago society had, besides human and animal energy, only three forms of energy at its disposal. All three of these forms could be traced to the sun. Hydropower was used to operate mills, which ground grain or wove cloth. Hydropower required water running downward to the sea from the uplands, where water fell as rain. Of course, the sun caused the water to evaporate in the first place. Wood was another source of energy; it was used for cooking, heating, and smelting iron. Again, the use of the sun's energy in the process of photosynthesis is responsible for the existence of wood. Also, wind was used to pump water for irrigation and to fill the sails of the great wooden ships. Wind is the result of temperature differences in the atmosphere caused by uneven heating by the sun. The giant windmills of Holland are vivid reminders of an early era when only such ingenuity and muscle were available to do our work.

In the past hundred years, our industrial society has

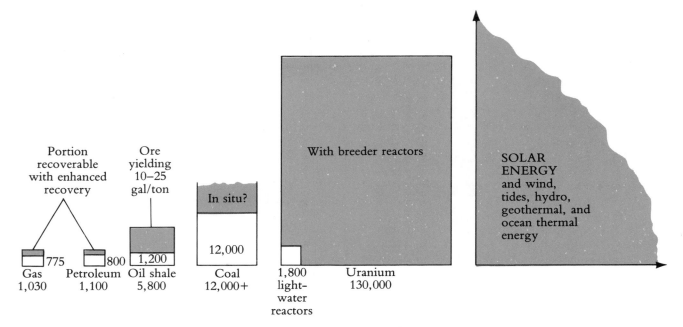

Figure A Available energy resources in the United States measured in quads. (1 quad = 1 quadrillion Btu = 1 million billion Btu.) The total U.S. energy consumption in 1974 was 73 quads. (Adapted from *A National Plan for Energy Research, Development and Demonstration: Creating Energy Choices for the Future*, Energy Research and Development Administration [now part of the Department of Energy], 1975.)

relied on the heat from fossil fuels, the heat from fission, and the energy from flowing water to power the enormous development that has taken place in our civilization. To only a minor extent has wind been utilized, heat withdrawn from the earth, or the sun's energy tapped directly. Yet these sources of energy, which stem from the earth's natural processes, are all around us.

In the last decade, we have increasingly been turning to these natural sources of energy because, in many respects, they provide energy without limit (Figure A). Furthermore, the pollution that comes from them is often minor compared with that from fossil fuels and fission. We have been turning to these sources for another fundamental reason, though. As fuel supplies become less secure and more expensive, these sources become more attractive and more economical. The rising prices of oil and gas have in a large measure been responsible for our renewed interest in water, wind, and sun. It seems strange to be saying "Thank you" to the oil cartel, but it has made the point of dwindling

energy resources more vividly than the arguments of any resource economist.

Thus, we are standing today on the edge of an exciting era: a frontier in much the same way that the western United States was a frontier (Figure B). People will be laying claim to economic territory as a solar industry is built. Challenges of exploration and discovery in energy generation and energy conservation lie before us. Once again, we have the opportunity of becoming self-sufficient in energy resources, this time by learning to use the natural energy sources that surround us. It is a period that will give new meaning to existence as we struggle to secure a stable, fruitful, and peaceful life in harmony with nature for generations to come.

In Chapter 22 we discuss how electrical energy is generated from falling water. We describe not only conventional hydroelectric generation but also the potential for small-scale hydropower installations and the development of power from the tides.

In Chapter 23 we discuss the generation of electric power from the wind. We also show how heat stored

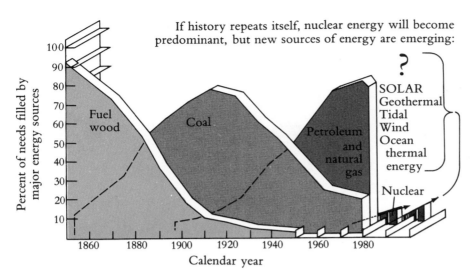

If history repeats itself, nuclear energy will become predominant, but new sources of energy are emerging:

Figure B U.S. energy consumption pattern through time. (Adapted from *A National Plan for Energy Research, Development and Demonstration: Creating Energy Choices for the Future*, Energy Research and Development Administration [now part of U.S. Department of Energy], 1975.)

in the interior of the earth can be used for both space heating and the generation of electric power. Environmental impacts are pointed out as well.

In Chapter 24 the many useful forms of solar energy are discussed. We include not only water heating and space heating but also solar electric generation. We also describe how solar energy can be converted to biomass fuels and how temperature differences in the layers of the ocean can be used to generate electric power.

Chapter 25 details methods of energy conservation. When energy was inexpensive, hot water from electricity was inexpensive, and large, heavy automobiles could still be operated at low cost. Homeowners

could afford not to have storm windows or adequate insulation. Electrical heating of homes was reasonable though expensive. Today, electrical heating of homes is an exorbitant option. People often choose lighter, more efficient autos to keep the cost of gasoline down. Public transportation is being installed or upgraded in many parts of the country and car pooling is being encouraged. Insulation of homes and other energy-saving improvements have been rewarded by tax incentives, as we reach out to save energy in many ways. The savings are turning out to be quite large. However, changing patterns of behavior takes time and requires a national commitment that does not change with temporary fluctuations in the price of oil.

CHAPTER TWENTY-TWO

Power from Falling Water

Hydroelectric Power

Conventional Hydroelectric Power Generation/Pumped Storage/Small-Scale and Low-Head Hydropower/The Redevelopment of Hydropower

Power from the Tides

Tidal Power Explained/Tidal Power in North America/ Biological and Physical Impacts of Tidal Power at Fundy

Hydroelectric Power

Water, an ancient source of power, today remains a good option for supplying electrical energy to our industrial civilization. The energy from falling water turning a waterwheel has been used directly to grind grain, cut lumber, and weave fabrics. But the gristmills and sawmills on our rivers began to fade when, in the 1880s, the generation of electric power began at waterfalls. In 1882 the first hydroelectric power plant in the United States was built on the Fox River in Appleton, Wisconsin. In 1895 the power of the mighty Niagara River was harnessed and used to produce electricity. This was the first large-scale effort for the production of hydropower in the United States. No one will say that hydroelectric power is without problems. (See Chapter 10 for a discussion of the environmental impacts of reservoirs.) Nonetheless, when we consider the pollution and the potential hazards of other forms of power generation, hydropower begins to look much less menacing.

Hydropower provides about 12% of the nation's electrical energy needs. A generating capacity of 81,000 megawatts was available in 1985 at some 1,200 hydropower plants across the country. By the mid-1990s, hydropower capacity is expected to reach about 90,000 megawatts. This renewed growth calls to mind the 1930s when hydropower furnished 30% of the nation's generating capacity. However, decreasing costs at fossil-fuel power stations gradually made hydropower generation at smaller sites (up to 25 megawatts) uneconomical. Many of the smaller hydropower plants in the eastern United States closed, and the newer, larger hydropower plants were constructed in the west—at Hoover Dam, at Grand Coulee Dam, and at other sites.

Conventional Hydroelectric Power Generation

The principle of hydropower generation is simple. The kinetic energy of falling water is used to turn a turbine, which is linked to an electrical generator.* Early hydroelectric plants were of the *run-of-river* type, in which water flowing in the river was not dammed but merely directed through a turbine. Large changes in river elevation were needed for these plants; the Niagara Falls development was of this type. Most modern hydropower installations, however, use dams to increase the volume of water that can be steadily discharged through the turbine (Figure 22.1). Dams do more than provide a reservoir of water from which to draw; they increase the height of the water surface. The increased pressure provided by this higher water surface gives the falling water a higher velocity and hence more kinetic energy. The amount of energy derived from this more powerful flow of water is, as a consequence, much larger.

In practice, water is drawn from the reservoir downward through a long smooth channel, called a

*Kinetic energy is the energy associated with the movement of matter. The energy is proportional to both the mass being moved and the velocity squared.

516

Figure 22.1 Shasta Dam and Reservoir on the Sacramento River, north of Redding, California. The dam is a curved concrete gravity structure, 602 feet (183 m) high, and the reservoir stores over 4.5 million acre-feet (556,000 hectares). Shasta Power Plant has five main generating units with a total capacity of 442,310 kilowatts. (Bureau of Reclamation)

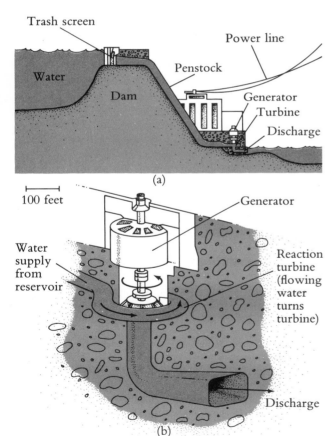

Figure 22.2 (a) Components of a hydropower system. (b) Turbine-generator unit. (Adapted from D. Kash et al., Energy Alternatives, *Report to the President's Council on Environmental Quality*, 1975.)

penstock, and is directed across turbine blades that rotate horizontally (Figure 22.2). The turbine shaft is directed upward into the generator unit. Many turbine/generator units are needed at a typical installation. Efficiency factors on the order of 60%–70% are common; that is, 60%–70% of the energy in the falling water is converted to electricity.

Hydropower units are costly to install and require maintenance, but they use a fuel that is free and not subject to inflation. The fuel is supplied by the sun, which evaporates water into the atmosphere from oceans, lakes, and rivers. The water vapor condenses as rain, which falls in the highlands and flows down to the sea. Hydropower units interrupt that flow to the sea and capture the energy in the moving water, energy that would otherwise be used up in carrying sediments to the sea.

Hydropower is not free of environmental impacts, however. Dams and reservoirs modify not only the land that is flooded but also the quality of the water that is stored and released. In addition, the stream channel downstream from the reservoir is affected. The loss of land habitat is an obvious effect. The decrease in the quality of the water is surprising, though. Water released from a reservoir, depending on the season, can be very low in dissolved oxygen and hence an unfavorable environment for fish and other normal aquatic species. Finally, the water released from a reservoir erodes and scours the stream channel to a greater extent than the undammed stream would have. (All these effects are discussed more completely in Chapter 10.)

It is not actually necessary to dam a free-flowing stream in order to create the benefits of an elevated water surface or a steady flow. Partial diversions of upstream water can be used to create an artificial lake off to the side of the river. Such a lake has the benefits of both elevation and steady availability. As you read about pumped storage in the next section, you will see that off-river storage lakes are actually quite common, but they have not been thought of as replacements for reservoirs.

In relation to other continental areas, the hydropower resources of North America are extensive (see Table 22.1). Although most sites in the United States with a large hydropower potential have already been put to use, much of the capacity at low-head sites remains to be exploited—as we shall discuss in a moment.

In Canada, on the other hand, huge hydropower resources are just now being tapped. Hydro-Quebec is building hydropower projects on a massive scale, including an installation on James Bay that will be the largest hydropower plant in North America. Power from these projects has already been sold to New York State and is likely to be available in other parts of the eastern United States as well. High-voltage transmission lines will carry the power down from Quebec to New York and elsewhere.

Table 22.1 Regional Distribution of Hydropower Resources

Location	Potential (1,000 MW)	Percent of world total	Developed (1,000 MW)
North America	313	11	59
South America	577	20	5
Western Europe	158	6	47
Africa	780	27	2
Middle East	21	1	—
Southeast Asia	455	16	2
Far East	42	1	19
Australia	45	2	2
USSR, China, etc.	466	16	16
	2857	100	152

Source: M. King Hubbert, *Resources and Man*, copyright 1969. Used with permission of the National Academy of Sciences, Washington, D.C.

Pumped Storage

The conventional method of hydropower generation on rivers is one way that falling water can be used. However, water can also provide supplemental power during times of peak electrical demand. Ordinary reservoirs can be used to save and then release water through turbines when the need arises, or water can be stored off-river for this purpose. **Pumped storage** is the name given to an off-river reservoir that is drawn from at the time of peak loads.

Pumped-storage reservoirs are created by pumping water up to high elevations during times when power demands are low, when an electric station, perhaps a coal-fired plant, is not being used to full capacity. (See Figure 22.3.) There the water remains until peak power loads occur. Then, the water is allowed to fall through the penstocks and turn the turbine pump to generate more electricity. The same pump used to force water uphill is now turned in the opposite direction by the falling water to generate electric power.

Essentially, a pumped-storage system uses spare generating capacity during periods of low demand. Electrical energy is converted into potential energy during these periods of low electrical demands by pumping water to a higher elevation. During peak demand times, the potential energy is converted back to electricity to supplement power from the base load plant, which is now taxed to its capacity.

There are two principal disadvantages of pumped storage. The first is that the high pool used for storage has daily changes in elevation, going from nearly full to nearly empty (Figure 22.4). The banks of the reservoir are exposed by this daily fluctuation and are quite unattractive, resembling dark desert mounds with an undulating shape. In addition, the pumped-storage system is not fully efficient. Of the quantity of electric energy used to raise the water to the storage pool, only two-thirds is recovered. If coal is used to generate the electricity operating the pumps, the efficiency of converting the combustion heat to electricity is decreased by the two-thirds factor.

Pumped storage can be costly to build. The unit in Ludington, Michigan, which pumps water from Lake Michigan 358 feet (110 m) up to a 1.6-square-mile (4.5 km²) storage pool, cost $340 million to construct. An alternative currently being investigated is underground water storage in a tunnel or other subsurface chamber. Cost and safety of such underground structures are issues that need to be resolved, but the under-

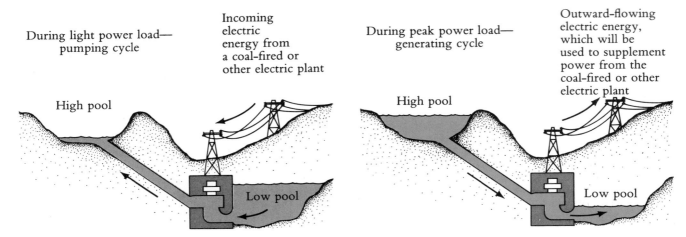

During light power load—
pumping cycle

Incoming
electric
energy from
a coal-fired or
other electric plant

High pool

Low pool

During peak power load—
generating cycle

Outward-flowing
electric energy,
which will be
used to supplement
power from the
coal-fired or other
electric plant

High pool

Low pool

Figure 22.3 Pumped-storage operation. (Adapted from D. Kash et al., Energy Alternatives, *Report to the President's Council on Environmental Quality,* 1975.)

Figure 22.4 The Blenheim-Gilboa Pumped-Storage Project. The high pool is atop Brown Mountain, and water falls 1,000 feet (300 m) when electricity is produced from the unit. The turbine-generators can produce power at a rate of 1 million kilowatts (1,000 megawatts). (Photo courtesy of Power Authority of the State of New York)

ground reservoir would obviously not have the usual environmental impacts of a pumped-storage project.

Small-Scale and Low-Head Hydropower

As a U.S. Corps of Engineers 1977 study noted, it is surprising that hundreds of hydropower sites, with dams already in existence, are no longer in use for electric power generation. Many of these sites are small in capacity, 5 megawatts or less, but their potential is real. Abandoned in favor of central electric power stations during an era of cheap fuel, the sites could nearly double our capacity for hydroelectric power generation, if they were put into service once again.

In the mid-1970s, about 57,000 megawatts of hydroelectric capacity was available in the United States. By 1980 this figure had expanded to 73,000 megawatts. The Corps of Engineers estimated that by restoring the

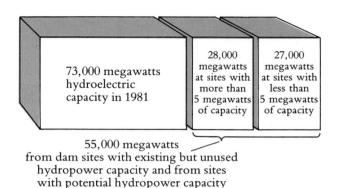

Figure 22.5 Used and unused hydropower capacity in the United States (not including bulb turbine run-of-river possibilities). By 1985, hydropower capacity had been expanded to 81,000 megawatts.

abandoned dams and adding generating equipment to those dam sites never equipped for hydropower, 55,000 megawatts of capacity could be added. About half this quantity would be drawn from the small dams (5 megawatts or less). Since a new thermal power plant is usually rated at about 1,000 megawatts, adding 55,000 megawatts is equivalent to building 55 major plants. Moreover, there would be little more disruption than has already occurred because the sites have been previously developed (Figure 22.5).

We mentioned earlier that the first hydropower developments were run-of-river plants; the typical sites chosen for these plants were waterfalls with sufficiently large changes in elevation to turn the turbines. When all such sites were exhausted, dams were built to create the needed changes in elevation. The types of turbines—the Francis, Kaplan, and Peyton wheels—all required large "heads" of water, 100–2,000 feet (30–610 m), depending on the type.

In the early 1950s, a French firm, Neyrpic, invented a new kind of turbine—the **bulb turbine**—that is so versatile it can generate power just from rapidly flowing water. Not even a dam is required. This development has dramatically changed the potential contribution of hydropower to electric power needs. The bulb turbine has made tidal power, with its low heads of 30–50 feet (9–15 m), technically feasible as well. The tidal power plant on the Rance River Estuary in France uses the Neyrpic bulb turbine. The bulb turbine can be installed at sites previously too small to use, and it can also be used to convert existing dams to hydropower installations. On existing dams, it could be installed

just downstream of the outlet works in such a way that reservoir releases could be channeled through the turbine.

Another development in small-scale hydropower has also taken place in France. The Leroy-Somer Company, a manufacturer of motors, has recently produced a miniature hydropower plant they call Hydrolec. Marketed for $7,800 in 1978, the smallest plant can produce power at a capacity of 4 kilowatts. A change in a stream's water level of as little as 3 feet (0.9 m) is sufficient to generate power, as long as the flow rate exceeds 70 gallons (266 L) per second. Operating full time, the smallest Hydrolec could produce power worth over $2,000 in a single year (at 5 cents per kilowatt-hour).

The American giant in turbine manufacture, Allis Chalmers, is now producing a standardized low-head turbine with a maximum power output of 6 megawatts. This low-cost unit was used in the redevelopment of power capacity at the Barker Mill Dam in Auburn, Maine, a project that began producing power again in 1980 after a 33-year interruption in power production at the site.

The Redevelopment of Hydropower

The number of applications to the Federal Energy Regulatory Commission for licenses to develop hydropower sites has increased steadily, from 28 in 1978 to more than 200 in 1982, and has remained at more than 200 each year since. Many of these applications have been submitted from municipalities and electrical cooperatives, who appear to be playing a major role in the redevelopment of hydropower.

The tax-free bonding power of municipalities may be one reason for the predominance of applications from the public sector. Another reason is that municipalities often own sites that can be developed for hydropower: their water-supply reservoirs. In 1981 in Litchfield, Connecticut, Northeast Utilities reopened a 320-kilowatt generator, originally built in 1905; their cost was only $30,000. Hydroelectric power generators are also being installed at the Boonton Reservoir, the water supply reservoir for Jersey City, New Jersey.

The New England River Basin Commission identified some 1,511 dam sites in the state of Connecticut; of these, 200 were believed to hold potential for hydropower generation. In New York State, the Commission counted 5,300 dam sites, 1,672 of which were suitable for electric power generation. The Energy

Master Plan of New York State calls for development of 1,050 megawatts of power capacity at these sites by 1994, the equivalent rating of a new coal-fired or nuclear power generating station. These dams are in the right place at the right time to relieve the high cost of power generation in New York State, where oil is often burned as a power-plant fuel to relieve extensive air pollution from sulfur compounds.

Even irrigation canals are now being trapped for electric power. In California, hydropower has been generated since 1982 by water flowing in the Richvale Irrigation Canal. Using a Schneider generator, the power plant generates a steady 75 kilowatts (0.075 megawatt) from water falling only 9.5 feet (2.8 m). A similar power plant is opening on the Turlock Irrigation Canal in California, as well. Also in California, the Metropolitan Water District, which supplies water to Los Angeles and San Diego, has been installing generating turbines in their water-supply system. Their water supply from Northern California and from the Colorado River in Arizona has to be pumped over mountains to reach the city. It arrives with heads of up to 1,500 feet (450 m), ripe for conversion to electricity. The Metropolitan Water District is investing about $100 million in 15 projects, and the power is being sold to Southern California Edison.

The redevelopment of hydropower is not without problems, however. Breaking through tangles of red tape at both the state and the federal levels is one such problem. Cost is another. Each new kilowatt of installed hydropower capacity at existing sites can cost between $300 and $2,000, depending on how much refurbishing and site development is necessary. With 1985 construction costs of nuclear and coal-fired power plants projected at $2,000 and $1,500 per kilowatt, respectively, the economics of installing hydropower do not always appear favorable—unless the fuel cost enters the equation. Receiving free fuel for most of the life of the structure makes hydropower quite competitive when the longer view is taken.

Finding a market for this power is a potential problem as well. In the past, utilities discouraged private development. They did this either by refusing to purchase surplus capacity or by paying discouragingly low prices for the electrical output. Federal law (1978) now requires utilities to purchase such power at fair prices. Utilities, however, have been levying a stand-by charge on hydropower producers. The stand-by charge is a fee paid by a hydropower developer who may occasionally

have to purchase power from the utility rather than sell his surplus power. Even with such stand-by charges, several manufacturers in New England have redeveloped existing hydropower installations for their own use.

One factor favoring the redevelopment of small-scale hydropower plants is the relatively short time needed to bring new capacity on line. It takes from 1 to perhaps 5 years to produce electricity once activity is begun. Compare this to the 10–12 years of lead time needed for nuclear plants.

Power from the Tides

Franklin Roosevelt, President of the United States from 1932 to 1945, made his summer home on Campobello Island, which lies along the western edge of the Bay of Fundy. From his home, he could see the rise and fall of the largest tides in the world, and he recognized and spoke of the potential of this region for power generation. A tidal power project was actually begun on Cobscook Bay in 1935 at his urging, but Congress suspended funding of the project in 1936. Though nearly half a century has passed since that era, Roosevelt's vision has not been fulfilled. Yet the promise in the tremendous Fundy tides remains.

Tidal Power Explained

Tides are the result of the gravitational pull of the moon and, to a lesser extent, of the sun on the great oceans. As the earth rotates, a portion of the ocean waters is lifted and held in position for a time by this gravitational pull. When the swell of water in the grip of the moon reaches the land, as it must because of the rotation of the earth, it appears as a high tide. Further rotation of the earth releases the grip of the moon on that portion of the ocean and the tide falls away. Tides rise and fall twice each day, although the times shift with the season and the moon's position (Figure 22.6).

The average height of the tidal swell is only a few feet (half a meter), *except* when the ocean tides move within relatively narrow bodies of water. In this case, an oscillation wave is set up that may be 10–20 times the normal height of the tidal swell. Bay of Fundy tides, the largest in the world, run up to 53 feet (16 m). Between England and the European coast (France, Belgium, and Holland), such large tides are also created. The highest tides of the year occur when the moon and

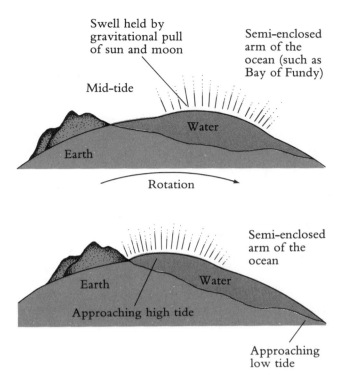

Figure 22.6 Tides. The swell of water that we call a tide is the result of the gravitational forces of the moon and, to a lesser extent, the sun. When the earth rotates, the swell seems to travel toward the shore. In reality, the swell is simply remaining in the same position relative to the moon and sun; the shore is approaching the swell.

Figure 22.7 La Rance Power Station lies across the estuary of the Rance River, which empties into the Atlantic Ocean between St. Malo and Dinard on the coast of Brittany. (Courtesy of Electricité de France)

sun are most nearly in line so that their combined gravitational pull increases the volume of water dragged across the sea.

The possibility of using the energy in the tides has been made a reality in France. In 1968 Electricité de France finished a tidal power station on the Rance River, which flows into the Atlantic Ocean. To understand how tidal power works, we can discuss the operation of the station on the Rance (Figure 22.7).

The Rance River experiences tides nearly as high as those in the Bay of Fundy, up to 44 feet (13.5 m). Although nearby coastal tides are lower, the Rance is a relatively narrow river, and when the tide moves up and down within its banks, the water moves at a very high speed. A half-mile-long dam has been built across the Rance and is used to store the waters of the arriving high tide. When tidal waters are receding, the stored water is released to the ocean through bulb turbines

below the dam, and electricity is generated. To many people, this is the way tidal power is expected to operate. At La Rance Power Station, there is more. Energy can be generated on both the falling tide and the rising tide.

The rising tidal waters are captured inland by opening a set of sluice gates, which allow the arriving tide to flow upstream in the direction of the river's source. The gates are closed when the tide reaches its highest stage; then, as the tide recedes, the water that was barricaded inside is allowed to flow seaward through turbines. At low tide, most of the water has been released. (See Figure 22.8.) As the tide builds again, it does so against the closed gates, so the water levels on the seaward side exceed those on the land side

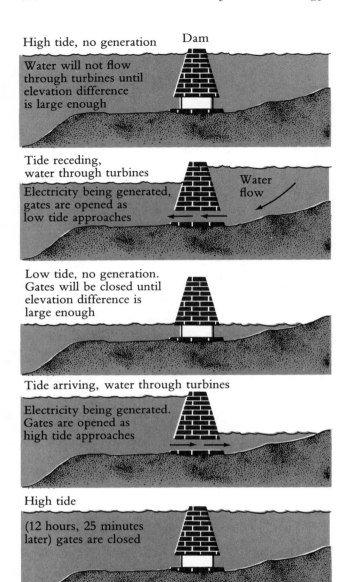

High tide, no generation

Dam

Water will not flow through turbines until elevation difference is large enough

Tide receding, water through turbines

Electricity being generated, gates are opened as low tide approaches

Water flow

Low tide, no generation. Gates will be closed until elevation difference is large enough

Tide arriving, water through turbines

Electricity being generated. Gates are opened as high tide approaches

High tide

(12 hours, 25 minutes later) gates are closed

Figure 22.8 Schematic diagram of the operation of a tidal power station.

of the dam. When a sufficient head is built up, the water is allowed to flow upstream through the turbines, again generating electricity. Thus, electricity is generated both on the receding tide and on the arriving tide. The generation of electrical energy from these low heads of water is made possible by the use of the bulb turbine, the device we described in our discussion of small–scale hydropower. (See Figure 22.9.)

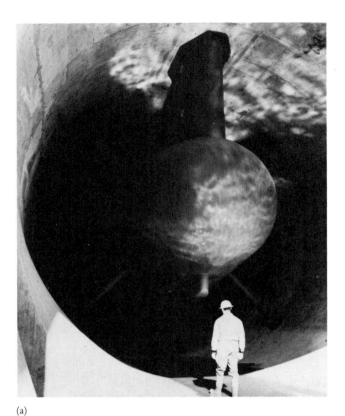

(a)

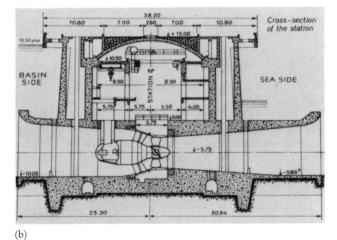

(b)

Figure 22.9 (a) The bulb turbine at La Rance Power Station, as viewed from the basin side. Note the size of the workman for scale. (b) A cutaway view of the power station, showing the bulb turbine. There are 24 bulb sets at the La Rance Power Station. (Courtesy of Electricité de France)

La Rance Power Station has 24 separate bulb tur-bine units, capable of generating a total of 320 mega-watts of electricity. Together, the units capture 25% of the energy possible to capture. Because the tide is mov-ing up a river rather than along a broad coastline, the turbine and dam rest about three miles inland and are thus protected from the ravages of the open sea. (Along the Bay of Fundy, the Passamaquoddy Bay on the east-ernmost coast of Maine could provide similar shelter for the dam and turbine. The potential for annual power production on the Passamaquoddy is more than five times that of the Rance project.)

A remarkable technique has been put to use at La Rance. During the final portion of the high tide, the difference in elevation between the water in the reser-voir and that in the ocean may be only a few feet. During this time, electricity from some other source can be used to pump ocean water (using the turbines) up into the tidal basin. The water is pumped up only a few feet, so not much energy is required. When the tide has receded, that extra water falls a distance of 20–30 feet, generating far more electric power than was used. The same idea works at low tide, except that the water is pumped out of the tidal basin into the ocean. The water level in the basin is thus reduced below sea level, and the incoming tide falls a greater distance.

Because of the enormous cost of these structures, governments have been reluctant to invest in tidal power. The Rance project cost two and one-half times the estimated cost of a river hydropower station of the same average power output, primarily because of the added cost of cofferdams both ahead of and behind the project. Yet, once the initial investment is made, the production of power requires no fuel. Only mainte-nance of the structure is required; thus, the costs of power are held down. The Rance station is proof that the concept of tidal power can work on a large scale, and it is already influencing the public's views on tidal power.

A number of sites where tides could be used to generate electricity are located around the world. Some of the most attractive are listed in Table 22.2.

Other drawbacks to tidal power exist besides cost. If the tidal project is a great distance from the nearest large load center, long and costly transmission lines will need to be built to provide the tide-generated elec-tricity an entry into the transmission network. On the other hand, such long-distance transmission is becom-ing more common as the new and more efficient 765-kilovolt lines are installed. New York City may soon be getting power from Hydro-Quebec via a 765-kilovolt line that runs several hundred miles. Tidal power reaching Boston from Passamaquoddy Bay or Cobscook Bay in Maine no longer seems so unthink-able. Furthermore, the distance of load centers may not be so important if industries that are heavy users of electricity find it desirable to locate near tidal power stations. Examples of such industries are aluminum

Table 22.2 Selected Tidal Resources

Location	Average difference between high and low tide		Average rate at which power could be produced (megawatts)
	Feet	Meters	
Severn, England	32.7	9.8	1,680
Mont St. Michel, France	28.0	8.4	9,700
White Sea, USSR	19.0	5.7	14,400
Mozen Estuary, USSR	22.0	6.6	1,370
Passamaquoddy, U.S., Canada[a]	18.3	5.5	1,800
Cobscook, U.S.[a]	18.3	5.5	722
Annapolis, Canada[a]	21.3	6.4	765
Minas-Cobequist, Canada[a]	35.7	10.7	19,900
Cumberland, Canada[a]	33.7	10.1	1,680
Petitcodiac, Canada[a]	35.7	10.7	794

Source: J. McMullan, R. Morgan, and R. Murray, *Energy Resources* (New York: Halstead Press, John Wiley and Sons, 1977).
a. All of the U.S. and Canadian sites listed are along the Bay of Fundy.

companies, steel companies using the electric arc furnace, and chemical companies producing chlorine and caustic soda via electrolysis.

Distance from the load center aside, there is another drawback to tidal power: The production of electricity is not steady. A moment's reflection on the nature of the tides should explain this. In the ordinary production of tidal power, electricity is generated only when the tide is receding—that is, when the height of water stored in the reservoir exceeds that of the receding water by a sufficient amount. As the tide drops and low tide approaches, the rate of power generation falls to zero, since there is no height difference. If the tidal project has reversible turbines, as La Rance does, power can be generated on the incoming tide as well, but still, power cannot be generated until the height of the rising tide exceeds the height behind the dam by a sufficient amount. As high tide is reached, the rate of power production again falls toward zero. Thus the curve of power production rises and falls twice each day, once for each of the two tidal cycles.

This cyclic production of electricity is unlikely to match the daily cycles of demand at load centers. Peak demands and peak production will occasionally occur together because the two tides shift in time as the seasons progress, but more often, the peaks of demand and production will occur at different times of the day. Somehow, the tidal power must be fed into the transmission network at the proper rate. This means that the power output of other central power stations must usually be phased down as the rate of tidal production reaches its maximum and phased up as the tidal production rate falls. A computer performs this task for Electricité de France at La Rance Power Station. The electricity from the tidal power station is, in effect, replacing on a fairly regular basis the electricity generated by other means. If the electricity replaced is from a coal-fired power plant, it is coal that is being saved.

Tidal Power in North America

Although Roosevelt and others envisioned the development of power on the Bay of Fundy, little effort was actually invested there—until this decade. The International Joint Commission of the United States and Canada studied a proposal for a 1,800-megawatt tidal power plant on the Passamaquoddy Bay in the late 1950s, but the investment was considered too great and the power uneconomical. In the early 1980s, a

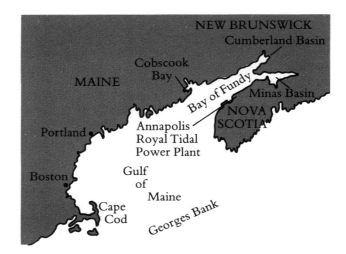

Figure 22.10 Tidal power projects and proposals on the Bay of Fundy.

12-megawatt tidal power plant on Half Moon Cove in Cobscook Bay was proposed by the Passamaquoddy Indian Tribal Council and is still under consideration. The first of these projects was to be a joint Canadian-U.S. effort, the second a solely U.S. undertaking. (See Figure 22.10.)

While the United States has studied, however, the Canadians have forged ahead with a 20-megawatt, simple turbine tidal power plant. Located on the Annapolis River near the town of Annapolis Royal on the Nova Scotia side of the Bay of Fundy, the plant was completed in 1984 at a cost of $55 million. The plant was built to demonstrate the feasibility of a new concept in tidal power production—the Straflo (for straight-flow) turbine. The Straflo turbine generates power only on the falling tide, in contrast to the bulb turbines at La Rance that generate on both the rising and falling tides. However, the Straflo turbine is both smaller and cheaper (by about 15%) than the bulb turbine. The Annapolis plant, built by the Nova Scotia Tidal Power Corporation, is the first operating tidal power plant in the Western Hemisphere and was built as a forerunner of a massive Canadian development 200 miles to the north and also on the Bay of Fundy.

The project to the north will also be built by the Nova Scotia Tidal Power Corporation and is most likely to be sited in the Minas Basin—where the world's highest tides sweep in and out. The Minas Basin power complex is forecast to cost $23 billion;

between 106 and 140 turbines would capture the energy in tides that reach up to 53 feet and average 39 feet. Depending on the number of turbines, between 3,800 and 5,300 megawatts would be generated by the Minas Basin power plant with about 90% of the power projected to be sold to the United States. In 1985, the Nova Scotia Tidal Power Corporation was lining up capital for the project with the hope that the undertaking could begin in the next few years and be completed by 1995— but $23 billion is not easy to raise. The construction cost per kilowatt of generating capacity is estimated at $1,500—not far different from the cost of conventional coal-fired capacity or nuclear capacity—and the fuel is free, unlike coal and enriched uranium. Another 1,000-megawatt tidal power plant on the nearby Cumberland Basin is also a possibility.

Cheap power? Yes. Environmentally sound? Not necessarily. Unfortunately, there are unknown and troublesome features of tidal power, some peculiar to the Bay of Fundy.

Biological and Physical Impacts of Tidal Power at Fundy

Physical impacts. As we look longingly at the tides and the awesome energy they carry, we must reflect as well on the environmental impacts of tidal reservoirs. We focus here on the physical changes that occur seaward of the tidal power plant. If the length of the Bay of Fundy is shortened by the tidal dam on the Minas Basin or on the Cumberland Basin, the natural cycle of Fundy tides will approach the oceanic tidal cycle more closely and be enhanced by it. The "push" of the ocean tide applied to the Fundy tide will increase the range of the tide from Fundy through the Gulf of Maine and down to Cape Cod, although the magnitude of the change will decrease as the distance from Fundy increases.

Although the tidal range might increase by only 1 foot (30 cm), even so small a change can have wide-ranging effects. The incoming tide could rise half a foot (15 cm) higher, which might cause salt-water intrusion into coastal wells and threaten structures built close to the high-tide mark. Beach erosion could be hastened, and low lying areas, including roads, could be flooded when storms and the increased tides combine forces. Shorelines will be physically unusable because of the higher tide. Estimates of shoreline lost to tidal flooding

range from 4,200 to 10,000 acres (17 km² to 40 km²). Of course, local shoreline loss depends on the slope and character of the shore. The receding tide, which could fall half a foot lower, could reduce the usefulness of docks in providing access to boats and water. An increased tidal range may swing saltier water into estuaries, changing the balance of aquatic species.

With the increased tidal range will come increased tidal currents, as much as 5% to 10% faster, which can scour up and move sandbars and fill present navigation channels. Navigation charts may need to be redrawn, but ships may get stuck in the short run as passages are altered by moving sand. These faster currents will make oil spills more difficult to contain, but on the other hand will disperse the oil more swiftly.

The tidal power plant on the Rance and the new plant at Annapolis Royal are built off the main basin; they do not shorten the basins that determine the tides. Hence, these tidal power developments do not have the impact on the tidal range that the Minas Basin and Cumberland Basin developments are expected to have.

Biological impacts. A large-scale tidal power plant on the Bay of Fundy is likely to have both local and distant biological impacts. In the basin behind the tidal dam, the operation of the power plant is likely to have an effect on the important biological area that extends along the shore of the ocean. This area, known as the intertidal zone, stretches from the point of the highest tide (or spray from the tides) to the lowest point exposed when the tide recedes. (Both of these points vary with the seasons, within limits.)

Communities in this region consist, first, of those organisms that spend all or most of their time there. On sandy beaches are burrowing creatures such as crabs, shrimp, worms, and clams. Rocky shores support organisms attached to the rocks, such as mussels, oysters, barnacles, and the larger algae. In the waters of the intertidal zone, another set of organisms occurs: the phytoplankton. These microscopic floating green plants include the diatoms, the dinoflagellates, and the microflagellates, organisms that are swept in and out with the tides. One of the species of dinoflagellates is responsible for the infamous *red tides*, which kill fish and sometimes make the flesh of shellfish poisonous to humans.

The phytoplankton, along with the attached larger red, brown, and green algae, or "seaweeds," are the "producers" of the intertidal region. A variety of "consumers" also live in the intertidal zone. Some consumers come in with the tide, among them zooplankton, which spend their entire lives as plankton, and the larval stages of crabs, jellyfish, sea urchins, snails, starfish, and other creatures. The intertidal zone is thus an important part of the sea's "nursery grounds." Still other consumers are present even at low tide, including the mature stages of crabs, clams, barnacles, snails, and starfish.

Tidal power has the potential to change the relative balance among species that make up the communities of the intertidal zone. We are not at all certain how the larval stages of marine species will survive passage through the turbine. Moreover, it is conceivable that nuisance species, such as those responsible for red tide, could be favored while the spawning of desirable species such as crabs or oysters could be harmed. In addition, we are not sure whether erosion could be accelerated or the deposit of sediments hastened by such projects.

Although local species may be altered by the presence of a tidal power plant on the Minas Basin, migrating species may be severely damaged. Shad, an important commercial species, migrate each summer to the top of the Bay of Fundy, feeding in the many small embayments there. Passage through the turbines of a tidal power plant is unlikely to do them any good. At the Annapolis Royal Tidal power plant, 15% of the shad are killed with each passage through the turbines. Screens may be used to block the entrance, but the usefulness of fish ladders to provide them a way around is still an open question. Migrating birds that feed on the tidal flats, such as sandpipers and plovers, are likely to find a reduced food supply in the tidal basin behind the power plant due to the mortality caused by the turbine. The birds stop for sustenance at Fundy each fall on the way from the arctic to their South American playgrounds. All of these effects take place locally, though their effects may be felt more widely.

The distant biological impact of tidal power development on Fundy is due to the increased strength of the tidal currents that accompany the increase in tidal range. Stronger tidal currents will upset and tend to mix the temperature layers in Fundy and the Gulf of Maine. The lower-lying cold layers are the richest in nutrients because nutrients gradually settle into these layers. Thus, more nutrients will be brought to the surface layers along with more cold water. Summer air and water temperatures may decrease by an average of 1°C. An increase in both fog and onshore winds is a likely consequence of these decreased temperatures, and the biological productivity of Fundy and the Gulf of Maine is likely to be increased. Algae and zooplankton will likely grow in numbers, as will the organisms that feed on these species, but we are not wise enough to know what particular species might be favored and what species diminished. Fisheries may be enhanced, but we cannot say for sure. The toxic dinoflagellates that cause red tide could also be favored by the combination of physical changes. The taking of clams, mussels, and oysters might need to be prohibited more frequently. The biological unknowns that could accompany tidal power development are large indeed.

Summary

In the generation of hydroelectric power, the kinetic energy of falling water is used to turn a turbine that is linked to an electric generator. The process is about 60%–70% efficient; that is, 60%–70% of the energy in the falling water is converted to electricity.

Early hydropower plants used the river in its natural form, but most hydropower installations now use a dam to increase the water height and to even out the supply of water. Hydropower installations are expensive to build, but they are relatively cheap to operate because the fuel is free. The energy used can be traced to the sun, which evaporates water that later falls as rain.

Environmental impacts of hydropower include loss of land, low oxygen in the water released from the dam, and increased downstream erosion. Pumped storage is a special use of water that provides supplemental power during times of peak demand. Water is pumped to an elevated storage area when demand for electricity is low, and it is released to generate power when demand is high. Only about two-thirds of the energy used to pump the water upward is regained later when power is generated.

The United States currently has some 73,000 megawatts of hydroelectric capacity around the country. This number could easily be expanded by 55,000 megawatts by restarting abandoned small hydropower stations and by utilizing existing dams not currently used for hydropower, such as those used for municipal water supply reservoirs.

Newly developed turbines allow power generation with much smaller water height differences, or heads, so power could be generated from tides, streams with only small elevation differences, irrigation canals, and streams below existing dams.

Tidal power is generated from the flow of water caused by the rotation of the earth and the gravitational pull of both the moon and the sun. La Rance Power Station in France uses bulb turbines and has demonstrated that tidal power can be produced on a commercial scale. Construction costs for tidal power are high, but operational costs tend to be lower because the fuel is free. Transmission of tidal electric power to the point of use must be considered, as must the fact that the amount of power generated oscillates with the tides.

A possible environmental impact of tidal power is an increase in the tidal range on the ocean side of the gates. This might cause flooding of land and structures either at high tides or during storms, and it might cause salt-water intrusion into estuaries and groundwater. Aquatic food chains and intertidal communities may be affected by changing water levels and stronger currents, both behind and in front of the dam, as well as by passage of the aquatic organisms through the turbines.

Questions

1. What do we mean by *hydropower*? Where does the energy actually come from?
2. What are the environmental advantages and disadvantages of hydropower schemes? Briefly compare these to the environmental disadvantages of power from the burning of coal.
3. Explain how tides could be used to generate electric power. Where does the energy come from?
4. Discuss the possible economic and environmental advantages and disadvantages of tidal power. Compare these to the disadvantages of generating power by burning oil.
5. Is there a net gain of electric energy when a pumped-storage system is used to generate electricity? Explain. What is the advantage of a pumped-storage generating system?

Further Reading

Hydroelectric Power

Erskine, G. "A Future for Hydropower," *Environment*, **20** (2), 33.
An engineer converts his expert knowledge of hydroelectric generation and the bulb turbine into public property. A rare article.

Kohler, J. "Home-Sized Hydro Power," *Solar Age* (September 1983), p. 34.
How one person built a private power plant to supply his home—with references on how to get started.

Loeb, W. "How Small Hydro Is Growing Big," *Technology Review* (August-September 1983), p. 51.
Readable account of the development of small hydropower stations, principally in New England.

Palmer, T. "What Price Free Energy," *Sierra* (July-August 1983), p. 40.
The down side of hydropower development is described, in which valuable areas and free-flowing streams are lost.

"Hydropower," *Civil Engineering/ASCE* (July 1984), p. 58.
The up side of hydropower development in which clever engineering provides hydropower without a dam.

"Hydro Recovery," *Civil Engineering/ASCE* (June 1983), p. 56.
Describes how the Metropolitan Water District of Southern California is generating electric power from the water flowing from its reservoirs to the points of use.

Tidal Power

Bernstein, Lev. "Russian Tidal Power Station," *Civil Engineering Magazine* (April 1974), p. 46.
A well-written article by the engineer in charge of a tidal power station built in Russia.

Britton, P. "Tapping the Tides for Power," *Popular Science* (January 1985), p. 56.
A thorough physical description of the tidal power plant at Annapolis Royal, Nova Scotia.

Larsen, P. "Potential Environmental Consequences of Tidal Power Development Seaward of Tidal Barrages," *Oceans* (September 1981).
An overview of biological and physical impacts of the Bay of Fundy tidal power plan.

Ryan, Paul. "Harnessing Power from the Tides," *Oceanus*, **22** (4) (Winter 1979–1980), 64.
A review of current activity in tidal power and a description of a new concept in tidal power generation, in which a huge reinforced plastic sheet replaces the conventional dam.

CHAPTER TWENTY-THREE

Power from the Wind and Power from the Heat in the Earth

Power from the Wind

A Brief History / Windmill Design / Wind Resources of the United States / Environmental Problems and Costs of Wind Power / The Development of Wind Power

Geothermal Energy

Origin of Geothermal Heat / Two Important Uses for Geothermal Heat / Resources of Geothermal Heat and Projections of Use / Cautions and Environmental Impacts

CONTROVERSY:

23.1 *Geothermal Energy Development—Should It Take Precedence over the Risk to a National Park?*

Power from the Wind

A Brief History

Wind has been in the service of humankind since primitive people first raised a sail above a fragile log canoe. The prevailing westerlies were the winds that powered the voyages of discovery to the New World, and wind carried the Spanish Armada to victory after victory. The trade winds caught the sails of the great clipper ships and opened India and China to commerce with the West.

Long before the wind was harnessed by the ingenious windmills of the durable Dutch people, the ancient Persians had captured the wind to grind their grain. The Persian windmill turned on a vertical shaft, a design reinvented in the modern age. The windmills in Holland were used not only to grind grain but also to pump water out of the low-lying lands (polders) so that the fertile delta of the Rhine River could be farmed.

The Dutch windmills, in contrast to the Persian windmills, had the familiar horizontal axis (Figure 23.1). The blades of these wooden windmills reached 80 feet (24 m) in diameter, and the windmill could operate in winds of 25 miles per hour (42 km/hr). Operators in the base of the mill oriented the blades to the direction and force of the wind. Many of the Dutch windmills, now more than 500 years old, are still in operating order, and courses are offered in Holland on windmill operation!

In the 1850s a new type of windmill was invented in the United States. This American innovation, the multivane windmill, became common in rural America in the following years and was first used to raise water from wells; it pumped water for the steam locomotives that began crossing the American continent in the 1870s. The multivaned windmill is still in production today, but steel blades are now used in place of the handmade wooden blades of early models. With blades up to 30 feet (9 m) in diameter, multivaned fans could produce up to 4 horsepower (3 kilowatts) in a wind of 15 miles per hour (25 km/hr) (Figure 23.2). In the 1930s, about 6 million multivaned windmills were in use, pumping water, across the United States.

A new invention from Denmark gave windmills another chore on the American farm. In 1890, the Danes became the first people to generate electricity from a windmill. This windmill used a propeller with two or three thin blades instead of the multivaned design, and it could capture the wind's energy more efficiently and at higher wind speeds. Windmills brought

Figure 23.1 The two basic Dutch windmill designs. The design at the top had a *cap*, which rotated so that the blades could be oriented to the wind. In the older design on the bottom, the entire building, except for the base, was rotated to set the blades in proper position relative to the wind.

electricity to the farms of America long before power lines were extended from central power stations. By the 1930s these 1-kilowatt windmill generators were helping rural America "tune in" to an exciting medium, the radio.

The windmill generators were used to charge automobile-type batteries, which could then be used for lights or radio, even when the wind was calm. As central station electric power reached more and more of rural America through public power programs like TVA, the use of such wind generators decreased. Until the present era, only relatively remote settings relied on small-scale wind power. Such sites may be rarely visited yet require small amounts of power on a steady basis. It is uneconomical to run power lines to such sites, so small-scale windmills are natural choices. Navigational beacons on islands at sea may often be powered by windmills.

Just for a moment, though, about 40 years ago, it looked as though electricity from the wind could compete with electricity from fossil-fuel generating sta-

tions. It was 1941, and a Smith–Putnam Generator rated at 1,250 kilowatts in a wind of 35 miles per hour (60 km/hr) was installed at Grandpa's Knob, near Rutland, Vermont. Hooked into a power grid, the windmill delivered commercial power for 3½ years. The tower was 110 feet (33 m) tall; its two blades weighed 8 tons apiece and were 175 feet (54 m) in diameter. But in 1945 a sudden strong wind broke one of the blades, hurling it 750 feet (283 m). A wartime shortage of materials prevented the windmill from ever being rebuilt, but the windmill at Grandpa's Knob proved that wind could be used to deliver commercial power.

(a)

(b)

(c)

Figure 23.2 The multivane windmill, invented in the United States in the 1850s, is still used on farms today to lift water from wells for livestock. Here an early model and two modern adaptations of the multivane concept are shown. (a) The self-regulating Perkins windmill as advertised in the *American Agriculturist*, December 1892. (b) The windmill supported by the pole with guy wires is from Chalk Wind Systems. (c) The windmill atop the tower is the product of Dempster Industries, Inc.

In the present era of high fuel prices, it appears that such windmills can become cost competitive and contribute to the electricity needs of the nation. In this chapter we focus on electrical energy from the wind, but the reader should note also that sail power, abandoned on commercial ocean vessels around 1910, is once again a viable option. Combining motor and sail on a single vessel and using durable new fabrics for sailcloth may return us to the era of the square-rigger, but with none of the delays and uncertainties of arrival.

Windmill Design

Windmills produce power when the wind pushes on the blades. The greater the *reach* of the blade, the more wind energy it can capture. Also, the greater the velocity of the wind, the greater the force on the blades and the greater the amount of energy captured.

The power response to the diameter of the blade and the speed of the wind is not one to one. The power produced goes up with the square of the diameter of the blade and with the cube of the velocity of the wind. Table 23.1 indicates how power output changes for a typical horizontal-axis windmill at different wind speeds and blade diameters.

Note that at a wind speed of 20 miles per hour (33 km/hr), multiplying the blade diameter by 4 (from 50 feet to 200 feet) multiplies the power output by 16. Observe also that at a blade diameter of 100 feet (30 m), a wind of 30 miles per hour (50 km/hr) generates 26

Table 23.1 Power Output at Various Wind Speeds

Wind speed		Power output in kilowatts		
(mi/hr)	(km/hr)	Diameter 50 feet (15 m)	Diameter 100 feet (30 m)	Diameter 200 feet (60 m)
10	(17)	3	14	54
15	(25)	11	46	182
20	(33)	27	108	432
25	(41)	53	211	844
30	(50)	91	365	1458

Source: Alternative Long Range Energy Strategies, Joint Hearing Before the Select Committee on Small Business and the Committee on Interior and Insular Affairs, U.S. Senate, December, 9, 1976.

times more power than a wind speed of 10 miles per hour (17 km/hr). This is why engineers lean toward big windmills and why they try to capture the higher winds.

Most large windmills now being built or already in use are designed to operate at wind speeds of 10–35 miles per hour (17–58 km/hr). Winds less than 10 miles per hour produce little useful energy; winds higher than 35 miles per hour could wreck the windmill. To appreciate more fully the wind speeds at which windmills operate, Table 23.2 lists events that occur at different wind speeds.

In a sense, the accident at Grandpa's Knob was an early indication of design problems in windmill tech-

Table 23.2 Observations and Events at Increasing Wind Speeds

Condition	Wind speed	
	Miles/hour	Kilometers/hour
Calm; smoke rises vertically.	1–3	2–5
Wind can now be felt on the face; leaves rustle.	4–7	6–11
Leaves and small twigs move constantly.	8–12	12–20
Dust is raised; loose paper blows; small branches move.	13–18	21–29
Small trees sway; the waves on water show crests.	19–24	30–39
Large branches are in motion; it is difficult to use an umbrella.	25–31	40–50
Entire trees are set to swaying in the wind; walking against the wind becomes difficult.	32–38	51–61
The wind snaps twigs off trees.	39–46	62–74
Structural damage starts to occur.	47–54	75–87
Trees may be uprooted; severe structural damage is possible.	55–63	88–101
Such velocities are unusual inland; they cause widespread damage.	64–72	102–115
Hurricane conditions.	73–132	116–212

Source: Industrial Instruments, Ltd., Stanley Road, Bromley, Kent, United Kingdom.

nology, especially in durability and safety. The serious accident indicated that the huge propeller blades could fatigue. *Fatigue* is a term from metallurgy, describing a metal that has undergone strong forces so that its structural strength is weakened. Metal fatigue caused the accident at Grandpa's Knob.

Windmills should not be designed to capture gale winds. Even though such winds deliver far more power than low-speed winds, they exert such a strong force on the blades that the machine itself may be destroyed. Furthermore, the proportion of time that gale winds blow is so small that the contribution of gale winds to total power output is extremely small, making such risks pointless. To combat the problem of gale winds, windmill blades are curved in such a way that they turn slightly to one side, out of the direct force of the wind, so that the full impact of large gusts does not damage the propeller. This old practice is known as *feathering*. Newer materials that can withstand higher forces are also being used to prevent the breaking of blades.

Other problems in windmill design occur simply because of the nature of the system needed to capture the power of the wind. Windmills typically stand on tall towers, enabling the blades to reach the stronger winds that occur at higher elevations. Near the ground, houses, trees, small hills, and the like interrupt and obstruct the wind. Thus, tall masts are needed. The heavy equipment—the propeller, gearbox, and generator—must sit atop the mast, however, and this requires a strong structure (Figure 23.3).

Another problem in using the power from windmills is the nature of wind itself. From little freshets to great gusts, the speed of the wind varies over a wide range. As a consequence, the cycles per second of the electrical output from a windmill varies. To correct this, the alternating current generated by the turning shaft is *rectified*; that is, it is converted into a steady, one-directional flow. For large windmills, this steady flow of current is fed to an *electronic inverter*, which produces a stable alternating current that can be fed into a power grid. Small windmills, such as those used on isolated farmsteads or on islands at sea, feed the rectified output to a large storage battery instead of to an inverter. The batteries are essential to store electrical energy for periods when the wind is too calm to produce any energy.

The problem of cycles per second can be corrected, but the adjustment of power output is more difficult. Just as with tidal power, there are times when wind-

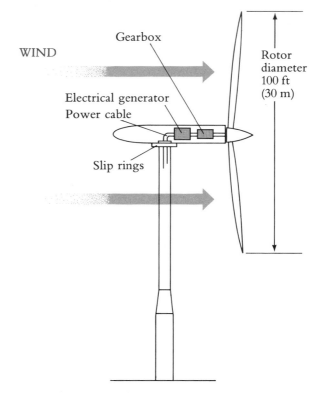

Figure 23.3 Typical wind rotor system. (Adapted from D. Kash et al., *Energy Alternatives*, Report to the President's Council on Environmental Quality, 1975.)

mills produce little or no power. At such times, conventional electric power will have to be increased elsewhere to meet power demands.

Wind Resources of the United States

Wind varies with the features of the natural landscape or urban terrain; it varies with the nearness to bodies of water, with the weather and the season, and with the height above ground. Winds over the land tend to be slower, on average, than those on the coast or over the ocean. Ocean winds are not only stronger, they are steadier as well.

Recognizing that some of our best wind resources are at sea, engineer William Heronemus of the University of Massachusetts proposed several banks of steel windmills off the coast of New England. He envisioned that the electricity developed by the mills would be converted to a steady, direct current that could be used to electrolyze water to hydrogen. The pure water to be electrolyzed would be provided by the distillation of sea water, which would also use the electrical energy developed. The hydrogen would come ashore through a pipeline and be recombined with oxygen in a **fuel cell**. Operation of the fuel cell would produce electricity and water. Heronemus's vision is, however, a long way off. For now, we need to look at the wind energy we can reach easily, the energy on the land.

Across the United States, the average annual wind speed is about 10 miles per hour (17 km/hr). It varies from a lowest average of 6.5 miles per hour up to 37 miles per hour, the average wind speed at the top of Mount Washington, New Hampshire. In fact, the winds on Mount Washington have been known to reach an incredible 150 miles per hour (250 km/hr).

Figure 23.4 indicates the wind resources available across the United States. Interestingly, there is a steady wind across the Great Plains, though the Plains are relatively free of obstructions.

The President's Council on Environmental Quality has estimated that wind resources, if rapidly developed, could be used to displace 4–8 quads of thermal energy by the year 2000. This is energy that would be consumed to generate electricity. Depending on how well we conserve energy, this could equal 5%–7% of annual U.S. energy needs. The Council quotes one study* that estimates the wind potential of the United

---

* *Solar Energy: Progress and Promise*, Council on Environmental Quality, April 1978.

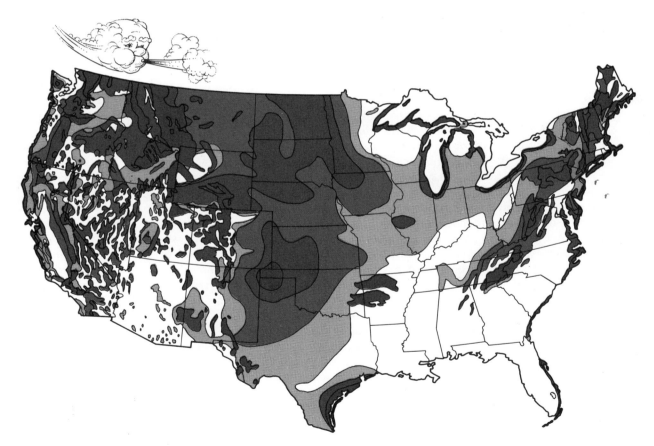

Figure 23.4 Wind resources of the United States. The darker the shading on the map, the more wind power is available (in watts per square meter, averaged over a year). The most wind energy is available along the ocean coasts, the lower Gulf coast, the coasts of the Great Lakes, the Great Plains, and the Eastern and Western Mountain Regions. (From U.S. Department of Energy.)

States at 1–2 trillion kilowatt-hours per year. U.S. consumption of electricity in 1985 was on the order of 2.5 trillion kilowatt-hours, indicating excellent potential for wind power.

Environmental Problems and Costs of Wind Power

Does wind power cause air pollution? No. Does it require water for cooling and cause thermal pollution? No. Does it consume fuels? No. It does cause noise, it does use land, and it does take materials to construct. It also has a visual impact, but the towers of long-distance electric lines have a height near that of the tallest windmill presently being considered, and cooling towers are even taller.

There is one other impact of wind power. Large windmills rotate at about 30 cycles per second. This is also the synchronization frequency of television in the United States. For a distance of up to 1 mile, these large windmills may interfere with television reception. With the use of fiberglass blades, which are proving to be lower in cost than metal ones, the distance falls to about a half mile. Again, this is for large windmills; it is not expected to be a problem for small windmills.

Birds may be hurt by windmill blades, but it is difficult to predict to what extent this would occur.

Certainly there are other environmental costs in the mining of ore, the manufacture of storage batteries, and the many more transmission wires and lines needed to collect electrical energy from multiple sources. But basically, when we count all the environmental costs, wind power comes out with a very low cost. However, what is the economic cost of wind power?

The first experimental windmills produced for the federal government were definitely not cost competitive with conventional power plants, but later test units are approaching a competitive status. The wind turbines at Goodnoe Hills, Washington, an installation built for the Department of Energy, produce power at about 8 cents per kilowatt-hour. Power produced at this cost would be competitive where high-priced oil is being used to generate electricity. When new blade designs have made the capture of wind energy even more efficient, costs on the order of 2–3 cents per kilowatt-hour will make wind energy fully competitive with conventional power.

The Development of Wind Power

Large-scale wind power. Since the "energy awakening" that began in 1973 and 1974, a growing number of large experimental wind generators have been built—principally by the U.S. Department of Energy but more recently by private groups. A 100-kilowatt generator was first tested by the federal government at Sandusky, Ohio in 1975. Although the station was beset by a number of difficulties, the Department of Energy went ahead with five additional wind turbines, each at a different site. Four of the wind turbines were rated at capacities of 200 kilowatts each. A fifth windmill was substantially larger. The four 200-kilowatt windmills are at Clayton, New Mexico; on Block Island, off the coast of Rhode Island; on the island of Culebra, a part of Puerto Rico; and on Oahu, the largest and most populous of the Hawaiian Islands. The generator at Clayton has been delivering power to the town's municipally owned utility since early 1978. The Culebra wind turbine began operation in July 1978 and is delivering electricity to about 150 island homes. The Block Island wind station began operation in 1979, and the Oahu wind turbine began to generate power in July 1980.

A significantly larger windmill began operation in July 1979 near Boone, North Carolina. The mill stands atop a mountain, and a steel tower reaches 140 feet (42 m) into the air. The 200-foot (61 m) diameter twin blades were able to produce 2,000 kilowatts of electric power from winds of 25–35 miles per hour (40–50 km/hr). The Boone wind station was shut down in 1981 after 20 months of successful testing. During this time, the wind turbine delivered power to a utility grid, the first time this had occurred since the ill-fated generator at Grandpa's Knob ceased operating in 1945.

In May 1981, a cluster of three still larger windmills (they are being called *wind turbines*) began operation at Goodnoe Hills, Washington. Built by Boeing Engineering for the U.S. Department of Energy, these machines, named the MOD-2, are rated at 2,500 kilowatts (2.5 megawatts) in a wind of 27.5 miles per hour (44 km/hr). The 200-foot (61 m) towers are built to survive winds of 125 miles per hour (200 km/hr). The rotor blades of each of the turbines are 300 feet (92 m) in diameter and weigh 80 tons apiece. The power from the wind farm at Goodnoe Hills, which is sufficient for 2,000–3,000 average homes, is fed into the northwest power grid by the Bonneville Power Administration.

Another MOD-2 is generating power at a U.S. Department of Interior installation at Medicine Bow, Wyoming. A fifth MOD-2 wind turbine is being operated in Solano County, California by the Pacific Gas & Electric Corporation.

Also at Medicine Bow is a 4-megawatt wind turbine designed by Hamilton Standard, a division of United Technologies, Inc.—but Hamilton Standard has withdrawn from the wind turbine business, as has the General Electric Co., slowing the pace of innovation. The Boeing Company was to provide a larger third-generation machine, but it scaled down its massive design. Boeing's machine was to demonstrate the capabilities of a commercial windmill and was to be in operation by late 1986, but funding from the Department of Energy was uncertain.

The slowdown in producing large wind turbines—those with power ratings of 2.5 megawatts and higher, with 300-foot blades, and with taller towers—is partly due to the presidential administration that took office in 1981. The Reagan Administration's philosophy of staying out of the private marketplace led to severe cuts in the funds for wind research. The slowdown is also influenced by problems with the large machines. At Goodnoe Hills, one of the windmills failed during high winds in 1982. A hydraulic valve failed in an emergency shutdown and damaged the generator; nine months were needed for repair, and the other MOD-2s were modified as well. The same wind machine failed in 1983; this time the horizontal steel shaft that supported the 100-ton rotor blade developed a crack. Support shafts on all MOD-2s were replaced with tougher designs and crack-detection systems.

The damage to confidence remained, however. In the utility industry, as represented by the Electric Power Research Institute, and in the U.S. Department of Energy, the concern is that the giant wind machines will not become commercially feasible; their size and complexity may make them too unreliable. One major manufacturer still remains committed to wind power; Westinghouse, who pioneered the 200-kilowawtt windmills of the mid-1970s, is using its experience to develop a 500-kilowatt machine that it hopes will be a commercial success.

Small- to moderate-scale wind power. Although wind power at the behemoth scale encouraged by the Department of Energy has not caught on, wind power at a more modest scale has been, and continues to be,

successful. Thousands of small windmills have been built; over 100 small companies were producing wind power in 1983. Private interest in wind power has been spurred by federal tax credits and state tax credits for renewable energy investments. California tax law has made such investments particularly attractive. In addition to tax incentives, a 1978 federal law has given small-scale wind power (and also water power) a real boost. Known as PURPA, the Public Utility Regulatory Policies Act requires public utility electric companies to purchase power from small producers at their avoided cost rate—that is, at the cost that would be incurred by the utility to produce the power itself. Until the act, utilities were free either to refuse to purchase small-scale power or to offer discouragingly low rates for it. PURPA and tax credits have lead to substantial wind power development.

In 1981, Windfarms Limited, a San Francisco–based firm, began a project to produce and sell electricity on the island of Oahu, Hawaii. The Hawaiian Electric Company will be purchasing up to 80,000 kilowatts (80 megawatts) of power from the cluster of about 30 windmills that the firm is building there. Windfarms Limited has also contracted to deliver power to Pacific Gas & Electric in California. In the Altamount Pass, 50 miles east of San Francisco, the firm has built some 600 relatively small windmills, each with a 50-kilowatt capacity under peak conditions, for a total power capacity of 30,000 kilowatts (30 megawatts).

Another firm, U.S. Wind Power, Inc. (USW), has also installed windmills in the Altamont Pass. Now the largest producer in the Pass, USW erected 200 wind turbines, each rated at 50 kilowatts, in 1982 and 400 more in 1983. By the end of 1984, USW was to have a total of 1,400 wind turbines operating in the Altamont Pass, enough capacity to supply 15,000 residential customers. USW designs and manufactures its own wind machines (see Figure 23.5).

All together, some 4,000 wind turbines have been built in the Altamont Pass. Another 5,500 are expected before the decade is over. The Pass was chosen because of its strong summer winds; summer is the time the power is needed most in California because of air-conditioning loads. Two other wind farm development areas have sprung up in southern California, in the Tehachapi Mountains north of Los Angeles and in San Gorgonio Pass near Palm Springs. Power from the Altamont Pass is sold to Pacific Gas & Electric. Power

Figure 23.5 The wind farm of U.S. Wind Power, Altamont Pass, California. By 1985 more than 4,000 windmills (with capacities of 50–200 kilowatts) had been erected in the Altamont Pass by 15 firms. U.S. Wind Power had installed over 1,300 of these windmills. Power is sold by these firms to Pacific Gas & Electric. Winds in the pass average 18–27 miles per hour from May to August. State and federal tax incentives helped to make these investments profitable. (Photo courtesy of U.S. Wind Power; photograph by Ed Linton)

produced in southern California is purchased by Southern California Edison. Contracts have been signed for the delivery of more than 1,000 megawatts from these California wind farms, enough to supply a city of a quarter million homes. About 500 megawatts of capacity was already installed in 1984. Interestingly, more than 10% of the windmill sales in 1983 were Danish machines. Denmark originated the windmill that produced electric power, and its machines are proving very reliable.

Not to be forgotten, however, are the small windmills that have served the countryside since the early 1900s. These modernized private wind systems include 8- to 40-kilowatt machines that could be used on farms, at rural and suburban residences, and even at city residences. According to one study, if these wind systems can be made reliable, can be given a 20- to 25-year life, and can be made cost competitive because of quantity

manufacture, a U.S. market for over 17 million machines awaits the industry. These systems have a long way to go, however, to become affordable. One manufacturer of 25-kilowatt windmills is charging $25,000 including installation for his machine, and that does not include delivery charges, which are on a mileage basis. Still, this only amounts to $1,000 per installed kilowatt, a figure one-third less than the installed cost of coal or nuclear power. Since most of us do not buy electric generating capacity, this interpretation does not make personal wind power more affordable, but it does suggest that it could be a good investment for electric utilities.

How fast could wind power grow? The 1981 Annual Report to Congress by the Energy Information Administration contained a projection made jointly by the Department of Energy and the Solar Energy Research Institute. Wind power produced in commercial quantity for utilities, as opposed to individual residences, was predicted to deliver 264 megawatts of electricity by 1985, a remarkable spurt from virtually no contribution just 10 years earlier. Hold onto your hats though. The actual capacity installed by the end of 1984 in California alone was 740 megawatts, almost three times the national prediction. The original 1981 projection called for generating capacities of 1,993 megawatts by 1990 and 18,400 megawatts by 1995, but the wind seems to be moving faster than predicted.

Geothermal Energy

"Healthful and Refreshing Warm Baths," the advertisements read for the resorts of Lake County, California. It was the late nineteenth century, and residents of the bustling city of San Francisco, some 75 miles to the south on the California coast, would travel to these geyser baths to be "restored." The odor of hydrogen sulfide gas was so noticeable in the area of the hot springs that the local stream became known as Big Sulphur Creek. Restorative though the baths must have been, no one saw any other commercial potential in the hot springs until the 1920s, when, perhaps inspired by success in Italy in the early 1900s, drilling for steam began in the area. But the drillers could not interest the electric companies in their steam.

Although Italy was able to produce electricity using steam from the earth in 1904, it was not until the 1950s that Pacific Gas & Electric became interested in using the steam that came bursting forth in northern California. By 1960, Pacific Gas & Electric was gener-

Figure 23.6 A geothermal power plant at the Geysers in Northern California. This is one of nearly 30 plants in the steam fields of Northern California. Plans call for the completion of nearly 2,000 megawatts of electrical capacity by 1990. Of this quantity, about 1,400 megawatts of capacity was already in place in 1983. A bank of mechanical-draft cooling towers in the foreground of the plant are a reminder that thermal discharges are a part of geothermal electric power. (Photo courtesy of The Marley Cooling Tower Company.)

ating electric power using geothermal steam at a capacity of 11 megawatts. The capacity of the sprawling plant had reached 1,400 megawatts in 1983 (Figure 23.6). Nearly 2,400 megawatts of capacity are likely to be available at the Geysers by 1988. The capacity of a modern coal-fired or nuclear power plant is about 1,000 megawatts. Wells are going ever deeper at the Geysers. Although the first steam was found at about 1,000 feet, recent wells have been drilled to more than 9,000 feet. Table 23.3 identifies existing geothermal power plants around the world.

What is geothermal heat? How can it be used, and where can we find it? Is it a clean and renewable source of power?

Origin of Geothermal Heat

Simply put, geothermal heat is the energy from the earth's interior. Of course, the eruption of a volcano, such as Mount St. Helens, is visible evidence of the enormous heat inside the earth. Scientists estimate the temperature in the core of the earth at thousands of degrees Celsius. From the intensely hot interior of the earth, where molten metal and molten rock are thought to be the only form possible, up to the surface of the earth, temperature decreases steadily.

Only a few miles beneath the surface, one can occasionally find molten rock at 1,000°C or more. The

more likely find, however, is hot solid rock at a temperature of perhaps 300°C. At a number of points on the earth's surface, usually in the areas of volcanic and earthquake activity, this immense heat comes bubbling to the surface in the form of water and steam at temperatures as high as 300°C. This water and steam comes

Table 23.3 Anticipated Worldwide Geothermal Power Production, 1984

Country	Generating capacity (in megawatts)
United States	1664
Philippines	891
Mexico	700
Italy	467
Japan	228
New Zealand	203
El Salvador	95
Iceland	41
Indonesia	32
Turkey	31
Kenya	30
Soviet Union	21
China	10
Azores	3

Source: R. DePippe, "Worldwide Geothermal Power Development," *Geothermal Resources Council Bulletin* (May 1983).

from groundwater that has wound its way from the surface down through porous rock and through rock fissures into a region of very hot rock. Heated by the rock, even boiled, this water comes bursting to the surface under pressure as steam and hot water. We call this erupting column of water and steam a *geyser*. Old Faithful in Yellowstone National Park is our best-known geyser because of its towering spray and its insistent punctuality.

Two Important Uses for Geothermal Heat

Geothermal heat has the potential to be used in two basic ways: in the production of electricity and in the heating of homes, offices, and factories. Whether the heat is used for electricity or for heating homes and factories depends on the form in which the resource makes its appearance. Sometimes the water comes billowing up from the earth as pure "dry" steam—that is, entirely vapor with no water droplets mixed in. This dry steam can be used directly to turn a turbine and generate electricity. Condensed water may be reinjected into the earth or, if the quality is good enough, disposed of in a nearby body of water.

The Geysers, the field in California where Pacific Gas & Electric has its plant, and the Larderello field in Italy are both examples of fields that produce dry steam. Japan has developed one small generating unit using geothermal steam in Matsukawa. Finding dry steam is still by chance, however; geyers that discharge steam at the surface led us to the California fields and to others as well.

In some fields, the geysers spew forth a mixture of steam and water droplets. Underground, these "wet" steam fields really have only water in them, but the water is at unusually high temperatures (180°–370°C) and is under very high pressure from the overlying rock.★ The release of pressure when the water comes to the surface causes about 10%–20% of the water to *flash* to steam; the rest remains as hot water.

This mixture of steam and droplets cannot be used directly for the generation of electricity; the impact of the droplets would damage the turbine. In addition, geothermal water contains corrosive salts that make the use of wet steam even less advisable. The mixture must be separated into dry steam and water by a centrifugal

separator, a device that whirls the heavier water droplets to the outer edge of the separator for collection. The steam is then directed to the turbine for the generation of electricity, while the hot water is disposed of in various ways. At Roosevelt Hot Springs, Utah, a device is being tested by Utah Power and Light that generates electricity with the water as well as the steam by directing the water droplets through a separate liquid turbine.

The steam that turns the turbine is condensed and must be gotten rid of. Although some novel ideas have been suggested for using the hot water that remains, reinjection into the earth appears to be favored because the water is often high in dissolved salts. The salts could damage any body of water receiving the condensed steam. The largest electric power plant in the world that uses wet steam is in Wairakei, New Zealand, where a plant with a power rating of 192 megawatts is in operation. The plant began operation in 1958. Whereas the Geysers plant and the plant at Larderello, Italy are the models for power production with dry steam, the New Zealand plant is the model for power production with wet steam (Figure 23.7).

The first plant in the United States to use wet steam began operation in 1980 at Brawley, California. Operated by Southern California Edison and the Union Oil Company, the plant will be producing electricity at a capacity of 10 megawatts. Another demonstration using water flashed to steam has been built in Valles Caldera, New Mexico, about 60 miles north of Albuquerque. The Union Oil Company teamed with the Public Service Company of New Mexico to build the 50-megawatt electric power plant there, which will tap geothermal water at 280°C.

There is a third type of water-containing geothermal field in addition to the dry-steam and wet-steam fields. This is a field that produces hot water only. Such fields are even more common than wet- and dry-steam fields. At first glance, hot water alone, especially if it has a high salt content, does not seem terribly desirable, but it does, in fact, have good uses. Reykjavik, the capital of Iceland, with a population of 85,000, is heated almost entirely by hot water drawn from deep geothermal wells under the city (Figure 23.8). The use of geothermal water in Iceland began only in 1943. After it has been used to heat the city, the water is sent to greenhouses that produce fresh vegetables for Icelanders. In the United States, geothermal heat has been used for several decades to heat homes in Boise, Idaho and Klamath Falls, Oregon.

★Under ordinary atmospheric pressure, water boils at 100°C.

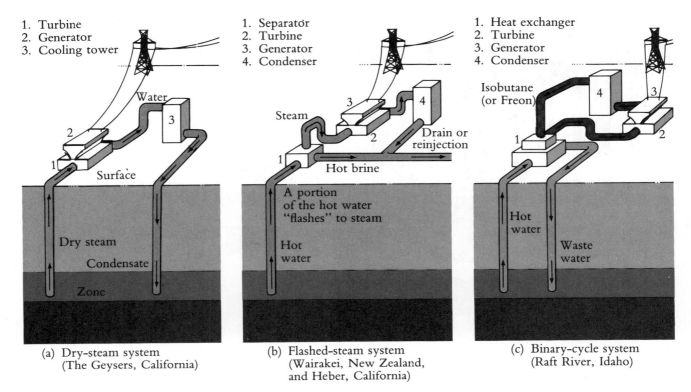

1. Turbine
2. Generator
3. Cooling tower

1. Separator
2. Turbine
3. Generator
4. Condenser

1. Heat exchanger
2. Turbine
3. Generator
4. Condenser

(a) Dry-steam system
(The Geysers, California)

(b) Flashed-steam system
(Wairakei, New Zealand,
and Heber, California)

(c) Binary-cycle system
(Raft River, Idaho)

Figure 23.7 Electricity can be produced in three ways from geothermal resources. (a) Dry steam, when it is available, can be used to turn a turbine directly for electric power generation. (b) When only hot water is available, a portion of the flow is allowed to flash to steam. The steam, after being separated from the water, is then used to turn the turbine. (c) Another alternative for electric power generation is to use the hot water to boil a fluid such as isobutane to a vapor. The isobutane steam is used to turn a turbine to generate electricity. (Adapted from "Western Energy Resources and the Environment: Geothermal Energy," Report to the Office of Energy Materials and Industry of the U.S. Environmental Protection Agency, April 1977.)

Hot water may be used for more than heating homes and offices. Geothermal water can be used to produce a steam, or vapor, from a *working fluid* that boils at a lower temperature than does water. Isobutane, one such fluid, might be boiled to an isobutane "steam" in a heat exchanger by using hot geothermal water as the heat source; this "steam" is then used to turn a turbine. The spent isobutane vapor is condensed and returned to the heat exchanger to be boiled again. (See Figure 23.7c.) The United States, Japan, and the USSR are experimenting with this approach, which is called *binary cycle* because it uses two loops of fluid: the water loop, which heats the working fluid, and the working fluid loop. In the working fluid loop, the fluid boils, turns the turbine, condenses, and reboils.

A significant demonstration of electric power production from a binary cycle plant is the 45-megawatt facility at Heber, California. The Heber Project in the Imperial Valley is a joint effort of numerous agencies and organizations, among which is Southern California Edison. The binary cycle plant went "on stream" in 1985. Its power was added to the power from a "wet-steam" geothermal plant that flashes hot water to steam, which was finished at the same site in 1982, another project of Southern California Edison.

The Heber binary cycle plant makes use of a moderate-temperature brine, 150°–210°C, not otherwise useful for efficient power production. Because such geothermal brines are four times more common than high-temperature brines, successful operation of the

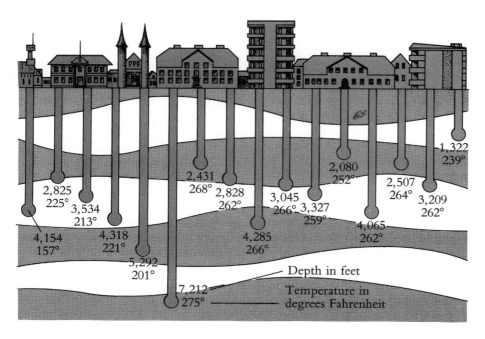

Figure 23.8 The city of Reykjavik, Iceland has been using geothermal water to heat its homes, offices, shops, and factories since 1943. To provide the city with pollution-free heat, 32 holes have been drilled through the underlying lava beds to tap reservoirs of extremely hot water. A number of the holes have been productive, as shown in this diagram. Nine holes are currently in use. (© 1974 by The New York Times Company. Reprinted by permission.)

Heber plant opens the way to use a new set of resources.

There are two additional ways to extract energy from the earth. One is not highly developed but shows promise. The other is only a concept at the moment. Geologists have noted that buried dry rock at temperatures as high as 300°C is about ten times more abundant than water-bearing hot rock. Experiments indicate that this dry rock can be fractured by pumping down water under very high pressure. This technique, known as *hydrofracturing*, originated in the petroleum industry to get at gas deposits that were cut off from one another. Once the rock is fractured, water can be forced through the cracks and returned to the surface at a much higher temperature. The water produced in this way may be as hot as ordinary geothermal hot water, depending on the temperature of the rock from which it was produced. There are, however, significant energy costs for pumping, depending on the depth at which the hot rock is found.

There are also deposits not far below the surface where molten rock (known as magma) resides. At least one such deposit lies in the United States. The deposit of molten rock in Marysville, Montana may be a little more than a mile below the surface. Apparently, it is the remainder of a magma flow that was pressing toward the surface to become a volcano but never made it. The temperature in the molten mass may be on the order of 1,000°C. Can the heat in the molten rock be tapped by drilling? We do not know. In many ways, the molten rock, or magma, resembles the interior of an active volcano. Thus the question is even broader: can we tap volcanic heat? Will it be possible to bring the heat to the surface in the form of steam? Sandia Labs in New Mexico has been charged by the Department of Energy with developing the answers to these questions, but at the moment, there is no known method of drilling into magma.

Resources of Geothermal Heat and Projections of Use

The U.S. Geological Survey estimates that our reserves of high-temperature water and steam (those useful for the generation of electricity) are at a level capable of producing 11,700 megawatts each year for the next 30 years. (These resources and reserves exclude those in national parks; see Controversy 23.1.) A plant of 1,000 megawatts, the size of a modern nuclear or coal-fired power plant, is needed to supply an average city of 1 million people. These reserves, then, are capable of supplying the electricity needs of about 12 million people, or about 5%–6% of current demands.

In contrast to these *known* reserves, the as yet undiscovered resources of high-temperature water and steam are estimated to be capable of producing 127,000 megawatts each year for the next 30 years (if they were immediately discovered and put to use). Some of these

CONTROVERSY 23.1

Geothermal Energy Development—Should It Take Precedence over the Risk to a National Park?

The grizzly bear can't survive if Yellowstone Park is its only refuge. It also needs portions of the five adjacent national forests.

John Townsley, *Superintendent of Yellowstone National Park*

Any activity that would tend to damage them (the geothermal features of Yellowstone Park) . . . would be an insult of the worst kind—insensitivity typical of nineteenth-century America.

Ralph Maughan

One must simply accept the conclusion that our geothermal resources should be developed promptly, regardless of their location.

Alan Buck, *Stewart Capital Corporation*

Union Oil has pioneered the development of geothermal resources in the United States. The Union Oil Company operates the Geysers geothermal plant for Pacific Gas & Electric and is a partner in the geothermal development at Brawley, California and at Valles Caldera, New Mexico. The company is now seeking to develop the Island Park Geothermal Area (IPGA) in Wyoming. The potential of this area is thought to rival that of the Geysers in California.

Unfortunately, the IPGA lies directly alongside Yellowstone National Park, the first (1872) National Park to be established in the United States. Yellowstone, as you know, is the home of Old Faithful, the world's most famous and most unique geyser. Yellowstone is also home to the grizzly bear, a threatened species in the lower 48 states, as well as home to bison, elk, and such western waterfowl as the trumpeter swan

and Canada goose. The IPGA is outside the park, but many are concerned that the proposed development will affect not only the geothermal features of the park but its wildlife as well.

The grizzly bear is the species thought to be most at risk from development of the IPGA. About 20% of the IPGA, which is in the Targhee National Forest, has been classified as land so critical to the grizzly that decisions about its use must first take into account the need to protect the bear. In fact, much more of the IPGA is crucial habitat for the bear, whose range includes both Yellowstone and parts of six adjoining National Forests. The presence of the additional people who would come to the area with geothermal development is thought to be the largest threat to the bear from geothermal-related activities.

The unique features of the 2.5-million-acre park

may also be at risk from development at the IPGA. Even though Old Faithful is inside Yellowstone and is some 13.5 miles from the IPGA, the possibility certainly exists of a deep underground connection between the thermal area and the geyser. Geologists have been unable to rule out the existence of such a connection, though they think it is unlikely. Could Old Faithful fall silent if development proceeds?

The opening of the Wairakei geothermal plant in New Zealand silenced the Great Geyser of Geyser Thermal Valley. Once the fifth most active geyser area in the world, this valley last had a geyser eruption in 1965. Similar declines in activity have occurred with development and drilling in Italy and Iceland, as well as at Beowaive and Steamboat, Nevada. Geysers in the

Boundary Creek area of Yellowstone are within a few miles of the proposed development; thus, an effect on this geothermal area would seem likely.

Can the United States take a chance on geothermal development at the IPGA? What will be saved or replaced by this development? Given that power from the development will still be almost as costly as conventional power, what would be the benefits of development? To whom would the benefits go? What would be the risks? Who would bear the risks? Should we preserve areas that would have economic value to the nation if they were developed?

Source: All quotes are from "The Incredible Shrinking Wilderness," *National Parks* (January/February 1982), p. 21.

resources, however, might not yet be profitable to recover.

The resource base we have described so far is only that portion useful for the generation of electricity. In addition, lower-temperature water at 90°–150°C is abundant. Such water is useful for heating, and the U.S. Geological Survey estimates the heating value of identified resources as equivalent to that of 14.3 billion barrels of oil. Given that the United States uses about 7 billion barrels of oil each year *for all purposes* and has only about 30 billion barrels of reserves, this amount of heat is staggering. Furthermore, the value does not yet include unidentified resources, which the Survey suggests may be three times the value of the identified resource base. We must be cautious, however, in interpreting these numbers, since population centers are not often on top of geothermal resources, as is Reykjavik, Iceland.

Two areas of the United States have unique geothermal resources. In the Gulf Coast region are huge deposits of extremely hot water that is under high pressure due to sediment layers above them. We refer to these deposits as *geopressured systems*. The Geological Survey places these resources (in regions they have assessed so far) at the equivalent of 11.2–43.3 billion barrels of oil. Perhaps three times these energy resources are thought to exist in areas as yet unassessed. In our discussion of natural gas, we pointed out that methane is in solution in geopressured water and that some geologists consider the quantity of recoverable methane

to be enormous, far larger than our current natural gas reserves. Thus, the geothermal deposits of the Gulf Coast are very attractive for exploration.

In California extensive deposits of very hot water are found in the Imperial Valley. An intriguing idea for this water-short region is to flash the hot water to steam, generate power, condense the steam, and use it for irrigation. Though the water is initially high in salts, the condensed steam has, in fact, been distilled; that is, it has first been made into vapor and then condensed. Whereas the briny water would have laced the soil with salt if used as irrigation water, the condensate will not.

The estimates we provide here are only for underground water and steam systems. There is also the hot rock beneath the surface, which may be useful for heating water pumped into it, and these resources are likely to be far more abundant than the water and steam systems. Finally, geothermal energy, in many cases, may be renewable. Where the rock contains the heat and not merely the water, the heat in the rock can continue to be extracted by circulating water through the formation. The water may be water originally drawn from the formation or water brought to the site for that specific purpose.

Projections for geothermal electric power are, in part, built on a survey of the expansion plans of companies. One company, Pacific Gas & Electric (PG&E) is planning significant expansion of geothermal power generation. By 1990, PG&E hopes to have expanded its facility at the Geysers to 2,000 megawatts from its 1983 level of 1,400 megawatts. The 2,000 megawatts should be able to be produced for 30 years, based on the steam resources existing there, and since most of the heat is in the rock, water injection may extend the resource still further. PG&E is lucky enough to have dry steam, and the company plans to make the most of its resources.

Most other sites will be using wet steam and binary cycle systems, and these systems are just coming out of the development phase. In all, if present growth rates continue, geothermal electric generation could expand to 16,000 megawatts by the year 2000. The facilities for this generation would probably be in the western states, where most geothermal resources exist.

In summary, the geothermal potential of the continental United States appears to be large and worth developing.

Cautions and Environmental Impacts

When geothermal energy is used, hot water must be disposed of. This is true even if dry steam is used to generate electricity. When hot water is "wasted" from wet steam plants, it often has considerable amounts of salts dissolved in it.

Brines from the Imperial Valley may be as much as 20% salt, or about six times the salt concentration in seawater. At a geothermal deposit in El Salvador, where salt content is about half that of seawater, the brine has been reinjected into a nearby well that is almost a kilometer deep. Up to 800 tons of water per hour are injected there, making it unnecessary to carry the water via pipeline to the sea, an earlier alternative. Nonetheless, we are still not certain that waste geothermal water can be reinjected for the long periods that may be necessary. The quantities of salt are astounding.

Reinjection may prove to be more than a way to dispose of brine. It may also prevent the surface above geothermal wells from subsiding as water is drained from the deposit. The water that is reinjected into the earth via a well not far from the drawing well will help hold the earth intact. Not only may it prevent subsidence, it may also provide a *recharge* to the geothermal deposit. That is, if the rate of withdrawal exceeds the rate at which surface water descends through the earth to the deposit, reinjection may prevent the well from

going dry. There is a possible drawback to reinjection, however: The water reinjected is typically more salty than that withdrawn, and this could lead to increasing levels of salt in the geothermal water.

Geothermal electric plants require a source of cool or cold water, as well as a source of heat for steam. The cool water is needed to condense the steam prior to disposal. In this regard, then, a geothermal electric power plant is no different from a coal-fired or nuclear power plant. The water used to condense the steam may come from a nearby freshwater source, such as a lake or river, or it may come from the ocean. If it comes from a freshwater source, cooling towers will be needed so that the heated fresh water from the condenser can be cooled and used again for condensing steam.

The volume of cooling water needed by a geothermal power plant is larger than that for a nuclear or coal-fired power plant of the same electric capacity. This is because the geothermal electric power plant is less efficient than its cousins, a defect often cited by critics of geothermal energy. Even the Geysers plant in Lake County, California, which uses dry steam, has this problem. The problem stems from the nature of the geothermal resource; the steam is under lower pressure and at a lower temperature than steam used in most electric plants. Because of this, the transfer of energy to the turbine blades is less efficient. At the Geysers, the thermal efficiency is about 22%. This compares with an efficiency of 30% for most new nuclear power plants and nearly 40% for most new coal-fired power plants. Accordingly, cooling water consumption is much larger for geothermal electric plants.

Geothermal water is also likely to contain hydrogen sulfide. About 25% of geothermal water sources have hydrogen sulfide present as a contaminant. Hydrogen sulfide not only smells bad (like rotten eggs), it may affect human health at high enough levels. Emissions from the Wairakei plant in New Zealand have caused silverware to blacken in a nearby village. The hydrogen sulfide content at the Geysers is high, about 200 ppm, but experiments at the Geysers show that more than 90% of the hydrogen sulfide can be removed.

To summarize briefly: geothermal energy may have a real, even dramatic contribution to make to our energy resource base. It is worth exploring further, even with its drawbacks.

Summary

The wind has served humankind for thousands of years, furnishing the power for sailing ships and the energy to grind grain and pump water. On the American farm, the multivane windmill pumped water for livestock and brought electricity to the rural homestead. The emphasis today is on wind-generated electric energy.

Windmills are built to capture only moderate winds (10 to 35 miles per hour); gale winds could destroy them. The devices require strong structures to support heavy blades as well as rectifiers and inverters to turn the wind's energy into a stable alternating current. The wind potential of the United States is 1–2 trillion kilowatt-hours per year, a figure that can be compared to 1985 U.S. electricity consumption of 2.5 trillion kilowatt-hours.

In 1975 the U.S. Department of Energy began a program to develop wind power, contracting for a succession of larger and larger machines (first 100 kilowatts, then 200 kilowatts, then 2,000 kilowatts, and finally, 2,500 kilowatts). Mechanical and structural problems beset the larger machines—as did the hands-off policy of the Reagan administration. Little activity on the larger windmills continues. The technology for smaller windmills (50 to 200 kilowatts) was already well advanced though; federal and state energy tax credits caused the wind power industry to grow quickly—especially in California. Wind-generated electric power seems to be coming along faster than anyone predicted.

Geothermal power is generated by using steam and hot water from the interior of the earth. The water reaches the interior of the earth as groundwater which winds its way down into the region of hot rock; there it is heated under great pressure. The heated water is forced up to the surface where it emerges from the ground as steam, as a steam-water mixture, or simply as very hot water.

The first geothermal development in the United States began at the Geysers in California, where Pacific Gas & Electric is projected to have 2,400 megawatts of generating capacity by 1988. Dry steam (without water droplets) is available at the Geysers to turn the turbines and generate electricity. At other sites where a water-steam mixture (wet steam) is all that is available, the steam is separated out and then used to turn a turbine; the water droplets would harm the turbine. At still other sites, the most common, only hot water is available, and power can be produced by using the hot water to boil isobutane to a vapor and then using this isobutane "steam" to turn a turbine. This process is the binary cycle system. The hot water can also be used directly to heat homes, commercial buildings, and factories. Such use is called *district heating*.

The U.S. Geological Survey has identified fairly extensive resources of geothermal energy, especially in the western United States and the Gulf Coast region. The agency estimates that far more remains to be discovered.

Geothermal development can have adverse results. The Great Geyser of New Zealand's Geyser Thermal Valley ceased to erupt after the geothermal development nearby. Old Faithful at Yellowstone could meet a similar fate if geothermal power production begins in the area surrounding the park. In geothermal power production, the steam that turns the turbine must be condensed, requiring a source of cooling water just as a coal-fired or nuclear power plant does. Thermal pollution can result from disposal of both the cooling water and the condensed hot water. In addition, where a water-steam mixture is drawn from the earth for a wet steam power plant, or where hot water is drawn for a binary cycle power plant, the water must be disposed of. The water can be incredibly salty, up to 20% salt, and may need to be pumped to the ocean or reinjected into the earth. Adding it to a river or lake could destroy the freshwater aquatic life. Geothermal water is also likely to contain significant quantities of hydrogen sulfide, a foul-smelling gas that is dangerous at high concentrations.

Questions

1. How can the wind be used to generate power? Where does the energy that is captured originate?
2. Why do wind and hydropower, together, make a good system?
3. What major environmental and economic disadvantages and advantages does wind power have? Compare these to the disadvantages of generating power from nuclear fuels.
4. Briefly describe how the energy in geothermal heat can be used as a substitute for other energy sources.
5. What environmental impacts are associated with the use of geothermal heat? Compare to the impacts of fossil-fuel (coal, oil) power plants.

Further Reading

Wind Power

Merriam, M. "Wind Energy for Human Needs," *Technology Review*, **79** (3) (January 1977), 28.
The author, an engineer, treats the subject of wind energy at the level of *Scientific American*. Although there are a few formulas and engineering graphs, most of the paper can be understood without these elements.

Putnam, Palmer C. *Power from the Wind*. New York: Van Nostrand, 1948; reprinted, Van Nostrand Reinhold, 1974.
This book is listed for historical as well as for practical interest. Note the author's name and recall the generator at Grandpa's Knob.

Marx, Wesley. "Seafarers Rethink Traditional Ways of Harnessing the Wind for Commerce," *Smithsonian*, **12** (9) (December 1981), 51.
Discusses the possibilities of combining sail and motor as the moving force of ships and boats. Easy reading.

Sorensen, Brent. "Turning to the Wind," *American Scientist*, **69** (September-October 1981), 500.
A reasonably nontechnical treatment of wind power with cost comparisons of electricity generated by the wind with electricity generated by coal-fired and nuclear power plants.

"Wind Power: A Question of Scale," *EPRI Journal* (May 1984), p. 6.
EPRI has contact with most wind power developments in the United States and abroad.

Kahn, R. "Harvesting the Wind," *Technology Review* (November-December 1984), p. 56.
Excellent writing distinguishes this up-to-date treatment of wind power prospects.

Johnson, M. "Wind—An Ancient Solution to a Modern Problem," *Electric Power Quarterly*, Third Quarter (1984), p. 3.
Describes the status of wind power including political and economic factors.

Scott, D. "Sail-Assist for Cargo Ships: Computer-Controlled," *Popular Science* (June 1985), p. 68.

Geothermal Energy

Keifer, Irene. "Earth Boils Below While We Scratch the Surface for Fuel," *Smithsonian*, **5** (8) (November 1974), 82–88.

Wheeler, Romney. "The Geothermal Option," *Panhandle Magazine*, No. 4. (1978), pp. 2–10.

"Geothermal: New Potential Underground," *EPRI Journal* (December 1981), p. 19.

"Tapping the Mainstream of Geothermal Energy," *EPRI Journal* (May 1980), p. 6.
These articles are clearly written without excessive technical jargon; they should be accessible to a wide audience.

Heiken, G., H. Murphy, G. Nunz, R. Potter, and C. Grigsby. "Hot Dry Rock Geothermal Energy," *American Scientist*, **69** (July-August 1981).
Explores the option of injecting water into otherwise dry rock that is heated by underlying magma. Fairly technical and scientific.

Sourcebook on the Production of Electricity from Geothermal Energy, U.S. Department of Energy (1980), DOE/RA/4051-1. Available from the U.S. Government Printing Office, Washington, D.C.
Comprehensive on its chosen subject.

Johnson, T. "Hot Water Power from the Earth," *Popular Science* (January 1983), p. 70.
Color illustrations and no mathematics make this article relatively easy to read. Probably the best "concept" pictures of geothermal power you will find.

Hamilton, B. "Geothermal Energy: Trouble Brews for the National Parks," *Sierra* (July-August 1983), p. 21.
Geothermal development can have unwanted effects. Examples where conflicts are in the offing are discussed.

Reed, M., ed. "Assessment of Low-Temperature Geothermal Resources of the United States—1982," U.S. Geological Survey, Circular 892 (1983).
A dry but informative treatment of a wet resource.

CHAPTER TWENTY-FOUR

Solar Energy

Solar Heat and Hot Water: Active Systems

The Collector/The Storage System/Air as the Collecting Medium

Passive Solar Energy

South-Facing Windows/Trombe Walls/The Solar Greenhouse/The Future of Passive Solar Heating/Interior Lighting from Sunlight

Conversion of Sunlight to Electricity

Photovoltaics/Central-Station Electricity from Solar Energy

Biomass: Biological Conversion of the Sun's Energy (Alias Photosynthesis)

Wood/Field Crops

Ocean Thermal Energy Conversion (OTEC)

Prospects for Solar Energy

CONTROVERSY:
24.1 *Should Energy Sources Be Centralized or Decentralized?*

Legend tells us that Archimedes saved his home city of Syracuse in Greece with solar energy. Ordering a thousand soldiers to turn their shields to the sun and lining them up in the shape of a parabola, Archimedes focused the sun's rays on the sails of the ships of an invading navy and burned them. No further practical applications of solar energy are recorded until the nineteenth century, when inventors began to experiment with the sun's ability to heat and even to boil water.

In the early decades of the twentieth century, it was still an open question as to what fuels would power society. Early and exciting efforts to harness solar energy took place in the late nineteenth and early twentieth centuries. In France, a solar steam engine was invented by Mouchot; his collector and engine were first displayed at the 1878 World's Fair (see Figure 24.1). By the early 1900s, farmers in California and Arizona had constructed solar irrigation pumps. By focusing the sun's rays on a boiler, they were able to produce steam, which was used to turn a pump. Such a steam engine was even introduced in Meadi, Egypt in 1912. Built by Shuman and Boys of Philadelphia, this engine was a long parabolic collector that could be turned to track the sun. The collector provided enough heat to boil steam for a 100-horsepower piston engine.

Such pumps and engines were only one aspect of a budding solar industry. Another dealt with solar hot water (Butti and Perlin, 1978). Around the turn of the century, residents of southern California found themselves paying dearly for the coal needed to make hot water. To reduce these costs, they painted water tanks black and placed them so that the tanks would absorb heat from the sun. On most days, the water would not be warm enough for showering until afternoon and then would cool quickly in the evenings. The Climax Solar Water Heater, invented and patented in 1891 by Clarence Kemp of Baltimore, gave better results. With tanks packed in an insulated box with a glass window to let the sun in, the Climax Water Heater was often roof-mounted or attached to a wall. The water in the

Figure 24.1 Solar collector exhibited at the 1889 Paris Exposition. The collector is driving a steam engine, which is furnishing the power to drive a printing press. The first solar steam engine was displayed at the 1878 World's Fair. (Courtesy of the Bettmann Archive)

black tanks warmed more quickly and retained its heat longer because of the decreased heat loss from the box. The Climax Water Heater sold for $25 in Pasadena around 1900.

An even more efficient design was invented by Frank Walker of Los Angeles in 1898. Although double the cost of the Climax, Walker's heater was nonetheless popular because it was hooked into a conventional water-heating system so that hot water could be drawn at all times, not just in late afternoon. Walker's design was mounted in the building so that the glass window cover was flush with the roof; water was drawn from the top of the tank, where the hottest water naturally collected. Even though hot water could be drawn at all times of the day, the need to use coal or gas for early-morning hot water made the system expensive to operate. The sun's energy had somehow to be stored if one wanted hot water early in the day without using coal or gas.

To accomplish this, William Bailey created the basic elements of the modern solar hot-water system—in 1909! Bailey's design, called the Day-and-Night Solar Water Heater, separated the component used for water heating from that used for heat storage (Figure 24.2). While a collector with copper tubes, much like a modern collector, was mounted on the roof, an insulated tank within the attic of the building stored the heated water. Water rose from the collector to the tank because water is warmer and lighter. Bailey's system could hold heated water through the night and into the next day. Although the Day-and-Night sold for $100, it could cut gas heating bills by 75%, or about $25 per year. In four years, the water heater paid for itself.

In 1913, however, a record freeze hit some parts of southern California. The water froze and cracked the metal tubes, destroying the plumbing of the system; water leaked into houses. Bailey was called upon to invent a system that would not freeze. His new design simply mixed alcohol with water, creating an antifreeze that was used as the collecting liquid, and the collector loop was separated from the hot-water loop. The collector liquid, now resistant to freezing, passed through a coil in the storage tank to heat the water. By the end of World War I, more than 4,000 units of Day-and-Night had been sold.

When natural gas discoveries in California made solar hot water less attractive in the 1920s, Day-and-Night built and sold gas water heaters. In 1923 Bailey sold the rights to manufacture the solar water heater to

Figure 24.2 Day-and-Night brochure of 1923. In the first two decades of the century, solar water heaters were popular items in southern California. In fact, the basic design still in use today was invented in California in that era. The arrival of inexpensive natural gas brought an end to the industry, however. Now, with fuel costs rising rapidly, solar hot water seems economical once again—this time throughout the country.

a Florida businessman, H. M. Carruthers. In 1932 an employee of Carruthers, Charles Ewald, added many new design features to the basic Day-and-Night model. One important change was to use soft copper tubing in the collector. The soft copper tubing, which proved resistant to cracking even in Florida freezes, allowed Ewald to avoid Bailey's alcohol-filled loop and to heat the water directly.

Because the basic patents on the Day-and-Night had expired by the 1930s, other companies were able to enter the business as well. In 1941, when America entered World War II, sales of solar water heaters in Miami were twice that of electric and gas water heaters. More than 15,000 solar hot-water units had been placed in service in Miami by that time. The war brought the business to a quick halt, though; civilian uses of copper were forbidden.

The industry did not recover after the war because electric rates were falling while costs in both labor and materials were shooting up. Although the industry has yet to recover, thousands of solar hot-water units are still in service in Miami. Today, as electric rates rise, solar hot water is once again looking attractive in Miami (see the section on "Prospects for Solar Energy" later in this chapter). From coast to coast in south Florida, new homes are once again being offered with solar hot water. In a very real sense, it is something of a "homecoming."

Solar Heat and Hot Water: Active Systems

When people talk about the promise of solar energy, they may be talking about any of a number of solar options. If they are pressed for the solar option that can deliver energy today, however, they will tell you that one option is available that does not require further research and development to be installed in homes today. It is the flat-plate collector that has been used for so many years to provide home heating and hot water. The systems we will discuss in this section, including the flat-plate collector, are termed *active* because they require assistance from conventional energy sources, such as the use of an electrically driven pump to move liquids.

The Collector

To collect the sun's rays efficiently for the production of heat, special solar collectors have been designed, much like the one that Bailey invented. The designs are meant to decrease the loss of heat energy. Typically, the collector is flat and stationary. This *flat-plate collector* is set on a roof at an angle roughly equal to the local latitude. At this angle, the rays of the sun fall nearly perpendicular to the surface of the collector.

Many different designs are possible for a flat-plate collector. The principles behind the designs, however, are much the same. We will discuss a design in which

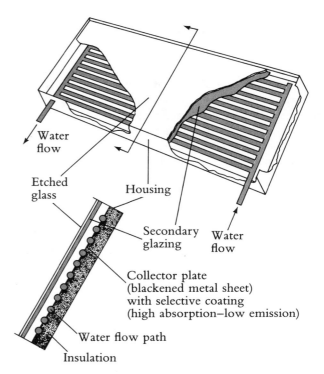

Figure 24.3 Flat-plate solar collector. Cold water is pumped in at the bottom of a row of parallel metal pipes. Warm water exits from the top pipe. The pipes are bonded to a blackened metal sheet. Below the sheet and around the outside of the box in which the pipes are housed may be glass-wool insulation. Two glass plates cover the top of the box. (Adapted from *Solar Energy Handbook*, Honeywell, Form 74-5436, p. 9.)

parallel metal tubes cross the collector; the metal tubes are welded to a metal plate, and either liquid or air flows through the tubes. Figure 24.3 illustrates the design of the collector with a liquid flow.

When the rays of the sun fall on the metal plate, the plate is warmed. The metal pipes, which are welded to the plate, are warmed in turn. The liquid pumped through the pipes receives this heat and carries it off. The pipes and backing plate are of metals such as aluminum, copper, and iron, chosen because they are good conductors of heat. Both pipes and plate may be blackened by a paint containing carbon black in order to decrease reflection of the sun's rays and increase their capacity to absorb heat.

Above the pipes are two panes of glass. Much of the light is absorbed by and passed through the glass, but a small portion (perhaps 10% of the sun's rays) is reflected back. The primary purpose of the glass is to prevent back radiation from the metal plate to the sky. Just as the glass panes of a greenhouse prevent the warmth inside from being radiated out of the structure, the glass cover of a solar collector traps heat within the device. The panes of glass do more besides. Because glass is a poor conductor of heat, it provides a barrier between the warm interior and colder air outside the collector. Thus, loss of heat by conduction to the outside air is decreased. Convection losses, as winds sweep by the collector, are also decreased. Plastics may be used over the heat-collecting surface in place of the glass.

The box in which the metal collector is housed is made of a low-conducting material, such as wood or plastic, which helps prevent heat loss to the surrounding air and structures. Insulation is placed in the bottom of the box, and the box itself is usually mounted on the roof in a stationary position, though it can be mounted on the ground.

The flat-plate collector we have just described can be incorporated into the heating systems of new structures, including private homes, by replacing portions of the roofing that would otherwise be required. Also, the collector can easily be mounted on roofs of existing homes. The cost of these collectors has been coming down as assembly-line production provides savings in manufacture.

The Storage System

The collector itself is the exterior symbol of a home or building that uses solar energy. Inside the building, attached to the collector, is a system of heat exchangers and a storage tank, which together store and transfer the sun's heat. We will describe two different solar hot-water systems. The first is the conventional design that has been popular in the past. The second is a newer design, gaining in popularity.

The conventional design uses an antifreeze-water mixture as the collector fluid, which circulates through the tubes of a flat-plate collector. The heated antifreeze mixture is pumped to and through a heat exchanger, which is usually a coil of pipe inside a large water-filled storage tank. The heat in the flowing liquid is transferred through the pipes to the water in the tank, which stores the heat for later use and for rainy days. (See Figure 24.4.) The antifreeze-water mixture then cycles back to the collector.

An antifreeze-water mixture is used as the collector fluid instead of water to prevent the fluid from freezing

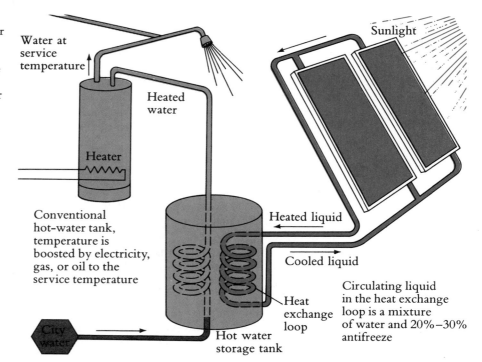

Figure 24.4 A typical arrangement of collector and storage tank for solar hot water. The water and antifreeze mixture in the collector loop circulates through the collector where the sun's energy is captured. The fluid then passes through a heat exchanger within a water-filled tank, transferring its heat to the water in the tank. The heat in the tank water is extracted by cold water, which flows through a second heat exchanger in the tank. The heated water in this service loop is now available for use in the house, either directly if it is hot enough, or with a boost from a conventional hot-water heater. The large water-filled tank is used to store heat from sunny days for use on cloudy days.

Water at service temperature

Heated water

Sunlight

Heater

Conventional hot-water tank, temperature is boosted by electricity, gas, or oil to the service temperature

Heated liquid

Cooled liquid

City water

Heat exchange loop

Circulating liquid in the heat exchange loop is a mixture of water and 20%–30% antifreeze

Hot water storage tank

when outdoor temperatures dropped to frigid levels. This is Bailey's basic design except that ethylene glycol (the antifreeze used in your car) has replaced the alcohol that Bailey used. This design (Figure 24.4) requires a heat exchanger placed in the tank in the collector loop.

The water in the storage tank serves as a heat reservoir. Heat can be extracted from the tank for space heating or, by passing a portion of the house water supply through a heat exchanger within the storage tank, hot water can be extracted for bathing, washing dishes, washing clothes, and the like. The water flowing through the heat exchanger is heated by the hot water in the tank. The heated water then passes to the conventional electric or gas water heater, where it can be given a temperature boost if needed.

The newer design for solar hot water is called the *drainback system.* It is increasing in popularity because it is less costly and more reliable than the conventional design. Total solar energy costs are keyed to the square footage of collector area on the roof. The conventional system has a total cost of more than $50 per square foot of collector area. The drainback system, in contrast, can be installed for as little as $25 per square foot of collector area. The reason the drainback system is both less costly and more reliable is that it is simpler, having

fewer components to install, repair, and replace. The drainback system is also relatively free from freezing problems.

In essence, the drainback system is Bailey's original design, with water as the fluid in the collector loop. The water that circulates is drawn directly from the storage tank, sent to the collector and then back to the storage tank with no intervening heat exchanger. Thus, the contents of the tank are heated directly as they circulate through the collector loop. Moreover, the drainback design has an added feature; when the temperature of the collector falls below the temperature of the circulating fluid, the fluid drains back into the storage tank leaving the collector loop empty, which prevents a loss of heat from the tank to the atmosphere and thereby prevents subsequent freezing. (See Figure 24.5.)

To use the heat in the storage tank to heat a home, several options exist, depending on what form of conventional heat the home uses. If home heating is supplied by forced air, as is so common now, water from the tank can be used in a water-to-air heat exchanger to warm the air flow to the house. If the air flow is not heated sufficiently, the conventional furnace turns on to provide a boost in temperature. The use of solar energy for space heating requires a far larger invest-

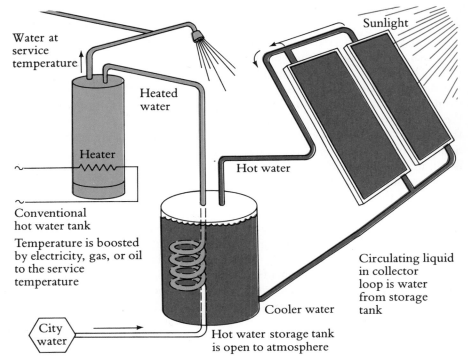

Figure 24.5 Schematic of a drainback solar hot water heater. The drainback system differs from the previous design only in the fact that the fluid in the collector loop is water that is drawn directly from the storage tank. The water "drains back" from the collector loop into the open storage tank to prevent freezing. This draining occurs when the temperature difference between collector and tank would cause a loss of heat from the tank contents to the outside.

Water at service temperature

Heated water

Sunlight

Heater

Hot water

Conventional hot water tank

Temperature is boosted by electricity, gas, or oil to the service temperature

City water

Circulating liquid in collector loop is water from storage tank

Cooler water

Hot water storage tank is open to atmosphere

(a)

(b)

Figure 24.6 Solar heating. (a) The Town Elementary School in Atlanta, Georgia. (Westinghouse photo) (b) A house in District Heights, Washington, D.C. (U.S. Department of Energy photo by Jack Schneider) On both structures, solar panels collect the sun's energy for use within the buildings. Provision is made in both buildings to store heat energy for cold, cloudy periods when sunlight is not available.

ment in collectors than solar hot water does. Nonetheless, structures using solar space heating are now being built. (See Figure 24.6.)

If one is going to use solar energy to heat a home or to produce hot water, the storage tank is an essential feature. The heated water can store the sun's energy from today until tomorrow or the day after or the day after that, depending on the volume of water in the tank. The storage tank frees the home from variable sunshine. It is much like a reservoir that stores water from periods of high flow until periods of need when the flow is low. The storage tank reserves solar energy from sunny days until days of need when there is little or no sunshine. When the stored heat runs out and there is still no sunshine, the conventional hot water heater or furnace supplies the needed energy.

One of the obstacles to installing solar hot-water or space-heating systems is the need for this large-volume tank; finding space for the tank may be difficult. In the late 1970s, a technology for heat storage, which was long thought possible, began to make its appearance on the solar components market. The technology consisted of a *phase-change salt* stored in a plastic or metal tray or tube. Glaubers salt (sodium sulfate decahydrate) melts in the 80°F range when air or water heated by radiant solar energy is passed over the tubes or trays. When the salt changes to a liquid, which it

does without a change in its temperature, a great deal of heat is stored. Thus heat can be stored in a very small volume. This heat can be recovered gradually as it is needed, simply by allowing the salt to solidify once again. The phase-change salt will provide heat storage, then, with a substantially reduced volume of the heat-storage medium.

Air as the Collecting Medium

A distinctly different system for solar heating relies on air as the collecting medium. Air passes through the collector and absorbs the sun's energy. Of course, this air cannot be stored the way water can be; the volume of hot air that would have to be stored would be enormous! An easy substitute for hot-air storage is a pebble or rock bed. The hot air from the collector is passed through a tank—a concrete or even a wooden structure—filled with rocks and pebbles. The rocks and pebbles are heated by this flow of warm air and store the heat until the dwelling requires it. When heat is required by the house, a blower forces house air through the warm pebble bed. The air that exits from the bed is heated and is distributed through the house. If the temperature in the house is not kept high enough by the solar heat, a conventional furnace will then turn on and assist in heating the air.

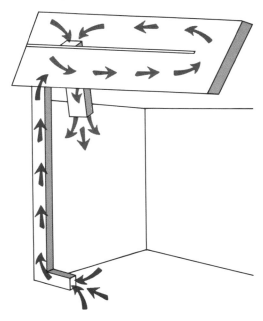

Figure 24.7 Solar hot air heater. This simple device, manufactured by the Solar Age Manufacturing Co. of Albuquerque, N.M., circulates air through the roof-mounted collector and back to the living space. Because air is the heat-collecting fluid, no freezing problems can occur.

A relatively simple and inexpensive device for solar house heating that uses air as the collecting medium is shown in Figure 24.7. A blower draws air up a duct to a roof-mounted solar collector. On its passage through the collector, the air is warmed and then is forced through the collector and back to the living space. Nothing is used for heat storage; the house and furnishings serve that function. Thermostatic controls close the ducts when the air would lose heat in the collector—that is, when the temperature of the collector falls below the temperature of the house air.

Passive Solar Energy

Unlike active solar-heating systems, some solar equipment uses little or no outside energy to assist in providing heat for the home. These systems are termed *passive*.

South-Facing Windows

One design for passive solar heating avoids collecting heat from the sun during the summer and captures the

sun's energy in the winter, when heat is needed. Large-paned windows are installed on the south wall of a building. These windows face the sun, and without further modification, the rays of the sun would be expected to enter the window both summer and winter. However, a roof overhang extends out over the windows (Figure 24.8). In the summer, when the sun is high, the overhang shades the windows from unwanted heat. But the winter sun hangs low in the southern sky throughout the day, and the roof overhang allows the sun's rays to enter and provide welcome heat for the dwelling.

The concept of passive solar-assisted heating has long been known. "In houses that look toward the south, the sun penetrates the portico in winter, while in summer the path of the sun is over our heads and above the roof so that there is shade." The speaker is Socrates, the philosopher of ancient Greece. The quote is drawn from *A Golden Thread: 2,500 Years of Solar Architecture and Technology* by Butti and Perlin (1980). As the title implies, people have been using the sun since the days of ancient Greece.

The overhang concept can be extended to multiple-floor office buildings. Here an overhang extends over the outer rim of every floor, blocking the sun in summer but allowing its entrance in winter. Interestingly, for the individual dwelling, a deciduous tree can serve

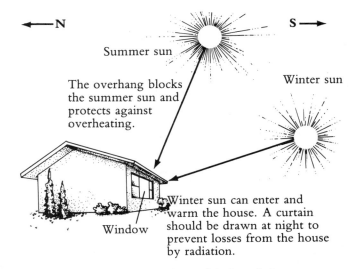

Figure 24.8 Proper orientation and design of a house can decrease heating and air-conditioning requirements. The south-facing house has an overhang that blocks the sun's rays in summer but does not block them in winter. (The house illustrated is in the Northern Hemisphere.)

much the same purpose as an overhang. Again, assume we have large windows on a south-facing wall. The leaves of a deciduous tree growing in front of the window will block out much of the searing summer sun, but in winter the sun's rays can penetrate through the bare branches to enter the window. Pine trees planted elsewhere around the house help to cut the winter wind and thus decrease the loss of heat from the house.

Although a large south-facing window admits and captures more sunlight than a smaller one, it also provides a wide surface that radiates heat into the night sky, an undesirable loss of energy. Opaque insulating curtains offer one solution to the problem of back radiation. Another novel solution to the problem is to reduce the size of the window and arrange mirrors to capture and reflect additional sunlight into the smaller window. Reasonable quantities of sunlight are captured in this way, and the potential for back radiation is reduced.

Still another way to use windows more fully as heat-capturing devices involves changing the transmitting properties of the windows themselves. Several manufacturers are now producing a clear polyester film coated with a thin layer of metal. The metal layer is so thin (only a few hundred atoms) that when the coated film is applied to a glass window, the window remains completely transparent. That is, light passes through the window unhindered. However, the metal layer blocks the passage of heat energy from the house to the outside. Windows with this extra glazing are ideal for passive solar heating because they retain the sun's energy as heat within the home. The special layer can be either added to a film, which is then applied to glass, or coated directly onto the glass in a factory process. Film for the homeowner to apply should be on the market.

Trombe Walls

A clever design of excellent potential that uses no moving parts is the Trombe (rhymes with prom) wall, a solar collector that is a part of the house itself. In a sense, it, too, is a large panel of windows facing south to catch the sun, but it is more. The large windows do allow entrance of the sun's rays, but a few inches behind the windows is a concrete wall. Behind the concrete wall is the living space of the home. The side of the wall that faces south is painted black so that the maximum amount of sunlight is absorbed as heat. The wall heats up, and the air space between the wall and the

window heats up as well. Openings near the top of the wall allow the warm, less-dense air in the air space to rise up and out to the living area. That is, natural convection carries warm air into the room and cool air out. The dense concrete structure acts as a heat storage device as well.

The Solar Greenhouse

Another architectural feature used for passive collection of the sun's energy is the solar greenhouse. The device is simply a greenhouse attached to a larger structure. Solar energy passes through glass and into the greenhouse. Drums painted black, which are filled with water, may be stored in the greenhouse (Figure 24.9). These drums and their contents are heated, and they store this solar heat for periods of no sunshine. A combination of a solar greenhouse using drum heat storage and a Trombe wall for further heat storage is shown in Figure 24.9.

The Future of Passive Solar Heating

These ideas for passive solar heating have been tried, and improvements have been suggested to correct some problems. The first problem is one of condensation. Moisture tends to accumulate in passive homes because of their tight construction. Ceiling fans to keep air moving should solve such problems. The second difficulty is the occurrence of local "hot spots" in the passive solar structures. Again, fans can be used to distribute heat from these hot spots through the rest of the house. Basically, passive heating seems to need a slight assist to make it fulfill its promise.

Passive solar architecture results in a house about 10% more expensive than a conventional home. The heating costs, however, may be 50%–80% less than those for the conventional home. It should come as no surprise, then, that in 1984, 3%–5% of new homes were employing some form of passive solar design. In addition, passive solar systems were installed on new homes in the northern United States climes and in Canada, not simply on homes in the southwest. Several sources predict tremendous gains in popularity of passive solar design. One set of architects predicted in *Solar Age* that passive solar heating would become standard practice in new home design in the decade 1985–1995 (Holtz et al., 1985). That may be industry hype, but if only half the predictions come true, a revolution

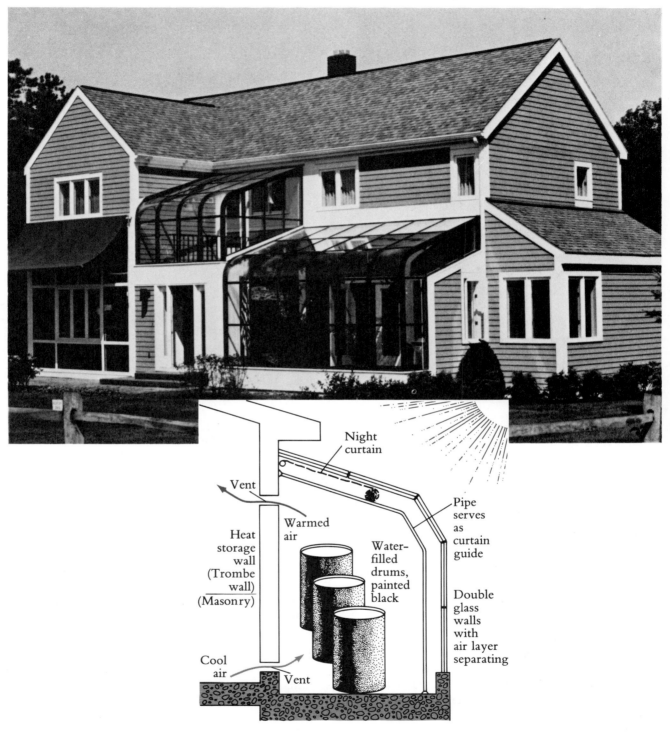

Figure 24.9 The solar greenhouse, or solarium, a passive solar system. This old idea from the Victorian era uses a greenhouse attached to and open to a residence. The furniture, floor, and perhaps drums of water are heated by the sun's rays. These fixtures radiate heat to the solar room and to the rest of the house when sunshine has ceased. (Photo courtesy of Four Seasons Solar Products Corp.)

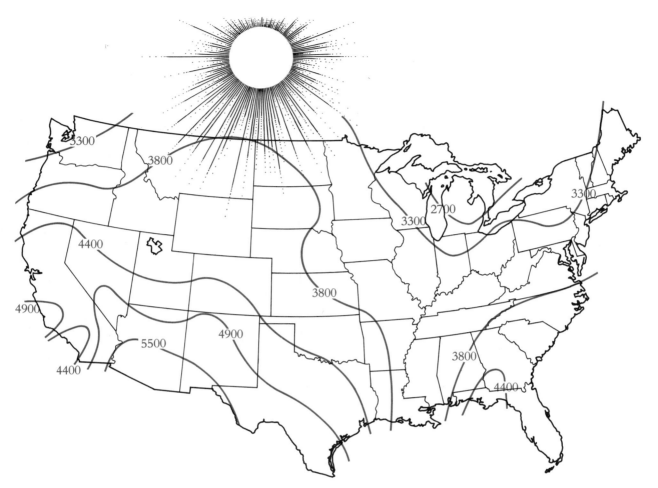

Figure 24.10 Distribution of solar energy over the United States. The figures give solar heat in Btu per square feet per day. Although some sections of the country have more sunshine than others, the attractiveness of solar energy depends also on the cost of alternate energy sources, such as electricity, oil, and so on. (From D. Kash et al., *Energy Alternatives*, Report to the President's Council on Environmental Quality, 1975.)

is in the offing. Solar energy has a distinct advantage in those portions of the country that have a high percentage of sunny days. The rates of solar heat reaching different portions of the United States are shown in Figure 24.10.

Interior Lighting from Sunlight

We have always known that the sun gives us light, and homes are often built with numerous windows to take advantage of this gift from nature. However, in most situations, interior lighting is provided only by electricity. In industrial and high-rise office buildings, interior rooms are often built without access to sunlight because windows are simply impossible to provide. Instead, incandescent and fluorescent bulbs furnish the light required for work. Artificial lighting in residential and commercial buildings has been estimated to account for 15% of U.S. electricity consumption. This fact motivates us to explore how we can use the sun's rays to light our way.

The idea of "piping" sunlight through a building

was tried in 1974 when the Hyatt Regency Hotel in Chicago brought in sunlight to decorate the glass ceiling in its lobby. In 1976 two physicists at Sandia Labs in New Mexico piped in sunlight to illuminate their windowless office. To their pleasure, they found the lighting scheme not only economical, but also pleasing in the quality of lighting they obtained. The faint image of the sun was reflected on the walls, and clouds could be seen passing over the image.

The system for lighting interior rooms with sunlight has only one moving part, a flat mirror controlled by a small computer to track the sun through the daytime hours.

Relative economics for solar versus electric lighting appear favorable because incandescent lights are only 10% efficient in converting electrical energy to light. Fluorescent bulbs, while more efficient, still convert only 20% of electrical energy into light. Sunlight directed into a building provides heat as well as light. Since lights furnish a portion of a building's heat requirement in winter, the use of sunlight for interior lighting can potentially save heating fuel as well.

Conversion of Sunlight to Electricity

Photovoltaics

The process of converting sunlight directly to electricity is known as **photovoltaic conversion**. The devices called *semiconductors*, made from silicon, that have so transformed the computer industry, are the basis of solar cells. The cells are constructed by layering two wafers of silicon crystal with junctions between them. Light falling on one of the wafers "boils" electrons out of the crystal and across the junction into the other wafer. This current is a direct current (dc), and so requires conversion to alternating current (ac) before it is used by a consumer.

The efficiency of converting solar energy to electricity is low; only about 12%–14% of the arriving solar energy is converted to electricity by the currently available silicon-based cells. Although silicon is the only material available commercially, cadmium sulfide and gallium arsenide can also be used in solar cells. Gallium arsenide, in particular, holds promise for higher conversion efficiencies, perhaps up to 30% of the arriving solar energy.

Solar cells were first used in dramatic applications

where cost was not a matter of concern. For instance, solar cells furnished the power for the Apollo moon rockets and the Viking space stations. In addition, the cells have been used for mountaintop radios, ocean signal buoys, highway call boxes, pipeline corrosion signal systems, and other applications. Typically, these uses require only modest amounts of power and are far removed from an electric transmission grid. A fledgling U.S. photovoltaics industry supplies these needs. Some 23 firms manufactured and shipped photovoltaic modules in 1984. Another 10 to 15 firms were producing solar cells for their own internal use, but on a very small scale.

In 1981, completed photovoltaic systems (as opposed to unconnected modules) cost between $15 and $60 per peak watt.* Because the 1981 cost for conventional central-station power was on the order of $1.20 per watt, the cost of power from solar cell arrays was still more expensive than conventional power. By 1985, costs per peak watt in assembled systems had fallen to $5, and the utility industry began to anticipate the dawning of central-station photovoltaic power. Numerous medium-scale applications (100–200 kilowatts) were underway in the mid-1980s.

Not only are medium-scale applications moving forward, solar cells are also now being linked into huge arrays to produce electric power at central stations. Arco Solar Industries is taking the lead in this activity. By 1981, the company had built and was operating a 1-megawatt (peak) photovoltaic power plant near Hespena, California; power was being sold to Southern California Edison. And in late 1983, Arco completed a 6.4-megawatt (peak) photovoltaic power plant near California City, California. Pacific Gas & Electric is buying the power from this plant. The power production of these plants reaches a peak when the sun is at its zenith; this is also near the time when power is most needed for air-conditioning loads in the southwest.

A remarkable concept in photovoltaics was made a reality in Frederick, Maryland. There a photovoltaic plant, fully powered by solar cells, has begun operation. The Solarex Corporation has named it the *solar breeder*, a name that recalls the breeder reactor, a plutonium-fueled nuclear power plant that the United States has decided not to build (see Chapter 18). The breeder

---

*The cost per peak watt is calculated by dividing the total cost by the number of watts produced during the portion of the day when the maximum solar energy is being received.

Figure 24.11 The solar breeder. This unique plant in Frederick, Maryland manufactures photovoltaic cells and panels using only energy from the sun. The 200-kilowatt array of solar cells that forms the roof of the facility provides enough electricity for all needs of the plant's production lines. A large battery storage system puts away energy on sunny days for continued production on cloudy days. Heating requirements are furnished by solar thermal energy, again with energy storage for sunless days. (Photo courtesy of the Solarex Corporation)

reactor would have produced tons of radioactive wastes; the solar breeder produces none. (See Figure 24.11.)

The manufacture of solar cells has increased. From about 4 megawatts of capacity in 1981, shipments increased to 7.9 megawatts in 1982, grew to 12.6 megawatts in 1983, and sagged slightly to 9.9 megawatts in 1984. About 50% of the capacity produced in 1984 was destined for use at central power station arrays.

Photovoltaic power has some way to go before it becomes economical for generating central-station power or for turning on our lights at home. But, in a world where technology advances at a rapid pace, there is hope that the cost of semiconductors can be markedly reduced, especially when they become mass produced.

If cost barriers can be overcome, and utility experts believe they can be, then photovoltaic electricity becomes very attractive indeed. No steam is generated; nothing is burned, heated, or fissioned; no fuel is mined and transported. No parts move and wear out. The plants function unattended, and the cells are expected to last a long time when enclosed in glass or plastic. Finally, the supply of fuel is inexhaustible.

Central-Station Electricity from Solar Energy

Power tower. Picture again the invasion of ancient Syracuse and envision Archimedes directing his solar assault on the approaching warships. This is the principle of the **power tower**; the sun's rays are concentrated on a single point by properly adjusted mirrors. The mirrors, known as *heliostats*, rotate through the day to follow the sun across the sky. They reflect and focus the sun's rays on the power tower, where the enormous concentration of energy boils water to steam. The steam, piped to a turbine on the ground, turns the turbine and generates electricity (Figure 24.12).

A decade ago, the power tower was only a concept on a drawing board. Today, it has been become a reality in the United States and overseas, and utilities see this new type of power plant as cost competitive with oil- and gas-fired electric plants.

The largest operating power tower is Solar One, a

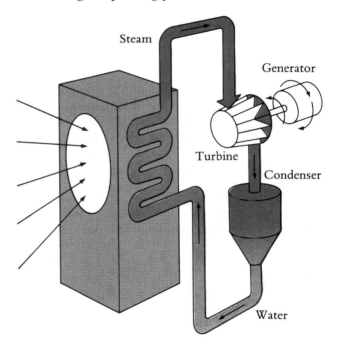

Figure 24.12 The power tower concept. Water is boiled to steam by the concentrated rays of the sun. The steam is used to turn a turbine. (Adapted from D. Kash et al., *Energy Alternatives*, Report to the President's Council on Environmental Quality, 1975.)

Figure 24.13 The power tower in operation. The Solar One facility, located near Barstow, California, can deliver up to 10 megawatts of power. A field of heliostats track the sun and focus the sun's rays on a boiler at the top of the power tower. Here water is boiled to steam, which is piped to a turbine generator unit at the base of the tower for the generation of electricity. (Photo courtesy of the Electric Power Research Institute)

10-megawatt (at peak) power plant occupying 130 acres in the Mojave Desert at Dagget, California near Barstow (Figure 24.13). Begun in 1978, the unit, which was built by McDonnel-Douglas, started delivering power to the grid in 1982. The $140 million facility is a joint effort of the Southern California Edison Company ($20 million) and the U.S. Department of Energy ($120 million). Although the construction cost was $14 per watt, about ten times the cost of conventional power, there is promise that future plants will be far less expensive because of the savings that come from mass production of the heliostats. Costs in the range of $1.50 to $4.00 per watt seem feasible if heliostat costs can be reduced to $250 per square meter.

How does Solar One work? Its 1,818 heliostats, which were furnished by Martin Marietta Corp., each have 12 facets, and each is 23 feet (7 m) on a side. The heliostats track the sun through the sky, reflecting the rays of the sun to the top of the power tower. There, at the top of the 310-foot (95 m) tower, the sun's rays focus on the boiler and produce steam at 950°F. The steam turns a turbine on the ground, generating enough power for a community of 7,000–10,000 people.

The power tower has provision for heat storage—a 1-million-gallon tank filled with rocks. The steam produced during periods of sunlight is used to heat oil, which circulates through the tank and heats the rocks. Using this stored heat to produce steam during hours of darkness, the facility can produce 7 megawatts for up to four hours. But heat storage for off-sun hours may not be necessary if the power tower is viewed only as a device to provide power at peak times of demand.

Solar One *works*. It was expensive to build, but this is because components were specially constructed. The information provided by its operation, such as frequency of breakdown of heliostats and the failure of construction materials, will prove valuable in the next generation of power towers. Southern California Edison is already planning for that next generation: a 100-megawatt unit to be producing power by 1988. In the meantime, it also contracted for a 12-megawatt power tower in San Bernardino County.

The power tower concept is being explored elsewhere in the United States, Europe, and Japan. There are units at Sandia Labs (Albuquerque, New Mexico), and solar power towers have been built at Almeria, Spain; Adrano, Sicily (Italy); Themis, France; and Nio Town, Japan, though none are as large as Solar One.

Solar Plant 1. The power tower is not the only concept for central-station power plants that would use the sun's energy to produce steam. Solar One and its cousins have real competition from a 5-megawatt station named Solar Plant 1 in Warner Springs, California. The plant, finished in 1985, is different from Solar One in two distinct ways. First, it does not use expensive glass mirrors; instead it uses a metal-coated plastic film stretched tightly over five-foot aluminum hoops. A vacuum pump pulls the film into the needed curved reflective shape, producing a collector for a mere $20.

The second way in which Solar Plant 1 differs is that it uses no power tower. Instead, each array of mirrors is aimed at its own attached local collector at the focal point of the array. Inside the collector, a salt-water mixture is heated by the sun's reflected rays. Water circulating in pipes through the mixture in the collector is boiled to steam, and after another step of heating in another collector associated with another array, the steam is gathered together to turn a central turbine.

The new concepts at Solar Plant 1 reduced the construction cost to $2.80 per watt, about one-sixth the cost of Solar One. The cost to produce power at any plant is the cost of operation, maintenance, and fuel. Solar Plant 1 is expected to produce power at costs comparable to a coal-fired power plant, 2 to 3 cents per kilowatt-hour. Power will be sold to San Diego Gas & Electric. The conclusion from such costs is that solar central-station electricity using steam turbines is almost ready to compete with conventional power.

Biomass: Biological Conversion of the Sun's Energy (Alias Photosynthesis)

Of all the conventional and widely used energy sources we have discussed, only nuclear power is not linked to the sun. Oil and gas are thought to be the buried remains of ocean plankton, species that derived their energy directly or indirectly from the sun. Coal was originally merely woody plants buried beneath the sediments in geological history. Hydropower can be traced to the sun's evaporation of water in the hydrologic cycle. Thus, our most common energy resources are linked closely to the sun.

Now, in an era of awakening to the finiteness of our energy resource base, we are turning to the sun to convert its rays directly to heat, hot water, electricity, and mechanical energy. That is, we are attempting to use our mechanical and engineering skills to apply the sun's rays directly to our purposes without any intermediate processing by nature.

It looks as though we will be reasonably successful at this conversion, but the fact that oil, gas, and coal all owe their energy to the sun reminds us strongly that nature already does an excellent job at converting the sun's rays to energy-laden molecules. We know the process as *photosynthesis*, the conversion of carbon dioxide and water to energy-rich organic molecules through the use of the sun's energy. Why not take advantage of the energy being captured by green plants here and now? It is an idea that was already used in earlier times.

Wood

Until the twentieth century, most American families cooked and heated their homes with wood-burning stoves. Some people never stopped heating their homes with wood; they are typically rural people largely con-

centrated in Maine, Vermont, New Hampshire, and parts of the rural South. Wood was always cheaply available to them, and it made sense to be economical. Now we are emulating them: About 11.5 million new wood-burning stoves were sold between 1974 and the end of 1981. By 1981 some 4.5 million households were using wood as their primary source of heat, and 10 million households were burning wood as a secondary source of heat. In 1982 and 1983, another 2.7 million stoves were sold nationally. Newer stoves are often equipped to burn wood or coal and may have catalytic combusters to make the wood burn more efficiently. It is estimated that 48 million tons of wood were burned to heat residences in 1981 in the United States, double the tonnage used in 1969. Wood is also the fuel used by traditional societies throughout the developing world.

The burning of wood in industry has grown almost as fast. From 59 million tons of wood in 1969, industry increased its use to 81 million tons in 1981. This wood is chiefly scrap material from the pulp and paper industry and from the wood products industry, and it is burned in the boilers of these industries.

The saying goes that wood heats twice: once when you cut it and again when you burn it. As opposed to drafty fireplaces that can cost additional heat, wood stoves are also a valuable backup to conventional heat. If an ice storm were to snap transmission lines, as one did in New England in the winter of 1973, the loss of electricity would mean little or no heat in forced-air heating systems because no fan would be available to blow heat through the house. A wood stove with fuel would be a welcome feature at such a time.

Wood is a fairly clean fuel; apparently sulfur does not become incorporated into wood until it has reached the stage of advanced peat formation. Its heat content is also high: A cord of wood (a stack 4 feet × 4 feet × 8 feet) has about 20,000,000 Btu on average; a ton of coal has 24,000,000 Btu. The price of a ton of coal and that of a cord of wood vary by region and season, but they are not far apart.

The fuel for wood stoves does not diminish our resource base if it is derived from scrap or from downed trees that otherwise would be wasted or would rot. Indeed, it takes pressure off the oil and natural gas markets. However, if the wood is drawn from woodlots that are not being replenished by the planting of new trees, there is an environmental effect. Natural regeneration of woodlots, depending on the locale, may take as little as 25 years or as long as 100 years. It is also possible to get wood from downed timber in state and national forests, where removal is often allowed. Two positive effects of this removal are that fire hazard may be reduced and that sunlight will reach seedlings more easily. But there is a negative effect as well. If left in place, the downed wood decays gradually, recycling its nutrients and minerals into the soil. Or fire may consume the downed timber, recycling the minerals quickly to the soil. When downed timber is removed, the cycle of growth and decay is interrupted.

If wood biomass were to become a major fuel, thousands upon thousands of new acres of trees would have to be planted; most of our present tree resources are already devoted to timbering and recreation. One proposal for producing wood for fuel in quantity is to grow sycamore trees for only 5 years to immature size, harvest them mechanically, and replant. Poplar is another fast-growing variety that may produce high yields. Although the burning of wood adds carbon dioxide to the atmosphere, just as the burning of oil, gas, or coal does, the growing trees would be withdrawing carbon dioxide from the air for photosynthesis. A rough balance of carbon dioxide addition and removal seems possible.

Field Crops

Sugar cane is another crop that gathers the sun's energy efficiently. In fact, sugar cane may be the best biological converter we have. Its energy can be made available for nonfood use by fermenting it to ethyl alcohol. Sugar cane does not grow as well in this country as other crops, particularly corn. Brazil, on the other hand, is well suited for sugar cane and is seriously pursuing the commercial production of fuel alcohol from sugar cane. By 1981 Brazil had 400,000 cars that ran on ethyl alcohol as fuel and 5,000 alcohol-based service stations. At that time, Brazil was also planning for a major increase in fuel alcohol production from sugar cane.

One unusual species of plant, the Hevea rubber plant, first found wild in Brazil, produces hydrocarbons that are used to make natural rubber. Grown almost exclusively in Indonesia and Malaya, the plant yields almost a ton per acre. Though this is half the yield of sugar cane, the growers think Hevea yields can be tripled. Although the Hevea plant has been our source of natural rubber for many years, enthusiasts see the hydrocarbons it produces as substitutes for oil.

Hevea has competition, however, from another plant that produces hydrocarbons.

Native to the deserts of Mexico and the American Southwest, the wild desert shrub known as guayule (pronounced gwy-oo'-lee) has been cultivated in the United States in the past. In the early 1900s, guayule was one of our principal sources of rubber, but rubber from the Hevea plant replaced it in the marketplace. During World War II, when Southeast Asia was dominated by the Japanese, we again established a rubber industry based on guayule. That industry collapsed, however, when the rubber-producing nations of Malaya and Indonesia were liberated. Now a number of scientists want to give guayule another try after first selecting the best strains for cultivation. Concern has been voiced, though, over the hybridized species of guayule that are being developed. The plant may cause severe skin rashes on the workers who harvest them, making mechanical harvesting a necessity. Still other desert plants have potential for cultivation as oilseed crops; these are jojoba, buffalo gourd, the gopher plant, guar, and devil's-claw.

Corn and corn wastes have been suggested as a source of biomass fuel. The burning of husks and stalks may be used in grain-drying operations, which presently burn propane and liquefied petroleum gas for heat. Corn can also be used to produce one component of the blended fuel popularly known as gasohol. Gasohol is a mixture of 90% unleaded gasoline and 10% grain alcohol. Corn carbohydrates can be fermented to produce the grain alcohol, which is then distilled to a high-purity alcohol. The technology of fermentation and distillation is well known and is the basis of the manufacture of whiskey from corn. The process is so simple that it has been done at home in illegal stills for many years.

The economics of gasohol are most uncertain. Gasohol is nearly competitive with gasoline, but the price of corn and the price of oil interact in a complex fashion to determine whether gasohol is profitable. A low price for corn and a high price for oil is definitely most favorable for gasohol, but year-to-year variations in the prices of corn and oil make investment in the fuel alcohol industry an uncertain proposition.

The potential for biomass fuels is very large, but the impact of this development is uncertain. When use is made of otherwise discarded materials, producing biomass fuels is only good conservation practice, reflecting the ethic "Waste not, want not." When new crops, new processes, or new acreages are proposed, though, we tread on less certain ground.

We do not know whether we can afford the acreage for energy crops at the expense of food crops. Turning cropland devoted to food plants into cropland devoted to energy plants could cause a rise in the price of basic foodstuffs. U.S. consumers aside, for a nation that serves as a breadbasket to the world, the choice of energy crops over food crops could have wide effects. The use of residues left after harvesting conventional crops, rather than going to an agriculture that grows plants for energy, might be a more sound course of action while we learn about biomass energy. We also do not know whether we can devote enough land to wood for burning because of the large demands for timber and paper. Also, many biomass processes do not have the advantage of some solar technologies because they require central processing. Solar house heating, in contrast to biomass energy plans, is a decentralized activity, carried out in each dwelling. (See Controversy 24.1.)

Of perhaps greatest concern, however, is the soundness of biomass proposals from a biological point of view. Planting only a single crop has a tendency to deplete the soil of particular nutrients, attract insect pests, and reduce the survival of a diversity of animals. Caution is needed lest tropical forests be destroyed to produce a source of energy. Field crop research on these ideas is needed to evaluate biomass further. And careful economic analysis of the effects on food prices is needed as well.

Ocean Thermal Energy Conversion (OTEC)

Another idea to exploit the sun's energy draws on the temperature differences that exist between the surface and the depths of tropical oceans. The temperature differences are very large in some places, especially in areas where the warm Gulf Stream flows. Between the surface and a depth of 2,000 feet (600 m), the temperature difference may be as great as 22°C (40°F).

The OTEC idea is not at all new. First proposed by the French physicist d'Arsonval in 1881, the concept was actually tested more than half a century ago in the Bay of Matanzas, Cuba. In 1929 d'Arsonval's pupil, Georges Claude, built and tested the forerunner of current OTEC concepts, but waves and currents destroyed his experiment before he could make the idea practical.

The principle of OTEC is to use these waters at different temperatures alternately to boil and condense a *working fluid*. In between, the vapor at high pressure is used to turn a turbine. The working fluids most studied and discussed are ammonia and propane. OTEC works this way: Warm surface waters are used to boil the *liquid* working fluid to a *vapor*. The vapor is used to turn a turbine and generate electricity. Cold waters are brought up from the ocean's depths to condense the vapor in another heat exchanger (Figure 24.14). The condensed fluid is returned to be vaporized again. Electricity would be generated continuously.

Of course, all of this is to be done at sea, so the

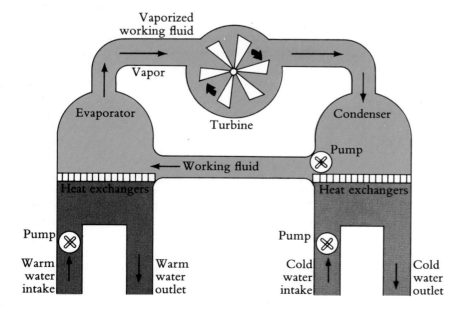

Figure 24.14 Schematic diagram of a closed-cycle ocean thermal power plant. (Adapted from John R. Justus, "Renewable Sources of Energy from the Ocean," in *Project Interdependence*, Committee Print 95-3, U.S. Congress, November 1977.)

CONTROVERSY 24.1

Should Energy Sources Be Centralized or Decentralized?

Amory Lovins, a noted independent energy policy analyst, has suggested that there are

> . . . two energy paths that the U.S. (and, by analogy, other countries) might follow over the next 50 years. The first, or "hard" path is . . . high technology, centralized, increasingly electrified, and reliant chiefly on depletable resources (coal and uranium). The second or "soft" path is . . . relatively low technology and decentralized, electrified only where essential, based on renewable resources, and fission free. . . . (*Alternate Long-Range Energy Strategies*, 1976)

Alvin Weinberg, director of the Institute for Policy Analysis of the Oak Ridge Associated Universities, takes issue with Lovins's suggestion that other forms of energy can be employed as effectively as electrical energy. He feels that

> Lovins ignores electricity's advantages . . . The question of whether the advantages of electricity—its convenience, its cleanliness (compared with decentralized fossil fuel systems) and its costs are worth sacrificing . . . [because of] its admittedly poorer thermodynamic match with its end use and its possible social consequences if it is generated by fission (Weinberg, 1978)

But David Lilienthal (1978), former Chairman of the Tennessee Valley Authority, one of the largest suppliers of electricity in the nation, disputes what was once accepted wisdom:

> We were persuaded to accept the fashionable idea that great new generating stations and huge, regionalized transmission systems would deliver electric energy more efficiently at lower cost to

the public than small, local decentralized ones. Nobody foresaw what would happen to the cost of oil and gas and coal, to the cost of transporting these fuels, or to the cost of constructing huge generating plants and mighty transmission lines. No one foresaw the rapid diminution of the world's reserves of oil. We placed major reliance on nuclear energy without being fully aware of its costs or its hazards.

Finally, Wilson Clark (1981), commenting on the risks to our energy supplies in time of war, points out:

> . . . the increasing centralization of . . . our energy systems is increasing our nation's vulnerability, in the name of decreasing it.

Alvin Weinberg favors central-station electric power; Amory Lovins advocates dispersed siting of power stations and individual energy generation. Their debate in many ways is *the* debate on centralized versus dispersed power sources; their exchanges are *the* exchanges of two communities. Both brilliant, both articulate, they have squared off, combating each other's views. To call one a conservative and the other a liberal, or one a traditionalist and the other a modernist, would probably be incorrect. Weinberg, however, has referred to Lovins and others as energy radicals. This is a most inappropriate label because of its political implications.

Both points of view have a long history. Individual energy generation is perhaps the oldest tradition, but the last half-century has been an era of growth in centralized power generation. Thus, the debate is not one between tradition and new ideas, but one between two points of view that were both valid at different times in history. Each point of view sees current problems being solved in a different way. Whose way do you think is most efficient and why? Or is it possibly a new time or a transition time when both points of view are valid? The issue of whose way is best should be approached by attempting to construct a list of advantages and disadvantages in each concept.

To answer a question such as this probably requires focusing on a single use of energy and answering specific questions. Since nearly half of America's homes heat water with electric energy, and since solar hot water is feasible and comparable in cost to electric hot water, it would be useful to compare these two energy uses. In making your comparison, recall that solar hot water may require electric energy to boost its temperature when the sun's rays are hidden for too long. Of course, burning oil or gas could be used to boost the temperature of the water as well.

Following are some questions you might use to make your comparison. Which is more reliable: hot water from electricity from a huge central power station, or hot water from solar energy with an occasional electric assist? Which is safer? Which is less subject to disruption? Which is less costly? Which is more convenient? Which is cleaner? Or, more accurately, where is the mess? If you add up the environmental problems, which method has less impact? Which uses more resources?

Are there intangibles in your comparison between central and decentralized power? That is, are there elements in the one you favor that you cannot quantify but that you simply prefer? What are those elements, if any? Can you identify a tradition that has the same point of view on what is preferred?

Sources:

Alternate Long-Range Energy Strategies. Joint Hearings Before the Select Committee on Small Business and the Committee on Interior and Insular Affairs, U.S. Senate, Interior Committee Serial No. (94–47) (92–137), December 9, 1976.

A. Weinberg. In a review of *Soft Energy Paths: Toward a Durable Peace*, by Amory Lovins (Ballinger, 1977), appearing in *Energy Policy* (March 1978), p. 85.

David Lilienthal, quoted in *National Journal* (April 29, 1978), p. 674.

Clark Wilson, quoted in *Science*, **21** (February 13, 1981), 683.

J. Ferguson, "Is Central Station Generation Becoming a White Elephant?" *Public Utilities Fortnightly* (March 21, 1985), p. 32.

electricity would require transport to land. Underwater cables would be used for this purpose. On the other hand, the electricity could be used at sea as it is generated, for such electric-intensive industries as aluminum or electric arc steel.

Both research and development on OTEC are being carried forward. The design of the huge heat exchangers, as well as of turbines, pumps, and other support structures, is being investigated. In addition, the pipe for a 100-megawatt plant must reach several thousand feet into the ocean depths; it would have a staggering 50-foot (15 m) diameter. Can such a pipe be made to withstand buffeting by waves and currents? Or will multiple cold-water pipes be needed?

Two pilot projects, both in the waters of Hawaii, are coming to grips with OTEC problems. A Department of Energy (DOE) facility, designated as OTEC-1, began operation in 1982 near Keahole Point, Hawaii; the facility, mounted on a former U.S. Navy tanker, is drawing cold water from 2,100 feet (640 m) below the surface. Ammonia is being vaporized in a heat exchanger by the warm water at the surface, and it is being condensed in a second heat exchanger by the cold water from the ocean depths. The ammonia "steam," however, is not used to generate electricity; the test is only of the heat exchangers—in particular, how to keep them effective and functioning in a marine environment.

The second major test facility, known as Mini-OTEC, was mounted on a barge off the Kona coast of Hawaii. In 1980, this 50-kilowatt plant, also using ammonia as a working fluid, was able to generate 12 kilowatts of net power after electricity needs for such operations as pumping were deducted. The plant was a joint venture of the State of Hawaii and three private corporations. Although a milestone in OTEC testing, the facility resolved few of the engineering problems faced by the OTEC concept.

On the drawing boards, DOE is planning a demonstration facility consisting of a cluster of four

10-megawatt OTEC units. Four separate units are used in order to ensure reliability; one unit going down will leave three others operating.

Many problems must still be solved. Corrosion by seawater is one problem area. Ocean organisms and slime may be foul and clog the heat exchanger, reducing its effectiveness. Large quantities of antifouling chemicals may be required. A novel idea is to bounce hard rubber balls back and forth through the heat exchanger pipes to keep them clear of scale.

Environmental questions such as impact on climate by diverting Gulf Stream waters still need to be answered. Nonetheless, vast quantities of electricity may be generated in this way. One estimate by the Department of Energy suggests that power equivalent to that from 200 electrical plants of a 1,000-megawatt rating could be produced from OTEC plants moored in U.S. waters (Figure 24.15).

Prospects for Solar Energy

Solar energy can benefit people in a number of basic ways. First, by replacing fossil fuels, air and water pollution are decreased. Second, the replacement of fossil fuels means a decrease in fuel imports, especially of oil, and this will help secure the value of the dollar. Third, by replacing nuclear fuels here and abroad, the threat of the spread of nuclear weapons is decreased. Finally, solar sources can provide us some protection by making our fuel supply less subject to interruption.

What portion of our energy needs can solar energy contribute? There are certainly optimistic views of its potential. Both the President's Council on Environmental Quality (1978) and the Project Independence Report (1974) suggested that under conditions of "accelerated development," solar power could contribute nearly 25% of the nation's energy requirements by the year 2000. Both reports had a wider definition of solar energy than we use here; solar included energy from wind and from hydropower. We have to remove the potential of these two energy sources from the solar category to arrive at our estimate of solar's potential contribution. The projection, then, for the year 2000 is that about 13% of the nation's energy needs could be supplied by solar energy in the form of heating, cooling, electricity generation at central stations, photovoltaic electric energy, and biomass. Thirteen percent is

Figure 24.15 Artist's conception of an OTEC factory ship. Ocean thermal energy conversion would exploit the temperature differences between the layers of tropical oceans. (The Johns Hopkins University Applied Physics Laboratory)

a very large number for an economy that is expected to be using 80–120 quadrillion Btu of energy in the year 2000.

It is good that such a major contribution is in sight, but what are the conditions of *accelerated development*? This term has no precise definition; it means many things. Basically, however, it means some form of government intervention in the market for solar technology. Without such involvement, the paths that the solar market can take will differ dramatically.

Government intervention in markets is nothing new and nothing radical for Democrats or Republicans. Before the OPEC era, oil from overseas was brought in on a quota system for many years. The quota system prevented the domestic oil market from being flooded with cheap imported oil and thus protected the American oil industry. Strategic metals have been stockpiled for many years, both to have a store of the metals for emergencies and to shore up the markets for these materials. To keep their incomes steady, farmers have received price supports on basic grains for many years. And in order to keep crop prices up, they have been paid to keep acreage out of production. In addition, a system of agricultural extension agents provide free advice to farmers on all aspects of farm operations. Why not have a system of energy extension agents who would provide free advice to homeowners on the most appropriate ways to heat homes and who would lay out choices that include solar options and energy conservation?

The federal government, through the Atomic Energy Commission, has done most of the research and development on nuclear power and still provides the insurance against a nuclear catastrophe. Several nuclear fuel reprocessing plants were heavily supported by government money. In fact, the entire nuclear industry has been backed by government involvement from the start. If nuclear power research and development merit government support, why not solar power research? The suggestion that the government stimulate the solar market by incentives and by bearing some development costs is not at all a radical proposal.

Incentives are necessary because active solar heating systems are expensive. A house with an area of 1,500 square feet (135 m²) and two or three days of heat storage may need 750 square feet (68 m²) of collector area. At a cost of $22 per square foot ($242/m²), the cost of the collectors alone is $16,500. The volume of the tank in gallons would be about twice the collector area. The 1,500-square-foot house will then need a tank with storage of 1,500 gallons.★ Installation and other equipment may push the cost to $20,000 and more for the entire solar house-heating system.

Basically, two kinds of costs need to be reduced to help people buy solar technology. The cost of collectors can be brought down by mass production—that is, by expansion of the market to many thousands, even millions, of units. The cost of installation, pipes, plumbing, and the instruments for solar heating can be brought down by having a number of experienced companies willing to compete with one another for jobs. Again, a mass market is the key. These two costs, collectors and installation, make up the bulk of the cost of a solar heating system.

★A cylindrical tank with a 3-foot diameter and 7-foot height would provide roughly this volume.

One mechanism the government has used to influence people to buy solar equipment is tax credits. Until 1986, 40% of the investment in solar energy, up to a total credit of $4,000, could be subtracted from taxes owed. Thus, a family that invested $5,000 in solar hot water was able to deduct from taxes owed 40% of $5,000, or $2,000, effectively reducing the investment in solar equipment from $5,000 to $3,000.

There are many more variations on these methods of stimulating the use of solar energy, but this is basically what the Presidential Council on Environmental Quality meant by the term *accelerated development*—a market for solar equipment that is given government support through low-interest loans, tax credits, deductions, or similar concepts. Since the precise effect of any of these programs is unknown, the future is difficult to predict. Unfortunately, federal solar tax credits were allowed to expire on January 1, 1986, making the future of solar power even more of a question mark.

A number of states still offer tax credits for solar installations. A 1981 survey conducted by the National Solar Heating and Cooling Institute found 28 of the 50 states offering some form of state income tax incentive for people who install solar equipment. Also, 11 of the 50 states offer either an exemption from state sales tax or a refund of any sales tax applied to residential solar equipment. Finally, 29 of the 50 states offer some form of exemption from the annual property tax that would otherwise be paid on the value of the solar installation.

All these incentives and the removal of barriers had a positive effect on the manufacture and sales of solar equipment for heat and hot water. Shipments of solar thermal collectors in 1982, 1983, and 1984 were 28.6, 16.8, and 16.4 million square feet. Of the 1984 manufacturing activity, about 85% of the collector area was destined for home use. Solar hot water installations accounted for about 55% of area sold in 1984; space heating, 15%; and pool heating, about 27%. The 55% figure for solar hot water translates into 40,000–50,000 installations in 1984.

Nonetheless, solar heat and solar hot water still have a long way to go. One of the reasons solar power has not taken hold even faster is that up-front cash is required to purchase solar equipment, and the interest rates on borrowed money were at historic highs in the early 1980s. To further stimulate solar purchases, President Carter proposed a Solar Development Bank, a repository of federal money that could be loaned at low interest rates for the purchase and installation of solar equipment. Although President Carter wanted to open the bank with $450 million to loan, President Reagan asked Congress to rescind its appropriations to the bank.

Another suggested incentive in addition to the Solar Bank would be used to increase the number of solar homes being built. Home builders do not like to install costly equipment that may make a home too expensive to sell, so they have shied away from building solar homes. To increase the number of solar homes being built, solar advocates proposed a $2,000-per-home tax credit to builders for each solar home erected. However, in a congressional mood of budget cutting, the proposal was defeated.

As important as government incentives are, communication to the homeowner about available solar technology is basic. Here, popular magazines are playing a significant role. One solar engineer has remarked that the regular feature articles on solar heating in *Popular Science* have done more to stimulate solar technology for the home than the government programs. It may be an overstatement, but it reflects the remarkably active interest of how-to magazines in solar technology. Not only are how-to magazines encouraging solar growth, but so are kit manufacturers. Heath-kit, probably the premier kit company in electronics and computing, has introduced a do-it-yourself solar water heater kit. Heath offers use of a computer program to size the unit for each home and the use of its technical consultation service for installation and operation.

One of the greatest potential areas for solar energy, of course, is in the heating of water and buildings. All of the preceding discussion on progress in solar energy has focused on solar heating of buildings and on solar hot water. The reason is that these uses of solar energy are ready now; their application is not held up by lack of fundamental research. The only barriers are economic ones, which can be overcome.

Where are solar heating and solar hot water economical? The answer depends on the package of tax incentives being offered by the federal government and by the state governments at the moment the question is answered. Several studies have attempted to answer the question of where solar energy is economical, but each was built on now slightly out-of-date information.

A study at the National Science Foundation (McGarity, 1976) examined a number of U.S. cities to see where solar heating with an electric backup was less costly than electric heat alone. The study assumed that

the costs of solar panels were reduced by 20% from their 1975 levels due to mass production and that installation costs were reduced by 30%. Using costs current in 1975 and a real rate of growth of electric prices of 5% per year, the study found a wide list of cities in which solar heat was more cost effective than electric heat. (See Table 24.1.) The study assumed no federal tax incentives since none were then in existence.

A study at the Department of Energy (Bezdeck et al., 1979) assessed both solar heat and solar hot water for single-family residences. The study assumed arbitrarily that the solar installation would provide 70% of the hot water and 50% of the heating needs over an average year. The study considered only four cities (cities also investigated in the NSF study): Los Angeles, Boston, Washington, D.C., and Grand Junction, Colorado. Solar equipment was assumed to be exempt from property taxes, and federal tax credits were counted in the analysis. (The tax credit figure, though, was at a lower level set by the 1978 National Energy Act, not at the 40% level later passed.)

The analysis compared solar heat plus electric backup to electric heat alone, just as the 1976 NSF study

Table 24.1 Where Solar Heat with Electric Backup Beats Electric Heat Alone

City	Solar fraction
Albuquerque	0.96
Boise	0.80
Boston	0.50
Charleston	0.99
Grand Junction, Colo.	0.80
Indianapolis	0.70
Los Angeles	0.80
Madison	0.74
Miami	1.00
New York City	0.80
Oklahoma City	0.90
Phoenix	0.96
Rapid City, S.D.	0.85
Santa Maria, Calif.	0.90
Washington, D.C.	0.80

Solar fractions are the fraction of annual heating loads met by solar energy. This chart was prepared for a 1,500-square-foot house with heating loads specific for the city.

Source: National Science Foundation, 1976.

Table 24.2 Time Required to Repay Solar Investments with Accumulated Savings

City	Solar hot water (years)	Solar heating plus solar hot water (years)
Boston, Mass.	13	17
Washington, D.C.	13	17
Grand Junction, Colo.	9	12
Los Angeles, Calif.	9	11

Source: Department of Energy, 1978

did. For solar hot water, the time for accumulated savings to repay the full cost of the system was less than or equal to 13 years (see Table 24.2). For combined solar heating and solar hot-water systems, the time for accumulated savings to repay fully the initial cost was less than or equal to 17 years in all four cities (see Table 24.2). Under all these assumptions, the study concluded that solar water heating is economical in most U.S. locations and that solar house heating is economical in many locations. Recall that these conclusions were drawn for the earlier and lower levels of federal tax credit and did not include state tax incentives.

Allowing the demise of tax credits for solar installations seems an unwise choice by Congress. The credits no more favored solar for the home than massive government research efforts favored the nuclear power industry. Solar and conservation remain keystones in the prevention of another energy crisis in the United States (see next chapter). To remove their support is to risk the stability of the structure.

Summary

An efficient solar hot water heater was invented in 1909. The devices first became popular in southern California and then in Florida. After World War II, gas and electric water heaters captured the market because of the availability of natural gas and cheap electricity.

In active solar systems, pumps powered by electricity move air or water from one part of the system to another. The system is composed of a collector and a storage facility. The collector is warmed by the sun's rays and transfers the heat to a collecting medium: water, water and antifreeze, or air. The heat is then stored in a tank of water, a phase-change salt, or a pebble or rock bed. Hot water for direct use or hot air

for heating is obtained by passing house air or water through or over this heat storage device.

Passive systems use little or no energy other than the sun's rays. Passive solar devices include large south-facing windows with overhangs, Trombe walls, and greenhouses. Passive solar homes are about 10% more expensive than conventional homes, but they cost 50%–80% less to heat, leading to predictions that some passive solar features will become standard in all new homes. The attractiveness of passive solar features depends on the costs of alternate heating sources in an area.

Sunlight can also be piped into a building to illuminate interior rooms and decrease electric use. Photovoltaic cells convert sunlight to electricity using semiconductors. Some industries already use them for electric power in specialized applications. If the devices become inexpensive enough, they could be used at central stations to provide additional needed electricity at times of peak demand. In the future it may be possible to provide all of the power at a central generating station by photovoltaic cells.

Steam electric generation at central stations using solar energy is already possible using the power tower concept; mirrors focus sunlight on a tower in which water is boiled to steam. In a related concept, mirrors made of metal-coated plastic film are used to reflect sunlight onto smaller focal points, and heat is then collected centrally (no power tower is utilized).

The sun's energy can be used indirectly by photosynthesis—in which sunlight converts carbon dioxide and water to biomass. Biomass products such as wood or field crops can be burned for heat or converted to alcohol for use in automobiles. Trees, sugar cane, guayule, and Hevea are all possible fuel crops.

Electricity may be generated by using the differences in ocean water temperatures at different depths. The concept is known as OTEC for ocean thermal energy conversion. In this concept, warm surface waters are used to vaporize a liquid such as ammonia, and the high-pressure vapor turns a turbine to generate electricity. Cold subsurface waters are used to condense the spent ammonia vapors to liquid again. Many problems remain to be solved in the OTEC concept.

By the year 2000, some 13% of the U.S. energy needs could be supplied by solar energy, but government programs are needed to help offset the high initial costs of installing solar technology and to help commercialize photovoltaic manufacture. Studies have concluded that solar water heating is economical over almost all of the United States, and solar space heating is economical in many areas.

Questions

1. Briefly, sketch how the sun's energy can be collected and used to heat water for an average house. Why might you need a conventional *backup* heating system?
2. Contrast active and passive solar energy systems.
3. Describe how the sun's energy might be used to generate electricity using solar cells. What environmental advantages does such a system have over generating electricity by burning coal or oil?
4. Suppose, instead, that electricity was generated by focusing the sun's rays to produce steam. Now compare the environmental advantages of solar versus conventional electric generation.
5. What environmental problems can you see arising if large tracts of land are devoted to growing trees for wood to be used as a fuel source? (Part of the answer is in Chapter 2, but think about it yourself, first.) What about problems associated with burning the wood?
6. Discuss the economics of solar energy from the point of view of a homeowner. What are the major costs? What are the savings? How can the government help homeowners?

References

General References on Solar Energy

Butti, K., and J. Perlin. *A Golden Thread: 2500 Years of Solar Architecture and Technology.* New York: Cheshire Books/Van Nostrand Reinhold, 1980.

Metz, W., and A. Hammond. *Solar Energy in America,* American Association for the Advancement of Science, 1978.

"Solar Energy: Progress and Promise." Council on Environmental Quality. Washington, D.C.: U.S. Government Printing Office, Stock No. 041-011-00036-0 (April 1977).

Lovins, A. As quoted in *Alternate Long-Range Energy Strategies.* Joint Hearings Before the Select Committee on Small Business and the Committee on Interior and Insular Affairs, U.S. Senate. Interior Committee Serial No. (94–47) (92–137). Washington, D.C.: U.S. Government Printing Office, December 9, 1976.

Weinberg, A. In a review of *Soft Energy Paths: Toward a Durable Peace,* by Amory Lovins (Ballinger, 1977), appearing in *Energy Policy* (March 1978), p. 85.

Lilienthal, D. As quoted in *National Journal* (April 29, 1978), p. 674.

Clark, Wilson. As quoted in *Science,* **21** (February 13, 1981), 683.

Ferguson, J. "Is Central-Station Generation Becoming a White Elephant?" *Public Utilities Fortnightly* (March 21, 1985), p. 32.

Ocean Thermal Energy Conversion

Metz, W. D. "Ocean Thermal Energy: The Biggest Gamble in Solar Power," *Science,* **198** (October 14, 1977), 178–180.

Cohen, R. "Energy from Ocean Thermal Gradients," *Oceanus,* **22** (4) (Winter 1979–80), 12.

Whitmore, W. "OTEC: Electricity from the Ocean," *Technology Review* (October 1978), pp. 58–63.

Office of Technology Assessment, "Renewable Ocean Energy Resources," Part I, *Ocean Thermal Energy Conversion,* 1978.

Biomass: Photosynthetic Conversion of Solar Energy

Broad, W. J. "Boon or Boondoggle: Bygone U.S. Rubber Shrub Is Bouncing Back," *Science,* **202** (October 27, 1978), 410–411.

Burwell, C. C. "Solar Biomass Energy: An Overview of U.S. Potential," *Science,* **199** (March 10, 1978), 1041–1048.

Griffin, D. "Natural rubber Has a Future After All," *Fortune* (April 27, 1978), p. 78.

Hammond, A. "Alcohol: A Brazilian Answer to the Energy Crisis," *Science,* **195** (February 11, 1977), 564–566.

Plotkin, S. "Energy from Biomass: The Environmental Effects," *Environment,* **22** (9) (November 1980).

Edelson, E. "The Great Gasohol Debate," *Popular Science* (July 1981), p. 53.

"Energy from Biological Processes, Summary." Office of Technology Assessment, Congress of the United States, July 1980.

Maugh II, T. H. "Guayule and Jojoba: Agriculture in Semiarid Regions," *Science,* **196** (June 10, 1977), 1189–1190.

Photovoltaic Conversion and Interior Lighting

Hammond, A. L. "Photovoltaic Cells: Direct Conversion of Solar Energy," *Science,* **178** (November 17, 1972), 732–733.

Kelly, H. "Photovoltaic Power Systems: A Tour Through the Alternatives," *Science,* **199** (February 10, 1978), 634–643.

Metz, W. D. "An Illuminating New Use for Solar Energy," *Science,* **194** (December 24, 1976), 1404.

Smith, J. "Photovoltaics," *Science,* **212** (June 26, 1981), 1472–1478.

Solar Heating

Butti, K., and J. Perlin. "Solar Water Heaters in Florida, 1923–1978," *Co-Evolution Quarterly* (Spring 1978), p. 74.

Butti, K., and J. Perlin. "Solar Water Heaters in California, 1891–1930," *Co-Evolution Quarterly* (Fall 1978), p. 4.

Dallaire, G. "Will This Low-Cost Solar Collector Revolutionize Home Heating?" *Civil Engineering* (February 1980), p. 38.

Hammond, A. L., and W. D. Metz. "Capturing Sunlight: A Revolution in Collector Design," *Science*, **201** (July 7, 1978), 36–39.

McGarity, A. "Solar Heating and Cooling: An Economic Assessment," National Science Foundation, Directorate for Scientific, Technological and International Affairs, Division of Policy Research and Analysis. Washington, D.C., 1977.

Bezdeck, R., A. Hirschberg, and W. Babcock. *Science*, **203** (March 23, 1979), 1214.

Lunde, P. *Solar Thermal Engineering*. New York: John Wiley and Sons, 1980.

"Solar Collector Manufacturing Activity." Energy Information Administration, U.S. Department of Energy, 1984.

Holtz, M. et al. "The Future of Passive Solar Design," *Solar Age* (October 1985), p. 49.

Arctander, E. "Heat Storing Salts," *Popular Science* (January 1983), p. 82.

Passive Solar Design Handbooks, Vols. I, II, and III. Washington, D.C.: U.S. Government Printing Office.

President's Council on Environmental Quality. *Solar Energy: Progress and Promise*. Washington, D.C.: U.S. Government Printing Office, April 1978.

Project Independence Blueprint, Final Task Force Report: Solar Energy. Federal Energy Administration, Washington, D.C.: U.S. Government Printing Office, 1974.

Solar Thermal Electric Power

Metz, W. "Solar Thermal Electricity: Power Tower Dominates Research," *Science*, **197** (July 22, 1977), 353.

"Spinning a Turbine with Sunlight," *EPRI Journal* (March 1978), p. 14.

Van Atta, D. "Solar-Thermal Electric: Focal Point for the Desert Sun," *EPRI Journal* (December 1981), p. 37.

Schefter, J. "Solar Power Cheaper than Coal, Oil, Gas," *Popular Science* (February 1985), p. 77.

CHAPTER TWENTY-FIVE

Energy Conservation

Is the Energy Crisis Over?
Energy Crisis Revisited/What Turned the Situation?/Is the Worst Over?

Looking at Two Possible Energy Futures

How We Heat Homes and Make Hot Water

The Manner of Transporting People and Goods
Passenger Transport/Freight Transport

CONTROVERSY:
25.1 *How Do We Get Fuel-Efficient Cars on the Road?*

Is the Energy Crisis Over?

Energy Crisis Revisited

You have heard people say the energy crisis is over. What do they mean, and are they correct? It is true that oil supplies are adequate for the moment. Further, supplies are more secure now than they were 10 years ago because the United States is importing less oil in total and a lower portion from the Middle East. Such is not the case for all the industrial countries, however. The Middle East still dominates world production and is a principal supplier to western Europe. Do those who claim the energy crisis is over mean that oil is as inexpensive as it once was? World oil will probably never again sell for $2 per barrel, as it did in the early 1970s. The price of oil did fall, however, from a high of $34 per barrel in 1979 to the range of $12–$18 per barrel in 1987.

Perhaps these people are talking about natural gas when they claim the energy crisis is past. From a low point in 1977, gas supplies seem to have recovered slightly as the Natural Gas Policy Act allowed prices to rise. It is not clear, though, whether there was a real resource shortage in natural gas or whether companies were simply withholding gas from the market. The voices that say the energy crisis is over also say that the free market solved the gas crisis. They claim, too, that a lack of a free market caused it.

It is worth our attention to examine these claims and come to our own conclusions because public policy is eventually formed on the basis of citizen understanding. In conducting this examination we must take care not to accept the simplistic claims of any one political ideology. The energy policy conflict that continues to the present is not simple.

We need to examine the dimensions of the event that was called the energy crisis and determine which fuels were involved. We also need to investigate the forces that shaped the conflict and that are still in place today. We should look for points of leverage to forestall or prevent another such crisis. One issue we would like answered is whether, by allowing the operation of a completely free market, we can prevent the return of another energy crisis. If that will not work, we would like to know whether there are other forms of energy self-defense.

The energy crisis, the popular name for the energy policy conflict, was first brought home to Americans when, in 1973 and 1974, the Organization of Petroleum Exporting Countries (OPEC) imposed an embargo on oil exportation to the United States. The oil embargo halted importation of some 10%–15% of the oil that entered the United States. Long lines formed at gas stations, odd–even rationing was imposed, and gasoline theft was common.

Subsequently, OPEC fixed new and higher prices for its oil. From a price of $2 per barrel, OPEC raised its prices in steps, feeling its way toward a price the market could carry. As the price of oil rose, price inflation occurred for goods that depended on oil for transport or for manufacture. Higher prices decreased demand for these goods, and people were thrown out of work. The government response was to force higher fuel efficiencies for the next generation of autos. Congress also created a strategic petroleum reserve. Money was authorized for the development of synthetic fuels from coal and oil shale. Legislation provided tax incentives for the home installation of renewable energy technologies and for the installation of energy conservation materials.

Congress fixed the price of oil from existing U.S. wells to prevent the oil companies from obtaining windfall profits during this time. Such profits would have occurred if the companies had sold their oil at the new, higher prices fixed by OPEC for its oil. Of course, the OPEC countries were getting windfall profits, but that could not then be prevented. In the late 1970s, all regulation of domestic oil prices by the government came to an end. These were the major events of the oil component of the energy crisis.

A second component of the energy policy conflict was the shortage of supplies of natural gas. By the winter of 1977–78, it seemed that gas was in such short supply that some users would have to be cut off and new users turned away. Gas had been regulated in price since the 1950s, but shortages were suddenly occurring. The reason was that industries in the states where the gas originated began offering higher prices than the natural gas companies could get on the interstate market. As a result, the gas was sold in the state in which it originated, but at the higher price.

The government's response was to modify the act that regulated the price of natural gas. The revised act allowed newly discovered gas to be sold at whatever price the gas companies could get. As more and more newly discovered gas came on the market from 1978 to 1985, the "old" price-fixed gas gradually decreased in volume and the price rose. By 1986, all controls had been removed from the price of natural gas. Whether steadily rising prices will lead to new controls is not clear. These were the many events of the natural gas component of the energy crisis.

Those who claim that the energy crisis has passed will generally point to several features of the energy landscape. First, OPEC's price of oil has fallen. From an official high of $34 per barrel it had fallen to the range of $12–$18 per barrel by 1987. Long-term prices in this range were widely expected as OPEC announced that it was abandoning its efforts to hold an official price. Instead of an official price, OPEC agreed to let individual members sell their oil for whatever price they could in order to protect their share of the market.

Second, there have been no recent oil- or gas-

supply interruptions, and U.S. demand for oil and gas is down. In addition, oil imports are down, and the portion of oil imports from the Middle East has fallen significantly. These facts all point to an easing of the crisis, but what factors *caused* the crisis to ease?

What Turned the Situation?

To answer this question, we first note that the higher prices for oil and gas caused people to buy more efficient cars, to insulate their houses, to install ceiling fans, to install wood stoves, to dial down thermostats, and to install active and passive solar heating systems. The high prices caused industries to use energy more efficiently as well. Thus conservation, spurred by high energy prices, was responsible for decreasing demand. The high prices for oil that stimulated the conservation efforts were the result of OPEC price fixing, not the result of a free market. The high prices of natural gas, in contrast, *were* the result of a free market. Thus a free market seems to have solved the gas shortage, but a free market did not cure the oil energy crisis. Higher fuel prices did help to turn both situations.

We also note that oil imports are down. Why? Once again, this is the result of conservation, but the reason that we are importing less oil from the Middle East is more complicated. The high OPEC oil prices caused more exploration for oil around the world. Mexico, in particular, became an important oil producer and supplier to the United States, replacing much of the oil imports from the Middle East.

The price of OPEC and world oil suddenly fell by 50% in 1986. Why? The decline in price, which took place over a four-month period, was the result of a combination of conditions. Again, discoveries of oil around the world have taken place in response to OPEC prices. This new oil, much of it non-OPEC, competes with OPEC oil and has pulled the price down as more oil chases a shrinking market. That is, the presence of more oil in a market diminished by conservation has caused producers to lower their price to sell their product.

A final factor in the easing of the energy crisis is that the United States continues to build a strategic petroleum reserve. The reserve consists of government-purchased oil stored up against a potential embargo. The reserve had reached 450 million barrels at the beginning of 1985, 60% of the way to the 750-million-barrel goal. Physical facilities limited the drawdown rate to 1.7 million barrels per day, about 10% of daily demand, but the withdrawal capacity is to be increased to 3 million barrels per day, a bit less than 20% of daily demand. The presence of the reserve warns OPEC that the United States is prepared to survive an oil embargo.

One action taken that has not been a factor in the easing of the energy crisis is the development of synthetic fuels. Though several projects were completed, nearly all have been abandoned because the government walked away from its commitment to subsidize the production of synthetic fuels. Those projects were needed, not for fuel but for experience. When an embargo and price increase threatens us again, we will not be ready with synthetic fuels—and we could have been. Another set of actions we pursued and then finally abandoned in 1986 were tax incentives for energy conservation and for alternate sources of energy. Tax incentives are still needed; they could still play a significant role in dampening demand.

Is the Worst Over?

Though matters now seem in relative control, we cannot afford to let slide the many measures put into effect during the crisis years. Take a careful look back at Figure 17.5; note the commanding resources of the Middle East. By the mid-1990s, as low-cost supplies from other areas of the world begin to dwindle, OPEC's enormous proved reserves will again place it in a dominant position. Once again, it will be capable of raising prices and controlling the oil market. The Middle East, OPEC's heartland, has greater proved reserves than all other regions of the globe combined. Of the world's 700 billion barrels of proved oil reserves, the resources of OPEC countries total about 475 billion barrels, or two-thirds of the total.

The original question was, "Is the energy crisis over?" The answer is that, as a cancer may sometimes do, the crisis has gone into remission for a period. We can confidently expect its return. Will we have conservation ready? Synthetic fuels? Solar and wind power?

Looking at Two Possible Energy Futures

"We must extend our resource base to replace the fuels now becoming scarce." That is the theme of the energy companies, the businesses that profit from selling fuel

and electricity. There is, nonetheless, another way to extend our energy resource base and to reduce the fuel imports that are both costly and risky. This other way does not make profits for the energy companies, however. In fact, they earn less and consumers save money by finding ways to save energy. The United States has Saudi-sized resources of energy that can be saved.

Energy conservation ranges from simple and inexpensive to costly and complex. It can be as simple as turning down the thermostat, turning out unnecessary lights, wearing a sweater, showering instead of bathing, using a fan instead of an air conditioner, keeping a car in tune, or riding the bus instead of driving. Conservation can be accomplished by such relatively simple purchases as insulation for an attic, storm windows, weatherstripping for a door, or a flow-restricting device for a shower. Investments can go still larger; high-efficiency appliances or fuel-efficient automobiles may be purchased.

Energy conservation saves money; the dollars we do not spend for energy remain in our pockets. Energy conservation can decrease the need to import oil and gas from overseas. Energy conservation postpones the days of scarcity and keeps fuel prices from rising too fast. Finally, energy conservation means less of the air pollutants that are by-products of combustion. It may also mean fewer hazardous products of nuclear fission to dispose of or to guard for centuries.

Energy conservation also becomes a focus for attention when we observe the average quantity of energy consumed per capita in the industrial world and in the less-developed countries. In 1950 the United States had a population of 152 million, which in that year consumed on average about 210 million Btu *per person.* By 1979 the U.S. population had grown to 230 million people, roughly a 50% increase. Energy consumed in that year reached 330 million Btu *per person,* roughly a 55% increase in per person consumption alone (Figure 25.1). Similar per person increases were noted in Canada. The combination of increases in both population and per person consumption of energy caused the total annual energy consumed in the United States to jump from 31 quadrillion Btu in 1950 to 79 quadrillion Btu in 1979. (One quadrillion is a million billion, or 1,000,000,000,000,000.)

These two aspects of energy use—the growth in per person consumption and the growth in population—reveal the two basic control points that could dampen the growth of energy use in the United States. Conservation can decrease the energy consumption per

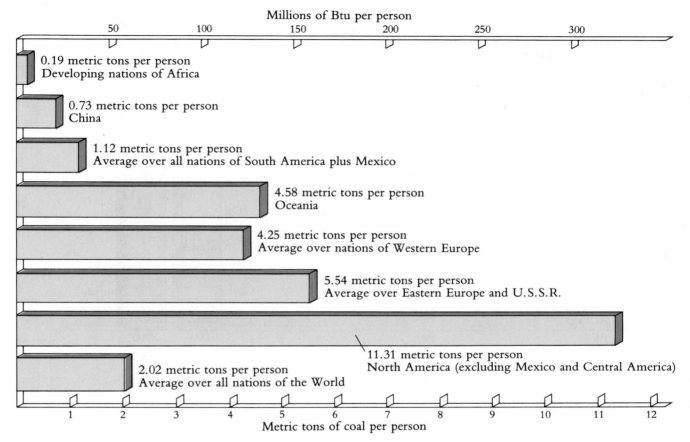

Millions of Btu per person

- 0.19 metric tons per person
  Developing nations of Africa
- 0.73 metric tons per person
  China
- 1.12 metric tons per person
  Average over all nations of South America plus Mexico
- 4.58 metric tons per person
  Oceania
- 4.25 metric tons per person
  Average over nations of Western Europe
- 5.54 metric tons per person
  Average over Eastern Europe and U.S.S.R.
- 11.31 metric tons per person
  North America (excluding Mexico and Central America)
- 2.02 metric tons per person
  Average over all nations of the World

Metric tons of coal per person

Figure 25.1 Annual energy consumption per capita by world regions, 1979. The people of North America are, far and away, the world's largest energy users on a per-person basis. The 330 million Btu consumed by the average person in 1979 in North America is equivalent to the energy in 11.3 metric tons of coal per person per year. (From 1979 Yearbook of World Energy Statistics, United Nations.)

person, and a stable population can limit the total energy consumption.

What specifically can energy conservation do? The Council on Environmental Quality has described two alternate futures (*The Good News About Energy*, 1979). Although both futures include a maximum commitment to solar energy, the first future assumes that conservation will be given sustained attention and emphasis. In Figure 25.2, Future I, *the conservation route*, is contrasted to Future II, *business as usual*. Following the conservation route will make it possible to decrease our reliance on oil and gas due to the combination of conservation and the development of solar energy. Our reliance on coal and nuclear power need increase only slightly. In stark contrast, the business-

as-usual future, which projects energy demands increasing at 1.9% per year, would require a sixfold expansion of nuclear power and more than a doubling of our reliance on coal, even with an expansion of solar energy to its maximum possible contribution.

The relative impacts of these two energy futures are shown in Figure 25.3. The annual requirements for coal mined in the business-as-usual future are about double those in the conservation route. And in the business-as-usual future, the number of coal-fired power plants is double that in the conservation route. The role of nuclear power in the conservation route, measured in terms of the number of nuclear electric plants, is less than half that in the business-as-usual future. Spent fuel generated is likewise about halved.

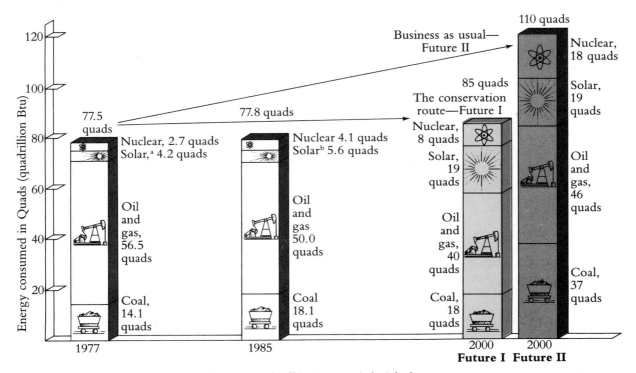

^aIncludes 1.8 quads of biomass; remainder is hydropower
^bIncludes 2.0 quads of biomass (est); remainder is hydropower

Figure 25.2 Two possible energy futures: the conservation route and
business as usual. Note that both futures include a maximum commitment to solar energy development. Energy use in 1985 reflects a pronounced slackening in demand growth, particularly in the oil and gas
sectors where use has gone down, implying the success of energy conservation measures. Growth in solar options, however, has not been so
successful. (Data for 1977 and Projections for 2000 are from *The Good
News About Energy*, Council on Environmental Quality, 1979. Data
for 1985 are from the Energy Information Administration.)

How did the Council on Environmental Quality
envision the conservation route being achieved? The
specific steps to achieve Future I include more insulation in homes, apartments, public buildings, and
factories; more efficient appliances and machines;
better-insulated refrigerators; gas stoves without pilot
lights; and fluorescent lamps instead of incandescent
bulbs. Total energy systems in industry may both generate electricity and produce process heat (heat for
industrial processes), as well as space heating. Automobiles can be made (and are becoming) far more fuel
efficient. In short, energy conservation, if we can
achieve it, holds as great an opportunity as discovering
new sources of energy.

As can been seen in Figure 25.2, total energy consumption from 1977 to 1985 was virtually flat. Nuclear

power grew about as anticipated in the conservation
route. Coal energy consumption grew faster than was
projected in the conservation route but a bit slower
than the projection in the business as usual future. One
reason for the growth in coal energy consumption was
that about 2.7 quads of electric energy from burning
oil at power plants was replaced by electricity from
burning coal. Oil and gas use has been declining, almost as fast as projected in the conservation route, reflecting both conservation and replacement by coal.
Solar power has made very little gain at all. The success
pictured in Figure 25.2 is, of course, the decline in oil
and gas consumption, but it is not clear whether further
gains in energy saving will be easily forthcoming.

We have selected two areas on which we shall focus
attention in our discussion of energy conservation.

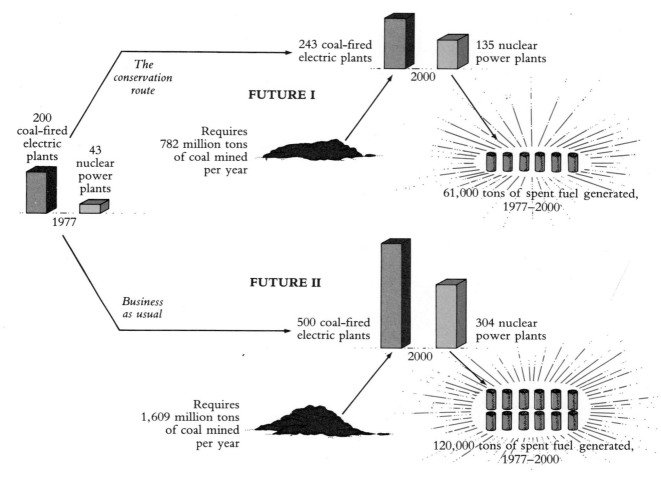

Figure 25.3 Some of the impacts of two possible energy futures. (Data from *The Good News About Energy*, Council on Environmental Quality, 1979.)

These are (1) the way we heat buildings and make hot water, and (2) the way we transport people and freight. Building heating and water heating together account for about 20% of our annual use of energy. Transportation consumes about 25% of the energy we use each year.

Choices in these two areas can be made now by the consumer or by the business person that can influence energy consumption. New technology is not necessary in these areas; methods presently available are sufficient to make dramatic differences both in energy consumption and in pollution. For instance, if the cars on U.S. roads in the mid-1970s matched the average efficiency of the cars on European roads at the same time, our annual travel nationwide could have been accomplished with 42% less gasoline. Because the

choices now available in these areas can make a big difference in our energy use, we discuss these two areas in detail.

Each of these two areas is what is called a *point-of-use*. Other improvements at points-of-use are possible. Some we have already mentioned, such as natural and fluorescent lighting in place of incandescent lighting. A very effective and even charming method for saving energy is to use old-fashioned ceiling fans for cooling instead of air conditioning. The breeze from such a fan, whose power usage is about that of a standard light bulb, produces a cooling effect that is about equal to a 6°F-drop in temperature.

More efficient appliances are possible. Spark-ignition gas stoves can replace pilot-light gas stoves; better-insulated refrigerators can replace current models; better-insulated water heaters are now available; and high-efficiency oil and gas furnaces are now on the market. The labeling of appliances with their energy consumption per unit of time will be a valuable aid to consumers, not only helping them save energy but helping them save money as well.

In addition to improved point-of-use efficiencies in energy consumption, the penalty of using the "electric middleman" can be softened as well; that is, improvements in the efficiency of electric power generation are achievable. Cogeneration uses a jet engine and a conventional boiler one after another to increase the efficiency of electric generation. Magnetohydrodynamic generation in conjunction with a conventional steam boiler can also, as discussed in an earlier chapter, increase the efficiency of electricity generation. *District heating*, in which waste heat from electric generation is put to use in heating buildings, makes the same fuel do two jobs. All three of these energy conservation methods are discussed earlier, although we did not then point to these methods as being ways to save energy. Thus, efficiency of conversion, in addition to efficiency at the point-of-use, offers opportunities for energy conservation.

Nonetheless, we return to the two areas mentioned earlier because of their importance in energy consumption and because of their potential for change.

How We Heat Homes and Make Hot Water

The choice about how to heat a home or how to heat water is simple to make in most cases, at least from the standpoint of what to reject. In the recent past, there have been only three options, but solar energy has now entered the picture as a real alternative. The three traditional options are natural gas, oil, and electricity. Let us look at these choices carefully.

Suppose you are in the market to purchase a home, and you have seen three houses; all of them are affordable and similar in other respects, except in the fuel used for heating. One heats with natural gas, another with distillate oil (home heating oil), and the third with electricity. In each house, hot water is produced in the same way that the home is heated; that is, the gas-heated home heats its water with natural gas, and so forth. Which home is preferable from the point of view of energy conservation, and which is preferable from the point of view of monthly energy costs?

Assuming that the price of natural gas is allowed to

continue to increase, heating a home with natural gas will become, in short order, about as costly as heating with oil. Gas had been much cheaper for some years, but the pattern is changing as the government allows the price of natural gas to rise in order to correct shortages. The energy efficiencies of an air-heating system that is oil-fired and one that is gas-fired are about the same. Either new system is 80% efficient. That is, about 80% of the energy derived from burning the fuel goes into raising the temperature of the house.

Electric heating, on the other hand, even when the house receives extra insulation as the electric company suggests, still costs more than heating by gas or oil. How much more depends on the local electric rates. Only in the Pacific Northwest have electric rates been low enough for electric heating to be an economical choice, and these rates are on the rise. In most cases, the electric home is considerably more expensive to heat than the home heated by gas or oil. The cost of creating hot water using electricity is large as well. A home heated with oil or gas but using an electric water heater has an electric bill about double that of a home that heats water with gas or oil. Solar energy is rapidly becoming an available alternative for heating homes and water in many sections of the country. Its economics relative to oil, gas, and electricity are discussed under solar energy in Chapter 24, where we show that the sun has quickly become a serious rival for the other methods of heating water and homes.

Not only is the electric home more costly to heat, but in several instances it is extremely wasteful of resources. In an earlier chapter we described the result of the second law of thermodynamics. We noted that as a consequence of the second law, there is an upper limit on the fraction of heat energy that can be converted to work or into some other form of energy. For our purposes, the other form of energy is electrical energy, and the impact of the second law is that little more than about 40% of the heat liberated by burning coal, oil, or natural gas can be converted to electrical energy. The remainder of the heat from burning the fuel, the portion not converted to electrical energy, is wasted by the electrical plant into the environment, either into the air or into the water. The latter is the phenomenon we know as *thermal pollution*.

Let us suppose that the electric generating station is oil- or gas-fired. If we assume that the plant is 33% efficient, a reasonable efficiency figure, then two-thirds of the heat energy released by burning the fuel is lost into the environment. The remaining one-third in the form of electrical energy can be used to heat a home. If there is a 5% loss in transmission and in heating at the home, we have about 31% or 32% of the original energy in the fuel left to heat the home.

In contrast, if oil or gas is burned in the home to provide heat, the efficiency is much greater. New oil and gas furnaces attain 80% efficiency; but after some use, the heat exchanger efficiency falls and the conversion of combustion heat to useful heat decreases to 60%–70%. It is reasonably clear in such circumstances that electric resistance heat, which is produced by a generating plant that burns oil or gas, is wasteful of that fuel. The same conclusion follows immediately for hot water. If the electric plant is coal-fired or nuclear, similar comparisons are not meaningful, since these fuels are not used to heat homes. (Coal use for home heating is very limited, though increasing.)

If their economics are so unfavorable, why are electric heat and electric hot-water heaters seen so frequently in new homes? Why do builders seem to favor these types of water heaters? It is a matter of *first costs.* An electric water heater may cost $150–$200 installed; an oil-fired heater, $750 installed; a solar water heater with collectors runs several thousand dollars. If gas is available, gas water heaters will probably be installed, since their purchase price for the builder is about that of electric water heaters. If gas is not available, a builder will almost always give customers electric water heaters because they are cheaper for the builder to buy. A smaller investment is needed by the builder, and the house can be sold at a lower price, increasing the number of people who can afford the product.

In the near future, however, the **heat pump** could change both the economics and energy cost of electric heating (Figure 25.4). This electrical device, a sort of refrigerator or air conditioner in reverse, is capable of heating a home during many, though not all, winter days; it exhausts cold air to the outside and warm air to the inside of a house. Measurements of the performance of these units have indicated that two units of heat energy can be produced for every one unit of energy (as electricity) delivered to the house. This cuts the cost of electric heating in half. Even better performances of heat pumps may be around the corner as more people try them as alternatives to conventional heating. One reason for their growing popularity is that the same unit that heats the home in winter can be used to cool the home in summer. Thus, one investment, not much more costly than central air conditioning alone, provides a unit that can both heat and cool.

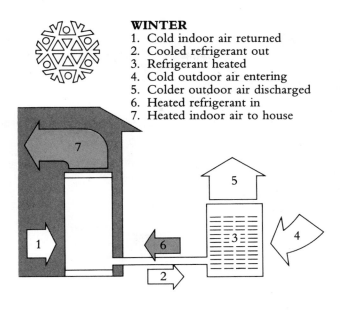

WINTER
1. Cold indoor air returned
2. Cooled refrigerant out
3. Refrigerant heated
4. Cold outdoor air entering
5. Colder outdoor air discharged
6. Heated refrigerant in
7. Heated indoor air to house

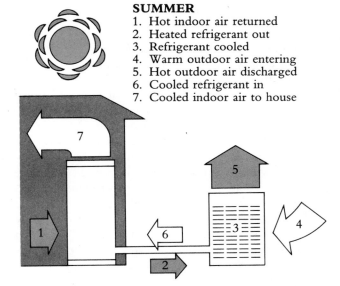

SUMMER
1. Hot indoor air returned
2. Heated refrigerant out
3. Refrigerant cooled
4. Warm outdoor air entering
5. Hot outdoor air discharged
6. Cooled refrigerant in
7. Cooled indoor air to house

Figure 25.4 How a heat pump works. (Adapted from GE Weathertron Heat Pump Brochure.)

Water heating can also be accomplished via heat pump, and two manufacturers are now offering heat-pump water heaters. Their electric requirements are expected to be about half that of conventional electric water heaters.

The means of heating and cooling a building and the means of heating water have a large influence on energy consumption and on cost. Nonetheless, there

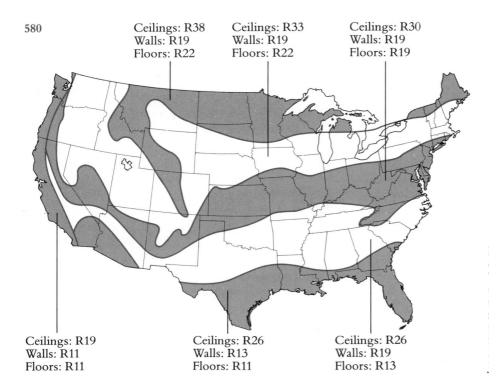

Ceilings: R38
Walls: R19
Floors: R22

Ceilings: R33
Walls: R19
Floors: R22

Ceilings: R30
Walls: R19
Floors: R19

Ceilings: R19
Walls: R11
Floors: R11

Ceilings: R26
Walls: R13
Floors: R11

Ceilings: R26
Walls: R19
Floors: R13

Figure 25.5 Insulation recommendations by area of the United States. To obtain approximate thickness of the needed fiberglass blanket, divide the R-value by 3 and round to the nearest integer. (From "Providing for Energy Efficiency in Homes and Small Buildings," Department of Energy, Office of Consumer Affairs, June 1980.)

are numerous additional steps that can be taken to reduce the energy requirements of buildings. Of course, many of them come under the heading of insulation, the addition of materials that prevent the exit of heat from the building.

Two statements of caution about insulation are in order, however. The first caution has to do with careful selection and application of insulating material. Urea-formaldehyde insulation foam was banned as unsafe and then reinstated by the courts. Most manufacturers have wisely left the business. If the foam can be found, it should not be used. The foam released extremely irritating gases into a number of homes. Styrofoam panels, despite manufacturers' claims, are highly flammable, as is cellulose. To show that styrofoam is flammable, put a match to a styrofoam cup, out of doors; AVOID BREATHING THE BLACK FUMES. Traditional fiberglass insulation seems a reasonable alternative. Even with this material, caution is needed in application. Insulation applied to homes with poorly insulated old electric wiring should be placed in a way that allows free exit of heat from the wires. If this is not done, heat buildup around the wiring could lead to fires. Thus, careful selection and application of insulating material is in order.

The second caution about insulation is indoor air pollution. Gases and particles from cooking and smok-

ing may accumulate in a house made too tight. Fresh filters on the furnace air flow, needed to allow dust-free circulation, become even more important to maintain. In addition, radon gas from the walls of homes built of stone can accumulate. Though formerly uncommon in U.S. homes, air-to-air heat exchangers are being used more frequently to bring in fresh outside air with little heat loss.

These cautions on insulation are not meant to dampen our enthusiasm for saving energy. Most houses can use added insulation, but the type and the extent should be carefully chosen. Figure 25.5 shows the insulating values recommended for homes and buildings by area of the United States. These values are rather standard figures, reflecting practice up through the 1970s. As the 1980s opened, however, new ideas in insulation were spreading. In particular, the notion of superinsulated houses was becoming a reality. Many of these superinsulated houses with minimal energy requirements have also made use of passive solar energy; that is, much of the reduced heating requirements could be met by the sun's energy.

Other possibilities for decreasing the heating and cooling requirements of buildings range from earth-sheltered housing to passive solar heating to the planting of windbreaks. Windbreaks of trees can be especially effective in reducing heat requirements.

Other energy-saving devices for the home include flow restrictors on faucets, ceiling fans, automatic setback thermostats, interior airtight shades on windows, awnings on the outside of windows, and jacket insulation on hot water heaters. To find out about such developments, all that is necessary is to open the pages of several issues of *Popular Science*.

The Manner of Transporting People and Goods

Passenger Transport

No nation has more automobiles per person than the United States, nor do other people drive as far as we do. We have had a speed limit of 55 miles per hour, however, and there are countries where people drive faster—much faster. On many stretches of the autobahn in Germany, no speed limits impede the motorist. Still, no society consumes as much energy per person in auto travel as ours. With so many cars and with so much space and sprawling development, we cover many more miles than do the people of other nations, and hence we consume far more energy in travel.

Americans are often accused of wastefulness in this regard, compared to Europeans who consume far less energy in travel. Is the accusation true or false? It is both. It is true that we consume far more energy, and part of the reason for this has been our tradition of larger cars. That tradition results from a low price of gasoline in the United States over many years. The taxes we placed on gasoline were always small in comparison to the taxes Europeans placed on their gasoline. So we hung on to our tradition of large cars while the Europeans typically manufactured smaller cars that used less gasoline. In that sense, because we were never willing to tax gasoline sufficiently to discourage larger cars, the accusation of wastefulness is true. In addition, there are about 56 cars for every 100 people in the United States. The figure in the countries of western Europe ranges from about 27 (Italy) to 40 (Switzerland) cars per 100 people. In another sense, however, the accusation of American wastefulness compared to that of Europeans is false.

If we compare Europe to the United States, we see a very different society. Population centers are much closer together in European countries; families are, in general, less widely separated by distance. People in Europe tend to stay in the nations of their birth, and the nations are small, on the order of the size of our states. The huge distances between cities and towns in the American countryside are not characteristic of Europe. Nor is car ownership anywhere near as extensive. In the United States, there is slightly more than one car for every two people; in the western European countries (the wealthier countries), there is about one car for every four people. In Europe, auto purchases are heavily taxed; special road taxes levied on car owners increase the burden still further. Furthermore, the price of gasoline in European countries is two and a half to three times that in the United States.

Public transportation developed in Europe before the era of the auto. Because of this, the networks of the bus, train, and streetcar systems are excellent; service, moreover, is frequent. Thus, Europeans generally have available to them a public transport system which, by appropriate transfers, can link almost any city address in western Europe with any other, nearly door-to-door. Such a journey would begin by walking to the end of the block to catch a subway, bus, or streetcar. From there, one would proceed to the central station, where long-distance buses and intercity trains are available. The need to use the automobile, therefore, is less frequent.

Even though such excellent systems are available, the importance of the auto to European travel is growing at a phenomenal rate. It is growing as fast as the growth of personal wealth allows it. In the Netherlands, a typical European country with regard to auto use, virtually all the growth in passenger kilometers is going into auto travel. Mass transit ridership is only holding even. Similar patterns are seen for other western European nations. The Europeans enjoy their cars every bit as much as Americans do. If they could afford to waste more energy in auto travel, they probably would.

Having defended Americans, we now must also admit the truth. We have come to "need" our automobiles. We need them not simply because we have built a dispersed society in which the auto is a necessary means of transportation. We need them for personal reasons as well. Kenneth Boulding, an economist, explained our relation with the auto in the following way:

The automobile, especially, is remarkably addictive. I have described it as a suit of armor with 200 horses inside, big enough to make love in. It is not surprising that it is popular. It turns its driver into a knight with the mobility of the aristocrat and perhaps some of his other vices. The pedestrian and the person who rides public transportation are, by comparison, peasants looking up with almost inevitable envy at the knights riding by in their mechanical steeds. Once having tasted the delights of a society in which almost everyone can be a knight, it is hard to go back to being peasants. I suspect, therefore, that there will be very strong technological pressures to preserve the automobile in some form, even if we have to go to nuclear fusion for the ultimate source of power and to liquid hydrogen for the gasoline substitute. The alternative would seem to be a society of contented peasants, each cultivating his own little garden and riding to work on the bus, or even on an electric streetcar. Somehow this outcome seems less plausible than a desperate attempt to find new sources of energy to sustain our knightly mobility. (Boulding, 1974)

Boulding may be suggesting that our "habit" is incurable, that the best we can hope for is a substitute fuel, such as a diesel or an electric "fix." He may be correct that the habit is beyond breaking. The growing use of the auto in Europe under such costly conditions gives us an idea of how enticing is the siren call of the automobile (Figure 25.6).

If the habit cannot be broken, what can we do? Try to imagine a society such as ours with 120 million private automobiles traveling as many miles as we do and using 25% less gasoline. Fuel efficiency is a trick the Europeans learned long ago, a trick we are only starting to learn. European cars are, on average, much more fuel efficient than American cars. An accepted figure for the efficiency of European autos, averaged across all cars on the road, is 10 km/L (22.5 miles per gallon). These high efficiencies are the response of motorists to the high prices of gasoline, since more efficient cars are less expensive to operate. The comparable figure, the average mileage for all cars in the United States, was 16.7 miles per gallon in 1983. If our fleet of cars had the average efficiency of European autos, each million miles our fleet logged would require 25% less fuel.

Put another way, a car that travels 16.7 miles on a gallon of gas requires 599 gallons to go 10,000 miles (an average yearly figure). The car that travels 22.5 miles per gallon needs only 444 gallons to go the same

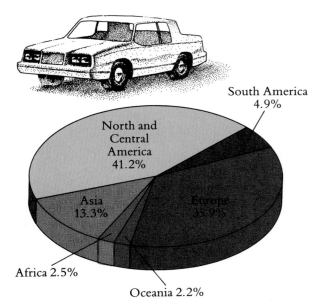

Figure 25.6 World distribution of motor vehicles in 1983. (From Motor Vehicles Manufacturers' Association, Facts and Figures, 1985.)

distance. At the current price of gasoline, the 155 gallons saved over a year amounts to a healthy sum. It seems that more fuel-efficient cars have a significant impact on our budgets as well as on energy conservation.

How do we arrive at an era of fuel-efficient automobiles? It is reasonably clear that high prices have encouraged the production of efficient cars in Europe. These high prices include, in general, taxes that amount to half or more of the total price of a liter of gas. Gasoline prices in the United States have risen rather sharply since the oil embargo and the rise of the cartel, from 30–35 cents per gallon in 1973 to their present levels. Gasoline prices in 1985 were nearly four times what they were before the embargo. Since the federal tax on gasoline has increased by only 5 cents per gallon from its original level of 6 cents per gallon during this period, most of the price increase pays for more expensive crude oil. This rise in prices, which has channeled so many dollars to OPEC members, has had a desired effect in terms of new automobile manufacture. More energy-saving cars are coming down the production line and are entering the fleet of cars on the road as consumers see the value of saving gasoline and dollars.

The American public's recognition of the importance of fuel economy is mirrored in legislation passed

Box 25.1 Daylight Savings Time Versus 64 Million Pounds of Candles Every Year

Benjamin Franklin was astonished. "An accidental sudden noise waked me about six in the morning, when I was surprised to find my room filled with light. I imagined at first that a number of lamps had been brought into the room; but rubbing my eyes I perceived the light came in at the windows," he said in a letter to the *Journal of Paris*.

> I looked at my watch, which goes very well, and found that it was but six o'clock; and still thinking it was something extraordinary that the sun should rise so early, I looked into the almanac where I found it to be the hour given for the sun's rising on that day. Those who with me have never seen any signs of sunshine before noon, and seldom regard the astronomical part of the almanac, will be as much astonished as I was, when they hear of its rising so early; and especially when I assure them *that it gives light as soon as it rises.* . . . And, having repeated this observation, the three following mornings, I found always precisely the same result. (Franklin, 1784; Prerau, 1977)

The discovery gave rise to "several serious and important reflections." Franklin considered that had he slept to his usual rising hour, he would have slept six hours by the light of the sun, and in exchange have been up six hours the following evening by candlelight. Since the latter was a much more expensive light, he was induced by his "love of economy" to estimate how much could be saved by using sunshine instead of candles. For Paris, where he was residing, he calculated a saving of over 64 million pounds of candles every year. "It is impossible that so sensible a people, under such circumstances, should have lived so long by the smokey, unwholesome, and enormously expensive light of candles, if they had really known that they might have had as much pure light of the sun for nothing." (Franklin, 1784)

Daylight Savings Time (DST) was first proposed in a serious way as a method by which to save energy in England in 1907. By 1918 the United States had adopted the procedure of setting the clocks ahead one hour during the spring, summer, and fall. The procedure was dropped, however, after the end of World War I. It was not until 1942, the year after the United States entered World War II, that the nation returned to Daylight Savings Time, except that this time it was called "wartime" and it was in effect year-round. Turning the clock ahead brought a decrease in the peak demand for electric power that occurred in early evening. The war over, the nation again abandoned its manipulation of the clock, but now Daylight Savings Time had become popular enough for a number of states to adopt the plan on their own.

By 1966, to correct the lack of uniformity in times across the country, Congress enacted a bill that put all states (with several exceptions) on Daylight Savings Time from the last Sunday in April to the last Sunday in October. In 1974, year-round Daylight Savings Times went into effect again, this time for about a year and a half as an energy-saving response to the oil embargo. Public opinion on the potential hazard to schoolchildren in the dark morning hours was apparently responsible for the abandonment of year-round Daylight Savings Time. We then went back to a DST that lasted from the last Sunday in April to the last Sunday in October.

However, an unfortunate error was apparently made when daylight time was first adopted. Congress made daylight time from one month after the Spring Equinox, March 21, to one month after the Fall Equinox, September 21. To begin and end daylight time on days of the same length, the law should have placed the beginning of daylight time from one month *before the Spring Equinox* to one month *after the Fall Equinox* (from the beginning of March to the end of October). This error was partially corrected in July of 1986 when DST was lengthened to the period from the first Sunday in April until the last Sunday in October.

Sources:
B. Franklin, "An Economical Project," Letter to *The Journal of Paris*, 1784.

David Prerau, "Changing Times: National Time Management Policy," *Technology Review*, **79**(5) (March/April 1977), 54.

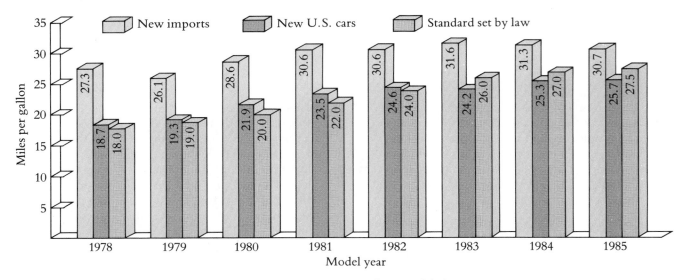

Figure 25.7 Fuel economy standards, fuel economy of cars, and fuel economy of imports, by model year. Fuel economy standards were set by Congress and the U.S. Department of Transportation. These standards are to be achieved "on average" by all manufacturers. That is, the average fuel economy of all vehicles sold by a particular manufacturer in a given year must meet the standard for that year. The standards for 1978, 1979, 1980, and 1985 were set by Congress as part of the Energy Policy and Conservation Act of 1975, while those for 1981, 1982, 1983, and 1984 were established in June 1977 under the authority of the Act. At first, U.S. manufacturers were able to exceed the standards. By 1983, however, General Motors and Ford had begun to fall short of the standards. In 1985, they won a reprieve—permission without a monetary penalty—to violate the 27.5-mile-per-gallon standard for 1986. Obviously, the technology was ready; it was the manufacturers' will to forgo profits on larger cars that failed.

by Congress in 1975, setting standards for the average fuel economy of autos manufactured and imported after 1977. The standards raised the required fuel economy in stages from 18 miles per gallon in 1978 to 27.5 miles per gallon in 1985. (See Figure 25.7.)

In the process of improving fuel economy, U.S. auto manufacturers, especially Ford and General Motors, complained of the costs of retooling to meet the standards and warned consumers of car price increases to come. They also argued that actual redesign of drive trains and new advanced engines would be necessary. In the first few years, they were able to increase the average mileage of their cars by selling smaller and hence lower-weight cars. They also achieved lower weights by substituting plastic and aluminum for steel in larger cars. General Motors introduced diesel engines on a number of its models in 1978 but has since given up production of diesels. It is of interest that the

new cars manufactured in 1981 only had to achieve a mileage standard that approaches the average of all cars, of all model years, already on the road in Europe. Obviously the technology already existed for that achievement.

In 1974, new cars manufactured in the United States had an average fuel economy of 13.2 miles per gallon. By 1980, new American-made cars had achieved a fuel economy of 21.9 miles per gallon, and by 1984, 25.3 miles per gallon. Beginning in 1985, Ford and General Motors failed to meet the fuel economy standard, not because the technology was not available but because these manufacturers chose to build and sell larger, less efficient cars—since profits on such cars were better. Chrysler, which spent billions to retool its factories, was able to meet the standards. It is of interest that new imported cars have exceeded the fuel economy standards since the very beginning of the process in 1977. Figure 25.7 shows the fuel economy standards year-by-year along with the achievements of U.S. and foreign manufacturers.

The automakers, specifically General Motors and Ford, argued that if they cut back on the manufacture of less-fuel-efficient cars, they would be "ignoring the needs of the consumer." In 1985, the National Highway Traffic Safety Administration (NHTSA) in the U.S. Department of Transportation granted the automakers a lower standard for 1986, namely 26 miles per gallon, and the opportunity to escape financial penalties. The possibility of extending the 26-mile-per-gallon standard beyond 1986 was hinted at as well. Lee Iacocca, chairman of Chrysler, noting that his company had spent $5 billion on retooling to meet the standards, said, ". . . it's turning out we were pretty damned stupid [to meet the standards]." The administrator of NHTSA told Congress that, ". . . market factors are the most efficient means of achieving . . . fuel economy over the long run." Do you agree? (See Controversy 25.1.)

Even though more-efficient new cars will be sold as time goes on, it will take time for energy-efficient cars to replace inefficient ones. The state of the nation's economy in terms of unemployment and inflation will influence how fast new cars are purchased and how fast older cars are retired. Table 25.1 shows the ago-

Table 25.1 Average Efficiencies of All Cars in the United States, Old and New

Year	U.S. passenger car registrations (millions)[a]	Average miles traveled per gallon of fuel consumed
1973	101.1	13.10
1974	104.8	13.43
1975	106.7	13.53
1976	110.1	13.72
1977	112.2	13.94
1978	116.5	14.06
1979	118.2	14.29
1980	121.7	15.15
1981	122.7	15.54
1982	123.7	16.33
1983	126.7	16.70

Source: U.S. Department of Energy, Energy Information Administration & Motor Vehicles Facts & Figures, 1984, 1985.

a. Federal Highway Administration data.

nizing slowness of improvements in the average fuel efficiency of automobiles in the United States. The near-depression level of the economy in 1981, 1982, and 1983 slowed even further the purchase of new automobiles and slowed severely the movement toward a fleet of fuel-efficient cars.

Even though energy-efficient cars save energy day by day, they cost more to buy initially, and because they are efficient, they depreciate in value more slowly. That means that lower-income individuals will hold their older cars longer before they can afford an efficient used car. Thus, both the economy and the higher cost of fuel-efficient cars could delay a transition to a fleet of cars with higher mileage per gallon.

What can we do in the meantime? Individual action and responsibility will help a great deal. First, and most simply, there is a speed at which automobiles and trucks are most efficient in terms of fuel consumption. For automobiles that speed is 35–40 miles per hour (56–64 km/hr); for trucks it is a bit higher. During World War II, a nationwide speed limit was imposed of 35 miles per hour. That speed limit was designed to conserve fuel for the war effort.

As a result of the energy crisis, the government mandated a 55-mile-per-hour (89 km/hr) speed limit. This represented a compromise between the speed that conserves the most energy and the speed that people apparently prefer to travel. In 1987, Congress backed away from the 55 mile-per-hour limit and passed a new law that allowed states to raise the speed limit to 65 miles per hour on rural portions of their interstate highways. By May 1987, twenty states had already done so. "Rural portions" refers to segments of the highway system outside cities of 50,000 or more people. Several states were raising the limit for autos but not for trucks. The results of this political compromise will be interesting to observe.

The 55-mile-per-hour limit, while initially planned to save energy, turned out to be a life-saving device as well. Motor vehicle fatality and accident rates have fallen significantly since the speed limit was reduced (Figure 25.8). A 1985 study by the National Research Council concluded that the 55-mile-per-hour limit saved 2,000 to 4,000 lives each year and reduced serious highway injuries by 2.6%. Fuel savings were estimated at $2 billion annually.

There are other features in the way we drive that influence mileage. Underinflated tires can decrease mileage by a mile per gallon, but radial tires can in-

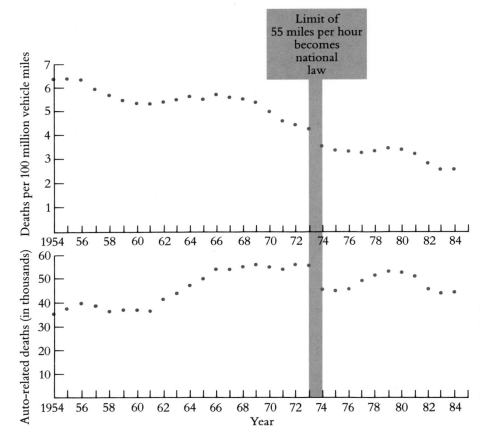

Figure 25.8 Traffic deaths and death rates. Between 1973 and 1974, the year the 55-mile-per-hour limit was first enforced, the annual death toll dropped by 9,000 and it has remained lower than the 1970–1972 levels every year since. In addition, the number of deaths per 100 million vehicle miles fell from 4.24 to 3.59, or by 15%, between 1973 and 1974, and it has not gone back up. (From National Safety Council, Accident Facts, 1984.)

crease mileage by a similar amount. Idling gasoline. If a car idles for a half hour, about a quart of gasoline is needlessly burned. Racing starts, uneven acceleration, and braking in city traffic can reduce the mileage a car obtains by up to 30%. Driving in lower gears takes more gasoline, as does an untuned engine. Short trips in cool and cold weather use more gallons per mile while the engine is warming up. Longer trips take advantage of the time already spent warming the engine.

When talk turns to saving gasoline, someone inevitably raises the possibility of the electric car. In the early days of the automotive industry, the electric vehicle (EV) was in competition with the auto that was powered by an internal combustion engine. But cheap gasoline for the internal combustion engine car and a limited range for the EV determined that the EV would not survive. In recent years, research on new and longer-lasting batteries has again raised people's hope for an electric vehicle.

A new concept has also emerged, the *hybrid car.* The hybrid vehicle, being developed by GE for the U.S. Department of Energy, would use an internal combustion engine for long trips and a battery for around-town trips. The battery would be charged by household current. The vehicle would definitely save petroleum, not a bad idea in a nation that is so dependent on oil imports. The electric energy would be likely to come from coal-fired and nuclear plants, so oil is conserved. Nonetheless, the energy savings of the hybrid vehicle would not be great, if indeed there are any savings. The reason is that electricity production is already so inefficient: 40% for the modern coal-fired plant, 30% for the nuclear plant.

After electrical transmission losses have been deducted, the hybrid vehicle and conventional auto are not far different in their efficiency of energy use. Suppose an electric vehicle goes 2 miles per kilowatt-hour and uses energy from a 40%-efficient coal-fired plant, which energy suffered a 5% loss in transmission. This

CONTROVERSY 25.1

How Do We Get Fuel-Efficient Cars on the Road?

We should adopt a surtax on all cars weighing more than 3,000 pounds escalated upward to as high as $1,000 tax on the big luxury cars weighing over 5,500.
John Quarles

I would like to describe a proposal for a stiff gasoline tax, specifically, an increase in the tax to $2 a gallon (with rebates on a per adult basis) . . . A $2 tax is not much higher than present taxes in countries like France and Italy.
Robert Williams

We must be cautious to see that a national policy of energy conservation does not unfairly burden the poor and those living in rural areas.
C. ReVelle

To stimulate oil conservation, we need a generation of high-mileage cars, but how do we get them? We can tax new car purchases by their weight or by their horsepower; we can tax gasoline; we can give income-tax rebates to individuals who purchase efficient new cars; we can ration gasoline, allowing one or two gallons per car per day. The many possible ways to direct people's attention to fuel-efficient cars are not all equal in their market effect; nor are they equal in their impact on upper-income, middle-income, and lower-income families.

There are four basic options listed above: (1) tax the purchases of inefficient new cars; (2) tax gasoline; (3) give rewards for purchasing efficient new cars; or (4) ration gasoline. We can assemble many plans from these basic options, but for each plan there is a set of questions to be answered:

1. How will it work? That is, what is the effect on the

rate of sales of different kinds of autos? On the sale of gasoline today and in the future?
2. (a) If the government pays, where should the money come from? From general tax revenues? From a special fund?
(b) If a tax raises money that goes to the government, what should be done with the money? Should it go into general tax revenues with no special earmarks? Or should it be set aside in a special-purpose fund? To give you an idea of what amounts of money are involved, each cent of federal gasoline tax raises $1 billion each year. That means that the five-cent-per-gallon tax that began in April 1983 will produce revenues of about $5 billion each year.*

(Continued on page 588)

---

*This small new tax is designated for road repair and jobs. It was not publicized as an energy-conservation measure prior to its passage.

3. Who is affected? Are the poor hurt more than the well-to-do? That is, is the policy progressive, taking dollars from those better able to afford the expense? Can you think of ways that the impact on the poor can be softened?
4. What are the political chances?
5. Is there an impact on the national economy? On local economies?

To focus the way you answer these questions, here are some hypothetical individuals, all of whom are very much concerned about such policies. Explain how the various conservation policies could affect the following people.

● Juan Lopez and his family live in Chicago; he maintains a set of apartments in the inner city in return for his own apartment. He just bought a gas-guzzling 1976 Zonker Grand Marquis.

● Arthur Williams and Sally Adam-Williams have moved back to their home town of Wayside, Nebraska, but the nearest job is in Omaha, some 70 miles away. They do not want to leave their rural environment for a city, but they need to make a living. They are about to buy their first car.

● Calvin and Nancy Beale own and run a diner in Ellsworth, Maine. Their diner, the Ellsworth Eatery, is well patronized and locally famous for homemade pies and the "Maineburger." The bulk of their business comes in the summer, when tourists flock to nearby Acadia National Park. Tourists with tents and motor homes come from all up and down the Eastern Seaboard, even from as far south as Virginia. No trains serve Ellsworth.

Can you think of other scenarios showing how people will be hurt by the various gas-conservation policies? Can you think of people who will hardly be affected at all by the listed methods?

Sources:
John Quarles, speech before Coal and the Environment Conference, Louisville, Ky., October 1974.
Robert Williams, in Daniel Yergin, ed., *The Dependence Dilemma* (Cambridge, Mass.: Harvard University Press, 1980).
C. ReVelle, "Public Transport and the Netherlands: Implications for Transport Policy in the U.S." Center for Metropolitan Planning and Design, John Hopkins University, 1977.

electric vehicle would require the consumption of 4,860 Btu at the power plant per mile of travel. A conventional auto getting 27.5 miles to the gallon (the 1985 new car standard) uses 4,727 Btu to go one mile. We use a small fuel-efficient car for this calculation because electric vehicles are expected to be small. Different assumptions about miles per gallon and generating efficiencies do alter these calculations, but not so significantly as to alter the general conclusion: electric vehicles are not significant energy savers. Instead, they are energy shifters.

Not only is energy efficiency about the same, but the pollution burden is shifted to other substances and to other sites—the sites of the electric power plants. A nation plagued with acid rain should think long and hard about a switch to electric vehicles.

So far, we have argued principally for changes in the automobile itself, not in the way our society transports itself. Public transportation (buses, train systems, and the like), if properly implemented, can have an important role in reducing energy consumption. Fig-

ure 25.9 illustrates the relative efficiency of various means of transportation. As can be seen, long-distance travel is typically most efficient because vehicles are more fully loaded and because fewer starts and stops save energy.

A vehicle not listed is the "auto in city traffic with a load equal to 75% of capacity." Such a vehicle is known as a car pool. People in car pools most often live near one another or near the line of travel. They also work near one another. The four-person car pool with an average load equal to 75% of capacity consumes only about 3,000 Btu per passenger mile, an efficiency level about that of a train traveling between cities.

Clearly, buses and trains are energy-efficient modes of travel, but how do we promote and encourage their use? To obtain a large proportion of people who will ride the bus, train, or subway, a number of factors must be positive. Routes must cover a large portion of the geographic area, so that access to public transport is not difficult. Stops must be frequent enough and

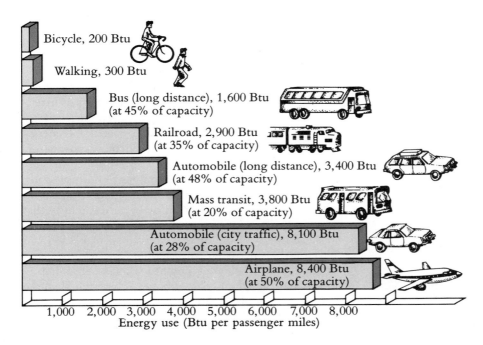

Bicycle, 200 Btu

Walking, 300 Btu

Bus (long distance), 1,600 Btu
(at 45% of capacity)

Railroad, 2,900 Btu
(at 35% of capacity)

Automobile (long distance), 3,400 Btu
(at 48% of capacity)

Mass transit, 3,800 Btu
(at 20% of capacity)

Automobile (city traffic), 8,100 Btu
(at 28% of capacity)

Airplane, 8,400 Btu
(at 50% of capacity)

1,000 2,000 3,000 4,000 5,000 6,000 7,000 8,000
Energy use (Btu per passenger miles)

Figure 25.9 Relative efficiency of various modes of travel. (Data from E. Hirst and J. Moyers, "Efficiency of Energy Use in the United States," *Science*, **179** [March 30, 1973], 1299. Copyright 1973 by AAAS. Capacity assumptions are those of Hirst and Moyers.)

spaced closely enough on the route so that the time to walk to a stop plus the time to wait for a bus is not too great. Equipment must be clean, conditions must be safe and comfortable, and costs must be reasonable when compared to use of the automobile. Routes cannot be too winding or the time from origin to destination will put off potential riders. In this regard, the use of express bus lanes on super highways has a positive effect in reducing travel time.

Even when all conditions are favorable, ridership on public transport may still be so low that the costs of providing the service (in equipment, personnel, gasoline, and maintenance) cannot be paid by the revenues generated from fares. Even in western Europe, where public transport runs quite full, governments make up the difference between revenues earned and the costs of actually operating and maintaining the system.

If the costs of public transport must be supported in part by the government, we need to justify this expenditure. It can be justified on the basis of energy savings, but, of equal importance, air pollution will be reduced by diminishing the number of cars on the road. Ozone, nitrogen oxides, carbon monoxide, and lead will all decrease as use of the auto decreases. These are positive benefits of public transportation. In addition, those who continue to drive have some benefits in terms of less-crowded streets and fewer accidents.

Where should the government find the money to make up the costs of public transport that are in excess of the system's revenues? Should income taxes be raised to pay this bill? Or should some other form of taxation be used to raise the funds to support public transportation? Should gasoline be taxed more heavily? The tax revenues could be used not only to help pay for public transport, but would simultaneously make public transport economically attractive to more people.

Freight Transport

About 30% of the energy we spend each year on transportation goes into the movement of goods; the remainder is expended on moving people. Of the various modes of freight transport, railroads, pipelines, and waterway transport all have comparable levels of efficiency. Trucks consume four to five times more energy per ton-mile than these other modes (Table 25.2).

The transport of goods, especially between cities, has evolved in much the same way as the transport of people. Whereas waterways and trains initially carried the bulk of goods moving between cities, trucks have been increasing their share of the load quite steadily. Thus, the energy-efficient modes of transport—barge and rail—are losing ground to an inefficient mode, the truck. Certain bulk goods such as coal, however, are

Table 25.2   Energy Cost of Freight Transport Modes

Mode	Btu/ton-mile
Pipeline	450
Railroad	670
Waterway	680
Truck	2,800
Airplane	42,000

Source: E. Hirst and J. Moyers, "Efficiency of Energy Use in the United States," *Science*, **179** (March 30, 1973), 1299.

unlikely ever to give way to truck transport. Instead, coal-slurry pipelines might capture a portion of the coal transportation market. (See Chapter 16 for a discussion of coal-slurry pipelines.)

The reasons for the growth in trucking are rooted in the door-to-door nature of truck service. Once goods are loaded, they often need not be reloaded before arriving at the final destination. If the particular goods are not moving in sufficient quantity to justify a rail track from door-to-door, truck transport, though costly, becomes a very attractive alternative. A routing that involves first a truck, then a train, and then a truck may be low in transport cost because of the rail savings, but the two transfers consume time and money and also increase the likelihood of damage to the shipment or of theft.

Summary

The oil crisis was caused by a rapid rise in the oil prices charged by members of OPEC. The crisis was eased by new suppliers entering the market from countries around the world and by industry and consumers adopting conservation measures—both events being responses to the price increases. Because the Middle East still possesses two-thirds of the world's proved petroleum resources, an oil crisis could occur again. In contrast, the natural gas shortage appears to have been caused by controls in the price of gas sold interstate and it was relieved by deregulation—the removal of price controls.

North Americans use much more energy per person than any other group of people in the world, so the potential for energy savings is great. Energy conservation measures can help postpone scarcity of fuel resources, save money for energy users, slow fuel price increases, and decrease pollution. Conservation mea-

sures could halve the predicted use of coal by the year 2000, as well as decrease sixfold the potential expansion of nuclear power. Conservation measures include more building insulation, more efficient appliances, more efficient automobiles and machines, manufacture of gas stoves without pilot lights, and replacement of incandescent bulbs with fluorescent lamps. Further, energy can be saved by increasing the efficiency of energy generation and by using, rather than disposing of, waste heat.

Using electricity to heat buildings and heat water is much less efficient and generally more expensive than using oil or natural gas. However, heat pumps that use electric energy may prove an efficient way both to heat and to cool buildings. Additional insulation can save most homeowners money, but care must be taken in the choice of materials and how they are installed.

Americans own more cars per 100 people than the citizens of any of the European nations. Further, people in the United States travel more miles per year than people from other countries do. The United States could save 25% of the energy used in automobile transportation if the cars on U.S. roads had the same average efficiency as cars in European countries. One way to encourage such efficiency is to increase gasoline taxes; another is to set fuel efficiency standards for new cars and enforce penalties for manufacturers who fail to meet the standards. Reducing the speed limit to 55 miles per hour saved fuel and also reduced traffic deaths and injuries. Electric cars would not save energy, but would shift the location of the pollution they cause from cities to rural areas where power plants are located. Car pools and public transportation systems (such as buses and trains) also promise energy savings. In addition, such systems reduce air pollution.

Questions

1. Compare heating water for the home by gas, electricity, and solar energy in terms of initial cost of the equipment, cost to run the system, and efficiency of energy use. Which system would you choose if you expect to live in your home 3 more years? 10 more years? The rest of your life?
2. What are the purposes of Daylight Savings Time?
3. How is it that people in some European countries manage to use 25 percent less gasoline while traveling almost as many miles by private auto as we do in the United States?

4. If you do not ride a bus to work or to school, what would it take for you to do so? If you do ride one, why do you choose this method of transportation?
5. Are electric vehicles energy savers? Explain.

References

General

The Good News About Energy. Washington, D.C.: U.S. Government Printing Office, 1979.

Boffey, P. M. "How the Swedes Live Well While Consuming Less Energy," *Science*, **196** (May 20, 1977), 856.

Franklin, B. "An Economical Project." Letter to *The Journal of Paris*, 1784.

Schipper, L. and A. J. Lichtenberg. "Efficient Energy Use and Well-Being: The Swedish Example," *Science*, **194** (December 3, 1976), 1001–1013.

Starr, C. "Is Sweden More Energy-Efficient?" *Science*, **196** (April 8, 1977), 121–124.

Prerau, David. "Changing Times: National Time Management Policy," *Technology Review*, **79** (5) (March-April 1977), 54.

"Energy Conservation: Spawning a Billion Dollar Business," *Business Week* (April 6, 1981), pp. 58–69.

Ross, M. and R. Williams. "The Potential for Fuel Conservation," *Technology Review* (February 1977), pp. 49–57.

Automobiles and Energy Conservation

Lave, L. "Conflicting Objectives in Regulating the Automobile," *Science*, **212** (May 22, 1981), 893–899.

Gray, C., and F. von Hippel. "The Fuel Economy of Light Vehicles," *Scientific American*, **244**, No. 5 (May 1981), 48–60.

Dallaire, G. "Transportation Innovations That Would Banish America's Energy Crisis," *Civil Engineering* (November 1981), pp. 47–50.

Boulding, Kenneth. "The Social System and the Energy Crisis," *Science*, **184** (April 19, 1974), 255.

Residential Energy Conservation

Dallaire, G. "Zero-Energy House: Bold Low-Cost Breakthrough That May Revolutionize Housing," *Civil Engineering*, **52**, No. 5 (May 1980), 47–59.

Bartos, Jr., M. "Underground Buildings: Energy Savers?" *Civil Engineering* (May 1979), p. 85.

"Tips for Energy Savers." U.S. Department of Energy, Assistant Secretary for Conservation and Solar Applications, DOE/CS-0020, March 1978.

Frieden, B. and K. Baker. "The Record of Home Energy Conservation: Saving Bucks, Not Btu's," *Technology Review* (October 1983), p. 23.

Schipper, L., and A. Ketoff. "The International Decline in Household Oil Use," *Science*, **230** (December 6, 1985), 1118.

Srnay, V. "Radon Exclusive," *Popular Science* (November 1985), p. 76.

Hileman, B. "Indoor Air Pollution," *Environmental Science and Technology*, **17** (10) (1983), 469A.

Srnay, V. "Heat-Saving Vents (Air-to-Air Heat Exchanger)," *Popular Science* (January 1983), p. 78.

Goldstein, D. "Refrigerator Reform: Guidelines for Energy Gluttons," *Technology Review* (February-March 1983), p. 36.

Nesbit, W. "Pumping Heat into Cold Water," *EPRI Journal* (February 1984), p. 16.

(National Archives/Documerica; Gene Daniels)

PART SEVEN

Human Health and the Environment

A startling and disturbing fact, emerging from the mass of scientific data on cancers, is that 60%–90% of human cancers are probably caused by environmental factors. "Environmental" is used, in this case, in the broadest sense. The term covers such things as eating and smoking habits as well as exposure to possible carcinogens (cancer-causing agents) in air, water, and manufactured goods.

About 400,000 Americans die each year from cancer, and some 700,000 new cases are detected (Figure A). Cancer is a worldwide problem as well. Around 2,900,000 new cases occur each year in the developed world and another 3,000,000 in developing nations.

A large portion of U.S. cancers must be attributed to smoking, and another smaller portion to working in certain industries. There is also evidence that the general public may be endangered by their chosen diet or by toxic substances that are widely distributed in the environment (Table A).

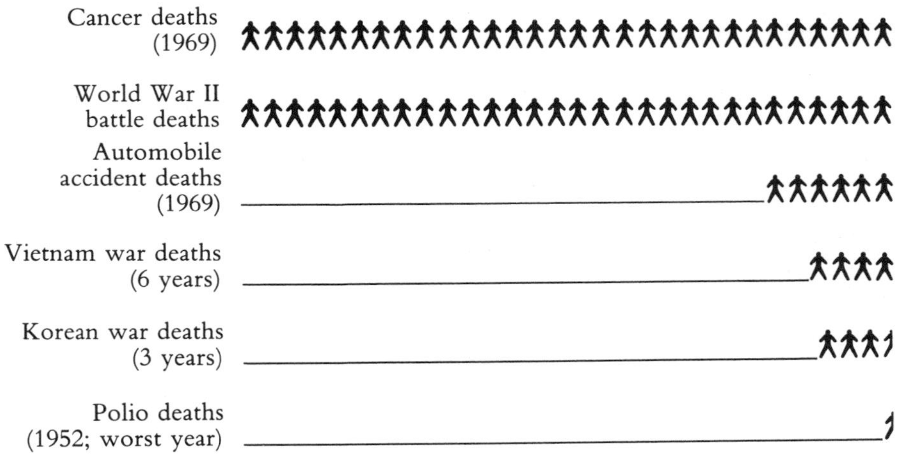

Cancer deaths (1969)
World War II battle deaths
Automobile accident deaths (1969)
Vietnam war deaths (6 years)
Korean war deaths (3 years)
Polio deaths (1952; worst year)

Figure A U.S. deaths from various causes. (Each figure represents 10,000 deaths.) (From *The Implications of Cancer-Causing Substances in Mississippi River Water* [Washington, D.C.: Environmental Defense Fund, 1974], p. 15.)

Table A   What Causes Cancer?

Cause	Estimated percent due to cause	Possible range
Smoking	30	25–40
Diet (not including food additives)	35	10–70
Workplace hazards	5	2–8
Alcohol	3	2–4
Environmental radiation (ultraviolet, natural x-rays, and cosmic rays)	3[a]	1–4
General air and water pollution	2	1–5
Medical treatment (drugs, x-rays)	1	.5–3
Food additives	1	2–minus 5 (protective effect)
Consumer products (for instance, Tris; urea-formaldehyde foam insulation; asbestos in patching compounds; trichloroethylene in aerosol cans; some hair dyes)	1	1–2
Nonenvironmental causes (infection, pregnancy, childbirth, sexual development)	17	2–?

Source: Adapted from R. Doll, and R. Petro, *Journal of the U.S. National Cancer Institute,* June 1981.
a. Does not include skin cancers other than melanomas.

In Parts Three, Four, and Five, we discussed a number of water and air pollutants, emphasizing where they come from and why they are found in our environment. These pollutants can have a variety of effects on human health. The health hazards of bacterial or viral water pollution were noted in Chapter 11. The health effects of chemicals such as mercury, cadmium, and nitrates, which sometimes contaminate drinking water, and the special case of trihalomethanes (carcinogens that may be formed from water purification procedures) were detailed in Chapter 14. In Chapters 20 and 21 we examined the harmful effects on human health of those air pollutants that result from power generation.

Now we take a closer look at certain other environmental factors and pollutants. Of particular concern are those suspected of contributing to, or causing, cancer deaths. Although there are a number of important health problems in the United States, cancer remains overall the greatest cause of death from disease in people under 55.

In Chapter 26 we discuss how the risks of health hazards such as carcinogens are evaluated. An interesting feature of much of the information used is that it comes from **epidemiology**—studies of the occurrence of diseases (such as cancer) in groups of people, or populations. In this chapter we also explore the difficulties in balancing risks against possible benefits of hazardous substances.

In the following chapter, a number of pollutants suspected of being contributors to the burden of carcinogens or otherwise harmful to human health are discussed. Most of these pollutants could be classified as "environmental" in the sense that individuals have little control over their exposure. Examples are asbestos, radiation, and workplace pollutants.

Chapter 28 covers three major contributors to the class of environmentally caused cancer: diet, drugs, and smoking. Although these factors are also called environmental, individuals can exercise a great deal of control over exposure to them. (There are exceptions; for instance, pesticide contamination of foods or smoke from a neighbor's cigarette.)

CHAPTER TWENTY-SIX

Science and Public Protection

Risks and Benefits

Lifestyle and Risks/A Basis for Judgments

The Epidemiologist as Detective

An Atlas of Cancer Deaths/A Mystery Case/Can Drinking Water Cause Cancer?/Latency Period of Cancers

Testing for Carcinogens

Using Animals to Test for Human Cancers/The Strengths of Carcinogens/Changes in DNA/New Tests

Risk–Benefit Analysis: Policy or Panacea?

Quantitative Risk Assessment/Risk Assessment and Limited Resources/Adding the Benefits to the Risk Decision/Risk or Risk-Versus-Benefit

Risks and Benefits

Lifestyle and Risks

In 1958, Congress passed a law that has become known as the Delaney Clause. This law specifies that no additive can be called safe (and thus used in food) if it has been shown to cause cancer in animals or humans. At the time this law was passed, it seemed a reasonable precaution. To many scientists, it still seems a good policy. To others, however, an outright ban on **carcinogens** appears naive. It is argued that consideration must be given to the possible benefits of a substance as well as to the possible risks. Basic to such a risk-benefit analysis is the idea that we can, in some way, measure both risks and benefits so that they can be compared.

What kinds of data do we have on cancer-causing substances and how good are the data? Who should make the final comparisons of risks versus benefits? These questions are important not only in legislating carcinogens but also in developing laws for many other environmental pollutants. Radiation standards, for instance, involve the same kinds of questions about public safety and how large a risk is tolerable. In fact, we all make risk-benefit decisions every day. Even such simple decisions as whether to cross the street, go for a ride in a car, or fly in an airplane involve this type of decision, although most of us hardly think about it. We are generally willing to accept a certain amount of risk as the price of the lifestyle we choose. But is such a procedure suitable for regulating carcinogens?

A Basis for Judgments

With respect to cancer-causing substances, we need to know how potent a carcinogen is, who is likely to be exposed to it, and how great that exposure is. We could then compare this risk to the possible benefits gained from allowing the use of the substance. Although this may sound like a perfectly reasonable and relatively straightforward procedure, a number of pitfalls exist.

Probably the greatest problem is that we do not have reliable scientific data on how carcinogenic most substances are to humans. Since we cannot experiment directly on humans, we must rely on three other kinds of data. The first is **epidemiological** data: data on disease patterns in large groups of people. The second kind is data from experiments in which animals are dosed with a test substance and then examined for cancer. A third type of evidence can be gained from tests on bacteria or on animal or human cells grown in test tubes. All of these methods have limitations, however, and none can be used directly to judge human cancer risk. The following sections examine more closely the various methods of testing for carcinogens.

The Epidemiologist as Detective

An Atlas of Cancer Deaths

In 1977, the U.S. Public Health Service published an atlas of cancer mortality for the 20-year period 1950–

1969. The atlas consists of a series of maps of the United States, showing the rates of death from various forms of cancer, county by county. Some striking variations can be found in the occurrence of various forms of cancer. For instance, in the northeast (New Jersey, southern New York, Connecticut, Rhode Island, and Massachusetts), there is a very high rate of mouth, throat, esophageal, laryngeal, and bladder cancers. The rates are high for males but not for females in this area, which suggests that the cancers may be related to the jobs at which the men work. In fact, this area is highly industrialized, with a large concentration of chemical industries. Several manufactured chemicals (2-naphthylamine is one) are known to cause bladder cancer. In Salem County, New Jersey, one-fourth of the workforce is employed in the chemical industry or related industries. The rate for bladder cancer in Salem County is the highest for any U.S. county with a reasonably large (10,000 or over) population. On the other hand, colon and rectal cancer is high for both males and females in this area. Although working in the chemical industries may eventually be shown to increase a person's chance of developing bladder cancer, some other explanation must be sought for the high incidence of colon cancer.

Examples of the occurrence of specific cancers in certain groups of people are found all over the world. Linxian County, in the People's Republic of China, has a very high incidence of cancer of the esophagus. Cancer of the stomach is very common in Japan, whereas in the United States it has steadily decreased to low levels since the beginning of this century. Liver cancer is a serious problem in Africa, Southeast Asia, and coastal China but is relatively rare elsewhere in the world. More examples could be given; however, it is clear that there are concentrations of various types of cancers.

Thus, there is a great deal of support for the idea that cancers are caused by some condition or some combination of conditions in the environments of different areas. This method of matching diseases in populations with factors that might have caused them is part of the science of **epidemiology**.

A Mystery Case

Such figures as those presented in the atlas cannot prove what causes a cancer, but they can provide clues as to where to look further. For example, on the Balti-

more census is an area where the cancer mortality rate is four-and-one-half times that of the city as a whole. The tract surrounds a former Allied Chemical plant that manufactured arsenic compounds for 100 years. Arsenic is a known carcinogen, and former plant employees have a lung cancer rate fourteen times higher than that of residents in other areas of the city. When investigators tried to determine whether emissions from the plant could have been responsible for the high neighborhood rate of lung cancer as well, they were at first puzzled by the pattern of where cancer victims lived with respect to the plant. Most victims seemed to live east and north of the plant, while prevailing winds blow north and northwest. If plant emissions contributed to the occurrence of lung cancer, one might have expected the pattern of cancers to follow the wind direction, the way pollutants would have blown. Eventually, former plant employees told researchers that there was once a railroad spur that ran to the plant from the north. Arsenic compounds were shipped to the plant along this line. If the cars were not sealed after unloading at the plant, arsenic could have blown out and settled along the rail tracks. In fact, a high concentration of arsenic was found in the soil along the old rail bed, the concentration decreasing as samplers moved away from the plant.

Actually, some 150–200 chemicals were manufactured at the plant at one time or another. Thus, it may never be possible to know for sure what specific substances caused the high cancer rate at the plant and in the neighborhood.

Can Drinking Water Cause Cancer?

A good example of the strengths and shortcomings of epidemiology is the controversy over carcinogens in the New Orleans water supply. In 1974, the Environmental Defense Fund (EDF) published a report noting that the number of cancer deaths in New Orleans was higher than expected. The report suggested this was due to carcinogens, both chlorinated and nonchlorinated organics, in the drinking water. (See Chapter 14 for a discussion of how these chemicals come to be in drinking water.)

Although no one argues with the fact that carcinogenic materials were found in New Orleans drinking water, it is not an easy matter to decide how much of a hazard this represents. According to a 1972 EPA study, various organic chemicals were present in the parts-per-billion range. In fact, except for a few chemicals,★ most were present at less than 1 part per billion. We do not know the effects of these very low concentrations of toxic chemicals, day after day, over a person's whole lifetime. Scientists from EDF attempted to solve this problem by looking at statistics on people drinking less-polluted water. Ten parishes in Louisiana and one-third of another draw their drinking water primarily from the Mississippi; the rest use groundwater (generally believed to be less polluted than surface water) or other surface water supplies. When these two populations are compared, it can be shown that there are more cancer deaths among those drinking Mississippi River water. However, this is not proof that the water is causing excess cancer deaths. Some other factor might be the true cause. The correlation with drinking water might be chance or might be associated in some way with the true cause (for instance, some food in the diet of people living in the affected area could be the cause). Researchers thus checked a variety of other possibilities.

Cancer deaths were divided into groups by race and sex. The deaths were still found to be associated with the source of drinking water for white and nonwhite males and for nonwhite females. Dividing people into groups by income made no difference, nor did dividing them by occupation. Cancer deaths were not correlated with living in southern versus northern parishes. (The southern parishes are different socioeconomically from the northern parishes and thus might be expected to have differences in diet and other factors.) Even elevation was checked, since air pollution could be expected to be worse at lower elevations than at higher ones. No correlation was found, however.

Cancer rates in the city itself might be falsely high if people have moved there for better medical care after the disease is diagnosed. However, subtracting the New Orleans data from that of all other Louisiana people drinking Mississippi River water did not change the positive relationship with cancer deaths.

The epidemiologist must be even more of a detective, however. Why are cancer deaths not correlated with the source of drinking water for white females? What about the role of smoking or alcohol consump-

tion? Are the excess cancers of a kind that could reasonably be expected to develop from contaminated drinking water (liver or digestive or urinary cancers), or are they more likely to have developed from air pollutants (lung cancers)?

More detailed studies, comparing cancer cases with the actual concentrations of carcinogens in the victims' drinking water, would come closer to proving a link between drinking water and cancer. Studies such as this are now under way. At the moment, most epidemiologists feel that the evidence linking chlorinated organics in drinking water to excess cancers is becoming stronger.

Although it is difficult to be sure that all possible interfering causes for a problem have been thought of, eventually the weight of epidemiological evidence becomes so great that most people consider the case proven beyond reasonable doubt. This is the case with evidence about cancer and smoking (Chapter 27).

Latency Period of Cancers

The possible involvement of arsenic in excess cancer deaths in a Baltimore area points out another problem epidemiologists face in determining the causes of cancers: A long time usually goes by between exposure to a carcinogen and diagnosis of cancer. For instance, it is now known that lung cancer caused by breathing asbestos dust does not appear for 20 years or more after exposure to the dust. In one study, an average of 39 years elapsed between the time workers breathed the asbestos and the time they developed lung cancer (Table 26.1). Thus, even methods that involve comparing groups of people who were unintentionally exposed to chemicals (pesticides workers, industrial employees, construction workers) with groups of people who were

★Chloroform was found at 113 ppb; four others were found at 1–8 ppb.

Table 26.1   Latency Period of Some Cancers

Type of cancer	Approximate number of years between exposure and diagnosis of cancer
Lymphomas	2–5
Bladder cancers (due to aromatic amines)	18
Mesotheliomas (asbestos-caused)	30
Lung cancers (asbestos-caused, bronchogenic)	20–40

not exposed are of limited value. Information can be gained this way, but it may come too late.

Because there is such a long time period between exposure to a carcinogen and the proof that harm has been done, many scientists and legislators feel that we must control substances released into the environment. Otherwise we may find out in 20 or 40 years, when it is too late, that some commonly used chemical is a carcinogen.

Testing for Carcinogens

Using Animals to Test for Human Cancers

Many different animals have been used to test for carcinogens (cancer-causing substances). Experiments on large animals such as monkeys or dogs are extremely expensive and also take years to produce useful results. Such experiments may be justified in cases where many people are being exposed to suspected carcinogens. Some of the information we have on smoking and lung cancer and on chemicals and bladder cancer comes from this type of research.

The accepted method of testing for carcinogens is generally to use two species of rodents: rats and mice or hamsters. The animals are given the test chemical in the highest doses possible without causing them to become sick or die (from some cause other than cancer). Periodically, the animals are examined to see whether more than the normal number of cancers can be found in the treated group of animals.

Very large numbers of animals may be needed to detect a cancer-causing substance. For instance, suppose a chemical caused cancer in 1 out of 10,000 humans exposed to it. If the entire U.S. population were exposed, 20,000 people would develop cancer, a fairly large number of cases. Yet to detect this hazard in an animal experiment, 10,000 rats would be needed to find one cancer. At least 30,000 would be needed to call the results significant. Further, humans may be more sensitive to some chemicals than rodents. For instance, humans are 60 times more sensitive to the effects of thalidomide than mice and 100 times more sensitive than rats.

The Strengths of Carcinogens

We know that, in general, some substances are able to cause cancer at lower concentrations than others. How-

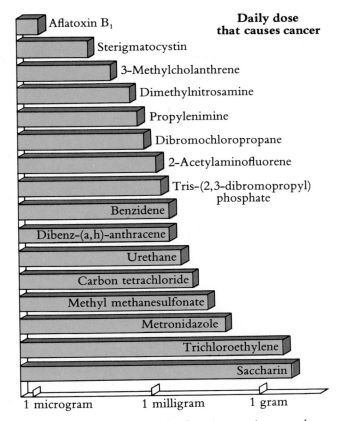

Figure 26.1 Relative strength of carcinogens in rats and mice. The doses shown are those that cause cancer in half of a group of test rats or mice when they receive it daily over their lifetime. As you can see, aflatoxin is over a million times more potent than saccharin. (From T. H. Maugh II, *Science*, **202** [October 6, 1978], p. 38. © 1978 by the American Association for the Advancement of Science.)

ever, we do not yet have the information that would make it possible to state the relative potency of most carcinogens for humans. In part, this is due to the fact that most of the information comes from animal experiments and cannot be directly applied to humans. The effects of carcinogenic materials on humans have been studied well enough in only six cases[*] to rank their relative strengths. In these six cases, the carcinogenic effect in humans is roughly comparable to that obtained in rat experiments. It is on this basis that results in rats are often applied to humans (Figure 26.1). Animal data, however, may not always be a valid basis to

---

[*]Benzidine, chlornaploazin, diethylstibestrol (DES), aflatoxin B_1, vinyl chloride, and cigarette smoke.

use in assigning human potency to carcinogens, because carcinogenic effects differ among animal species. To give two examples: aflatoxin is not carcinogenic in adult mice, although it is in rats; 2-naphthylamine is not a rat carcinogen, although it is carcinogenic to humans.

Changes in DNA

We are just beginning to understand how a carcinogen affects cells. There are many different carcinogens and there is probably more than one way that cancers arise. However, a large group of chemicals appear to cause some permanent change in a cell by reacting with its DNA. The **DNA** in a cell is in the form of one or more large molecules, composed of nucleic acids, phosphate groups, and the sugar deoxyribose linked in a long chain. Sections of the DNA chain, containing specific sequences of nucleic acids, are now recognized as genes, those units of cellular material that control heredity and the normal workings of a cell. A permanent change in one of these genes is called a **mutation**. Most, although not all, chemical carcinogens can be shown to cause mutations in specially designed experiments. Of course, not all mutations cause cancer (Figure 26.2). A mutation involving genes that control a cell's reproductive mechanisms is an example of the kind of change believed to lead to cancers. This first stage of carcinogenesis is called *initiation*.

The second stage is called *preneoplasia*. This is the latent period. Little is known about what happens during this long, apparently quiet period, but there is evidence that many cells are repaired or revert to normal during this time. Vitamin A may play a role in repairing cell damage, as do a variety of enzymes in cells. Cells that are not repaired go on to stage three, or *transformation*. Now they show the characteristics of cancer cells and begin to proliferate. Much research is being directed toward finding substances that can help cells repair themselves during preneoplasia in order to undo the effects of exposure to carcinogens.

In some cases, factors that are not carcinogenic themselves seem to cause cancer in combination with other substances. Some of these substances are called **promoters**. They seem to "turn on" cells that have been initiated by a carcinogen but are in a latent period. The artificial sweeteners saccharin and sodium cyclamate are both examples of compounds that are either very weak carcinogens or not carcinogens at all, but

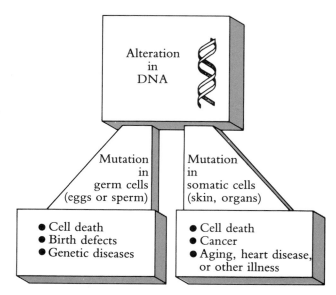

Figure 26.2 The effects of changes in DNA. (From *Short-Term Tests for Carcinogens, Mutagens and Other Genotoxic Agents*, Technology Transfer Report # EPA–625/9–79–003, U.S. Environmental Protection Agency, July 1979, p. 5.)

promoters. **Cocarcinogens** are substances that cause cancer in combination but not by themselves. In animal experiments, sulfur dioxide in combination with benzo(a)pyrene causes cancer, while neither of the two chemicals alone does so. Further, there is evidence that small amounts of carcinogens add together to increase the total carcinogenic effect. Finally infection with certain viruses appears to be a cause of some human cancers, but only if other factors, such as chemical carcinogens, are also present.

Of the almost 2 million chemicals known, only about 6,000 have been tested in animal experiments to determine whether they are carcinogenic. Of these, 1,000 have given some evidence that they do cause cancer. Of the 1,000, a few hundred are considered proven carcinogens.

New Tests

Animal welfare groups are pressing for new methods of testing. They want to see the use of tests that do not require animals (especially dogs, cats, and monkeys) to be killed or injured in order to determine the safety of chemicals for humans.

Several new methods are being developed to provide faster, cheaper screening for carcinogenicity. One group of methods uses animal cells grown in test tubes. Changes in the **DNA** of cells exposed to chemicals can be correlated to the development of cancers. Results are promising, but much work remains to be done on this group of methods. Ideally, the animal cells used should be human cells, but this is not yet possible.

Another method, called the Ames assay, is already in use. It involves adding the chemical to cultures of the bacterium *Salmonella typhimurium*. The bacteria are then examined to see whether the chemical has caused a *mutation*. In this particular test, researchers look to see whether bacteria that require the nutrient histidine to grow have changed, or mutated, so they can grow without it. (This is called a *back-mutation*.)

Although the Ames test detects mutagens (substances that cause mutations), not carcinogens, a good correlation has been found between the two. That is to say, 80% of the carcinogens tested have been found to be mutagens. Furthermore, most known human carcinogens give a positive result in the Ames assay. Of the noncarcinogenic substances tested, 87% were negative in the assay, while 13% gave a false positive result. These percentages may improve even more as the method is refined. For instance, it has been shown that certain substances undergo chemical changes in the body that turn them into carcinogens. Thus, benzo(a)pyrene is a carcinogen after conversion to an active form by the cell components called microsomes. If benzo(a)pyrene is used in the Ames assay, it is not mutagenic. However, if a preparation of liver microsomal enzymes is also added to the test system, mutations occur (Figure 26.3). Further refinements such as this will undoubtedly increase the accuracy of the Ames assay and of the other "quick tests," such as those using cell cultures. The Ames assay is quick (about three days to run a test) and inexpensive (about $200 per chemical, compared to $100,000 per chemical in the conventional rat and mouse assays). It is the possibility of false negatives, however, that limits the usefulness of the test. That is, up to 20% of carcinogenic compounds may not be mutagenic in the Ames assay. Thus, no chemical could be given a clean bill of health on the basis of the Ames assay alone.

A final group of methods uses mathematical modeling and computers to determine, by looking at the structure of a chemical, whether it is likely to be carcinogenic or perhaps a good anticancer drug.

Some combination of testing methods will even-

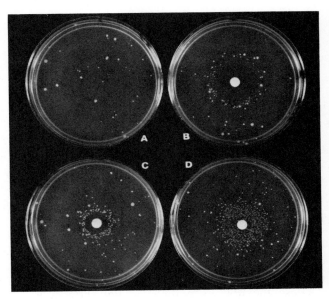

Figure 26.3 The "spot test" for mutagen-induced revertants. Each petri plate contains, in a thin overlay of top agar, the tester stain TA98 and, in the cases of plates C and D, a liver microsomal activation system (S-9 Mix). (Plate B did not require the liver system.) Mutagens were applied to 6 mm filter-paper discs, which were then placed in the center of each plate. (A) Control plate. (B) Revertant colonies produced by the Japanese food additive furylfuramide (AF-2) (1 µg). (C) (Revertant colonies produced by the mold carcinogen aflatoxin B$_2$ (1 µg). (D) Revertant colonies produced by 2-aminofluorene (10 µg). Mutagen-induced revertants appear as a circle of revertant colonies around each disc. (From Bruce N. Ames, Joyce McCann, and Edith Yamasaki, "Methods for Detecting Carcinogens and Mutagens with the Salmonella/Mammalian-Microsome Mutagenicity Test," *Mutation Research*, **31** [1975], 347–364. Reprinted by permission. Copyright © 1975 by the Elsevier Scientific Publishing Company, Amsterdam.)

tually have to be worked out in order to catch all potential carcinogens before they can be distributed in the environment.

Risk-Benefit Analysis: Policy or Panacea?

Quantitative Risk Assessment

There are many uncertainties in determining the carcinogenic risk to humans when using data from animal or test-tube experiments. Nonetheless, many reputable scientists are pressing for a policy of quantitative risk assessment. This means assigning actual numbers to the risks of contracting cancer from particular substances or treatments.

We have already discussed some of the problems involved in using animals or bacterial cells to measure the potency of carcinogens to humans. This part of risk assessment is sometimes called *biological extrapolation*.

Another part of quantitative risk assessment involves *numerical extrapolation*. This means extending the results of animal experiments at high doses to the lower doses involved in workplace or environmental exposure by a mathematical calculation. Animals in experiments receive much larger doses of test compounds or of radiation than those to which workers or the general public are likely to be exposed. This is done on the assumption of some positive relationship between the dose of a harmful substance and its effect: The larger the dose, the more likely the effect will show up in an experiment. By using these high doses, scientists can use reasonable numbers of animals and still find evidence for even weak carcinogens.

Probable effects are then calculated for the exposure levels of workers or the general public. The question is often posed, however, whether this is reasonable; whether there is not, in fact, a level below which no effect occurs (the threshold level). Thresholds could be the result of such processes as repair mechanisms by which the body repairs damage caused by carcinogens. There is evidence for such repair, but at present the more conservative policy is to assume that there is no threshold. In other words, we assume that smaller and smaller doses result in fewer, but still measurable, numbers of cancers.

Risk Assessment and Limited Resources

Those who advocate quantitative risk assessment point out that we actually do not have the resources to properly test the 9,000 synthetic chemicals manufactured on a large scale today, much less the 500–1,000 new ones introduced each year. They argue that, flawed as it may be, such analysis would at least give a basis for deciding how to spend limited resources. Substances would be ranked both on how many people would be exposed and how toxic or carcinogenic they appear in a combination of animal and other tests. Substances judged to present higher risks would be tested first and most extensively.

Adding the Benefits to the Risk Decision

Where opinion divides most widely is on whether to use quantitative risk assessment to determine the cost to society of a hazard and then compare this cost to the possible benefits obtained. Some scientists, such as Arthur Upton, director of the National Cancer Institute, feel that the risk evidence is too flimsy to make such comparisons the basis for regulating carcinogens (for setting exposure standards in the workplace, for instance).

In addition to uncertainties in risk assessment, critics of cost-benefit analysis point out several drawbacks. Such analysis is intended to reduce all factors to a common denominator for purposes of comparison. Dollar values must therefore be put on such things as a human life or a pleasant environment. Many people feel such things are not quantifiable.

The analysis also cannot consider intergenerational effects. That is, many substances are both mutagens and carcinogens. People exposed may develop cancer and their germ cells may undergo mutations. The effects of these mutations must be borne by later generations.

Also, there may well be inequities in the distribution of costs versus the benefits. The benefits of a regulation go to the people protected by it, while the costs may be borne by another group (such as industry). On the other hand, without regulation, the health risk is borne by people who may not benefit at all from the process causing the risk. Cost-benefit analysis does not deal well with this sort of problem. Even determining the cost of regulating a substance, which may seem straightforward, involves extrapolating the probable cost of regulation into the future. The future value of the dollar is important to such analyses but not easy to agree upon, especially if inflation is a likely complication.

Once risk assessment is an accepted technique, however, the temptation will certainly exist for lawmakers and regulators to use cost-benefit analysis to regulate hazardous substances. It will be all too easy for regulators and the general public to lose sight of the uncertainties and limitations of this apparently simple but possibly treacherous tool. (See Controversy 26.1.)

Risk or Risk-Versus-Benefit

It is interesting to note the basis for regulation of the various laws dealing with toxic substances. Even when

CONTROVERSY 26.1

The Murder of the Statistical Person: Risk-Benefit Analysis Applied to Carcinogens

Public policy is fundamentally an economic exercise. It cannot evade the balancing of risks and benefits . . .

Gio Batta Gori, deputy director, Division of Cancer Cause and Prevention, National Cancer Institute

[Risk-benefit analysis] sets out to answer the question: What is *efficient* for society? But this is not what the public asks about environmental issues nowadays; they ask: What is *good* for society? And the most precious kinds of good cannot be quantified . . .

Eric Ashby

The use of risk-benefit analysis to regulate carcinogens is shot-through with controversy. One of the most serious limitations is that the data for estimating risk must come from animal or test-tube experiments, and we are not yet sure how well this data predicts human risk. Risk-benefit analysis also involves such arguable features as assigning value to human life or a pleasant environment and the probable cost of regulation at some date in the distant future. Further, it does not take into account who bears the costs of regulation or nonregulation, compared to who derives the benefits.

Another problem arises when costs and risks have been determined as well as possible and are ready to be compared. If risks are not to be reduced as close as possible to zero (because of the high cost), some judgment must then be made as to what is an acceptable risk. For instance, one might decide that since 20,000 out of every 100,000 Americans die of cancer, increasing the number to 20,001 out of 100,000 would be an acceptable risk. This represents 2,200 excess cancer cases (assuming a U.S. population of 220 million people), or 30 more deaths per year due to cancer in the U.S. (2,200 deaths spread over a 70-year life span).

Now, if these 30 people each year were actually known to us, the risk might not seem so negligible. This is called the "murder of the statistical person."

Do you favor regulation of carcinogenic substances by risk-benefit analysis, or by a policy of reducing the risk to as low a value as possible, regardless of cost? Does the kind of substance to be regulated make a difference? What if it were a food additive? What if it were a cosmetic? What if it were an air pollutant? Does it make a difference whether exposure is voluntary (food additive) or nonvoluntary (air pollutant)? Does it make a difference who is exposed (children, pregnant women, workers)? Who should make these kinds of decisions? Congress, the judiciary, or regulatory agencies? (All of these bodies have actually made such decisions—can you think of examples?) Or perhaps a special court that includes members of the general public or special-interest groups?

Sources:
Gio Batta Gori, quoted in *Science,* **208** (April 18, 1980), 256.
Eric Ashby, *Environmental Science and Technology,* **14** (10) (October 1980), 1178.

dealing with carcinogens, laws often allow consideration of the possible benefits of a substance or the cost of regulating it (Table 26.2).

At the present time, only part of the Food, Drug, and Cosmetic Act (the Delaney Amendment), part of the Clean Air Act, and the Resources Conservation and Recovery Act are risk based; that is, once a substance is shown to be carcinogenic, agencies are directed to reduce human exposure as close as possible to zero. No consideration of benefits is allowed. Other laws, such as the Clean Water Act and part of the Clean Air Act, direct regulators to use the best available technology or best practicable technology to reduce exposures. There is, of course, cost-balancing involved in deciding what technology is most practicable or even best available.

All the rest of the laws allow benefits such as increased food production or the cure of a disease to be considered when considering risk. None of these laws, however, specifies that a formal cost-benefit analysis be done. In a formal analysis, values might have to be attached to human life or happiness. Congress seems willing to accept more general value judgments from regulatory agencies about the magnitude of risks and benefits. (See Controversy 26.2 for a comparison of U.S. and Soviet approaches to regulating toxic substances.)

Summary

Risk-benefit analysis means comparing the possible risks of a substance or action with the possible benefits to be gained from it. Carcinogenic risk is judged from epidemiological data, from animal experiments, and from test-tube type evidence. Epidemiological data may be difficult to interpret because of interfering factors and because of the long latency period of cancers.

Carcinogens differ in their ability to cause cancers. Some substances do not cause cancer alone but act with other substances to cause cancers. Cancers seem to develop in three stages: initiation, preneoplasia, and transformation.

Animals are often used to determine the potency

Table 26.2 Regulatory Basis of Laws Governing Toxic Substances[a]

Law[b]	What is regulated	Basis of regulation
Food, Drug and Cosmetic Act (1938, 1958, 1960)	Food additives	Risk to health
	Drugs	Risk versus benefit
	Cosmetics	Risk to health
Occupational Safety and Health Act (OSHA, 1970)	20 substances in workplace	Risk versus benefit, considering technology available
Clean Air Act (1970)	Stationary sources of some 20 substances, including asbestos, mercury, and arsenic	Risk
	Vehicles; diesel particulates and control attempts	Technology available
	Fuel additives	Risk versus benefit
Clean Water Act (1972)	Toxic discharges	Technology available
Pesticides Control Act (1972)	Pesticides	Risk versus benefit
Resource Conservation and Recovery Act (RCRA, 1976)	Hazardous wastes	Risk to human health, environment
Safe Drinking Water Act (1974)	Substances in drinking water: pesticides, trihalomethanes, metals	Risk versus benefit
Consumer Product Safety Act (1972)	Hazards in consumer products (e.g., tris, asbestos, benzene, vinyl chloride)	Risk versus benefit
Toxic Substances Control Act (TOSCA, 1976)	Anything not covered in above laws (e.g., PCBs in environment)	Risk versus benefit

a. For more discussion of these regulations, see Clean Air Act, Chapters 20, 21; Clean Water Act, Chapter 13; RCRA, Chapter 14; Safe Drinking Water Act, Chapter 14; Pesticides Control Act, Chapter 7; Occupational Safety and Health Act, Chapter 27; Food, Drug, and Cosmetic Act, Chapter 28.

b. Dates in parentheses are the date the act was first passed and those dates on which major amendments were added.

CONTROVERSY 26.2

Toxic Substance Control in the Soviet Union

Even a brief comparison of the standards for exposure to workplace air pollutants in the United States and the Soviet Union shows an astonishing difference in the standards (Table C.1). Of the 100 substances for which both countries set standards, the Soviet standards are more strict 80% of the time, often by a significant amount. In only two cases are American standards more strict than Soviet standards. From the standards alone, it would appear that the Soviets are providing more protection for their workers than we do in the United States.

This appearance is partly due to the methods of testing used. Russian toxicologists give a great deal of weight to psychological and neurophysiological effects of toxic substances—that is, effects on the sense of smell, on electroencephalogram responses, and on animal motor responses. These effects can be subtle. They occur at much lower levels than those causing the more visible physical effects that concern U.S. toxicologists, such as production of tumors, decreased respiratory function, and so on. Further, not all scientists would agree that such responses prove any harm has been done.

Besides differences of opinion about the kinds of experiments to use, there seems to be a major difference in the philosophy of setting standards in the two countries. One Soviet author identified four possible bases for regulation decisions (Riazanov, 1976):

1. A normal unpolluted environment is the natural condition and any change caused by a pollutant should be considered not permissible.
2. The major consideration should be what levels of pollution control are technologically feasible.
3. Economic considerations should be the most important. That is, the allowed pollution level should be set so that the cost of controlling a pollutant is equal to the cost to society of the damage that will result if the pollutant is not controlled.

Table C.1 A Comparison of Representative American and Soviet Workplace Standards for Air Pollutants

Substance	American standard[a] (mg/m³)	Soviet standard[a] (mg/m³)
Aldrin	0.25	0.01
Aniline	19.0	0.1
Carbon monoxide	55.0	20.0
Dioxane	360.0	10.0
Ethyl alcohol	1900.0	1000.0
Ethyl mercaptan	25.0	1.0
Ethylene oxide	90.0	1.0
Heptachlor	0.5	0.01
Hydrogen cyanide	11.0	0.3
Methylchloroform	1900.0	20.0
Vinyl chloride	1300.0	30.0
Acrolein	0.25	0.7
Anisidine	0.5	1.0

Source: George J. Ekel and Warren H. Teichner, *An Analysis and Critique of Behavioral Toxicology in the USSR*. Washington, D.C.: U.S. Government Printing Office, 1976.

a. U.S. standards are 8-hour weighted average concentrations, while Soviet standards are ceiling values never to be exceeded. Thus, Soviet standards are actually more strict than appears from the numbers alone.

4. Standards should be set at levels for which there are no direct or indirect harmful or unpleasant effects on humans, in terms of effects on their ability to work as well as their physical and emotional well-being. (It is worthwhile noting that any measurable effect is felt to be potentially harmful; that is, an odor in the workplace, even if generally considered pleasant, would not be permissible, in the same way that adding sugar to the water supply would be undesirable.)

The author then goes on to state that the first

option of allowing no change in the environment is unnecessarily strict, because not all changes lead to measurable effects in humans. The third option, which considers only the costs of controlling pollution versus the cost of not controlling it, he feels is unnecessarily commercial, since it involves setting a price on human life, health, and well-being. He feels the second option is not strict enough, since in most cases the level of control technically feasible at the present would still allow harmful effects. Thus, standards set at such levels tend to hinder development of better pollution control. The fourth option he considered the only proper approach.

However, it should be noted that while standards in the Soviet Union are often set much lower than in the United States, the methods for achieving these standards are not always defined. In fact, it is likely that these levels are often exceeded, at least where they are lower than what is technologically feasible or where achieving those levels is extremely costly. The Soviet standards are thus ideals, or "targets" to work toward, rather than limits that must not be exceeded. It is expected that decisions will be made at the local level on whether or not standards will be met.

In the United States, although we do not seem to have a uniform philosophy about setting standards for toxic pollutants, once standards are set, it is expected that they will be enforced. There are certainly environmental protection groups in the United States who feel that the first option, no change in the environment, is the safest, most desirable course. The Delaney Amendment, which prohibits any substance in foods once it has been shown to be a carcinogen, is an example of this most restrictive category of U.S. law. The law

allows no consideration of possible costs or benefits involved. But it is also, in the strictest sense, unenforceable. Some carcinogenic substances such as pesticides do find their way into foods through indirect means. Congress was even forced by public pressure to exempt from control the food additive saccharin, a weak but known carcinogen.

A number of U.S. standards are set according to the second option: what is technologically feasible at the moment. Some water-pollution discharge standards, for instance, are based on the best available technology. So also are OSHA workplace standards, which require that worker exposure to carcinogens be lowered to the lowest possible levels. In other cases, Congress has set future standards below levels that were currently technically possible in order to stimulate industry to develop the necessary technology; standards for emissions from automobiles are an example. In many U.S. standards, economic considerations are given weight.

Which of the four possible bases for regulatory action do you feel are reasonable? Should one of them be the major basis for U.S. decisions, or does the present patchwork serve better? Why?

Do you agree or disagree with the Soviet basis of standard setting and decision making? Why?

Is the third option, which allows the cost of regulation to be compared to the cost of damage to society, the same as risk-benefit analysis? Can it be made to work? Why or why not?

Source: V. A. Riazanov, in George J. Ekel and Warren H. Teichner, *An Analysis and Critique of Behavioral Toxicology in the USSR.* Washington, D.C.: U.S. Government Printing Office, 1976.

of carcinogens. Other tests include the Ames assay, tests using human or animal cells, and mathematical modeling using computers.

Quantitative risk assessment involves both biological extrapolation of human risk from animal data and mathematical extrapolation of the effect of low doses of carcinogens from experiments using high doses. There may be a threshold for the effects of carcinogens, but it is prudent to assume there is not. The assessment of both the risk and the benefit halves of a risk-benefit equation has many difficulties and pitfalls.

Questions

1. Epidemiological studies, such as the study of the occurrence of cancer according to locality, are useful but cannot prove that certain substances or occupations lead to cancer. One of the reasons is that people are exposed to many factors at the same time. Thus, it is not easy to sort out which one or ones actually cause a disease. Give examples of factors or habits that could confuse a study on whether the smoke from a factory is causing lung cancer in a nearby neighborhood.

2. Why do we have to use animal studies to determine

whether substances are carcinogenic? What problems does this cause?

3. What is the Ames assay?
4. What is meant by quantitative risk assessment? How is it related to risk-benefit or cost-benefit analysis?

Further Reading

Searle, C. E. "Chemical Carcinogens and Cancer Prevention," *Chemistry in Britain* (March 1986), p. 211.

A good summary of current knowledge about carcinogenic chemicals in the human environment.

Drake, J., B. Glickman, and L. Ripley. "Updating the Theory of Mutation," *American Scientist*, **71** (November–December 1983), 621.

If you have some basic biological knowledge, you'll appreciate this paper. It explains more about mutations, how the body repairs them and what happens if they aren't repaired.

Assessment of Technologies for Determining Cancer Risk from the Environment. Congress of the United States, Office of Technology Assessment, June 1981.

A well-written summary covering the major points in this chapter. The almost 400 references enable the reader to delve further into the literature.

Barker, Brent. "Cancer and the Problems of Risk Assessment," *Epri Journal* (December 1984), p. 26.

A variety of opinions on the value of risk assessment are included in this article.

Yunis, Jorge. "The Chromosomal Basis of Human Neoplasia," *Science*, **221** (July 15, 1983), 227.

Explains how changes in chromosomes may lead to cancer.

Land, H., L. Parada and R. Weinberg. "Cellular Oncogenes and Multistep Carcinogenesis," *Science*, **222** (November 18, 1983), 771.

Provides details on the cellular steps that lead to cancer formation.

References

Weisburger, John H., and Gary M. Williams. "Carcinogen Testing: Current Problems and New Approaches," *Science*, **214** (October 23, 1981).

Squire, R. A. "Ranking Animal Carcinogens: A Proposed Regulatory Approach," *Science*, **214** (November 20, 1981).

Mary, J. L. "Viruses and Cancers," *Science*, **231** (February 28, 1986), 919.

Drinking Water and Health, Vol. 3. Washington, D.C.: National Academy of Sciences, 1980.

Ekel, George J., and Warren H. Teichner. *An Analysis and Critique of Behavioral Toxicology in the USSR*. Washington, D.C.: U.S. Government Printing Office, 1976.

Hattis, D., and Kennedy, D., "Assessing Risks from Health Hazards: An Imperfect Science," *Technology Review* (May/June 1986), p. 60.

CHAPTER TWENTY-SEVEN

Environmental Carcinogens

I n this chapter we look at some of the areas in which humans have little personal control over their exposure to environmental carcinogens. Exposure to asbestos, ionizing radiation, and a variety of workplace carcinogens is for the most part not affected by individual actions. Several air pollutants that may contribute to the burden of cancer (such as particulates) and some water pollutants (such as trihalomethanes) also fall into this "involuntary exposure" category. They were discussed in Chapters 13, 14, and 20. In the next chapter we discuss some of the factors over which individuals do have control (such as smoking and diet).

Asbestos

Diseases Caused by Breathing Asbestos

It took a lot of time and effort to discover that the microscopic creatures we call "germs" cause diseases like pneumonia and scarlet fever. Imagine how hard it would be to find the reason for a sickness if the symptoms did not show up until 20 or 40 years after a person was exposed to the causative agent. This is the case, however, with diseases brought on by breathing asbestos.

Asbestosis is a disease in which breathing is made difficult by the presence of asbestos fibers in the lungs. The tissue around the fibers becomes tough; oxygen cannot be transferred to the blood by such tissues. Asbestosis usually shows up 20 years or more after a person starts working with this substance.

Breathing asbestos dust can also cause cancer, both of the lungs and of the membranes covering the lungs. It seems that even a short exposure to asbestos can cause cancer 20 to 40 years later. In one case, a woman went to school for a short while near an asbestos field in South Africa. She and her classmates used to slide down piles of asbestos wastes on the way home from school. After her family moved away, she was never exposed to asbestos again. Fifty years later she died from a rare type of lung cancer (mesothelioma) caused by the dust she breathed when she was 5 years old.

Men who have worked in shipyards (where a great deal of asbestos was used for insulating ships) are at an increased risk of developing mesothelioma. This is true even if they worked there for periods as short as a few weeks and even if they did not directly handle the asbestos. Although most asbestos-related disease is seen among people who have worked in industries where asbestos is produced or used, there seems to be a risk also to the families of people who work with asbestos. Asbestos brought home on work clothes appears to have caused mesotheliomas in the children or spouses of asbestos workers. Because such a small amount of asbestos has been shown to cause disease, experts feel that no level of asbestos can be called safe.

Uses of Asbestos

The mineral nature of asbestos ensures that it does not burn. For this reason, asbestos is useful in fireproofing materials. It has been used in firemen's clothing, in heatproof gloves and mats, for oven linings, and for furnace ducts. Asbestos is also included in brake lin-

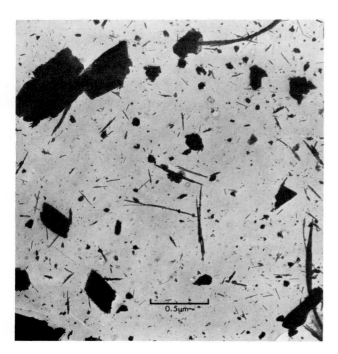

Figure 27.1 Asbestos fibers are found in several products used by the home handyman. This photograph, taken with an electron microscope, shows asbestos fibers in a consumer spackling product. Asbestos can contaminate the air when the spackling is mixed or when the dried spackling is sanded. Old floor tiles and linoleum should not be sanded off floors because of the risk of releasing asbestos fibers. (From A. N. Rohl et al., *Science*, **189** [August 15, 1975], 551. © 1975 by the American Association for the Advancement of Science)

Figure 27.2 Asbestos-containing coatings have been used on ceilings in a number of schools and public buildings. In some cases, the coating has begun to flake off, scattering asbestos fibers into the air. In this photograph, parents at Ramtown School in Hopewell, New Jersey, examine the flaking ceiling that has scattered asbestos onto the hall runner (lower right). A similar situation in a building at Yale University was improved when the entire building interior was scraped and repainted at great cost. In New Jersey, where the problem has been studied most thoroughly, 10% of the schools were found to be painted with the asbestos coating. In 1982, the EPA ruled that all schools must be inspected for asbestos coatings, and in 1986 Congress passed a law that all schools found to have potentially hazardous asbestos must begin cleanup procedures. (Frank Dougherty/NYT Pictures)

ings, ceiling and floor tiles, and roofing tar and cement. In some products, the asbestos is firmly bound so that it is unlikely to contaminate the environment. Floor tiles containing asbestos fall into this category. In other cases, asbestos is used in ways that permit its release into air and water (see Figures 27.1 and 27.2).

In some cases, asbestos is found where it is not meant to be. Several prescription drugs, as well as some brands of beer and gin, have been found to contain asbestos fibers. Most likely, the asbestos fibers have been washed into these products when the liquid passed through filters made of asbestos. For over 20 years, asbestos fibers contaminated the drinking water in Duluth, Minnesota, and Superior, Wisconsin. Both cities draw their water from Lake Superior. For 22 years, Reserve Mining Company dumped 67,000 tons of mining wastes into Lake Superior every day. A major com-

ponent in the waste was asbestos. No excess of gastrointestinal cancer has so far been demonstrated in the area. However, asbestos fibers have been found in the urine of people drinking water from Lake Superior. Although asbestos is a known carcinogen when inhaled, the effects of drinking it are not clear. Twenty

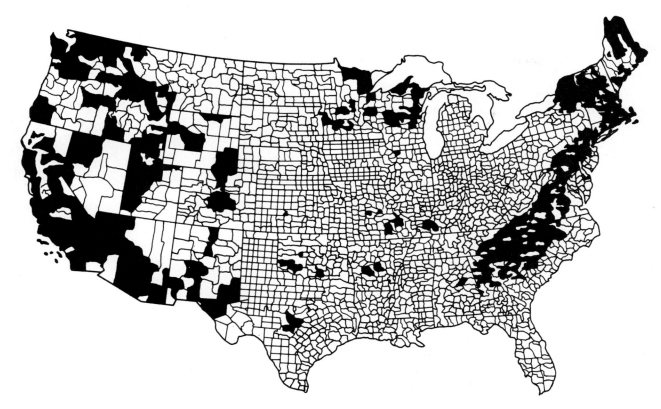

Figure 27.3 The mineralogy of the blackened counties and regions is such that quarries found in them may produce asbestos-bearing rock. The map was prepared by the Environmental Defense Fund from information derived by the Mining Enforcement and Safety Administration from reports by Batelle-Columbus and the U.S. Geological Survey. (From L. J. Carter, "Asbestos: Trouble in the Air from Maryland Rock Quarry," *Science*, **197** [July 15, 1977], 237. © 1977 by the American Association for the Advancement of Science.)

years is not long enough to tell whether an excess of cancers will develop among residents who drank contaminated water.

Estimating the Risk

Low levels of asbestos fibers are definitely present in the air in cities. The main cause appears to be construction or demolition of buildings built with asbestos fireproofing and asbestos wallboard. Fibers are also found in the air where asbestos is mined, milled, or made into various products. Unfortunately, there is no good method for determining low levels of airborne asbestos fibers. In some areas of the country, natural deposits of

asbestos-bearing rock can cause unsafe levels of asbestos fibers in the air (Figure 27.3). Researchers recently recommended closure of dirt-bike trails in the Clear Creek Recreational Area in San Benito County, California. Dust samples along the trail were shown to be 90% asbestos.

We do know that almost all of us—not just people who work with asbestos—have asbestos bodies in our lungs. (Asbestos bodies are fibers of asbestos that the body coats with a special protein after they are breathed into the lungs.) Unfortunately, we do not know whether the presence of asbestos bodies means that cancers or other asbestos-related diseases will someday develop. Remember, cancers caused by asbestos may not become visible for 20 to 40 years. Because we know so little at the moment about the effects of small amounts of asbestos, it seems wise to prevent as much asbestos as possible from getting into air, water, and food.

With this in mind, in 1986 the Environmental Protection Agency proposed a ban on all future production of asbestos products. If the proposal becomes a final ruling (after a period of public comment and agency reconsideration), all manufacture of asbestos roofing felts, asbestos floor tiles, asbestos cement pipes, and asbestos clothing must cease immediately. The manufacture of all other asbestos-containing products as well as mining, milling, and importation of asbestos would be phased out over a 10-year period. There is an adequate substitute for most asbestos-containing products. The main exception is brake and clutch linings for heavy vehicles. The EPA estimates that the use of more costly substitutes will cost consumers about $1.8 billion over the next 15 years. The agency also estimates the ruling will prevent 1,900 deaths from asbestos-related disease (mostly in asbestos workers) over the same period (see Controversy 27.1).

Toxic Substances in the Workplace

Types of Toxic-Substance Hazards on the Job

To most people, the term *job safety* brings to mind safeguards against immediate hazards to life and limb: machinery accidents or chemical burns. These are hazards workers rightfully expect their employers to help them guard against. However, other, less visible, but equally crippling job hazards exist, and the extent of these problems is only now becoming known. A study published in 1977 by the National Institute of Occupational Safety and Health (NIOSH) stated that one out of every four U.S. workers may be exposed during their working lifetimes to a toxic substance that can cause disease or death (Maugh, 1977). This means that almost 22 million workers have been or are being exposed to toxic substances: solvents, mercury, lead, pesticides, and so on (Figure 27.4). Furthermore, some 880,000 of these workers, or 1% of the entire workforce, are currently being exposed to substances known to be carcinogenic, such as asbestos, chromate compounds, arsenic, or chloroform (Table 27.1). Often, the workers or even employers are not aware of

Figure 27.4 At one time, miners were concerned mainly about hazards to life and limb. However, breathing dust is now recognized as a serious health hazard. Among coal miners, it leads to black lung disease. Workers in the cotton industry may contract brown lung disease from breathing cotton dust, while uranium miners are at an increased risk of lung cancer from breathing uranium dust. (U.S. Department of Energy)

Table 27.1 The Most Hazardous Industries and Some of the Carcinogens Used in Them

Industry	Carcinogens
Industrial and scientific instruments	Solder, asbestos, thallium
Fabricated metal products	Lead, nickel, solvents, chromic acid, asbestos
Electrical equipment and supplies	Lead, mercury, solvents, chlorohydrocarbons, solders
Machinery	Cutting oils, quench oils, lube oils
Transportation equipment	Formaldehyde, phenol, isocyanates, amines
Petroleum and products	Benzene, naphthalene, polycyclic aromatics
Leather products	Chrome salts, tanning organics
Pipeline transportation	Petroleum derivatives, welding metals

Source: T. H. Maugh, "Carcinogens in the Workplace," *Science*, **197** (September 23, 1977), 1268.

Note: When industries are ranked according to how likely workers are to be exposed to a carcinogenic material, some surprises are found. Industries commonly thought to be "clean" show up near the top of the list. For instance, manufacturing of industrial and scientific instruments requires very little in the way of carcinogenic materials. However, those hazardous substances that are used are involved in hand assembly of machines; thus worker exposure is great. In contrast, the chemical industry, which may manufacture tons of carcinogenic materials, may employ only a few people to carry out the operation. Thus total worker exposure is relatively small.

CONTROVERSY 27.1

The Economics of Protecting Workers

> When the pollution-oriented health administrators and the public alike begin to focus clearly on the enormity of the bill that would be required to reduce pollution to meet unnecessarily severe standards . . . then will come the day of reckoning . . .
>
> **H. E. Stokinger, chief, Laboratory of Toxicology and Pathology, National Institute for Occupational Safety and Health**

> . . . liability for asbestos-related deaths of workers will exceed $38 billion . . . more than the combined book value of the major asbestos defendants and 51 insurance companies involved . . .
>
> **The New York Times, *March 9, 1982***

> But we paid dear and we will pay dear.
>
> **Billie Walker, *widow of asbestos worker***

The price of environmental protection is a recurring question. How much are we willing to pay? H. E. Stokinger (1971) presents one view:

> When the pollution-oriented health administrators and the public alike begin to focus clearly on the enormity of the bill that would be required to reduce pollution to meet unnecessarily severe standards . . . precipitously prepared from undigested, dubiously related facts . . . on which the public has been ill-advised or misled . . . then will come the day of reckoning and rude awakening to the folly of past antipollution actions. Already industry has felt the bite; shortly, the public will. Hardest hit are the mineral and chemical industries. On top of multimillion-dollar outlays for air pollution control, and sums of similar magnitude for water, are multibillion-dollar legal suits that stagger the imagination, cripple large industry, and eliminate small industries. Two consequences of profound economic importance are the increased price of basic chemicals and the loss of employment. Already a number of small manufacturing plants have been forced to close, unable to bear the burden of meeting pollution standards. Heavy industry, unable to survive on repeated annual financial losses or to continue on less than a 4 to 6 percent profit margin, will ultimately pass the needless charge on to the consumer.

It thus should be evident that such actions, with their unbearable consequences, should only be taken when it is clear beyond a shadow of scientific doubt that human health is in imminent danger.

Until a few years ago, there was an asbestos plant in Tyler, Texas. Workers there were exposed to asbestos dust from 1954 to 1972. Although public health officials told plant managers that the plant violated health standards for working with asbestos, the management did not install the necessary safety equipment. Neither the public health officials nor the plant management told the workers of the dangers involved. Some 25–40 of the 900 workers at the plant have died from breathing the asbestos dust, and the death toll may reach 200. The remaining workers sued the plant owners and the government. A $20 million settlement was agreed upon, to be provided by the owners, the government, and the asbestos suppliers.

> William Morris, a 50-year-old former asbestos worker, is afraid that he is dying. And he is angry, because for years neither his employer nor visiting Federal health inspectors warned him that clouds of dust he was sucking into his lungs at work caused cancer.
>
> "I may live for another six months," he said, panting heavily as he sat in his darkened living room, a shoebox full of pain killers and other drugs at his feet.
>
> "It was pretty damned dirty of them not to let us know" . . .
>
> Reports of the Tyler settlement have reached many of the workers and their families. "It sounds like a whole lot of money," said Billie Walker, a 51-year-old widow whose husband worked at the plant for almost 11 years before dying in November of 1973. "But we paid dear and we will pay dear," she added as she sat at her kitchen table in Tyler with her two teenage sons. (*The New York Times*, December 20, 1977)

Suits brought by former asbestos workers have been increasing yearly. Some 10,000–12,000 cases are pending on behalf of over 20,000 workers. Dr. Irving Selikoff, a medical expert on the effects of asbestos exposure, estimates that as many as 9 million workers have been exposed to significant levels of asbestos. The liability as a result of asbestos-caused deaths could total more than $38 billion over the next 20 years. Beleaguered asbestos and insurance companies question whether court suits are the right way to solve such an enormous problem. In addition, because of normal delays in the judicial system, many, if not most, asbestos workers die as did Billie Walker's husband before their cases are settled.

One major asbestos company, Johns-Manville, faced with potentially billions of dollars worth of suits by current and former workers, declared bankruptcy. This protected the company from suits while a plan was worked out for a more than $2 billion trust fund from which all asbestos-related claims would be paid.

As Dennis Connolly, senior counselor for the American Insurance Association, summarized the situation:

> [The] tort recovery system is running a little wild, enormously large businesses are threatened by it and a tremendously large number of people must be compensated. That may be a societal problem rather than a legal problem. (1981, p. 166)

The experience of asbestos companies points out that, at least in some cases, the cost of *not* protecting workers may eventually far exceed the cost of protecting them.

Do you feel that this argument alone proves that workers must be protected from possibly harmful substances in their work environment? What should be society's course of action if a company does not fulfill this responsibility, as happened in the case of asbestos?

Do you agree that private companies cannot be expected to handle a problem this large? Should Congress provide a remedy? If so, of what sort?

Sources:
H. E. Stokinger, *Science,* **174** (November 12, 1971), 662.
Dennis Connolly, quoted in *Business Week* (April 13, 1981), p. 166.

----

the hazards because the toxic substances are in products known by trade name only.

White-collar workers are not exempt. Secretaries and management employees may be exposed to toxic air pollutants at plants, as may the families of asbestos, lead, or pesticide workers exposed to these substances from the workers' clothing. Anesthetists and other operating-room personnel are more likely than the gen-

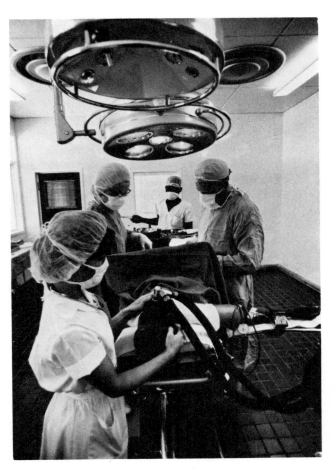

Figure 27.5 White-collar workers are exposed to hazards in industry, research laboratories, and hospitals. People who work as anesthetists in hospitals have an increased risk of developing cancer and other serious diseases, apparently due to exposure to chemicals in their work environment. (© Marc and Evelyne Bernheim/Woodfin Camp & Associates)

eral population to suffer kidney and liver diseases or cancers or to have babies with birth defects (Figure 27.5).

Matching Diseases with Toxic Substances

How could a problem of this size have been ignored for so long? One reason is the use of trade names and the absence of labels that list ingredients by common or chemical names. The 1977 National Institute of Occu-

pational Safety and Health (NIOSH) study took two years to complete, partly because 70% of the substances found at workpaces were identified only by trade name. NIOSH had to contact 10,000 manufacturers to find out the ingredients.

As of May 1986, the Occupational Safety and Health Administration (OSHA) began to require all chemical companies to provide workers with sheets of detailed information on the hazardous chemicals they handle. Unions and environmental groups have objected to the provision that companies need not publish such information if a trade secret is involved. Manufacturers can also decide whether to label chemicals with their generic name (for instance, halogenated hydrocarbons) or with their chemical or trade names (vinyl chloride).

A more difficult problem is to determine which workplace materials are toxic. As was explained in Chapter 26, epidemiological studies of workers' health are necessarily indirect investigations, since the information must be taken as it is found. Neat experiments cannot be planned to test interesting theories. As part of the National Toxicology Program (NTP), chemicals are being tested using animal experiments in a priority system. The system is based on both the suspected cancer-causing ability of the chemicals and on how many workers or members of the general public are likely to be exposed to the chemicals.

In the past, workers have often been accidental guinea pigs for toxic materials. Asbestos, arsenic, and vinyl chloride were all recognized as carcinogens after studies showed that workers handling these materials developed more cancers than comparable groups in the population. However, some diseases, such as cancer, do not appear for many years after exposure to toxic material. Even if a company keeps careful records on employees, it is easy to lose track of workers who are no longer employed and who may become sick after they leave. In some cases, companies have been reluctant to release data on employee health, for a variety of reasons. Fears of being sued by employees or confined by additional government regulations are probably among these reasons.

Laws requiring that death certificates give information about a person's occupation, and his or her parents' occupations as well, would help to pinpoint hazardous jobs and ones involving materials that can cause birth defects.

Laws Protecting Workers

The Occupational Safety and Health Administration is by law responsible for protecting the health of workers. It sets limits for exposure to various toxic materials or determines what safety equipment is needed by workers. Furthermore, OSHA inspects workplaces to see that employers are following regulations.

The problems involved in proving that certain materials are hazardous to workers have made it difficult to set standards. A great many regulations are challenged by manufacturers in hearings or in court. This has resulted in long and costly fights among OSHA, manufacturers, and workers' rights groups. During its first 9 years of existence, OSHA managed to regulate only 20 substances. In the past, in order to justify regulations, OSHA estimated the risk of hazardous substances to workers in a general way, best described as qualitative risk assessment. However, a court decision has directed OSHA to make a stricter justification for its regulation, rather than assuming that reduction of exposure is always a worthwhile course of action. That is, OSHA must present the actual benefits of measures to reduce worker exposure compared to the cost of those measures.

Economics of Controls

Understandably, manufacturers feel that worker protection devices will increase their production costs. We can expect these costs to be passed on to consumers in one way or another. However, society already pays costs comparable to these in medical care, lost productivity, and welfare payments to miners and millers crippled with lung diseases, to cancer victims, and to those poisoned by lead, mercury, or pesticides. (See Controversies 27.1 and 27.2.)

Unions also worry about new restrictions. They fear that increased manufacturing costs will cause plants to close, especially small ones, with the result that many jobs will be lost. In reply to this concern, Douglas Costle, then administrator of the Environmental Protection Agency, wrote in the December 1977 issue of the *EPA Journal*:

> It must also be recognized that environmental regulations create jobs. The facts show that more people have been employed now than would have been without the major pollution control programs. Approximately 19,000 job losses have been attributed to pollution control compared to perhaps half a million jobs that were generated because of cleanup efforts. Such jobs are generated in three ways. First, construction of equipment and plants required by environmental programs create the largest number of jobs . . .
>
> The second way in which jobs are created is in the pollution control equipment manufacturing industry.
>
> Finally, many more indirect jobs are stimulated by these expenditures.
>
> I want to stress that when we talk about the issue of jobs versus the environment we are caught in the old mindset of looking at pollution controls as unproductive, profit-decreasing expenditures. Rather we need to explore calculating productivity in a larger and more meaningful perspective, one that includes protection of workers' health. (p. 4)

Industrial Hazards and the Community

Workers may not be the only people exposed to hazardous substances in the workplace. Residents in the surrounding community may suffer from plant emissions of toxic chemicals on a regular basis or if an accident occurs. The Bhopal disaster, in which more than 2,000 people were killed when isocyanate leaked from a plant in India, was the worst industrial disaster in history.

In order to discover whether the possibility for similar incidents exists in the United States, the EPA conducted a survey of hazardous chemicals used and stored in the United States. A list of 403 chemicals that pose a hazard to public health was compiled. The EPA has suggested that local governments or citizens' groups check industry sourcebooks that list chemicals and who produces them to see whether any of the 403 hazardous chemicals are produced in their community. Then the agency suggests that local officials ask the companies involved about their handling procedures to ensure that there is no risk to the community. The EPA has written a guide for communities or groups that want to undertake such a project (Chemical Emergency Preparedness Program, Interim Guidance, U.S. EPA, November 1985).

CONTROVERSY 27.2

The Cost of Cancer Surveillance

. . . There is now the urgent problem of surveillance and management and treatment of people who [are] at increased risk of developing cancer.
Irving Selikoff

. . . It could cost as much as $54 billion to provide warnings and health surveillance services . . .
The New York Times, *October 3, 1977*

An area that still needs attention is the notification of workers who were exposed to toxic materials in the past but may not know it. According to Irving J. Selikoff (1977):

> Both in the workplace, and in the environment in general, there is now the urgent problem of surveillance and management and treatment of high risk groups—people who inadvertently were exposed in the past to agents which we now know places them at increased risk of developing cancer in the future. At present, there is little surveillance or care for them. I consider this a social lapse and I strongly urge that attention be devoted to this as rapidly as is possible. (p. 8)

However, this could cost a great deal of money. According to an article in *The New York Times*, October 3, 1977:

> In discussing the costs involved, according to one government analysis circulating within the Carter Administration and Congress, it could cost as much as $54 billion to provide warnings and

health surveillance services to the 21 million Americans now believed to be exposed to harmful conditions while working. (p. 1)

Yet this cost must be offset against the cost to society of treating occupational disease. Each year, $3–$5 billion is spent on treatment for cancer victims and about $12 billion is lost in wages. Early diagnosis can cure or prevent the spread of many cancers.

Do you feel that the government should attempt to ferret out and warn all workers who have been exposed to toxic materials in the past and who may develop a disease from this in the future? (Remember how much this will cost you in increased taxes. Remember, too, that there are savings to be gained from decreased health and welfare costs.) Do we owe these people treatment as well as warnings?

Can you think of benefits to yourself from keeping track of the health of these people?

Source: Irving Selikoff, quoted in *EPA Journal* (November–December 1977), p. 8.

CONTROVERSY 27.3

Women's Rights and the Workplace

If American Cyanamid . . . can get away with removing women of childbearing age from these jobs, we will have established the principle of altering the worker to the configuration of the workplace instead of altering the configuration of the workplace to protect the worker.
Anthony Mazzocchi, vice president, Oil, Chemical, and Atomic Workers Union

A number of companies have started to exclude or remove women workers of childbearing age from jobs where they might be exposed to teratogenic substances (substances that cause birth defects). In one case, four women workers at American Cyanamid claimed they had themselves sterilized because talks with company officials led them to believe they might lose their jobs otherwise. Although some companies are transferring women to less hazardous positions with salaries equal to their former jobs, in other cases women allege they have been discharged or offered only lower-paying positions.

Anthony Mazzocchi, vice president of the Oil, Chemical, and Atomic Workers Union, notes that the reproductive capacity of men as well as women can be affected by many hazardous substances (such as the pesticide DBCP, which can cause sterility in males). Furthermore, lead workers, asbestos workers, and men who worked at a Kepone plant have brought home enough dust in their work clothes to affect other members of their families.

Dr. Karrh, corporate medical director at DuPont, agrees that cleaning up workplaces to make them safe for everyone would be best, but:

. . . first of all it is not technically and economi-

cally feasible to clean it up to a safe level. And, second, we don't have the data to know what is a safe level.

Sue Nelson, director of the Office of Policy Analysis for the Occupational Safety and Health Administration, adds:

What this is really about is that the employers are trying to save themselves from expensive lawsuits . . . It is easier for a woman worker than a man to bring a lawsuit against a company on behalf of a fetus.

Do you think women able to bear children should be excluded from jobs involving exposure to hazardous chemicals? What about men able to father children? Supposing there is some doubt about whether a chemical is teratogenic, who should make decisions (the company, the government, the worker), or should everyone be protected in any case? Suppose protective measures would cost so much that the company might not produce the product?

Source: All quotes are from *The New York Times*, January 15, 1979, p. A-1.

Radiation: Microwaves, Radiowaves, High-Voltage Power Lines

During the years 1953–1976, the U.S. Embassy in Moscow was subjected to low-level microwave irradiation. Why the Soviets chose to irradiate the embassy is still unclear. Possibly the intent was to cause some sort of neurophysiological condition. Certainly there was an unsettling mental effect on embassy personnel. However, as far as current medical methods have been able to determine, the effect was due more to a fear of what microwaves might do than an actual physical effect. We know that radiation can affect human health. However, we also know that the kinds of effects seen depend on what sort of radiation is involved and how high the dose is.

The Electromagnetic Spectrum

The whole electromagnetic spectrum is shown in Figure 27.6. As you can see from this diagram, there are many kinds of radiation. They range from radiation of very long wavelengths, such as that used in power generation, all the way to radiation of very short wavelengths, such as x-rays and cosmic rays. The wavelengths visible to the human eye are also part of the electromagnetic spectrum, but they are only a small portion of the entire range.

The health effects of radiation depend on the wavelength involved. The effects most commonly associated with the term *radiation*—that is, radiation poisoning and the various x-ray- or atomic-bomb-caused cancers—are caused only by the shorter wavelengths. These types of radiation are known as **ionizing radiation**. In contrast, the longer wavelengths, from the ultraviolet region to the power generation region, are called *nonionizing radiation* and their health effects are quite different.

Microwaves are in this nonionizing range, while x-rays, gamma rays, and cosmic rays are ionizing. The effects of nonionizing radiation will be covered first and the effects of ionizing radiation after that. Finally, the health effects of ultraviolet rays, mostly skin cancers, are detailed in Global Perspective VI.

Biological Effects of Nonionizing Radiation

Nonionizing radiation can cause thermal motion of the molecules in living tissue. This results in a temperature rise in the tissue and may lead to harmful effects such as burns, cataracts, and birth defects. There is also the possibility that complex biological systems, such as those existing in cell membranes, may be disrupted. Systems such as cell membranes depend on an orderly arrangement of molecules for proper functioning. Thus, nonionizing radiation could cause effects over and above those caused by a simple increase in temperature. The experimental evidence for such effects, however, is incomplete.

Most experimental results on nonionizing radiation concern radio-frequency wavelengths. These results show that doses above 100 milliwatts per square centimeter (cm^2) produce definite thermal damage, as well as cataracts in the eye. Between 10–100 milliwatts/cm^2, changes due to heat stress, including birth defects, are seen. From 1–10 milliwatts/cm^2, changes have been noted in the immune system and the blood-brain barrier. In the range of 100 microwatts/cm^2 and 1 milliwatt/cm^2, almost no confirmed effects have been found.

In general, it appears that the immediate effects of nonionizing radiation, such as the effects of excess heat on a body tissue, are the important ones (although there are some new and incomplete data suggesting that workers exposed to microwaves as well as people living very close to high-voltage power lines may suf-

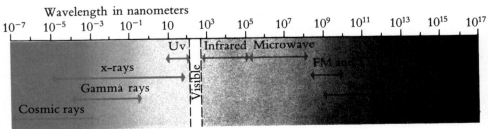

Figure 27.6 The electromagnetic spectrum. Wavelength is measured in nanometers. A nanometer is one one-trillionth of a meter.

Wavelength in nanometers

10^{-7} 10^{-5} 10^{-3} 10^{-1} 10 10^3 10^5 10^7 10^9 10^{11} 10^{13} 10^{15} 10^{17}

Uv   Infrared   Microwave

x-rays

Gamma rays

Visible

FM an

Cosmic rays

fer increased cancer rates). In the Moscow embassy, microwave concentrations reached a maximum of 18 microwatts/cm² and no effects on the personnel there could be found.

Microwaves and Radio-Frequency Radiation

Against this lack of apparent effect due to low-level microwave exposure must be set the fact that the growth in use of microwaves is at least 15% per year. Besides their use in microwave ovens, microwaves are used in radar and as transmission links for TV, telephone, and telegraph. The United States does not have a standard for exposure to nonionizing radiation, although OSHA has recommended that workers be exposed to no more than 10 milliwatts/cm². The Soviet Union has set a standard of 1 microwatt/cm² for the general public.

Industrial workers involved with heating, drying, and laminating processes may be at some risk, as may technicians working in live broadcasting, radar, and relay towers and military service personnel. Worker compensation suits have been filed charging that microwaves have contributed to disabilities, and in at least one case, the board ruled in the worker's favor.

Meanwhile, as the number of sources of microwave exposure increases, concern about the exposure of the general public has been growing. Construction of a microwave TV transmission antenna atop New York's World Trade Center was halted when engineers realized that it would subject some office workers, as well as tourists atop the building, to nonionizing radiation in the range of 360 microwatts/cm². The Coast Guard was denied permission to construct a microwave transmission tower as part of a vessel-traffic monitoring system in New York Harbor due to public concern about the safety of microwaves.

Government agencies and public interest groups would like to see standards set for the various forms of nonionizing radiation. Many industrial groups would like to see federal standards both to help them in designing equipment and to prevent the growth of a patchwork of local ordinances.

High-Voltage Power Lines

The second area of concern about nonionizing radiation, after microwave and radio-frequency radiation, is radiation from high-voltage power lines. These are electric power lines designed to carry large quantities of electric power from generating plants to large population centers. The largest such lines now in use are 765,000 volts (765 kV), one of which could carry enough power for the cities of Boston and Baltimore combined. For the future, lines carrying up to 2,200 kV are planned. Power lines this size have both electric and magnetic fields surrounding them (Figure 27.7).

An ordinary kitchen has an electric field of about 3 volt/meter due to the electric appliances in it. Directly under a 765 kV transmission line, the field at ground level is about 10 kV/m. However, if one moves 500 feet (152 m) away from the line, the field decreases to 0.1 kV/m. Thus the possible problems center on effects that might occur directly around or under the lines. These effects include electric shocks, biological effects induced by electric and magnetic fields, and the effects of corona.

Electric shocks. High-voltage power lines cause electric shocks to people or animals moving below them. For a distance of several feet around the line itself, there can be a "flashover," or breakdown of the air between the line and a conducting object, allowing a dangerous current to flow. Power lines must be set high enough so that no objects passing underneath (such as a boat with a high mast) come into range to allow such flashover.

However, the electric field surrounding the power line can also cause a shock hazard. The reason is that objects in an electric field collect an electric current. For instance, a large tractor under a 765 kV power line can collect up to 4–5 milliamps. Such a current is still not a hazard unless someone touches the tractor while he is grounded (for instance, standing on wet ground), thus allowing current to flow from the tractor through the person and into the ground. The shock hazard in this example is probably just at the limit of one that would be very painful but not otherwise harmful to a child. Higher-voltage power lines than this, however, could lead to more serious results.

Electric field effects. Besides the possibility of shocks, electric fields can have other effects on living matter. The external power-line field causes an internal electric field to form in living tissue. For humans, the internal current density caused by an external electric field of 10 kV/m is still 10–100 times less than the cur-

Figure 27.7 High-voltage power lines. Lines such as these carry up to 765,000 volts. Possibly harmful environmental effects may be caused by electric shocks or by electric and magnetic fields generated directly around or under the lines. (U.S. Department of Energy)

rent density that acts on the membrane of a muscle or nerve cell and causes it to react, or "fire." Whether electric current densities of this magnitude can cause other, more subtle, effects on cells is still hotly debated. None are yet proven, but several experiments are in progress.

At the surface of a body or tip of a pointed leaf, the local field can be much higher than the internal field. This leads to a tingling sensation in humans caused by vibration of hairs on the skin. In addition, pointed leaves may show burned leaf tips (round leaves are not affected). Neither of these effects seems to have any harmful results for the organism as a whole, although some people find the tingling sensation unpleasant. Other effects, such as excess fatigue, have been reported from Russian experiments but could not be repeated by U.S. investigators.

The electric field under 765 kV power lines can definitely affect certain types of pacemakers. Although such pacemakers have a fail-safe mechanism that takes over, farmers or other workers with pacemakers who have to spend time under high-voltage power lines should discuss the problem with their doctors. People with pacemakers driving under such power lines are not at risk because the metal car body shields them from external electric fields.

Magnetic field effects. At ground level under a 765 kV power line, the magnetic field is about 0.56 Gauss, but it decreases rapidly to 0.016 G at 500 feet (152 m) from the line. Migrating birds appear able to detect the 0.4 G magnetic fields generated by large-scale antennas such as the Navy's Project Sanguine. However, the birds appeared able to compensate by using other cues (such as sun and star positions). No actual disturbances of migration patterns have been seen.

Other harmful biological effects of magnetic fields have not been confirmed at the levels found under currently operating power lines.

Corona effects. Corona, which occurs mainly in bad weather, is the breakdown of air directly surrounding a power line. It is most notable for the noise it generates—a crackling or frying sound. Although well below levels that cause hearing damage, this noise can be annoying. Corona can also cause interference with radio and TV signals, which could be a severe problem in fringe areas of reception. In addition, ozone and nitrogen oxides may be formed. Levels, however, seem too low, compared to other sources, to be of concern.

The outlook. In summary, the electric and magnetic fields generated under high-voltage power lines have not been proven to cause serious biological effects. If transmission voltages are increased, however, problems, especially with electric shocks, could result. Power companies will need to introduce safety devices to shield people, plants, and animals from the higher electric fields that might be generated.

Radiation: X-rays, Gamma Rays, and Particles

Kinds of Ionizing Radiation

Looking at the electromagnetic spectrum again (Figure 27.6), we can see that the short-wavelength end of the spectrum is made up of x-rays, gamma rays, and cosmic rays. These rays possess enough energy to free an electron from the atom of which it is a part. This leads to the formation of ions (which is why these types of radiation are called ionizing). The eventual effects of ionizing radiation on living cells are due to the formation of these ions. Certain types of particles, such as those given off by radioactive materials, also produce these ions.

The nuclear decay of unstable elements gives rise to both ionizing particles and ionizing rays. The unstable elements, which we refer to as **radioactive**, emit alpha particles, beta particles, and gamma rays. Gamma rays have the greatest penetrating ability of any radiation from radioactive decay. They may pass through several centimeters or more of lead without a significant weakening of energy. Those who work near substances that emit gamma radiation must exercise the greatest caution to limit their exposure.

Like gamma rays, x-rays are highly penetrating; several centimeters or more of lead are necessary to block these beams. Cosmic rays, which consist of both particles and electromagnetic radiation, are constantly bombarding us from outer space. Although a portion of cosmic rays can be blocked by several thicknesses of lead, another portion penetrates even into the deepest mines. The intensity of cosmic rays increases at higher altitudes, so much so that jet crews might one day conceivably be classified as radiation workers. Cosmic-ray intensity also increases as one moves toward the polar latitudes.

Sources of Human Exposure to Ionizing Radiation

Human exposure to ionizing radiation comes via one or more of these routes: decay of radioactive elements,

x-rays, and cosmic rays. We measure exposure to radiation most commonly in "rems" and "millirems" (one one-thousandth of a rem), units reflecting both the intensity of the radiation and its effect on human tissues. Radiation standards are also set in rems and millirems. If we exclude radiation from medical x-rays and other human-made sources, we have what is referred to as *natural background radiation*. This is the radiation we would receive if there were no exposure to radiation from human sources.

Background radiation exposure in the United States is most commonly in the range of 100–150 millirems per year. Leadville, Colorado, at 2 miles (3.3 km) above sea level, has one of the highest annual radiation exposures in the United States at 160 millirems, due to high intensity of cosmic rays reaching the city. Radioactive elements in the earth's crust, such as potassium-40 and radium, are also contributors to background radiation.

X-rays ordered by physicians and dentists are a common source of exposure to ionizing radiation. Although such exposure has been estimated at 90 millirems per year, this is an average figure. Many individuals receive far higher doses and some none at all. The annual chest x-ray for tuberculosis is now thought to be not such a good idea. Older x-ray equipment is still in use in many places and delivers radiation doses far higher than necessary.

During the era when atomic and hydrogen bombs were tested in the atmosphere, radioactive elements were scattered around the globe in clouds of particles. Rain washed these particles from the atmosphere in *fallout*. The fallout in areas very near atomic explosions, such as in the Pacific Ocean islands, was large enough to leave measurable levels of radiation in the soil, but the minor amount of continuing fallout is negligible when compared to background radiation exposures. The various exposure sources are summarized in Figure 27.8.

Radiation Exposure and the Nuclear Power Industry

Exposure to radiation from radioactive elements is a concern that has grown with the arrival of the nuclear electric power industry. Routine discharges of low levels of radiation from electric power plants were allowed in the first 15 years of the nuclear industry. The older nuclear plants were to be operated so that no individual in the general population would receive more than

Human Health and the Environment

a Ship Vulcanus, burning Agent Orange at sea

c Spraying pesticides

b

Spraying pesticides without protective clothing

d Indian child playing on asbestos pile

Modern technology can carry a high price if human health is risked by its use. In developed nations, such as the U.S., improper disposal of hazardous waste has endangered water supplies and made whole communities uninhabitable. (a) Some chemical wastes, for instance Agent Orange, are so toxic that they can only be disposed of by incineration at sea in specially designed ships such as the Dutch vessel, Vulcanus. In developing countries there is the additional problem that workers may be unaware of the hazards of the materials they use. (b) Here a barefoot worker in Thailand rests after applying pesticides without wearing protective gear, and (c) an unprotected worker in Ethiopia sprays pesticides in an attempt to save crops from threatening locust swarms. (d) This child, playing on a mound of asbestos-containing waste in India, cannot know the health risks such exposure poses.

Land Resources

b Lake Manyara, Tanzania

a Saguaro cacti, Arizona

In the U.S. large areas have been set aside as parks or preserves because they are scenic or because they are needed for the preservation of plants and animals. (a) Saguaro National Monument in Tucson, Arizona, preserves giant cacti that need hundreds of years to reach this size. In developing countries, as well, there is a realization that many species will be lost unless enough of their habitat is set aside. (b) Lake Manyara National Park in Tanzania. And yet in both developing and developed countries, growing populations threaten the success of plans to set aside recreational or refuge lands. (c) In Shenandoah National Park rangers recently had to begin a permit system to limit use of the trails lest park flora and fauna suffer. In Africa, hard choices have had to be made between the needs of wildlife populations and growing human populations (who require ever-increasing amounts of land to farm, for grazing or for wood-gathering).

c Hiking trail, Shenandoah

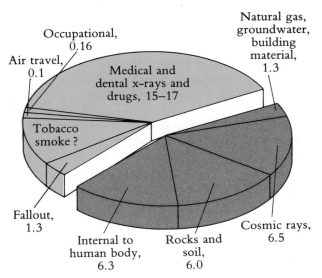

Figure 27.8 Estimates of exposure to ionizing radiation (measured in million person-rems per year, where 1 rem to 1 million people in a year equals 1 million person-rems per year). On the average, about half of human exposure to radiation is due to natural environmental sources. One-third of this natural background is from cosmic rays, one-third from natural radioactive materials in soils and rocks, and one-third from radioactive materials (such as potassium-40) that are incorporated into the human body. People using groundwater for drinking or using natural gas are exposed to additional natural radon from these sources. Further, certain building materials (notably stone, brick, and plaster board made from phosphogypsum) emit some radiation from naturally occurring radioactive materials. Of the other sources of radiation, medical x-rays and radioactive drugs are the most significant, nearly equalling natural background. Air travel, which increases cosmic ray exposure, is significant for some people, as are the radioactive particles in tobacco smoke. (Data from *Pilot National Environmental Profile*, US EPA, 1980.)

500 millirems of radiation per year, and the average exposure of people away from the site would be kept below 170 millirems per year. Such operation reflected the then-current standards for exposure of people in the general population. In the middle 1970s, the Atomic Energy Commission came under attack for these standards and responded with a tightening of the allowed emissions of radioactive elements from nuclear plants. The largest annual allowable dose was reduced from 500 millirems to 5 millirems and the average dose to less than 1% of natural background, or 1 millirem.

Under these new exposure levels, the normal operations of nuclear power plants are quite clearly not a

significant concern; yet radiation remains a problem. One remaining source of radiation exposure involves fuel reprocessing plants (see Chapter 18). Another source is the worked and discarded uranium ore. Of greater concern, however, is the possibility that terrorists or warring nations could hijack spent reactor fuel or fissionable material and build nuclear weapons—an atom bomb from fissionable material such as plutonium, or a dispersal device that spreads the deadly contents of a spent fuel rod. The detonation of these weapons could bring exposure to radiation on a massive scale.

Finally, mention should be made of the fact that a number of radioactive elements can be concentrated in food chains (**biologically magnified**). An example is phosphorus-32, which was found to be concentrated as much as 5,000 times above levels in the water in whitefish downstream from the Hanford Atomic Power Plant on the Columbia River. Blue gills and crappies in the same river were found to contain 20,000–30,000 times the concentration in water of phosphorus-32, while filamentous algae were found with up to 100,000 times the water level of phosphorus-32. Zinc-65, iron-59, and iodine-131 are other examples of radioactive elements concentrated by living organisms. These are sources of radiation exposure to the general public when certain foods are eaten. Fish may contain phosphorus-32; oysters and clams, zinc-65; and milk, iodine-131. Unfortunately, not enough is known about the paths such elements take through food chains to estimate the hazards involved.

Biological Effects of Ionizing Radiation

The effects of radiation on the health of people alive today can be divided into two categories. The acute symptoms that result from intense short-term exposures occur within days or weeks. Such exposures are very unlikely except as a result of a nuclear war or a severe, accidental occupational exposure. The effects of long-term, low-level radiation exposures are not likely to be seen for a number of years. These delayed symptoms cannot be distinguished from the common diseases of aging, especially cancer. We know that ionizing radiation can cause breast cancer, thyroid cancer, leukemia, lung cancer, gastrointestinal cancer, and bone cancer. These diseases have been observed in individuals exposed to radiation levels of 100 rems or more in an accident or catastrophe such as the atomic bomb explosions at Nagasaki and Hiroshima.

Although radiation can induce cancer, a person who has cancer cannot ordinarily point to radiation as the cause. Radiation hazards to which the general public or workers are exposed are generally low-level, 0.1–5 rems. The cancers caused by radiation are not caused solely by radiation but are caused by other agents as well. For these reasons, in order to prove by ordinary statistical methods whether certain cancers are caused by these low levels of radiation, very large numbers of exposed people would have to be studied. These numbers would have to be on the order of hundreds of thousands of people. It is not usually possible to find enough exposed people or enough cancers of a specific type to use epidemiological studies as proof of the harm caused by low-level radiation. Some investigators have examined the available information with more complicated statistical methods (see Land, 1980), but their results are not accepted by most other scientists.

Instead, the results of exposure at doses over 100 rems are used to estimate the probable effect of lower doses. There is a heated argument among radiation specialists about how to extend the data from high doses to predict low-dose effects, and whether there is a threshold below which no effects occur. This is similar to the arguments about the effect of low doses of chemical carcinogens explained in Chapter 26. Most scientists agree that there is no proof of a threshold level, especially for certain types of radiation such as alpha particles.

In addition to cancers, radiation might be expected to cause genetic damage—that is, mutations that can be passed on to future generations. Evidence from animal and cell culture experiments lead scientists to believe such damage is likely. However, although damaged chromosomes have been found in survivors of the atomic bomb and their children, no significant increase in birth defects has been found. This is probably due to the low average dose to the gonads of parents (50 rems) and also to the small population (78,000 children to parents who survived Hiroshima and Nagasaki). It is estimated that natural background radiation (about 100–150 millirems) causes perhaps 0.1%–2% of genetic disease.

Should Radiation Standards Be Made More Strict?

On the basis of population studies of atomic energy workers and others exposed to low-level radiation, some experts argue that the standard for occupational exposure of 5 rems per year should be cut to 0.5 rems, or 500 millirems, per year. A recommendation for such a standard was made in 1978 by a Committee of the National Academy of Sciences. The recommended level is only about twice the level of a chest x-ray exam.

Nearly 5,000 of the 70,000 atomic energy workers were exposed to levels greater than 2 rems in 1977. In addition to the employees of nuclear power plants, the nation has about 170,000 x-ray technicians. Although they should be limited to less than the occupational standard of 5 rems per year, their actual exposures are largely unknown.

Uranium Mine Tailings: An Unnecessary Hazard

Uranium miners are exposed to radiation on their jobs and as a result have experienced a high rate of lung cancer. The radiation to which the miners are exposed comes from the natural radioactive decay of uranium. After a number of steps, beginning with the initial uranium decay, a radioactive gas known as radon is produced and dispersed thoroughly in the air of the mines.

The decay of radon results in new radioactive elements in particle form. Since the particles are only single atoms, they remain suspended in the air for a long while. This contaminated air is inhaled by the miners, and the radioactive "daughters" of radon are deposited in the miners' lungs. Masks or filters are ineffective in capturing particles consisting of a single atom. Hence, the only way to control exposure of the miners is to replace the air in the mine with fresh air on a frequent basis.

Miners are not the only people exposed at this step in the atomic fuel cycle. Wastes, called **tailings**, remain after the uranium oxide has been extracted from uranium ore. This rubble still contains some uranium that could not be removed. Since radon is a product of uranium decay reactions, the air in the vicinity of the rubble also contains radon gas at levels up to 500 times the natural background. The air near the tailings is likewise contaminated with radon "daughters."

Through oversight and a lack of awareness of the hazards, tailings have been used to make concrete for homes and buildings. Rubble from the Old Climax Mill in Colorado is about 0.03% uranium; it was used as building material and as landfill for homes in the town of Grand Junction, Colorado. The tailings were

Figure 27.9 Radiation physicist standing on a 250-foot (76 m) pile of uranium tailings. The city of Durango, Colorado, is seen clearly on the other side of the pile. (Photo courtesy of Oak Ridge National Laboratory Review)

not only used to level land surfaces, they were also used in mortar, concrete, and backfill around basement walls.

Levels of radon decay products in the air of these homes and buildings are up to five times the concentration permitted for uranium miners. Hence, the exposure of people who live in these homes to a potential cancer-causing agent has been substantial. A number of homes have had foundation materials, even walls and chimneys, replaced and backfill removed and replaced; the federal government and the state of Colorado assisted with the costs of these repairs.

Most of the deposits that are rich enough in uranium to be mined are found in the western portion of the United States. The states of New Mexico, Wyoming, Colorado, and Utah are the principal suppliers of ore, and here mounds of tailings may still be found (Figure 27.9). When milling operations moved on to more radioactive pastures, the piles of tailings were simply abandoned. In this way, piles of tailings fell into private hands. Often, the new owners were ignorant of the dangers of the material. In eight states, such piles still exist, some very near large communities. The smallest such pile is only 2 acres (0.8 hectares) in area; the largest is 107 acres (43 hectares). One pile, next to an abandoned mill, is within 30 blocks of downtown Salt Lake City, Utah. In all, the Western states are known to have 23 uranium tailings piles over which no control is exercised. The mills have closed; the piles remain.

Controlling the problem is difficult. Several feet of soil can be deposited over the piles and vegetation established. This prevents the wind from lifting the dust-like particles into the air. However, it is thought that 10–20 feet (3–6 m) of earth will be necessary to prevent the escape of radon gas. This suggests the immediate need to bury the tailings that are currently being produced from continuing mining and milling operations. In 1978, Congress passed legislation stating that the government would cover the costs of cleanup for mill tailings generated before 1978, on the basis that much of this was produced by companies fulfilling government contracts. However, the law required industry to clean up any tailings produced after 1978. Nonetheless, very little cleanup has been done by the industry, which claims the costs to be excessive.

Summary

Environmental carcinogens can be divided into two groups: those over which the individual has little control, such as workplace exposure to toxic chemicals or radiation, and those over which the individual has a great deal of control, such as smoking and diet. Asbestos falls into the first group. Breathing even small amounts of asbestos dust can lead after 20 to 30 years to development of asbestosis, lung cancer, or mesothelioma. Asbestos is, or has been, used for insulation, in brake linings, in fireproof clothing, tile, cement, and roof shingles, and (unintentionally) as paving material or in drugs and cosmetics. Most of the asbestos to which the general public is exposed is airborne, from wear on brake linings or demolition of asbestos-containing buildings. All uses of asbestos will be phased out over the next 10 years to reduce asbestos-caused diseases in workers and, eventually, to reduce the exposure of the general public.

Many workers, both blue collar and white collar, are exposed to carcinogenic substances. Because of scientific uncertainty and industry pressures, the Occupational Safety and Health Administration has regulated only a few of the many hazardous substances found in the workplace. Residents in surrounding communities may also be subject to hazardous emissions from manufacturing plants.

Radiation hazards are of two kinds: ionizing and nonionizing. Ionizing radiation has sufficient energy to free electrons from atoms. X-rays, gamma rays, and cosmic rays fall in this category. Nonionizing radiation, such as microwave, radio waves, and power waves, cannot free electrons. These waves can still cause thermal damage to tissues, however, and may also disrupt cellular structure or cause cancers. We do not know at present whether current exposures to nonionizing radiation from transmitters or high-voltage power lines causes any hazard to the general public. There is some evidence that workers exposed to such radiation are at risk.

Ionizing radiation can cause acute effects such as radiation sickness and death, or chronic effects such as various cancers and genetic damage. Radiation-induced cancers are also caused by other hazards, so particular victims cannot usually prove their cancer was caused by radiation. Occupational exposure to radiation from industrial or electric power sources is limited to 5 rems per year, while the average exposure of the general public is limited to 1 millirem, or 1% of natural background radiation.

Uranium mine tailings pose a severe hazard, up to 500 times natural background radiation. Although the government is involved in cleaning up mine tailings produced before 1978, the cleanup of tailings produced since then is the subject of various court fights.

Questions

1. Should the benefits of certain products be balanced against risks to the workers who produce them? Give an example. Where does the concept of personal freedom fit in with your answer?
2. Explain the differences between ionizing and nonionizing radiation. Note which parts of the electromagnetic spectrum are ionizing and which are nonionizing, and what different biological effects they cause.
3. What is meant by natural background radiation?
4. Suppose the frequency of leukemia in male workers in a particular industry with radiation exposures were three times that for males of the same age in the general population. Could a worker who has cancer prove that the cancer was the result of his exposure on the job? Why?
5. What are the major causes of concern with respect to transmission of electric power by high-voltage power lines (765 kV and greater) and microwaves?

Further Reading

Hill, D., et al. "Management of High-Level Waste Repository Siting," *Science*, **218** (November 26, 1982), 859.
Radioactive waste disposal is part of hazardous waste disposal but poses extra problems that must be considered. These papers provide an entry into the literature.

"OSHA on the Move," *Environmental Science and Technology*, **11**(13) (December 1977).
A good history of the problems of regulating toxic substances in the workplace up to this date.

Selikoff, I. J., and D. H. Lee. *Asbestos and Disease*. New York: Academic Press, 1978.
A comprehensive and readable treatment of the asbestos problem written by a public health authority who is probably the foremost expert in the field.

"Asbestos and Home Improvements," *Environmental News*, EPA Region I, June–July 1981.

"Asbestos Guidance to Schools," *Environmental News*. Washington, D.C.: U.S. EPA, Office of Public Awareness, March 16, 1979.
These two EPA reports cover asbestos problems most likely to affect the general public.

Upton, Arthur C. "The Biological Effects of Low-Level Ionizing Radiation," *Scientific American*, **246** (2) (February 1982), 41.
A clear explanation of what is known about ionizing radiation and its biological effects. Very good if you would like to know more about the actual cellular events leading to radiation-caused cancers.

Marshall, Eliot. "New A-Bomb Studies Alter Radiation Estimates," *Science*, **212** (May 22, 1981), 900.

Marshall, Eliot. "Japanese A-Bomb Data Will Be Revised," *Science*, **214** (October 2, 1981), 31.
These two articles detail some of the ongoing controversy about radiation standards and their safety.

Eckholm, Erik. "Study Finds Genetic Damage in Plants After Atomic Blast," *The New York Times*, August 8, 1985, p. A8.
Studies in Japan show the type of radiation damage that occurs in plants.

References

Higginson, John. "Proportion of Cancers Due to Environment," *Preventive Medicine*, **9** (1980), 180.

Identification, Classification and Regulation of Potential Occupational Carcinogens. Federal Register, January 22, 1980.

Selikoff, I. Quoted in *EPA Journal* (November–December 1977), p. 8.

Workplace Exposure to Asbestos. U.S. Department of Health and Human Services Publication #81-103, November 1980.

Sun, Marjorie. "EPA Proposes Ban on Asbestos," *Science*, **231** (February 7, 1986), 231.

Nonoccupational Health Risks of Asbestiform Fibers. Board on Toxicology and Environmental Health, National Academy of Sciences, National Academy Press, 1984.

Stokinger, H. E. Quoted in *Science*, **174** (November 12, 1971), 662.

Costle, D. *EPA Journal* (December 1977), p. 4.

Koslov, Samuel. "Radiophobia: The Great American Syndrome," *Johns Hopkins Applied Physics Laboratory Technical Digest*, **2** (2) (1981), 102.

Land, Charles E. "Estimating Cancer Risks from Low Doses of Ionizing Radiation," *Science*, **209** (September 12, 1980), 1197.

"Ionizing Radiation and Its Health Impact: 1980." *Pilot National Environmental Profile*, U.S. Environmental Protection Agency, 1980.

Genetic and Cellular Effects of Microwave Radiations. U.S. EPA, Health Effects Research Laboratory, May 1980, EPA-600/1-80-027.

LaDou, J. "The Not So Clean Business of Making Chips," *Technology Review* (May/June 1984), p. 23.

Miller, M. W. and G. E. Kaufman. "High Voltage Overhead," *Environment* (January/February 1978), p. 6.

Kennedy, D. M. "Microwaves and Cancer: New Evidence," *Technology Review* (October 1985), p. 77.

Crawford, M. "Mill Tailings: A $4-Billion Problem," *Science*, **229** (August 9, 1985), 537.

Maugh, T. H. "Carcinogens in the Workplace," *Science*, **197** (September 23, 1977), 1268.

CHAPTER TWENTY-EIGHT

The Voluntary Factors: Diet, Drugs, and Smoking

Smoking: A Personal Form of Air Pollution

The Problem / What's in a Puff? / The Health of Smokers / The Effects of Smoking on Nonsmokers / Smoking Laws: Should There Be More? / Smokeless Tobacco / Smoking and Fires

Diet, Disease, and Food Additives

Intentional Food Additives / Food, Drug, and Cosmetic Laws / Artificial Sweeteners and the Delaney Clause / Artificial Colors in Foods / The Special Case of Nitrates and Nitrites / Added Vitamins and Minerals / Unintentional Food Additives / A Healthful Diet

Toxic Substances in Drugs and Cosmetics

Hair Dyes / The Case for Testing Before Marketing / Alcoholic Beverages / Weighing Risks and Benefits of Drugs

ancer has been labeled by some experts an "environmentally caused disease." However, they are quick to point out that we have a great deal of control over some of the environmental factors that cause cancer. The factor over which we have the greatest control is probably smoking. Fully 30% of U.S. cancer deaths are due to smoking. An enormous improvement in public health would thus result if no one smoked.

The role of diet in cancer is also believed to be large. Thirty-five percent of cancers are probably due to the kinds of food people choose, and a smaller percentage, perhaps 2%–4%, are due to food additives. However, the role of diet in cancer is not as well understood as that of smoking. The following sections examine more closely the role of smoking, diet, and drugs in relation to health in general and cancer in particular.

Smoking: A Personal Form of Air Pollution

The Problem

Scientists studying the health effects of air pollution often decide that they must separate the people in the study into two groups: those who smoke and those who do not. The reason is that people who smoke are adding a variety of pollutants to the air they breathe.

Some of these pollutants are already present in city air; some are unique to cigarette smoke. It has been said that smoking is a personal form of air pollution. While the effects of breathing in some of the substances in tobacco smoke are known, the effects of others are not. Still, there is a growing acceptance in this country of the fact that smoking is hazardous to health. Over the past 15 years, surveys by the National Clearinghouse for Smoking and Health have noted that a smaller and smaller percentage of adult Americans smoke. In 1965, 51% of American men smoked. The rate has dropped to 30%. For women, the rate has declined from 32% in 1965 to 28.2% now.

A 1975 survey also noted that 61% of those who smoke have tried seriously, at least once, to stop smoking. In fact, 9 out of 10 people say they would stop if it were easier to do.

Scientific studies on the effects of smoking fall into two general groups. First, there are studies in which people who smoke are compared with people who do not smoke. People who smoke are found to suffer not only more lung diseases, such as lung cancer and emphysema, but also more heart disease.

Second, there are studies in which tobacco smoke is examined to see what chemicals it contains. These chemicals are then often given to animals to determine whether they are harmful. An examination of the makeup of tobacco smoke shows that it contains a number of chemicals known or suspected to cause cancer. Other harmful chemicals, such as lead and carbon monoxide, are also found in cigarette smoke.

Table 28.1 Major Toxic and Tumor-Producing Agents in Cigarette Smoke

Component	Effects
In vapor phase	
Carbon monoxide	toxic (heart disease)
Nitrogen oxides	toxic
Hydrogen cyanide	toxic
Acrolein	toxic
Acetaldehyde	toxic
Formaldehyde	carcinogenic
Hydrazine	carcinogenic
Vinyl chloride	carcinogenic
Urethane	carcinogenic
2-Nitropropane	carcinogenic
Quinoline	carcinogenic
Nitrosoamines	carcinogenic
Nickel carbonyl	carcinogenic
In particulate phase	
Benzo(a)pyrene	carcinogenic
5-Methylchrysene	carcinogenic
Polonium-210 (radioactive)	carcinogenic
Cadmium	toxic

Source: 1982 Surgeon General's Report on the Health Consequences of Smoking.

What's in a Puff?

Major toxic components. Table 28.1 lists some of the major toxic components of cigarette smoke. The table is not complete, however. Many other compounds present in smaller amounts may eventually be found to have serious effects.

Carbon monoxide. People who smoke expose themselves to rather high levels of carbon monoxide. Carbon monoxide, which is also a pollutant in city air (Chapter 21), binds to the hemoglobin in blood. Hemoglobin so bound is incapable of carrying oxygen to tissues in the body. Hemoglobin binds 200 times more tightly to a molecule of carbon monoxide than to a molecule of oxygen. Thus, the body tissues in a smoker receive a lower-than-normal supply of oxygen.

The air-quality standard for carbon monoxide is determined by averaging carbon monoxide concentrations over an 8-hour period. This standard, 10 milligrams of carbon monoxide per cubic meter of air, is the level that should not be exceeded more than once a year.

When we live in such a concentration year round, about 2% of our hemoglobin is tied up as **carboxyhemoglobin** (the combination of carbon monoxide with hemoglobin). Compare this percentage to that of a pack-a-day smoker, who may have further tied up about 6% of his hemoglobin. Thus, we cay say smokers have greatly decreased the quality of the air they breathe.

How much of a problem is this? It probably depends on where they live and what they eat. Most cities already have a high concentration of carbon monoxide in the air. Concentrations are highest in traffic jams, underground garages, and tunnels. Levels in these areas can reach 70 mg/m³ or more. Someone breathing air at the 70 mg level for 8 hours would have a blood concentration of about 10% carboxyhemoglobin. At this level, tests have shown a decreased driving ability. People cannot respond as quickly to such stimuli as brake lights and changes in the speeds of cars. Since smokers have inactivated 6% of their hemoglobin due to the carbon monoxide in cigarette smoke, we can see that in certain situations they may well have further decreased their ability to act quickly when they need to be alert and quick.

Increased blood levels of carbon monoxide may also increase the risk of fatal heart attacks. This is explained later.

What does a smoker's diet have to do with the problem? Nitrites, which are used to cure meats such as hot dogs and corned beef, also tie up hemoglobin. (Nitrites in food are covered later in this chapter.) Nitrites react with hemoglobin to form methemoglobin. Methemoglobin, like hemoglobin when it is tied to carbon monoxide, is incapable of carrying oxygen. A large corned-beef sandwich—about one-quarter pound (110 g) of meat—could inactivate 1.5%–5.7% of an adult's hemoglobin by converting it to methemoglobin. Thus, the effects of air pollution, a food additive, and smoking may all combine to inactivate a significant portion of a person's hemoglobin. The combined, or synergistic, effects of smoking and various air, food, and water pollutants are just beginning to be uncovered.

Cigarettes also contribute particles of nickel, arsenic, cadmium, and lead to the lungs of a smoker, in addition to a variety of gaseous and particulate organic materials. Individually and in high quantities, many of these compounds have been known to have harmful effects on humans. Together and in lower quantities, their effects are not clear.

Arsenic and lead. In an earlier era, the pesticide lead arsenate was used on tobacco. Because neither lead nor arsenic is broken down in the soil, both are still found wherever lead arsenate was once used. Tobacco plants grown in that soil today absorb lead and arsenic. The lead level in tobacco leaves is variable, depending on where the tobacco is grown. One study, however, estimates the lead in a cigarette at an average of 13 µg. Of this quantity, about 1.5 µg of lead appears in the smoke. This is the quantity inhaled by an individual while smoking one cigarette. Twenty cigarettes in a day mean 30 µg of lead inhaled. About one-third of this quantity is absorbed into the blood.

Lead has many adverse effects on health (Chapter 21). Smoking alone does not lead to levels that cause apathy, sluggishness, and brain damage. However, lead is also found in food and in water. It is also in air, due to lead additives in gasoline. Smoking a pack a day adds about 50% more lead to the daily respiratory intake for an individual who lives in a polluted city.

Arsenic is a cumulative poison; that is, many small doses can accumulate over a period of time until a poisonous level is reached. Tobacco smoke does not contribute enough arsenic to kill anyone in the classic manner. However, it is suspected that at lower levels, arsenic is carcinogenic.

Cadmium and nickel. A pack of cigarettes contains about 30–40 µg of cadmium and 85–150 µg of nickel. In large enough quantities, cadmium has several effects in the body. It interferes with the body's use of calcium, and may also contribute to high blood pressure and heart disease. A smoker inhales about 2 µg of cadmium in a pack of cigarettes, and about .5–1.5 µg of nickel. Smokers have, on average, about twice the amount of cadmium in their kidneys and liver as nonsmokers. The effect of cadmium and nickel on human lungs is not clear. Studies in animals, however, have shown that both nickel and cadmium can, under certain circumstances, be carcinogenic.

Flavorings. A number of flavoring materials are added to cigarettes. Experts have voiced concern over cocoa, coumarin, angelica root, and triethylene glycol, which are carcinogens or give rise to carcinogens when burned. See Controversy 28.1 for another nontobacco product that causes problems when smoked.

The Health of Smokers

Comparisons of groups of smokers to groups of nonsmokers reveals a number of ways in which smokers appear less healthy than their counterparts who do not smoke.

Heart and circulatory problems. According to a 1979 mortality study by State Mutual Life Assurance Company of America, people who smoke suffer death rates double those of nonsmokers at any age. Smokers are at greater risk with respect to both heart disease and lung disease. For instance, people who smoke are approximately twice as likely to suffer an immediately fatal heart attack as those who do not smoke. When a person stops smoking, the risk decreases. In 10 years, the risk to former smokers is about the same as for nonsmokers. Researchers suspect that there is a relationship between the carbon monoxide in cigarette smoke and an increased risk of sudden death. Possibly, carbon monoxide, because it ties up hemoglobin in blood, reduces the amount of oxygen available to the heart muscle itself. This puts a strain on the heart and makes it pump harder. This strain on the heart may explain the higher likelihood of sudden death from heart attacks in groups of smokers.

Smoking also seems to increase the possibility of cerebral hemorrhages. It further increases the occurrence of peptic ulcers and the chance of dying from them.

Effects on unborn babies. Women who smoke tend to have smaller babies. There are more spontaneous abortions, more stillbirths, and up to one-third more deaths of newborns among babies born to mothers who smoke during pregnancy. In general, it can be said that there are more frequent complications of pregnancy and labor among women who smoke. These effects seem to be due to a shortage of oxygen in the blood during pregnancy. Again, carbon monoxide in cigarette smoke, which ties up the oxygen carrier, hemoglobin, is the culprit.

Effects on lungs. It is not surprising that many adverse effects of smoking involve the lungs. Smokers

(a)

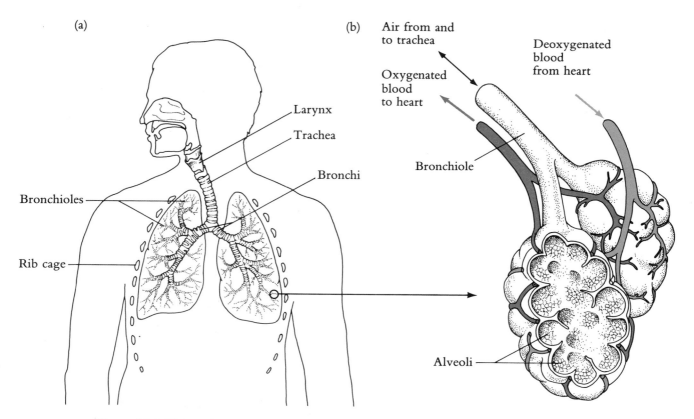

Larynx

Trachea

Bronchi

Bronchioles

Rib cage

(b) Air from and to trachea

Deoxygenated blood from heart

Oxygenated blood to heart

Bronchiole

Alveoli

Figure 28.1 The respiratory tree. (a) Air is breathed in through the nose and mouth, moves down the trachea into the bronchi and bronchioles, and finally enters the alveoli. (b) The alveoli are tiny sacs at the end of the bronchioles. Oxygen enters the bloodstream at this point. Capillaries carry blood around the alveoli, allowing oxygen to diffuse from the alveoli into the blood. (Adapted from *Life: The Science of Biology*, copyright 1983 by William K. Purves and Gordon H. Orians. Used with permission of Sinauer Associates and Willard Grant Press.)

suffer more ordinary respiratory illnesses and have been shown to have more complications, such as pneumonia, after surgery than nonsmokers. Smoking is a major cause of emphysema. This is a disease in which the alveoli, tiny sacs at the end of each bronchiole, do not expel air as they normally should in exhaling (Figure 28.1). People with emphysema have trouble getting the used air out of their lungs so that they can take a fresh breath. Those who stop smoking usually improve.

The disease most commonly associated with cigarette smoking is lung cancer, although smokers also have more of a chance than nonsmokers of contracting cancer of the larynx, esophagus, mouth, bladder, kidney, and pancreas.

Lung cancer accounts for 25% of cancer deaths in the United States. Cigarette smoking is believed responsible for 85% of lung cancer cases. (Despite attempts at early diagnosis, the survival rate for lung cancer is only about 10%, a rate that has not changed much over the past 15 years.)

The rate of lung cancer for women, which has historically been lower than that for men, has increased significantly in recent years. This probably reflects the fact that women did not begin smoking in large numbers until after World War II. Because of the long la-

CONTROVERSY 28.1

Government Protection for an Illegal Activity? Marijuana Smoking

The United States Government has a responsibility to insure that its actions do not foreseeably endanger the health and safety of its citizens.
U.S. Senator Charles Percy of Illinois

They [marijuana smokers] don't have to smoke.
Lee Dogoloff, Office of Drug Abuse Policy

In 1977, many marijuana smokers were jolted by the news that the pot they were smoking could be contaminated with a poisonous herbicide, paraquat. This came about because the Mexican government, supported and encouraged by the United States, began a new program to stamp out marijuana farming in Mexico. Scouts located fields of marijuana by plane and then destroyed them by spraying with the herbicide paraquat, a program for which the U.S. government helped pay. Paraquat is absorbed by marijuana plants and kills them by preventing photosynthesis. However, paraquat does not always act immediately. It can take up to 2–3 days to kill plants. Thus, some enterprising farmers managed to harvest their crops and get them on the market even after they were sprayed. Paraquat levels in some samples of marijuana were found to be as high as 2,264 parts per million. At this level, one to three cigarettes a day for a few months could cause irreversible lung scarring.

Public reaction to the problem varied. Marijuana is, after all, an illegal substance. An official in the White House Office of Drug Abuse Policy stated:

The government does feel some responsibility to smokers, but individuals do have some responsibility and choice in the matter—they don't have to smoke.

In 1983 the U.S. government continued its eradication program by spraying paraquat on marijuana fields in national forests in Georgia and Kentucky. Environmental groups protested the use of a dangerous pesticide on national forests, while other groups voiced concern about health effects on marijuana smokers.

What do you think? Does the government have a responsibility to protect the health of its citizens even if they are doing something illegal? The fact that the U.S. government helped pay for the spraying program can be viewed as causing harm. Does the government have the responsibility to at least not harm its citizens' health directly? Does the fact that the government was attempting to stamp out trade in an illegal drug have any bearing on this case?

Source: All quotes are from *Science*, **200** (April 28, 1978), 418.

tency period before cancers are diagnosed, women's lung cancer rates are only now increasing. Lung cancer now surpasses breast cancer as the leading cause of cancer deaths in women.

When people stop smoking, their risk of developing lung cancer drops to that of nonsmokers over a period of 10–15 years. Thus, cigarette smoking appears to be the largest single *preventable* cause of cancer in the United States today.

The choice of a filter-tip cigarette or a low-tar brand does not appear to reduce the risk of lung cancer significantly. In addition, these cigarettes do not necessarily reduce the amount of carbon monoxide the smoker inhales. For this reason, low-tar cigarettes do not protect a smoker from smoking-related heart disease.

Tobacco smoke can increase a person's risk of developing cancer from other causes. For instance, a person who both smokes and works with asbestos has a nine times greater chance of developing asbestos-caused lung cancer than an asbestos worker who does not smoke.

The Effects of Smoking on Nonsmokers

Two-thirds of the smoke from a cigarette is not inhaled by smokers but rather goes into the environment around them. Actually, the larger portions of both cadmium and nickel are found in the "side-stream smoke," that is, the smoke not inhaled by the smoker. Thus, about 9–13 μg of cadmium and 12–21 μg of nickel per pack of cigarettes are added to the environment around the smoker. In addition to nickel and cadmium, this side-stream smoke contains twice the tar and nicotine, three times the 3,4-benzopyrene, and five times as much carbon monoxide as the smoke the smoker inhales. Studies on "passive smoking" (simply breathing in a room where people are smoking) show that in a very smoky room, nonsmokers can inhale as much nicotine and carbon monoxide in an hour as if they smoked one cigarette themselves. Recent studies appear to show that working in smoky offices can have harmful effects on lung function in nonsmokers. Further, studies show that the nonsmoking wives of men who smoke may be at a higher risk for lung cancer than wives of men who do not smoke.

Children exposed to tobacco smoke at home suffer more respiratory illness and may have a higher risk of developing cancer when they are adults. Some doctors

are recommending that people with heart disease avoid smoky places because they could inhale enough smoke to injure their health.

Smoking Laws: Should There Be More?

In public vehicles transporting people between states, federal regulation requires separate sections for smokers and nonsmokers. Some 70% of the people queried in a 1975 study by the National Clearing House on Smoking and Health felt that smoking should be prohibited in more places, although only 51% of smokers felt this way. A total of 78% felt that smoking should be prohibited in places of business if the management wishes. And three out of four people believe that doctors, teachers, nurses, and others in the health professions should set us all an example by not smoking themselves.

Smokeless Tobacco

It is a common misperception that the use of smokeless tobacco, snuff, or chewing tobacco, carries no risk. Snuff is powdered tobacco and chewing tobacco is roughly cut. Both are usually held between the cheek and gum and both can cause oral cancers. There is a growing use of these types of tobacco among teenaged boys, partly because it is felt that the dangers of actually smoking tobacco can thus be avoided. However, since these products appear to raise the probability of oral cancers four times over nonusers, they are definitely not harmless.

Smoking and Fires

A dropped cigarette will burn for 20–45 minutes. If it falls on upholstered furniture, it will commonly smolder and cause flames in 15 minutes. Nationwide, almost 10% of fires are caused by smoking and more than one-fourth of fire deaths are related to smoking. Possible solutions would involve either cigarettes that go out quickly when dropped or fire-retardant upholstery and mattress fabrics. Cigarette manufacturers claim to have had little success developing a self-extinguishing cigarette that tastes good.

Should the government take a more active role in trying to stop people from smoking? See Controversy 28.2.

CONTROVERSY 28.2

What Is the Government's Role in Controlling Smoking?

If we have to make a mistake we should make a mistake in the direction of protecting public health.
Gus Speth

. . . he should first remove the products about which there is no mistake.
Rodney Adair

I would have thought that today, fifty years after the repeal of the great ignoble experiment of Prohibition, that we as a society had learned something about the failures of government intervention into people's lives.
J. Michael Jablons

The New York Times quoted Gus Speth, the chairman of the Toxic Substances Strategy Committee, commenting on a proposed ban of the artificial sweetener, saccharin:

> So far, we have no scientific basis for setting a safe threshold dose for a carcinogen. Until such a scientific basis is demonstrated, if we have to make a mistake we should make a mistake in the direction of protecting public health. And that means we assume, for the present, that there is no safe level for a carcinogen.

In a letter to the editor of *The New York Times*, Rodney Adair commented:

> How can Mr. Speth and the F.D.A. expect anything more than cynical laughter toward a Government that subsidizes the tobacco industry and at the same time threatens the removal of other products which are often only marginally suspect and on which there is no statistical evidence of any undue incidence of cancer among its human users?

If he truly believes his own statement in reference to removal of carcinogenic materials from the market, i.e., "if we have to make a mistake, we should make a mistake in the direction of public health," he should first remove the products about which there is no mistake.

At the present time, the government's role in preventing smoking is limited to a requirement for warning labels on cigarette packs, a few prohibitions on smoking in public places, and warnings about the dangers of smoking by the Surgeon General.

What is the responsibility of the government to its citizens when there is a proven health hazard such as smoking? Do you agree with Mr. Adair that the government should remove tobacco products from the market?

(Continued on page 636)

Other people have suggested a ban on cigarette advertising:

> With one-and-a-half-billion dollars spent on advertising cigarettes, we get positive reinforcement, communicating "hey it's really OK to smoke, look at all the nice folks who are doing it," everywhere we turn: from magazines and newspapers to the tops of taxis and fronts of grocery shopping carts. There are literally thousands of such positive messages communicated to us

each month for every one depicting the frightening medical truth. (Whelan, 1984)

Is this a reasonable proposal or does it, as J. Michael Jablons (1985) suggests, smack of prohibition?

Sources:
Gus Speth, quoted in *The New York Times* (March 9, 1978), p. A-21.
Rodney Adair, Letter to the Editor, *The New York Times* (March 29, 1978), p. A-26.
Elizabeth Whelan, *ACSH News & Views* (September/October 1984).
J. Michael Jablons, *ACSH News & Views* (January/February 1985).

Diet, Disease, and Food Additives

Many nutritionists now agree that the foods and drinks Americans choose may be contributing to the development of disease. As much as 40% of cancer deaths may be related to eating or cooking American-style. Broiling, frying, and charcoal grilling of meats may cause carcinogenic substances to form. High-fat diets may cause the production of excess bile and lead to intestinal cancers. Excess fat can also stimulate the production of hormones that could promote breast cancer.

Alcohol is suspected of causing birth defects and also of increasing the risk of respiratory and digestive tract cancers. Excessive amounts of salt can contribute to high blood pressure, while excess sugar leads to tooth decay. Furthermore, we are all concerned about various other additives or contaminants that are found in small amounts in foods, drugs, and cosmetics and which may cause diseases such as cancer.

Later in this chapter, we will examine a diet that is recommended as a possible means of reducing the risk of cancer. First, however, we take a closer look at chemicals that, intentionally or unintentionally, become part of the food supply.

Intentional Food Additives

Why do food manufacturers use additives? A trip to the supermarket should convince anyone that many additives are used to color, preserve, or otherwise "improve" food, drugs, and cosmetics. Over 2,000 different chemicals are added to foods alone. The term *additive* covers a wide variety of materials. In foods,

additives fall into three main groups. The first includes natural substances such as sugar, salt, and vitamin C. A second group consists of laboratory copies of natural substances, such as vanillin, the chemical that is the main flavoring in natural vanilla bean extract. There are also the substances that are entirely synthetic, or invented in the laboratory: BHA, EDTA, and saccharin, among others (Figure 28.2).

Additives are used for many reasons, all of them understandable. However, some are more justifiable than others. Many are used to increase a product's appeal to consumers. Drugs are flavored to help hide bitterness or other unpleasant tastes. Foods may be colored to help identify flavors (yellow for lemon candies, pink for strawberry ice cream). But colors and flavors are also used as substitutes for expensive ingredients that are left out of cosmetics or foods. For instance, real fruit is often missing from artificially colored and flavored soft drinks.

Modern methods of selling food have made certain additives necessary. Chemicals that kill mold and keep foods moist mean that bakery products and candies can be shipped across the country and still remain fresh tasting for long periods of time. Antioxidants, which prevent fats from becoming rancid, make convenience foods such as boxed cake mixes possible. In fact, the whole group of convenience foods and special diet foods probably could not exist without the food additives that flavor, color, and stabilize them. In some cases, additives make a wider variety of foods possible. Certain foods could not be canned, frozen, or packaged for shipment or out-of-season selling without additives.

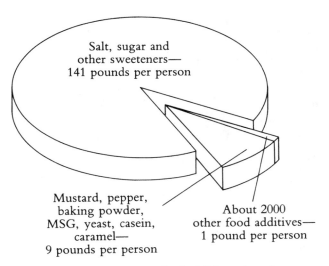

Salt, sugar and
other sweeteners—
141 pounds per person

Mustard, pepper,
baking powder,
MSG, yeast, casein,
caramel—
9 pounds per person

About 2000
other food additives—
1 pound per person

Figure 28.2 Americans and food additives. Americans consume about 150 pounds (68 kg) of food additives per person each year. Most of these are salt, sugar, and other sweeteners.

There is not complete agreement, of course, that all or even most additives are necessary. For instance, it has been argued that the consumer does not gain from the use of preservatives:

> Even assuming that these chemicals are harmless, the advantage in selling bread that does not go stale for a week or more, to take an example, seems to lie more with the baker and retailer than with the consumer. (Lijinsky, 1972)

It should be noted, however, that a shorter shelf life would mean higher costs to manufacturers. In a competitive market, this would be passed on as higher food prices for the consumer.

In another instance, at Senate hearings in 1972 Michael Jacobson, a co-leader of a consumer action group, demonstrated several dozen pairs of similar food items, one of which had no chemical additives while the other had preservatives, emulsifiers, dyes, or antifoaming agents. He stated, "This shows that such chemicals are rarely, if ever, needed in the production of quality foods if manufacturers use good manufacturing practices."

However good the reasons for using food, drug, or cosmetic additives may seem, we would like to be sure that the additives are safe.

Of special concern are the effects of additives on children and young adults. These members of the pop-ulation may be more susceptible to additive chemicals because their bodies are still growing or maturing. In addition, because they have most of their life-span ahead of them, their potential exposure to chemical additives in foods is greater than for other age groups. (See Figure 28.3.)

Food, Drug, and Cosmetic Laws

The safety of food, drug, and cosmetic additives is governed by the Federal Food, Drug, and Cosmetic Act of 1938 and its amendments. Adulteration of food is prohibited. This means harmful substances cannot be added and that unexpected ingredients (such as low-cost fillers) cannot be substituted for materials the consumer would normally expect to find in a product.

In 1958, a food additives amendment was passed. Under this law, the manufacturer is required to prove the safety of an additive before it can be marketed. It is estimated that it costs a manufacturer $200,000 to

Figure 28.3 In some schools, parents who are concerned about the amount of artificial flavors, colors, and preservatives that children eat and drink have banned the sale of foods containing additives in the school lunch room and vending machines. Soft drinks are replaced by milk and fruit juices. Other snack foods are replaced by fruit or additive-free cookies and potato chips.

$3,000,000 to test a new food additive and prove it safe enough to market.

Also in the 1958 amendment is the Delaney Clause, which forbids the use of an additive, in any amount, if it has been shown to cause cancer in humans or animals. This clause has aroused a great deal of argument because some people feel it prevents reasonable judgments about risks compared to benefits. (See Chapter 26 for a discussion of risk-benefit comparisons.) Some of the arguments for and against the Delaney Clause are detailed in the later section on artificial sweeteners.

These are federal laws, which apply to foods that are sold across state boundaries. Foods sold only within a state are subject to the laws of that state.

The GRAS list. When Congress passed the 1958 Food Additive Amendment, which required manufacturers to prove the safety of food additive chemicals they wished to manufacture, a list of some 600 chemicals was prepared and designated the GRAS (Generally Regarded As Safe) list. These chemicals were exempted from the amendment. That is, they were allowed to be added to food even though they had not undergone a great deal of testing to prove their safety, since they had been in use for a number of years. Chemicals were put on the list if the opinions of experts in nutrition and toxicology were that no unhealthful effects had ever been traced to these additives. Since then, some chemicals have proved to be less safe than was believed. In 1969, the Food and Drug Administration began an extensive testing program to determine whether the items on the GRAS list were indeed safe.

About 450 items on the list have been tested so far. Of these, 80% have been declared safe. Another 14% are probably safe but are undergoing further study. About 4% have been declared uncertain and are also being studied.★

Artificial Sweeteners and the Delaney Clause

Substitutes for sugar have come to seem not only desirable but necessary to a nation beset by weight prob-

Figure 28.4 Artificially sweetened products. Saccharin is used not only in products designed for weight control but also in products where sugar would be too bulky, such as in pills to make them easier to swallow. Saccharin is also used to flavor many toothpastes because sugar is felt to contribute to the formation of dental caries (cavities). The product on the extreme right does not contain saccharin. It uses the new artificial sweetener aspartame.

lems. In the United States, almost 5 million pounds of the artificial sweetener saccharin were sold in 1976 (Figure 28.4). Thus, it came as unpleasant news in March 1977 that the FDA was considering a total ban on the use of saccharin in foods. New studies had shown that saccharin caused bladder cancer in rats.

Saccharin was not the first artificial sweetener to be declared unsafe. In 1969, the FDA banned cyclamates. Sales of this group of artificial sweeteners had reached around $1 billion per year when it was found to cause bladder cancer in rats. At the time of the cyclamate ban, manufacturers of diet foods were upset, but not desperate. After all, they had saccharin to fall back on. Diet soda recipes were changed to leave out cyclamates and rely on saccharin.

When the saccharin ban was proposed, reactions were much stronger. Mail flooded congressional offices. One Representative said he had received more mail about saccharin than about any other subject during the three years he was in Congress. Most of the mail opposed the saccharin ban. In addition, soft-drink manufacturers began a vigorous lobbying campaign. Three-quarters of the saccharin used in 1976 went into diet soft drinks. The controversy that followed highlights many of the problems involved in regulating food additives and drugs.

★In 1978, the Center for Science in the Public Interest published a table in which common food additives are grouped according to estimates of their safety. The table can be ordered as a poster from C.S.P.I., 1755 S Street, N.W., Washington, D.C. 20009.

Experimental evidence. In the first place, experiments showing that saccharin causes cancer were challenged. The experiments were done on rats, a common experimental animal. However, because rats concentrate their urine more than other species before excreting it, their bladders were exposed to higher concentrations of saccharin than might be the case in other animals.

Saccharin does not cause mutations in bacterial cells (the Ames assay), but it has been shown to cause mutations in test-tube cultures of hamster cells and human cells. Although this does not prove that it is carcinogenic, it is cause for suspicion. (These tests are described in Chapter 26.)

In the rat experiments, very high doses of saccharin were used, much higher than humans would be expected to consume. The rats were fed saccharin as 5% of their diet for two generations. This prompted a number of people to calculate that a human would have to chew 6,700 pieces of artificially sweetened bubble gum or drink 800 diet drinks each day to receive the same dose the rats did. Are these reasonable comparisons? Not really. Rats live for only 2½ years compared to humans' 70- to 80-year lifetime. Thus, researchers using high doses are trying to compensate for the longer exposure time humans have (and thus re-create the higher accumulated dose). Furthermore, we are interested in finding out whether a chemical is carcinogenic even if it only causes one case in 100,000 people per year. (That would be 2,000 cases of cancer in the United States each year caused by saccharin.) In order to detect a carcinogen of this strength at normal human levels of exposure, scientists would have to run experiments on hundreds of thousands of rats. The expense and work involved would be enormous. Thus, higher doses are used in the belief that there is a relationship between dose and rate of cancer causation: The higher the dose, the more cases of cancer will be found. There is evidence that this is a reasonable assumption.

Leaving aside questions about the experiments themselves, many people felt that a saccharin ban smacked of the Prohibition era. They argued that if there was a risk, they would rather be told about it and left to make their own decision.

The Delaney Clause. The FDA, however, was on firm legal grounds in proposing a saccharin ban. The 1958 amendment to the Food, Drug, and Cosmetic Act states "no additive shall be deemed to be safe if it is found, after tests which are appropriate for the evaluation of the safety of food additives, to induce cancer in man or animal" [Section 409(c)(3)(A)]. This is known as the Delaney Clause after the Representative who proposed it.

There has been much criticism of the Delaney Clause, mainly because it allows administrators no flexibility. That is, they are not given the opportunity to decide whether a very small amount of a carcinogenic substance might either cause no cancers or be worth the risk. Second, foods such as meats and dairy products usually contain detectable amounts of pesticides such as DDT. This is a carcinogenic chemical that has become widespread in the environment. It is incorporated into foods without being intentionally used in their production. The Delaney Clause could be used to justify stopping the interstate shipment of such foods. In practice, the clause is not used this way, but it is felt by some to be a law that could not be enforced and, therefore, one that is invalid.

Those in favor of retaining the Delaney Clause point out that we do not know whether there are threshold (or "no-effect") levels of carcinogens. The safest policy is to assume that any amount of a carcinogen can cause some cancers. Further, since even weak carcinogens can act together to cause greater hazards, the Delaney Clause provides a safeguard against the additive effects of small amounts of carcinogens in the environment.

Another important feature of the Delaney Clause is that it allows decisions to be made on black-and-white issues. Either a substance causes cancer in experimental animals or it does not. This prevents the kind of delays in removing carcinogens from the marketplace that would occur if regulatory agencies had to fight long, drawn-out court battles over risks versus benefits. Nonetheless, the saccharin issue has provided the strongest challenge to the Delaney Clause since it was passed in 1958.

Is a saccharin ban the right solution? In the end, Congress bowed to strong public pressure and directed that FDA hold off on imposing a saccharin ban. Meanwhile, the FDA required stores selling saccharin-containing products to display a warning sign stating that saccharin has been shown to cause cancer in rats. This warning was also required on the packages of foods containing saccharin. The original House version of the bill did not include a warning label, which prompted Representative Andrew Maguire to jokingly propose that manufacturers should include an "assurance" label stating: "Assurance: Saccharin does not cause cancer, in the opinion of your Congressman, despite all scientific evidence that it does."

A study committee of the National Academy of Sciences and the Institute of Medicine concluded in 1979 that saccharin was a "moderate carcinogen." Subsequent studies have confirmed that saccharin is, at least, a weak carcinogen in rats. Nonetheless, Congress has repeatedly extended the moratorium on banning saccharin.

In contrast, the Canadian government did ban saccharin in foods in 1977. The Canadian diet soda industry has developed two new types of diet sodas for the Canadian market. One is a group of low-sugar sodas that have about 60 calories per can, about half that in the regular product. The second type consists of no-sugar sodas that can be sweetened with saccharin tablets, which are still available to consumers. (See Controversy 28.3.)

Aspartame. The situation changed again in 1981 when a new sweetener, aspartame, was approved by the FDA. This is marketed under the trade names Nutrasweet and Equal. Aspartame is composed of two amino acids that are present in the human body: aspartic acid and phenylalanine.

There are still some scientists who feel this compound should be tested more thoroughly. There is incomplete evidence from studies that it may cause brain tumors in rats or a variety of subtle effects on brain function. However, other studies seem to show no effect. The FDA commissioner, when approving aspartame, noted that more studies would be helpful, but he felt the weight of evidence was on the side of no serious effects at the levels at which aspartame was expected to be consumed. However, none of the regulators seems to have judged accurately the market success aspartame would have or how much would eventually be consumed by the average user.

In Canada, where aspartame is the only allowed artificial sweetener, diet drinks sweetened with aspartame captured 22% of the soda market in 3 years.

In the United States, the FDA set an allowable use level of 50 mg per day, expecting the average use to be about 8–10 mg. By 1984 however, average use was 19 mg per day in the United States, and by 1985 that rose to 30 mg, with many consumers above the 50 mg limit. Aspartame's success may thus promote new concerns about its safety.

Artificial Colors in Foods

A fact known to every food manufacturer is that grape soda tastes better when it has a rich purple color. Ice cream seems creamier if its slight yellow color hints at thick cream and egg yolks. And no one wants to buy oranges that are pale orange or slightly greenish. These and other problems and effects call for the use of food colors: substances added to food to cover up or improve natural colors. Any material added to food to color it is called *artificial* color, although the material used might be called *natural*. For instance, carotene, the substance that makes carrots orange, is often used to color other foods orange. The food must then be labeled artificially colored.

Food colors have a somewhat unpleasant history. Before strict regulations were enforced, it was not uncommon for colors to be added to cover up spoiled or adulterated food (that is, food to which some unlabeled, cheaper ingredient had been added). Furthermore, the food colors used were sometimes poisonous. Mercuric sulfide and red lead were used on occasion to color cheese. Candies were often colored with lead chromate, red lead, or white lead. Forty-six percent of the candy sampled in Boston in 1880 contained one or more of these poisonous coloring agents.

There are two general groups of food colors in use today. The first group, called uncertified colors, consists of colors that are for the most part natural products. That is, they are usually extracts of various plants. The second group of colors is called Food, Drug and Cosmetic Act Certified (FD&C) colors. These are synthesized or extracted from nonplant materials, such as coal tar, and *certified* or tested for purity by the FDA.

The industry using the largest amount of certified color is the beverage industry. About 30% of all soft drinks contain certified color. Sixty percent are colored with caramel coloring (cola drinks), while 10% are uncolored (club soda and lemon-lime drinks).

In 1963, Congress directed the FDA to check the safety of the certified and uncertified dyes in use. Nine certified food colors were subsequently taken off the approved list because it was decided they were unsafe or that their safety was not proven. Since then, a total of 63 dyes have been removed from the original list of 200. A final decision is yet to be made on 10 more. Six

CONTROVERSY 28.3

Protecting People from Themselves: Saccharin

All day long we've been taking calls from people, some of them in tears, demanding that we leave saccharin alone.
FDA Official, The New York Times, *March 11, 1977*

. . . put a warning label but don't ban. Let people make their own choice.
Representative James Martin

Surely the argument that people should be allowed to make their own decisions about using possibly harmful materials is not serious, particularly when those subject to the greatest risk are children.
William Lijinsky

Establishing "low-level" residue tolerances for any one carcinogen in food would, by precedent, permit unlimited numbers of other carcinogens also to be added.
Samuel S. Epstein

Morris Crammer, head of the National Center for Toxicological Research, has written about saccharin:

> I think most of us are unprepared, in the absence of persuasive evidence, to replace all those items of human convenience, even though an ultraconservative mathematical hypothesis may imply a relatively high level of risk.

Do you feel that the FDA should ban saccharin on the basis of the Delaney Clause (no substance that has been shown to cause cancer may be added to food), or should people simply be warned about the hazard and left to make their own decision? This is the way alcoholic beverages (known to cause birth defects) and cigarettes (known to cause lung cancer) are treated. Are there basic differences between alcohol, tobacco, and foods that mean they should be regulated differently?

Sources:
James Martin and William Lijinsky, quoted in *Chemical and Engineering News* (June 27, 1977).
Morris Crammer, quoted in *The New York Times* (July 20, 1978), p. A17.
Samuel Epstein, testimony before the Senate Select Committee on Nutrition and Human Needs, September 20, 1972.

of these are animal carcinogens, including the widely used red #3.★

Three successive FDA commissioners have recommended that these dyes also be banned, but all have withdrawn their recommendation under pressure from industry, senior officials, or the Office of Management and Budget. Allowing the use of the 6 dyes shown to be animal carcinogens would clearly be a violation of the Delaney Clause. No FDA administrator has done that yet, although all have expressed discomfort with the idea of banning chemicals that are used in such small quantities that they may pose a negligible risk. Nonetheless, administrative delay (a ban on the 6 dyes in question has been under deliberation for 25 years) is equivalent to an outright violation of the Delaney Clause, according to several consumer activist groups. (See Controversy 28.4.)

The Special Case of Nitrates and Nitrites

Nitrites and methemoglobin. Nitrates (NO_3^-) and nitrites (NO_2^-) are a special case among food preservatives. They were originally added to meats and fish to prevent them from spoiling before refrigeration was available. Over the years, however, people have come to like the cured flavor these additives give to meat. Thus, we now eat cured meats not only because they are a safe way to keep meat but because we like the flavor.

The combination of sodium nitrate and sodium nitrite has three effects on meats: (1) It prevents the growth of bacteria that cause various kinds of food poisoning, such as botulism; (2) it creates the typical pink color of cured meats, such as ham; and (3) it gives meat that special "cured" flavor. Meat was originally cured by adding only potassium nitrate. It was later found that bacteria were converting some of the nitrate to nitrite. The nitrite is actually the chemical that prevents the growth of bacteria and gives the meat a pink color.

Nitrites are not harmless compounds. Hemoglobin in the blood reacts with nitrites to form methemoglobin. As noted earlier, methemoglobin cannot carry oxygen. When 70% of the hemoglobin in the blood is converted to methemoglobin, death results from suffocation. At lower levels of methemoglobin, there may

---

★The other five are reds #8, 9, 19, and 37 and orange #17.

be problems such as dizziness or difficulty in breathing. For this reason, legal limits exist to the amount of nitrite that can be added to meat or fish to cure it. Cases have occurred in which children have been poisoned by bologna and frankfurters containing more nitrite than the legal limit.

Although a large portion (40%) of the nitrite we eat is from cured meats, the nitrate we eat comes mostly from vegetables. Only about 2% is from cured meats. Spinach, beets, eggplant, radishes, celery, lettuce, and greens all can have high levels of nitrates. Since nitrates are not as poisonous as nitrites because they do not oxidize hemoglobin, problems do not usually arise unless bacteria change the nitrates to nitrites. This has happened in some cases, for instance, when opened jars of baby-food spinach have been left unrefrigerated.

Nitrites and cancer. Besides the formation of methemoglobin, there is another problem with adding nitrites to foods. Nitrites can react with certain amines, chemicals that are also present in foods, to form nitrosamines. In one study, 75% of a variety of nitrosamines tested were carcinogenic; that is, they caused cancer in test animals. Cancers were produced in all species of animals tested (dogs, monkeys, rats, parakeets, hamsters, guinea pigs, mice, and trout). Nitrosamines also seem to act with weak carcinogens to make them stronger.

Nitrosamines are found as air pollutants, especially in cities. It is believed they are formed from nitrogen oxide pollutants in urban air. Further, nitrosamines are found in cutting oils (used in industry where metals are cut or ground) and in some pesticides, drugs, and cosmetics. Nitrosamines are also known to be components of tobacco smoke and certain alcoholic beverages. Beer and whiskey manufacturers were able to significantly reduce the amounts of nitrosamines in their products after this was discovered. The total effect of nitrosamines in food and the environment is not yet known. Certain scientists feel that the effect, in terms of causing cancer, will eventually be found to be very large.

Should nitrites be banned as food additives? Before 1978, it was illegal for cured meats such as sausage, ham, or bacon to contain more than 200 ppm of nitrites. This limit was determined by considering the effects of nitrites on an average person. At this level, 100 g of meat (1/4 pound of corned beef on rye) would

CONTROVERSY 28.4

Does Everything Cause Cancer?

FDA's absolute zero risk standard—if generally and sincerely applied to all synthetic and natural chemicals—would ban most of the food supply, most industrial jobs, going outdoors and staying indoors, and much of the rest of the Universe.

Congressman James G. Martin, Letter to Science, June 6, 1980

. . . the public wants to live in an environment free from the risk of cancer, asking, and rightly so, that food, air and water be clean and wholesome . . .

World Health, *September–October 1981*

It is argued that any substance in large enough amounts is carcinogenic. For instance, Representative James Martin, speaking before the Manufacturing Chemists Association, said:

> The rapidly growing abundance of over 2,000 suspect chemicals seems limited only by the resources of rat breeders and feeders. It seems probable that for every chemical there is a dose above which the metabolic defenses and DNA repair and immune systems of some cancer-prone rat can be overwhelmed, especially if concentrated *in utero* by a transplacental effect. Ironically, the only chemicals unlikely to be overdosed to a carcinogenic level are those that are lethal poisons. If cyanide were a sweetener, it would pass.★

★Martin is also a Ph.D. chemist. This passage is taken from a speech in October 1978.

This type of argument has been used to protest bans on artificial sweeteners, hair dyes, and food colors, but experiments do not really bear out the claims made. Although researchers have tested 7,000 chemicals (by feeding them to animals at the maximum amount that does not kill or poison them directly), less than 7% of the chemicals have caused cancers. If this small a proportion of tested chemicals causes cancer, it seems reasonable to assume that the ability of a chemical to cause cancer is a relatively rare phenomenon.

Suppose you were debating Representative Martin on the floor of the House as he attacked regulations banning a popular food additive suspected of being a weak carcinogen. Could you explain the arguments that not everything causes cancer and that dosing animals with high levels of suspected carcinogens is a reasonable way to determine whether or not these chemicals are carcinogenic?

convert 1.5%–5.7% of the consumer's hemoglobin to methemoglobin. If the person were also a heavy smoker living in a city, a further amount of hemoglobin would be inactivated by the carbon monoxide in the cigarette smoke and urban air. The combined effects of the air pollution, smoking, and corned beef sandwich would be just short of causing the first sign of oxygen starvation: headaches.

In 1975 it was shown that when bacon containing legal levels of nitrites was cooked until it was crisp, nitrosamines formed at the parts per billion level. There is some evidence that vitamin C can prevent this reaction. The FDA considered a ban on the use of nitrites in cured meats such as bacon. However, the question of safety from food poisoning was also involved. That is, would people be in more danger of dying from food poisoning if nitrites were not allowed in cured meats than they would be in danger of suffering a nitrosamine-caused cancer if nitrites were allowed to be added?

The main problem appeared to involve cooked bacon because it is usually cooked at high temperatures and has a very high fat content, conditions that encourage nitrosamines to develop. For this reason, the FDA lowered the amount of nitrite allowed in bacon and required the use of vitamin C to reduce nitrosamine formation. You can see from the ingredient list on the packages that some manufacturers of sausage meat and bacon no longer add nitrites to their product.

Added Vitamins and Minerals

A number of manufactured food products contain added vitamins and minerals. In some cases, this is required by law to replace nutrients lost during processing. Thus, white bread is enriched with thiamine and riboflavin, vitamins lost when wheat is milled to white flour. In other cases, vitamins are added to commonly used foods to ensure good nutrition. For instance, ready-to-eat breakfast cereals, which are of little nutritional value but are eaten for breakfast by millions of children, are enriched with vitamins. Artificial orange breakfast drinks are enriched with vitamin C because they are widely used as a substitute for orange juice, a natural source of the vitamin.

Such supplements are not without hazard, however. Certain vitamins, notably A and D, can accumulate to toxic levels in the body. Children given large doses of vitamin A have developed a deformity in

which one leg is shorter than the other. Some pregnant women who received calcium and vitamin D supplements, drank vitamin-D enriched milk, and sunbathed (a natural source of the vitamin, which is formed in the skin by the action of sunlight), have borne babies with excess calcium deposits in the skull. Some scientists have voiced concern that excess iron may cause people to become susceptible to disease. Such an effect has been seen in animal experiments.

The FDA is responsible for ensuring that people do not receive excess doses of vitamins and minerals from foods to which they have been added. In the face of the uncertainty that exists over what is and is not an excess dose, however, the consumer would probably be wise to avoid foods that must be enriched to be nutritious. Instead, it would be safer to rely on less processed foods, traditional safe sources of the vitamins and minerals necessary for good health. (See Controversy 28.5.)

Unintentional Food Additives

Four groups of toxic substances in foods are not added to them on purpose. One is naturally occurring toxins. A second group includes traces of pesticides (residues) left from the spraying of crops. Third, there may be drug residues in meat, left from chemicals added to animal feeds. The fourth category includes additions that occur because of some accident during the manufacture or shipment of food. The PBB disaster, discussed later in this chapter, provides an example of such an accidental contaminant.

Pesticide residues in food. In 1978, the House Commerce Committee began hearings on cancer-causing chemicals in foods. Many problems in regulating carcinogenic substances, such as pesticides in foods, were brought out in testimony before the committee.★

How do pesticides, toxic chemicals intended to kill pests, get into food? In many cases, small amounts of pesticides used to treat crops or animals remain on produce or in meat when they are marketed. These small amounts are called **residues**. Certain government regulations state exactly how high residues can legally be

---

★Pesticides and the problem of growing enough food for the world's people are discussed in Chapters 7 and 8. Pesticides as water pollutants are noted in Chapter 14, and pesticide use in developing countries is covered in Global Perspective VI.

CONTROVERSY 28.5

Lifestyle and Disease: Health Foods

Don't let any food faddist or organic gardener tell you there is any difference between the vitamin C in an orange and that made in a chemical factory.

Dr. Frederick J. Stare, chairman of Department of Nutrition, Harvard School of Public Health

The guru of food faddism is not Adelle Davis, but Betty Crocker.

Nutrition Action

The American diet has become a controversial subject. One group of people, sometimes labeled "health food faddists," argue that we should eat "natural foods," grown without fertilizer or pesticides and packaged without additives or fortifiers of any kind. Others reply that there is no difference between natural foods and those grown with fertilizers and pesticides, or between vitamin C from rosehips and that produced in a laboratory.

The composition of our bodies can be stated in terms of water, protein, fat, carbohydrates, vitamins, and minerals. And so can the composition of an orange or a loaf of bread. And don't let any food faddist or organic gardener tell you there is any difference between the vitamin C in an orange and that made in a chemical factory and added to grape or apple juice. (Stare, 1971)

Furthermore, says another group of scientists led by Bruce Ames (1983), "the human intake of 'nature's pesticides' (natural toxins found in foods) is likely to be several grams per day—probably at least 10,000 times higher than the dietary intake of man-made pesticides."

The other side of this argument can be seen in the following discussion:

Food faddism is indeed a serious problem. But we have to recognize that the guru of food faddism is not Adelle Davis, but Betty Crocker. The true food faddists are not those who eat raw broccoli, wheat germ, and yogurt, but those who start the day on Breakfast Squares, gulp down bottle after bottle of soda pop, and snack on candy and Twinkies.

Food faddism is promoted from birth. Sugar is a major ingredient in baby food desserts. Then come the artificially flavored and colored breakfast cereals loaded with sugar, followed by soda pop and hotdogs. Meat marbled with fat and alcoholic beverages dominate the diets of many middle-aged people. And, of course, white bread is standard fare throughout life.

This diet—high in fat, sugar, cholesterol, and refined grains—is the prescription for illness; it can contribute to obesity, tooth decay, heart disease, intestinal cancer, and diabetes. And these diseases are, in fact, America's major health problems. So if any diet should be considered faddist, it is the standard one. Our far-out diet—almost 20 percent refined sugar and 45 percent fat—is new to human experience and foreign to all other animal life. . . .

It is incredible that people who eat a junk food diet constitute the norm while individuals whose diets resemble those of our great-grandparents are labeled deviants. . . . (*Nutrition Action*, 1975)

How do you feel about "health foods"? What does make up a good diet?

Sources:
Frederick Stare, *Food Additives: What They Are, How They Are Used.* Washington, D.C.: Manufacturing Chemists Association, 1971, p. 16.
B. Ames, "Dietary Carcinogens and Anticarcinogens," *Science,* **221** (September 23, 1983), 1258.
Editorial, *Nutrition Action*, Center for Science in the Public Interest, April 1975.

and list procedures that must be followed to keep residues below the legal limit.

In other cases, however, foods become contaminated with pesticides that were not even used on food crops. This is a process of secondary contamination in which pesticides in dust or rain land on food crops. For instance, most food products from animals (milk, meat, cheese) contain measurable amounts of the pesticide DDT. This comes about because of widespread contamination of the environment with DDT, which is picked up and blown around the world on dust particles. Rain washes DDT out of the air and onto pastureland. Grazing cattle eat DDT-contaminated grass and incorporate the DDT into their meat and milk.

In the United States, the Environmental Protection Agency is responsible for protecting society from the risks involved in using pesticides. The most recent major pesticides law, passed in 1972, is the Federal Environmental Pesticide Control Act. Under this law, manufacturers must supply the EPA with data showing that a pesticide they wish to sell is effective against pests and will not harm humans or the environment. If the EPA is satisfied with the data, it will register the pesticide. In some cases a **tolerance** is allowed. That is, a certain amount of pesticide residue may safely remain on food or feed when it is marketed.

Regulation of pesticides has proven to be a major problem for the EPA. By 1976, $2.4 billion in pesticides were sold every year. By 1984 this had grown to $3.3 billion. Industries on this scale have political muscle. The EPA itself estimates that one-third of the 1,500 active ingredients in pesticides are toxic and up to one-fourth may be carcinogenic.

Critics point out that although a number of pesticides are restricted to use by certified operators (Figure

Figure 28.5 Farmer studying a manual on pesticide use during a pesticide applicator training course. A number of pesticides have been classified for restricted use only. This means that only persons who have passed a state test and received a permit may use them. Almost all commercial operators and about half of the nation's farmers are certified to use restricted pesticides. Training sessions for applicators stress the hazards to people and to the environment of certain pesticides. Safe use, storage, and disposal of these pesticides are taught. Examples of pesticides in the restricted-use class are endrin, paraquat, sodium cyanide, and strychnine. (USDA Photo)

28.5), only a few pesticides have so far been banned except for emergency use (for instance, DDT, heptachlor/chlordane, Mirex, DBCP, and aldrin/dieldrin). Part of the reason for this record may be that industry and other special-interest groups are almost certain to challenge regulations through lawsuits and by stimulating congressional pressure to reverse decisions. For instance, when aldrin/dieldrin was suspended, both Shell Oil (the manufacturer) and the Environmental Defense Fund filed suit, because neither group was happy with the decision (although for exactly opposite reasons). In the same vein, when Mirex was banned for control of fire ants, congressmen from the Southern states where the ants are a problem pressured EPA into allowing the use of ferramicide, a mixture of Mirex and two other chemicals. Ferramicide was alleged to break down more quickly in soil than Mirex, but no hard data existed on whether it actually did so. The EDF filed a successful suit to delay the use of ferramicide until further tests were made. In a climate such as this, knowing a long legal battle will result, officials may be reluctant to propose new restrictions.

This sort of pressure has, to some extent, caused a situation in which the EPA seems to make regulations only when it is forced into complying with toxic substances laws by lawsuits brought by public interest groups.

The House Commerce Committee has been critical of the EPA's tolerance-setting program. In many cases, pesticide tolerances have been set without all the needed information. On pesticides registered more than 10 years ago, we need information on carcinogenicity and data on birth defects and mutations. Furthermore, the EPA relies on safety data supplied by the pesticide manufacturer. The committee, noting that the manufacturer is hardly a disinterested party, suggests that a better program to check the accuracy of such data is needed.

In addition, problems arise when the EPA tries to calculate how much of a pesticide people are exposed to. When setting tolerances for pesticide residues in foods, the EPA usually divides the total amount of a particular food sold in the United States by the total U.S. population. This is the yearly per capita consumption. An obvious problem with this method is that it averages consumers with nonconsumers, making no allowance for the wide variation in dietary habits in the United States. The procedure leads to underestimates of the amounts of certain foods consumed. For example, a 1965–1966 study showed that only 1 in 20 families eats fresh pears. However, the tolerance for pesticides used on pears is set using the assumption that everyone in the United States eats pears. What this means is that, since those who do eat pears are eating more of them than the tolerance calculation assumes, they are exposed to higher residue levels than calculated (Table 28.2). A better system for estimating consumption of various foodstuffs is needed.

The U.S. Department of Agriculture is responsible for checking that meat and poultry do not contain more than the legally allowed residue of hazardous chemicals such as pesticides, drugs (such as antibiotics or hormones),* and other substances (such as lead or mercury).† The Food and Drug Administration checks all foods except meat and poultry. The FDA is also the agency that is supposed to actually take contaminated foods off the market if the USDA or FDA monitoring programs find excessive residues. In addition, the FDA is the agency that can prosecute growers for sending contaminated foods to market.

*More about why these drugs are used and the resulting problems is on page 648.

†The human health hazard of mercury is discussed in Chapter 14, that of lead in Chapter 21.

Table 28.2   Foods Consumed in Small Amounts—Less than 7.5 Ounces (213 g) per Year per Person

Artichokes	Eggplant	Muskmelons	Swiss chard
Avocadoes	Figs	Nectarines	Tangelos
Barley	Honeydew melon	Okra	Tangerines
Black-eyed peas	Hops	Plums	Turnips
Blueberries	Horseradish	Radishes	Walnuts
Brussels sprouts	Kale	Raspberries	Winter squash
Coconut	Mangoes	Rye	
Cranberries	Molasses	Safflower	
Dates	Mushrooms	Summer squash	

Note: It is easy to see that, in many of these cases, if someone eats the food at all, he or she will eat more than 7.5 ounces (213 g) per year.

However, neither the FDA nor USDA monitors all shipments of food or all toxic substances known to occur as residues in food. This is partly a matter of economics. Both agencies rely mainly on tests that detect whole groups of chemicals at one time. The tests for some chemicals are difficult and take a great deal of time to carry out. Furthermore, in some cases there are no good tests for a particular chemical. Thus a number of hazardous chemicals are not tested for at all. (See Global Perspective VI for a discussion of the additional problems in monitoring pesticides in imported foods.)

Even if the USDA finds high levels of a chemical in a meat sample, the meat may not be taken off the market. The reason is that meat leaves the slaughterhouse within 24 hours, while test results are not received for up to a week. Since meat is not tagged after leaving the slaughterhouse, there is no way to retrieve a contaminated carcass by the time the test results come back. In 1976, the USDA sampling program found that 5%–15% of meat and poultry samples contained excess chemical residues. Since almost none of the products from which the samples came were removed from the market, it is estimated that 1.9 million tons (1.7 million metric tons) of beef and 1.1 million tons (1 million metric tons) of pork were sold in 1976 with illegal chemical residue levels. The FDA has similar problems in locating and removing contaminated shipments of other foods.

How serious a health problem is this apparent pattern of tolerance violations and of contaminated food reaching the marketplace? It is probably impossible to say at the present time. We are dealing with quantities that are almost always too small to make people sick immediately. Rather, we are concerned that some of the chemicals may be carcinogenic; that they may have additive effects with other carcinogens in the environment; and that we probably will not know the final effects until after the long time it takes for cancers to develop.

The Commerce Committee recommended more personnel and better monitoring procedures for sampling and detecting illegal chemical residues. In addition, a method of retrieving contaminated meat and produce from the market is needed.

Drugs in animal feeds. A wide variety of substances is added to poultry and cattle feed to keep the animals healthy and to cause them to grow faster. Farmers are encouraged to look for growth-promoting additives because of the high cost of animal feed. Anything that causes livestock to put on weight faster or that shortens the time to market means higher profits. This is a matter of concern because small amounts (or residues) might be left in the meat and thus be eaten by humans. The hormone diethylstilbestrol (DES) has been used as a growth promoter in cattle. This drug is known to have caused cancer in children born to women who took it while they were pregnant. There is also evidence that it has increased the risk of cancer to the women themselves. Whether DES should be allowed as a feed additive depends in a large part on whether any remains in the meat when it is marketed. Arsenic, another known carcinogen, is approved as a feed additive on the basis that it is all excreted by the cattle before they are slaughtered.

A second reason for concern about drugs in animal feed involves the possibility of drug-resistant bacteria developing in animals fed antibiotics on a routine basis (Figure 28.6). In the United States, nearly 100% of poultry, 90% of pigs and veal calves, and 60% of cattle are given antibiotics in their feed. Animals raised in the close quarters of feed lots or large chicken ranches respond to antibiotics by producing more meat per pound of feed consumed, although healthy, well-nourished animals raised in exceptionally clean quarters show no such response. It is known, however, that adding antibiotics to feed increases the possibility of developing strains of bacteria resistant to the antibiotics used. It now seems clear that such antibiotic-resistant organisms can cause disease in humans. There was a case in England in which the injection of large doses of antibiotics into veal calves appeared to start a human epidemic of antibiotic-resistant salmonellosis. Since then, antibiotics for animals in England have been available only by prescription. A similar epidemic in the United States in 1984 led to calls for restrictions on the use of at least some antibiotics in animal feeds. At a minimum, those antibiotics especially useful in treating humans should not be allowed as feed additives.

PBBs: Story of an accidental disaster. During the summer of 1973, 10 to 20 bags of the flame retardant Firemaster were mixed in with a shipment of the feed additive Nutrimaster. Firemaster is the Michigan Chemical Company's trade name for PBB (polybrominated biphenyl). The contaminated feed was eaten by animals on dozens of Michigan farms in what is now viewed as a classic example of how a toxic chemical can

Figure 28.6 Feeding beef cattle in Missouri. Farmers found antibiotics sharply cut feed bills. (USDA Photo)

cause an accidental environmental disaster. Fredrick Halbert, a dairy farmer whose farm is near Battle Creek, Michigan, received a shipment of the PBB-contaminated feed in late summer, 1973. His cows ate the feed for about 16 days, each consuming up to a half pound (about .25 kg) of PBBs in the feed. During this time they began to show signs of serious illness: loss of weight, runny eyes and noses, overgrowths of the hoofs, and sharply decreased milk production.

Halbert and the veterinarians he called in were unable to determine what was wrong. They sent all sorts of samples out for analysis, but no one could find anything unusual in the cattle or in their feed. By this time, Halbert had changed his cattle feed because he was still suspicious that there might be something toxic in it. All told, Halbert estimates that he spent $5,000 tracking down experts and sending out samples. Finally, he located George Fries, a scientist at the USDA's Agricultural Research Center in Maryland, who, by lucky chance, was familiar with PBBs. Fries was able to determine that the feed was contaminated with large amounts of PBBs. The men also realized that Michigan Chemical Corporation manufactured both a fire retardant containing PBBs and Nutrimaster, a feed additive with a similar appearance. In the investigation that followed, it was found that Firemaster was usually packaged in bags with red lettering. However, due to a

shortage of preprinted bags, for a time in 1973 both chemicals were packaged in plain brown bags with the names stenciled on in black. The final link was the finding of partially used bags of Firemaster at the Farm Bureau Service mill where Halbert's feed had been mixed.

Although the authorities were now aware that there was a serious problem, some 11 months had gone by since Halbert had received his contaminated shipment of grain. Meanwhile, other farmers had received some of the contaminated feed and had fed it to their animals. State agricultural investigators visited farms suspected of receiving contaminated feed and quarantined those animals that had high PBB levels.

By this time, several thousand Michigan farm families and their neighbors had eaten contaminated meat, eggs, and milk. An undetermined but presumably lesser amount of PBBs had entered the food supply of the country as a whole.

Although the effects on the most heavily contaminated cattle were obvious, it is still not clear what effects the PBBs have had on people who ate contaminated food. Researchers are having difficulty sorting out symptoms and levels of exposure. In part, this is because the investigations are being carried out years after people were exposed. It is also true that individual susceptibility to PBB poisoning may vary. Because related chemicals, PCBs, are suspected carcinogens, groups of individuals exposed to PBBs are being followed for the next 20 years to determine the long-term effects of the exposure.

In the PBB episode 30,000 cattle, 6,000 hogs, 1,500 sheep, and 1.5 million chickens died or were condemned. Further, 18,000 pounds (8,000 kg) of cheese, 2,500 pounds (1,100 kg) of butter, 34,000 pounds (1,500 kg) of dry milk products, and 5 million eggs had to be destroyed.★ Many farmers whose animals tested out below the legal limits reported sickness among their animals and their families. They were not able to destroy sick animals and receive compensation from the state because their animals were below the legal PBB limit. Some state officials have blamed these particular problems on poor animal husbandry. Some officials also point out that no one has been able to find a correlation between blood levels of PBBs and symptoms in humans. The farmers claim the state did not want to lower the PBB limit because the government didn't want to pay the additional compensation to farmers whose herds would be condemned under the new limit.

Gerald Woltjer, 39, who had a $1 million investment in his dairy farm, lost everything at auction and still owes $500,000. Unable to work because of a wide range of health problems, including painful joints, dizzy spells, extreme fatigue and blurred vision, Mr. Woltjer, his wife and five children are forced to live on welfare.

A man with a big smile who was once, according to his wife, jolly and even-tempered, Mr. Woltjer says he is now very moody, anxious, irritable and is also extremely impatient with the children.

Mr. Woltjer and a number of others, examined by Dr. Sidney P. Diamond, the neurologist on a team of investigators sent by Mount Sinai School of Medicine, were suffering from a loss of confidence and ambition and diminished sexual activity.

"How much of the problems are organic and how much is emotional overlay may be a valid scientific question, but in terms of the lives of these people, it's really not a relevant issue," Dr. Diamond remarked.

"These people's lives are destroyed, and that is as important as pain in their joints," he continued. "You can't take fluid out of the soul and show PBB levels." (*The New York Times*, November 8, 1976, p. 1.)

Almost all of the PBBs have probably disappeared from the country's food supply by now, although many farm families still live with the disastrous effects on their lives and health. According to *The New York Times* (August 12, 1976):

This disaster we have seen was caused by at the most only 2,000 pounds of PBB. . . . But Michigan Chemical made and sold 12 million pounds of Firemaster, and they aren't saying where it went.

Most of the chemical is believed bound up with plastics to reduce flammability. But, Dr. Selikoff asks: "How much PBB escapes into the general environment? Will we discover 40 years from now that we have another problem like PCBs on our hands?"

In a similar incident, a piece of heavy machinery knocked a PCB (polychlorinated biphenyl)-containing transformer off a utility pole at a feed manufacturing plant in Billings, Montana. The transformer fell into an open storage facility containing meat meal. Figure 28.7 illustrates how this accident contaminated meat, eggs, bakery goods, and soup with PCBs in 17 states.

Better coordination of FDA, USDA, and EPA efforts at monitoring for and responding to such incidents, or perhaps the designation of one agency to deal with these events, could improve the response to incidents of accidental contamination.

Natural toxins. Many plants, including those eaten by humans, synthesize toxic chemicals, apparently as a defense against insects and other predators. Tests show that a large variety of these chemicals are *carcinogenic, teratogenic,* or *mutagenic.* Further, toxic materials are formed in foods by natural processes during storage. Mycotoxins are produced by molds, especially under damp storage conditions. Aflatoxins, produced by the mold *Aspergillus fluvus,* are potent carcinogens that

---

★The state of Michigan set the limit for PBB in meat at 0.3 ppm. In October 1974, levels of PBBs were found as high as 595 ppm in milk; in poultry, as high as 4,600 ppm; in eggs, up to 59.7 ppm; and in meat, up to 2,700 ppm.

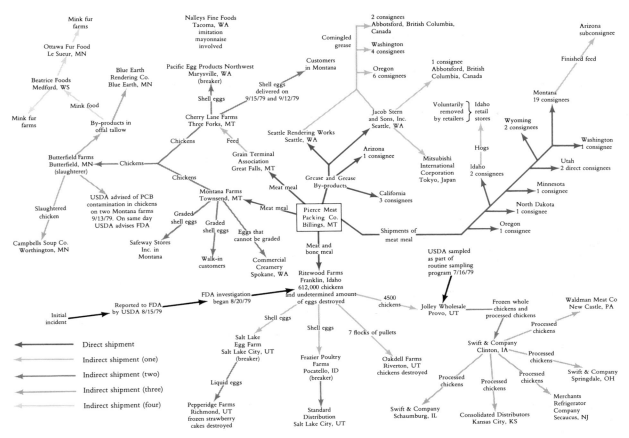

Figure 28.7 Possible spread of PCBs through the food distribution system after accidental contamination of feed at Billings, Montana. This chart illustrates the scope of the federal government investigation of the PCB contamination incident in the western states. Not all material identified contains PCB. (Source: Food and Drug Administration.)

have been found in rice, peanuts, corn, wheat, and food products from Asia, Africa, and North and South America. (See Figure 26.2 for an estimate of the carcinogenicity of aflatoxins.) In Georgia in 1977, a drought was followed by an aflatoxin plague that destroyed the value of the remaining corn crop. Corn cannot be sold in interstate commerce if it contains more than 20 ppb of aflatoxin.

Charcoal broiling of meats may even be hazardous. Two types of possibly carcinogenic materials are produced. One is related to the tars in cigarette smoke, the other is a breakdown product of proteins in the meat. (See the reference by Ames 1983 for more on this subject.)

A Healthful Diet

Up to this point, we have mainly discussed additives and contaminants that may be harmful in foods. The other side of the coin is also important, however. If, as has been claimed, dietary patterns are responsible for as much as 35%–40% of cancer cases, what *should* we eat to prevent disease and maintain health?

In 1982, the Committee on Diet, Nutrition and Cancer of the National Research Council issued a report on dietary patterns and cancer risk. The committee noted that the understanding of how diet is related to cancer is about at the stage that the understanding of how smoking is related to cancer was about 20 years ago. Nonetheless, the committee stated that several general agreements seem to be emerging from the mass of experimental evidence. The evidence is most persuasive that high fat and high alcohol intake increase cancer risk. On the other hand, vitamins A and C, as well as vegetables in the cruciferous family (broccoli, cabbage, cauliflower, and brussels sprouts), appear to decrease cancer risk. The committee recommended that:

1. A one-quarter reduction in the amount of fats in the average diet (from 40% to 30%) would be wise.

2. The diet should include vegetables, fruits, and whole-grain cereals, especially citrus fruits, which are high in vitamin C; green leafy vegetables and deep yellow vegetables, which are high in β-carotene that the body can convert to vitamin A; and vegetables of the broccoli–cabbage group.

3. The consumption of salt-cured, salt-pickled, or smoked foods should be minimized.

4. Alcoholic beverages should be consumed only "in moderation," particularly by smokers, both because of cancer risks and other health problems (cirrhosis of the liver, fetal alcohol syndrome, high blood pressure, and so on).

The committee also stated that although there is no evidence as yet that food additives and contaminants are causing significant numbers of cancers, neither have these groups of substances been found in foods for long enough time to say that they are safe. Thus, efforts should be made to minimize intentional or unintentional contamination of the food supply with possible carcinogens. Care should also be taken to minimize food contamination with substances found to be mutagenic in bacterial tests (which implies that they could be carcinogenic). Evidence was not felt to be strong for dietary fiber as a cancer preventive or for the need for more vitamin E than Americans already consume.

Most of these diet recommendations are not new. They are, in fact, similar to those on the "prudent diet" recommended by the American Heart Association and to the "Dietary Goals for the United States" issued by the Senate Select Committee on Nutrition and Human Needs in 1977. (Also, they appear similar to what mothers and grandmothers have been urging children to eat for years.)

Undoubtedly, further recommendations will be made as further research uncovers more associations between diet and disease. Nonetheless, according to Clifford Grobstein (1982), chairman of the panel that prepared the report, "it is time to further spread the message that cancer is not as inevitable as death and taxes."

Toxic Substances in Drugs and Cosmetics

Hair Dyes

In the past, cosmetics were viewed simply as materials that stay on the outside of the body. Cosmetics were thus felt to be much less able to cause harm than foods and drugs. We now know that this view is too simplistic. Many substances are, in fact, absorbed through the skin.

A case in point involves hair dyes. When the 1938 Food, Drug, and Cosmetic Law was passed, hair dyes were specifically not included in the regulations. The reason was that they were known to cause allergic reactions in some people and it was felt that this might, under the wording of the law, be enough to ban them from the market. So they were exempted, with the provision that warning labels be put on them to tell consumers about possible hazards. This action set the stage for a later problem. In 1975, certain hair dyes were found to cause mutations (changes in the genetic material) in bacteria (the Ames assay). This indicated that dyes might be carcinogenic. Studies by the National Cancer Institute showed that at least one dye did indeed cause cancer when fed to laboratory animals. The FDA was already aware that the dyes were absorbed through the scalp. In fact, there are many complaints on file that hair dyes have caused a temporary color to appear in the urine of consumers. The industry, however, argued that the feeding studies were irrelevant since hair dyes are not eaten but are applied to the skin. They further pointed out that studies of hairdressers showed no excess occurrence of cancers. Nonetheless, the FDA probably would have had enough evidence to ban certain dyes if they had not been specifically exempted from the law. The agency had to settle for warning labels on products containing 2,4-DAA (2,4-diaminoanisole), 4-MMPD (4-methoxy-*m*-phenylenediamine), or 4-amino-2-nitrophenol.

CONTROVERSY 28.6

Pitfalls of Regulation: Tris, Pajamas, and Children's Safety

The use of an untested chemical [tris-BP] as an additive to pajamas is unacceptable.
Arlene Blum and Bruce Ames

Sleepwear treated with tris-BP was being sold in late 1977 (after the ban).
Marian Gold and co-authors

There are two problems a regulatory agency must face when it considers taking a potentially dangerous product off the market. In the first place, something just as dangerous may be substituted. Second, the product may be sold anyway but not labeled. Both problems occurred when tris-BP was banned as a flame retardant in children's pajamas.

In 1975, the Consumer Products Safety Commission established standards for the flammability of children's pajamas. Arguing that as many as 3,000 injuries and 100 deaths each year resulted from flammable children's sleepwear, the commission required all fabrics used in children's sleepwear to pass a standard test proving the fabric would not burn.

Since making a piece of clothing fabric flame-retardant adds 10%–30% to its cost, it was felt that many people, especially the poor, would not, by choice, buy flameproof garments. In large part to protect children in families with low incomes, then, the commission did not simply require a label stating whether a fabric was flammable.

Most manufacturers met the new standards by treating sleepwear with tris-BP. This was the most effective and least expensive way to meet the standards.

All treated garments were to be labeled fire-retardant and the material used was to be specified. Up to 5% of a fabric's weight consisted of tris to make it fire-retardant. However, in 1977 it was discovered that tris-BP was mutagenic in bacterial assays (the Ames assay). This indicated strongly, although it did not prove, that tris-BP might be carcinogenic. Further, tris painted on rabbit skin at doses of 2.27 g/kg caused male rabbits' testicles to atrophy. As a result, tris-BP was banned for use in children's sleepwear and all treated garments were recalled.

This is not the end of the story, however. Researchers checking in late 1977 found tris-BP-treated garments still being sold—unlabeled. There is evidence that a replacement flame retardant, Fyrol FR2, may also be mutagenic. (The manufacturer has since removed this substance from the market.)

Thus, we have a dangerous situation that may be partly caused by a regulation. That is to say, sleepwear is required by law to be flameproof, but the safety of the chemicals being used to meet the regulation are in doubt.

Adding flame-retardant chemicals to almost all

children's pajamas, as a consequence of the Consumer Product Safety Commission's standards, most probably is reducing the number of burns and deaths due to children's nightwear catching fire, although statistics are unavailable. As we have indicated, there are also other ways of reducing fire injuries.

The risk of the exposure of tens of millions of children to a large amount of a chemical must be balanced against the risk of fire. A calculation . . . suggests that the risk from cancer might be very much higher than the risk from being burned. Flame retardants (and most other large volume industrial chemicals) either have not been tested or have not been adequately tested for carcinogenic-

ity. The use of an untested chemical as an additive to pajamas is unacceptable in view of the enormous possible risks. (Blum and Ames, 1977)

Can you think of other ways to help solve the problem of fire-related injuries? Do you think children's pajamas should be required by law to meet flammability standards? (You may want to read the two articles listed below for more information.)

Sources:
A. Blum and B. Ames, "Flame Retardant Attitudes as Possible Cancer Hazard," *Science*, **195** (January 7, 1977), 21.
M. Gold et al., "Another Flame Retardant: Tris-(1,3)-dichloro-2-propylphosphate and Its Expected Metabolites Are Mutagens," *Science*, **200** (May 19, 1978), 785.

The Case for Testing Before Marketing

In past years, the FDA has been able to ban several cosmetic ingredients, including hexachlorophene, an antibacterial agent used in soaps, deodorants, and skin creams. Hexachlorophene was found to cause brain damage in infant monkeys. Another material banned was vinyl chloride, now known to cause birth defects and otherwise rare liver cancer. Vinyl chloride was once widely used as a propellant in aerosol cans of hair spray and pesticides.

However, even premarket testing by manufacturers may not assure the safety of a product. Such testing depends in part on good faith by the manufacturer. Cases have been uncovered in which a manufacturer has submitted false data or incorrect conclusions. A case in 1976 involving the Searle Company led to the tightening of FDA regulations on animal testing procedures, as well as a more careful review of data submitted to the FDA.

Alcoholic Beverages

Another toxic material, sometimes classified as a drug, that is exempt from certain provisions of the Food, Drug, and Cosmetic Act is alcohol. The authority to label alcoholic beverages is given by law not to the FDA but to the Bureau of Alcohol, Tobacco, and Firearms (BATF) in the Treasury Department.

This has caused two recent problems. In the first

case, the FDA wanted to require the listing of ingredients on beer, wine, and liquor labels. This would make it easier to recall products containing an ingredient found to be harmful (a food coloring, for instance). In addition, some people are allergic to a few of the ingredients used in alcoholic beverages. Ingredients that might cause such reactions include yeast, fruit, malt, molasses, spices, sulfites, egg white, and fish glue (the last two are used as clarifying agents in wine). Labels listing ingredients would make it possible for people to avoid those products that might cause an allergic reaction.

Winery owners and distillers opposed the labeling requirements. In part, they opposed listing ingredients because, although they know what goes into making any particular batch of beverage, without expensive tests they cannot be sure what is left in the final product or what new compounds have formed during fermentation or aging. The FDA lost a federal court suit in 1976 in which it was confirmed that only the BATF was in charge of labeling alcoholic beverages. BATF had earlier refused to require ingredient listing.

The problem again came up when new data were found on the effects of alcohol on unborn children. As little as 3 ounces (90 ml) of alcohol a day, or one drinking binge, during pregnancy can cause fetal alcohol syndrome. Babies with the syndrome may show reduced growth, mental retardation, smaller head size, and defects in other organs. In addition, it is not known whether lesser amounts of alcohol might cause smaller but still unwanted effects. Again the FDA was unable to require labeling, this time warning of the dangers of drinking alcohol during pregnancy.

In addition to its teratogenic effects, alcohol acts synergistically with tobacco smoke. That is, alcohol and smoking act together to greatly increase the risk of cancers of the mouth, larynx, and esophagus.

Weighing Risks and Benefits of Drugs

As a final caution, we should note that even if all the regulations and procedures work properly, the consumer can never be completely sure that approved drugs and cosmetics are safe. There is uncertainty in the testing procedures themselves, and certain effects, notably the production of cancers, take many years to become apparent.

New data comes to light periodically on the hazards of drugs once thought to be without serious side effects. For instance, 15 years after use of "the pill" became common, benign liver tumors (which grow but do not spread to other organs) were found to be a (small) risk to women taking certain oral contraceptives. (Those containing mestranol are suspect. This includes Enovid, Ovulen, Ortho-Novum, Norinyl, and Norquen.) As other examples, birth defects are now believed to result from the use of the common tranquilizers Valium, Miltown, and Librium if they are taken in early pregnancy. Phenacetin, an ingredient in several common analgesics, is suspected of causing renal tumors in humans. The use of diethylstilbestrol (DES) was found to cause vaginal cancer 15–20 years later in a percentage of daughters born to mothers who had taken it to prevent miscarriage during pregnancy.

It seems only reasonable that consumers should know as much as possible about the benefits and possible harmful effects of the drugs or cosmetics they are using. In this way they can take some part in the decision about long-term risks compared to present benefits.

Summary

Smoking, choice of diet, and use of drugs may all have an effect on a person's risk of developing cancer. Tobacco smoke contains many harmful and carcinogenic substances, including carbon monoxide, arsenic, lead, nickel, cadmium, and carcinogenic flavorings. Smok-

ers suffer more heart and circulatory problems than nonsmokers as well as more respiratory illness (lung cancer, emphysema, and so on). Pregnant women who smoke increase their risk of miscarriage, stillbirth, and other complications. When people stop smoking, health risks almost all decrease over time. Nonsmokers living or working with smokers may also be at risk for diseases caused by cigarette smoke. Smokeless tobacco, snuff, and chewing tobacco are not harmless. Their use also carries a cancer risk.

Diet may be responsible for as much as 40% of cancers. However, this figure includes cancers due to cooking and storage methods, fat content of the diet, alcohol use, and natural toxins in foods as well as the effects of food additives.

According to the Federal Food, Drug, and Cosmetics Act, the manufacturer is required to prove the safety of any food additive before it can be marketed. This provision does not apply to additives on the GRAS list. The Delaney Amendment to this law states that no chemical known to cause cancer in humans or animals may be added to foods.

The artificial sweetener saccharin, a proven animal carcinogen, provides the strongest challenge to this part of the law. A special law passed by Congress allows its use. The new artificial sweetener aspartame was judged safe by various experts at levels it was expected to be used. However, its enormous success has led to its being used at much greater than expected levels, prompting new concern over its safety.

Of the original 200 food colors approved for use at the time the GRAS list was established, 63 have since been banned. Ten more are still under review, including 6 known to be carcinogenic in animals.

Nitrates and nitrites are used in foods as preservatives, flavoring agents, and color preservers. Forty percent of the nitrite eaten is in cured meats, but most of the nitrates are natural components of vegetables. Nitrites can cause methemoglobinemia and may combine with other chemicals in foods or in the body to form carcinogenic nitrosamines.

A number of substances in foods are unintentional additives. These include pesticide residues, drug residues from drugs given to animals to cure disease or in their feed, contaminants from accidents (such as PBBs), and natural carcinogens and toxins.

The Committee on Diet of the National Research Council has recommended that Americans reduce their fat, cured-food, and alcohol intake and increase consumption of yellow and green leafy vegetables, fruits, and whole grains in order to remain healthy and decrease their risk of developing cancer.

The Food, Drug, and Cosmetics Act gives the FDA the right to regulate drugs and cosmetics with two important exceptions: hair dyes and alcoholic beverages. Alcoholic beverages are regulated by the Bureau of Alcohol, Tobacco and Firearms. This led to problems when the FDA wished to require a label listing of ingredients to which some people may be allergic.

The scientific uncertainties about the safety and future effects of a variety of food additives and drugs appear to make it appropriate to allow the consumer enough information to make decisions about the risks and benefits of these substances.

Questions

1. Suppose a friend asked you whether you thought smoking was really dangerous. What would you tell him? Why?
2. How would you feel about the government's banning all but natural plant extracts as flavoring agents and food colors? Flavorings and colors would be available, but their cost might be so high that many foods (soft drinks, candies, cake icings, fruit drinks, etc.) could not be economically produced. Think about the artificially colored or flavored products you normally use, which might not be available. Would this bother you?
3. What is the Delaney Clause? Are you in favor of letting it stand as is, or would you rather see agencies allowed to balance the risks from carcinogens against the possible benefits from their use?
4. What sorts of toxic substances in foods could be called unintentional additives?
5. How do pesticide residues get into foods?
6. Do you believe the Department of Agriculture and the FDA should try to monitor all possible toxic residues in foods? What about the cost of this?
7. Why might the EPA's method of calculating the average daily consumption of foods be especially hazardous for milk?
8. What rules would you make about toxic chemical manufacturing to protect the public against a disaster similar to that which occurred with PBBs? What kinds of information and employees would you want on hand so that you could determine quickly if a dangerous accident had occurred?
9. Do you think alcoholic beverages should carry a label warning about the dangers of fetal alcohol syndrome or

synergistic effects with tobacco smoke? Can you think of any problems this might cause?
10. Would you find it useful to have the ingredients listed on all cosmetics and alcoholic beverages? Why?

Further Reading

Introduction to Lung Diseases. American Lung Association, 1975.

    The American Lung Association has a number of booklets, pamphlets, and posters on smoking. Call or write to the association in your state for a listing of what is available.

Schilling, R. F. II, et al. "Smoking in the Workplace: Review of Critical Issues." *Public Health Reports,* **100** (5) (September–October 1985), 473.

    Smoking subjects many workers to special hazards because of interaction between tobacco smoke and other chemicals in the workplace. In addition, the freedom to smoke while working may infringe on others' rights to a smoke-free place to work. Such issues are considered in this paper.

Adult Use of Tobacco 1975. Center for Disease Control, National Clearing House for Smoking and Health, Bureau of Health Education, Atlanta, Georgia.

    The results of an extensive study of smoking habits of U.S. citizens is published in this booklet.

Fenstermaker, C. "Can Our Food Supply Ever Be Completely Risk-Free?" *GAO Review,* Spring 1982.

    A good summary of the problems involved in regulating possible carcinogens in the food supply.

Sun, M. "Food Dyes Fuel Debate Over Delaney," *Science,* **229** (August 23, 1985), 739.

    The current status of the Delaney Clause barring carcinogens in the food supply is summarized here.

Food Additives, Who Needs Them? Manufacturing Chemists Association, Inc., 1825 Connecticut Avenue, Washington, D.C.

    Industry's side of the argument is presented in this well-written booklet.

Marijuana and Health. Washington, D.C.: Department of Health and Human Services, 1982.

    Current views on the hazards of marijuana use are summarized.

Searle, C. E. "Chemical Carcinogens and Cancer Prevention," *Chemistry in Britain* (March 1986), p. 211.

    A readable summary of the lifestyle-causes-cancer viewpoint.

Sun, M. "In Search of Salmonella's Smoking Gun," *Science,* **226** (October 5, 1984), 30.

    Fascinating details of the epidemiological investigation

that uncovered the link between antibiotics in animal feed and a human epidemic.

Evaluation of Cyclamate for Carcinogenicity. Commission on Life Sciences, National Academy Press, 1985.

A new review of cyclamate data leads to the conclusion that it is probably not a carcinogen but may be a promoter or co-carcinogen. However, its other effects, such as possible testicular atrophy, may be even more influential in keeping cyclamate off the market.

"Nitrate Levels in Bacon," *Federal Register*, **50** (72) (April 15, 1985), 14711.

The whole issue of nitrites in cured meats is reviewed here, along with new proposals for curing bacon.

References

The Health Consequences of Smoking: Cancer. Report of the Surgeon General, U.S. Department of Health and Human Services, 1982.

"The Changing Cigarette," *Mortality and Morbidity Weekly Review*, U.S. Department of Health and Human Services, February 6, 1981.

The Health Consequences of Involuntary Smoking: A Report to the Surgeon General. Office on Smoking and Health, U.S. Public Health Service, Rockville, Md., 1986.

Brody, J. E. "Food Additives, Do They Hurt?" *The New York Times*, July 12, 1978, p. C-10.

Human Food Safety and the Regulation of Animal Drugs. 27th report, Committee on Government Operations. U.S. Government Printing Office, December 1985.

Ames, Bruce N. "Dietary Carcinogens and Anticarcinogens," *Science*, **221** (September 23, 1983), 1256.

Diet, Nutrition and Cancer. Committee on Diet, Nutrition and Cancer, National Research Council, National Academy Press, 1982.

Reduced Tar and Nicotine Cigarettes: Smoking Behavior and Health. National Research Council, National Academy Press, 1982.

Lijinsky, W. Testimony at Senate hearings on food additives. *The New York Times* (September 22, 1972), p. 1.

Jacobson, M. Quoted in *The New York Times* (September 22, 1972), p. 1.

Marshall, E. "USDA Under Fire," *Science*, **217** (July 1982), 37.

Stare, F. *Food Additives: What They Are, How They Are Used.* Washington, D.C.: Manufacturing Chemists Association, 1971.

Blum, A., and B. Ames. "Flame Retardant Attitudes as Possible Cancer Cause," *Science*, **195** (January 7, 1977), 21.

Gold, M., et al. "Another Flame Retardant: Tris-(1,3)-dichloro-2-propylphosphate and Its Expected Metabolites Are Mutagens," *Science*, **200** (May 19, 1978), 785.

Norman, Colin. "Regulating Pesticides: The Delaney Paradox," *Science*, **236** (May 29, 1987), 1054.

Regulating Pesticides in Food: The Delaney Paradox. Committee on Scientific and Regulatory Issues, Board on Agriculture, National Academy of Science, 1987.

A GLOBAL PERSPECTIVE VI

Worldwide Effects of Pesticides and Threats to the Ozone Layer

Pesticide Use in Developing Nations

*Effects of Pesticide Use in Developing Countries/
Some Solutions*

The Ozone Layer

*Ozone: An Essential Gas/Threats to the Ozone Layer/
Effects of Decreasing the Ozone Layer/Global Regulation*

CONTROVERSY:
VI.1 ***Should the United States Sell Dangerous
Pesticides Overseas?***

Pesticide Use in Developing Nations

One chemical Americans don't expect to find in their food and drink is DDT. After all, its use has been banned in the United States. But the unhappy fact is that residues of DDT, as well as of other toxic pesticides such as lindane, aldrin, and chlordane, can be found in the American diet because these chemicals are still used in developing countries. They are sprayed on food crops such as coffee, fruit, and vegetables that are then shipped to the United States. Ninety-eight percent of the coffee, 25% of the fruit, and 6% of the vegetables consumed in the United States are grown elsewhere, usually in developing nations.

Although the U.S. Food and Drug Administration (FDA) has a monitoring program to detect harmful substances in imported food, the program is too small to sample more than a tiny percentage of food imports. In 1982, the FDA took one sample out of every 200 million pounds of imported oranges, every 8 million pounds of imported grapes, every 4 million pounds of imported tomatoes, and every 5 million pounds of imported coffee.

Even then, FDA tests do not check for the presence of some commonly used pesticides that are banned in the United States but used overseas, such as the suspected carcinogens EDB and DBCP. In general, tests detect only half of the kinds of pesticides that may be used on imported foods.

Ironically, the source of much of the pesticides used in developing countries is the United States itself, because it is not illegal for U.S. manufacturers to ship banned pesticides overseas. Tougher laws on pesticide use in the United States have led to new higher-priced, low-toxicity pesticides being sold in the United States while the older, lower-priced, more toxic and persistent pesticides are exported. The United States supplies 30% of the pesticides used worldwide, approximately one-third of which consist of products that are not approved for use in the United States.

Effects of Pesticide Use in Developing Countries

Effects on citizens. In 1980, it was reported that 99% of Americans had detectable levels of DDT in their blood and fatty tissue. Studies now show that these levels are being matched or exceeded by levels in the bodies of citizens in developing nations (see Table VI.1). In Costa Rica in 1980, there were 593 reported cases of pesticide poisoning, and many more cases are believed to go undiagnosed or unreported. Even the number of reported cases, however, is comparable to the reported cases of measles, whooping cough, or tuberculosis. Between 1970 and 1980, the number of cases of conventional diseases, such as measles and whooping cough, decreased 70% in Costa Rica, while the number of cases of pesticide poisoning increased 250%. In Choluteca, Honduras, a city located near a major rice-producing area, 10% of the residents were found to show evidence of pesticide poisoning in 1981.

Not only are pesticides that are considered too dan-

Table VI.1   Pesticides in Humans

Pesticide	Detected level (parts per million)			
	Americans[a]	Mexicans[b]	Nigerians[c]	Hondurans[d]
Pentachlorophenol (in urine)	0.007		0.025–0.23	
DDT (in blood)		0.003–0.068	0.07–14.9	
DDT (in fatty tissue)	3.6		6.5	19–89
Dieldrin (in fatty tissue)	0.12		0.02–0.18	

Sources:
a. National Center for Health Statistics, *Health and Nutrition Examination Survey and Human Adipose Tissue Survey*, Pesticides Monitoring Report, USEPA, Exposure Evaluation Division, Volume 1, Nos. 1 and 2, 1980.
b. K. Radetzke and A. Gonzalez, *Journal of Environmental Health* (January–February 1985).
c. S. Atuma and D. Okor, "Pesticide Usage in Nigeria," *Ambio*, **14** (6) (December 1985).
d. "Surveillance of Intoxications by Pesticides in Central America," *Human Ecology and Health*, **3** (3) (1984), 3.

gerous for use in the United States being used in developing countries, but the pesticides are often misused or misapplied by workers who are unaware of the risks involved. In Nigeria, large drums of pesticides are delivered to fields where they are often left in the open, under intense sun and torrential rains. "Some or all of the labels, including instructions for use, are washed off. Some containers get rusty and leak pesticides that may leach down to the groundwater . . ." (Atuma and Okor, 1985). In Ecuador, pesticides are usually sold without instructions for use or with instructions that farm workers are not likely to understand. "Many farm workers are poorly educated, and often illiterate. They commonly think of pesticides as a medicine which is used to cure plants of the diseases caused by insects or other pests. As a result there is a total lack of awareness that pesticides are poisons which can, if improperly used, harm their health and their environment" (Sevilla, 1984).

Protective devices such as waterproof clothing and respirators are often not available or are not affordable. In tropical climates, workers find it difficult to work while wearing heavy clothing or goggles and so forgo such protection (Figure VI.1). And when pesticide workers do wear heavy clothing, it may be all they own and, as is the case with the heavy wool clothing worn by workers in the mountains of Ecuador, it may not be washed after being worn to spray pesticides. Finally, in the small one- or two-room houses where workers live they often have no place to store pesticides out of the reach of children or animals.

In Nigeria, empty pesticide containers are dumped into rivers and streams or, worse, are used to store drinking water. In some cases pesticides are even used

as medicines. In rural areas, γ-HCH, a member of the DDT family, is used to cure people of intestinal worms.

Development of pest resistance. Just as pests have developed resistance to pesticides in the United States, pesticide resistance is showing up in developing nations. In Central America, cotton that was once sprayed five to seven times a growing season must now be sprayed over thirty times. This increased spraying has led to much higher costs for growing cotton. In addition, it has contributed to the development of pesticide resistance in the mosquitoes that carry malaria, leading to an increase in cases of malaria.

Effects on wildlife. Wildlife, especially migrating birds that winter in the tropical habitats of many developing nations, are exposed to toxic pesticides such as DDT. In the United States the main reason DDT was banned in the early 1970s was its lethal effects on wild bird species. Wildlife experts are concerned that our migrating birds saved by the U.S. law banning DDT may succumb to DDT exposure in developing nations.

Some Solutions

Techniques for diagnosing, reporting, and treating pesticide poisoning must improve before a realistic picture of the extent of the problem can be obtained. However, it is already clear that public education and training of pesticide applicators are serious needs in developing countries. In April 1985, the World Bank and the U.S. Agency for International Development, two agencies that fund a major portion of agricultural

Figure VI.1 This barefoot farmer in Indonesia is unaware of the hazards of using pesticides without protective clothing. After the day's work is over he will allow the water buffalo to cool off in the nearby canal, possibly contaminated with pesticides. (Photo courtesy of Dr. James Perry)

development projects in developing nations, adopted strict new guidelines for pesticide use, handling, and storage.

In the long run, the real solution to pesticide problems in developing countries and in the United States lies in decreased use of pesticides and increased use of other methods of pest control such as the techniques of integrated pest management (IPM). For example, in Ibadan, Nigeria, tiny South American wasps in capsules are dropped to prey on mealybugs that threaten the cassava, an especially important crop that withstands the drought now affecting Africa.

This type of solution offers hope for both the victims of pesticide use in developing nations and the unsuspecting consumers in the United States. But in light of low literacy levels and the relative poverty in developing countries, it may take decades or longer for such integrated pest management techniques to be adopted. (See Controversy VI.1.)

Pesticide use in developing nations is one global issue directly concerned with human health. Another such issue is the destruction of the earth's protective ozone layer by various air pollutants.

The Ozone Layer

Ozone: An Essential Gas

High above the earth, in that part of the atmosphere called the stratosphere, is a relatively little known gas essential to life on earth. That gas is **ozone**. Each molecule of ozone is made up of three atoms of oxygen. Ozone in the stratosphere absorbs over 99% of the ultraviolet light that comes from the sun. Ultraviolet rays are often welcomed because they cause skin to tan. They can also cause it to burn and, over long periods of time, ultraviolet rays cause skin cancers.

High concentrations of ultraviolet rays are harmful to many forms of plant and animal life. They are sometimes used to sterilize objects because, in high concentrations, they kill bacteria. Scientists generally believe that life on land did not develop until the earth's protective ozone layer formed. Because ozone in the stratosphere is essential to life, reports that a number of human activities can destroy ozone are understandably alarming.

Threats to the Ozone Layer

Fears about the possible effects of supersonic transport planes (SSTs) on the ozone layer were a strong factor in the decision to stop development of a U.S. fleet of these planes. The planes fly in the stratosphere and release two contaminants, water and nitrogen oxides, that can destroy ozone. However, high costs have slowed the growth of supersonic travel to a point where it is not now considered a serious hazard to the ozone layer.

In 1974, a more serious threat was uncovered. Ozone is known to be destroyed by chlorine, and a major source of chlorine in the atmosphere is chloro-

CONTROVERSY VI.1

Should the United States Sell Dangerous Pesticides Overseas?

In a world of growing food interdependence, we cannot export our hazards and then forget them. There is no refuge.

D. Weir and M. Shapiro

It is simply ridiculous to suggest that my country can be or should be able to undertake the extensive testing analysis and reviews which are carried out in the U.S. in order to decide whether or not to allow the use of a pesticide. . . . To a Latin American our concern about the export of banned pesticides is not much different than the United States' concern about exports from Latin America of cocaine and marijuana.

Roque Sevilla, president of an environmental protection organization in Ecuador

For Nigeria and all developing countries, pesticide use is indispensable to the struggle against hunger and disease. Emphasis should be placed on judicious and safe use rather than the barring and/or restriction of pesticides.

Samuel Atuma, chemistry professor, University of Benin, Nigeria

Risks and benefits must be balanced differently in a nation on the edge of famine.

John E. Davies, Department of Epidemiology and Public Health, University of Miami

The United States currently sells overseas some pesticides that are banned in the United States (except for public health emergencies), including DDT, aldrin, dieldrin, heptachlor, and chlordane. In 1976, several environmental groups successfully sued to prevent any sale of these unregistered pesticides that involved the use of money from the U.S. Agency for International Development. Chemical companies can still ship unregistered pesticides and other countries can still buy them, however, as long as they don't use U.S.A.I.D.

money. Ten percent of U.S. pesticide production consists of unregistered or banned products destined for overseas markets.

What do you think? Should we allow sales to other countries of pesticides we restrict severely in our own country? Do we have the right to make this decision for other people? Do we have the responsibility? What about the argument that if we do not sell pesticides to other countries, someone else will?

Sources:

D. Weir and M. Shapiro., "The Circle of Poison," *The Nation* (November 15, 1980).

R. Sevilla, "Pesticides South of the Border," *The Amicus Journal* (Winter 1984) p. 12.

S. Atuma, "Pesticide Usage in Nigeria," *Ambio* **14** (6) (December 1985), 340.

J. Davies, *Technology Review* (November/December 1984), p. 72.

---

fluorocarbons (or fluorocarbons), widely used as propellants in spray cans and in refrigerators and air conditioners. Because these compounds do not react easily with other materials, they are ideal for use in spray cans. After they are released into the environment however, they diffuse slowly into the stratosphere where they are broken down by ultraviolet light, with the release of chlorine. This chlorine can then break down ozone.

How much ozone are the chlorofluorocarbons destroying? Unfortunately, there are many things we still don't know about ozone and the stratosphere. A 1982 report by the National Academy of Sciences predicted a 5%–9% decrease in stratospheric ozone by late in the next century if fluorocarbons are released at about the rate they were in 1977. In 1984, the prediction was lowered to only a 2%–4% decrease in ozone over the same time period. But then in 1986, scientists discovered a 40% decrease in the ozone column over Antarctica. (The ozone column is the total quantity of ozone that ultraviolet radiation would have to penetrate from the upper atmosphere down to the earth's surface at any given location.) Thus, there seems to be an unexplained "hole" in the ozone layer over Antarctica. Further, scientists have now realized that chlorofluorocarbons and SSTs are not the only substances affecting the ozone layer.

Nuclear explosions in the stratosphere destroy ozone by the release of nitrogen oxides. Thus, in the event of a nuclear war, the resulting increase in ultraviolet rays could be as serious a problem as nuclear fallout. Automobile exhausts and fertilizers in soil also release nitrogen oxide, which can react with ozone.

Bromine is known to destroy ozone in the laboratory. As methyl bromide, it is widely used as a fumigant in agriculture. How much bromine escapes to the stratosphere is not yet known. Similarly, the effect of quantities of industrial chemicals such as carbon tetrachloride and methyl chloroform is not clear, although it is felt that they could release significant amounts of ozone-destroying chlorine.

Effects of Decreasing the Ozone Layer

The average concentration of ozone in the stratosphere is probably about 3 ppm. However, ozone concentrations vary geographically. In addition, ozone concentration can vary, normally, by as much as 30% from day to day. Over a period of years, the variation averages out to 10% due to what seems to be a natural balance between the formation and destruction of ozone.

The normal distribution of ozone in the earth's atmospheric layers is shown in Figure VI.2. There are two major concerns about the decreases caused by human activities: health effects and climate effects.

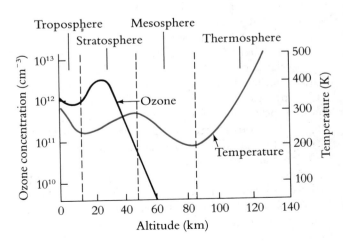

Figure VI.2 Normal temperature profile and ozone distribution in the atmosphere.

Health effects. A decrease in the ozone layer would allow more of the damaging ultraviolet rays to reach the earth, which would almost certainly be harmful to living organisms. It seems quite certain that a large proportion of skin cancers are caused by exposure to sunlight over long periods of time. The occurrence of the two most common types of skin cancer, basal cell and squamous cell carcinoma, appears to be strongly related to exposure to sunlight, especially its ultraviolet-B component. (Ultraviolet-B includes those wavelengths from 290 to 320 nm.) Fair-skinned people are much more apt to develop skin cancers, since the pigments in dark skin help to screen out ultraviolet rays. People living in tropical climates are more subject to skin cancer, apparently because they are exposed to more ultraviolet light.

The rate of skin cancer in the United States is reported to be 300,000–400,000 cases of basal cell carcinoma and 100,000 cases of squamous cell carcinoma per year. Each 1% decrease in ozone is believed to cause a 2% increase in ultraviolet radiation and as much as a 2%–5% increase in skin cancer. Proportionately, then, each 1% decrease might cause 10,000–30,000 extra cases of skin cancer each year in the United States.

Ultraviolet-caused skin cancer is almost always curable, although the treatment is not pleasant. The rate of cure is 95% or greater by chemotherapy (using drugs such as fluorouracil), surgery, x-rays, or cauterization. Basal and squamous cell cancers rarely spread to other parts of the body.

The third type of skin cancer, melanoma, does spread rapidly to other parts of the body. It is a much less common type, however, and is not as strongly linked to exposure to sunlight. For instance, the rate of occurrence of melanoma is 75% greater in southern states than in states on the Canadian border, while the rate of the other two types of cancer is 250% higher.

A related concern is the effect of ultraviolet-B radiation on the immune system of humans and animals. This type of radiation appears to prevent normal immune responses not only in the skin but also in other parts of the body. Scientists speculate that this is how ultraviolet-B radiation causes skin cancers to form.

Plants and animals also show sensitivity to ultraviolet-B rays. An increase in ultraviolet light could be expected to cause the extinction of some microscopic life forms and to damage other species or decrease the living space available to them.

Climate effects. The second major concern about changes in the stratosphere is that they might cause changes in climate. Chlorofluorocarbons, in addition to effects they have on ozone, absorb infrared radiation. This could trap heat in the atmosphere and cause a warming of the earth. Such a warming would add to the greenhouse effect, in which carbon dioxide accumulating in the atmosphere may lead to a warming of the atmosphere near the earth's surface (see Global Perspective V). Despite warming the atmosphere close to the earth, the greenhouse effect is expected to lead to a cooling of the stratosphere. This would, in turn, slow the rate of ozone destruction in the stratosphere.

You may recall from Chapter 21 that methane and nitrogen oxides (all produced when fossil fuels are burned) contribute to the production of ozone. Thus, we have one complex of factors producing ozone or at least slowing its destruction (the greenhouse effect, fossil fuel pollutants) and another complex of factors (chlorofluorocarbon release and production of nitrous oxides, methyl chloroform, and carbon tetrachloride) leading to the destruction of ozone. An important point needs to be made before we begin to speculate on whether the two opposing factors might to some extent cancel: The destruction of ozone takes place mainly in the *stratosphere*, while the production of ozone occurs in the *troposphere*. At the very least, we are engineering a transfer of ozone from one layer of the atmosphere to another (with unknown results). At worst, we are allowing the destruction of the earth's protective ozone covering (Figure VI.3).

At the present time there is a great deal of uncertainty about the final effect of chlorofluorocarbons and changes in the ozone layer on climate. Predictions actually range from a new ice age to temperature increases of up to 1°C. What should be remembered is that life on earth has evolved under the present set of conditions. Lasting changes in one direction or another will have profound effects on life as we know it.

Global Regulation

Because chlorofluorocarbons are believed to be a serious threat to the ozone layer and because the United States was responsible for half of the total world fluorocarbon releases, U.S. government agencies decided to prohibit unnecessary uses.* The use of chlorofluo-

---

*A few drug products, certain pesticides, some aircraft, and some electrical maintenance products are exempt.

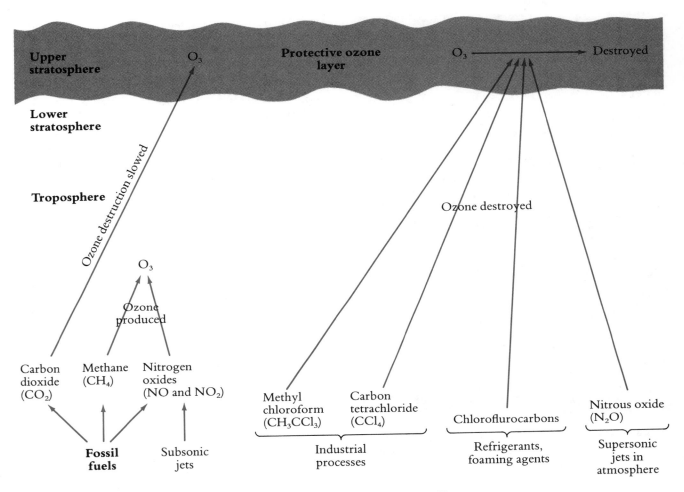

Figure VI.3 Factors affecting the ozone layer. Ozone is destroyed by chlorofluorocarbons and by N_2O from supersonic jets flying in the stratosphere. It is also destroyed by methyl chloroform and carbon tetrachloride. Ozone production is accelerated by pollutants generated by the burning of fossil fuels (CO_2, CH_4, NO, and NO_2) and by subsonic jets flying below the stratosphere. The destruction of ozone is slowed by stratospheric temperature decreases that can result from absorption of heat by carbon dioxide in the troposphere. Note that the ozone produced by fossil fuels and subsonic jets is mainly in the troposphere or the lower stratosphere, while the destruction of ozone by chlorofluorocarbons takes place mainly in the upper stratosphere. Thus, even if destruction and formation rates balance, there may still be an imbalance in where the ozone is located. The effect of such an imbalance is unknown.

rocarbons as spray-can propellants was outlawed in the United States after 1979.

As a result of the regulations, the United States now releases only one-third of the world total. However, the United States is not the only nation to use fluorocarbons. Thus, any solution to the problem must involve other countries as well as the United States.

There are no good substitutes for fluorocarbons as refrigerants and foaming agents for polymers. Increasing use of these compounds in refrigerators and air

Figure VI.4 Balloons such as this one are used to carry aloft equipment for measuring ozone concentrations in the stratosphere. (Photo courtesy of NASA)

conditioners in the United States and other nations threatens to offset gains from the ban on fluorocarbons as spray-can propellants. Reflecting this, world production of chlorofluorocarbons is rising again, after falling in the late 1970s.

In March of 1985, 45 countries agreed on a treaty to protect the ozone layer. This first treaty did not dictate any limitations on chlorofluorocarbons or other substances that destroy ozone. Rather, the treaty was an agreement to share monitoring data and to continue research on ozone (Figure VI.4).

Despite strong initial reluctance to controlling the use of chlorofluorocarbons, delegates to a later conference (perhaps spurred by discovery of the ozone hole over Antarctica) were finally able to agree on a plan to decrease use of these compounds. The plan drawn up in 1987 calls for a freeze on world production of chlorofluorocarbons at 1986 levels. A series of cutbacks would then reduce production by about 50% by the end of this century. American industry is not happy with the plan, which may force the use of substitutes and manufacturing processes that are untried on a large scale. It seems likely that such controls would lead to small price increases for such items as air conditioners, refrigerators, and foam insulation.

Questions

1. Why can traces of pesticides barred for use in the United States still be found in the U.S. food supply?
2. What problems does pesticide use cause in developing countries? In what ways are such problems different from problems in the United States? In what ways are they the same?
3. What solutions do you see to this problem?
4. Why is the ozone layer important?
5. What are the major threats to the ozone layer?
6. What are the two main concerns about a decrease in the ozone layer?

Further Reading

"Stratospheric Ozone Protection Plan," *Federal Register*, **51** (7) (January 10, 1986).
    The U.S. position on and intentions about the ozone problem are summarized here.
Maugh, T., II. "What Is the Risk from Chlorofluorocarbons?" *Science*, **223** (March 9, 1984), 1051.
    A good assessment of the assumptions underlying predictions about the ozone layer.

References

Atuma, S. and Okor, D. "Pesticide Usage in Nigeria," *Ambio*, **14** (6) (December 1985), 340.
Human Ecology and Health, (Newsletter of the Pan American Center for Ecology and Health), **2** (3) (1983) and **3** (3) (1984).
Bull, B. "Bringing Home the Poison." New York: Natural Resources Defense Council, 1985.

Leonard, H. J. *Are Environmental Regulations Driving U.S. Industry Overseas?* Washington, D.C.: Conservation Foundation Issue Report, 1984.

Sevilla, R. "Pesticides South of the Border," *The Amicus Journal* (Winter 1984).

Shaikh, R. A. "The Dilemmas of Advanced Technology for the Third World," *Technology Review* (April 1986), p. 57.

Causes and Effects of Stratospheric Ozone Reduction: An Update. Washington, D.C.: National Academy Press, 1982.

"Review of Atmospheric Science Issues Related to Ozone Modification." National Aeronautics and Space Administration, World Meteorological Organization, United Nations Environment Program, 1986.

Kerr, R. A. "Has Stratospheric Ozone Started to Disappear?" *Science*, **237** (July 10, 1987), 131.

Cicerone, R. J. "Changes in Stratospheric Ozone," *Science*, **237** (July 3, 1987), 35.

(U.S. Department of the Interior, National Park Service photo)

PART EIGHT

Land Resource Issues

I conceived that the land belongs to a vast family. Of this family, many are dead, few are living, and countless members are still unborn.

A Nigerian Chief*

An often-heard phrase is "the quality of life." It refers to society's and an individual's well-being. One kind of well-being is economic; another is social. Elements of economic well-being include jobs, pay, and the cost of food, housing, clothing, transportation, medical care, and other basics. An assessment of social well-being asks how civilized and stable a society is—that is, whether there is widespread crime, alienation, or prejudices. Well-being also takes in the concept of freedom to move, change jobs, and travel. Beyond the economic and social components of the quality of life is still another concept, less distinct and less measurable, but nonetheless real and important. It has to do with the quality of our environment.

The quality of the environment does not lend itself to neat characterization. Instead, a set of questions might be used to grope toward the meaning of quality of the environment. One might ask, for instance, whether there is unspoiled forest nearby that is open to the public to hike or backpack and observe wildlife. Are there parks where families can picnic? Are there coastal beaches within reach that have not been locked up in private ownership? Are historic buildings preserved so that people can see the origins of their settlement? Are farms preserved nearby, both for the food they supply and their contribution to our need for green? Or does surburbia with shopping centers and strip development sprawl to infinity? Is there a place where children can buy pumpkins from the farmer who grew them? Or better yet, where the children can grow pumpkins themselves? Each individual is likely to have a different set of questions to reveal the quality of the environment. What we ask obviously reveals our own personal measures of environmental quality.

One element stands out as a key in all these questions and touches every one of these concerns: how we use the land. If we use the land badly, the questions will

---

*Quoted in *Land Use and the Environment*, Virginia Curtis, ed., Environmental Protection Agency, Office of Research and Monitoring, Environmental Studies Division, 1973.

be answered in a negative way. If we use the land wisely, the questions can be answered in a positive way.

The subject of land use is an unusual blend of issues. To improve the use of the land, we need to know how we can influence the numerous private decisions that go into creating a pattern of land use. We also need to know where private decisions are likely to fail to bring desired patterns of use; that is, where it is necessary for the public to acquire and preserve the land and its qualities. And when the public comes to own such lands, how can their qualities best be protected?

The following chapters discuss how humans use land areas and how we are attempting to preserve some of these areas and the natural communities living there. Chapter 29 treats private land-use decisions, the numerous microscale decisions that go into the formation of a pattern of land use. We discuss the concept and practice of zoning, a tool to enforce the preservation of established land-use patterns in urban and suburban areas. We also discuss new methods that are evolving to control the loss of farmland to development. In Chapter 30 we take up the issue of public lands and their purposes. There we explore the historical development of public ownership of land and the various forms it currently takes.

CHAPTER TWENTY-NINE

Private Land-Use Decisions

A Changing Tradition

Enforcing Urban/Suburban Land-Use Plans
Zoning by Use Classification/Zoning as a Tool for Environmental Improvement/Zoning to Protect the Public/ Growth Management Plans

Preserving Rural Land
Farmlands Are Becoming Suburbs/Taxation as a Root Cause of Farmland Loss/Ways of Preserving Rural Agricultural Land/Preserving Land That May Not Have Agricultural Value

A Changing Tradition

Traditionally, decisions about land use made by private individuals and companies have been "let alone" to a large degree. This tradition is changing, however, as we discover that our land resource is limited and that the impacts of many private decisions have a way of adding up. In the past, a new factory meant jobs and purchasing power. Stores clustering along a major highway meant more access to goods and services. Now we also recognize that certain kinds of industry and commercial development may bring air and water pollution or high volumes of traffic.

Stores, shopping centers, and the like are beginning to face controls on their freedom of action as citizens express their concern about the noise, the traffic, and the visual impact caused by new commercial development. We are faced with difficult choices about what is desirable or needed and what is undesirable or can be done without.

Finally, we now realize that the steady development of suburbs is consuming prime farmland at an agonizing rate. The consumption of prime agricultural land for homes is a trend that somehow must be checked.

Enforcing Urban/Suburban Land-Use Plans

Zoning by Use Classification

The most common form of land-use control is that practiced at the local level by community and county governments. Referred to as **zoning**, it is another way of saying that land has been classified as to its appropriate uses. By and large, such zoning is not concerned with air and water pollution but with the uses of neighboring land. "Is the proposed activity a reasonable 'neighbor' to the activities that already exist in the area?" is the question asked by zoning boards.

The idea of zoning is young. The first comprehensive municipal zoning law in the United States was established in 1916 by New York City. The appeal of the zoning idea was enormous and immediate. In only 10 years, more than 400 American cities had adopted zoning laws. In 1926 the U.S. Supreme Court upheld zoning for the Village of Euclid, Ohio as a legitimate means of controlling nuisances. The legal justification of zoning is that it protects the health, safety, and welfare of citizens. While zoning may, in fact, serve these purposes, it is used most often to protect the values of privately held land.

Zoning laws. Zoning begins when a state government passes legislation that "enables" local and county governments to create land-use plans if they wish to do so. The local government then draws a map that labels each area or zone of the locality as to its allowable uses. Typically, the uses fall into one of four categories, in addition to agriculture: industrial, commercial, multifamily residences, and single-family residences. These four categories may be further divided as well.

Figure 29.1 shows the four categories in an arrangement that has been termed "cumulative zoning." Within this planning framework, a given category of use can accept all uses above it in the chart. That is,

them in the first place. More often, though, those who are putting land to "nonconforming" uses may apply for "variances" or "special exceptions." Such exceptions are considered by a planning board or zoning commission on a case–by–case basis.

The idea of having a procedure to consider uses different from those normally allowed is quite reasonable. However, the results of the procedure are often not so reasonable. In the past, and probably in the future as well, commercial interests, apartment developers, and the like often have obtained exceptions for their activities. They do this simply by pleading hardship or by pointing to existing nonconforming uses. A common technique is to begin construction without a variance but with the excuse that a variance is expected. When the variance is not forthcoming, the developer pleads hardship because of his already "sunk" costs.

When a sufficient number of exceptions have been made, the original labeling of an area loses meaning, and further exceptions become easy to obtain. Zoning as a barrier to certain kinds of development then resembles a windbreak. It can help shelter an area from the winds of change, but it cannot always protect it.

Enforcement of zoning laws. Although zoning is interpreted and administered by a planning board or zoning commission, it is enforced by the courts. Those who violate the zoning statutes can be enjoined from doing so; that is, they may be given a court order to prevent the violation. The court's muscle is its ability to fine those who disobey its orders.

We noted, though, that zoning sometimes fails. Failures may occur as individual instances or they may accumulate to produce wholesale changes in a neighborhood. Such failures may be due in large part to the way variances have traditionally been granted.

Requests to rezone a given parcel of land are heard, as we mentioned, by a zoning board or commission; the meetings are generally open to the public. The commissioners or board members take note of community members who come to protest the exception that has been requested. A lack of protest by citizens is often interpreted as meaning that the residents of an area do not object to the proposed exception in use of the land. In fact, citizens may not know the meeting is being held, may have prior commitments, or may be unable to attend. If only a few people appear at the meeting to protest a proposed change in land use, the board may grant the exception to the person or firm

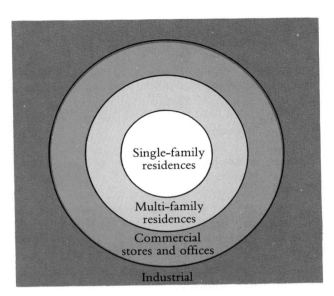

Figure 29.1 The Cumulative zoning concept. Single-family and multi-family residences and commercial activities are allowed within an area zoned "industrial." Single-family and multi-family residences are allowed within an area zoned for "commercial" activities. Single-family residences are allowed in an area zoned for "multi-family" units.

single-family residences, multi-family residences, and commercial activities are all allowed within an area zoned "industrial." In a similar fashion, land zoned in the "commercial" category can be used for its designated use (stores and offices), but single- and multi-family dwellings are also allowed. On the other hand, land earmarked as being in the "multi-family" zone cannot be used for either commercial or industrial purposes. The "single-family residences" category is often referred to as the "highest," or most restrictive, grouping because no other uses are allowed.

Zoning used to protect a land-use pattern. In addition to stating the allowed uses of the land, zoning may be utilized to enforce a uniform pattern of use. For example, in the single-family residence category, the minimum size of the lot may be specified, or the distance of the dwelling from a sidewalk (setback) may be given, or the minimum space between dwellings may be spelled out, as well as allowable construction materials and building heights.

Zoning is typically not cast in concrete. Changes in the ordinances can be made by the body that enacted

requesting it. Thus, making zoning work requires alertness and involvement on the part of citizens.

The publicly proclaimed purpose of zoning is to maintain stable patterns of settlement, but the goal of zoning in practice is quite different. People use zoning primarily to protect the value of the residential property they own. Nearby factories and traffic are seen as threats to property value, and zoning is used to exclude uses that could diminish property value. Zoning, then, has not generally been an instrument to protect the general environment—only the environment of some people.

Shortcomings of zoning. It has been argued that some forms of zoning amount to economic discrimination. Consider, for instance, the common requirement that lot sizes be an acre or more. Such a requirement will cause builders to erect expensive homes in order to earn enough profit on their investment in the land. Expensive homes are bought by people with upper levels of income; lower-income people are unable to afford these dwellings. (While the zoning of acre lots does seem to foster upper-income ownership, the original reason for acre lots may have been the protection of groundwater from septic tank discharges.)

It has also been argued that zoning can lead to a monotonous conformity in the landscape. The diversity of having shops, playgrounds, and public buildings near residential sites can create an atmosphere of interest and appeal. Those who have visited Europe will have seen neighborhoods with a wider variety of activities than in United States suburbs. In Europe, each neighborhood, in addition to its homes and apartments, is likely to have a school, a bakery, a vegetable store, a meat market, a drug store, a candy store, a cigar store, and so on.

Suburbs in the United States, in contrast, are often simply miles of homes with services very distant. Residential areas in the United States that exclude commercial shops tend to waste energy. A purchase at the grocery store often requires a trip by car. Most Europeans can find a sufficient selection of food within a few blocks of their homes or apartments. They not only save on gasoline by walking to market but get useful exercise as well. Thus, the exclusion of any kind of commerce in a residential area results in an added use of energy.

Zoning as a Tool for Environmental Improvement

However, for all its shortcomings, zoning is firmly in place in the United States. And despite its flaws, zoning can be used in clearly positive and beneficial ways. New zoning concepts are providing these improved possibilities for zoning.

Natural area zoning. One new concept is *natural area zoning.* First adopted in New York City in 1974, natural area zoning can be used to protect wetlands such as swamps, bogs, ponds, and creeks, as well as forested areas and rugged landscapes. The greenbelt area of Staten Island, a borough of New York City, was the first area to be classified by the New York City Planning Commission as a protected natural area and hence not open to development.

Cluster zoning. We noted that zoning has been criticized for the economic barrier it often sets up. A concept that would not set up such economic barriers is *cluster zoning.* This concept allows homes to be built much closer together than conventional zoning allows, as long as the total ratio of dwellings to total land area of the development remains at the desired level. The homes may be built with shared exterior walls, common walkways, and small yards. Such common wall construction requires fewer materials, shorter lengths of water and sewer lines, less sidewalk area, less pavement, and less labor. As a result, the dwellings can be sold at a relatively moderate price.

As an example, suppose a community insisted on one-half acre per home (or two homes per acre). The community might be willing to accept 40 homes built on a total of 20 acres, with the 40 homes clustered onto just 10 of those acres. To get the savings of clustered construction, the remaining 10 acres surrounding the clustered development would have to be left undeveloped to meet the requirement of no more than two homes per acre. For the entire area of 20 acres, the ratio of homes to acres would be the required two to one. The added open space could enhance the community and provide resources for recreation and leisure to the development itself. In addition, the clustered homes might be less expensive than individual dwellings, thereby providing more access to home ownership for more people.

Zoning to Protect the Public

Zoning has also been used to protect the public from hazards. Although hazard zoning is used much less often than general land-use zoning, its most common example is flood-plain zoning. The flood plain is the area outside a river's normal channel over which its flood waters may flow. In the past, it has been common to see dwellings built in a flood plain. They have been built there because memories of floods may have faded and because the homes have access to the water and an attractive natural setting. The arrival of a flood, however, refreshes memories and makes the setting less attractive.

When an area is classified by zoning as a flood plain, certain land uses may be banned. These prohibitions make a great deal of sense because local, state, and federal governments bear much of the burden of flood damages. Such burdens include rescue, cleaning up, and restoration of homes, shops, and factories devastated by flood. Governments may even go to the extent of building flood-control structures such as levees and costly dams in order to protect dwellings that ought never to have been located on the flood plain in the first place.

Some years ago, Professor Gilbert White recognized that the traditional approach to the problem of flood control was much like closing the barn door after the horse has escaped. It would be far better, reasoned White, to avoid the misery, hardships, and enormous expense that were caused by homes and commerce built in a flood plain. The expenses of restoration and of flood-control structures could be avoided by simply barring new development in the path of a potential flood. Hardships and expense could also be avoided by removing flood-threatened dwellings from their current sites.

White's ideas were translated into federal law in 1968. Flood insurance was provided under the National Flood Insurance Program enacted in that year, and this insurance was used to entice local communities to adopt flood-plain zoning. This desirable low-cost insurance that the act made available is paid for, in part, by the federal government. However, the insurance can be made available only to residents of those communities that have taken active measures to prevent further development in the flood plain. By the mid-1970s, more than 13,000 communities had indicated a desire to participate in the insurance programs and hence prevent further development in the flood plain.

Growth Management Plans

There are still other ways to influence private decisions on the development of land. Several of these fall under the heading of "growth management," and their history is quite recent. While growth management plans appear legitimate for the moment, they are, nevertheless, controversial. Although general land-use zoning has been approved in principle through decisions of the U.S. Supreme Court, challenges to growth management plans have, to this time, never been heard by the Court. Their present status could, therefore, change.

First in Ramapo, New York and then in Petaluma, California, communities attempted to control the rate at which they grew. Ramapo was then being developed as a suburb of the New York metropolitan area. To slow and control its development, Ramapo drew up an 18-year plan that included an orderly expansion of its road network and of its water and sewer utilities. If developers had been allowed to build homes and shopping centers wherever they wanted, the network of roads and water and sewer lines would have had to grow rapidly in many directions to meet new needs. The costs of rapidly expanding this network to go wherever developers chose to build would have been prohibitive to the community. Hence, the Ramapo plan was not only aimed at preserving the rural environment of the community but also at preventing the explosion of local taxes. Developers who wished to build in areas as yet unimproved by the community had to bear the costs of roads and of water and sewer lines themselves. Challenged in court by the developers, the town's plan was upheld in 1972.

Petaluma, California, under pressure for housing from the San Francisco metropolitan area, adopted in that same year (1972) a plan that limited the construction of new houses to 500 per year. This figure was far lower than the growth rate of nearly 2,000 homes per year in the recent past. Petaluma's plan was at first challenged in Federal District Court, where the plan was ruled unconstitutional for infringing peoples' "right to travel" where they wished. When the case was brought to the Court of Appeals, however, the ruling was reversed, and Petaluma's authority to limit its growth was upheld.

Two features have been common to all methods of influencing private land-use decisions that we have discussed so far. First, they have been aimed at urban and suburban land-use decisions, as opposed to decisions on rural land. Second, the methods have all involved using the police power of government (the power of the courts) to enforce a set of rules. Private individuals are bound by law to follow these rules, unless exceptions are granted in response to specific requests.

Preserving Rural Land

In contrast to the methods that use government enforcement power are techniques that use economic means to convince or require private owners to make better decisions on land use. These economic methods fall short of actual purchase of the land, but in terms of accomplishing their more limited objectives, they are very powerful. Moreover, they are generally more suitable for use on rural land, especially farmland, than the enforcement power of government.

Farmlands Are Becoming Suburbs

The rural land surrounding most of our cities is in genuine danger of extinction as the suburban fringe expands, foundation by foundation, out into the countryside. We value and admire the farmland that surrounds us because it reminds us of an earlier era in which certain aspects of the quality of life were better—when traffic, noise, and congestion were less and when the proportion of our population that worked the land was far larger than today. The suggestion of "a ride in the country" is a very common reference to our feelings for rural land. The tranquil, less hurried atmosphere and the visual pleasure it may impart touch a responsive chord in many of us.

There is more, however, beyond our feelings for rural land. The rural lands on the fringe of the suburban front contain among them the richest agricultural lands near our cities. Some refer to such land as "prime." It must be surprising to be told that the rural farmland being gobbled up by suburban sprawl is among the best and most productive in the country. It seems almost perverse that the rural land most in need of protection for our own economic well-being is also the land most threatened. The explanation is simple.

When our cities and towns were first established, dirt roads connected them one to another, and dirt roads reached out into the surrounding countryside where vegetables, fruit, and animal stock were grown. Naturally, these roads went to the richest farmlands, for these were the roads farmers would use to transport their animals and produce to the town and city markets. This fact of historical geography has brought us to the present, in which those same roads are now widened and paved. These roads, which once carried produce-laden wagons into the cities, now carry commuters in steel-clad vehicles back and forth between homes and commerce and industry. The result is that the rich farmland along these roads is taken for new housing developments. Thus, more and more farm products come from farther away at greater expense. This increased expense is due to the increased distance of shipment and to the fact that the produce may possibly be grown on less productive land. It is as though we were eating not merely the bread but the breadbasket as well.

How severe is this problem? How fast is prime farmland being gobbled up? While the statistics are patchy, the fact of farmland development can be observed by simply taking a car ride to the city fringes.

In 1981, an estimate of the rate of consumption of farmland was prepared by the National Agricultural Lands Study, a cooperative study by twelve U.S. government agencies. The study estimated that during the 10-year period from 1967 to 1977, about 30.8 million acres of farmland were taken; these acres were used for suburbs, for transportation, for shopping centers, for water-resource development, and for production of minerals such as coal. That is a loss of some 3 million acres per year or 12 square miles per day (1.2 million hectares per year or 32 km² per day).

An estimate in the late 1970s by the President's Council on Environmental Quality placed the agricultural land consumed by suburban "sprawl" at 1,000,000 acres (400,000 hectares) per year. This is the equivalent of 1,520 square miles (4,000 km²). California, at about the same time, estimated its annual rate of agricultural land conversion at 21,000 acres (8,400 hectares), or about 3% of the annual loss in the entire country. These figures, however, must be viewed in the perspective of how much land we have and how it is currently being used.

The land area of the United States, excluding Alaska and Hawaii, is 3,522,000 square miles (9,100,000 km²). This is equivalent to 2.27 billion acres (0.91 billion hectares). Of this quantity, about 2.7% of

Figure 29.2 Farmlands are being converted to commercial buildings, dwellings, and pavement. Prime producing farmlands are being consumed as the suburbs of the cities sprawl across the countryside. In Massachusetts, for instance, if present trends continue, more than half of its prime farmland existing in 1977 will be consumed by the year 2000. (Photo courtesy of USDA)

this land area is built over with homes, commerce, and industry. About 18% (413 million acres) of our total land area is actually in use as farmland, according to the U.S. Department of Agriculture. Another 127 million acres is available for farmland but is presently used as pastureland, rangeland, or forests.

We noted earlier that about 1,000,000 acres of agricultural land was being converted to urban use each year. A million acres out of the 413 million acres of producing farmland amounts to about 0.25% of our producing farmland lost each year. This loss is the more significant because of its location. Cities are the obvious focal point for urban expansion. Suburban development spreads in much the same manner as the ripples caused by a stone falling into a pool of water. Cities are also consumers of agricultural products. Continuing loss of farmland in the urbanized east means higher expenditures for produce because of the increased transportation costs and the higher consumption of energy. It also means greater dependence on other portions of the country for food (Figure 29.2).

Taxation as a Root Cause of Farmland Loss

We indicated that the roads of agricultural commerce are the paths along which people from the city move to suburbia; this is but one reason for the threat to farm-lands. Another and more powerful reason is the way we tax land. While the states and the federal government derive most of their income from taxes on industrial profits, taxes on wages, and taxes on certain luxuries, such as alcohol, the counties and local governments draw much of their income from taxing the land.

To tax the land, the local government must first attach a value to the land. The attachment of value is called *assessment* of the property; the value is known as the *assessed value*. Practices differ from locality to locality, but the basis of the assessed value is frequently the price the land would bring if sold on the market. Some local governments may assess a property at an amount equal to the value in the marketplace; others may assess a property at 50% of the market value; and so on. The tax that the owner pays on the property is determined by multiplying the assessed value by the tax rate. The tax rate is stated in dollars per $1,000 of assessed value and is a uniform value throughout the taxing locality.

As housing sprawls across the landscape, land once distant from the suburbs comes closer and closer to spreading development, and the value of that land may soar. Land once of value only for farming may become many times more valuable for homesites. The increase in the value of rural land is a natural outcome of suburban expansion.

Since the market value of the land has increased, the assessed value increases and the taxes on the land

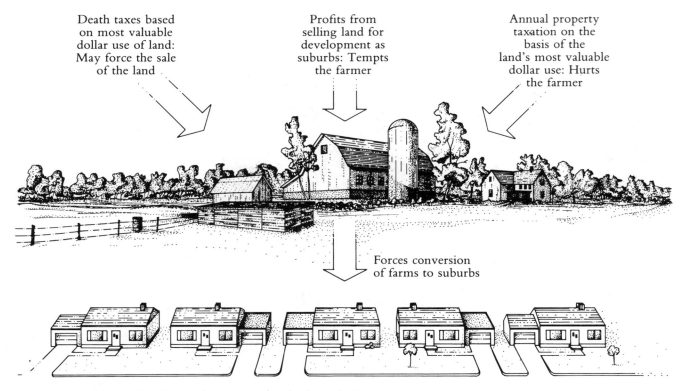

Death taxes based
on most valuable
dollar use of land:
May force the sale
of the land

Profits from
selling land for
development as
suburbs: Tempts
the farmer

Annual property
taxation on the
basis of the
land's most valuable
dollar use: Hurts
the farmer

Forces conversion
of farms to suburbs

Figure 29.3 Forces that convert farmland to suburbs.

rise. Farming, therefore, becomes less profitable be-cause of the farmer's obligation to pay higher taxes. At the same time that farming is becoming less profitable because of higher taxes, the opportunity for profits in the real estate market begins to tempt the farmer who lives on the suburban fringe.

Land in the family for generations or purchased decades earlier can become a source of wealth for a farm family. Land is often now a farmer's money in the bank for retirement. A farmer would have to love the land a great deal not to be tempted. At first, he might sell only a lot or two on the main road and then some less-used acreage. Each sale makes his land more valuable for homesites, makes his taxes higher, and increases the temptation to retire from farming. Some people jest-ingly call farmland "the farmer's last cash crop."

Still another force pushing farmlands into suburbs is federal inheritance taxes. These taxes are levied on the value the estate would realize if it were sold on the open market. Even if the heirs to the property wish to continue to farm the land, the higher dollar value re-flecting the property's use for development purposes

causes a very high tax to be levied on the heirs. To pay such a tax, portions or even all of the farm must be sold (Figure 29.3).

Ways of Preserving Rural Agricultural Land

Assessing land on the basis of use. The vicious circle of increasing temptation for the farmer has been ob-served for some time, and efforts have been made to interrupt the economic forces that bear down on farm-land. Many states have decided not to value rural land on the basis of its most valuable use; instead, they are assessing it on its actual use. That is, even if a parcel of farmland could be sold for a high price, its assessed value reflects its use as farmland. Property worth $4,000 an acre for homesites but only $1,000 an acre as a farm will be taxed as a farm as long as that use contin-ues. The tax threat to the farmer's pocketbook from an unexpected rise in value of the land is thereby eliminated.

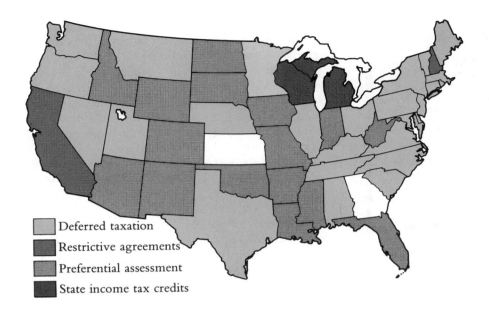

Deferred taxation

Restrictive agreements

Preferential assessment

State income tax credits

Figure 29.4 States where farmland is assessed on the basis of current use. (Note: Michigan and Wisconsin allow farmers to credit property tax against state income tax.) (From Natural Agricultural Lands Study.)

By the end of 1980, all states except Georgia and Kansas had passed some form of tax legislation to decrease the pressures on farmers. Seventeen states allowed taxation of farmland at its value for agricultural use. This *preferential assessment* clearly reduces a farmer's taxes.

Another 28 states have opted for a program of *deferred taxation*. This form of taxation involves a combination of lower taxes for farmland with a penalty if the land is sold. If the land is sold for development purposes, the penalty the owner must pay usually depends on the taxes that were avoided. The taxes that were avoided is the difference between the taxes that would have been owed if the land had been valued at its market price and the taxes actually paid while the property was valued as farmland. This amount, added up over the years during which the special tax was applied, might have to be paid if such a sale is made (Figure 29.4).

Not all states add up the taxes avoided; seven states levy a "land-use change" tax equal to a portion of the taxes avoided or as a percentage of the fair market value at the time the land is developed. Special requirements exist in New Hampshire and California. In these states, a farmer must sign a long-term agreement not to develop the land, in order to obtain preferential assessment of his land.

Deferred taxation and long-term contract programs are designed to prevent speculation and slow still further the conversion of farmland. However, these added requirements and penalties are expected to decrease the number of farmers willing to participate in the programs.

Although the tax threat is decreased by such systems of taxation, the temptation to the farmer from the increased value of the land remains. If profits from the sale of the land are banked, they draw interest. When the annual interest on the proceeds of the sale begins to approach the annual profit the farm family earns through their labor, farming may be abandoned, whether there is a lower tax or not. Furthermore, the method of putting lower taxes on land used for farming can have unintended results. The farmer could sell his land to a speculator who might simply hold it or rent it out for farming while the price of the land continues to rise. The speculator may be rewarded with low taxes while he waits to make a killing on the real estate market. The speculator may, in fact, be attracted to the property because it will cost so little to hold it. Special property tax treatment for the farmer, it is now generally agreed, cannot by itself halt the expansion of suburbia.

Incorporating family farms. Another reason that farms are declining in number on the suburban fringe is also rooted in taxes, but the taxes are not on land; they are on inheritances. When a farmer dies, estate taxes must be paid on the value of the estate. Since the farmlands in the path of advancing suburbs become

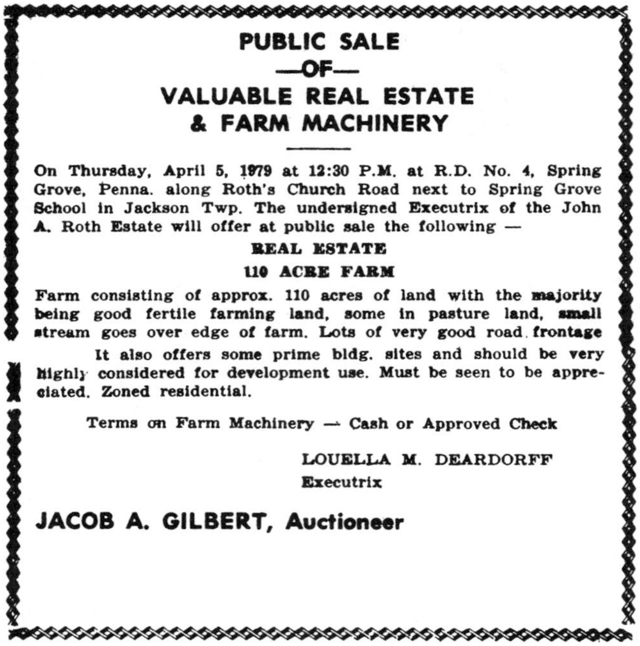

PUBLIC SALE
—OF—
**VALUABLE REAL ESTATE
& FARM MACHINERY**

On Thursday, April 5, 1979 at 12:30 P.M. at R.D. No. 4, Spring Grove, Penna. along Roth's Church Road next to Spring Grove School in Jackson Twp. The undersigned Executrix of the John A. Roth Estate will offer at public sale the following —
**REAL ESTATE
110 ACRE FARM**

Farm consisting of approx. 110 acres of land with the majority being good fertile farming land, some in pasture land, small stream goes over edge of farm. Lots of very good road frontage

It also offers some prime bldg. sites and should be very highly considered for development use. Must be seen to be appreciated. Zoned residential.

Terms on Farm Machinery — Cash or Approved Check

LOUELLA M. DEARDORFF
Executrix

JACOB A. GILBERT, Auctioneer

Figure 29.5 A farm falls to suburbia. Note the advice to developers that this fertile farmland is likely to be useful for a subdivision. Advertisements like this are common in many newspapers.

increasingly valuable, the value of the estate soars. Federal taxes on the estates of deceased farmers can be considerable, amounting to as much as 25% of the value of the estate. The sons and daughters who inherit the farm have to pay such taxes to receive the inheritance. A farm in Cutchogue, New York, valued at only $800 per acre for agriculture, was valued by the Internal Revenue Service at $3,500 per acre. The tax came to about $1,200 an acre. As a neighbor described the situation, "they [the federal government] forced its sale, very simply."★ (See Figure 29.5.) Although the value of an estate that can be inherited tax-free was increased in 1981, the problem of farm inheritance remains because of the high value of farm estates.

To preserve their family farms, farmers have increasingly incorporated. As the owners of their corporations, they are able to make stock gifts each year of up to $3,000 to individuals, in this case their sons and

★"Estate Taxes Drive Farmers Off the Land," *The New York Times,* May 14, 1972.

daughters. These gifts can be made in the form of stocks, and no taxes need be paid as long as the value of the gift is less than $3,000 annually. If they were simply to give deeds for the land to their children, the decreased value of their property would affect their credit at the bank. The banks would see the farm decreasing in value and would be less likely to make loans the farmers need for seed, fertilizer, and equipment. Giving stock does not dilute the corporation, which remains a single entity despite having many owners. Incorporation obviously does not answer all the farmer's estate tax problems. Clearly some change in federal taxation of farm estates is needed.

The federal government has begun to recognize its role in hastening the conversion of farmland by estate taxes. Congress, in 1976, passed a tax reform act that attempted to address the issue. Nevertheless, the eligibility requirements and the penalties for farmland conversion by the heirs to the property are thought to be so strict that very few farms would be saved from development by this law. Further changes in estate tax are needed. Some states have also adopted laws to value farm estates at the farm-use value (Table 29.1).

Purchasing and transferring of development rights by the public. The problem of the loss of farmland is occurring nationwide. Around nearly every major city, suburbs lick greedily at the farms on

Table 29.1 States with Laws That Value a Farmer's Estate on the Basis of Its Current Use

Alabama	Mississippi
Alaska	Missouri
Arizona	Montana
Arkansas	New Mexico
California	New York
Colorado	North Dakota
Connecticut	Oregon
Florida	South Carolina
Georgia	Tennessee
Illinois	Utah
Iowa	Vermont
Kansas	Virginia
Kentucky	Washington
Michigan	Wisconsin
Minnesota	

Source: National Agricultural Lands Study, 1981.

the rural fringe. At a few places in the nation, however, local governments are attempting to preserve their rural character by an interesting and powerful method. The method involves separating the right to develop the land from the land itself.

The idea of outright public purchase of development rights was first put into practice on Long Island, which feeds commuters by the hundreds of thousands into the nation's largest city, New York. Suffolk County, one of several counties on Long Island, was once noted for fine agriculture, but through the decades of the 1960s and 1970s, it came under increasing pressure from real estate developers. Farms were gobbled up by the apparently unstoppable movement of New York City's suburbs.

In 1972, John Klein, county executive, proposed that it would be in the citizens' interest if the county government could somehow halt the conversion of farms to homes. He argued persuasively on economic grounds that the loss of farmland would bring higher tax rates to the community. Because of the increased school-age population that would result, he argued, new schools would be required, and taxes would have to be raised to pay for them. It took the county legislature until 1977 to pass Klein's program.

In September of that year, John Klein sat down at a card table in a potato field with Nathaniel Talmage and signed an agreement guaranteeing that Talmage's 133-acre farm would remain undeveloped forever (Figure 29.6). Klein's program is aimed at preserving 15,000 acres in Suffolk County and is the model that counties, states, and even nations are watching.

When Klein and Talmage sat down at the card table, Talmage transferred to Suffolk County the ownership not of his land but of the "development rights" to his land. He was paid a substantial amount for giving up those rights to Suffolk County, but not the amount he would have received if he had sold the land itself to real estate developers. With the signing of the agreement, ownership of the Talmage property was now split into two pieces. Nathaniel Talmage still possessed the land and all dwellings and improvements on the land. He retained the right to farm it or even to let it go wild. He may not subdivide it to build homes on the lots, but he may, if he wishes, sell off parcels of land. The buyers of those parcels, however, will not be able to build new dwellings because the right to develop the land has passed to Suffolk County. As a consequence of the transfer of the development rights, the market

Figure 29.6 Saving a farm. John V. N. Klein (right), Suffolk County executive, and Nathaniel A. Talmage signing land agreement in a potato field on Mr. Talmage's farm. (Louis Manna/NYT Pictures)

value of the land has decreased because of the restriction on how the land may be used.

There are distinct benefits for Talmage, in addition to the money he was paid—benefits that go far to compensate him for the decreased market value of his land. First, as he clearly planned to continue farming, he has lost no privilege that he needs. His choice of crops, of whether or not to sow, of equipment—these items remain his decisions. Second, the loss in market value of the land means lower real estate taxes to pay, since taxes are calculated by multiplying a tax rate times some portion of the market value.

Third, Talmage's children are more likely to be able to continue farming this land after his death because the federal inheritance taxes on the Talmage estate will be much lower than on land with the potential for development. The estate or inheritance taxes are essentially determined by a tax rate multiplied by the market value of the estate. The reduced market value of the land, now shorn of its development rights, will lead to lower estate taxes. A tax burden that could have brought the family to dispose of the farm is now lightened, and it is more likely that the family will be able to continue farming.

It cost Suffolk County $357,000 to obtain the development rights to the Talmage land; the county had to raise the money by selling bonds, but they still have a bargain. Their payment to Talmage was 80% of the market value of his land. That is, the possibility for developing the land is by far the most important factor in determining its market value. With development rights for the new land in the hands of the county, the value of the land is expected to fall to a bare 20% of the earlier market price. Put another way, the value of the land when sold for homesites is five times its value as farmland only.

As time goes by, that purchase will become more of a bargain. Pressures for new homes in the county will only increase with time as the suburbs continue to expand. Such pressures will drive the market value of the land that can still be developed for homes ever higher; the value of the same land used only for agriculture will rise more slowly.

By 1981, nine local or state governments had established programs for the purchase of development rights. The state of New Jersey passed such a law as well, but terminated the program before any development rights were purchased. The experience of these programs and their funding levels are indicated in Table 29.2. Apparently, little consideration has been given to the compactness of the lands preserved by these programs. The notion here is that if such properties are clustered, the equivalent of an agricultural district is created. Farm activities, such as manure spreading, are much more likely to be tolerated by farms than by neighboring homes.

Agricultural districts, right-to-farm laws, and agricultural zoning. Farming is not always a perfectly compatible activity with the suburban pattern of land use. Noise from equipment operation, odors from livestock wastes, pesticide spraying, manure spreading, and the like can put a suburbanite's nose out of joint. A farmer may find himself sued as a "public nuisance" if the suburbanite leans toward that form of coercion. Alternatively, homeowners may force the passage of local laws to control the farmer's operations. Such actions make a farmer's profession that much more difficult. It may seem especially unfair to the farmer, since his farming activity is likely to have preceded the suburban homes in the area.

State governments have responded to the farmer's plight in two basic ways. One is the creation of agricultural districts within the states. By the end of 1980, California, New York, Virginia, Maryland, Illinois, and Minnesota had each created programs to allow establishment of *agricultural districts*; no two programs were the same.

Three features of these agricultural district programs seem to be most important. The first is the allowance of differential assessment of farms within the district. That is, the farms in such a district are taxed on the basis of their current use, not on the basis of their market value as homesites. Four of the six states had such tax laws for agricultural districts. A second important feature of these districts is that local governments in the districts are prevented from establishing laws that interfere with a farmer's activities as long as the farmer's operations do not threaten the public health or safety. Five of the six states had such a feature incorporated in their programs for agricultural districting. A third common feature of these laws is that public investment in an agricultural district is limited by law as a means to deter the attraction of development; four of the six state programs included this concept.

Parallel in intent to the creation of agricultural districts is the passage in a number of states of general right-to-farm laws. By 1980, sixteen states had passed such laws as a means of protecting farmers from nuisance suits and having to plead their cases in court. The laws shield farmers from both the expense and annoyance that come with such proceedings.

Table 29.2 Programs for the Purchase of Development Rights

Jurisdiction	Year first funded	Acreage under easement as of 1981	Total authorized funding ($ million)
Suffolk County, N.Y.	1976	3,214	21.0
Maryland	1977	2,400	6.3
Massachusetts	1977	1,349	15.0
Connecticut	1978	2,585	9.0
Howard County, Md.	1978	0	1.5
Burlington County, N.J.	1979	810	3.0
King County, Wash.	1979	0	50.0
New Hampshire	1979	0	3.0
Southampton, N.Y.	1980	0	6.0

Source: National Agricultural Lands Study, 1981.

The laws differed from state to state, but two basic themes show up. One theme is the prevention of the passage of local laws that unreasonably interfere with farming activities. Of course, laws designed to protect either the public health or public safety would always be in order. Five of the eleven states with laws controlling local government statutes on farming chose to apply the law only in agricultural districts. The other major theme of state right-to-farm laws is the protection of farmers from private law suits by limiting their liability.

Right-to-farm laws are new; the earliest was passed in 1971 in New York State. Thus, the impact of their enforcement on preserving farming is as yet unknown and possibly unknowable. Further, the boundary between a nuisance and endangering of the public health can be faint.

Not only have state governments taken up the cause of the beleaguered farmer; but in a number of cases local government has been sympathetic to the farmer's exposed situation as well. In these cases, the zoning tool, a local government power, has been used to protect farming. Agriculture zones have been added to the land-use categories shown in Figure 29.1. Such agriculture zones can either permit nonagricultural uses or exclude them.

Can any of the techniques we have discussed reverse the trend toward farmland loss? The complete results will not be in for several decades. In the meantime, the programs should be watched carefully for early signs of success or failure.

Preserving Land That May Not Have Agricultural Value

Purchase and transfer of development rights by private corporations. The purchase of development rights is not restricted to governments, although many states are watching the Suffolk County experiment. Such rights may also be purchased by private individuals, by corporations, and by nonprofit institutions. In fact, any legal entity that can hold property can hold development rights as well.

Of what use could these development rights be to corporations? To understand how a home-building corporation could make use of purchased development rights, we need to reemphasize the nature of the transaction on Long Island. Suffolk County purchased the development rights on certain farms; those develop-

ment rights now in the hands of the county are likely to *never* be exercised. It is possible, however, that the development rights purchased by a corporation on a parcel of farmland in one place can be used by that company *elsewhere* than on that parcel. Once used elsewhere, the land from which they were withdrawn can no longer be developed and is thereafter taxed only as agricultural or undeveloped land.

The first use of this concept to preserve rural lands took place in the small town of Saint George, Vermont. This town, which then had no commercial development whatsoever (no store, gas station, post office, etc.), was in the path of the spreading suburbs of Burlington, and its population had gone from about 100 in 1960 to 500 in 1970. To guide its future growth, the town purchased 48 acres of land to be used as the nucleus for future homes and commercial enterprises. These 48 acres were made available to developers for building, but only if the developers agreed to keep land open elsewhere in the town.

A description of how the town will use this 48-acre project area to guide the expected development reads:

> To achieve the objective of concentrating settlement and preserving the rural character of most of the rest of Saint George, the town may oblige a developer to transfer to the town development rights purchased from owners outside the project area in exchange for the opportunity to develop in the core village area. For example, a developer wishing to construct twenty units of housing in the village area would have to purchase twenty acres of land zoned at one family to the acre elsewhere in Saint George and transfer his acquired right to twenty units of housing to the project area. The twenty acres from which the rights were transferred will remain open land in perpetuity or until the town releases it to meet future needs. The land will be taxed only at its value as undeveloped land. (Wilson, 1974, p. 51)

It should be noted that those 20 acres purchased elsewhere need not have been land suitable for building. It could be swamp, ravine, mountaintop, or hillside. The point is that a one-for-one trade of preservation area for housing development is taking place.

Conservation easements. State and local governments have a number of options available to them, short of outright purchase, to prevent the development of agricultural or wild land. One such option is to secure a conservation easement of the land, a statement attached to the deed specifying the uses to which a parcel of land may (or may not) be put. These uses may include whether or not buildings may be erected, the maximum height of buildings, the cutting of trees, the use of billboards, and the like.

How can a government go about attaching such statements to deeds? The simplest way is for the community to buy the property outright when it comes up for sale. The local government attaches the statement on allowable uses and then puts the land back on the market. The land will probably sell for less than the community paid for it because its use is now restricted. The loss in value, however, may be viewed as the amount the community had to pay to put the restriction in the deed.

Although the method of purchase/resale is the easiest way to attach a deed restriction to the land, desired parcels are not likely to come on the market just at the time when a community is ready to act. More than likely, the local government will have to convince the owners that donating the restrictions will bring them benefits. The benefits may be direct payment for the attachment by the government or may be tax advantages in terms of federal estate taxes, federal income-tax deductions, or reduction in assessed valuation and hence in property tax.

Summary

Zoning involves classifying land in terms of its appropriate uses. The most common standard is whether such uses are compatible with neighboring land uses. Common categories are industrial, commercial, multifamily residential, single-family residential, and agricultural. Zoning usually acts to protect real estate investments. It can lead to lack of diversity in community development. Certain types of zoning, enforced by governments, can be used to protect the urban and suburban environment; such types include natural area zoning, cluster zoning, hazard zoning, and growth management plans.

Other types of zoning, using economic incentives as the "carrot" for their adoption, are suitable for preserving rural land. Prime farmland is being lost to development at the rate of 1 million acres per year. One reason is tax policies in which agricultural land is taxed

at its market value. Another is the burden of inheritance taxes. A third is the rising land values that occur as development nears agricultural land. To avoid these problems, land can be taxed on current use rather than market value; farmers can incorporate; and local governments can buy the development rights to rural land, but not the land itself, from the owners. The designation of agricultural districts, an agricultural zoning category, and the passage of right-to-farm laws can allow farmers to carry on their needed farm activities and also act to prevent suburban development in agricultural areas.

Rural land that may or may not have agricultural value can be preserved by transfer of development rights, by private corporations, and by conservation easements.

Questions

1. What is the purpose of zoning? Contrast the stated purpose and the actual purpose. How can zoning be used to protect the environment and people?
2. Explain how taxation and the growth of suburbs are "eating up" prime farmlands in the United States.
3. Describe some ways of keeping privately held land rural. What advantages do such methods have? That is, why not just have the government buy the land outright?

Further Reading

Leopold, A. *Sand County Almanac.* Oxford University Press, 1949. Reprinted, Sierra Club/Ballantine, 1970.
Leopold observed nature and humans through 12 months on his Wisconsin farm. This highly readable, classic work has a fine essay, "The Land Ethic," in which the author describes the relationship that ought to exist between people and the land.
National Agricultural Lands Study (a report in four volumes from 12 federal agencies). Washington, D.C.: U.S. Superintendent of Documents, 1981.
Very readable and nontechnical; a comprehensive study on the transition of farmland to other uses.
International Regional Science Review, **7** (3) (December 1982). Special Issue on Regional Development and the Preservation of Agricultural Land.
The issue contains five articles on the problems of urban sprawl and its consumption of farmland.

References

Platt, R. *Land Use Control: The Interface of Law and Geography.* Resource Paper 75–1. Washington, D.C.: Association of American Geographers.

Stover, E., ed. *Protecting Nature's Estate.* Bureau of Outdoor Recreation, U.S. Department of the Interior. Washington, D.C.: U.S. Government Printing Office, 1976.

President's Council on Environmental Quality. *Untaxing Open Space.* Regional Science Research Institute. Washington, D.C.: U.S. Government Printing Office, 1976.

Whyte, W. H. *The Last Landscape.* Garden City, N.J.: Doubleday, 1968.

Wilson, L. "Precedent-Setting Swap in Vermont," *Journal of the American Institute of Architects,* **61** (3) (March 1974), 51.

Miner, D. "Land Banking in Canada: A New Approach to Land Tenure," *Journal of Soil and Water Conservation* (July/August 1977), p. 158.

America's Soil and Water: Conditions and Trends. Washington, D.C.: Soil Conservation Service, U.S. Department of Agriculture, 1980 (booklet).

Pease, J., and P. Jackson. "Farmland Preservation in Oregon," *Journal of Soil and Water Conservation* (November/December 1979), p. 256.

Dunford, R. "Saving Farmland: The King County Program," *Journal of Soil and Water Conservation* (January/February 1981), p. 19.

Ognibene, P. "Vanishing Farmlands," *Saturday Review* (May 1980), p. 29.

Pierce, J. "Conversion of Rural Land to Urban: A Canadian Profile," *Professional Geographer,* **33** (2) (1981), 163.

CHAPTER THIRTY

Preserving Public Natural Areas

The Public Preservation Movement

Why Should Lands Be in Public Hands?/A Brief History of the Federal Preservation Effort

Federal Natural Areas

A Fragmented System/National Wildlife Refuges/National Forests/The National Park System/Wilderness/National Wild Rivers and National Trails

Problems of Accessibility

Reviving an Old Tradition of Land Preservation

Expanding the System of Natural Areas

The Land and Water Conservation Fund/Donations and Tax Benefits/Conservancies or Land Trusts/Coordinated Federal and Private Actions

CONTROVERSIES:
30.1 *The Baxter Fire—Would You Let It Burn?*
30.2 *Wilderness*
30.3 *Should Wilderness Be Accessible?*
30.4 *The ORV Fight*

The Public Preservation Movement

Why Should Lands Be in Public Hands?

State and national parks, state and national forests, national wildlife refuges, national monuments, national seashores, national lake shores, wild and scenic rivers—the list of ways that the public sets aside land for recreation and preservation is longer still.

Wetlands may be rescued from development because they serve migrating birds on their annual journeys up and down the continent. Wildfowl may winter at such places or simply stop at them en route. A forest or desert may be set aside for its unique vegetation. The Redwood National Park and numerous California state parks are devoted to preserving the two species of redwood tree, the largest living species. The Joshua Tree National Monument and the Saguaro (cactus) National Monument are used to preserve remarkable desert plants. Some parks are used to preserve stunning natural landscapes. Death Valley National Monument, Yellowstone National Park, and Grand Canyon National Park are examples that come immediately to mind. Other parks, such as the Everglades, are chosen to preserve whole ecosystems of plant and animal species. And for every preserved area, it would not be an exaggeration to say that there is another area yet to save.

Save from what? From timbering in the case of the redwoods; from vacation homes in the case of the wetlands; from mining in the case of Death Valley; from commercial exploitation in the case of Grand Canyon and Yellowstone; from being drained or made into a jetport in the case of the Everglades. The threats are abundant. Although sometimes innocent, often they are from those who are ignorant of the values that other people place on such natural resources. Threats to these landscapes and ecosystems continue to occur.

The road to preservation of our landscape, our species, and our ecosystems has been marked by several disasters, not all in the distant past. Our first spectacle was Niagara Falls, which fell quickly to commercial development around 1820–1840. Mills were built at the foot of the falls, the surrounding forests were cut, and tourist operators set up at the rim of the falls. Visits to the falls were highly commercialized. One pair of English visitors who observed the commercial activities at and around the falls during this period called for protection of the falls as "the property of civilized mankind." In California in the 1850s, one of the largest of the giant sequoia trees in the mountains of southern California was cut down and transported to New York and to Britain for exhibition. Such acts led to pressure to put the Yosemite region in public hands.

As late as the 1960s, over 100 years later, the awesome and beautiful rock walls of Glen Canyon were sealed finally from view by construction of a massive dam. And even into the 1980s, we find mining of existing claims officially approved in Death Valley by act of Congress. Preservation seems to require not only the initial act of setting aside, but constant watchfulness as well. As if to prove the point, in the early 1980s the Reagan administration and, in particular, the Secretary of the Interior, began an evaluation of wilderness and even of national parks for their mineral potential.

A Brief History of the Federal Preservation Effort*

Creation of the federal parks. In 1864, the tradition of federal creation of park lands began. The Yosemite Act gave 44 square miles (113 km²), which had been in the possession of the United States simply as public land, to the State of California to preserve for the public. The act was introduced by a senator from California who justified his unusual proposal by noting that the lands were "for all public purposes worthless," by which he meant that no mineral resources or usable waterpower could then be derived from the land. The preserve included the Yosemite Valley, the towering mountains that formed its sides, and one other tract. The giant sequoia was at last protected. In 1890, the park was taken into federal possession by an act of Congress.

In 1872, Congress created a park in the Yellowstone region of Wyoming. The park preserved not only the fabulous falls and gorge of the Yellowstone River but the unique geysers of the region, of which Old Faithful is the most famous example. This time the park was put directly under federal direction, presumably because Wyoming was still only a territory. Only a year before the creation of the park at Yellowstone, the fabulous geysers had been in danger of being claimed as private property, apparently because of their value as a tourist attraction. Nevertheless, the oratory in Congress that supported the act stressed the "worthlessness" of this park land. Again, the lack of value referred to a lack of mineral resources or of potential agricultural use. In the 1980s, the geyser region was again threatened, this time by the possible development of geothermal power in the area directly adjacent to the park. (See Controversy 23.1.)

An 1895 act of Congress prohibited hunting in Yellowstone National Park. However, the thrust of wildlife protection by setting aside preserves devoted only to animals was still in the future. Indeed, the protection provided by National Wildlife Refuges evolved in a tradition entirely separate from that of the National Park System.

The low commercial value of the lands committed to parks was a theme heard again and again. Another theme was that the areas represented one aspect of America's claim to greatness. The beauty of the parks was often compared to the architecture and art of Europe. Though we did not have the wealth of history of the European continent, we nonetheless had our own natural history treasure that was worth preserving. In its day, this viewpoint served us well, for the notion of preserving an ecosystem (then still undefined) could never have occurred to people in that era.

No one was concerned then with preserving a forest, for we had at the time what seemed an abundance of land, species, timber, and mineral resources. Wilderness was not far from anyone's doorstep. Indeed, in those early years, the nation saw its business as taming and subduing the land rather than preserving it intact. The realization that land and forest were limited resources dawned only gradually; and when it did occur to people, it was in the perspective of making profit. Mark Twain poked fun at the madcap commercial atmosphere of land grabbing that occurred after the Civil War. "Buy Land; they ain't making it any more!" one of his characters suggested.

Forest preserves and the U.S. Forest Service. In the period 1880–1910, a genuine appreciation of the role of natural resources and the public land began to surface. Individuals like Franklin Hough, Bernhard Fernow, and Gifford Pinchot helped to shape a policy of *forest protection* as opposed to scenery or wildlife protection. Thus began a tradition different from the national parks movement and different from the movement to protect wildlife. Their efforts, which came out of the Division of Forestry within the Department of Agriculture, were directed at forest protection and at soil and water conservation.

With the passage of the Forest Reserve Act of 1891, the president was empowered to set aside areas of land in the public domain as public reservations to ensure that adequate timber would be available to the nation in the future. Although President Harrison created, within two years, 15 forest preserves with over 13 million total acres (5.2 million hectares), no means was actually provided to protect these reserves until 1897. In that year, Congress passed an amendment to an unrelated appropriations bill that did offer protection for the reserves. The Pettigrew Amendment, as it is known, directed the Secretary of the Interior to make

*This discussion of the history of the national parks movement is drawn from "The National Park Idea" by Alfred Runte, *Journal of Forest History,* **21**(2) (April 1977). The article is insightful, exciting, and beautifully illustrated. It reaches back and recaptures an early era in the environmental movement.

rules for protection of the reserves and authorized the sale only of mature or dead timber within the reserves. The timber was to be marked for removal before cutting could begin. The amendment of 1897 shaped the nation's forest management policies for 63 years, until the Multiple Use Act of 1960 expanded its provisions.

Effective protection, however, did not exist until 1905, when the administration of the forest preserves was transferred from the Department of the Interior to the Bureau of Forestry within the Department of Agriculture. The man who was given the responsibility for the 63 million acres (25 million hectares) now in the preserves was Gifford Pinchot. He renamed his agency the U.S. Forest Service, the name still in use today. The preserves were renamed as well; they became the National Forests. By 1907, there were over 150 million acres (61 million hectares) of preserves in the National Forests.

Wildlife protection and the refuge system. Still another tradition of land preservation was growing at about the same time. Whereas the purpose of the National Parks was to preserve magnificent scenery and the purpose of the National Forests was to preserve timberland, the purpose of wildlife refuges was the preservation of animals by protection of their habitat.

The federal government did not become active in the preservation of wildlife until the turn of the century. In 1900, Congress passed a law instructing the Secretary of Agriculture to take steps to protect and restore game and wild birds. This was the time at which the nation was witnessing with sadness the disappearance of the passenger pigeon, and this loss undoubtedly influenced passage of the law (see Figure 30.1).

In 1903, the United States created its first preserve for animals, a refuge on Pelican Island off the Florida coast. This was soon followed by the establishment of wildlife ranges in the Wichita National Forest (1905) and the Grand Canyon National Forest (1906). In 1906, Congress prohibited hunting on all preserves that were now to be set aside for the protection of wildlife. In 1908, Congress set aside a National Bison Range in Montana for the threatened buffalo.

The fate of the passenger pigeon helped move Congress to pass the Migratory Bird Treaty Act in 1918, which placed migratory game under the protection of the federal government. Two years later, the Supreme Court declared that the federal government's power to regulate migratory wildfowl came before that of the states. Justice Holmes, writing for the majority, noted that without federal regulation, "there soon

Figure 30.1 Long before the last passenger pigeon died, it was clear that their numbers were declining every year. Yet, despite many efforts to save it, the passenger pigeon became extinct. At some unrecognized moment, a crucial point was passed, the opportunity to save this species lost forever. This etching of the 1870s depicts hunters in Northern Louisiana shooting for sport at a migrating flock of passenger pigeons. (From S. Bennett, *Illustrated Sporting and Dramatic News*, London, July 3, 1875, p. 332.)

might be no birds for any powers to deal with . . . It is not sufficient to rely upon the states. The reliance is vain."★

The 1918 Act, however, had not provided for the acquisition of bird habitats. This defect was corrected by the 1929 Migratory Bird Conservation Act. The act authorized the Secretary of the Interior to create refuges for migratory wildfowl and to operate these refuges as "inviolate sanctuaries" (Figure 30.2). A portion of the funds to purchase wildfowl sanctuaries from states and private owners has come from the sale of hunting stamps. Revenue for types of refuges other than wildfowl refuges has not been so readily available.

Federal Natural Areas

A Fragmented System

These three traditions—the National Parks, the National Forests, and the National Wildlife Refuges—all evolved over the years. Uses and purposes have changed and blended as new needs and new pressures have emerged. Today these three systems are still managed by different agencies, in keeping with their differing purposes at the outset. The National Park Service administers the National Park System within the Department of the Interior; the U.S. Forest Service manages the National Forests in the Department of Agriculture; and the U.S. Fish and Wildlife Service is responsible for the system of National Wildlife Refuges in the Department of the Interior. Table 30.1 lists the land areas these agencies administer, as well as the supply of state parks and forests.

Not all the land the federal government holds for the public is in the National Parks, the National Forests, or the National Wildlife Refuges. A considerable portion of the government's holdings are in the National Resource Lands, some 174 million acres (70 million hectares) of public land in the lower 48 states and another 70 million acres (28 million hectares) in Alaska. These lands, which are supervised by the Bureau of Land Management (BLM), are the remainder of the "public domain." The bulk of the lands we refer to as the public domain are those that remain from the great

Figure 30.2 The fate of wild birds, as seen from the 1930s. This series of cartoons was the work of J. N. Darling, a national syndicated cartoonist with the *Des Moines Register*. In the middle 1930s, because of his interest in wildlife protection, Darling was appointed Chief of the U.S. Bureau of Biological Survey. His efforts were instrumental in building and strengthening the National Wildlife Refuge System. (Courtesy of the *Des Moines Register*.)

land purchases of the past, such as the Louisiana Purchase. Much of these lands were granted to the railroads, and settlers acquired a portion of them by homesteading. The Forest Preserves and the National Parks and Wildlife Refuges withdrew other lands from the public domain. But the land that has never been sold and that remains in the hands of the BLM is vast and often trackless.

Some of the land held by the BLM is timberland. By a quirk of history, some 2.4 million acres (about 1 million hectares) of the most productive timberland in the United States fell into the hands of the Bureau of Land Management. Two large tracts of land granted to

---

★Quoted from *Evolution of National Wildlife Law*, prepared for the Council on Environmental Quality by the Environmental Law Institute, 1977.

Table 30.1　Land Areas Administered by State and Federal Agencies

Agency/office	Name of land holding	"Lower 48"	Alaska	Total
		Millions of acres		
Bureau of Land Management	National Resource Lands	174	70	244
U.S. Forest Service	National Forests	168	23	191
National Park Service	National Park System	27	52	79
U.S. Fish and Wildlife Service	National Wildlife Refuges	13	76	89
State and local parks and forests	——	about 25	——	——

railroads in Oregon reverted back to the Bureau of Land Management when a violation of the grant occurred. These are the 2.4 million acres of highly productive timberland that the BLM manages today. They apparently are administered by the BLM from much the same viewpoint as the Forest Service.

The National Resource Lands, as BLM lands are called, are available for mining, grazing, and timbering. Within the last decade, the recognition of valuable wilderness in the National Resource Lands has prompted an evaluation of these lands for their wilderness and recreation potential. The management of the National Resource Lands for mineral purposes has been a long-term issue. In the pages to come, we will survey the management of natural areas by the federal agencies involved in their protection.

National Wildlife Refuges

Although originally set aside as "inviolate sanctuaries," the National Wildlife Refuges have evolved over the years so that their use includes other purposes. The Migratory Bird Hunting Stamp Act of 1934 set up the sale of "stamps," or permits to hunters. Revenue from the sale of hunting stamps has been used to finance in part the purchase of refuges for wildfowl. The use of these revenues from hunters may have influenced decisions by Congress in 1948 and 1959 to permit the Secretary of the Interior to authorize hunting on the refuges.

Although hunting was originally limited to no more than 25% of the refuge area by the 1948 legislation, the limit was raised to 40% in 1959. The Secretary was allowed to use his or her judgment to decide whether hunting on a particular refuge was compatible with the purposes of the refuge. No challenge to this use has yet been successful in a court of law. In 1962, the Refuge Recreation Act authorized the Fish and

Wildlife Service to open wildlife refuges to public recreation. The authorization gave to the Secretary of the Interior the responsibility to determine whether recreation at a particular site would interfere with the basic purpose of the refuge.

In 1966, the wildlife system, which consisted of game ranges, wildlife management areas, wildlife refuges, and other units, was consolidated into the National Wildlife Refuge System under the administration of the Fish and Wildlife Service (Figure 30.3). The allowable uses of these areas were expanded as well. Now in addition to bird hunting and public recreation, the uses, when found to be compatible with the central purpose of the refuge, could include hunting of other species, fishing, and public accommodations. Just as the National Forests were expanded to become multiple-use lands in 1960 (see next section), the National Wildlife Refuges were becoming multiple-use, with the added proviso that their original purposes as species habitats should not be compromised. This notion of accompanying but secondary uses of the refuges has led to their being referred to as "dominant-use" lands, as opposed to "multiple-use."

National Forests

The multiple-use concept. The term **multiple-use** has come to be applied most frequently to the use and management of the National Forests. Indeed, an act of Congress known as the Multiple-Use-Sustained Yield Act was passed in 1960 to define in a formal way the meaning of multiple-use management in the National Forests.

The act stated that the National Forests were to "be administered for outdoor recreation, range, timber, watershed, and wildlife and fish purposes." It further stated that "The establishment and maintenance of

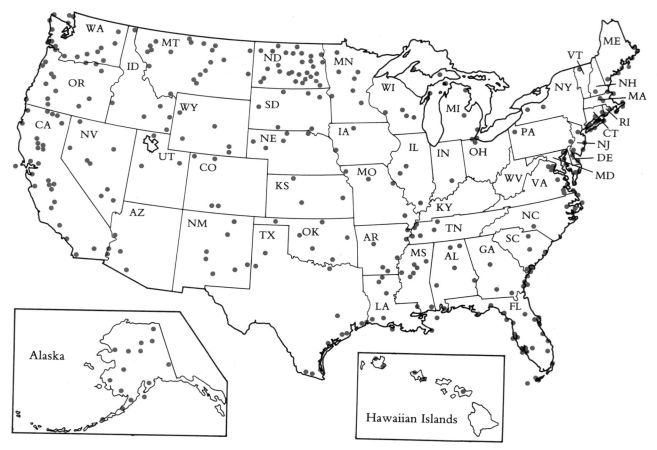

Figure 30.3 Map of the National Wildlife Refuge System, September 30, 1981. Most, but not all, of the wildlife refuges managed by the Fish and Wildlife Service are included in this map. The Fish and Wildlife Service publishes a directory of the nearly 400 refuges. The directory lists each refuge by state and county, noting its size, date established, and primary species protected. (Map courtesy of the Fish and Wildlife Service, U.S. Department of the Interior.)

areas of wilderness are consistent with the purposes and provisions of this act." The National Forests were also to be managed for a "sustained yield" of forest products. **Sustained yield** was defined as "the maintenance in perpetuity of a high-level annual or regular periodic output . . . without impairment of the productivity of the land."

Surprisingly, the act, which is now the principal policy directive of the Forest Service, was opposed both by conservation organizations *and* by the timber industry. In fact, the Forest Service saw the law as a means of making legitimate the management practices

it had followed since the earliest days of the agency. The National Forests had always been administered for all these purposes, and for mining, in addition. The opposition by both conservationists and by timber people stemmed from their suspicion of the motives of the Forest Service.

Conservationists were concerned that the act would be the tool by which the Forest Service could authorize extensive new cutting in the National Forests. Because of pressure from the Sierra Club, a nationwide organization of conservationists and environmentalists, the statement on wilderness was added to the act. The Sierra Club was deeply distrustful of the motives of the Forest Service in supporting the act and never supported the bill, though the organization had a hand in shaping it. The distrust of the Forest Service is explained by Steen (1976):

> Several events coincided during the 1950s which cost the Forest Service the support of many conservationists. Since Pinchot's time, the national forests had been held in reserve to supply lumbermen when privately owned timberlands had been cut over and were in the regrowth cycle. Year after year and chief after chief reaffirmed the policy to log a substantial portion of the national forest system—at the appropriate time. The appropriate time followed World War II, when increases in population and affluence concurrently multiplied recreational pressures on all public lands. Couched in terms of timber famine and devastation, decades of propaganda by the Forest Service to justify federal regulation of logging had created the popular image of rangers protecting forests against rampaging, greedy lumbermen. Now the postwar public came to the forests—camping, fishing, hiking—and "discovered" logging in the national forests. To many of the public, national forests and national parks were the same, and logging either was bad. Implementation of long-term timber management plans collided with the results of an extremely effective public relations program against "destructive" logging. As a result, the public felt deceived. (p. 302)

Whereas conservationists were concerned with excessive logging being allowed in the National Forests, industry, in contrast, saw the bill as giving weight to public uses. Such uses, they felt, could only work against the interests of the timber companies.

Four years later, in 1964, conservation interests were finally successful in passing the Wilderness Act. This act made it possible for areas within the National Forests to be specifically set aside as wilderness areas only. The act was essentially a restatement of distrust of the intentions of the Forest Service. That distrust, which surfaced first in 1960, reflected a concern that the Forest Service would not protect wilderness unless specifically ordered to do so.

The Multiple-Use Act of 1960, while it affirmed wilderness as a legitimate "use," did not order the establishment of such areas. The Wilderness Act gave Congress the right to designate areas of the National Forest as wilderness and hence protect them from logging. By 1982, 19.7 million acres (7.9 million hectares) of the National Forests had been included in the National Wilderness Preservation System from national forest land in the "lower 48" states.

Timber management in the National Forests. In addition to the order from Congress to the Forest Service to set aside wilderness areas, court cases arose over the Forest Service's application of the Multiple-Use Act. Specifically, these cases claimed that the Forest Service was not giving due consideration to wilderness in its timber sales.

Controversy on the way in which the Forest Service has administered sales of timber has focused not only on where timber sales have been allowed but also on the manner in which trees have been cut. One specific method has been attacked again and again: the practice known as **clear-cutting**. Clear-cutting means simply the cutting of all trees in an area, regardless of size, quality, or age. The practice of clear-cutting in National Forests has grown through the years. Lumber companies prefer it because it is obviously more economical than selective cutting of mature and dead trees. Since there is a cost of moving lumberjacks and equipment to each site, the companies prefer to take all the timber they can get before moving on to a new area.

Nonetheless, there are costs to the public when clear-cutting takes place. In addition to the visual impact of a forest cut to the ground, clear-cutting can affect rainfall retention in the area, and it can affect erosion. More rapid runoff over wide areas can lead to widely fluctuating stream flows, including damaging flood flows. High runoff rates can speed the erosion process as well. Furthermore, species habitats are completely wiped out by clear-cutting, and wilderness values are completely destroyed in the region of the clear-cut (Figure 30.4).

The timber companies argue for clear-cutting for reasons other than economics. One commercial timber species, the Douglas fir, is thought to grow best in direct sunlight. Company foresters point to clear-cutting as the way to assure optimum growth of a new forest of Douglas fir, but the claims that direct sunlight is necessary for the regrowth of the Douglas fir have been disputed by forest scientists.

One aspect of clear-cutting deserves mention be-

Figure 30.4 A poorly designed clear-cut can ruin the view of a forested mountain area. This photo was provided by the Bureau of Land Management, but does not necessarily depict activities on BLM lands. Clear-cutting has been a recent practice on Forest Service lands.

cause of its bearing on the ecological system of the harvested site. Recall that the clear-cut virtually wipes out the entire habitat. The species that return to an area regrown from a clear-cut are expected to be somewhat different from the species that existed prior to the clear-cut. The reason is that the regrowing trees will all be about the same age. A natural forest—one that has had time to mature through several growing cycles—has trees of many different ages. The forest regrown from a clear-cut is referred to as an even-aged stand; the forest with trees of different ages is called an uneven-aged stand. The uneven-aged stand is thought to be ecologically superior in terms of the balance of species present.

Certain areas are particularly harmed by clear-cutting. A clear-cut on a steep slope may lead to erosion of the best soil off the slope. With the best soil eroded away, a long time is required for the forest to return. An area that includes a stream channel can also be damaged if the trees are cut too close to the banks. Erosion of the stream banks damages the forest, but the stream is damaged as well by the heavy load of soil particles that it carries. Removal of trees also opens the stream to sunlight, which increases the temperature of the flowing water (Figure 30.5).

Other methods of cutting can be and are used by the timber industry. These include selection cutting, seed-tree cutting, and shelterwood cutting. **Selection cutting** removes mature timber, usually the oldest and

Figure 30.5 Construction of a logging road near a creek. This exposure of loose, unanchored soil causes rapid erosion from the land. The particles will enter the creek, producing soil-choked waters. Sediment in the creek can decrease the ability of the stream to carry high flows, causing flooding downstream. (This photo was provided by the Bureau of Land Management but does not necessarily depict activities on BLM lands.)

Figure 30.6 Seed-tree cutting prior to log removal. Some trees are left to seed and regrow the cut-over area. (Bureau of Land Management)

largest trees. The cutting cycle involves a return to the area for timber cutting every 5–20 years to remove the newly generated mature timber. This method of cutting produces an uneven-aged stand and has been widely practiced.

Seed-tree cutting is a poor name for this logging method, which actually leaves seed trees uncut (Figure 30.6). The seed trees left to regenerate the stand may be left singly or in groups. Although seed-tree cutting has been used for commercial harvesting in Montana, it is still relatively experimental.

Shelterwood cutting proceeds in three stages. In the first, or *preparatory* stage, dying trees, defective trees, diseased trees, and trees of unwanted species are removed, leaving space for new trees to grow. About 10–15 years later, a stage called *seed cutting* opens the stand further so that seedlings can receive adequate sunlight and heat. In the final phase, called *shelterwood removal cutting*, when seedlings have been established, the remainder of the mature trees are cut.

It is important to note that without supervision, any of these logging systems can produce unwanted results. As an example, selection cutting, if unsupervised, can be used not to select mature trees as it is supposed to do but to select the best or most marketable trees.

In 1973, clear-cutting was taken to court, and it was found that the 1897 law that formed the basis for management of the forests still has an impact in the present. In *Izaak Walton League* vs *Butz*,★ a district court examined the timber harvesting that the U.S. Forest Service was allowing in the Monongahela National Forest in West Virginia. The court declared that clear-cutting in the National Forest, the removal of all trees in an area regardless of age, was in violation of the 1897 law. The 4th Circuit Court of Appeals upheld that decision in 1975, citing the rules of the 1897 act, which limited cutting to the trees individually marked and trees of mature growth or dead trees.

These interpretations of the original law, however, were hollow victories. Industry and the Forest Service wanted other options, and Congress responded to the pressure. In 1976, the Forest Management Act finally replaced the original law, putting in the hands of the Forest Service the responsibility to manage the national forests as it judged best. Although the Forest Service was instructed by the act to consider topography, type of forest, size of cut, climate, and the like, the result has so far been a return to the practice of clear-cutting so clearly prohibited in the 1897 law. (See Box 30.1.)

How much timber comes from public lands? In the 1900s, virtually all the lumber produced in the United States was cut from private forestlands. After World War II, housing pressures forced the government to open the National Forests to companies in a serious way, although private logging had been going on in the National Forests to a minor extent since the establishment of the preserves. The share of the annual production of lumber from public lands had reached 15% by 1950. By the early 1970s, the lumber from public lands accounted for 40% of annual timber production. In the early 1980s, that percentage dipped substantially to about 20%, as a severe recession and historically unprecedented high interest rates slashed home building activity.

The Forest Service has obviously "upped the cut" from the period before World War II, and the explanation lies in part in the initial mission of the service. The preserves, now the National Forests, were set aside for the day when timber supplies on private lands fell so low that the preserves would be needed to meet de-

---

★Earl Butz was then Secretary of Agriculture and thus was in charge of the U.S. Forest Service, which is housed within the Department of Agriculture.

Box 30.1   The Tragedy of One-Shot Forestry

The developing pattern of forest management in the 50 eastern national forests is like a personal tragedy to me. I am puzzled and dismayed by the five new management plans I have seen for national forests from Vermont to South Carolina. They call for clear-cutting and even-aged management for most of the mixed hardwood types of forest area close to heavily populated areas. The plans show a relative insensitivity to nontimber values and propose an almost exploitative, primitive silviculture, with the objectives of short-term financial returns to the timber operator and ease of administration and future management by the Forest Service. . . . Our bureaucrats will designate a boundary of timber, and the operator will simply harvest all the trees, large and small, regardless of quality or species, and the next clear-cut will be 80–120 years hence.

How can this be? The Forest Management Act of 1976 has noble and true words about the importance of environmental values, about the relative unimportance of dollar values alone, and says in so many words that clear-cutting shall not be used unless it is the "optimum" method. But there is a catch. The discretion and the judgment are left to the professionals in the Forest Service. . . . The "discretion" has been exercised all in one direction—cheap timber production on the bulk of the forests. How can any professional agency prescribe *one* silvicultural system for the vast diversity of types, topography and climates that constitute eastern United States forests? . . . now the

Forest Service seems to believe it has a legal basis for almost universal clear-cutting, and it is acting accordingly. . . .

What are the consequences of large-scale block clear-cutting of eastern hardwoods? . . . The consequences . . . are greater waste of timber, lower environmental values and greater danger of damage to soil, site and water. Let me explain. Most of the individual trees in eastern hardwood forests are now below mature saw-timber size. But clear-cutting harvests these smaller trees, just when they are growing fastest. This is a waste. . . .

This crisis in forestry has been caused by failure to reconcile and balance the values of the timber as a commodity, the forest as an environment, and the integrity of the forest-site-soil-water ecosystem. Such a balance cannot be achieved by the one-shot system of clear-cutting and even-aged management now being proposed for the eastern national forests. . . . If the Forest Service will not or cannot reconcile and balance forest values, then Congress or the courts will have to act again.

Leon S. Minckler★

Source: Sierra Club Bulletin, July/August 1978.

★Leon S. Minckler is now retired from the U.S. Forest Service; he has taught environmental science and forestry at Virginia Polytechnic Institute and the State University of New York. He is considered one of the shapers of forestry opinion.

mands. Rising demand for lumber in the decades after World War II was the signal to open the forests. Now, with the prices of lumber at near record levels, there is pressure from many directions to keep the National Forests producing. Lumber companies, home builders, construction workers, and consumers want lumber at relatively low prices. Management of the forests is a very large problem in the national economy.

If we listen to the debate between conservationists and the timber companies, two terms are heard again and again: *sustained yield* and *allowable cut.* A **sustained yield** cut is a quantity of timber less than or equal to the natural increase (in cubic feet of lumber) occurring in the area in the past time period. The allowable cut is

the quantity of timber that a company is given the right to harvest from an area over a set period. A forest area that has an allowable cut equal to the new growth will, when the cut is complete, still have the same quantity of standing timber it had before the new growth took place. That quantity of new growth can be expected again, and the yield from the forest can be sustained.

Should the rate of cutting in the National Forests be allowed to approach the new growth? Remember that the National Forests are our final reserve of timber, to be cut when the yield from private forests is insufficient to meet demands. Although we are now cutting extensively in the National Forests, we must ask whether it is because of high demand or because private forests have been poorly managed. Private forestry efforts must be encouraged so that the use of the National Forests for meeting timber demands can be decreased. There is, in addition, the issue of whether we should allow timber cut from the National Forests to be sold in export to foreign nations, as it now is.

If the cut in the National Forests is equal to sustained yield, the stock of standing timber will not decrease, but it will be standing still in the face of growing demands. Unless private forests are substantially regrown, the pressure for cutting in the National Forests can only increase. Allowing the cut to approach sustained yield in the National Forests makes sense only if the private forests are being regrown. It may be that decreasing the allowable cut in the National Forests could help encourage adequate private forestry efforts. The problem is that people want houses now, and forests take a minimum of 25 years to grow.

Fire—No stranger to the forest. A careless match drops glowing on the forest floor amid fallen and dried needles. The tiny embers slowly ignite nearby needles; at first there are only embers, but dried needles are everywhere thick beneath the match; the embers grow hotter and tiny flames lick at the jumble of dead twigs. The twigs ignite and a flame leaps higher to catch fallen and dead branches. For a moment it looks as though the fire will go nowhere; then the bark on a dry branch crackles and flames encircle it. The branch itself begins to burn.

Now the fire spreads rapidly on many fronts; the small dry materials on the ground spread it quickly to more branches. At first, the living trees and fallen logs do not catch fire, but as the heat grows, the leaves, twigs, and stems of the live trees start to ignite. As the intensity of the fire increases further, fallen logs catch fire and even the trunks of live trees begin to burn. Though the live trees could not catch fire by themselves, in the heat of the forest fire, they will burn.

The fire front moves quickly through the dry underbrush, but fallen logs continue to burn long after the fire front has passed. Thus, though the fire may be stopped at a fire break, the fire within the forest continues. The embers from these burning logs may later feed a new phase of the fire.

Fire is no stranger to the forests. Lightning may set a forest fire as well as a match. When rain from high clouds evaporates before reaching the earth, a lightning bolt that strikes dry ground may ignite a blaze. Forest ecologists have been able to study the frequency of fires by observing fire scars embedded in the annual growth rings of trees. By careful counting of both rings and scars, they have been able to show a remarkable phenomenon. A study in a California forest found that forest fires reoccurred in about 8-year cycles as far back as 1685.

A study in the Boundary Waters Canoe Area of Minnesota, now part of the Wilderness System, found through a study of lake sediments that fires have reoccurred again and again in these forests for thousands of years. In the jack-pine forests of this region, fire is thought to be the agent that regenerates the stand, since the cones on the jack-pine are opened by intense heat. Once the cones are opened, the seeds fall on a ground where competing plants have been burned away and where sunlight finds better entry. The fire has made it possible for new jack-pine seedlings to grow. Fire, then, can be an agent to regenerate a forest.

Fire can also be a means of protecting the forest from a holocaust. It does so by consuming the litter and underbrush in the forest, thereby preventing a major buildup of deadwood that could lead to a truly deadly fire. If a light surface fire burns quickly through a small buildup of underbrush, the trees in such a forest may be only scarred, not destroyed. These are the scars ecologists counted in the tree rings to establish the frequency of fires. If the buildup of underbrush is reduced by light surface fires, any individual fire may not reach the searing intensity that consumes the living trees themselves. Thus, fire can be an agent of forest survival and regeneration—an idea that, after years of warnings from Smoky the Bear, may seem hard to accept.

Smoky's warnings are sincere nonetheless. For a fire to be of value to a forest, the forest must be in a condition that will not lead to complete destruction. If large amounts of dead fuel have accumulated over decades of careful protection, the hazard to the forest is great, as the fire is likely to reach an intensity that will consume live trees as well as underbrush. No one but the professional forester or forest ecologist can know the hazard or value of fire to a forest. Smoky is still telling the truth.

How is the tool of fire to be utilized? Prescribed or controlled burns have been used in some western forests to prevent buildup of dead underbrush, but in other forests, the careful protection of the Forest Service since the turn of the century has made prescribed burning impossible to use. Any attempt at burning in such areas would lead not to surface fires but to major fires, which are difficult to control. Such major fires would occur because of the large quantities of dead underbrush accumulated at ground level. In an area where a major fire has already occurred recently, however, controlled burning might be used in following years because of low levels of deadwood buildup on the forest floor. (See Controversy 30.1.)

The National Park System

The National Park System, now grown to more than 79 million acres (33 million hectares), is administered by the National Park Service. The Park Service was created in 1916 to administer a system of 15 National Parks and 22 National Monuments already in existence. The Park Service was charged with conserving:

> the scenery and the natural historic objects and the wildlife therein and to provide for the enjoyment of the same in such manner and by such means as will leave them unimpaired for the enjoyment of future generations.★

By the end of World War I, the Park Service was able to take over protection of the parks from the U.S. Army, whose troops had been used to build roads, fight fires, and generally protect these federally owned areas.

Although the Park System lists 287 units, the Park Service administers other areas as well. The more familiar responsibilities of the Park Service include the 38 National Parks, 61 National Monuments, 52 National Historic Sites, 16 National Historical Parks, 16 National Recreation Areas, and 14 National Seashores and Lake Shores. In 1982, the Park Service recorded 245 million visits to its units, over a 100% increase from the 1965 level of use (Figure 30.7). The less familiar responsibilities of the Park Service include areas that the Service does not even consider part of the Park System: the National Natural Landmarks, National Environmental Education Landmarks, and National Historic Landmarks.

Over the years, the Park Service has accepted other tasks in addition to marshalling people carefully through and past its holdings. One of its important

★*Preserving Our Natural Heritage*, prepared for the U.S. Department of the Interior by the Nature Conservancy, 1977, p. 32.

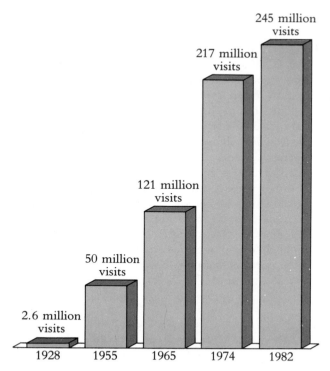

Figure 30.7 Visits to units of the National Park System. Though the National Park System has been growing in size, the growth in visitors is principally at the larger and more famous members of the system. (From Statistical Unit of the National Park Service.)

functions is the preservation of historical landmarks, both natural and man-made. The Service not only preserves but also restores early settlements and interprets early events for visitors. For instance, at DeSoto National Memorial, a relatively out-of-the-way site in Florida, a camp much like that of the early Spanish explorers has been set up. Weapons such as the crossbow and musket are displayed and demonstrated by the staff, and a movie re-reates the DeSoto expedition. Such educational efforts are common in other units of the National Park System.

The Park Service also conducts surveys of historic buildings and engineering works and maintains a register of sites significant for historical and educational reasons. In addition, the Service provides grants for surveys of sites possibly worthy of preservation. Under agreements with other agencies, the Park Service supervises recreational use of land not in their system.

New lands enter the National Park System by a number of methods, the most basic of which are (a) pur-

CONTROVERSY 30.1

The Baxter Fire—Would You Let It Burn?

Just think if you can, but don't strain yourself, do-gooder, of all the comfortable homes that could have been built with that wasted lumber.

E. D. Chasse, **Bangor Daily News,** *July 27, 1977*

If you could look at the park as something besides a giant woodlot you might not be so concerned about the burn. This beautiful park has burned many times in the past, and it will burn in the future.

Melvin Ames, **Bangor Daily News,** *August 19, 1977*

As we have seen, fire can play two important functions in the forest ecosystem. First, light surface fires clear away dead underbrush, which, if it were to accumulate, could cause massive and intense forest fires. In this century, our very success in preventing forest fires has led to conditions in many forests that could lead to terribly severe fires, fires far more severe than might otherwise have occurred. *Prescribed burning* has been used by professional foresters in some forests to prevent such buildups. Prescribed burning is undertaken with caution; days chosen are moist and windless; all forest uses are curtailed.

The second function of fire is replacement or renewal of forest stock. Fire may act as nature's instrument of forest renewal and has acted in this fashion for hundreds and thousands of years without the aid of the forester. Only recently have we discovered the significance of fire to the forest ecosystem. Such knowledge of the role of fire will increasingly be put to use in "managing" forests, but it is not yet clear precisely how the knowledge will enter into future decisions.

In the summer of 1977, a massive fire occurred in Baxter State Park in Maine. The fire was started by lightning; its intensity was fed by downed timber. Let

us, for a moment, reenact the scene: Prime recreation land is being destroyed, but the wilderness is renewing itself. Will you let the fire burn? You will want to know all the circumstances of the fire, how it has spread and how it is being fought, what resources may be destroyed.

The story began many decades ago when the will of Percival Baxter, former Governor of Maine, was opened by his lawyers. To the people of Maine, Baxter bequeathed a 200,000-acre (81,000 hectares) forested tract with 46 mountain peaks and ridges. The jewel of the tract was towering Mount Katahdin. This mountain, at 5,240 feet (1,600 m), is the northernmost point of the Appalachian Trail, the wilderness footpath that runs from Georgia to Maine. Governor Baxter's will instructed that the land is

> forever to be held by the State of Maine in trust for public park, public forest, public recreational purposes, and scientific forestry, the same also forever shall be held in its natural wild state and except for a small area forever shall be held as a sanctuary for wild beasts and birds.

Over the years, the park had been developed, so

that by 1977 it included an automobile road and seven campgrounds. The camps had space for trailers and tents, as well as bunkhouses and shelters. Because of the popularity of the park, reservations were suggested for those who needed to be sure of space. When the fire struck Baxter Park in 1977, it was a foregone conclusion that the park authority would fight it. The battle, however, was exceedingly difficult because of the rough terrain and the presence of numerous downed trees. The trees were in the aftermath of a fierce winter storm that had struck the southwest slope of Mt. Katahdin in 1974. Wind and ice had combined to snap tall trees in two, leaving a pileup of dead logs on the slopes. The Great Northern Paper Company had been allowed into the park to harvest the blowdown for use as pulpwood in the hopes of lessening the risk of fire, but the harvest had never been completed.

Legal action had been taken by the Baxter Park Defense Fund, and a judge's order brought the harvest to a halt in August 1976. While the judge did not forbid the removal of the downed trees, he directed that the heavy equipment being used had to be removed from the park. The heavy equipment was cutting into the forest floor, leaving gouge marks that might be used as roads or that could reroute streams into the depressions. Increased erosion from scarred areas would be likely as well. The equipment was removed by the paper company and the harvest of the blowdown came to an abrupt end. The following summer, when lightning ignited the blaze, much of the tangle of deadwood still lay on the ground.

The primary means chosen to fight the fire was to scrape out a 16-mile (27 km) fire line around the perimeter of the fire. Bulldozers and skidders were borrowed from the Great Northern Paper Company to cut the 12-foot (3.6 m) fire line; the company, which owns forest adjacent to the park, sent firefighters as well. Beaver seaplanes were also used in an effort to control the fire. The planes loaded water from nearby ponds and dropped the water in 150-gallon (564 L) loads on fire spots. On the ground, firefighters carried hoses and pumps into the forest. The pumps were set up on streams and ponds; the hoses, connected to the pumps, were dragged to the fire front.

In the heat of the blaze, the Baxter Defense Fund, which had prevented removal of the blowdown, let it be known that it planned once again to go to court. The group's aim, as before, was to force out the heavy pieces of equipment that were scarring the slopes of the park and crisscrossing the mountain streams in an effort to create the fire line. Although public reaction to the threatened suit was intense, the Defense Fund would have gone ahead with the legal action had the group not seen that the fire was in fact being brought under control faster than court action could be taken. The legal steps to cause removal of the machinery were not undertaken, and hence no injunction against the firefighters was obtained.

The fire line was completed by the eighth day of the blaze, and the fire was brought under control within the line. Nonetheless, weeks of watching were required to be certain that the fire was truly out, since fire can smolder in the dry litter that makes up the forest floor. The Baxter Park Authority, in an effort to slow erosion, built water bars in the deep gouges left by the bulldozers. These miniature log dams slowed the flow of water in the depressions and allowed earth particles to settle rather than be swept away by flowing water.

Although the fire was out, the controversy raged on. Letters and editorials in the newspapers of Maine insulted the environmentalists. Letters also appeared expressing sympathy for the concerns of the Baxter Defense Fund. The quotations that began this discussion were from such letters. There were also aftershocks, including court action by the Defense Fund to ensure a natural reclamation of the scars left from the firefighting; but eventually the controversy, too, appeared to go out. Like the remains of a forest fire itself, however, it is only smoldering beneath the surface. If the fire rages again, would you let it burn?

chase; (b) condemnation; (c) gift; and (d) exchange or transfer. Lands have had to be purchased frequently in recent years as the system of National Seashores has grown. Often, where no purchase can be negotiated, the Park Service has had to resort to condemnation. In legal terms, condemnation is made possible by the power of *eminent domain*, the privilege of the government to take lands for its use. Though lands may be condemned, they must still be purchased. *Just compensation*, to use the legal term, or *fair market value*, must be paid for lands so taken.

Over the years, the National Park System has benefitted significantly from gifts from states and from private individuals. Two gifts from the Rockefeller family have been especially important. One formed the nucleus of Acadia National Park in Maine, the first National Park in the east. The other gift provided land on the Island of St. John in the Virgin Islands (a U.S. territory) for the Virgin Islands National Park, a treasure of beach, ocean, and coral reef.

Lands may enter the system by what might be called "barter," in which land in the park system is traded for land outside. More important, however, is the transfer of land from the jurisdiction of another federal agency to the park system. Voyageur's National Park in Minnesota was acquired by gifts, transfer, purchase, and condemnation. Minnesota is donating about 50% of the 220,000 acres (89,000 hectares) of land and water in the park; the Forest Service, which controls 10% of the proposed park, is transferring another component to the Service; and the Park Service will have to purchase the remaining 40% of the area. Those purchases may be willing sales or may require condemnation.

The National Park System still bears the imprint of the early laws with which it began. Justified as the preservation of spectacles and grandeur, the parks were to serve people in a way that would enable them to see the most memorable scenes within the park. Since the parks were initially remote, this required development at the park itself. Such development included accommodations, food, information, and the like. Initially, such development was not intrusive, since relatively few people had the means to visit the parks. Now, however, the Park Service is increasingly called on to usher vast crowds past its most important possessions. Balancing its need to serve and its need to preserve is becoming increasingly difficult for the Park Service.

Wilderness

The Wilderness Act of 1964 cut across agency boundaries and called for the preservation of wilderness in all public land holdings of the government. Units that entered the wilderness system could be areas in the National Parks, National Forests, National Wildlife Refuges, and National Resource Lands (Figure 30.8).

Private holdings could be acquired as well. The language of the Wilderness Act explained its purposes in this way:

> In order to assure that an increasing population accompanied by expanding settlement and growing mechanization does not occupy and modify all areas within the United States and its possessions, leaving no lands designated for preservation and protection in their natural condition, it is hereby declared to be the policy of Congress to secure for the American people of present and future generations the benefits of an enduring resource of wilderness. For this purpose there is hereby established a National Wilderness Preservation System to be composed of Federally owned areas administered for the use and enjoyment of the American people in such manner as will leave them unimpaired for future use and enjoyment as wilderness, and so as to provide for the protection of these areas, the preservation of their wilderness character, and for the gathering and dissemination of information regarding their use and enjoyment as wilderness. . . .

Congress, to guide recommendations on what to include in the Wilderness Preservation System, attempted to define wilderness in the Act:

> A wilderness, in contrast with those areas where man and his own works dominate the landscape, is hereby recognized as an area where the earth and its community of life are untrammeled by man, where man himself is a visitor who does not remain. An area of wilderness is further defined to mean in this chapter an area of undeveloped Federal land retaining its primitive character and influence, without permanent improvements or human habitation, which is protected and managed so as to preserve its natural conditions and which (1) generally appears to have been affected primarily by the forces of nature, with the imprint of man's work substantially unnoticeable; (2) has outstanding opportunities for solitude or a primitive and unconfined type of recreation; (3) has at least five thousand acres of land or is of sufficient size as to make practicable its preservation and use in an unimpaired condition; and (4) may also contain ecological, geological, or other features of scientific, educational, scenic, or historical value.

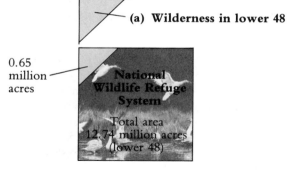

(a) Wilderness in lower 48

0.65
million
acres

National Wildlife Refuge System

Total area
12.74 million acres
(lower 48)

(b) Alaska Wilderness

Agency	Total Alaska Holdings (million acres)	Area of Wilderness (million acres)
National Wildlife Refuge System	76.1	18.6
National Park System	52.1	32.4
National Forests	23.2	5.45

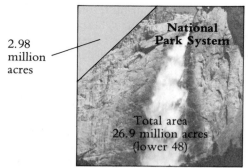

2.98
million
acres

National Park System

Total area
26.9 million acres
(lower 48)

Figure 30.8 (a) The boxes represent the Wilderness System as a component of federal parkland, refuges, and forests in the lower 48 states. (b) The table shows that in recent years Alaska has added dramatically to our national parks, refuges, and forests. Much area classified as wilderness is included in these additions. Unfortunately, the Alaska wilderness is not accessible to most people. (Photos: top, National Park Service photo by Richard Frear; middle, American Airlines; bottom, Appalachian Mountain Club photo by Roger Chapman.)

19.70
million
acres

National Forests

Total area
167.5 million acres
(lower 48)

The 1964 Act first instructed both the Department of Interior and the Department of Agriculture to study their holdings for suitability as units in the National Wilderness Preservation System. The Act then quickly designated 54 areas, totalling over 9 million acres (3.6 million hectares), as the first components of the Wilderness System. In 1975, the Eastern Wilderness Areas Act supplemented the first act by focusing attention on Eastern wild areas that had recovered from earlier impacts of human activity. A number of eastern areas scarred by fire or logging and now largely regrown or reforested were placed in the Wilderness System. Other eastern areas were designated for study for possible inclusion.

The Wilderness Act, in historical perspective, ranks in importance with the setting aside of Yosemite and Yellowstone, the Forest Reserve Act of 1891, and the establishment of Wildlife Refuges. Each of these earlier acts responded to an emerging awareness. The first of the National Parks set aside grandeur. The Forest Preserves were aimed at preserving our timber resources. The Wildlife Refuges were created to protect species from extinction. The Wilderness Act went further; it demonstrated both how far our knowledge of our environment had come and the concern we had to preserve not one item but nature in its entirety. Grandeur was not enough; timber resources were not enough; species alone were not enough. The preservation of nature's intricate pattern as seen through wilderness was required. (But see Controversy 30.2.)

CONTROVERSY 30.2

Wilderness

All this land is being set aside for non-use at a time when public interest in the maintenance of single-use wilderness . . . is evaporating.
R. L. Brown

The U.S. Forest Service should think in terms of generations if not of centuries.
Editorial, **Business Week,** *January 29, 1979*

Additions to the National Wilderness Preservation System are surrounded by controversy. A study by the U.S. Forest Service made public in 1979, considered the potential uses of 62 million acres of roadless land in the National Forests. Known as RARE II, for the second Roadless Area Review and Evaluation, the study concluded that the areas in question should be largely given over to other uses than wilderness. Of the 62 million acres, only 15 million acres were proposed for wilderness, while 36 million acres were to be "open"

for multiple-use activities. These activities include "mineral entry" (prospecting and mining), grazing of stock, and timber cutting. Eleven of the 62 million acres were not designated either open or for wilderness but were recommended for further study. The next step after the Forest Service proposal was for Congress to consider the study and then allocate areas to wilderness as they saw fit.

Some people thought the area of land that the Forest Service had recommended for wilderness to be

704 *Part Eight: Land Resource Issues*

quite ample and perhaps even too generous. In a letter to *The New York Times* (May 9, 1979), R. L. Brown asserted:

All this land is being set aside for non-use at a time when public interest in the maintenance of a single-use wilderness, which appears to have been a fad of some three or four years' duration, is evaporating. Entry into wilderness areas last year totaled only seven million man-days, a fraction of the entry by Americans into national parks, and onto other types of Federal land. Sporting-equipment manufacturers report that sales of tents, sleeping bags and other equipment used by hikers for treks into the wilderness are decreasing. . . .

It is not true that all use of land results in permanent loss of wilderness characteristics. Most exploration for mineral deposits, for an example, fails. All signs of preliminary exploration are usually obliterated by new growth in a very short time. Even clear-cut log areas reforestate, and wilderness groups now suggest that many reforested areas are in fact wilderness and should be so designated. (p. A24)

(To put some of Brown's figures in perspective, the National Parks include about 27 million acres in the lower 48 states. Visitor days to the parks in 1982 reached a total of about 245 million.)

Other people felt the Forest Service had recommended too little wilderness:

The U.S. Forest Service should think in terms of generations if not of centuries. But continuing pressure from lumbermen, mining companies, and enterprising recreational developers makes it hard for Forest Service officials to think beyond the day's schedule of appointments. The result is a built-in bias in favor of early utilization, which shows in the recommendations the Forest Service has just drawn up for classifying some 62 million acres of undeveloped land in the national forests. (Editorial, *Business Week*, January 29, 1979, p. 28)

They [the recommendations] would designate only 15 million acres as wilderness and 11 million for future study, while opening 36 million for various kinds of development. There are strong arguments for at least reversing these proportions, designating sufficient wilderness areas besides barren ice and rock, and being sure possibilities are not overlooked for both preserving wilderness and attaining necessary development.★

Is Brown right that wilderness use is a passing fad? Brown implies that wilderness is only set aside for human uses such as hiking and camping. Is that a correct interpretation? What criteria should be used in deciding how much public land should be set aside as wilderness compared to the amount of land on which multiple uses are allowed? Are visitor-use days a good measure of the value of parks or of wilderness?

★Editorial, *Christian Science Monitor* (January 8, 1979) p. 28. Reprinted by permission from the *Christian Science Monitor*, © 1979 The Christian Science Publishing Society. All rights reserved.

National Wild Rivers and National Trails

Four years after the Wilderness Act of 1964, the preservation of wilderness was again before the Congress, this time in two related acts. The wilderness under consideration in 1968 was not simply land and mountains. The remainder of the nation's free-flowing rivers whose banks were still free of development were the subject of the Wild and Scenic Rivers Act. The historic wilderness footpaths—the Appalachian Trail, the Pacific Crest Trail, and others—were the subject of the National Trails Systems Act. The passage of these two acts, while they had defects, strengthened the wall that preservationists were building around the wilderness.

National wild and scenic rivers. The 1968 Act began the process of setting aside the yet unspoiled rivers of the nation. Eight rivers were tagged for immediate entry into the system:

1. Clearwater River, Middle Fork, Idaho
2. Eleven Point River, Missouri
3. Feather River, California
4. Rio Grande River, New Mexico

5. Rogue River, Oregon
6. Saint Croix River, Minnesota and Wisconsin
7. Salmon River, Middle Fork, Idaho
8. Wolf River, Wisconsin

By 1978, the system had been expanded by Congressional, state, and federal agency action to include 28 rivers. Taken together, these 28 rivers cover some 2,300 miles (3,800 km). The Park Service has sole administration of 5 of the 28 and shares two others with the Forest Service. The Forest Service administers 10 rivers as the single guardian and shares other rivers with the Bureau of Land Management. Five of the rivers are administered solely by the states and one by the Bureau of Land Management. A river may be recommended for inclusion by individual state legislatures. Such a river, if approved by the Secretary of the Interior, becomes part of the system without further Congressional action. The Little Beaver River in Ohio entered the system in this way and is administered by the state of Ohio. In all, five state-administered wild rivers have joined the system: the Allagash in Maine; the Little Miami and Little Beaver Rivers in Ohio; the New River in North Carolina; and the Lower St. Croix in Minnesota and Wisconsin.

The rivers or portions of rivers included in the system are to be kept free of dams or other structures; that is, their free-flowing character is to be preserved. Further, a corridor on both sides of the river is to be kept free of all development, thereby maintaining the wilderness setting of the river. The rivers placed in the system are classified into one of three categories, according to the legislation that created the system:

(1) Wild river areas—Those rivers or sections of rivers that are free of impoundments and generally inaccessible except by trail, with watersheds or shorelines essentially primitive and waters unpolluted. These represent vestiges of primitive America.
(2) Scenic river areas—Those rivers or sections of rivers that are free of impoundments, with shorelines or watersheds still largely primitive and shorelines largely undeveloped, but accessible in places by roads.
(3) Recreational river areas—Those rivers or sections of rivers accessible by road or railroad, that may have some development along their shore-

lines, and that may have undergone some impoundment or diversion in the past.★

The protection provided a river in the national system is of several kinds. First, the Federal Power Commission is prohibited from licensing new power dams on the protected portion of the river. Rivers under study for inclusion are also protected in this way while they are being studied. Power developments above and below the protected portions of the rivers are not prohibited, but if they conflict with the flow in the protected stretch they may still be prevented.

In addition to the ban on dams and diversions, the U.S. government can no longer sell or exchange public lands within the boundaries specified for the rivers. Although mineral claims in these areas that were established before the act were not voided, new mineral-related activities are under the regulation of the Secretary of the Interior to prevent either pollution or other degradation of the area.

National scenic and national historic trails. Although the Appalachian Trail is today under the primary administration of the Secretary of the Interior and the Park Service, and the Pacific Crest Trail is under the primary administration of the Secretary of Agriculture and the Forest Service, it would be a mistake to assume that the federal presence is the reason these trails exist today. In particular, the bulk of the Appalachian Trail is not a result of federal effort. The federal presence does provide some added protection to the trails and helps to ensure their preservation, but at the time the trails were assembled, ordinary citizens and private organizations furnished the creative force.

The original proposal for the Appalachian Trail came in 1921 in an article by Benton MacKaye, a forester and regional planner, in the *Journal of the American Institute of Architects*. MacKaye suggested linking a number of trails, as well as building new trails. Hiking and outdoor clubs expressed great interest in MacKaye's idea, which initially was for a trail without end, and they undertook to build MacKaye's dream. Within a year of the proposal, the first section of the trail was opened in the Palisades Interstate Park in New York and New Jersey. Of the more than 2,000 miles ultimately assembled, only 350 miles of trails existed at the

---

★Title 16, Section 1273 of the U.S. code.

Figure 30.9 Part of the Appalachian Trail. Pinnacles Overlook, Shenendoah National Park. (National Park Service photo)

outset. All of the trails that already existed were in New England and New York. The formation of the Appalachian Trail Conference, a federation of clubs, government agencies, and individuals, in 1925 gave hope that the trail could become a swift reality. The complexity of assembling the trail, however, slowed its progress, especially south of Pennsylvania where hiking clubs were still to be organized.

Completed in 1937, the trail runs along the mountain backbone of the east, the Appalachian Mountains, from Mount Katahdin, Maine to Springer Mountain, Georgia (Figure 30.9). Public land in the National Forests and in the National Parks from Virginia south helped to make the final assembly possible (Table 30.2). To protect the portion of the trail through public land, the conference signed an agreement with the Park Service and Forest Service, as well as most of the state governments involved in 1938. Under its terms, the federal government promised not to construct incompatible projects within a 1-mile corridor on either side of the trail; the states agreed to a quarter-mile strip on each side.

Nevertheless, establishing the connections between such sections proved to be a large task, as the trail often wound its way across private lands. Permissions had to be requested of owners; at the time, simply an oral agreement was usually needed. As the pressure from use grew, however, individual owners became

Table 30.2 Approximate Division of National Scenic Trail Jurisdictions

Agencies with jurisdiction	Appalachian Trail		Pacific Crest Trail	
	Miles	**Kilometers**	**Miles**	**Kilometers**
U.S. Forest Service	719	1158	1856	2989
National Park Service	215	346	249	401
Bureau of Land Management	—	—	204	329
States	289	465	43	69
Private[a]	805	1296	106	170

a. No cooperative agreements for use exist on 75% of the 805 miles (1,296 km) in private ownership.

more reluctant to extend their hospitality. The trail was threatened by its very popularity.

At about the same time the Appalachian Trail was beginning to be assembled, Clinton Clark of California proposed (1932) a similar footpath extending from Canada to Mexico on the West Coast (Figure 30.10). The Pacific Crest Trail was also planned as a mountain range path. Its 2,600 miles (4,300 km) includes the Cascade Crest Trail and the Oregon Skyline Trail in Washington and Oregon. The Pacific Crest Trail follows the Lava Crest Trail, the Tahoe-Yosemite Trail, and the John Muir Trail, among others, in California. Fully 85% of the Pacific Crest Trail is on federal lands, such as national park or national forest, so the trail is less threatened by development than the Appalachian Trail.

After several years of effort, Congress passed and President Lyndon Johnson signed the National Trail Systems Act in 1968. The Act created a system of scenic and recreational trails across the country and made the Appalachian Trail and its sister trail, the Pacific Crest Trail, the first members of the system. Federal purchase of the needed right-of-ways for the trails was authorized by the 1968 Act, but funds were not appropriated

for another 10 years. If negotiations with private land owners proved unsuccessful, the government was given the right to acquire the land by "condemnation."

The Land and Water Conservation Fund administered by the National Park Service was also supposed to assist states in purchasing portions of the corridor needed for the trail, but this federal effort was not a part of the National Trail Systems Act. Individual trail clubs and the Appalachian Trail Conference still maintain funds to buy trail right-of-way, and they continue to make such purchases to the present time. The struggle to create the trail and to preserve it still falls largely to individual citizens and the trail clubs.

The National Trail Systems Act did more than promise to stabilize the historic mountain backbone trails of the East and West Coasts. While the Appalachian and Pacific Crest Trails were designated the first National Scenic Trails, the Act directed that other trails be studied to see whether they would qualify for inclusion in the Scenic Trail System. The Continental Divide Trail has since been designated as a National Scenic Trail (1978). From Glacier National Park in Montana, the trail follows the continental divide down to New

Figure 30.10 Along the Pacific Crest Trail. (U.S. Forest Service)

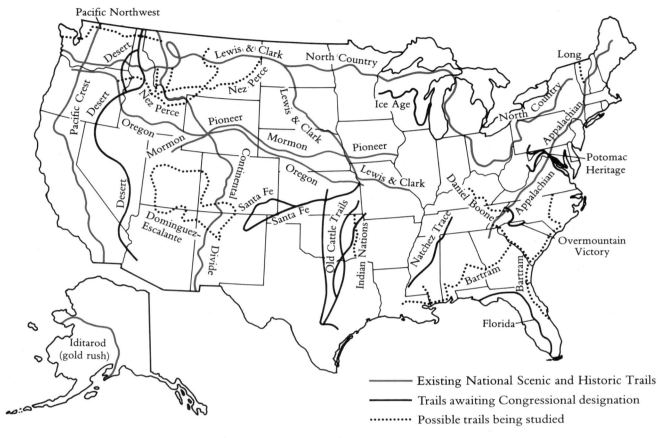

Figure 30.11 The National Trail System: national scenic and historic trails, 1980. (Reprinted by permission from *National Parks & Conservation Magazine*, October, 1980. © 1980 by National Parks & Conservation Association.)

Mexico and the Mexican border (Figure 30.11). In 1980, the 3,200-mile-long North Country Trail was added to the Scenic Trail System.

Recognizing the historical significance of a number of the proposed trails, in 1978 Congress created a new category of trail, the National Historic Trail. Four trails have been taken into the system as National Historic Trails. The Oregon Trail (2,000 miles from Missouri to Washington), the Lewis and Clark Trail (3,700 miles from Illinois to Oregon), the Mormon Pioneer Trail (1,300 miles from Illinois to Utah), and the Iditarod (an Alaskan Gold Rush trail, 2,000 miles from Seward to Nome) thus entered the National Trail System. Other routes are still under study. The trails do not necessarily follow the original routes; development has in some

cases wiped out entire sections of the original trail. Thus, some rerouting is necessary to achieve unbroken trails. In addition, federal protection is provided only to those portions of the trails that are on federal lands.

For a number of years, the promise of the National Trail Systems Act to preserve the nation's trails has been a hollow commitment. The hope that the Appalachian Trail could be finally secured by federal purchase of threatened areas has not yet been fulfilled. Congress finally appropriated some money under the act for such purchases in 1978, 10 years after the initial commitment to acquire the needed land. During those 10 years, development marched closer to the trail in many places, and the cost of acquiring right-of-way rose considerably. Now, as before, the trail's continued existence as an unbroken footpath is the result of the pooled private efforts of individuals and hiking organizations. There is a lesson in the history of the Appalachian Trail.

Problems of Accessibility

A stone-and-mortar walkway with wooden handrails guides visitors away from the picnic area down along a steeply sided stream. The visitors witness a quickly changing scene of rock cliffs and evergreen-covered earth banks. Then the valley opens with a thunderous roar; a magnificent waterfall confronts the visitors. It is a spectacle worth remembering—a vivid demonstration of the remarkable power of flowing water, cutting through the rock over countless centuries. The visitors, an elderly couple, would have been unable to see it had the walkway and handrails not been present to guide and steady their steps. The walkway was the legacy of a work crew from the Civilian Conservation Corps who, almost 50 years ago, helped to make parks more accessible to less athletic people. It was an effort of noble proportions and one of benefit to millions of people.

Yet such development at park after park, year after year, leaves its mark on the wilderness setting. Construction of roads, walkways, picnic grounds, swimming areas, restaurants, campgrounds, shops, service stations, and even hotels begins to take on a pattern. Nature is being enclosed in an amphitheater. A supporting staff of ticket takers, ushers, waiters, and waitresses is needed for an orderly display. Each "improvement" makes the park more accessible and enables it to accommodate more and more people. And the crush of people justifies the improvements; the park authorities are simply attempting to meet "demand" for public recreation.

Many argue that such developments, which are designed to meet the need for public recreation, degrade wilderness values to an unacceptable extent. Indeed, the word *recreation*, when broken into its parts, means "to be created again." Only by usage has its meaning evolved to include amusement, exercise, or educational activities.

The measurement of success in wilderness preservation is difficult. It is far easier to count visitors and use the annual visitor count as the measure of success of a park operation than to somehow assess how well the land has been preserved. At a famous midwestern site managed by the National Park Service, one of the goals of management is to keep the average time of a visit to less than 50 minutes. If the average time a party takes there is much more than 50 minutes, backups on the access road to the site will occur and motor homes will boil over. The Park Service needs to move people in and move them out.

Although an increased visitor count is a possible measure of success in a park, in a wilderness area an increased visitor count may be a danger signal. Congested trails in the John Muir Wilderness of California led the U.S. Forest Service to go to a permit system for trail camping there. The number of hikers on the trail to Mount Whitney was limited to no more than 75 per day. Since two days are usually needed to hike the 8-mile trail, the number of overnight campers is generally kept to less than 150 on any given evening. In the early 1970s, before the permit system was begun, more than 1,200 hikers had been on the trail during one day of the Labor Day weekend. (See Controversy 30.3.)

Often the development of tourist facilities at the site of the spectacle itself diminishes the experience of visiting. Old Faithful in Yellowstone, with its parking lot and visitor center, is often mentioned as a natural area that has been overdeveloped. The development in the Yosemite Valley is also seen as overdone. Such developments began in a era when visitors to the parks were few. The modest facilities needed for the visitors did not intrude on their appreciation of the park's values, even if the facilities were built at the site itself. But as the number of visitors grew, the development required to serve them had to expand as well. It was easiest, apparently, simply to "add on."

The alternative to development at the site, worthy of careful consideration, is a visitor center at or near the park gate. Such a visitor center, remote from the scenes of the park itself, was wisely built in the early 1970s at Acadia National Park in Maine. At other parks, however, existing facilities are unlikely to be torn down to be rebuilt at less scenic locations; the expense is considered too great. Nonetheless, the concept is attractive because not only development but also the litter that accompanies tourism can, to a degree, be kept from the park interior.

Wilderness and recreation come in conflict in more ways than numbers of people and development. Certain human activities allowed in a natural area can also diminish wilderness values. Prime examples are motor boating, off-road vehicle operation, and snowmobile use. These activities are supported, however, not only by a large public but by business interests who organize lobbying efforts so that their products can continue to be used. (See Controversy 30.4.)

Reviving an Old Tradition of Land Preservation

When the Forest Preserves were first created in concept at the federal level in 1891, Congress was responding to growing conservationist sentiment throughout the nation. This sentiment for preservation of our forests was expressed in conservation efforts by the states as well as by the federal government. The most ambitious of these efforts, the Adirondack Mountain Preserve, predates the nation's Forest Preserves by six years.

Created in 1885, the Adirondack Mountain Preserve was meant to protect the exquisite wilderness of mountains, forests, lakes, rivers, and wild species that the region possesses. Threatened by lumbering and mining interests, the land was snatched from exploitation by the preserve. A portion of the Constitution of the State of New York calls for the region to "forever be kept as wild forest lands."

In 1894, the New York legislature penciled a blue line around a six-million-acre region of the Adirondacks. The area enclosed by this line became known as a greenline park, an area within which wilderness would be preserved. The movement to designate such areas for needed protection became known as the "greenline movement." (Color blindness has been suggested for the shift in colors in naming the movement.)

Yet the Adirondack Forest Preserve is not a pristine and totally undeveloped wilderness. About 120,000 people are permanent inhabitants of the region, scattered in 106 towns and villages, and many more people come to spend the summer in the Adirondacks in cottages and homes that dot its lakes. Even more people come to vacation at the numerous public campgrounds that the state runs and at the many private resorts in the region. Although the state holds 2.3 million acres within the greenline area, private interests hold 3.7 million acres within the preserves. Some 62% of the land area of the preserves, then, is in private hands.

Commerce of all sorts proceeds in the preserves. Businesses open and grow. Developers speculate and subdivide throughout the region in search of profits. All these activities coexist with the state-owned

wilderness. One must ask how this can constitute "protection."

The answer is that protection comes from an elaborate set of controls on the use and development of privately held land. An example is found in the rules for using privately held land that fronts on lakes. Shoreline modifications and the building of structures require permits. The cutting of live trees within 50 feet of the shoreline is strictly limited. Septic tanks must be kept suitably distant from shoreline. Such controls as these are placed on land in private hands when the condition of the lands directly influences the visual aspect of the preserve. Development within the business districts of incorporated towns is governed by the zoning within those towns, but elsewhere development is controlled by the Adirondack Park Agency, the preserver/enforcer of the preserve. In areas still wild, though privately owned, the agency allows no more than one building for every 42.7 acres.

To many who make their home in the region, the regulations are "big government," unnecessary interference with people's natural rights. To millions of others in New York State, though, the rules are the means by which the Adirondacks were preserved for their use and the use of their children. The preservation of the Adirondacks without actual purchase of the land creates inevitable conflicts between people and the administering agency. Without these conflicts and tensions, however, the land could only be preserved at enormous expense—by the actual purchase of the land outright. The Adirondacks are being preserved, even in some cases against the will of its residents. The land-use controls are working, and working, in fact, to the long-term benefit of the residents of the region, whose income derives largely from tourism.

The success in preserving the Adirondacks is now leading to parallel efforts elsewhere in the United States. As in the Adirondacks, we will find that preservation by controlling land uses meets with opposition from commercial interests.

In 1978, the nation's second greenline park was created, this time in the unique pitch-pine forests of southern New Jersey. The action did not come a moment too soon. Under pressure for rapid vacation-home development from the nearby metropolitan areas in New York, New Jersey, and Pennsylvania, the pine barrens, as the forests are called, had already shrunk from 2,000 to 1,500 square miles. The sandy soil of the region had largely prevented farming and subsequent

CONTROVERSY 30.3

Should Wilderness Be Accessible?

> To be precious, the heritage of wilderness must be open to those who can earn it again for themselves. The rest, since they cannot gain the genuine treasure by their own efforts, must relinquish the shadow of it. We must earn again for ourselves what we have inherited.
>
> *Garrett Hardin (1971)*

> As a young forestry student in 1939 . . . I roamed the high country west of the Three Sisters, mingling with deer, elk, and bear. . . . Thirty-five years later I returned. It was only three months after open-heart surgery so I could not climb very much. . . . It was a logging road that brought a middle-aged college professor within walking distance of his goal.
>
> *Stephen H. Spurr (1976)*

The views of Garrett Hardin and Stephen Spurr are in conflict over the issue of whether wilderness ought or ought not be easily accessible to people. The sentiments expressed by Hardin suggest that the definition of wilderness as offered by Congress in the Wilderness Act is somehow inadequate.

Hardin is suggesting that wilderness should not be easily accessible. There are, of course, practical reasons to limit accessibility. You should be able to provide such practical reasons. Hardin's point of view, however, is one of principle. The practical implications are not of concern to him, at least not in this passage.

The sentiments of Spurr, on the other hand, suggest that wilderness should be used by people and that accessibility provides the wilderness values to many, rather than to the few.

You may have thought about these points of view before when you were on the trail and looking to escape crowds of people. You may consciously have chosen less-traveled paths. Your opinion on the accessibility of wilderness, however, may have something to do with your age and condition. No doubt age and condition influenced Stephen Spurr's opinion. You should examine your views on this issue because the shape of future wilderness will be influenced by them. Is there a middle ground or compromise between inaccessibility and accessibility?

Sources:
G. Hardin, quoted from the Foreword by David Brower of *Earth and the Great Weather: The Brooks Range* (San Francisco: Friends of the Earth, 1971).
Stephen H. Spurr, *American Forest Policy in Development* (Seattle: University of Washington Press), 1976.

CONTROVERSY 30.4

The ORV Fight

Just put your gang on Suzuki's DS trail bikes. And head for the boonies. Doesn't matter where you go. Peaks or valleys, it's all the same to those rugged off-road machines.
Suzuki advertizing copy, Cycle World 17(3) (1978), 110–111

St. Francis of Assissi himself while driving an off-road vehicle on wild land could not avoid diminishing the recreational experience of many non-ORVers in the same area.
CEQ report, 1979

One of the most bitter controversies over the use of public lands involves off-road vehicles (ORVs): trail bikes, snowmobiles, dune buggies, and four-wheel-drive vehicles. Some 25% of the American public enjoy the use of these vehicles. ORVs allow their users the physical thrill of conquering rugged terrain and also allow them to penetrate far into wilderness areas for hunting, fishing, and camping. But ORVs can destroy that very wilderness, as well as ruin the wilderness experience for hikers, backpackers, and other non-motorized users.

Off-road vehicles destroy the vegetation they ride over, leaving soil exposed to erosion from wind and rain. They compact soil, killing plant roots and decreasing the soil's ability to absorb water, which further increases erosion. In the arctic, ORV use leads to the melting of permafrost and the formation of water-filled tracks. The noise from ORVs can drive away wildlife or even damage vital hearing ability in certain species.

Non-ORV users are not only worried about these environmental effects but also extremely angry at the intrusive noise ORVs produce:

[O]ne ORV operator can effectively restrict a large public area to his own use through the emission of loud engine noise, obnoxious smoke, gas and oil odors and dangerously high speeds. Whereas previously many persons of all ages and wealth could observe the beauty of unspoiled land, now a single ORV can reign supreme. (Rosenberg, 1976)

Consider the snowmobile . . . and the often-heard argument that this machine makes it possible to "get way back in there, away from it all." There are, of course, several other ways to "get back in there," including snowshoes and skis. Maybe if you need an engine to get there, you don't belong there in the first place! To my mind, "getting away from it all" means, foremost, getting away from our society's overdependence on the combustion engine. (Buerer, 1975)

[We need to] provide some place on God's green earth for man to spend some time without hearing a damned motor. (Huffman, 1974)

I hope there is some way we could outlaw all off-road vehicles, including snowmobiles, motorcycles, etc., which are doing more damage to our forests and deserts than anything man has ever created. I don't think the Forest Service should encourage the use of these vehicles by even suggesting areas they can travel in. . . . I have often felt that these vehicles have been Japan's way of getting even with us [for World War II]. (Goldwater, 1973)

In some areas, attempts have been made to restrict ORVs to particular areas or trails. TVA, for instance, has tried this on an experimental basis. TVA ranger Scott Seber (1979) says,

I used to hate ORVs. Now I feel they can be worked with. We have demonstrated that they don't have to be running amuck everywhere.

This has not satisfied everyone, however:

[We] object to the continual enhancement of non-

ORV recreation at the expense of the off-road vehicle enthusiast. We do not feel that all compromises should be made at the expense of off-road motorcyclists. (Rasor, 1976)

How do you feel about ORV use? Do you think all public lands should be open to ORV users? Or do you feel their use should be forbidden on public lands, perhaps restricted to privately run ORV parks?

Sources:
Gary A. Rosenberg, *Environmental Affairs*, 1976.
Jerry Buerer, professor of sociology, Marquette University, and organizer of a group to protect the rights of non-ORVers, 1975.
Ben Huffman, Vermont Department of Forests and Parks, 1974.
Senator Barry Goldwater, letter to William D. Hurst, Regional Forester, Region 3, U.S. Forest Service, Albuquerque, New Mexico, March 23, 1973.
Scott Serber, in *Off Road Vehicles on Public Land*, Council on Environmental Quality, 1979, p. 14.
Robert Rasor, AMA, 1976, Environmental Impact Statement, supra note 19, at 260.

development in the area, so that much virgin forest and many unspoiled rivers still remained. The pine barrens region harbors rare plant and animal life, such as Pickering's morning glory, the Karner blue butterfly, and the carpenter frog.

The region was too large to purchase outright, so Congress created the Pinelands National Reserve, again an area defined by a "greenline" in which federal and state land would coexist with private land. Development on the privately held land would be carefully controlled. Just as in New York, where the Adirondack Park Agency administered the land-use controls, an intergovernmental commission has been established to oversee the regulation of development in the Pinelands. Just as in New York, the plan is to acquire the most ecologically sensitive lands and limit further development in the remainder of the million-acre area. At the outset, some 20% of the area was in state hands and another 5% of the area was to be purchased by the federal government.

As proof that the pinelands were under intense pressure and needed rescuing, would-be developers have filed suit to have the land-management plan set aside.

Greenlining may well be used to protect other natural areas of the United States. These areas would be ones where some development is already in place and cannot be removed, and where unique features still remain to be preserved. These are areas also where the cost of outright government purchase of land would be too great. The areas surrounding a number of National Parks fall in this category; these areas have developed to accommodate vacationers and often possess landscape and wildlife qualities worth preserving. Further, the development has often increased land values and made purchase at a fair price very difficult.

The National Parks and Conservation Association has identified a number of areas where greenlining, a combination of land purchases and land-use controls, could make an impact on the preservation effort. One such area is the Big Sur coastal region of Northern California. Rich in redwood forests, this area boasts a rugged seascape of towering cliffs and misty rock-encrusted shores. Its exquisite beauty is as haunting as its ocean waters are cold.

The Jackson Hole region of Wyoming is another area where greenlining could make a difference. The rugged Grand Teton Mountains look over this area where exciting western history was made. Tourist developments have pushed up property values and taxes so that ranchers are looking for relief from high taxes. Taxation based on current use rather than on value for development is one possibility (see Chapter 29), but care must be taken that such taxation does not simply reward speculators with higher profits. The purchase of scenic easements is another possibility. In this case payment is made for a promise not to develop. A question arises here, however. Should the ranchers be compensated by purchase of easements in this way? Land owners in the Pinelands and the Adirondacks do not receive compensation for leaving the land undeveloped; they simply *cannot* develop it.

A third area identified by the National Parks and Conservation Association is the Gorge region of the Columbia River. The Columbia's majesty has been reduced by numerous hydropower developments, but many waterfalls remain, and a number of unusual plant species are found in the Gorge. Although development pressures come from nearby urban areas, only a patchwork of state parks and national forests now protect the region. Because the Gorge region takes in portions of both Washington and Oregon, effective protection has been elusive. In 1982, legislation lay before the U.S. Congress to establish the nation's third greenline park around the Gorge of the Columbia.

Expanding the System of Natural Areas

The Land and Water Conservation Fund

The Land and Water Conservation Fund (LWCF) is a recent and powerful means to expand public natural areas at the local level and at the national level. Created by Congress in 1965, the Land and Water Conservation Fund has in the past furnished federal money to match local money for the purchase of natural areas. The Fund is presently administered by the Division of State, Local, and Urban Programs of the National Park Service, but was formerly administered by the Heritage Conservation and Recreation Service (HCRS), housed in the Department of the Interior. The original administering office was begun as the Bureau of Outdoor Recreation and was created in 1962 by the Secretary of the Interior at the request of President Kennedy.

The Land and Water Conservation Fund (LWCF) Act of 1965 gave clout to the mission of the office. The

Act established a fund of money to be allocated to state and federal agencies for the purchase of land. The fund is fed by revenues from the sale of surplus properties, from the leasing of the outer continental shelf for oil and gas, and from other sources, including direct appropriations by Congress. Of the fund's revenues, 60% have been for the use of the states. Three activities were supported by this portion of the fund: the planning related to a specific project; the acquisition of the area; and the development of recreation facilities and resources at the site. The fund has supplied up to one-half of the total cost of a specific project. For a state to be eligible to apply for money, it must have already developed a comprehensive outdoor recreation plan that includes an assessment of outdoor recreation needs and recreation resources in the state. The remaining 40% of the fund is for the use of federal agencies involved in outdoor recreation.

Contributions of the fund to purchases are often made in cooperation with the efforts of state governments and conservancies. Federal laws giving tax benefits for donations of land are often involved in the transactions.

After describing the functions of conservancies, we will illustrate the acquisition process with several examples. These examples will show how conservancies and state governments, using the Land and Water Conservation Fund, have been combining their talents and using the benefits of the federal tax law to enrich the public with new natural areas.

Donations and Tax Benefits

We mentioned that the National Park System has received a number of gifts, but we did not describe in any detail how gifts are made. To someone interested in giving a valuable property to the public, the monetary advantages that can be obtained from the gift are important.

The simplest type of gift, called an outright donation, transfers the land directly from owner to recipient at the time of agreement. Depending on whether the recipient is a conservancy (a nonprofit land trust) or a government agency, different but nonetheless attractive tax benefits to the donor are available. Land may also be given in a will and thus be transferred after the death of the owner. In still another type of gift, the transfer can be made in such a way that the donor retains for his or her lifetime the privilege of living on the land, even though the land has already changed hands. Still another popular form of gift is the bargain sale, in which the owner sells the land to a conservancy or government agency for less than its fair market value. The difference between what would have been obtained by sale on the open market and what actually was obtained in the bargain sale represents a gift or donation, and the value of the difference determines the tax benefits to the donor.

It is this bargain sale procedure that we will mention further in describing how the LWCF, conservancies, and state governments have been functioning together to expand the system of natural areas. First, however, we need to dwell for a moment on the tax benefits of donation, because these benefits are a part of the argument that convinces would-be donors to become the generous individuals they wish to be. The value of the land is first established through an appraisal by a professional appraiser. The appraiser normally considers what similar tracts of land in the geographical vicinity have been sold for on a per-acre basis and applies that to the land in question, with adjustments for the specific factors of its location, such as road frontage, distance from water and sewer, and so on. The appraised value reflects the price the property would bring if placed on the open market.

Specific federal tax benefits differ, depending on whether the property is given to a conservancy, a local government, or an agency of the federal government. If the donation is made to a conservancy or land trust, the donor may deduct from his taxable income either the value of the gift or 30% of his adjusted gross income for that year, if the gift is larger than this amount. If the value of the gift exceeds 30% of his adjusted gross income, the donor may carry over the remainder and deduct that remainder the following year, subject to the same percent limitation. Any excess may be carried over to the following year for a total of six years.

Here is an example.

Mr. Sandune donates a beach property worth $75,000 to the Nature Conservancy. His adjusted gross income is $130,000 per year. He can deduct from his taxable income up to $0.3 \times (130,000) = $39,000 of the $75,000 gift. The income on which he must pay taxes is now reduced to $130,000 − $39,000, or $91,000. His tax savings are considerable. The following year, if he has the same level of income, he can deduct up to $75,000 − 39,000$

$= 36,000$, the remainder of his contribution. Again, he obtains large tax savings.

Tax benefits are even greater if the gift is to a local government or government agency. The full value of the gift may be deducted from the adjusted gross income without any limitation calculated as above. In our simplified example, Mr. Sandune can deduct $75,000 from his adjusted gross income; that is, he will not need to pay any tax on $75,000 of his income because of the gift. Corporations may also donate land, but special rules apply for the tax treatment of their donations.

Conservancies or Land Trusts

A number of private nonprofit organizations assist in the process of obtaining land for the public and obtain lands for future transfer themselves. Most prominent among these is the Nature Conservancy, a private, nonprofit organization that buys ecologically and geologically valuable land. The Conservancy has purchased nearly 1,800 parcels of land, totalling 1.1 million acres, since it was founded in 1950. About two-thirds of these purchases have been transferred to government agencies, mostly to state governments, for their permanent protection as natural areas. The remainder are managed by the Conservancy itself.

The Conservancy, which has many state branches, does more than buy and transfer land. It is active in identifying and inventorying natural areas across the country. Because of the Conservancy's stature in preservation activities and because of the tax advantages that come to donors of land, the Nature Conservancy is often successful at obtaining land by simply telling potential donors about its purposes and the tax advantages from gifts.

A 138-acre parcel at Stevens Creek, South Carolina, was obtained by the Conservancy in this way from the Continental Group. The land, which harbors several endangered species, was then transferred to the protection of the State of South Carolina. In a similar fashion, the Conservancy was able to obtain a portion of Jupiter Island off the east coast of Florida. The manatee, a remarkable sea mammal resembling nothing so much as a large cigar, lives at Jupiter Island. This distant relative of the elephant, known more commonly as the "sea cow," is one of America's endangered species. The 500-acre tract on Jupiter Island will aid in protecting the manatee.

Another example of the Conservancy's work is the preservation of the 200-acre farm of Dr. and Mrs. Wright near Albany, New York. Given to the Conservancy outright by Dr. and Mrs. Wright, the area is now maintained as a nature preserve. One of the most significant of the Conservancy's purchases is the Dewey Ranch in Kansas. The 7,200-acre tract preserves a portion of the tall-grass prairie. The prairie that drew so many pioneers west for settlement is now virtually gone. Only small pockets of this unique ecological landscape still exist, so the Conservancy's purchase is an especially important one. The area is to be leased to Kansas State University for ecological research. (See Chapter 6 for a discussion of the tall-grass prairie.)

Coordinated Federal and Private Actions

Federal tax laws, as we pointed out, often make it economically attractive for people to donate land for natural areas. The Nature Conservancy helps explain these advantages to donors and may assist in the transactions themselves.

Take the case of a 3,100-acre parcel in Rutland County, Vermont, which was prime land for a vacation-home development. This parcel, which adjoins a state forest, was put up for sale in 1971. Its appraised value was $610,000. Vermont's Agency of Environmental Conservation had no money to acquire the property—not even enough to go halves with the Land and Water Conservation Fund in purchasing the land. The agency proposed to the owners that they sell the property to the state for half of its market value, or $305,000, money they felt they could obtain from the Land and Water Conservation Fund, whose guidelines allow them to contribute up to half of the market value for a tract of land. The difference between the full market value of $610,000 and the bargain sale price of $305,000 could then be treated as a donation for federal tax purposes. Although interested, time was important to the owners. They wished to move faster than the grant from the Land and Water Conservation Fund could be received. The Nature Conservancy stepped in to protect the potential sale.

They offered the owners $305,000 in cash and pointed out the tax advantages of donating the remainder of the property's market value. The owners could deduct up to 30% of their individual shares of the donation from their taxable incomes. Furthermore, they would save realtor fees and save taxes on a portion of

their investment profits as well. All of these advantages were theirs, said the Conservancy, in addition to public recognition of their gift.

The owners sold the property to the Conservancy and the Conservancy held the land in its name until the state agency obtained its grant from the Land and Water Conservation Fund. With the grant in hand, the Agency purchased the land from the Conservancy. The price was the original $305,000 plus miscellaneous interest and legal fees. Here the Conservancy both acted as an influential spokesman for land preservation and actually took possession of the land on a temporary basis to ensure that it would not be sold to developers.

The linkage and the coordination of the Nature Conservancy, the states, the Land and Water Conservation Fund, and federal tax laws has been in the past a very powerful tool in saving land. Although the LWCF fell on lean times under the Reagan presidency, the Division of State, Local, and Urban Programs continued to allocate small amounts of LWCF money in the early 1980s. This was money that had been obligated for state projects in prior years but for which actual purchases had fallen through.

The Reagan Administration was recommending zero funding for the office during this time, but Congress was trying to respond to local sentiment and attempted to provide funds to be channeled to the states. Each new session of Congress brought new attempts to restore the money and functions of the office. Although these changes took place in the early 1980s, the successes of the LWCF in the 1970s make it clear that the LWCF was an effective tool for bringing new lands into public ownership at low cost to the states. As such, the expectation is that its functions will one day be completely restored so that the state-federal-private partnership can be renewed.

Summary

The protection of land and wildlife resources evolved as three separate traditions in the United States. The first national parkland, Yosemite, was set aside in 1864 to protect part of the young country's natural wonders. In 1891, Congress began to protect areas of forestland as national forest reserves, later the National Forests. In 1900 the first steps were taken to protect native American wildlife on refuges and sanctuaries. The Department of the Interior administers the National Park

System and the National Wildlife Refuge System (as part of the Fish & Wildlife Service), while the Department of Agriculture manages the National Forests. The Park Service administers not only the National Parks but also national monuments, historic sites, recreation areas, seashores, and lake shores. The National Park System can acquire lands by purchase, gift, trade, or condemnation.

Other public lands are in the charge of the Bureau of Land Management. These National Resource Lands can be mined, grazed, and logged.

Some hunting and recreation are allowed on National Wildlife Refuge lands as long as wildlife protection is not compromised. This is similar to the multiple-use concept by which the National Forests are administered. Logging, fishing, and recreation, as well as wilderness, watershed, and wildlife protection are all allowable uses for national forest land.

Clear-cutting is the practice of cutting all trees in an area regardless of size or quality. It is a cheaper method of logging but leads to erosion, flooding, loss of wildlife habitats, and loss of scenic values. In selection cutting, only mature trees are harvested. In seed tree cutting, some mature trees are left to reseed an area. Shelterwood cutting is a three-step process in which first unwanted or unhealthy trees are cut, then, in 10–15 years, a portion of the mature trees are cut. Finally, the last of the mature trees are logged.

Since World War II, a significant portion of U.S. timber use has come from the National Forests. Some factions believe this portion should equal the possible sustained yield (or amount that is replaced each year by new growth). Others feel the timber should be kept in reserve for future needs and to encourage replanting in privately held lands.

Fires are a natural event in most forests. Often fires help regenerate certain forest species. Frequent fires also prevent the buildup of underbrush and dead material, preventing fires that burn not only dead material but also living trees.

The Wilderness Act of 1964 called for preservation of wilderness in all types of government-owned lands and represents a fourth type of wild resource to be saved (along with timber lands, spectacular scenery, and wildlife). Congress next acted to save wild, scenic, and recreational rivers in their natural state and wilderness trails such as the Appalachian Trail, the Pacific Crest Trail, and the Continental Divide Trail. Almost all of the Pacific Crest Trail is publicly owned, but parts of the Appalachian Trail are still in private hands. In addition to these scenic trails there are a variety of national historic trails, such as the Lewis and Clark Trail. Conflict exists over the purpose of wilderness preservation. Should wilderness be accessible to as many people as possible, or should it merely "be there"?

Where wilderness areas are too large to purchase and set aside, such as in New York State's huge Adirondack Forest Preserve, development can be controlled by land-use laws. This is known as a greenline park, an area where people live but where development is limited to prevent deterioration of wilderness. Another such region is the New Jersey pine barrens. Scenic easements can also protect those large, mainly wild, areas where some development has already taken place but that should be preserved in some way.

Additional public lands can be bought using funds from the Land and Water Conservation Fund. Various tax benefits are available to private donors, which makes giving land to the public park or reserve systems more attractive. Several private, nonprofit organizations, such as the Nature Conservancy, are dedicated to making it easier for individuals or corporations to donate lands for public use.

Questions

1. Why are so many different agencies involved in administering public lands? Briefly outline the agencies and their responsibilities.
2. Is fire always a bad thing in a forest? Explain.
3. What is a greenline park? How does it preserve the natural characteristics of the land?

Further Reading

Dodge, M. "Forest Fuel Accumulation: A Growing Problem," *Science,* **177** (July 14, 1972), 139.
This well-written article can lead you into the literature of this controversial area.

Fitzsimmons, A. "National Parks: The Dilemma of Development," *Science,* **191** (February 6, 1976), 440.
A nontechnical article with details of park development problems and methods to deal with intrusions on the setting.

Krieger, M. "What's Wrong with Plastic Trees?" *Science,* **179** (February 2, 1973), 446. See also Letters and Reply in *Science,* **179** (May 25, 1973), 813.
We don't all think alike, and this article proves it. But

can you argue with the point of view? The letters show you how to begin.

Marsh, G. P. *Man and Nature*. New York: Scribner's, 1864. (Reprinted by Harvard University Press, 1965.)

From a historical standpoint, this is the first warning of how badly we were using the land. Marsh points to overgrazing of land and overcutting of forests as having the potential to "destroy the balance which nature has established."

Moore, W. "Fire!" *National Wildlife Magazine* (August 1976), p. 4.

A readable account of how fire is used in forest management.

Preserving Our Natural Heritage: Volume I, Federal Activities. Prepared for U.S. Department of the Interior by The Nature Conservancy. Washington, D.C.: U.S. Government Printing Office, 1975.

No other book we have found brings together so much information on the role, holdings, and management practices of the federal agencies concerned with land preservation. Because it is so well organized and devoid of pictures and maps (these are referred to but never reproduced), there may be a problem with maintaining an adequate attention span. If you can remember your question on federal activities in conservation long enough, the answer is quite likely to be in this book, or a reference on the subject may be cited.

Journals on land preservation in which you may wish to browse are the *Journal of Forest History; American Forests; Journal of Forest Ecology; National Parks and Conservation Magazine;* and *Living Wilderness.*

References

Steen, H. K. *The U.S. Forest Service: A History.* Seattle: University of Washington Press, 1976.

Wonderfully detailed historical account of the beginnings and evolution of forest management from the post-Civil War Era to the early 1970s. Recommended for those keenly interested in the past of the forestry movement and its influence on the shape of the present.

Waver, R., and W. Supernaugh. "Wildlife Management in the National Parks," *National Parks* (July–August 1983), p. 12.

Bartlett, R. A. *Yellowstone—A Wilderness Besieged.* Tucson: University of Arizona Press, 1985.

The Environment—
A World Concern

Many Voices

*Our Narrow Niche in Time/Our Narrow Niche in Space/
A Thank You Across the Generations/Concern with the
"Here and Now"*

**Restoring the Environment—Personal
Choices**

A Choice of Methods/Environmental Decisions

Many Voices

In the introductory chapter of this book we told the stories of six Americans who cared enough about their environment to devote a substantial part of their life's energy to preserving it. Some people, for example John Van Dyke and Rachel Carson, seemed to view their efforts as a vital task—one of educating their fellow citizens to a danger or to the loss of something precious. Others, such as Lois Gibbs, eventually turned their efforts into a career. For still others, like William Green, environmental improvement became more of a crusade, into which money, time, and a tremendous force of personal will were poured. However we view these American pioneers of a better environment, though, it would be wrong to assume their concerns to be exclusive to Americans.

In countries all over the world, and among people of all levels of education and wealth, we find a similar understanding of the human need for a healthy environment, as well as a recognition of the value and rights of other species.

The American Joseph Wood Krutch wrote about wilderness:

These are the things which other nations can never recover. Should we lose them, we could not recover them either. The generation now living may very well be that which will make the irrevocable decision whether or not America will continue to be for centuries to come the one great nation which had the foresight to preserve an important part of its heritage. If we do not preserve it, then we shall have diminished by just that much the unique privilege of being an American.

No less eloquently, Renatas, a young Tanzanian park ranger, said:

After I be dead, others will follow. If people be killing, killing, there will be no more buffalo, no rhino. If they be cutting, cutting, there will be no more trees, no oxygen, no rain. Like a desert. What will my daughters think? They will come and there will be nothing. "Our father was stupid," they will say. (1985, p. 67)

The American naturalist Henry Beston (1962), writing on Cape Cod in 1928, shared these thoughts on the relationship of humans and other species:

For the animal shall not be measured by man. In a world older and more complete than ours they move finished and complete, gifted with extensions of the senses we have lost or never attained, living by voices we shall never hear. They are not brethren, they are not underlings; they are other nations, caught with ourselves in the net of life and time, fellow prisoners of the splendor and travail of the earth. (p. 25)

But in 1854, Chief Seattle, an Indian chief of the Washington Territory, had already expressed similar thoughts:

This we know. The earth does not belong to man: man belongs to the earth . . . all things are connected, like the blood which unites one family. . . . Man did not weave the web of life; he is merely a strand in it. Whatever he does to the web, he does to himself.★

Many quotations from American Indians show their respect for the environment and their concern for other species besides humans. A Hopi prayer is found in a story translated by Natalie Curtis:

> "When Lololomai, the chief, prays, how does he pray? Will you tell me?"
>
> He goes to the edge of the cliff and turns his face to the rising sun, and scatters the sacred cornmeal. Then he prays for all the people. He asks that we may have rain and corn and melons, and that our fields may bring us plenty. But these are not the only things he prays for. He prays that all the people may have health and long life and be happy and good in their hearts. And Hopis are not the only people he prays for. He prays for everybody in the whole world . . . everybody. And not people alone; Lololomai prays for all the plants. He prays for everything that has life. That is how Lololomai prays.

In Germany today, a new political party, die Grunen, or the Green Party, has developed partly out of the environmental movement. Although the concerns of the party are much more sweeping than just environmental protection, one of its basic tenets is that humans cannot continue to exploit their ecosystem but must learn to live within natural boundaries imposed by their environment. "Human life is bound up in the cycle of ecosystems: our actions affect nature, which in turn has consequences for us. The exploitation of nature as well as that of human beings must be opposed . . ."

Germany is a developed, or industrialized country, one that can well afford to spend time and money on improving the environment in which its citizens live and work. But it is not only the developed countries today that are concerned with their environment. An article in *China Reconstructs* (April 1986) argues that economic development is not the only factor that should be considered by officials who plan the country's future:

Poor in the past, [the farmers of Ghongxian County] thirst for a better life. However, some county leaders' limited knowledge of science and lack of foresight blind them to the way in which unchecked industrial wastes can destroy the environment and ruin life for themselves and their posterity.

For some developing countries respect for the environment has a long and honorable history. In an article in *Ambio* the Indian author Vandana Shiva noted that forest settlements, not urban areas, are recognized in India "as the highest form of cultural evolution, providing society with both intellectual guidance and material sustenance." In Part Eight, we noted the words of a Nigerian chieftain who, sensing the briefness of our own span of years compared to the infinity of years belonging to our ancestors and descendants, said, "I conceive that the land belongs to a vast family, many are dead, few are living and countless members are still to be born."★

We quote all of these people not only for their eloquence but also for the wisdom in their messages, a wisdom that recognizes the narrow niche we occupy in both time and space, a wisdom that is shared by all peoples, worldwide.

Our Narrow Niche in Time

The concept of the Nigerian chieftain places our years on earth in some perspective. Being aware of the finiteness of our time span makes us sensitive to the fact that depletable resources, such as the fossil fuels that we use and consume, are lost to future generations. The land we consume—by building houses, by paving, by surface mining, by allowing unchecked erosion—can only be restored over centuries. The species destroyed by human activities are lost forever. It is true that technology is constantly finding new resources, but this is not a process upon which we can depend. Nor *should* we depend on it, since each new application of technology seems to carry new environmental burdens. Shale oil may one day be produced in large quantities to power the engines of society, but the upheaval of the land and pollution of the water from producing shale oil make it

---

★Quoted in *Conservation News*, **38** (22) (November 15, 1973).

★Quoted in V. Curtis, ed., *Land Use and the Environment*, EPA Office of Research and Monitoring, Environmental Studies Division, 1973.

a potentially damaging technology. Commercial actions that maximize our well-being today may lead to degraded conditions for future generations. The sacrifices that parents make for their children are an example of how the human species instinctively looks to the protection of future generations.

Our Narrow Niche in Space

We are not alone, as Beston points out, nor can we survive alone. We are part of a wider ecological system. We cannot chart all the interrelationships in this system. Sadly, because we cannot find all the species of plants and animals upon which we depend, and upon which these species in turn depend, some of us are tempted to believe that what we can see is the whole of the system. Europeans of the fifteenth century felt that way about the extent of the earth, until Columbus sailed beyond the limits of contemporary knowledge. In the same way, new Columbuses show us, every day, that our knowledge of the environmental relationships on which we depend is incomplete. We are not alone, nor are we free to be independent of the environment that surrounds us.

A Thank You Across the Generations

These ideas on sharing time and sharing space lead us to consider actions that have long-term benefits. That is, the benefits of the actions we may take today are distant in time and are to be enjoyed principally by future generations, rather than by us. These ideas also lead us to consider the prevention of actions that will have long-run negative effects, even though the effects may fall outside of our own time span. We sense the necessity of protecting the welfare of future generations, but we have no way to measure the resulting satisfaction and comfort of those future generations. Few of us will be present 100 years from now to hear a "thank you," though it be a thunder that echoes from every succeeding generation. Yet, if you have ever seen the redwoods in California, or hiked the Pacific Crest Trail or the Appalachian Trail, or viewed the waterfall and geysers of Yellowstone, you may yourself have said a thank you to generations past. The wisdom that inspired the setting aside of these lands is both an inspiration and an instruction.

Long-run benefits stem from other actions than setting land aside. If population growth is slowed and

finally halted, our future impact on the earth can be slowed and controlled. When we turn away from the fossil fuels and nuclear power to limitless sources of energy such as the sun and wind, we preserve petroleum and coal for future uses. These steps also limit the degradation of the land from surface mining and help check oil pollution of the oceans. Turning away from nuclear power can lessen for future generations the burden of pollution by radioactive elements, and it can slow the spread of nuclear weapons, perhaps the most awesome threat to the future of life on earth.

Concern with the "Here and Now"

Concern about the narrowness of our niche in time and in space is complemented by a third concern that also motivates us to improve the environment: the "here and now." In some cases, human health and lives are threatened by pollution and toxic substances in the environment. In other cases, the welfare of plants and animals is threatened by pollution and human activity. To remove these threats requires action in the present. Such actions as controlling automobile emissions or scrubbing sulfur oxides from the stack gases of power plants have present benefits in decreasing the frequency of respiratory illnesses such as bronchitis. Removing such hazardous substances as asbestos and vinyl chloride from the workplace will lower the number of cancer deaths among the work force. Preventing oil spills in the oceans will protect waterfowl and aquatic species from being destroyed in local environments. Treating wastewater on a particular river to remove organic wastes will protect the oxygen content of the river and thereby ensure the survival of fish and other aquatic species.

This concern for the present is one point of focus for improving the environment. Indeed, a large portion of this book has been devoted to the subject of cleaning up pollution in water, in the air, and in the workplace. Fortunately, at the same time that we are making improvements in the environment, we can also make long-term changes for the better. Reducing the pollution burden today is plainly a commitment to maintain the quality of the air and water and workplace for future generations. Furthermore, since many toxic chemicals can also cause birth defects and changes in the genetic information, the removal of toxic substances from the workplace has long-term as well as short-term benefits.

Because of this implied promise to maintain the quality of the environment once it has been improved, our short-term interest in improving our surroundings merges with our long-term concern for the quality of the environment of future generations.

Restoring the Environment— Personal Choices

A Choice of Methods

It would be far too simplistic for us to offer you a formula for environmental progress. What works in one time and place may be terribly inappropriate in another. And it would be arrogant for us to advise you in what manner to act. As you can see in the statements that follow, people respond to the challenge in ways that differ not only in their impact, but also in their intent:

My contribution to the Sierra Club is helping to influence government action on environmental/ conservation problems. (Sierra Club member, *Sierra Club Bulletin*, March/April 1979).

Fix it up
Wear it out
Make it do
Do without
(Old Yankee saying)

Save the Whales, Don't Buy Japanese Products (Bumper Sticker)

We want to build a self-sufficient community, complete with organic gardens, solar heat, wind power, and—especially—the chance to provide an environment God intended for raising healthy families. . . . (From the advertising section of *Mother Earth News*, No. 36, November 1975)

We should applaud and look up to those who adopt life styles that are modest in terms of the amount of space they monopolize or the amount of materials and energy they consume. (Maurice Strong, in *Mazingira*, No. 3/4, 1977)

In 1976, we tried to save [the harp seal] pups by spraying their coats with a harmless green dye. . . . In 1977, we were back again, placing our bodies over the pups to save their lives. (Greenpeace letter, 1980)

Political action groups work within the system and, if they are successful, become a part of the system. Their implied purpose is to influence the government's position on environment and conservation issues and possibly, therefore, influence the lives of many other people. The choice of an alternate lifestyle, on the other hand, is an act that usually influences only one or a few lives. Although it does not touch the system, it has its special satisfactions. Civil disobedience, in contrast, threatens or blocks the system itself. It draws attention, but requires large sacrifices. Still another way to influence the environment is to choose a profession that is concerned with environmental improvement. These are the ways people have chosen to influence the environment:

political action groups
a conserving ethic
consumer boycotts
alternate lifestyles
civil disobedience
an environmental profession

Environmental Decisions

How you decide to influence environmental progress is up to you. What is important to remember, however, is that decisions about the environment are being made right now. Decisions are being made by politicians, administrators, company executives, and individual citizens. Often, these decisions require inputs from experts on matters such as the biology and technology of a situation; the effects of pollutants at various levels; the cost of cleaning up; and the people who gain or lose benefits when a substance or technology is changed or set aside.

Nevertheless, we must all decide the level of costs we are willing to bear, and who should bear them; what benefits we can give up, and what risks we are willing to endure. Experts can only lay out the choices; you must help make the decisions. No one is more qualified to decide social, moral, and economic issues—which come up again and again—than you, the individual citizen. This privilege and this burden are yours in a democratic society.

We have exposed you to the clash of opinions over values, over risk, and over science to show you where and how your own opinions and views are needed and are valuable in the environmental debate. And we have tried to give you as much information as possible about present-day environmental problems. In the future, however, you will have to face new problems, problems now only dimly seen or, perhaps, problems not yet dreamed of, even by the most far-seeing environmental expert. We can only leave you with a plea to enter the decision process because of the importance of your views.

References

Renatas. Quoted in B. McBride, *Sierra* (March/April 1985), p. 67.
Beston, Henry. *The Outermost House.* New York: Viking Press, 1962.
Shiva, V. "Reforestation in India," *Ambio*, **14** (6) (1985).

I know no safe depository of the ultimate powers of the society but the people themselves; and if we think them not enlightened enough to exercise their control with a wholesome discretion, the remedy is not to take it from them, but to inform their discretion.
Thomas Jefferson

Glossary

abiotic Nonliving; referring to the nonliving components of ecosystems (water, light, etc.).

acclimation The biochemical changes that enable an organism to withstand changed temperatures (either higher or lower).

acid mine drainage Water that has dissolved iron pyrites (ferrous sulfide), which were left behind from coal-mining operations. The water becomes acidic and deposits ferric hydroxide (yellow boy) on stream bottoms. The acid makes the water undrinkable and is harmful to aquatic life. The acid waters are also low in dissolved oxygen, a substance needed by aquatic life for survival.

acid rain Rainfall with a high acid content falling downwind of major fuel-burning areas. Sulfur oxides from the burning of fossil fuels are the culprit. Acid rain may stunt the growth of plants and turn lakes and streams acidic, driving out the normal aquatic species.

activated sludge plant A device for removing dissolved organics from wastewater. The plant is a well-aerated tank in which microbes using oxygen convert the dissolved organics to simpler substances.

adaptation A characteristic that helps an organism survive in a particular environment.

aerobic Referring to an environment in which oxygen is present.

algae Simple, often microscopic, plants that live in water or very moist land environments.

amino acids Chemical compounds from which proteins are made.

anaerobic Referring to lack of oxygen. In the context of water pollution, a condition of water in which all the dissolved oxygen has been used up or removed. Only a few specially adapted species can survive in anaerobic (oxygen-depleted) waters.

aquifer An underground body of water whose precise dimensions are unknown.

arable Able to be farmed.

area mining of coal A surface-mining method in which the overburden is removed and laid up in successive parallel rows; the technique is used on relatively flat lands.

bacteria Single-celled organisms visible with a microscope. Some bacteria are harmful and cause plant and animal diseases. Many more are harmless or even useful. Bacteria aid in the decay and recycling of organic matter and are used industrially (e.g., to produce drugs or ferment dairy products).

ballast Weight used in ships in certain areas of their cargo space to make them stable and easy to steer. Oil tankers use as ballast either the oil they carry or seawater.

barrel of oil One barrel of oil equals about 43 gallons. A ton of oil varies in volume according to what kind of oil it is (e.g., gasolines are lighter than diesel oils), but as a general statement, one ton of oil is about seven barrels or 300 gallons.

biochemical oxygen demand (BOD) The amount of oxygen that would be consumed if all the organics in one liter of polluted water were oxidized by bacteria and protozoa; it is reported in milligrams per liter. The BOD number is useful in predicting how low the levels of oxygen in a stream or river may be forced to go when organic wastes are oxidized by species in the stream.

biodegradable Able to be broken down by living organisms.

biological controls Pest-control methods that use natural predators, parasites, or diseases or that rely on the use of naturally produced chemicals such as insect pheromones.

biological magnification The process by which certain, often toxic, materials become more concentrated as they move up food chains. That is, organisms at the top of the food chain contain more of the substance than do organisms on the bottom of the food chain, or than does the environment itself.

biomass The weight of the living creatures in a given area.

biomass energy Energy derived from crops (trees, sugar cane, corn, etc.) by either direct burning or by conversion to an intermediate fuel such as alcohol.

biomes Climax communities characteristic of given regions of the world.

biota The living organisms, plant and animal, in a region.

birth rate The number of babies born each year per thousand people in a population.

bituminous coal   The most plentiful form of coal in the United States. Its high heating value and abundance also make it the most widely used coal. Its principal use is in steam electric power plants.

black lung disease   A condition in which the elasticity of the lung is destroyed, caused by the inhalation of coal dust over a relatively long period. Many coal miners are permanently disabled by the disease.

bloom   A rapid overgrowth of algae.

blowdown   The water removed from that circulating in a cooling tower to prevent solids from building up in the tower water. This water is replaced by fresh water called "make-up."

blowout   Explosive release of gas and/or oil from an oil well.

BOD   *See* biochemical oxygen demand.

boiling-water reactor (BWR)   A nuclear reactor in which the water passing through the core is heated by the fission process and is converted directly to steam, which turns a turbine. The BWR is one of the two major types of nuclear reactors used for electric production. *See also* pressurized water reactor.

breeder reactor   A nuclear reactor in which the coolant and heat-transfer medium is molten sodium. The fuel is plutonium-239, which is continuously bred from uranium-238.

Btu   The British thermal unit, the quantity of heat needed to raise the temperature of 1 pound of water by 1°F. One Btu is equivalent to 1.054 kilojoules (a common metric energy unit).

bulb turbine   A turbine whose blade face is oriented perpendicular to the direction of water flow. In contrast to the Peyton, Francis, and Kaplan wheels, which require water elevations of 100 feet or more, the bulb turbine can generate electricity from only rapidly moving water. It thereby makes possible both tidal power and hydro development at many sites not previously feasible.

Bureau of Land Management   The managing agency of the National Resource Lands. The Bureau is within the Department of Interior.

BWR   *See* boiling-water reactor.

calorie   A measure of the heat or energy content in food. Human food requirements are usually measured in kilocalories (the amount of heat needed to raise the temperature of 1,000 grams of water by 1°C), and the term is written with a capital c, "Calorie."

carbon cycle   The cycling of carbon in nature. Carbon dioxide is produced when plants and animals respire and is consumed by green plants during photosynthesis. Carbon dioxide also dissolves in the oceans and precipitates as limestone or dolomite.

carbon dioxide   A colorless, odorless gas at normal temperatures, composed of one atom of carbon and two of oxygen. It makes up about 0.03% of the atmosphere by weight. Carbon dioxide is consumed in photosynthesis by green plants and produced by the respiration of plants and animals and the burning of fossil fuel.

carbon monoxide   A gas consisting of one atom of carbon bonded to one of oxygen. Its action on human health is a result of its "tying up" of hemoglobin, the protein that carries oxygen to the cells. The principal source of this major air pollutant, formed from the incomplete combustion of carbon, is the automobile.

carboxyhemoglobin   Carbon monoxide combines with the oxygen-carrying blood protein hemoglobin to form carboxyhemoglobin. This combined form cannot transport oxygen in the bloodstream.

carcinogen   Anything that causes cancer.

carnivores   Meat-eaters.

carrying capacity   The largest population a particular environment can support indefinitely.

catalytic converter   The device used on automobile exhaust gases to convert carbon monoxide to carbon dioxide and hydrocarbons to carbon dioxide and water. An additional converter may be used to convert oxides of nitrogen back to oxygen and nitrogen.

chlorinated hydrocarbons   Chemicals composed mainly of carbon and hydrogen plus one or more atoms of chlorine. Examples are the pesticides DDT, aldrin, dieldrin, chlordane, and heptachlor.

chlorinated organics   Chemical compounds composed of carbon, hydrogen, oxygen, and one or more atoms of chlorine. These chemicals can be formed in drinking water by the action of the disinfectant chlorine on organic chemicals found in some water supplies.

chlorination   The controlled addition of chlorine to water destined for drinking and to wastewaters being discharged into receiving bodies. Chlorine kills bacteria and viruses and so renders the water safe for human consumption and use. The process is also known as disinfection.

cholera   An intestinal disease caused by specific bacteria. The disease can be spread by water polluted by human wastes.

clear-cutting   The practice of cutting all trees in an area, regardless of size, quality, or age. The practice hastens erosion, is visually displeasing, and leads to a loss of species habitat. Better practices are selection cutting, seed-tree cutting, and shelterwood cutting.

climate   A complex of factors affecting the environment. Climate includes temperature, humidity, amount of precipitation, rate of evaporation, amount of sunlight, and winds.

climax community   The characteristic and relatively stable community for a particular area.

coagulation   A process for the removal of suspended material from drinking water. A "floc" of insoluble material is created by the addition of alum or ferrous sulfate. When the floc settles in a detention basin, suspended material is captured in the floc and is settled out as well.

coal cleaning The removal of sulfur from coal by washing and chemical steps. Cleaned coal is less likely to require "scrubbers" to remove sulfur oxides from the stack gases.

coal gasification The conversion of the carbon in coal to a gas that can burn. Coal is burned in the presence of oxygen and steam to produce carbon monoxide and hydrogen gases. This low-Btu gas can be burned directly or can be converted to a high-quality methane by the addition of hydrogen in further reactions. The latter product can be substituted for natural gas.

coal liquefaction The conversion of coal to a hydrocarbon liquid. Pyrolysis is one method for this conversion. A modification of coal gasification will also produce a hydrocarbon liquid. Solvent refining of coal is a third possible process.

coal-slurry pipeline A technology in which coal is transported via pipeline as a mixture (about 50/50) of coal particles and water.

cocarcinogen A substance that does not, by itself, cause cancer but that can cause cancer in combination with some other substance.

coliform bacteria A category of bacteria largely derived from fecal wastes. The presence of these bacteria in a river or stream is taken as evidence of fecal pollution and indicates the possibility that pathogenic (disease-causing) bacteria may be present.

combined cycle power plant A power plant in which oil or gas is first burned in a turbine (jet) engine, generating turning power in the engine for electric generation. The hot gases are then used to boil water to steam to turn a conventional turbine for power generation. The efficiency of conversion of heat to electricity is increased by this two-stage process.

combined sewers A sewer system that carries both domestic wastewater and the water from rainstorms. During storms, the total of the flows is too large to be treated and hence sewage enters the water body virtually without treatment.

community All of the living creatures, plant and animal, interacting in a particular environment.

competition In ecological terms, the struggle between individuals or populations for a limited resource.

condense To change from a gas to a liquid, as when steam is condensed to water in a power plant condenser.

consumers Organisms who eat other organisms. Primary consumers eat producers, secondary consumers eat primary consumers, and so on.

contour mining of coal A surface-mining method in which L-shaped cuts are made into the hillside in long curving arcs that follow the contour of the hill.

contraceptive A device, chemical, or action that prevents pregnancy.

cooling tower A structure designed to cool the water that was used to condense steam at a power plant.

core (of a nuclear reactor) The concrete and metal shielded structure that houses the nuclear fuel in a reactor; the place where the fission process and heat production take place.

crude oil Oil as it comes from the ground, in its natural state. Crude oil is a mixture of many chemical compounds.

DDT A member of the chlorinated hydrocarbon family of pesticides. DDT was banned in the United States in 1972 because it was interfering with reproduction in certain bird species.

death rate The number of people who die each year per thousand people in a population.

deciduous forest biome A biome characterized by trees that lose their leaves each fall.

decomposers Organisms that take part in the decay of organic materials to simple compounds (e.g., bacteria and fungi).

demographic transition Pattern of change in which birth rates fall as, or after, death rates fall. After the transition, a country's birth rate is closer to its death rate and the population does not grow rapidly.

demography The study of populations.

desertification The severe degradation of a land environment to the point where it resembles a desert.

detritus Dead organic matter, composed of plant and animal remains.

developing nation A term applied to countries that have little or no technological development.

development rights The rights, most often accompanying ownership of the land, that enable the owner to construct buildings, build roads and sewers, and otherwise alter the land. Such rights can be sold or transferred to other parties, separating the rights from the land itself.

digester A water pollution control device used to further degrade and stabilize the organic solids that arise from primary and secondary wastewater treatment.

dissolved oxygen The amount of oxygen dissolved in water, reported in milligrams per liter. Levels of 5 mg/l or above indicate a relatively healthy stream. The maximum level of dissolved oxygen is ordinarily 8–9 mg/l, depending on the water temperature.

district heating System by which the spent steam from a steam-electric power plant is carried in underground pipes to homes, offices, factories, and so on for heating during the winter months.

diversity A measure of the number of species in a given area. The more species per square meter, the higher the diversity.

DNA (deoxyribonucleic acid) The hereditary material contained within cells that determines the characteristics of an organism.

drift Particles of liquid water that escape from cooling towers and cause corrosion or other environmental problems.

ecology   The study of the interaction between organisms and their environment.

ecosystem   All of the living organisms in a particular environment plus the nonliving factors in that environment. The nonliving factors include such things as soil type, rainfall, and the amount of sunlight.

effluent   A liquid or gaseous waste material produced from a physical or chemical process.

electric power plant (thermal)   Power plant in which the heat (hence the word "thermal") from a burning fossil fuel or from the fissioning of uranium is used to boil water to steam and the steam is then used to turn a turbine, generating electricity.

electrostatic precipitator   A device that removes particulate matter from stack gases; the particles are given an electric charge and then attracted to a collecting electrode. The precipitator is highly efficient and is widely used on electric power plants.

emergency core cooling system   A spray system designed to inject water rapidly into the core of a nuclear reactor that has experienced a loss-of-coolant accident.

emigration   The movement of organisms out of a population.

endangered species   A species with so few living members that it will soon become extinct unless measures are begun to slow its loss.

enrichment   The process of converting uranium from 0.7% uranium-235 to 3% uranium-235. The higher concentration is needed for the fuel rods of the reactor of a nuclear electric power plant.

entrainment   Entrapment of organisms in the condenser water pipes of a power plant.

epidemiology   The study of disease in groups of people or populations.

erosion   The loss of soil due to wind or as a result of washing away by water.

estuary   A coastal body of water partly surrounded by land but having a free connection with the ocean.

eutrophic   Water that has a high concentration of plant nutrients.

evolution   Change in the frequency of occurrence of various genes in a population over a period of time.

fauna   The animal life of a particular region.

fertility rate   The total number of live births a woman in a particular country is expected to have during her reproductive years. (Also called total fertility rate.)

fertilizer   A material that promotes the growth of plants. It can be natural or artificial.

filtration   In the context of water supply, the practice of forcing water through beds of sand to trap the bacteria that cause disease and thereby prevent their presence in drinking water.

first law of thermodynamics   Law stating that energy can be changed from one form to another, but it cannot be created and it cannot be destroyed.

fission   Process that occurs when the nucleus of certain heavy atoms is struck by neutrons, breaking into two or more fragments (fission products) with the production of heat and more neutrons; fission means "breaking apart."

food chain   A picture of the relationship between the predators in an area and their prey (i.e., who is eating whom). This term is applied when the relationships are simple and few creatures are involved.

food web   Interconnected food chains made up of many organisms, with many interrelationships.

fossil fuels   Fuels like coal, oil, and natural gas, derived from the remains of organic matter deposited long ago.

fuel cell   A device in which oxygen combines with hydrogen or carbon monoxide, producing direct current electricity. Fuel cells are expected to be relatively clean producers of electricity, unless coal is burned to provide the carbon monoxide.

fuel reprocessing   Process in which spent nuclear fuel rods are broken apart, and the highly radioactive fission products are separated and uranium and plutonium are recovered to be reused.

fusion   The combination of the deuterium atom (a form of hydrogen) with either another deuterium atom or a tritium atom (still another hydrogen form). The combination releases enormous heat, which scientists hope some day to capture to produce electrical energy. Fusion is also the basis for the hydrogen bomb.

gamma ray   A highly penetrating form of ionizing electromagnetic radiation that is produced by one type of radioactive decay.

geopressured methane   Natural gas known to be dissolved in salt water in deep caverns at extremely high pressures and temperatures. It is not known whether geopressured methane can be recovered profitably.

geothermal heat   Heat from the earth's interior carried to the surface as hot water or steam. The steam and hot water can be used to heat homes, offices, and factories or can be used to generate electricity.

geothermal power plant   An electric generating station that uses hot water or steam from the earth's interior as the energy source. Releases of steam from the earth can be used directly to turn a turbine. Alternatively, hot water releases can be "flashed" to steam for such uses. Or hot water can be used to vaporize a fluid such as isobutane, which will then be used to turn a turbine.

greenhouse effect   Atmospheric heating that occurs when outward heat radiation is blocked by carbon dioxide molecules that absorb the energy. The greenhouse effect suggests the earth will undergo a warming trend as carbon dioxide from fossil fuel combustion accumulates in the atmosphere.

green revolution   The term given to the new developments in farming, including the use of high-yielding grains, that promise to enable farmers to grow much more food on the same number of acres than with conventional techniques and older crop varieties.

groundwater   Water beneath the earth's surface.

growth promoter   A substance that makes an animal grow better or more quickly.

habitat   The physical surroundings in which an organism lives.

half-life   The time required for one-half of a given quantity of a chemical to disappear from the environment (or to be excreted by the body, if it is a chemical absorbed by a living organism).

hazardous wastes   Any waste materials that could be a serious threat to human health or the environment when disposed of, transported, treated, or stored.

heat pump   A device that draws in cold air and exhausts this air at an even colder temperature. The heat captured is transferred via a refrigerant liquid to the indoor air. The heat pump has the potential to cut in half electric requirements for home or hot-water heating.

hectare   A metric unit of area. One hectare is 100 meters by 100 meters and equals 2.47 acres.

hepatitis   A viral disease in which the liver becomes inflamed. Epidemics of hepatitis have been traced to contaminated water supplies.

herbicides   Chemicals used to kill weeds.

herbivores   Plant-eaters.

highwall   The wall of overburden and coal left behind after contour mining of hillsides.

high-yielding grains   New varieties of corn, wheat, and rice developed by agricultural research. These varieties produce much more grain per acre than older varieties. They also require more water, pesticides, and fertilizers.

humus   Large, stable organic molecules formed in the soil from the breakdown of organic waste materials. Humus contributes to soil fertility by helping to retain water and keeping the soil loose.

hydrocarbons   A class of air pollutants derived principally from the operation of internal combustion engines of motor vehicles. These pollutants contribute to photochemical pollution by the reactions they undergo.

hydroelectric energy   Electric energy derived from falling or moving water. The water is commonly stored behind a dam and released through penstocks to turn a turbine and generate electricity.

hydrogen economy   An energy system that uses hydrogen to store and produce energy. Hydrogen from chemical processing of fossil fuels or from the electrolysis of seawater is used in fuel cells, which in turn generate electricity by combining the hydrogen with oxygen or carbon monoxide.

hydrologic cycle   The cycling of water in the environment, from rainfall to runoff to evaporation and back again.

immigration   The movement of individuals into a population.

indicated and inferred reserves (of oil or gas)   Oil or gas known to exist and likely to be recoverable with the application of additional technology (indicated) and oil or gas for which we already have some limited evidence of existence (inferred). *See also* proved reserves.

inversion   A weather phenomenon in which cold air lies close to the earth's surface, trapped by a warm air mass above it. This is the inverse of the normal situation in which temperature decreases with increasing distance from earth.

ionizing radiation   Rays or particles that possess enough energy to separate an electron from its atom. Some ionizing radiation is electromagnetic, for instance, x-rays and gamma rays. Other forms are particles such as electrons (beta radiation), neutrons, and the nuclei of helium atoms (alpha radiation).

IPM   Integrated pest management, a combination of techniques designed to control pests using a minimum of chemical sprays.

irrigation scheme   To supply water, other than natural rainfall, to farmland.

kilowatt   A rate of providing electrical energy; equal to 1,000 watts.

kwashiorkor   A children's disease caused by protein deficiency.

laeterization   The process by which certain tropical soils containing iron harden into a material called laeterite when vegetative cover is removed. The resulting soil is hard enough to cut into blocks for building and is no longer suitable for plant growth.

land application of wastewater   A set of three different processes (overland flow, spray irrigation, and infiltration-percolation) in which wastewater is applied to the land as a means of treatment. The wastewater should previously have undergone primary treatment.

leaching   The movement of a chemical through the soil.

leaching field   A system of underground tiles designed to channel the flow from a septic tank into porous soil.

lead   A cumulative poison that can cause brain damage and death; it is present in air (from the lead in gasoline), food, and water. It was once in paint as well, and the eating of paint chips by children is a frequent cause of lead poisoning.

lignite   A lower form of coal with a little over half the heating value of bituminous or anthracite coal; it constitutes about 4 percent of the U.S. coal energy resource.

limiting factor   Whatever nutrient is in shortest supply compared to the amount needed for growth. This factor limits the growth of plants in a particular environment.

liquefied natural gas (LNG)   Natural gas made liquid at very low temperatures ($-162°C$) in order to transport it economically via special tankers.

liquid metal fast breeder reactor (LMFBR)   A nuclear reactor in which a liquid metal (sodium) circulates through the core, removing the heat from the fission of plutonium. The heat in the liquid metal is eventually used to boil water to steam to turn a turbine. The fuel is plutonium-239 created by neutrons striking the nucleus of uranium-238. The scheme is designed to use uranium-238 rather than uranium-235 because uranium-238 is so much more abundant.

longwall mining   A relatively new method of mining coal underground. The mine roof is allowed to collapse in a controlled way as the mine "room" moves across the coal seam. Greater coal removal and prevention of acid mine drainage are claimed advantages.

loss-of-coolant accident   What would happen in a nuclear reactor if the water in the core were lost because of a pipe break. The temperature in the core would rise rapidly as the heat from fission builds up. The fuel rods could melt and radioactive substances would be released through the pipe break if the core were not cooled quickly by the emergency core cooling system.

low-head hydropower   Energy generated using water elevations of fifty feet or less. The bulb turbine, a new technology, makes possible relatively efficient capture of energy from low-head dam sites not currently in use for electric generation.

magnetohydrodynamics (MHD)   Process in which hot gases from the burning of a fossil fuel are seeded with potassium, which then ionizes. Electric current is extracted from the hot gases, which are then used to boil water to steam for conventional electric generation.

malnourished   Referring to someone who does not get enough of the various nutrients needed for good health.

marasmus   A form of malnutrition caused by feeding infants overdiluted formula using unsterile water.

mariculture   The "farming" of marine organisms. This may be done in pens or rafts in coastal waters or in artificially maintained salt-water environments.

marine   Having to do with the oceans or salt waters of the earth.

megawatt   A term describing the rate at which electricity can be generated by a power plant. One megawatt is 1,000 kilowatts.

metabolism   The chemical reactions that take place within a living organism or cell. These reactions include those yielding energy for life processes and those synthesizing new biological materials.

methemoglobin   A form of hemoglobin in which the iron is oxidized. Methemoglobin is not able to carry oxygen in the bloodstream.

metric ton (or long ton)   1,000 kilograms or about 2,200 pounds, 10% larger than the (short) ton of English measure.

MHD   *See* magnetohydrodynamics.

middle distillates   That portion of crude oil refined to diesel fuel, kerosene (jet fuel), and home heating oil.

migration   The periodic movement of organisms into or out of an area.

monoculture   The cultivation of a single species of plant as opposed to mixtures of species, as is usually found in nature.

multiple-use   The concept applied to the management of lands in the national forests. The concept allows timbering, mining, recreation, grazing, watershed protection, fishing, wildlife protection, and wilderness as legitimate uses of the same land. The official designation of wilderness, however, most often excludes timbering, mining, and grazing.

mutagen   Substance that causes an inheritable change in a cell's genetic material.

mutation   An inheritable change in the genetic material of an organism.

National Park Service   The managing agency of the National Park System. The National Park Service is within the Department of Interior.

natural gas   A gas consisting mainly of methane, a simple hydrocarbon gas. It is derived from chemical and physical processes operating on buried ocean plankton. Nitrogen may be present in the gas as well.

natural selection   A difference in reproduction whereby organisms having more advantageous genetic characteristics reproduce more successfully than other organisms. This leads to an increased frequency of those favorable genes or gene combinations in the population.

NEPA   National Environmental Policy Act. One of the most important parts of this act is that it requires environmental impact statements. These statements are reports based on studies of how a proposed government project will affect the environment.

neutron   A fundamental particle, without any charge, found in the nucleus of an atom. In the fission process, neutrons are unleashed. They strike the nuclei of atoms of uranium-235 and cause these atoms to break apart.

niche   Where an organism lives and how it functions in this environment (i.e., what it eats, who its predators are, what activities it carries out).

nitrate (NO_3^-)   A salt of nitric acid. Nitrate is a major nutrient for higher plants and also a food additive and water pollutant.

nitrite (NO_2^-)   A salt of nitrous acid. Nitrite is a food additive but is very toxic above certain concentrations.

nitrogen oxides (NO_x)   A contributor to photochemical air pollution. Nitrogen oxides are produced during high-temperature combustion of fossil fuels. Oxygen and nitrogen from the air produce the pollutant gas. The oxides of nitrogen (nitric oxide and nitrogen dioxide) are both air pollutants, and nitrogen dioxide has been linked to an increase in respiratory illnesses.

non-point-source water pollution Polluted water, arising typically from rural areas, that enters a receiving body from many small widely scattered sources.

nucleic acids Chemical compounds from which important biological materials (such as the hereditary materials DNA and RNA) are made.

ocean thermal energy conversion (OTEC) A concept now being tested on a pilot scale to use the temperature difference between the warm surface waters of the ocean and the cold deeper waters to produce electrical energy. The warmer waters would boil a working fluid to a "steam," which would turn a turbine. The colder water would condense the vapor for reuse.

oil A substance derived from chemical and physical processes operating on buried ocean plankton. It consists mainly of liquid hydrocarbons (compounds of hydrogen and carbon). Nitrogen and sulfur are other elements that may be present.

oil shale Shale rock containing oil; *see* shale oil.

oligotrophic Referring to water that has low concentrations of plant nutrients.

organic chemical A chemical composed mainly of carbon, hydrogen, and oxygen.

organic wastes A class of water pollutants composed of organic substances. When these substances are oxidized by bacteria and other species, oxygen is removed from the water. When high concentrations of organic wastes are present, aquatic species may be deprived of the oxygen necessary for survival.

organophosphates A major group of synthetic pesticides consisting of organic molecules containing the element phosphorus. A number of them are extremely toxic to humans; however, they do not persist in the environment for long periods of time.

oxidant A chemical compound that can oxidize substances that oxygen in the air cannot. Ozone, a prominent photochemical air pollutant, is an oxidant, as are nitrogen dioxide, PAN compounds, and aldehydes. Levels of photochemical pollution are often reported as oxidant levels.

ozonation The process of treating water intended for drinking with ozone gas in order to kill microorganisms. Although this process is used widely in Europe, most water engineers in the United States prefer chlorination because of the simple test for free chlorine.

ozone A compound made up of three atoms of oxygen (O_3). The oxygen we need to breathe is O_2.

particulate matter A class of air pollutants consisting of solid particles and liquid droplets of many different chemical types. Examples are fly ash, compounds from photochemical reactions, metal sulfates, sulfuric acid droplets, and lead oxides. Particles will soil and corrode materials; they may also soil and react chemically with the leaves of plants. Particles are linked firmly to increases in human respiratory illnesses, and some particle types are suspected of causing human cancer.

pathogens Disease-producing microorganisms, including bacteria, viruses, and protozoa.

PCBs Polychlorinated biphenyls; a family of chemicals similar in structure to the pesticide DDT and having a variable number of chlorine atoms attached to a double ring structure.

peak-load pricing The practice of raising the price of electricity during the hours (or season) of peak demand and lowering the price during times of slack demand. The goal is to level the rate of electric usage through time and decrease the need for bringing into service new electrical-generating capacity to meet peak demands.

peat A low-heating-value fossil fuel derived from wood that decayed while immersed in water.

permafrost Frozen layer of soil underlying the arctic tundra.

persistence The length of time a pesticide remains in the soil or on crops after it is applied.

pesticide A substance that kills pests such as insects or rats.

petroleum A mixture of organic compounds formed from the bodies of organisms that died in prehistoric times. This complex mixture is separated into less complex fractions, such as gasoline, heating oil, asphalt, and so on.

pH A measure of the acidity or alkalinity of solutions. Solutions with a pH of 8 or above are alkaline or basic; solutions with a pH of 6 or below are acidic.

photochemical air pollution Air pollutants such as nitrogen dioxide, ozone, aldehydes, and PAN compounds produced as a consequence of sunlight-stimulated reactions involving nitrogen oxides and hydrocarbons. The various compounds have differing effects, but plants are damaged by ozone and PAN compounds. Eye and throat irritation and respiratory illness are common effects of these substances on people.

photosynthesis The process by which green plants use carbon dioxide, water, and the energy in sunlight to synthesize organic materials.

photovoltaic conversion A process in which silicon cells are used to convert sunlight directly to electricity. Although costs of cells are high, attempts to make photovoltaic conversion economical are underway.

phthalates Chemical compounds that are esters of phthalic acid and various alcohols; for example, diethyl phthalate.

physical-chemical wastewater treatment A relatively new set of methods aimed at replacing conventional primary and secondary wastewater treatment processes. Physical-chemical treatment is designed to remove phosphorus by precipitation and settling and to remove dissolved organics by adsorption on carbon particles.

phytoplankton Microscopic, drifting plant species.

pica   The habit of eating nonfood items; it occurs among about 50% of all children, independent of social or economic class, beginning at about one year of age. Such items as paper, string, dirt, and paint chips may be eaten.

plankton   Microscopic plants and animals that drift in water, mostly at the mercy of currents and tides.

plutonium   A radioactive element of high atomic weight, produced in nuclear reactors. Plutonium can be used to make atomic weapons or as fuel for nuclear reactors.

point-source water pollution   Water pollutants that arise in urban areas or from industries and that enter a receiving body from a single pipe.

population (natural)   The members of a species living together in a particular locality.

population profile   A bar graph showing the number of people in each age group in a population.

power tower   A method to generate central station electricity using the sun as the energy source. The sun's rays are focused by banks of mirrors on a tower to provide the heat to boil water to steam. The steam turns a turbine for conventional electric generation.

precipitation   In ecological terms, the amount of water that falls as rain or snow on a given area.

predator   A creature that eats another.

pressurized-water reactor (PWR)   A reactor in which the water passing through the core is heated by the fission process but does not boil because it is under great pressure. This extremely hot water is used to boil to steam a parallel but separate stream of water in a steam generator. The steam in this second loop is used to turn a turbine for electric generation. *See also* boiling water reactor.

prey   A creature that is eaten by another.

primary production   The energy captured by plants in photosynthesis. Gross primary production measures the amount of energy stored as organic materials, as well as that used in respiration by the plant. Net production includes only the amount stored.

primary recovery   The quantity of oil that flows out of a well by natural pressure alone; it averages about 20% of the oil in place.

primary treatment of wastewater   The first major wastewater treatment process in the typical set of processes used in American cities. Primary treatment consists of allowing the organic particles that were in suspension to settle out of the flowing water.

producers   Organisms who produce organic materials by photosynthesis.

productivity   Amount of living tissue (plant or animal) produced by a population in a given period of time.

promoter   A substance that does not cause cancer itself but that can act to cause another material to be carcinogenic.

pronatalist program   A government program that encourages families to have more children.

proved reserves (of oil or gas)   Oil or gas *known* to be contained in the portions of fields that have already been drilled and which is profitable to recover.

pumped storage   The pumping of water to a high elevation at times when spare electric capacity is available. The water can then be released from its high pool through the turbine pump that raised it in order to generate electricity. Such releases are made when the existing basic generating capacity is insufficient to meet electrical demand.

PWR   *See* pressurized-water reactor.

quad   One quadrillion (a million billion) Btu. The annual energy use of an entire nation is often given in quads.

radioactivity   Particles or rays emitted by the decay of unstable elements. Examples are alpha, beta, and gamma radiation (*see also* ionizing radiation).

rate of growth   A country's growth rate is the rate of increase (birth rate minus death rate) plus the rates of change due to immigration (people arriving) and emigration (people leaving).

rate of increase   The birth rate minus the death rate of a population.

recharge   Applied to groundwater, the replacement of water that has been withdrawn from the ground. The replacement can occur from surface or other underground sources.

reclamation of surface-mined land   In the mining of coal, the replacement of overburden and topsoil and revegetation of the land.

refining (oil)   Separation of crude oil into less complex mixtures of substances (i.e., gasolines, kerosenes, heating oils, waxes, tars, asphalts).

refuse banks (or "gob" piles)   The wastes from a coal preparation plant, which washes and grades coal. The wastes include low-grade coal, shale, slate, and coal dust.

reservoir   The body of water backed up behind a dam; also refers to an accumulation of underground water.

residue   The amount of a chemical left in a food by the time it reaches the consumer.

resistance   Condition that occurs when a particular germ or pest is no longer killed by a drug or pesticide.

room-and-pillar coal mining   A method of underground coal mining in which pillars are left in the mine in a regular pattern in order to support the mine roof while the coal mining is going on.

runoff   Water that comes to the earth as rain and runs off the land into lakes, rivers, and oceans.

salinity   Salt content.

salinization   The accumulation of salts in soil. Eventually, the salt buildup prevents plant growth.

sand filtration   *See* filtration.

satellite solar power station   The idea of installing photovoltaic cells in a satellite orbiting earth. Electricity generated by the cells would be beamed to earth via microwave. The system would be enormously costly if ever undertaken.

scrubbers   A class of devices (differing widely in their chemical process steps) that remove sulfur oxides from the stack gases of coal-burning power plants.

secondary recovery   Techniques such as gas injection or water injection to place increased pressure on oil in a deep reservoir and thus force more of it to the surface.

secondary treatment of wastewater   The second major wastewater treatment process in the typical series of processes used in American cities. Two processes, the trickling filter and the activated sludge plant, are used to remove dissolved organic material from wastewater.

second law of thermodynamics   Law stating that when energy is changed from one form into another, some energy is always unavailable or lost as heat. Another way of stating this explains the inefficiency of electric power generation—namely, that a natural limit exists on the extent to which heat can be converted to work and hence to electricity. The limit in the conversion of heat to work is about 45%–50% conversion.

sediment   The fine particles of soil washed off land by the erosion of water. These are the particles that become suspended in flowing water and ultimately settle to stream or lake bottoms or form river deltas.

seed-tree cutting   A logging practice that leaves behind trees to seed the area for regrowth.

selection cutting   Cutting only mature timber from an area.

septic tank   A device used in rural areas for partial treatment of wastewater. The device is a concrete or metal tank that detains wastes from a home for up to 3 days, providing settling of solids and partial treatment of dissolved organics.

shale oil   A hydrocarbon liquid derived by retorting (cooking) crushed oil-bearing rock. The hydrocarbon liquid is known as kerogen. The United States may have the equivalent of 600 billion barrels of oil in the shale rock of Colorado and neighboring states.

shelterwood cutting   A three-stage cutting plan in which (1) defective trees are first removed; (2) the stand is "opened" by further cutting 10–15 years later; and (3) the mature trees are cut after seedlings are well established.

siltation   The dropping or settling of sediments to the bottom of a body of water. Water brought to a halt behind a dam drops much of its load of sediment into the reservoir; the effectiveness of the reservoir for water storage is thus decreased.

smog   The term originally coined to describe the "fog" of photochemical air pollution.

solvent-refined coal   *See* coal liquefaction.

species   All those organisms that are able to interbreed successfully (if they are given the opportunity to do so), that share ties of common parentage, and that share a common pool of hereditary material.

stratification (of water)   Situation that occurs when the water at different levels of a lake or reservoir is at different temperatures and little mixing occurs between layers.

stratosphere   The layer of the earth's atmosphere directly above the troposphere, which is the lowest layer. Scientists are concerned because ozone levels in the stratosphere are declining and carbon dioxide levels are increasing.

subsidence   Collapse of land as a result of the empty spaces left underground by coal mining. Roads may buckle and sewer lines and gas mains may crack as a result of severe subsidence if it occurs in developed areas.

succession   A natural process in which the species found in a given area change conditions to make the area less suitable for themselves and more suitable for other species. This continues until the climax vegetation for the area grows up.

sulfur oxides   A class of air pollutants from the burning of fossil fuels (mainly coal and oil) containing sulfur. The sulfur is oxidized to sulfur dioxide and sulfur trioxide when the fuel is burned. The oxides and the acids they form with water vapor will damage building materials like marble, mortar, and metals and they damage plants and the health of people. Respiratory illnesses are increased during times when the levels of sulfur oxides are elevated.

surface mining (or strip mining)   The practice of removing coal by excavation of the surface without an underground mine. Area strip mining is practiced on flat lands. Contour strip mining is used on hillsides.

sustained yield   A quantity of timber less than or equal to the natural increase in that area occurring over the past time period. Increase is measured in volume units, such as cubic feet.

synthetic pesticide   A pesticide that was invented in a laboratory and not produced by any natural system.

taiga   Northern coniferous forest biome, characterized by spruce and other coniferous trees.

tailings   The remaining waste materials after a substance such as uranium or iron has been extracted from ore.

tar sand   A sand coated with bitumen, an oily black hydrocarbon liquid. Extensive tar sand deposits exist in Alberta, Canada, and smaller deposits are in Utah. In the Alberta sands, about two barrels of hydrocarbon liquid can be recovered from three tons of sand; the liquid can be refined to all grades of petroleum.

teratogen   A substance that causes a defect during prenatal development.

tertiary treatment of wastewater   The third major component of wastewater treatment plants. In tertiary treatment, which consists of a number of processes, nitrogen in its various chemical forms is removed; phosphorus is removed by precipitation and settling; and resistant organics are removed by passage through towers containing activated carbon.

tetraethyl lead   An organic lead compound that has been added to gasoline to increase octane and decrease "engine knocking."

thermal pollution   Pollution of the environment with heat.

threatened species   A species that is not yet endangered but whose populations are heading in that direction.

threshold   The level or concentration at which an effect can be detected.

tidal power   The capturing of the energy of the tides as electrical energy by the use of dams and a turbine/generator unit. Both the receding and arriving tides can be channeled through the bulb turbine to generate electricity.

tolerance   The amount or residue of a pesticide or drug legally allowed in food.

topsoil   The top few inches of soil, which are rich in organic matter and plant nutrients.

toxic substance   A material harmful to life.

toxin   A naturally produced poisonous material secreted by certain organisms.

trickling filter   A device for removing dissolved organics from wastewater. In the device, the wastewater is distributed across a bed of stones on which a microbial slime is growing. The microbes remove the dissolved organics by converting them to simpler substances.

trophic structure   How the various organisms in a community obtain their nourishment.

tundra   Arctic biome characterized by a permanently frozen subsoil and low-growing plants such as mosses and lichens.

turbidity   A measure of how clear water is. It depends on the amount of suspended solid materials or organisms in the water.

typhoid   An intestinal disease caused by specific bacteria. The disease can be spread by water polluted by human wastes.

ultraviolet light   A part of the electromagnetic spectrum comprising light waves that are shorter than visible violet light rays but longer than x-rays. These light rays can contribute to the development of skin cancer but are normally absorbed in the upper atmosphere by ozone.

undernourished   Referring to a person who does not get enough Calories to maintain body weight with normal activity.

undeveloped nation   A country with little technological development.

undiscovered resources (of oil and gas)   Oil or gas not yet discovered by drilling but that, on the basis of geological and statistical evidence, is expected to be found eventually.

upwelling   A result of offshore winds that "push" surface waters away from shore and allow nutrient-rich bottom waters to rise from the deeper oceans.

uranium-235   An isotope of uranium normally found in a mixture with uranium-238. Uranium-235 serves as the fuel for light-water nuclear reactors. It can also be used for making atomic bombs.

U.S. Fish and Wildlife Service   The managing agency of the National Wildlife Refuge System. The service is within the Department of Interior.

U.S. Forest Service   The managing agency of the National Forests. The Forest Service is in the Department of Agriculture.

waste stabilization lagoon   A large shallow pond into which wastewater is discharged for biological treatment. Solids settle out, and dissolved organics are removed by microbes in the pond. The device is used primarily where waste loads are small and when land with no conflicting uses is available inexpensively.

watershed   The land area that drains into a particular river, lake, or reservoir.

weather   The day-to-day variation in temperature, humidity, air pressure, cloudiness, and amount of precipitation. Thus, weather describes the state of the atmosphere at a certain moment.

windmill or wind turbine   A machine whose blades are rotated by the wind and that converts the wind energy into electricity or work (for example, pumping water).

zero population growth (ZPG)   The situation in which birth rates equal death rates. Assuming immigration and emigration are not significant, the population does not grow.

zoning   The practice by which local governments designate the allowable uses of a tract or tracts of land as a means of preventing incompatible land uses. Typical zoning classifications are single-family residential, multifamily residential, commercial, and industrial.

zooplankton   Microscopic forms of animal life that drift about in water, moving mainly with the currents.

Index

hunting, 111
hunting stamps, 691
hybrid car, 586
hybrid crops, 186–188
hydrocarbons, 386, 463, 478–479, 482, 485, 501
   liquid, 458, 460
hydroelectric power, 515–520
hydrofracturing, 540
hydrogen, 355, 356, 494
hydrogen sulfide, 260, 269, 394, 543
Hydrolec, 519
hydrologic cycle, 33, 35, 36, 213
hydropower, 341, 346, 352, 511, 515–526
Hydro-Quebec, 517, 523
hydroxyl ions, 494

I
Iacocca, Lee, 585
Iditarod, 708
Idso, S. G., 506
Igorots, 179–180, 190
illegal aliens, 74–75, 77
immigration, 74–76
   to U.S., 74–76, 77
immigration rate, 56–57
Imperial Dam, 241
Imperial Valley, California, 542
incineration, 310, 460
India, 721
   family in, 59
   family planning in, 96–97
   food production in, 63, 64, 186
   populatation problems in, 10
   population control in, 94, 96–97
   population growth in, 63, 64
   population profiles for, 60, 61
   sidewalk dwellers of, 73
Indonesia, 73–74
infant formula, 176–177
infant mortality, 9, 85
infectious diseases, 56
infertility, in Africa, 97
infiltration-percolation, 296
inheritance tax, 678
initiation stage, 600
insect pests, 161, 163–164, 168–169
insecticides, 163–164
in-stream aeration, 295–296
insulation, 580
integrated pest management (IPM), 168–172, 661
International Atomic Energy Agency (IAEA), 420
International Drinking Water and Sanitation Decade, 331, 333–334
International Whaling Commission, 127–128, 129
intertidal zone, 222, 525–526
Inuit. *See* Eskimos
inversions, 444, 449–450
iodine-131, 623
ionizing radiation, 619, 622–624
IPM, 168–172, 661
iron, 315

iron-59, 623
iron oxide, 454
iron pyrite, 374, 456
irrigation, 183–184, 188
   in California, 243
   in High Plains, 234, 235
   trickle, 235–236
   use of groundwater for, 233
   with wastewater, 296, 297
   water for, 230
irrigation canals, use in hydropower, 520
island ecosystems, 112, 119
Island Park Geothermal Area, 541–542
isobutane, 539
Israel, 297, 315
IUDs, 86, 87
Izaak Walton League v. *Butz*, 695

J
Jablons, J. Michael, 635, 636
jack-pine forests, 697
Jackson Hole, 713–714
Jacobson, Michael, 637
Japanese beetles, 169
Jefferson, Thomas, 724
job safety, 612–618
John Muir Wilderness, 709
Johns-Manville, 614
Johnson, Arthur, 499
Joseph, 192
Joshua Tree National Monument, 687
Johns-Manville, 614
Johnson, Arthur, 499

K
K selection, 47
Keahole Point, Hawaii, 564
Keepin, G. R., 419
Kemp, Clarence, 547
Kennedy, David, 246
Kenya, 130–131
Kerala (India), 87, 94
kerogen, 396
Kilbey, B. J., 279
Klamath weed, 169
Klein, John, 681
Korea, 100–101
Krakatoa, 437
krill, 185, 222
Krutch, Joseph Wood, 720
Kuck, David, 377
kwashiorkor, 176

L
La Rance Power Station, 521–523
labor force
   in developing nations, 65, 68
   illegal aliens in, 75
ladybugs, 163
lake ecosystems, 214–216
Lake Erie, 281–282
Lake Michigan, 252
Lake Nakuru (Kenya), 17, 23
Lake Ontario, 322
Lake Superior, 610

Lake Tahoe, 292
Lake Trummen, 281
Lake Washington, 280
lakes
   acid rain and, 498–499, 502
   artificial, 517
   eutrophication of, 275–282
   mercury pollution of, 316, 319
   oligotrophic, 276
   restoration of, 280–281
   succession in, 141
   temperature stratification in, 18–19, 214
Lamb, H. H., 438
land
   arable, 184–185
   arid, 201–207
   ownership of, 188
   reclamation of, 366–372
Land and Water Conservation Fund, 714, 716
land trust, 715
land use
   multiple-use concept of, 691–692
   private, 672
   public, 687–716
   rural, 676–684
   taxes and, 677–678, 679
   urban/suburban, 672–676
landfills, 291, 311–313
   methane from, 394
Larderello, Italy, 538
laterite, 149
Law of the Sea Treaty, 433
laws governing toxic substances, 604
laws of thermodynamics, 342–343, 578
leaching, 37
leaching field, 300
lead, 315, 319
   in air, 488–489
   in cigarettes, 631
   in drinking water, 497
lead arsenate, 630
lead poisoning, 488, 489
Leadville, Colorado, 622
legumes, 35–36
Leopold, Aldo, 121
leptospirosis, 332
Leroy-Somer Company, 519
Lewis and Clark Trail, 708
life, quality of, 669
lifeboat ethic, 193–194
lighting, solar, 556–557
light-water nuclear reactors, 403–409
lignite, 361
Lijinsky, William, 641
Lilienthal, David, 563
lime-limestone scrubbers, 456–457
limestone, 502, 503
limiting factors, 276, 277
Lin, Shen Dar, 257
lindane, 260
liquefied natural gas, 392
liquid-metal fast breeder reactors, 421–423
Little Tennessee River, 106, 115
Littleton, Colorado, 376
liver cancer, 597

Credits

(for full color illustrations)

Endangered Species

a. © Zoological Society of San Diego. All rights reserved.
b. FPG/A. Schmidecker
c. R. Perdue, Jr.
d. U.S. Fish and Wildlife Service/C. J. Henry
e. T. Goettel

Food Production

a. UN/K. Singh
b. M. Ritter
c. UN Photo Library
d. UN Photo Library

Water Resources

a. WHO/E. Scheidegger
b. UN/R. Witlin
c. J. Perry

Energy-Related Pollutants

a. U.S. DOE
b. UN/Exxon Photo, Rentmeester
c. U.S. DOE
d. SOVFOTO/EASTFOTO

Air Pollution

a. UN/B. Wolff
b. UN/J. Robaton
c. P. Norton
d. O. Bricker

Natural Sources of Power

a. U.S. DOE
b. U.S. Windpower/E. Linton
c. Solar Energy Research Institute
d. UN/S. Metalitsa

Human Health and the Environment

a. Greenpeace/McAllister
b. J. Perry
c. UN/G. Tortoli
d. B. Castleman

Land Resources

a. FPG/W. McKinney
b. FPG/G. Rowell
c. National Park Service/R. Frear

How Much Is a Part per Billion?

A number of environmental problems are defined in terms of parts per million, parts per billion, or even parts per trillion of pollutant chemicals or particles in a given quantity of the resource. These are very small numbers, indeed. But are the numbers so small they can be safely ignored? This, of course, depends on the frame of reference.

The December 1976 issue of *ChemEcology* carried a table compiled by Dr. Warren B. Crumett of the Dow Chemical Company that was designed to show just how small trace concentrations are (Table 1).

Table 1 Trace Concentration Units

Unit	1 part per million (ppm)	1 part per billion (ppb)	1 part per trillion (ppt)
Length	1 inch/16 miles	1 inch/16,000 miles	1 inch/16,000,000 miles (A 6-inch leap on a journey to the sun)
Time	1 minute/2 years	1 second/32 years	1 second/320 centuries
Money	1¢/$10,000	1¢/$10,000,000	1¢/$10,000,000,000
Weight	1 oz salt/32 tons potato chips	1 pinch salt/10 tons potato chips	1 pinch salt/ 10,000 tons potato chips
Volume	1 drop vermouth/ 80 fifths of gin	1 drop vermouth/ 500 barrels of gin	1 drop vermouth/ 25,000 hogsheads of gin
Area	1 sq ft/ 23 acres	1 sq ft/36 sq miles	1 sq in./250 sq miles
Quality	1 bad apple/ 2,000 barrels	1 bad apple/ 2,000,000 barrels	1 bad apple/ 2,000,000,000 barrels